DEEP-MARINE ENVIRONMENTS

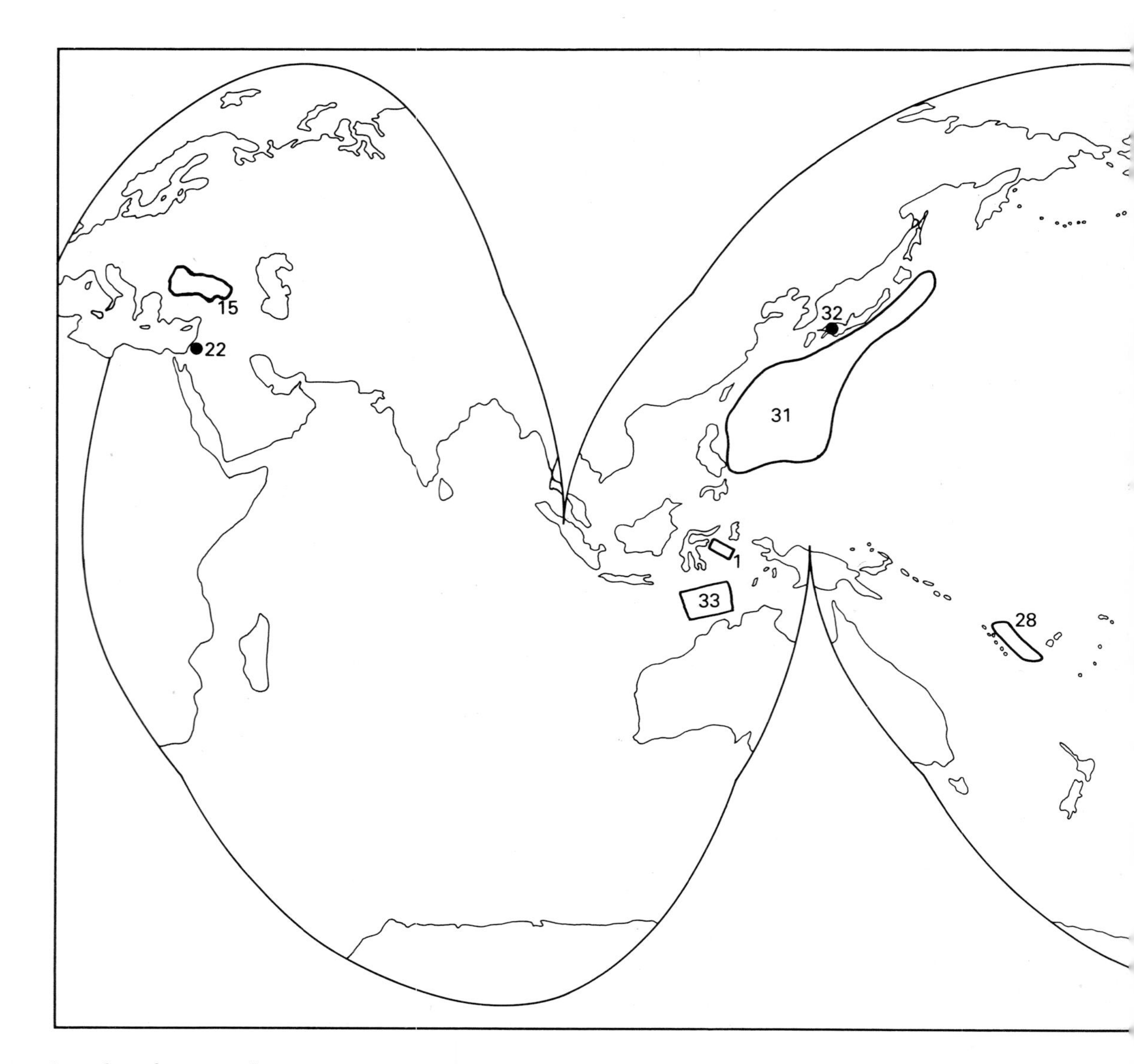

Location of case studies

Chapter 5

1 Seram Trough, Indonesia
2 Orca Basin, Gulf of Mexico
3 Jason Deposit, Yukon
4 Jameson Land, East Greenland
5 Gowganda Formation
6 Cow Head Group, Newfoundland

Chapter 6

7 Quinault Canyon, Washington Coast

Chapter 7

8 Navy Fan
9 Rhône Fan
10 Amazon Fan
11 Tyee Formation
12 Milliners Arm Formation, Newfoundland

Chapter 8

13 Sohm Abyssal Plain and Hatteras Abyssal Plain
14 Aleutian Abyssal Plain
15 Black Sea Euxine Abyssal Plain
16 Marnoso–Arenacea, Italian Apennines
17 Cretaceous Flysch, E. Alps

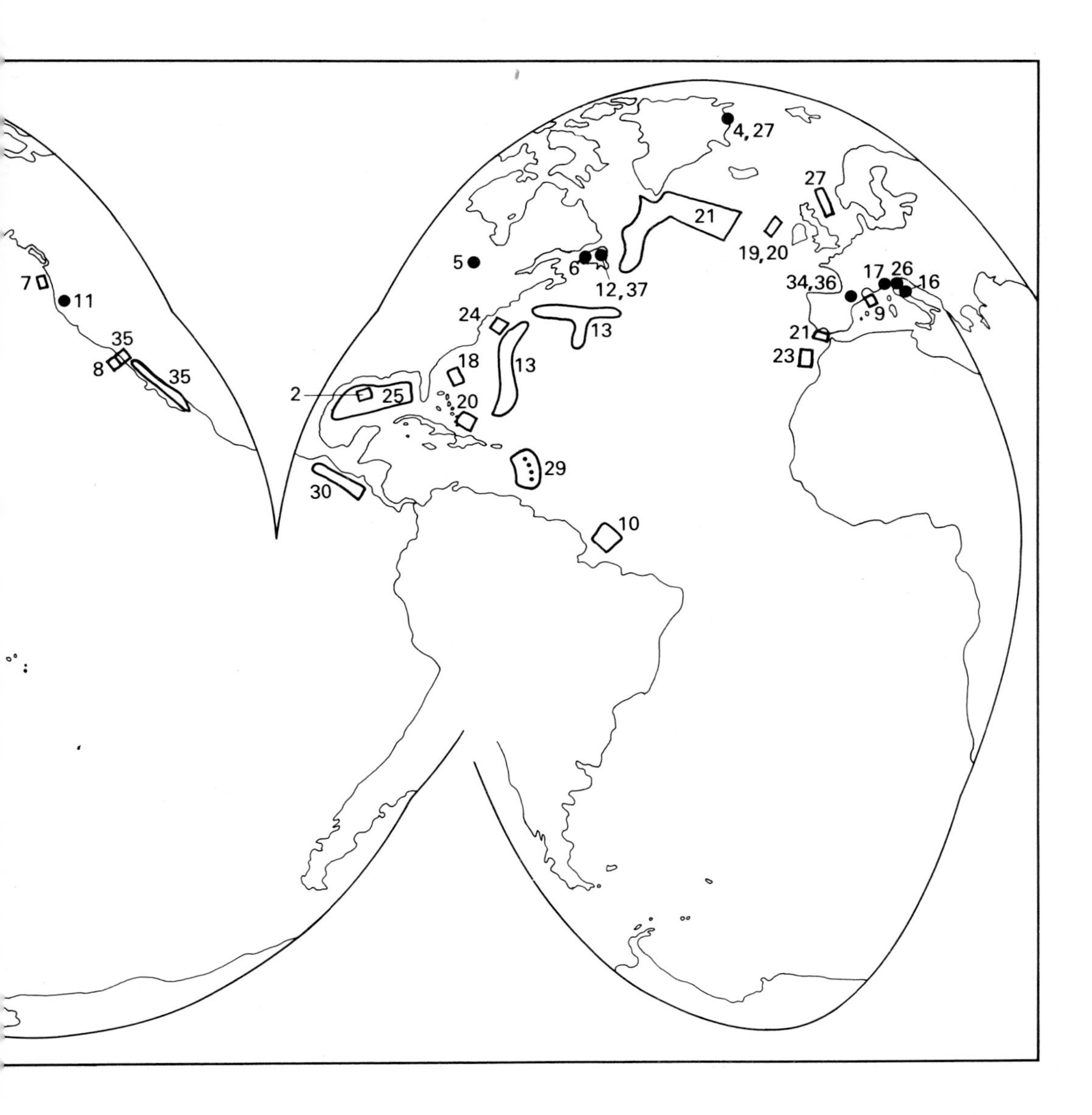

apter 9

Furrows, Blake–Bahama Outer Ridge
Rockall Trough
Mudwaves Rockall and Blake–
Bahama Outer Ridge
Central and North Atlantic Ocean
Talme Yafe Formation, Israel

apter 10

Mazagan margin, NW Africa
U.S. Atlantic margin, New Jersey
Northern Gulf of Mexico
Jurassic southern Alps
Late Jurassic/Early Cretaceous/Palaeogene E.
Greenland & Viking Graben

Chapter 11

28 N. Fiji Basin
29 Lesser Antilles
30 Middle America margin
31 Western Pacific 10°–45° N
32 Cretaceous–Neogene, S. Japan
33 Neogene–Quaternary, S. Banda Arc
34 Tertiary S. Pyrenean Foreland Basin

Chapter 12

35 California Continental Margin and Gulf of California
36 Mesozoic Pyrenees
37 Lower Palaeozoic, north-central Newfoundland

TITLES OF RELATED INTEREST

Aeolian geomorphology
W. G. Nickling (ed.)

Biogenic textures in sedimentary rocks
R. Bromley

Cathodoluminescence of geological materials
D. J. Marshall

Chemical fundamentals of geology
R. Gill

A dynamic stratigraphy of the British Isles
R. Anderton *et al.*

Experiments in physical sedimentology
J. R. L. Allen

Geomorphology and soils
K. Richards *et al.* (eds)

Geomorphology in arid regions
D. O. Doehring (ed.)

Geomorphological field manual
V. Gardiner & R. Dackombe

Geomorphological techniques
A. S. Goudie (ed.)

Glacial geomorphology
D. R. Coates (ed.)

Hillslope processes
A. D. Abrahams (eds)

The history of geomorphology
K. J. Tinkler (ed.)

Image interpretation in geology
S. Drury

Introduction to theoretical geomorphology
C. Thorn

Karst geomorphology and hydrology
D. C. Ford & P. W. Williams

Marine geochemistry
R. Chester

Mathematics in geology
J. Ferguson

Pedology
P. Duchaufour (translated by T. R. Paton)

Perspectives on a dynamic Earth
T. R. Paton

A practical approach to sedimentology
R. C. Lindholm (eds)

Principles of physical sedimentology
J. R. L. Allen

Rock glaciers
J. Giardino *et al.* (eds)

Rocks and landforms
J. Gerrard

Sedimentology: process and product
M. R. Leeder

Soils of the past
G. Retallack

Tectonic geomorphology
M. Morisawa & J. T. Hack (eds)

Volcanic successions
R. A. F. Cas & J. V. Wright

DEEP-MARINE ENVIRONMENTS

CLASTIC SEDIMENTATION AND TECTONICS

K. T. PICKERING

University of Leicester, UK

R. N. HISCOTT

Memorial University of Newfoundland, Canada

F. J. HEIN

Dalhousie University, Canada

London
UNWIN HYMAN
Boston Sydney Wellington

Published by the Academic Division of
Unwin Hyman Ltd
15/17 Broadwick Street, London W1V 1FP, UK

Unwin Hyman Inc.,
8 Winchester Place, Winchester, Mass. 01890, USA

Allen & Unwin (Australia) Ltd,
8 Napier Street, North Sydney, NSW 2060, Australia

Allen & Unwin (New Zealand) Ltd in association with the
Port Nicholson Press Ltd, Compusales Building, 75 Ghuznee Street,
Wellington 1, New Zealand

First published in 1989

British Library Cataloguing in Publication Data

Pickering, K.
Deep-Marine environments.
1. Clastic marine sediments.
I. Title II. Hiscott, R. III. Hein, F.
551.46′083
ISBN 0–04–551122–5

Library of Congress Cataloging-in-Publication Data

Pickering, K.
Deep-marine environments: clastic sedimentation and tectonics/
K. Pickering, R. Hiscott, F. Hein.
p. cm
Bibliography: p.
Includes index.
ISBN 0–04–551122–5
1. Sedimentation and deposition. 2. Marine sediments. 3. Plate tectonics. I. Hiscott, R. II. Hein, Frances J. III. Title.
IV. Title: Clastic sedimentation and tectonics.
QE571.P55 1988
551.46′083—dc19 88–17222
CIP

Typeset by Oxprint Ltd, Oxford
and printed in Great Britain by Butler & Tanner Ltd,
Frome and London

Preface

This book grew out of our desire to integrate process-based, environmental and large-scale plate-tectonic aspects of both modern and ancient deep-marine sedimentation into a unified and comprehensive text. Edited volumes that cover parts of this rapidly expanding area of knowledge are available, but they are unable to provide the broad ranging and consistent conceptual approach of an integrated text. We have endeavoured to produce an up-to-date, discursive and well-balanced text on deep-marine sedimentology. We have focused on clastic sedimentology, and in particular siliciclastic sediments, although much of the text is equally applicable to carbonate, volcaniclastic, or lithologically mixed environments. The geochemical, petrographic and palaeontological aspects of deep-marine environments are, however, beyond the scope of this book.

There are many ways in which to tackle a book of this type. We divided the text into three parts. In the first part, we discuss the fundamental building blocks of deep-marine environments and the nature of sediment transport and deposition in the deep sea. After considering the quantitative and semiquantitative aspects of deep-water sedimentation, we provide a rigorous classification of the range of sediment deposits or facies. There is a general section on the way in which facies can be vertically stacked as sequences and a discussion of how sequences can be recognized using objective, statistical criteria. The first part of the book ends with a chapter which considers the main factors controlling the nature of deep-marine sedimentation.

The second part of the book is concerned with specific deep-marine environments. Individual chapters focus on: slope aprons and slope basins; submarine canyons, gullies and valleys; submarine fans; sheet systems, such as abyssal plains; and contourite drifts. Our approach has been to introduce each of these chapters with a review of the important ideas and models associated with the environments, followed by selected modern and ancient case studies. We have carefully chosen the case studies to reflect the major developments in understanding a particular environment. The advantage of our 'case study' approach is that the reader has an opportunity to understand the important concepts and models for deep-marine sedimentation within an easily digestible context.

The third part of the book, tectonics and sedimentation, integrates the various deep-marine environments discussed in the preceding part of the text into a plate-tectonic framework. The approach, as in the second part of the book, is to follow the introductory sections with modern and ancient case studies. We have adopted a traditional three-fold division of plate settings into: evolving and mature passive continental margins; active convergent margins; and oblique-slip margins.

One of our principal aims has been to produce a truly integrated, co-authored, book with standardised nomenclature and conceptual approach to each subject. Naturally, in writing the chapters we had to delegate the responsibility for a first draft to an individual, and this was done as follows: Chapters 5 & 9 by Frances Hein; Chapters 1, 2, 3 & 7 by Richard Hiscott; and Chapters 4, 6, 8, 10, 11 & 12 by Kevin Pickering. Having written the first drafts for each chapter, they were then reviewed and modified by the other two co-authors. Many colleagues kindly read the manuscript at various stages of preparation; their constructive comment and advice have substantially improved the scientific content and readability of the text. In this regard we extend our sincere gratitude to A. E. Aksu, D. A. Cacchione, D. J. Cant, D. J. W. Cooper, D. S.

Gorsline, M. A. Hampton, C. W. Harper Jr., A. E. Hay, P. R. Hill, J. D. Hudson, M. R. Leeder, G. V. Middleton, P. M. Myrow, Y. Ogawa, D. J. W. Piper, A. J. Rex, A. D. Saunders, C. T. Schafer, R. D. A. Smith, D. A. V. Stow, J. P. M. Syvitski and J. H. McD. Whitaker. We sincerely thank Clare Stanga and Chris Mawby for retyping parts of the book after a file 'backup' command resulted in an instantaneous but temporary loss of three chapters! Many of the diagrams were kindly drafted by John Taylor, Gordon Hodge and Louise Pickering. A large proportion of the photographic work was undertaken at Leicester University by Colin Brooks, Ken Kowalski, Ian Paterson and Ken Garfield.

We are very grateful for the help and professionalism of the staff at Unwin Hyman Limited, without whose commitment and encouragement this book would not have been possible, particularly Andy Oppenheimer and Roger Jones. Roger is also thanked for his early involvement in the book at the stage where we had to transform a contract and book plan into preliminary chapters.

Finally, we thank our respective spouses and families for their multifarious support over the last two years when a substantial part of our 'leisure' time was taken up by 'the book'; thanks Louise Pickering, Paula Flynn and Doug Cant.

Kevin T. Pickering
Richard N. Hiscott
Frances J. Hein

Acknowledgements

We are grateful to the following individuals and organisations who have kindly given permission for the reproduction of copyright material (figure numbers in parentheses):

J. W. Beck, Ocean Drilling Program (1.3); H. M. Pantin (1.4); Figures 2.1, 3.4, 4.7, 5.4, 5.26, 5.27, 5.28, 5.29, 5.30, 6.14, 6.18, 7.10, 7.24, 9.5, 10.19, 12.3, 12.11 and Tables 10.1, 10.2, 12.1 by permission of the American Association of Petroleum Geologists; G. V. Middleton and the *Canadian Journal of Earth Sciences* (2.2); the Editor, *Journal Sed. Petrol.* (2.6, 2.12, 2.13, 2.15, 2.20); Blackwell Scientific Publications (2.11, Table 5.1); D. B. Prior (2.14); C. Hollister, reproduced by permission from *Nature* **309**, Copyright © 1984 Macmillan Journals Ltd (2.16); I. McCave (2.17); J. S. Booth (2.20); R. G. Walker (3.4); Academic Press (4.2a); Prentice-Hall (4.4); Figures 4.5 and 4.6 © 1985 Springer-Verlag New York; H. A. Karl (4.10); A. Robertson (4.11); Figures 5.1, 5.5 and 5.14 reproduced from Brown and Fisher in *Seismic Stratigraphy* (ed. C. E. Payton) 1977, by permission of the American Association of Petroleum Geologists; S. Thornton, reproduced from Stow and Piper, *Fine Grained Sediments* 1984, by permission of the Geological Society (Table 5.2, Figures 5.7, 5.8, 9.14, 9.15); P. Homewood and Academic Press (5.2); T. H. Nilsen and the American Geological Institute (5.3); C. D. Winker and the Society of Economic Paleontologists and Mineralogists (5.6, 5.9, 5.13); J. P. Syvitski, reproduced from Syvitski and Farrow, *Fjord Sedimentation as an Analogue for Small Hydrocarbon Bearing Submarine Fans* 1988, by permission of the Geological Society (5.10, 5.11, 5.12); J. C. Ingle and the Society of Economic Paleontologists and Mineralogists (5.15); D. Jongsma and the Netherlands National Institute for Sea Research (5.17); A. Leventer, reproduced from Leventer et al, *Marine Geology* **53**, by permission of Elsevier (5.18, 5.19); Geological Society of America (5.21–25, 6.5, 7.22, 11.9, 11.15, 11.18a, 12.2); A. D. Miall and Blackwell Scientific Publications (5.31–37); M. Coniglio and the *Canadian Journal of Earth Sciences* (5.38–44); D. C. Twichell and the Geological Society of America (6.1); R. M. Carter (Table 6.1, Figure 6.19); Figure 6.4 © 1981 Springer-Verlag New York; Society of Economic Paleontologists and Mineralogists (6.6, 6.16); Elsevier (6.7, 6.15a, 8.11, 11.4a–b, 12.12, 12.13, 12.17, 12.18); G. H. Keller (6.8d, 6.12); N. H. Kenyon and Elsevier (6.13); B. Carson and Elsevier (6.15b); W. R. Normark (7.1, 7.11); E. Mutti and the Editor, *Mem. Soc. Geol. Ital.* (7.2); E. Mutti and the International Association of Sedimentologists (7.4); W. R. Normark and Blackwell Scientific Publications (7.5, 7.12, 7.13, 7.15); E. Mutti and D. Reidel Publishing (7.6, 7.7, 7.8); W. R. Normark and Springer Verlag (Table 7.1, Figures 7.11, 7.16); L. Droz, reproduced from Droz and Bellaiche, *Bull. Am. Assoc. Petrol. Geol.* **71**, by permission of the American Association of Petroleum Geologists (7.17–20); Springer Verlag (7.21, 7.23); Figure 7.22 from Damuth et al, *Geology* **11**, 470–73, by permission of the Geological Society of America; M. A. Chan, reproduced from Chan and Dott, *Bull. Am. Assoc. Petrol. Geol.* **62**, by permission of the American Association of Petroleum Geologists (7.25, 7.26); P. P. Weaver (8.1, 8.2); E. L. Hamilton and Elsevier (Table 8.1, 8.8); R. Hesse and Blackwell Scientific Publications (Table 8.2, Figure 8.13); D. J. Stanley and Elsevier (8.4, 8.5); R. D. Elmore and the Geological Society of America (8.6, 8.7); D. A. Ross (8.9); F. Ricchi-Lucchi and Blackwell Scientific Publications (8.12); Figure 9.1 reproduced from Herzen et al, *Science* **152**, 502–508, Copyright © 1966 the American Association for the Advancement of Science; D. A. Stow (Tables 9.1, 9.2); R. H. Dott and the Society of Economic Paleontologists and

Mineralogists (9.2); Table 9.3 and Figure 9.19 reproduced from McCave and Tucholke in *Deep current-controlled sedimentation in the Western North Atlantic* (ed. P. R. Vogt and B. E. Tucholke) 1986, by permission of the Geological Society of America; R. Myers (9.4); D. Tolmazin (9.6, 9.7); C. D. Hollister and Plenum Publishing (9.9); P. C. Richards (9.11); J. R. Allen, reproduced from Allen, *Sedimentology* **30**, by permission of Elsevier; E. Gonthier (9.15); Esso Resources Canada Ltd (9.20–23); A. Bein and Blackwell Scientific Publications (9.24–30); J. C. Sibuet (10.1); D. R. Kingston and the American Association of Petroleum Geologists (10.2); B. G. Williams, reproduced by permission of Butterworth & Co. (Publishers) Ltd © 1984 (10.3–5); G. Deroo (10.6); C. W. Poag and the Geological Society of America (10.7); E. L. Winterer (10.8, 10.10); T. Steiger (10.9); R. M. Leckie (10.11); W. B. Ryan (10.12, 10.16); A. H. Bouma, reproduced from Martin and Bouma in *Framework, Facies, and Oil-Trapping Characteristics of the Upper Continental Margin* (ed Bouma et al) 1978, by permission of the American Association of Petroleum Geologists (10.17); F. J. Shaub, R. T. Buffler and the American Association of Petroleum Geologists (10.18); E. L. Winterer and the American Association of Petroleum Geologists (10.24); H. C. Jenkyns and Blackwell Scientific Publications (10.25, 10.26); F. Surlyk (10.27); C. C. Turner and Graham & Trotman Ltd (10.29, 10.30); D. R. Boote and Graham & Trotman Ltd (10.31); D. S. Hastings and Graham & Trotman Ltd (10.32); I. J. Stewart and Graham & Trotman Ltd (10.33, 19.34); W. R. Dickinson and the American Association of Petroleum Geologists (11.1); X. Le Pichon, reproduced from Le Pichon et al, *Earth Planet. Sci. Letts* **83**, by permission of Elsevier (11.3, 11.6); T. M. Thornburg and the Geological Society of America (11.5); S. H. Stevens and Elsevier (11.7); G. K. Westbrook and the American Geophysical Union (11.13b, 11.20); S. N. Carey (11.16); E. C. Leitch (11.17); J.-Y. Collot and Elsevier (11.18b-c); M. Baltuck (11.21, 11.22); J. Aubouin (11.23); J. Letouzey, reproduced by permission of Butterworth & Co (Publishers) Ltd © 1985 (11.24, 11.25); Japex Geoscience Institute (11.26); Figure 11.27 reproduced by permission of Butterworth & Co (Publishers) Ltd © 1985; T. H. Nilsen (11.28); A. Taira (11.29, 11.30, 11.32); A. Taira and Blackwell Scientific Publications (11.31); Y. Ogawa and Elsevier (11.33); P. Huchon and Elsevier (11.34); Figure 11.35 reproduced from Audley-Charles, *Journal Geol. Soc.* **143**, by permission of the Geological Society; P. Labaume, © 1985 Springer-Verlag New York (11.37, 11.39); P. Labaume and the American Geophysical Union (11.38); Figure 12.1 reproduced from Harding, *Bull. Am. Assoc. Petrol. Geol.* **58**, by permission of the American Association of Petroleum Geologists; T. Harding and the Society of Economic Paleontologists and Mineralogists (12.4–6); F. J. Davey and the Geological Society of America (12.7); K. B. Lewis (12.8); A. D. Saunders (12.9); D. G. Howell (18.10); K. Kelts (12.16).

Contents

Part 2: Deep-water basin elements

Part 3: Plate tectonics and sedimentation

List of tables

Ocean Drilling Program's JOIDES *Resolution* (SEDCO/BP 471) in Antarctic waters. The boat is 143 m long, up to 21 m wide and with a 62 m high drilling derrick. The boat accommodates 68 crew and 50 scientific and technical personnel. Operations at sea began in January 1985. Photograph by John W. Beck of ODP office, Texas A&M University.

CHAPTER 1

Introduction

Introduction

A broad understanding of deep-marine sedimentary processes and environments has only been acquired since about 1970. This book summarizes as much as possible of this explosion in information in an integrated approach. It is perhaps appropriate that the three authors of this book were all educated in geology and developed interests in deep-marine sedimentation in the 1970s at the same time as the science was rapidly expanding.

We take a rather liberal view of the range of sedimentary environments that can be classified as 'deep'. As would be expected, we include all areas of the oceans beyond the local shelf-slope break. In most cases, the shelf-slope break occurs at depths of about 100–200 m, but in some regions the break is either deeper or shallower than this range. The minimum depth at the edge of the shelf in modern oceans is *c*. 20 m (Emery 1981), well within the influence of wave action, but this shallowness is so rare that we will generalize our discussion to deal almost exclusively with sedimentation below storm wave base (*c*. 200 m).

Deep, nearshore, semi-enclosed basins such as the central parts of fjords are also sites of sedimentation below wave base, and fast rates of supply of sands and muds in such settings result in facies that are very similar to those beyond the shelf-slope break in more open parts of the oceans. As an example, Postma (1984) describes sediment-gravity-flow deposits from the fronts of marine fan deltas that show many similarities with deposits described in Chapters 2 and 3. The deposits of deep fjords and fan-delta fronts are within the scope of this book, although fan deltas are not treated specifically.

The *Challenger* expedition (1872–6) marks the birth of modern marine sedimentology. HMS *Challenger* (Fig. 1.1) collected 133 dredge samples, from which the character of deep-marine pelagic sediments, particularly the biogenic oozes, was determined and related to both water depth and latitude (Murray & Renard 1891). After this expedition, noteworthy advances in understanding of deep-marine sedimentation are largely restricted to the period after 1950. An interesting exception is the description by Sheldon (1928) of the essential features of classic turbidities,

Figure 1.1 Drawing of HMS *Challenger*. From Thompson (1878).

including the sequence of sedimentary structures summarized 34 years later by Bouma (1962). An excerpt of Sheldon's description of graded beds in the Devonian of New York State is given in Walker (1967a).

Critical observations, in the period 1950–70, that provided fundamental insight into deep-sea processes and sedimentary facies are listed in Table 1.1. From this list, we wish to highlight: (a) the revolutionary documentation that coarse-grained, graded sands are carried into deep parts of the oceans by turbidity currents (Kuenen & Migliorini 1950, Natland & Kuenen 1951, Heezen & Ewing 1952, Heezen *et al.* 1954) – this idea was anticipated by Daly's (1936) suggestion that turbidity currents were responsible for the cutting of submarine canyons; (b) the careful documentation of a preferred sequence of sedimentary structures in graded turbidites (Bouma 1962) and explanation of this sequence by deposition from decelerating turbulent currents (Harms & Fahnestock 1965, Walker 1965, Allen 1969); (c) publication of superbly illustrated photographic monographs of the variety of sedimentary structures, including sole markings, that characterize turbidites (Pettijohn & Potter 1964, Dzułynski & Walton 1965); (d) experimental modelling of deposition from turbidity currents (Middleton 1967); and (e) the discovery that water in the deep oceans is not still and clear, but locally contains significant concentrations of suspended mud (Ewing & Thorndike 1965).

Post-1970 publications form the major reference material that we attempt to synthesize in this book, so to a large extent we prefer to leave discussion of contributions during this time period to the separate chapters that follow. We wish, instead, to draw the reader's attention here to particularly influential volumes of collected papers that have appeared since 1970, and that provide comprehensive syntheses of various aspects of deep-marine sedimentation.

Turbidites and deep-water sedimentation, edited by Middleton & Bouma (1973), provided the first real review of theoretical, oceanographic, and field data related to deep-water, essentially siliciclastic deposits. The landmark paper in this volume by Middleton &

Table 1.1 Benchmark papers in deep-marine sedimentation, 1950–70.

Year	Author(s)	Contribution
1950	Kuenen & Migliorini	Synthesis of experimental and field studies resulting in conclusion that graded beds in flysch are deposited in deep water by turbidity currents
1951	Dorreen	Documentation of association of chaotic submarine mudflows and slides with turbidites in a deep-water setting, with the coarser deposits linked genetically to active tectonics
1951	Natland & Kuenen	Explanation of controversial alternation of shallow-water and deep-water faunas in graded sand–mud couplets of Ventura Basin as a consequence of turbidity–current transport of sands into deep water
1951	Shepard	Description of contemporary mass movements in heads of submarine canyons
1952	Heezen & Ewing	First indirect observation of passage of a large turbidity current down a continental slope (Grand Banks of Newfoundland, 1929) by analysis of timing of cable breaks. Apparent maximum velocity *c.* 28 m/s
1954	Heezen, Ericson & Ewing	Core description of the 1929 Grand Banks turbidite
1955	Renz, Lakeman & van der Meulen	Restored palaeogeographic link between major submarine slides and a tectonically active basin margin, including a map of slide geometry
1959	Gorsline & Emery	First detailed description of turbidites of the offshore California Borderland, and distinction of canyon–fan systems from slope–apron systems
1960	Sullwold	Comprehensive description of a relatively intact Miocene submarine fan, California
1961	Dott	Recognition that thick, coarse grained, poorly organised successions, particularly in areas of high sediment yield from glaciers, could be deposited in deep water by high-concentration flows
1962	Bouma	Field synthesis of internal structures in turbidites leading to formulation of the 'Bouma sequence'
1962	Whitaker	Description of sedimentology/palaeoecology of Silurian submarine canyons, Welsh Borderland
1963	Dott	Early thoughtful discussion of the various mechanisms for transport of coarse detritus into deep water
1963	Unrug	First thorough description of 'fluxoturbidites' and emphasis that many deep-water clastics are very much unlike Bouma turbidites. These other deposits received more consideration after this time
1964	Pettijohn & Potter	First photographic atlas of turbidite sole markings
1964	Seilacher	Summary of the diagnostic biogenic traces of deep-water siliciclastic sediments, specifically in flysch
1965	Ewing & Thorndike	Discovery of nepheloid layers containing suspended sediment deep in the modern oceans
1965	Dzułynski & Walton	Classic photographic atlas dedicated entirely to turbidites and flysch
1965	Harms & Fahnestock	Hydrodynamic model for Bouma turbidites
1965	Walker	Hydrodynamic model for Bouma turbidites
1966	Heezen, Hollister & Ruddiman	Discussion of the importance of contour currents in reworking sediments on the continental rise of eastern North America
1966	Middleton (a,b,c)	First geologically oriented experimental studies of flow of turbidity-current head and body regions
1966	Shepard	First description of meanders in submarine-fan channels
1966	Shepard & Dill	Major compilation of data on submarine canyons and valleys
1967	Middleton	Experimental analysis of deposition from turbidity currents of high and low concentration
1969	Enos	Exhaustive description of flysch turbidites, including specification of geometry and extent of single beds
1969	Hilde, Wageman & Hammond	Early description of subduction-related sedimentation with active deformation in the Nankai Trough area near Japan. This region is still providing fundamental ideas on the link between deep-marine sedimentation and tectonics
1969	Ricci Lucchi	First major English-language extended paper on Italian submarine-fan deposits, heralding the major contributions to fan models of Ricci Lucchi & Mutti in the 1970s
1970	Middleton	Comprehensive discussion of the hydrodynamics of sediment gravity flows, primarily turbidity currents
1970	Normark	Presentation of a model for development of modern sand-rich submarine fans, California Borderland

Hampton (1973) combined the results of Middleton's experimental work on turbidites (1966a & b, 1967) with Hampton's study of debris flows (1972) into a theoretical assessment of the features to be expected in subaqueous deposits from turbidity currents, debris flows, grain flows, and 'fluidized flows'. Never again would geologists be able to describe deep-water flysch successions as 'monotonous' alternations of graded sandstone beds and shales – it was now clear that the coarser deposits could be transported and deposited by a variety of processes, resulting in complex deposits. Also in this volume, Bouma & Hollister (1973) raised the continuing problem of the distinction between the deposits of turbidity currents (turbidites) and those of contour-following bottom currents (contourites – Chapter 2).

Pelagic sediments: on land and under the sea, edited by Hsü & Jenkyns (1974), is still the authoritative volume on this widespread class of deep-marine sediments, and is particularly recommended for its discussions of diagenesis of biogenic siliceous sediments to form bedded cherts. A clear link is also made in this volume between pelagic sediments, plate tectonic setting, and volcanism.

Fine-grained sediments: deep-water processes and facies, edited by Stow & Piper (1984a), provides a comprehensive treatment of those sediments that might be loosely called 'slope/rise deposits', or 'hemipelagic deposits'. Details on the transport of muddy sediments by turbidity currents, bottom currents, surface and mid-water plumes, and within sediment slides contribute to an overall impression of the complexity of sediment transport and deposition in the deep sea. The emphasis on fine grained silts and muds helps counter the bias that one often encounters in the literature in favour of the coarser grained deep-water facies.

Scientific drilling

A major turning point in understanding the distribution and genesis of deep-marine facies was the launching of the Deep Sea Drilling Project (DSDP) in 1968. The *Glomar Challenger* (Fig. 1.2) drilled 96 legs, both the first and last in the Gulf of Mexico, with a total core recovery from the deep oceans of *c*. 100 km. These first long cores from the oceans provided the raw data to study palaeoceanographic history since the Mesozoic, and to link facies to particular geographic, tectonic, and plate-tectonic settings. Fundamental advances achieved during the first decade of DSDP are summarized in Warme *et al*. (1981), and in the famous 'blue books' that have become essential reference texts for the professional marine geologist.

Scientific drilling has continued under the Ocean Drilling Program (ODP), using the drill-ship *JOIDES Resolution* (Fig. 1.3), registered as the *SEDCO/BP 471*. As with DSDP, this programme is destined to add substantially to our understanding of the deep ocean basins.

Tectonic influence on sedimentation

The link between large-scale tectonic processes and sedimentary processes/facies has developed considerably from a relatively primitive view based only on observations in mountain belts before the advent of plate-tectonic theory. These earlier ideas are of mainly historical interest – the reader is referred to Dott (1978) for a thorough review.

Only in the last 5–10 years has the wide availability of the following data been sufficient to allow a good understanding of the interaction between active tectonics and sedimentation in ocean basins: (a) long marine seismic-reflection profiles from the petroleum industry and government surveys (Bally 1983), (b) data from regional programmes of piston-core sampling and side-scan sonar, and (c) DSDP and more recently ODP results. This important link between sedimentation and tectonics is given considerable emphasis in this book (Ch. 10, 11, 12); the reader will quickly appreciate the importance of seismic data and deep drilling when reading case studies in these chapters. References in such case studies are generally less than five years old. We believe that these chapters could not have been written in anything like their present form, even a few years ago, due to a lack of well documented case studies, particularly in complex convergent-margin settings where trenches, forearc basins, marginal basins (including backarc basins) and small, ephemeral slope basins form a collage of deep-marine environments in an area where processes of thrusting, mud diapirism and mélange formation all complicate the stratigraphic record.

The relationship between tectonic setting and composition of siliciclastic sediments is not discussed in this book because of our emphasis on facies, not petrology. Readers interested in the petrographic aspects are referred to papers by Dickinson & Suczek (1979), Dickinson & Valloni (1980), Maynard *et al*. (1982), Bhatia (1983), Valloni & Mezzadri (1984), van de Kamp & Leake (1985) and Bhatia & Crook (1986).

Figure 1.2 The *Glomar Challenger*, drill-ship of the Deep Sea Drilling Project.

Figure 1.3 The *Joides Resolution* (*SEDCO/BP 471*), drill-ship of the Ocean Drilling Program.

A view to the future

We expect that the following three areas of study will feature strongly in the next 5–10 years of research on deep-water sedimentation: (a) scientific drilling (ODP) of specific deep-marine settings; (b) experimental and theoretical modelling of unusual or high-concentration, coarse to mud-rich/fine-grained, sediment gravity flows, and (c) development of objective criteria for recognition of cycles or sequences of various thicknesses in deep-marine sediments, together with explanations for the cyclicity. Undoubtedly, there will be other areas of research which will lead to significant advances in our understanding of deep-marine systems.

DRILLING OF MODERN DEPOSITIONAL SETTINGS

The last leg of DSDP, Leg 96 (Bouma, Coleman *et al.* 1986), was the first leg dedicated specifically to understanding the stratigraphy and growth of a submarine fan – the Mississippi Fan. Other DSDP and ODP sites have drilled into a variety of deep-marine sediments, but the general objectives were essentially related to palaeoceanographic, stratigraphic, or tectonic problems. If marine sedimentologists endeavour to write and defend strong proposals for HPC (Hydraulic Piston Corer) sampling of fans or other poorly understood modern settings in the oceans, specifically for the purpose of relating morphological features to associated facies and facies sequences, then our understanding of these deposits and modern–ancient analogues will increase considerably.

MODELLING OF SEDIMENT TRANSPORT PROCESSES

Turbidity currents, and to some extent cohesive debris flows, have been experimentally studied and theoretically modelled to the point that equations of flow and constraints on deposition are available (Ch. 2). This is not the case for complex flows that deposit most of the disorganized sands and gravels that are associated with ancient submarine fans – support mechanisms for clasts in these flows may include turbulence, buoyancy, matrix strength, particle interaction and pore-fluid pressure (Ch. 2). Experiments with such flows are very difficult to devise, but would probably add significantly to our understanding of deep-water sedimentation.

Thick megaturbidites, or 'megabeds' (Ch. 2, 8), may contain basal parts that resemble debris-flow deposits (Stanley 1982), or may show evidence for flow reversal due to reflection of a single large flow during deposition (Pickering & Hiscott 1985). The hydrodynamics of such flows must be exceedingly complex, but preliminary experiments and theoretical modelling (Pantin & Leeder 1987) have raised new possibilities for behaviour of large turbidity currents (Fig. 1.4).

SEQUENCES AND CYCLICITY IN DEEP-MARINE DEPOSITS

A powerful concept that developed during the 1970s was that thinning-upward and thickening-upward sequences in submarine-fan successions could be used to specify the original depositional setting (e.g. fan channels *vs* fan lobes). Recently, this hypothesis has been questioned both because the mechanisms of sequence generation are quite hypothetical, and because the sequences have not been rigorously defined using statistical techniques and are subjectively identified to the extent that there is little agreement among researchers as to their validity (Ch. 4). We expect that there will be a renewed effort to statistically demonstrate the presence (or absence) of asymmetric sequences in deep-marine deposits (e.g. Waldron 1987), so that fan environments, in particular, may be more easily identified in ancient successions. This will be most successful if 'modern' successions can be analysed in the same way, perhaps from ODP cores obtained from future fan drilling. We strongly support any initiative to drill thick clastic successions, particularly submarine fans, in order to test the depositional models based on vertical sequence analysis. A critical question in both ancient and modern systems will be to determine the relative importance of intrabasinal controls (e.g. switching of channels) versus extrabasinal controls (e.g. tectonic pulses, eustatic sea-level changes, palaeoclimatic and palaeoceanographic changes) in the development of cyclicity.

Sequences, or cycles, in hemipelagic and pelagic successions deserve a more rigorous examination in the coming decade as a part of a developing major programme to study changing global climates, perhaps driven by Milankovitch cyclicity (Ch. 4). This topic is to be studied by Working Group 3 of the recently conceived Global Sedimentary Geology Program.

Organization of the book

We have divided this book philosophically into three parts. In the first part (Ch. 2, 3, 4), we provide a physical basis for the transport and deposition of sedi-

Figure 1.4 Sequential (2 s) time-lapse photographs showing development of soliton waves after reflection of an experimental turbidity current from a sloping ramp at the end of a flume tank. The reflected flow is moving from right to left. The triangular optical effect in the right-centre of each frame is a reflection of part of the flume. Large turbidity currents in restricted basins may show this behaviour.

ments in the deep oceans, followed by a comprehensive facies classification scheme, with examples, and a survey of the main controls on deep-marine sedimentation. The second part (Ch. 5, 6, 7, 8, 9) concerns discussions of the main depositional systems or environments in which siliciclastic sediments accumulate in the oceans, and is illustrated by both modern and ancient case studies to show their variability. The last part of the book (Ch. 10, 11, 12) focuses on deep-marine sediments at passive, convergent, and oblique-slip continental margins and plate boundaries, again illustrated by case studies based in part on recent ocean drilling programmes and marine seismic-reflection profiles.

DEEP-MARINE ENVIRONMENTS

PART 1

Facies, processes, sequences and controls

Lateral snout of carbonate debris flows, Cambro-Ordovician Cow Head Group, western Newfoundland. 5 cm scale marked. Measurement of snout geometry and the extent of boulder projection above the mean bed surface allow calculation of debris static strength (cf. James & Hiscott 1985).

CHAPTER 2

Sediment transport and deposition in the deep sea

Introduction

This chapter provides a quantitative background for the main processes of long-distance transport and deposition in the deep sea of (a) particulate sediments and (b) masses of semi-consolidated sediment (slides). The processes are referred to in Chapter 3 to explain deposition of a variety of deep-sea facies.

Particulate sediments, composed of mixtures of terrigenous and biogenic grains from clay size to boulder size, are transported beyond the shelf-slope break by three main mechanisms: (a) bottom-hugging sediment gravity flows, (b) thermohaline bottom currents, and (c) surface wind-driven currents or river plumes that carry suspended sediment off the shelf. Tides and surface waves appear to be only locally important as transport agents on the upper parts of slopes and in the heads of some submarine canyons (Ch. 6), and will not be discussed here. A theoretical treatment of the grain settling that results in deposition of some pelagic sediments, including particles dropped from floating icebergs at high latitudes (Anderson *et al.* 1979, Srivastava, Arthur *et al.* 1987), will be omitted because unhindered settling is not a process particular to the deep sea, and because the subject of settling is well treated in the literature (Baba & Komar 1981). In situations where settled particles become incorporated into advecting nepheloid layers, or are reworked after initial deposition, the important process is transport by density currents or thermohaline flows, both of which are covered in this chapter.

Deposition of particulate sediments occurs either a few grains at a time – bed load traction, settling from suspension, rolling down an incline – or *en masse* where decreasing shear stress becomes insufficient to overcome intergranular friction or cohesion of a collection of grains. A variety of depositional processes may leave their imprint on a deposit, but are not unique to specific transport mechanisms. For this reason, a clear distinction must be made between long-distance transport agents and local depositional mechanisms in explaining the origin of various deep-sea deposits (Fig. 2.1). For example, the end-member deposits of sediment gravity flows, as recognized by Middleton & Hampton (1973), include grain flow deposits; however, pure grain flows cannot move on the gentle slopes that characterize

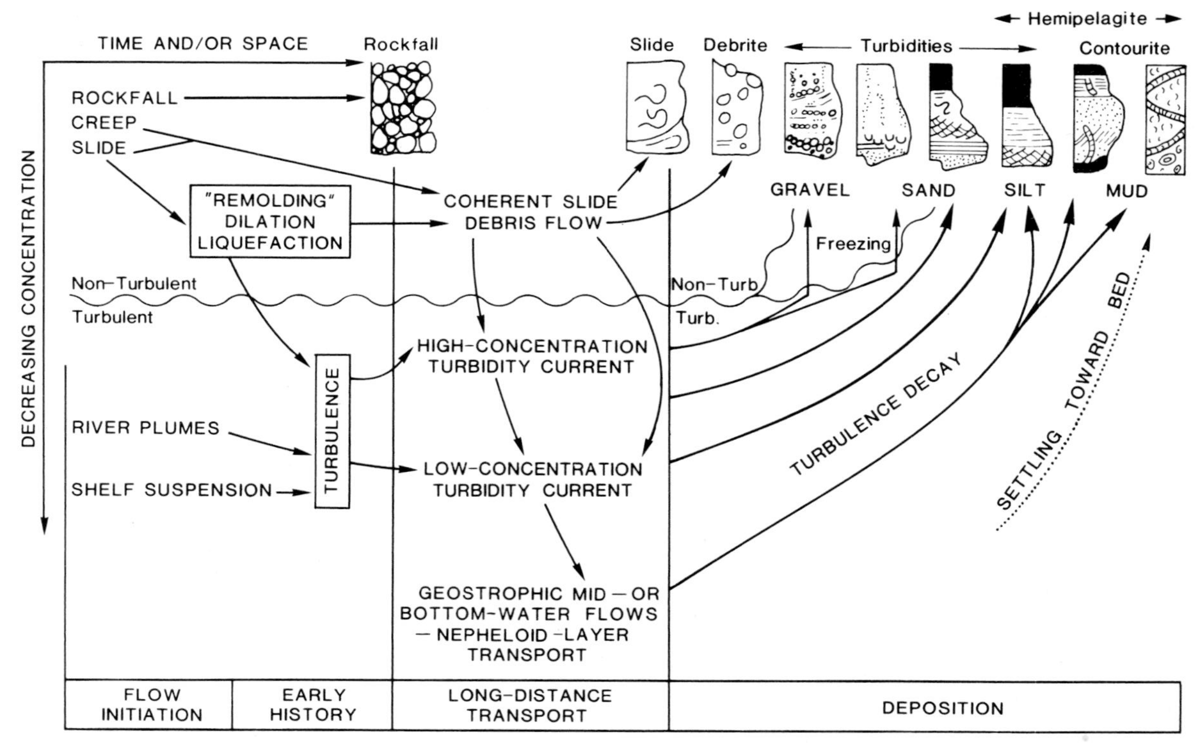

Figure 2.1 Summary of physical processes, involving flows of variable concentration, that are believed to be responsible for initiation, transport, and deposition of deep-sea facies shown as stylized beds in the upper right. Late-stage depositional processes are discussed in the text. Note that during long-distance transport, flows may be diluted, resulting in pronounced changes in suspension mechanism. Nepheloid layers obtain suspended load both from dilute turbulent flows and from settling of terrigenous and biogenic grains from the sea surface (aeolian and fine ice-rafted input, primary biological production). We are not suggesting that contourites form a subset of hemipelagites, but in ancient successions, it may be impossible to successfully distinguish silt/mud turbidites from hemipelagites and muddy contourites. Diagram modified from Walker (1978), Pickering *et al.* (1986).

ocean-basin margins (Middleton 1970, Middleton & Hampton 1973). Even on steeper slopes, the grain flow mechanism can only emplace sand layers a few centimetres thick (Lowe 1976a, Hein 1982). Thicker beds that have in the literature been referred to as 'grain flows' were probably deposited from decelerating high concentration turbidity currents, with transport in turbulent suspension followed by rapid settling from suspension, particle collisions and near collisions above the bed under continued shear (Bagnold 1956, Rees 1968) and *en masse* deposition due to intergranular friction, resulting in a structureless sand bed.

Most geologists working on land associate deep-sea sedimentation with turbidity currents and other sediment gravity flows. For this reason, we will begin with this familiar topic, and then proceed with a discussion of bottom currents, off-shelf transport of sediment suspended by river input or by storms, and finally slides.

Sediment gravity flows

Sediment gravity flows (Middleton & Hampton 1973) are mixtures of particles and water that move down slopes because the mixtures have a density greater than that of the ambient fluid, sea water. Gravity acts on the solid particles in the mixture, inducing downslope flow; the admixed water is a passive partner in this process. The flow will continue to move if the following conditions are satisfied: (a) the shear stress generated by the downslope gravity component acting on the excess density of the mixture exceeds frictional resistance to flow, and (b) the grains are inhibited from settling by one of several support mechanisms. These support mechanisms are listed in Table 2.1, along with their relative importance in supporting particles in the three flow types capable of transporting sediment on slopes of a few degrees or less.

The only sediment gravity flows capable of long-distance transport of particulate sediments into the deep sea on relatively gentle slopes (<10°) are debris flows and turbidity currents (Fig. 2.1). Liquefied flows (Lowe 1976b, Middleton & Southard 1977) can transport sands and silts on slopes of as little as a few degrees, but are probably ephemeral on these gentle slopes due to pore-fluid escape; on steeper basin-margin slopes, liquefied flows probably quickly accelerate and become turbulent (Middleton & Southard 1984). Debris flow deposits have been recorded on slopes less than 1° (Prior & Coleman 1982, Damuth & Flood 1984, Simm & Kidd 1984, Thornton 1984); the flows are capable of travelling for hundreds of kilometres from upper continental slopes to abyssal plains (Embley 1980). Turbidity currents can flow long distances on flat basin floors or even upslope (Hersey 1965a, Komar 1977, Elmore *et al.*1979, Stanley 1981, Pickering & Hiscott 1985). For example, Pleistocene turbidity currents carried distinctive coal fragments at least 1800 km from the eastern Canadian continental margin to the Sohm Abyssal Plain (Hacquebard *et al.* 1981), and Chough & Hesse (1976) suggest that turbidity currents flow for 4000 km in the Northwest Atlantic Mid-Ocean Channel.

The most popular classification scheme for sediment gravity flows is that of Middleton & Hampton (1973), most recently outlined by Middleton & Southard (1984). This scheme is based on the recognition of four process end members: turbidity currents, grain flows, debris flows, and liquefied flows (originally called fluidized flows). We prefer a more general classification based on the typical range of flow characteristics

Table 2.1 Support mechanisms for sediment gravity flows of significance in transport of marine sediment. 1 = turbulence, 2 = buoyancy, 3 = grain collisions (dispersive pressure), 4 = trapped or escaping pore fluids, 5 = frictional strength, and 6 = cohesive strength.

Flow type	Major suport	Minor support
Turbidity current	1	3,4 (high-concentrations)
Liquefied flow	4	3,1
Debris flow[a]	5,2,6,4,3	1 (very large flows)

[a]Debris flow mechanisms from Pierson (1981).

found in nature, and not restricted to strictly defined end members. For example, natural debris flows are characterized by a complex interaction of particle support mechanisms (Shultz 1984) (Table 2.1), one of which, dispersive pressure, is linked in end-member classification schemes to the process of grain flow. Lowe (1976a, 1982) calls all flows that are partly facilitated by dispersive pressure 'grain flows', in spite of field evidence and modelling that suggests that natural debris flows may owe much of their clast support to dispersive pressure (Takahashi 1981, Shultz 1984). Our classification is not as strict as that of Middleton & Hampton (1973) or Lowe (1982), and attempts to outline from the outset the complexity of flow and support mechanisms in the natural environment. We do not recognize separate categories of grain flow and liquefied flow, but instead indicate how pore fluid escape, elevated pore pressures, and dispersive pressure influence transport and deposition in turbidity currents and debris flows. We will begin our discussion with turbidity currents and their deposits.

Turbidity currents and turbidites

DEFINITION AND EQUATIONS OF FLOW

Turbidity currents are density currents (Simpson 1982) in which the denser fluid is a grain suspension, with particle support from upward velocity fluctuations associated with turbulent eddies (Bagnold 1966, Leeder 1983). Entrained sediment diffuses throughout the flow thickness. Depending on the manner in which the flow is initiated, the turbidity current may be (a) relatively short, quickly passing an observation point on the seafloor (surge-type flow), or (b) relatively long with steady discharge due to prolonged input from a long-lived source (steady- and uniform-type flow, generally river-fed). Modelling of deposition from surge-type flows is difficult because equilibrium velocities and sediment concentrations may never be attained. Experimental modelling of density surges by Laval *et al.* (1988) shows that: (a) the velocity of the surge is effectively proportional to the square root of the initial volume; (b) the surge velocity increases with increasing initial density of the flow, proportionally to the square root of the ratio of the excess density to ambient fluid density, and (c) surge speed increases again with an increase of slope. For continuously-fed flows or unusually large-scale surge-type flows on a constant slope, a condition of effective equilibrium may exist in the long body of the flow, in which the finest grain sizes have an essentially uniform concentration throughout the flow, whereas the coarsest grain sizes are most concentrated near the base of the flow (Rouse 1937). The surface of maximum velocity is below the top of the flow but in its upper part (Middleton & Southard 1984). Concentration of the coarsest load near the base of the body of the current, where time-averaged velocities are lowest, results in more rapid movement of the fine size fractions relative to the coarser fractions.

As shown by Middleton (1966b), the velocity of the body of a turbidity current, U_B, is given by a Chezy-type equation,

$$U_B^2 = \left[\frac{8g}{(f_0 + f_i)}\right]\left[\frac{\Delta\rho}{(\rho + \Delta\rho)}\right] d_B \tan\alpha \quad (2.1)$$

where $\Delta\rho$ = density difference between the flow and seawater, ρ = density of seawater, d_B = body thickness, α = bottom slope in degrees, f_0 = dimensionless Darcy-Weisbach friction coefficient for bed friction, and f_i = dimensionless friction coefficient for interfacial friction at the top of the flow. Friction coefficients have been determined empirically for rivers and in flumes, but turbidity currents differ from rivers in that there is significant friction between the flow and the overlying water. According to Middleton & Southard (1984), for large natural turbidity currents $f_o + f_i$ is likely to be about 0.01. For subcritical flow conditions (Froude number, $F < 1.0$), $f_0 >> f_i$, but for supercritical flows ($F > 1.0$), $f_i > f_o$ due to intense mixing at the upper interface of the flow.

The velocity of the head of the current, U_H, does not appear to depend significantly on bottom slope for turbidity currents moving over low slopes; Middleton (1966a) gives:

$$U_H^2 = 0.56\, g d_H \left[\frac{\Delta\rho}{(\rho + \Delta\rho)}\right] \quad (2.2)$$

On steeper slopes, from perhaps 2–10°, a more general result (Hay 1983) includes a dependence of head velocity on bottom slope:

$$U_H^2 = g d_H \left[\frac{\Delta\rho}{(\rho + \Delta\rho)}\right](0.50\cos\alpha + t\sin\alpha) \quad (2.3)$$

Experiments on small surges suggest that the empirical factor t is in the range 1.6 to 4.0 (Hay 1983). Note that for α = 1° and t = 3.3, Equation 2.3 reduces to Equation 2.2, but for α = 10°, the numerical constant is 1.1, not 0.56. The ratio of the head velocity to the body velocity (U_H/U_B) is approximately 1.0 on gentle slopes but is less than 1.0 on steeper slopes (Fig. 2.2). The head does not grow unchecked as suspension from the body of the current overtakes it. Instead, an equilibrium is developed. The rate of body flow into the head region is balanced by loss of suspension from a region of intense turbulence and flow separation at the back of the head (Simpson 1982, Fig. 2.3). This ejected material settles back into the flow top according to fall velocity, with the coarsest grains returning to the body nearest the head and the finest grains returning far behind the head. The result is lateral grading in the flow from a coarser suspension near the head to a finer suspension near the tail (Walker 1965). The finer sizes are recirculated at a faster rate, and therefore achieve a greater degree of lateral size sorting.

CONCENTRATION OF TURBIDITY CURRENTS

Flow concentration, commonly expressed as the volume fraction of solids in the suspension, C, probably depends largely on the nature of the process that generated the turbidity current. Unfortunately, generating mechanisms are still poorly understood, although several have been suggested: (a) direct deltaic input of sediment suspensions (Heezen *et al.* 1964); (b) spontaneous or cyclic shock-induced liquefaction (Seed & Lee 1966, Andresen & Bjerrum 1967) with transition to turbulence, perhaps as a result of movements of sediment slides or debris flows (Stanley 1982); (c) erosion of the front of a moving debris flow (Hampton 1972);

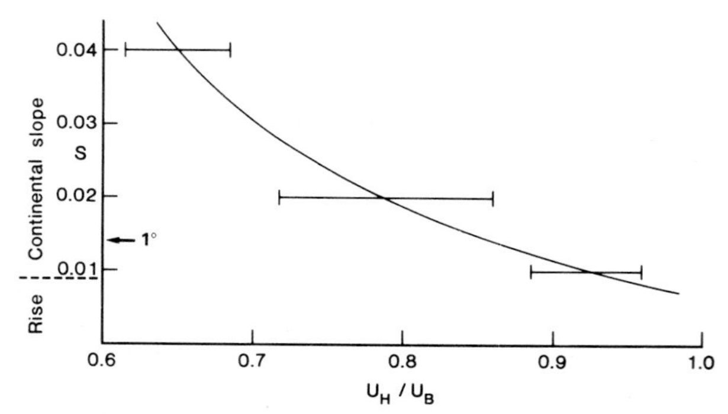

Figure 2.2 Ratio of head velocity to body velocity (U_H/U_B) plotted against bottom slope for experimental turbidity currents (5 m flume) of Middleton (1966a). Horizontal bars show extent of scatter in data. Ranges of bottom slope for continental margins are superimposed.

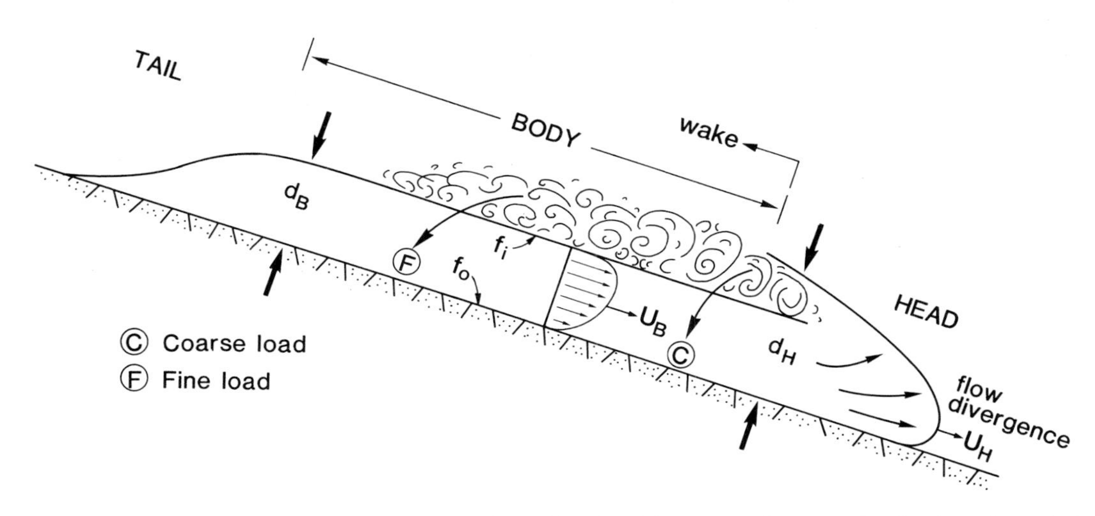

Figure 2.3 Simplified sketch of a large turbidity current, divided into head, body and tail regions. Settling from the wake behind the head produces a lateral size grading in the flow.

(d) dilution of a large, mobile sediment slide (Ricci Lucchi 1975b, Cita *et al.* 1984, Pickering & Hiscott 1985); (e) suspension of sediments in canyon heads by edge waves associated with storms (Inman *et al.* 1976, Fukushima *et al.* 1985); and (f) thickening and dilution of a debris flow as it undergoes an hydraulic jump (Weirich 1988).

Most workers recognize high-concentration turbidity currents and low-concentration flows, realizing that in nature there are all gradations in between. Middleton (1970), based on the work of Kuenen (1966), suggests that the boundary between high and low concentration be placed at a density of 1.1 g/cm^3 ($C = 0.06$), although there is no consensus on this value. Concentration has a bearing on flow velocity (Eqns 2.1–2.3), and therefore competence on basin margin slopes. It is not necessarily true, however, that high-concentration flows have a coarse-grained load and that low-concentration flows have a fine-grained load. Low concentrations of coarse material can be transported on steep slopes, whereas high concentrations of fine material can be transported on both steep and gentle slopes. On gentle slopes, coarse sizes can only be transported by high-concentration flows (Lowe 1982). Concentration influences flow density and viscosity, but has no direct effect on the mechanics of flow until particle concentrations become so high that (a) interparticle collisions become an important component of grain support, or (b) turbulence, particularly near the bed, becomes damped.

SUPERCRITICAL FLOW OF TURBIDITY CURRENTS

A fundamental distinction can be made between turbidity currents that are subcritical (Froude Number, $F < 1.0$) and those that are supercritical ($F > 1.0$). For a reasonable friction factor, $f = 0.02$, Komar (1971) concluded that turbidity currents would be supercritical on slopes $>0.5°$, a value exceeded on many basin-margin slopes and on the upper parts of submarine fans (Table 2.2). The transition to subcritical flow occurs as

Table 2.2 Typical gradients of submarine fans and basin-margin slopes.

Location	Bottom gradient in degrees	Sources
Upper continental slope	3–6	Heezen *et al.* 1959
Lower continental slope	1.5–3	Heezen *et al.* 1959
Continental rise	0.1–1	Heezen *et al.* 1959
Submarine canyons	1–3	Nelson & Kulm 1973
Upper fan channels	0.2–0.5	Barnes & Normark 1984
Middle fan lobes	0.1–0.4	Barnes & Normark 1984
Lower fan	0.1–0.2	Barnes & Normark 1984
Carbonate bank slopes	4–40	Mullins & Neumann 1979

bottom gradient declines, and may involve an hydraulic jump with intense turbulence and flow homogenization (Middleton 1970, Komar 1971).

AUTOSUSPENSION IN TURBIDITY CURRENTS

It is frequently claimed that turbidity currents can effectively carry sand-sized detritus across basin-margin slopes without leaving a deposit: there may even be net erosion in these areas. This 'bypassing' requires that the turbidity current be at the least self-sustaining on the slope, or while flowing through slope channels. This process of 'self maintenance' (Southard & Mackintosh 1981) was named *autosuspension* by Bagnold (1962), and is summarized in Figure 2.4. In words, the excess density of the grain suspension, combined with the downslope component of gravitational acceleration, induces basinward flow. The turbulence generated by the flow maintains the grains in suspension (Middleton 1976, Leeder 1983), and the suspension maintains its density contrast with the overlying seawater, allowing continued flow, turbulence generation, and effective grain suspension. According to Middleton (1966c), Allen (1982, II, p. 399) and Pantin (1979), true autosuspension is probably not common in nature, except for thick turbidity currents carrying fine particles on steep slopes. For all other cases, some of the suspended load settles through the flow and is deposited.

Southard & Mackintosh (1981) and Middleton & Southard (1984) claim that Bagnold's (1962) mathematical formulation of the autosuspension criterion fails to find support in experiments. This shortcoming is ascribed to flaws in Bagnold's (1962) energy-balance equation, resulting in a 'fallacious system of energy bookkeeping' (Paola & Southard 1983). As outlined by Middleton & Southard (1984), Bagnold's (1962) equations do not account for the fact that only a small percentage of a flow's power, about 2%, is available to suspend sediment, the rest being expended to overcome frictional resistance at flow boundaries; and to produce turbulence and heat, most of which does not contribute to grain suspension. This overestimation of available power by Bagnold (1962) was remedied by Pantin (1979), who introduced an efficiency factor, e, into the formulation of an autosuspension criterion. This criterion is:

$$e \alpha U_s > w \tag{2.4}$$

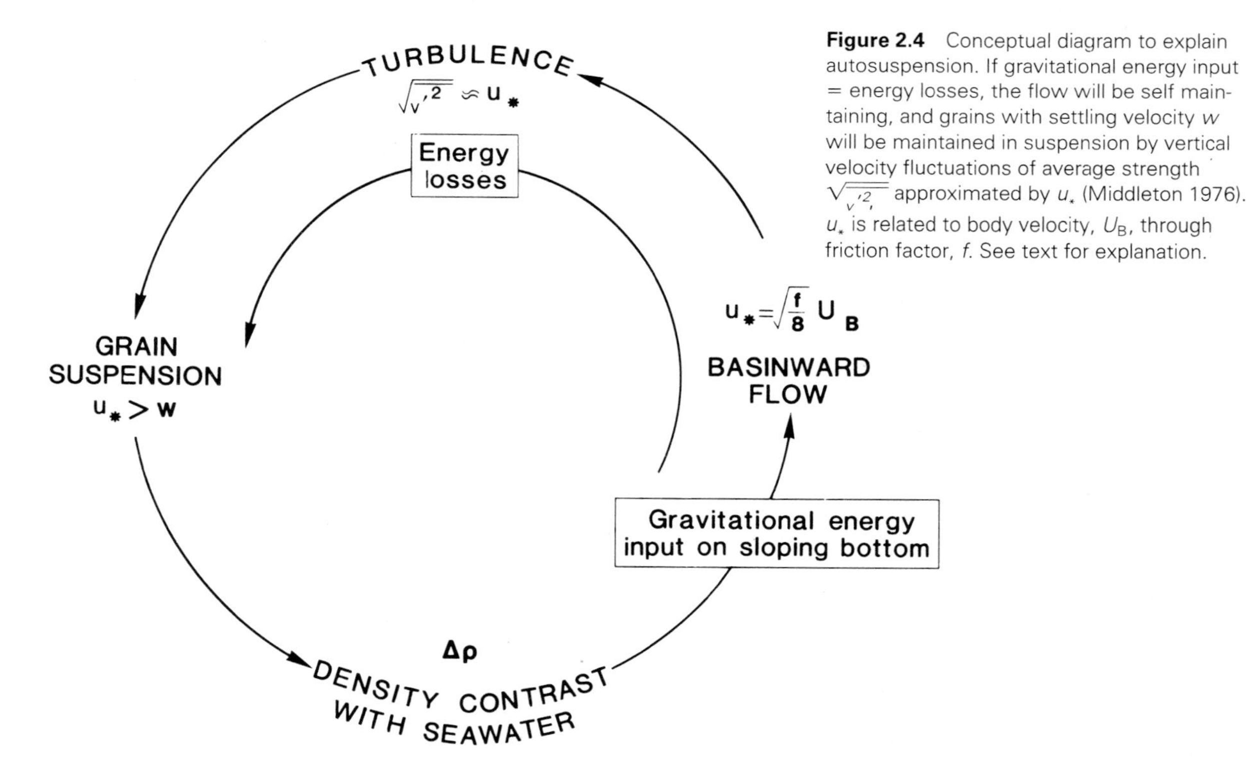

Figure 2.4 Conceptual diagram to explain autosuspension. If gravitational energy input = energy losses, the flow will be self maintaining, and grains with settling velocity w will be maintained in suspension by vertical velocity fluctuations of average strength $\sqrt{v'^2}$ approximated by u_* (Middleton 1976). u_* is related to body velocity, U_B, through friction factor, f. See text for explanation.

where α = slope angle, U_s = transport velocity of the suspended sediment, and w = grain settling velocity. In a worked example, Pantin (1979) sets $e = 0.01$. In general, Pantin (1979) shows that flow density is the main control on whether autosuspension will occur. Below a critical density, which varies with slope, flow thickness, grain size and drag coefficient, a turbidity current will 'subside' and deposit its sediment. Above the critical density, the flow will 'explode' – increase in density and velocity – and will achieve autosuspension. This latter condition is only predicted for sediment finer than fine sand. Application of the theory of Pantin (1979) and Parker (1982) to conditions for initiation of erosive turbidity currents in Scripps Submarine Canyon, California, suggests that 'ignition' of turbidity currents will occur for initial downchannel velocities in excess of about 0.5 m/s (Fukushima *et al.* 1985), in general agreement with field observations on the conditions required for initiation of flows in this canyon by Inman *et al.* (1976).

NATURAL TURBIDITY CURRENTS

Natural marine turbidity currents have been studied only superficially because of the inherent difficulties in devising and deploying monitoring systems. Man-made turbidity currents formed by the dumping of mine tailings have been studied by Normark & Dickson (1976) and Hay (1987a & b, see also Hay *et al.* 1982). Nascent turbidity currents in Scripps Submarine Canyon have been monitored up to the point at which current meters were lost, giving minimum flow speeds of 1.9 m/s (Inman *et al.* 1976). Larger natural flows have only been studied indirectly, by examining records of submarine cable breaks (Heezen & Ewing 1952, Heezen *et al.* 1964, Gennesseaux *et al.* 1980), variation in levee heights of submarine channels at channel bends (Komar 1969), and the height of flow deposits and deposit characteristics (Bowen *et al.* 1984). Quantitative results of some of these studies are summarized in Table 2.3.

The best documented large natural flow was generated by a 7.2 magnitude earthquake in the Laurentian Channel, eastern Canada, in 1929. The so-called '1929 Grand Banks turbidity current' broke a succession of submarine cables, apparently travelled as fast as about 20 m/s, and deposited a graded fine sand to silt layer about 1 m thick in a water depth of about 5200 m (Heezen & Ewing 1952, Heezen *et al.* 1954, Fruth 1965). According to Piper *et al.* (1985), the turbidity current initially suspended gravel 2 cm in diameter, deposited the gravel in water depths of 1600–4500 m, and subsequently reworked the gravel into large bedforms that probably are internally cross-stratified.

Natural turbulent density currents are not restricted to the marine environment, but also occur in lakes, in the atmosphere (Simpson 1982), and as powder snow avalanches (Hopfinger 1983).

TURBIDITES

Deposits of turbidity currents are called *turbidites*. The character of these deposits depends as much on the mode of deposition as on the long-distance transport mechanism (Fig. 2.1). Depositional process is mainly a function of (a) flow concentration, or density, and (b) rate of deposition.

Rate of deposition is dependent on the rate of decrease

Table 2.3 Data from natural turbidity currents.

Locality	Primary data	U (m/s)	D (mm)	d (m)	Sources
Scripps Canyon	Current meters	1.9	0.15	3–10	Inman *et al.* 1976
Monterey Fan	Levee heights	4–20	0.1–0.2	100	Komar 1969
Navy Fan	Grain size, cross-flow slope	0.1–0.8	0.5	20–70	Bowen *et al.* 1984
Laurentian Fan	Cable breaks	20	20	<300	Heezen & Ewing 1952, Piper *et al.* 1985
Laurentian Fan	Grain size	0.1–0.15	0.04–0.1	300–800	Stow & Bowen 1980

of both flow *competence* – the coarsest particle that can be transported, and flow *capacity* – the sediment discharge, in volume or mass per unit time, integrated over the cross-section of the flow. Competence and capacity both decrease as mean velocity decreases. Velocity may decrease for any of the following reasons: (a) decreasing bottom slope, (b) flow divergence, (c) increased bed friction, (d) increased particle interaction (intergranular friction), (e) decreasing flow density due to deposition, or (f) deflection of slow, mud-rich flows by contour currents or by the Coriolis effect so that they are constrained to move roughly parallel to the slope contours rather than down the slope (Hill 1984a). Van Andel & Komar (1969) provide mathematical expressions for the momentum losses in a turbidity current due to grain-to-grain friction, bottom and interfacial friction, and sediment loss due to deposition. It is noteworthy that flows with a high proportion of suspended mud can flow for a greater distance than mud-poor flows without suffering a crippling degree of sediment loss, even on quite gentle slopes. Such flows are therefore more 'efficient' in moving both their mud- and sand-size loads into the basin. The relative 'efficiencies' of turbidity currents have been used by some workers to characterize types of submarine fans, a discussion of which appears in Chapter 7.

Experimental monitoring of deposition from turbidity currents has only rarely been attempted. These laboratory studies are inadequate analogues for deposition from much larger natural flows, but provide some insight into depositional processes. The most well known experiments are those of Middleton (1967), who monitored deposition of graded beds from both low- and high-concentration flows. For the high-concentration flows, deposition was rapid, *en masse*, and resulted in formation of a 'quick bed' that was easily deformed by Helmholtz waves at the top of the deposit. The deposit was characterized by coarse-tail grading. Low-concentration flows deposited beds with good distribution grading; because of the nature of the experiments, no traction transport took place, and no lamination nor bedforms were generated.

Lüthi (1981) studied deposition from more dilute, unconfined flows that were free to expand laterally after leaving a narrow entry slot. With increasing distance from the entry slot, velocity and grain size both decreased, and sedimentary structures changed from parallel lamination, to climbing-ripple lamination, to fine parallel lamination. These structures are the same as those that are superimposed in vertical sections through ancient turbidites, and correspond respectively to Bouma (1962) divisions T_b, T_c and T_d. The structureless or graded division that corresponds to Bouma division T_a is probably more akin to the deposits of high-concentration flows produced by Middleton (1967).

By far the most detailed observations of the internal characteristics of turbidites come from study of ancient sediments. The Bouma (1962) sequence (Fig. 2.5) represents a summary of the transitions observed in 1061 beds in the Grès de Peïra-Cava in the Maritime Alps of southern France. The sequence was soon interpreted in relation to the progression of bedforms observed during deceleration of flow in flumes (Harms & Fahnestock 1965, Walker 1965), but with omission of megaripples. The Bouma sequence is a good model for medium-grained turbidites that result from deceleration of low-concentration turbidity currents, but is inappropriate for either coarse beds emplaced by high-concentration flows, and too general for fine-grained turbidites that consist entirely of Bouma's T_d and T_e divisions.

CROSS STRATIFICATION IN BOUMA-TYPE TURBIDITES

The general absence of megaripple-scale cross stratification in Bouma-type turbidites can be explained in at least four ways. Walker (1965) suggested that deposition from turbidity currents is too rapid for bedforms to equilibrate with the decelerating flow. The result is that megaripples, which require a significant time to develop because of their size, only begin to form by the time that velocity decreases to values consistent with ripple, not megaripple, stability. Allen (1969) used a similar argument, i.e. for grain sizes available in most turbidity currents 'the range of flow power appropriate to these forms was traversed too quickly, being narrow, to permit their growth' (Allen 1982, II, p. 414).

As an alternative, Walker (1965) suggested that many turbidity currents may be too thin to allow formation of megaripples, which only form if the ratio of flow depth to bedform height is about 5 : 1. For natural flows with velocity maxima below the flow top, and perhaps with strong internal density stratification, Walker (1965) suggested that flow thicknesses of 5–10 m might be necessary for the growth of megaripples 50 cm high. A third possible explanation for lack of megaripple-scale cross stratification is that for the appropriate flow conditions, most turbidity currents are depositing sediment too fine for production of megaripples (Walton 1967, Allen 1982, II, p. 414). Experiments show that megaripples are absent in sediment finer than 0.1 mm (Allen 1982, I, p. 339). This explanation gains strong support from the observation

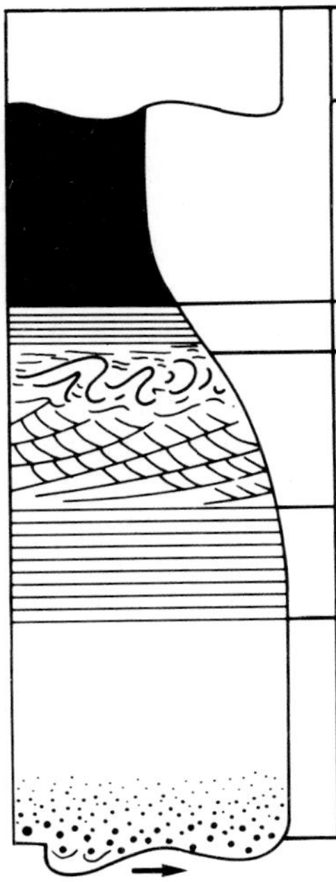

GRAIN SIZE	BOUMA (1962) DIVISIONS	INTERPRETATION
Mud	E Laminated to homogeneous mud	Deposition from low-density tail of turbidity current ± settling of pelagic or hemipelagic particles
Silt	D Upper mud/silt laminae	Shear sorting of grains & flocs
Sand	C Ripples, climbing ripples, wavy or convolute laminae	Lower part of lower flow regime of Simons *et al* (1965)
	B Plane laminae	Upper flow regime plane bed
Coarse Sand	A Structureless or graded sand to granule	Rapid deposition with no traction transport, possible quick (liquefied) bed

Figure 2.5 Ideal sequence of sedimentary structures in a turbidite bed (after Bouma 1962, with interpretation after Harms & Fahnestock 1965, Walker 1965, Middleton 1967, Walton 1967, Stow & Bowen 1980).

that coarse-grained bioclastic turbidites differ from generally finer grained terrigenous turbidites in that they commonly contain megaripple cross stratification in association with the Bouma T_b division (Hubert 1966a, Thompson & Thomasson 1969, Allen 1970).

Finally, cross stratification may be absent due to the occurrence of an hydraulic jump between deposition of the Bouma T_b and T_c divisions. Although upper flow regime parallel lamination does not necessarily indicate supercritical flow, it may do so. Even long-wavelength antidune lamination (Hand *et al.* 1972) may be so subdued as to appear flat. Downstream of an hydraulic jump, flow depth increases and velocity decreases sharply. This rapid velocity drop could cause complete omission of the megaripple stability field. Upstream migration of the hydraulic jump would result in superposition of ripples on a previously flat sediment bed. This mechanism is restricted to submarine slopes or the upper parts of submarine fans, where hydraulic jumps are believed to occur (Komar 1971).

There are no data on the mechanics of deposition from large natural turbidity currents, but it is not improbable that all four explanations for lack of megaripples in turbidites are valid in particular cases. In most cases, we favour the grain-size argument.

TURBIDITES FROM HIGH-CONCENTRATION FLOWS

High-concentration flows carrying gravel or coarse sand leave deposits quite different from those described by Bouma (1962). Important depositional processes are: (a) rapid *en masse* deposition of a quick bed (Middleton 1967) due to an increase of intergranular friction; (b) generation of inverse grading at the base of the flow because of grain collisions and dispersive pressure (Bagnold 1956), or because of a lag between velocity of the head and transport velocity of the coarse part of the size distribution (Hand & Ellison 1985); (c) formation and sequential deposition by 'freezing' of traction carpets driven along by shear from the overriding flow (Carter 1975, Hiscott & Middleton 1979, 1980, Hein 1982); (d) grain-by-grain deposition with little subsequent traction transport (Middleton 1967); (e) formation of irregular stratification by alternate deposition of bedload and suspended load (Walker 1975); and (f) syn- or post-depositional expulsion of pore fluids when unstable, open-grain packings collapse, generating dish and pillar structures (Lowe & LoPiccolo 1974).

The term 'traction carpet' (Dzułynski & Sanders 1962, p. 88) has gained acceptance (Lowe 1982), although it is strictly a misnomer. The clasts in these carpets are not under true tractional bedload transport, but are sheared as a dense dispersion just above the bed. The dispersion is maintained by dispersive pressure (Bagnold 1956), which varies directly with the applied shear stress. The relatively high shear stress needed to transport gravel generates strong dispersive pressure, capable of maintaining a relatively thick traction carpet (Lowe 1982). Generally, only one traction carpet is generated in gravel, giving an inversely graded basal division to the deposit, which may be overlain by normally graded gravel and pebbly sand, or by stratified gravel and sand (Walker 1975). In fine pebbly sands and coarse to medium sands, traction carpets are thinner, probably due to less effective dispersive pressure (Hein 1982, Lowe 1982) and weaker shear stress at the top of the carpet. During sequential 'freezing' to the bed of several traction carpets, the maximum grain size and thickness of each unit decrease upward through the deposit (Hiscott & Middleton 1979). Within the deposit of each traction carpet (Fig. 2.6), intense grain interaction produces both a basal inverse grading, and a strong alignment of grains with imbrication angles commonly >20° (Hiscott & Middleton 1979).

Sedimentary structure sequences for the deposits of high-concentration turbidity currents have been presented by Aalto (1976), Hiscott & Middleton (1979), Hein (1982) and Lowe (1982). The analysis of Hiscott & Middleton (1979) is based on a transition matrix (9 states, Table 2.4) for 214 medium- to coarse-grained sandstone beds (mean thickness 380 cm) from inferred submarine-fan deposits (Hiscott 1980) of the Tourelle Formation of Quebec, Canada. The preferred transitions, inferred from Markov Chain Analysis, are presented in Figure 2.7 with process interpretations. These transitions differ somewhat from those presented by Hiscott & Middleton (1979), due to a change in the method of calculation of the independent trials matrix (modified procedure includes iterative fitting of row and column totals; Powers & Easterling 1982).

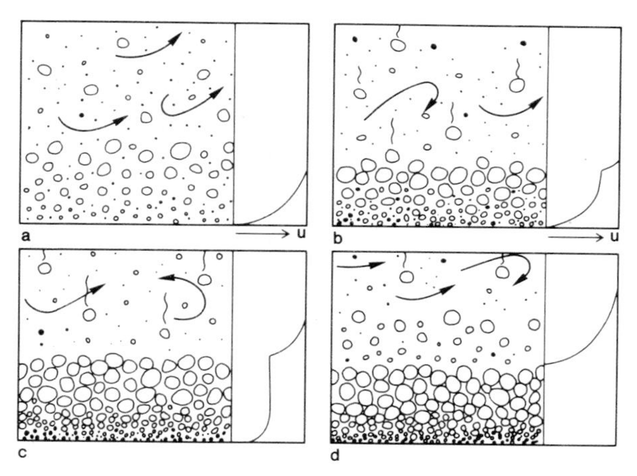

Figure 2.6 Origin of traction carpet layers, with sketches from Lowe (1982) and explanation based on Hiscott & Middleton (1979, 1980). (a) Grain interaction at the base of a high-concentration turbidity current produces inverse grading in the flow. (b) Continued fallout from suspension produces a dense traction carpet that is supported by dispersive pressure and sheared by the main body of the flow above. (c) The upper part of the traction carpet freezes (vertical segment in the velocity profile), with shear being concentrated at the base of the carpet. (d) The entire sheared layer ceases to move and becomes welded to the bed. Repetition of this process forms a division of inversely-graded stratification.

Except for some basal scour and fill structures, the first deposits of the high-concentration flows described by Hiscott & Middleton (1979) accumulated by rapid mass deposition, producing either a division of structureless coarse- to medium-grained sand, or, in the case of intense grain interaction and formation of traction carpets, a division of widely-spaced stratification. There is no evidence of grain-by-grain traction on the bed during the early stages of deposition. At the top of the deposit, deceleration of the more dilute tail of the current commonly produced a sequence of structures like the Bouma sequence. In other examples, the dilute tail of the current reworked the sandy top of the initial deposit into large ripples or megaripples, and then continued down the fan surface to deposit its fine load at more distal sites (cf. Lowe 1982).

Hein (1982) documents sequences of sedimentary structures in channelized conglomerates, pebbly sandstones and coarse sandstones that are similar to those recognized by Walker (1975) for conglomerates

Table 2.4 Transition matrix for 214 thick beds, Tourelle Formation.

State	Number of upward transitions 1	2	3	4	5	6	7	8	9
1 Basal scour	0	155	20	2	0	0	0	0	37
2 Massive or graded	135	0	10	13	21	10	9	12	23
3 Internal scours	4	27	0	0	0	0	0	0	1
4 Parallel lamination	7	1	0	0	5	1	2	0	0
5 Ripple lamination	25	0	0	0	0	0	0	3	0
6 Convolution	12	0	0	0	0	0	1	0	0
7 Crossbedding	9	0	0	2	1	0	0	0	0
8 Muddy lamination	16	0	0	0	0	0	0	0	0
9 Traction carpets	6	50	2	0	1	2	0	1	0

and Hiscott & Middleton (1979) for sandstones. Important additional divisions recognized by Hein (1982) are those bearing the imprint of syn- and post-depositional fluid escape; dish structures are particularly abundant in these rocks. Also, the pebbly sandstones of Hein (1982) contain more cross stratification than that observed in finer deposits by Hiscott & Middleton (1979). Hein (1982) interprets some of this cross stratification as the product of reworking of the tops of the deposits by dilute flows that spilled into the submarine channel from other, nearby channels (see Ch. 7, flow stripping process of Piper & Normark 1983).

A detailed study of turbidity current deposits of mud, from Madeira Abyssal Plain, has led McCave & Jones (1988) to propose a model for these deposits up to 10 m thick as due to deposition of ungraded mud from high-concentration (~ 50–100 kg m^{-3}), non-turbulent, turbidity currents. McCave & Jones (1988) believe that the turbidity currents were transformed from fully turbulent flow to a decelerating flow with suppressed turbulence and increasing density. With turbulence severely damped, the dense suspension of fine material consolidates like a static suspension while continuing to flow downslope with inter-particle forces preventing the differential settling of coarser grains. An ungraded mud turbidite results.

ANTIDUNES IN TURBIDITES

It is unclear whether supercritical turbidity currents ($F > 1.0$) deposit any diagnostic sedimentary structures. In rivers and flumes, supercritical flows mould antidunes on the bed. These antidunes may migrate either upcurrent or downcurrent to produce internal lamination with low angles of dip (Middleton 1965). Skipper (1971) and Skipper & Bhattacharjee (1978) described crossbed sets with wavy upper profiles at

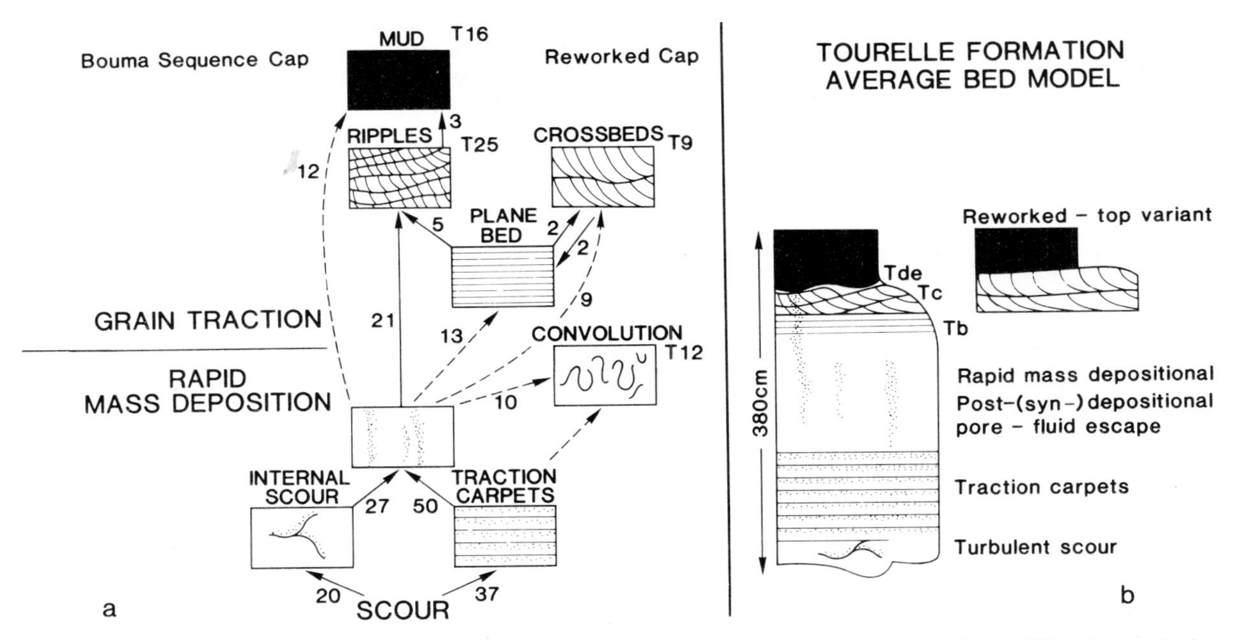

Figure 2.7 (a) Preferred transitions between nine 'states' in 214 thick sandstone beds (average thickness 380 cm) deposited from high-concentration turbidity currents, Tourelle Formation, Quebec, Canada. Numbers indicate the number of transitions between states; T25 signifies that a rippled division is the highest division in 25 beds. Complete beds (top preserved) only occur 62 times – the other 152 beds had their tops bevelled during amalgamation. Structural divisions were produced either by rapid mass deposition ('freezing') or by grain-by-grain deposition with traction transport. In some cases, post-depositional reworking of the top of the bed is indicated by medium-scale crossbedding. Arrow weight indicates the statistical significance of transitions between states (based on method of Powers & Easterling 1982, p. 922), with the levels of significance being > 98% (solid arrow), and > 93% (dashed arrow). No arrows are shown in cases for which there is only one transition between two states, regardless of significance level. (b) Generalized bed model based on the transition diagram, with interpretation for those parts of the bed emplaced by rapid mass deposition. Where appropriate, structural divisions are labelled with short notation for Bouma (1962) divisions (Fig. 2.5).

the base of some thick turbidites in the Ordovician Cloridorme Formation of Quebec, Canada, and interpreted these bedforms as short-wavelength antidunes (wavelength ≈ 65 cm), based primarily on the observation that the foreset dip directions opposed the local palaeoflow as deduced from flutes. The hydrodynamic interpretation of these bedforms proved to be difficult (Hand *et al.* 1972), because antidune wavelength beneath turbidity currents should be in the order of 12 times the flow depth. Recently, Pickering & Hiscott (1985) have re-interpreted this part of the Cloridorme Formation as a basin-floor sequence deposited in a constricted foreland basin, in which large turbidity currents were repeatedly reflected and deflected from marginal slopes. Many individual graded beds have sole markings and divisions of cross stratification or ripple lamination indicating flow reversals during deposition. Grain fabric data obtained by Pickering & Hiscott (1985) from the bedforms described by Skipper (1971) indicate that the depositing current flowed in a direction opposite to that indicated by most flutes in the sequence. These bedforms are therefore not antidunes, but instead record the migration of megaripples under subcritical flow conditions (Fig. 2.8).

Given that no other convincing examples of antidune lamination in turbidites have been described, we suspect that such structures are either exceedingly rare or are not produced by natural flows. Alternatively, lamination produced by long-wavelength, low-amplitude antidunes may appear flat, given the theoretical relationship between wavelength and flow depth. Indeed, some exceedingly large mud waves on the levees of the Monterey Fan (wave height 2–37 m, wavelength 0.3–2.1 km) have been interpreted as antidunes formed beneath low-concentration, low-velocity turbidity currents ($U \approx 10$ cm/s) approximately 100–800 m thick (Normark *et al.* 1980). The layering in these mud waves would appear flat in both deep-sea cores and outcrops.

TURBIDITES FROM LOW-CONCENTRATION FLOWS

The deposits of low-concentration turbidity currents transporting fine-grained sediment – silt and clay – have only recently been described in detail (Piper 1978, Stow 1979, Stow & Bowen 1980, Stow & Shanmugam 1980, Stow & Piper 1984). The Bouma (1962) divisions for sandy turbidites are too general for mud turbidites;

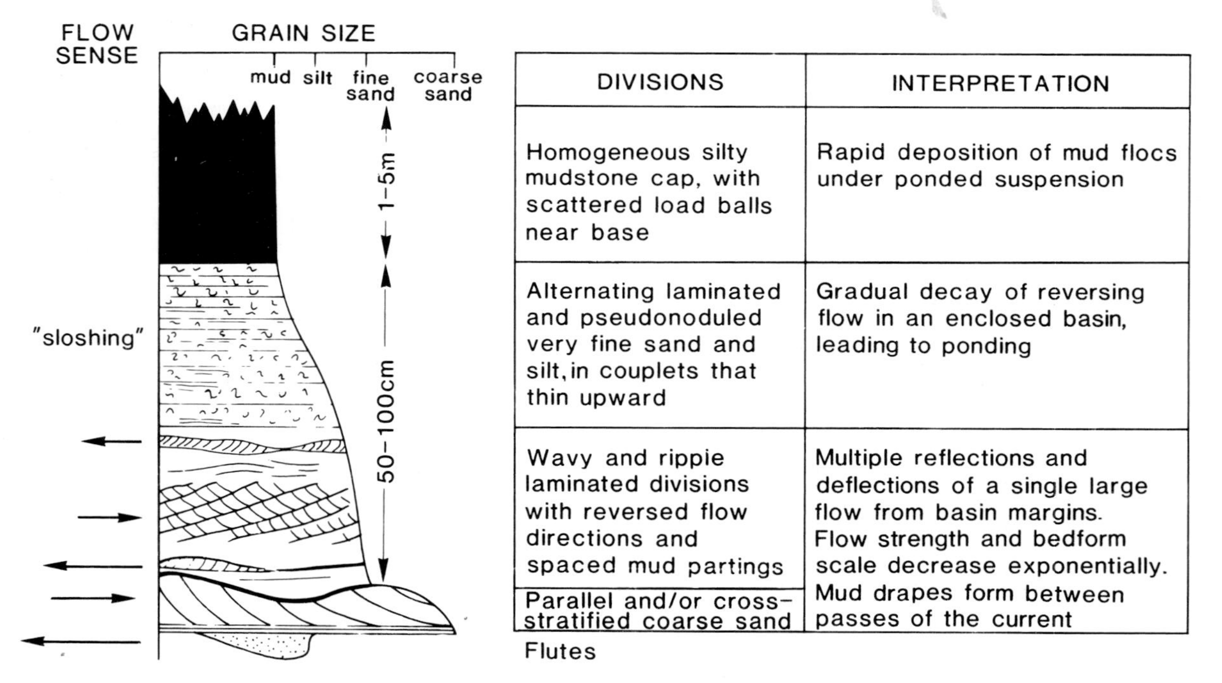

DIVISIONS	INTERPRETATION
Homogeneous silty mudstone cap, with scattered load balls near base	Rapid deposition of mud flocs under ponded suspension
Alternating laminated and pseudonoduled very fine sand and silt, in couplets that thin upward	Gradual decay of reversing flow in an enclosed basin, leading to ponding
Wavy and ripple laminated divisions with reversed flow directions and spaced mud partings	Multiple reflections and deflections of a single large flow from basin margins. Flow strength and bedform scale decrease exponentially. Mud drapes form between passes of the current
Parallel and/or cross-stratified coarse sand	

Figure 2.8 Generalized sequence of internal structures produced beneath large reflected turbidity currents (Pickering & Hiscott 1985). The thick mud cap is deposited after the flow becomes ponded within a confined basin. Single beds of this type may be in excess of 10 m thick.

many beds contain only Bouma's T_d and T_e divisions. Several schemes to further subdivide mud turbidites have been proposed, and are outlined in Figure 2.9. Stow & Shanmugam (1980) recognize nine divisions, numbered T_0 to T_8. The physical structures in these nine divisions are believed to result from suspension fall-out and traction (T_0–T_2), shear sorting of silt grains and clay flocs in the bottom boundary layer (T_3–T_5), and suspension fall-out with no traction (T_6–T_8). Division T_0 corresponds to Bouma's division T_c. As with Bouma-type turbidites, complete structure sequences are unusual; top-absent, base-absent and middle-absent sequences are common (Stow & Piper 1984). The very thin, regular laminations of division T_3 have been attributed to shear sorting near the base of the flow and then alternating deposition of silt grains and clay particles by settling through the viscous sublayer (Stow & Bowen 1980, Kranck 1984). According to this model, as the silt grains and clay flocs fall toward the bed, the increased shear in the boundary layer causes the clay flocs to break up (Fig. 2.10a). The silt grains then settle through the viscous sublayer to form a silt lamina (Fig. 2.10b). As more sediment is supplied to the top of the boundary layer, the mud concentration builds up, and some reflocculation may occur (Fig. 2.10c). At some critical concentration, the clays are able to form sufficiently large aggregates that they escape disaggregation by shear stresses and are able to settle rapidly through the viscous sublayer to form a mud lamina (Fig. 2.10d). The cycle of silt and mud deposition is then repeated for successively finer grain sizes.

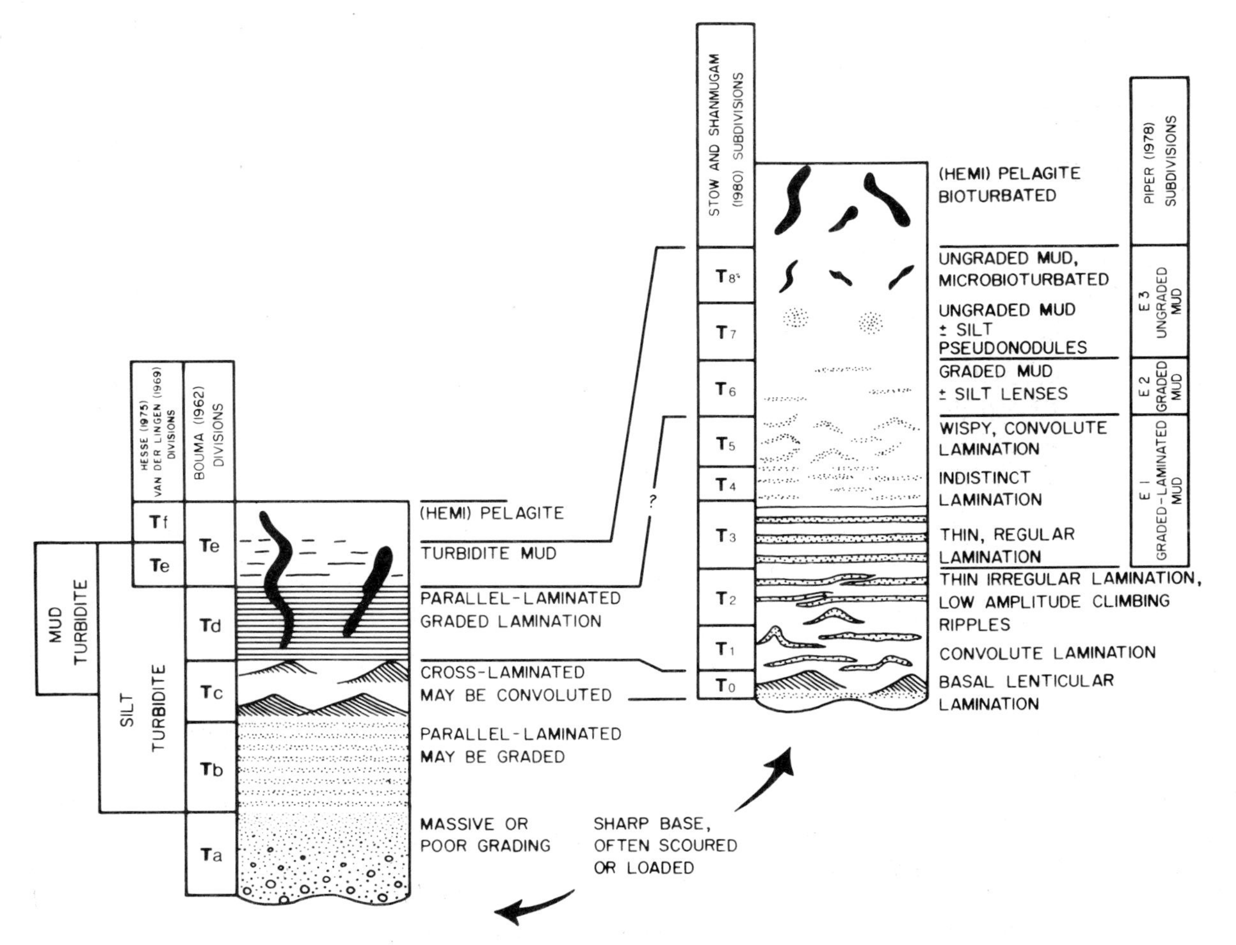

Figure 2.9 Summary of schemes for subdivision of fine-grained turbidites, based on Hesse (1975), Piper (1978), van der Lingen (1969), and Stow & Shanmugam (1980).

A second interpretation for fine-scale lamination in fine-grained turbidites has been proposed by Hesse & Chough (1980). The silt-rich laminae are believed to form during periods when bursts and sweeps repeatedly disturb the viscous sublayer of the flow, resulting in transport of clay-sized particles away from the boundary. The composite nature of individual silt laminae is used as support for this model. Clay-rich laminae are attributed to periods when bursts and sweeps are suppressed, perhaps by passage of large eddies leading to repeated establishment of ephemeral negative pressure gradients.

BOUMA-TYPE TURBIDITES

High- and low-concentration flows are end members, and many natural flows are intermediate in character, or evolve from high concentrations to low concentrations during deceleration and deposition. For a complete Bouma-type turbidite (Fig. 2.5), the T_a division was probably deposited from the high-concentration head region of a flow that later continued to deposit finer sediment from its more dilute body and tail, eventually dropping fine muds onto the seafloor. Bouma-type turbidites have been described in detail in the literature – further analysis of such structures here is not warranted.

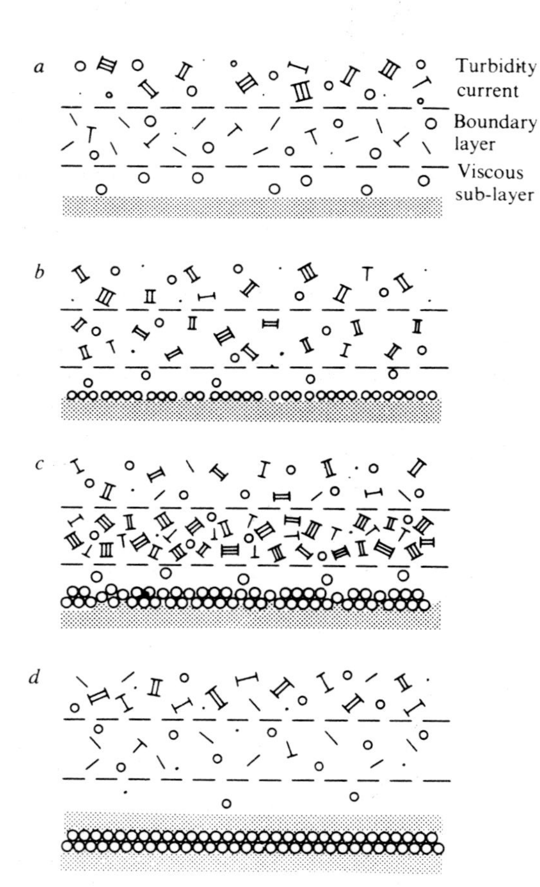

Figure 2.10 Schematic representation of the four stages of silt and mud deposition through the boundary layer of a turbidity current to form silt and mud laminae. After Stow & Bowen (1980).

GRAIN SIZE – BED THICKNESS TRENDS IN TURBIDITES

Turbidites deposited from both high-concentration and low-concentration flows in a single basin, with a consistent range of grain sizes in the source, exhibit a well defined relationship between bed thickness and grain size (Fig. 2.11 a & b) (Sadler 1982). Sadler's curves indicate a downslope trend from (a) proximal, relatively thin, medium-grained beds, to (b) intermediate, relatively thick, somewhat coarser-grained beds, to (c) progressively finer grained and thinner beds with increasing distal transport. The 'distal limb' of this trend is relatively easy to explain. Sadler (1982) demonstrates that 'for a given flow density and slope and with a uniform grain-size distribution, bed thickness becomes proportional to the square root of bed shear stress cubed'. Hence the decrease of both competence and capacity as turbidity currents decelerate leads to a related decrease of both grain size and bed thickness. The 'proximal limb' of the trend involves a temporary downslope increase in both grain size and bed thickness, and requires a special explanation. Sadler points to experiments of Kuenen (1951) that show a longitudinal decline of density and competence from the head to the tail of a turbidity current. As a flow that is in a state of autosuspension decelerates, deposition will begin beneath the tail of the current where competence and flow density are relatively low. 'The most proximal part of the deposit will therefore thicken and coarsen downcurrent until the point at which all of the flow, except a small region near the head, will be depositing sediment' (Sadler 1982, p. 48).

Debris flows and their deposits

DEFINITION AND EQUATIONS OF FLOW

The name *debris flow* commonly is used for both a flow process and a deposit. This usage generally does not create problems because the context makes the meaning clear. For clarity, however, we will use the term debris flow for the process only, and *debrite* for the deposit.

Debris flow is only well described from modern subaerial settings (Johnson 1970, 1984, Pierson 1981, Takahashi 1981). According to Takahashi (1981, p. 58), debris flows are flows 'in which the grains are dispersed in a water or clay slurry with the concentration a little thinner than in a stable sediment accumulation . . . [and] in which all particles as well as the interstitial fluid are moved by gravity'. Subaerial debris flows are capable of transporting boulders up to 2.7 million kg (Takahashi 1981), have bulk specific gravities in the range of 2.0–2.5, and may move at speeds of 20 m/s. Catastrophic submarine debris flows may carry (or push) enormous slabs weighing up to about 2300 million kg (immersed weight, Marjanac 1985). Experiments and theory (Hampton 1975, 1979, Rodine & Johnson 1976) suggest that only a small amount, about 5%, of interstitial matrix – mud + water slurry – is required to allow flow on surprisingly gentle slopes. The matrix serves several functions. It (a) lubricates the larger clasts so that they are able to slide past one another, (b) provides buoyant support for clasts of only slightly higher density, and (c) commonly exhibits elevated pore pressures that increase buoyancy and lower frictional resistance to flow (Pierson 1981).

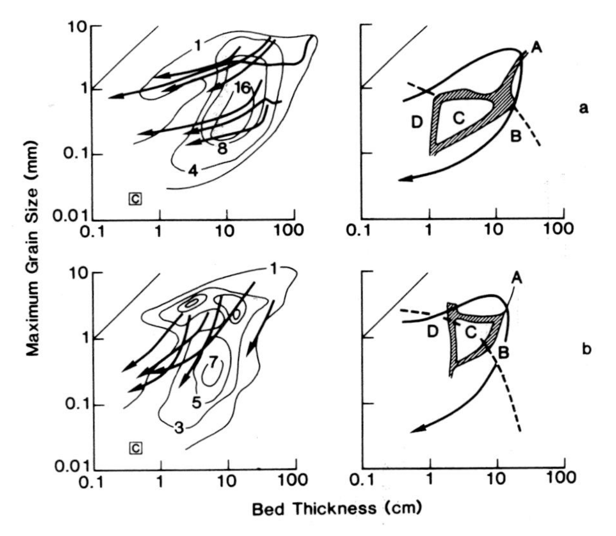

Figure 2.11 Maximum grain size plotted against bed thickness for Lower Carboniferous turbidites of (a) the Rhenaer Kalk (498 beds), and (b) the Posidonienkalk (235 beds), from Sadler (1982). The left-hand diagrams show contours of density of data points representing total bed thickness and corresponding maximum grain size. Arrows are selected vertical grading curves using the x-axis scale as 'distance to top of bed'. The small box labelled 'c' is the size of the counting square used in contouring. The right-hand diagrams show fields for Bouma divisions A through D found at the base of the beds for which maximum size was determined. Diagonal ruled zones = overlap of fields for divisions B, C and D. Dashed line = lower limit of field A, which has extensive overlap with the other fields. Arrow = modal horizontal grading curve, based on contour pattern in left-hand diagram.

In terms of rheology, debris flow resembles wet concrete and exhibits *strength*, which can be divided into two components: cohesive strength due to electrostatic attractions between clay-size particles, and frictional strength due to interlocking and surface contacts between clasts and between the debris flow and its bed. According to Pierson (1981), frictional strength far exceeds cohesive strength as a mechanism of clast support. Strength allows deposited debris to stand up in relief above its surroundings, with steep meniscus-like margins called *snouts*. Strength, and buoyancy of clasts in the matrix, permit the flow to carry clasts above the bed that are more dense than the bulk density of the flow itself.

Materials with strength will not deform until a critical yield strength is exceeded. Once deformation (flow) begins, laminar flow generally prevails in those parts of the debris flow where the critical yield strength is less than the shear stress. Elsewhere, friction and/or cohesion resist deformation. Particle support during flow comes from a combination of (a) frictional resistance to settling through finer matrix, similar in explanation to the kinetic sieve process described by Middleton (1970, p. 267), (b) matrix cohesion or strength that is not exceeded by the downward-directed force exerted by dispersed clasts, (c) buoyancy (partial support only), (d) elevated pore pressures of the slurry matrix, and (e) dispersive pressure (Bagnold 1956, 1973, Pierson 1981). Lowe (1982) believes that, in many cases, the largest clasts are not actually suspended, but remain in contact through rolling, sliding and intermittent bouncing downslope.

Debris flows move as a result of deformation in a basal zone of high shear stress. Lower shear stresses high in the flow, or at the flow margins, do not always exceed the yield strength of the debris, so that the upper part of the flow may be rafted along as a semi-rigid plug (Johnson 1970). As total shear stress decreases (i.e. as bottom slope decreases), or as intergranular friction increases, the semi-rigid plug thickens by downward growth until the entire mass ceases to move ('freezes') when bed shear stress declines to a value lower than the yield strength. Likewise, low shear stresses at flow margins result in marginal 'freezing' and construction of debris levees. Johnson (1970) favoured either a Bingham plastic or Coulomb-viscous rheological model for debris flow. A mathematical model for Bingham plastics is:

$$\tau = k + \mu \left[\frac{du}{dy} \right] ; \tau_{crit} = k \tag{2.5}$$

where k = strength of the debris = critical shear stress for movement (τ_{crit}), μ = dynamic viscosity of the debris after movement begins, and du/dy = the rate of change of velocity at any level, y, in the flow. Note that for $k = 0$, this equation reduces to Newton's law of viscosity. A similar model for Coulomb-viscous materials is:

$$\tau = C + \sigma_n \tan \phi + \mu \left[\frac{du}{dy} \right] ; \tau_{crit} = C + \sigma_n \tan \phi \tag{2.6}$$

where C = the cohesive strength component, due to electrostatic attractions between clay particles; $\sigma_n \tan \phi$ = the frictional strength component, due to intergranular friction; σ_n = normal stress; and ϕ = angle of internal friction. In contrast, Takahashi (1981) favours a 'dilatant-fluid' rheological model based entirely on dispersive pressure (Bagnold 1956). Dispersive pressure, P, is related to shear stress and internal friction angle by the following equation:

$$\tau = P \tan \phi \tag{2.7}$$

where friction angle is approximately equal to the static angle of repose. Note that, by convention, normal stress in Equation 2.6 is directed downward, but dispersive pressure in Equation 2.7 is directed upward. Clearly, when $|P| \geq |\sigma_n|$, dispersed grains in cohesionless flows behave as though weightless, and can freely move on the gentlest of slopes.

Instead of advocating a strict cohesive model (Lowe 1982) or a strict dispersive-pressure model (Takahashi 1981) for debris flows, we believe that natural flows exhibit a wide variation in the relative importance of available support mechanisms. This approach is the same as that taken by Shultz (1984), who recognizes a wide range in rheological properties of debris, with some 'debris flows' being relatively weak and fluidal in behaviour and some having little or no cohesion but being sustained almost entirely by dispersive pressure (Fig. 2.12).

TURBULENCE OF DEBRIS FLOWS

Most debris flows are laminar, with no fluid mixing across streamlines. Large flows may be turbulent (Enos 1977, Middleton & Southard 1984). For Bingham plastics, the criterion for turbulence is based on both the *Reynolds Number*, R, and the *Bingham Number*, B, where

$$R = \frac{Ud\rho}{\mu} \tag{2.8}$$

and

$$B = \frac{\tau_{crit} d}{\mu U} \tag{2.9}$$

In these equations, U = mean flow velocity, d = flow thickness, ρ = flow density, and μ = dynamic vis-

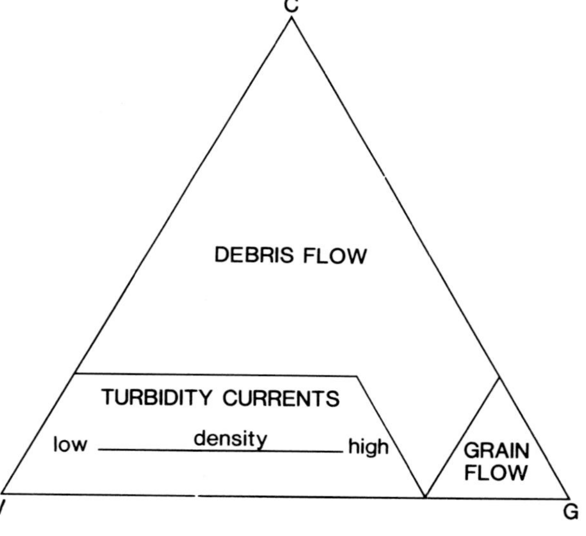

Figure 2.12 Ternary classification diagram for sediment gravity flows, based on relative importance of types of flow behaviour, where C = cohesive–plastic behaviour, V = viscous–fluid behaviour, and G = granular–collisional behaviour (simplified from Shultz 1984, who subdivided the debris-flow field). Flows approach V with increasing water content, C with increasing clay content, and G with increasing clast content and shear rate.

cosity after flow begins. Experimental data (Fig. 2.13), originally plotted by Hampton (1972), indicate that for large values of either R or B, a conservative criterion for turbulence is:

$$R \geq 1000\ B \qquad (2.10)$$

which is equivalent to

$$\frac{\rho\ U^2}{\tau_{crit}} \geq 1000 \qquad (2.11)$$

This last dimensionless product was named the *Hampton Number* by Hiscott & Middleton (1979), who used reasonable values for flow strengths and densities to show that even large, fast debris flows probably would not be turbulent.

Basal scour beneath fully freighted laminar debris flows is insignificant (Takahashi 1981), but underlying sediments may be plucked up and incorporated into the flow. Nevertheless, debrites may occur in channels cut by other processes, as debris flows seek out bathymetric lows. Even for laminar flows, secondary circulations due to clast rotations and encounters, or due to flow meandering, may cause internal mixing.

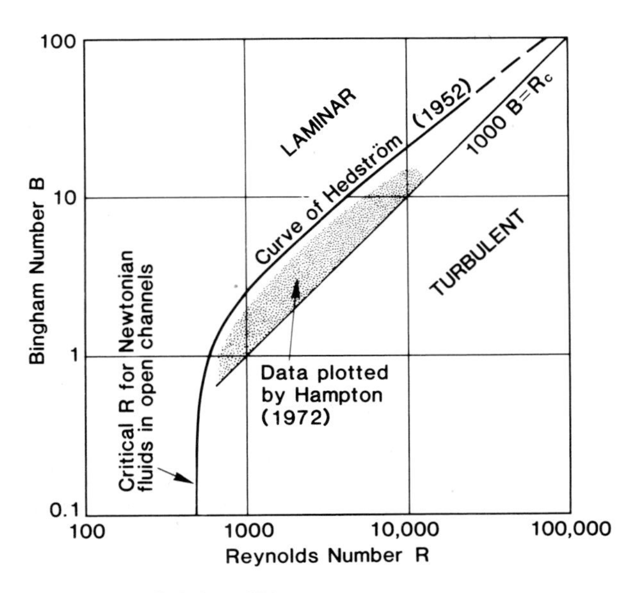

Figure 2.13 Relation of Bingham Number to critical Reynolds Number for turbulence in a Bingham plastic. Experimental data are for pipe flow but scales have been adjusted to the correct values for two dimensional turbidity currents, using the thickness of the flow as the length scale. From Hampton (1972).

COMPETENCE OF DEBRIS FLOWS

The competence – largest clast that can be carried – of fine-grained debris flows was determined by Hampton (1975) to be

$$D_{max} = \frac{8.8\ k}{\mathbf{g}\int(\rho_{clast} - \rho_{matrix})} \qquad (2.12)$$

where k = strength, $\mathbf{g}$ = gravitational acceleration, and ρ = density. Using kaolinite-water slurries as matrix, Hampton (1975) determined that (a) competence decreases approximately exponentially with increasing water content from about 20 cm at 40 weight per cent water, to about 2 cm at 60 weight per cent water, to about 0.5 mm at 85 weight per cent water, (b) competence of slurries is less after shear than before shear – by about 0.5 mm at 60 weight per cent water content, (c) competence decreases with flow duration for durations less than about one hour, and (d) competence is independent of flow velocity. Hampton's data indicate that quite dilute slurries can carry a fine sand load, resulting in a deposit with only a small amount of clay matrix. It should be noted that, had Hampton (1975) used montmorillonite-water slurries instead of kaolinite-water slurries, the competence values would have been even greater because of the greater strengths of montmorillonite-water slurries (Hampton 1972).

For debris flow with large amounts – > 20% by volume – of coarse sand and gravel, the competence increases dramatically with increasing concentration of coarse clasts (Hampton 1979). This is because the large clasts load the matrix and produce elevated pore-fluid pressures that counteract the tendency of the clasts to settle. These excess pore pressures (above hydrostatic) also reduce the strength of the debris flow by reducing normal stress (Eqn 2.6). At volume concentrations ≥50%, dispersive pressure due to clast collisions provides additional support (Rodine & Johnson 1976), and competence continues to increase dramatically above that predicted by Equation 2.8 (Hampton 1979). For these concentrations, dispersive pressure may be the dominant support mechanism (Pierson 1981, Takahashi 1981).

DEBRITES

Because of the mode of deposition, debrites are poorly sorted, lack distinct internal layering but may have crude stratification due to non-uniform migration through the flow of the base of the semi-rigid plug

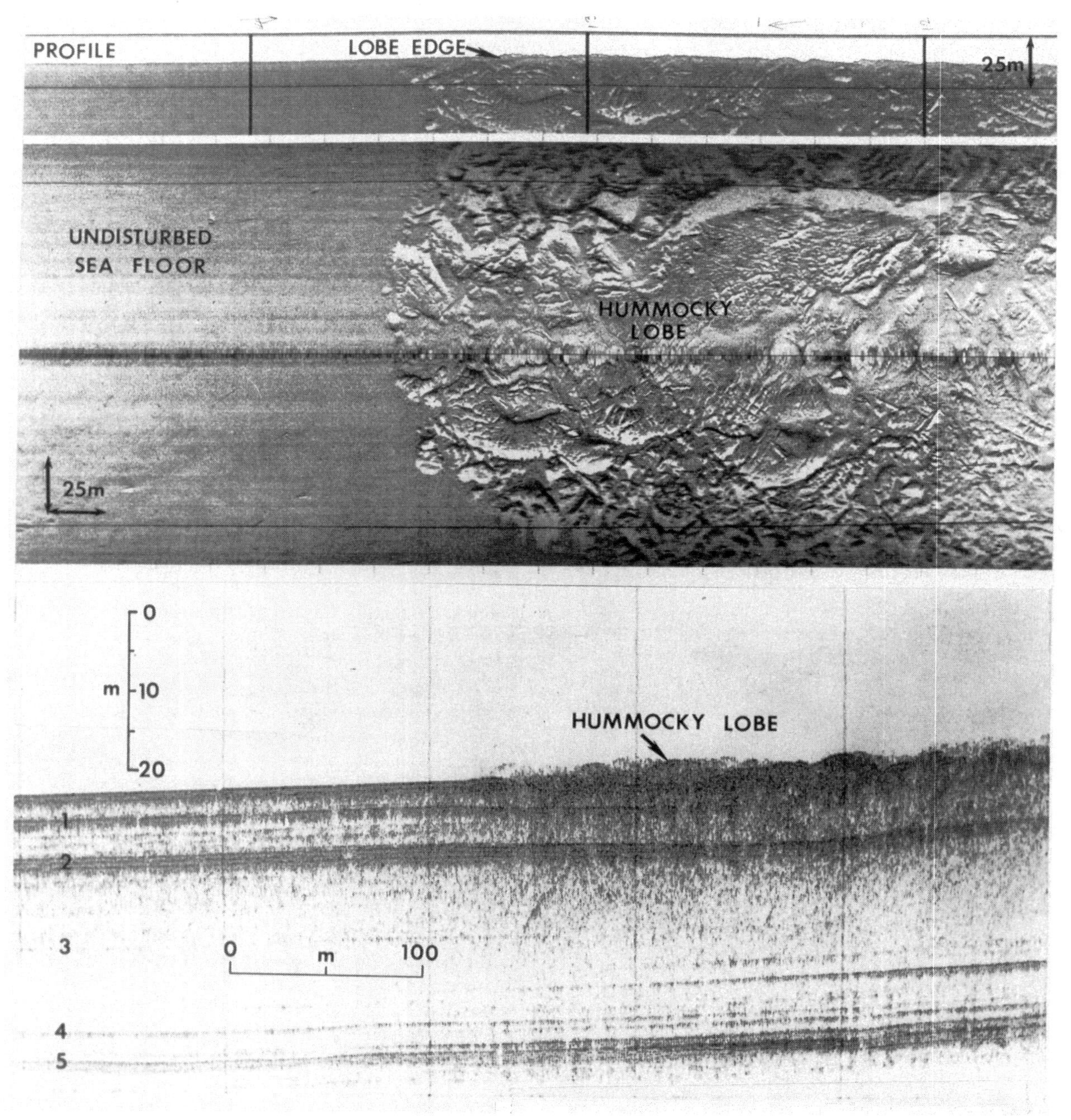

Figure 2.14 100–kHz side-scan sonar image (and relief profile), and matching subbottom profile of hummocky debrite lobe, showing low-relief hummocks; shear-bounded, irregularly shaped segments; and flat-lying fjord floor.

during deceleration (Hampton 1975, Thornton 1984, Aksu 1985), have a poorly developed clast fabric (Lindsay 1968, Aksu 1985, Hiscott & James 1985), irregular mounded tops (Fig. 2.14) (Prior *et al.* 1984), and tapered flow margins, or snouts. Grading is generally poor, but both normal and inverse grading may occur (Naylor 1980, Aksu 1984, Shultz 1984). Sandy debrites may be inversely graded because of the influence of dispersive pressure or because the base of the flow undergoes greatest and most prolonged shear, leading to reduced competence (Hampton 1975). Individual debris flow events may deposit separate tongues or lobes of material that have quite different textural characteristics (for subaerial example, see Johnson 1984, p. 266–74), making it difficult to distinguish separate flows in the geological record.

Broster & Hicock (1985) propose a novel explanation for inverse grading at the base of an ice-marginal subaqueous debrite that involves (a) formation of a normally graded cap on the moving flow due to fluid admixing and clast settling, followed by (b) overturning of the normally graded cap by movement around the frontal nose of the flow, leading to burial beneath the deposit after flow ceases.

Shultz (1984) attributes grading style and volume concentration of matrix in debrites to the relative importance of cohesion, clast interaction (dispersive pressure) and fluid behaviour (Fig. 2.12). The result is a continuum of deposit characteristics (Fig. 2.15) intermediate in character between four distinct end-member facies: *Dmm* = massive (structureless) matrix-supported diamictite, *Dmg* = graded matrix-supported diamictite, *Dci* = inversely graded clast-supported diamictite, and *Dcm* = massive (structureless) clast-supported diamictite (Shultz 1984). This facies classification is entirely non-genetic, but could be made more specific by replacing the term 'diamictite' with 'debrite'.

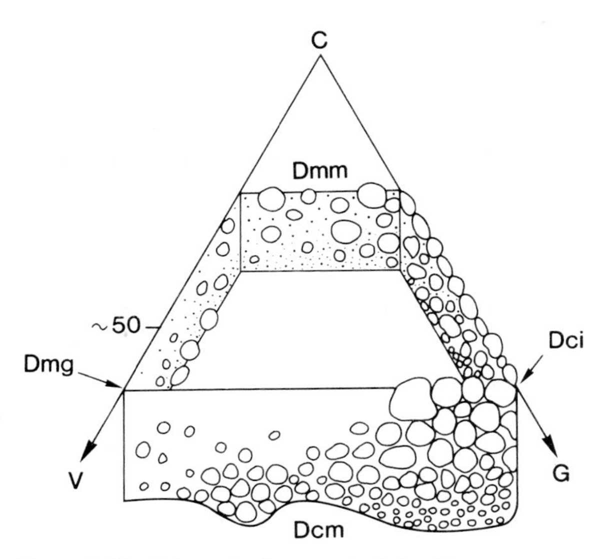

Figure 2.15 Schematic diagram of relationships among debrite types, with variables as in Fig. 2.12. See text for abbreviations.

The larger clasts – cobbles and boulders – of some debrites are concentrated near the base or in the middle of the deposit. Large boulders may, however, project from the tops of the beds, allowing a quantitative assessment of debris strength using a Bingham plastic model (Johnson 1970, p. 487).

Bingham plastic strength can also be calculated from the shape of debris snouts and from the thickness, T_{crit}, at which flow ceased (Johnson 1970, p. 488), according to the following equation:

$$T_{crit} = \frac{k}{\gamma_d \sin \alpha} \qquad (2.13)$$

where γ_d = specific weight of the debris and α = slope angle. Note that $\gamma_d = \mathbf{g}'\rho_{debris}$, where $\mathbf{g}' = \mathbf{g}(\rho_{debris} - \rho_{water})/\rho_{debris}$.

Calculations of debris flow strength based on either the extent of clast projection or snout shape are not generally available for ancient examples. Hiscott & James (1985) and Kessler & Moorhouse (1984) calculated strengths in the range 10^3–10^5 dynes/cm^2 for Cambro–Ordovician and Jurassic deposits, respectively. These estimates are in the same range as values calculated for subaerial flows by Johnson (1970).

SUBMARINE *VS* SUBAERIAL DEBRIS FLOWS

All evidence suggests that submarine debris flows and debrites differ only in minor ways from subaerial equivalents, although associated facies and processes of post-depositional modification are clearly different. Shear stress is dependent on the density difference between the debris and the ambient fluid, so that one might predict that somewhat higher slopes would be needed to permit subaqueous *vs* subaerial debris flow, given similar yield strengths. Recall, however, that submarine debris flows have been documented from very low slopes. In general, submarine flows probably have lower yield strengths than subaerial flows as a result of entrainment of sea water and wet mud, lack of downward percolation of water into the substrate, and elevated pore fluid pressures due to greater amounts of interstitial fluid (Pierson 1981). This suggestion is supported by the observation that, for deposits of equal thickness, subaqueous debrites contain smaller boulders than subaerial debrites (Nemec *et al.* 1980, Gloppen & Steel 1981, Nemec & Steel 1984), i.e. subaqueous debris flow is weaker than subaerial debris flow.

Deep, thermohaline, clear-water currents

There is growing evidence that the deep ocean is not a realm of great stillness, with little or no significant sediment transport except during invasion by sediment gravity flows. Large parts of the deep ocean basins, especially the Atlantic Ocean, are characterized by geostrophic currents moving at mean speeds of 10–30 cm/s (McCave *et al.* 1980, Hollister & McCave 1984), with short 'gusts' reaching about 70 cm/s (Richardson *et al.* 1981). Deep circulation in the oceans is the result of thermohaline effects. In the North Atlantic Ocean, for example, dense cold water sinks off the coast of Greenland and in the Norwegian Sea and moves southward as a bottom current (Worthington 1976); in the South Atlantic, ice formation in the Weddell Sea causes an increase in salinity and hence density, the dense sea water sinks and flows northward along the bottom (Stommel 1958, Pond & Picard 1978, p.134). Other regions of the world's oceans that are characterized by spreading cold bottom water are outlined by Mantyla & Reid (1983). The deep ocean bottom currents are deflected to the right in the northern hemisphere and to the left in the southern hemisphere by the Coriolis effect, with the result that they flow along the continental slopes and rise on the western sides of ocean basins. Two examples are the *Western Boundary Undercurrent* (WBU), which sweeps along the continental rise of eastern North America (Fig. 2.16) at depths of about 2000–3000 m and at peak velocities of about 25–70 cm/s (Stow & Lovell 1979),

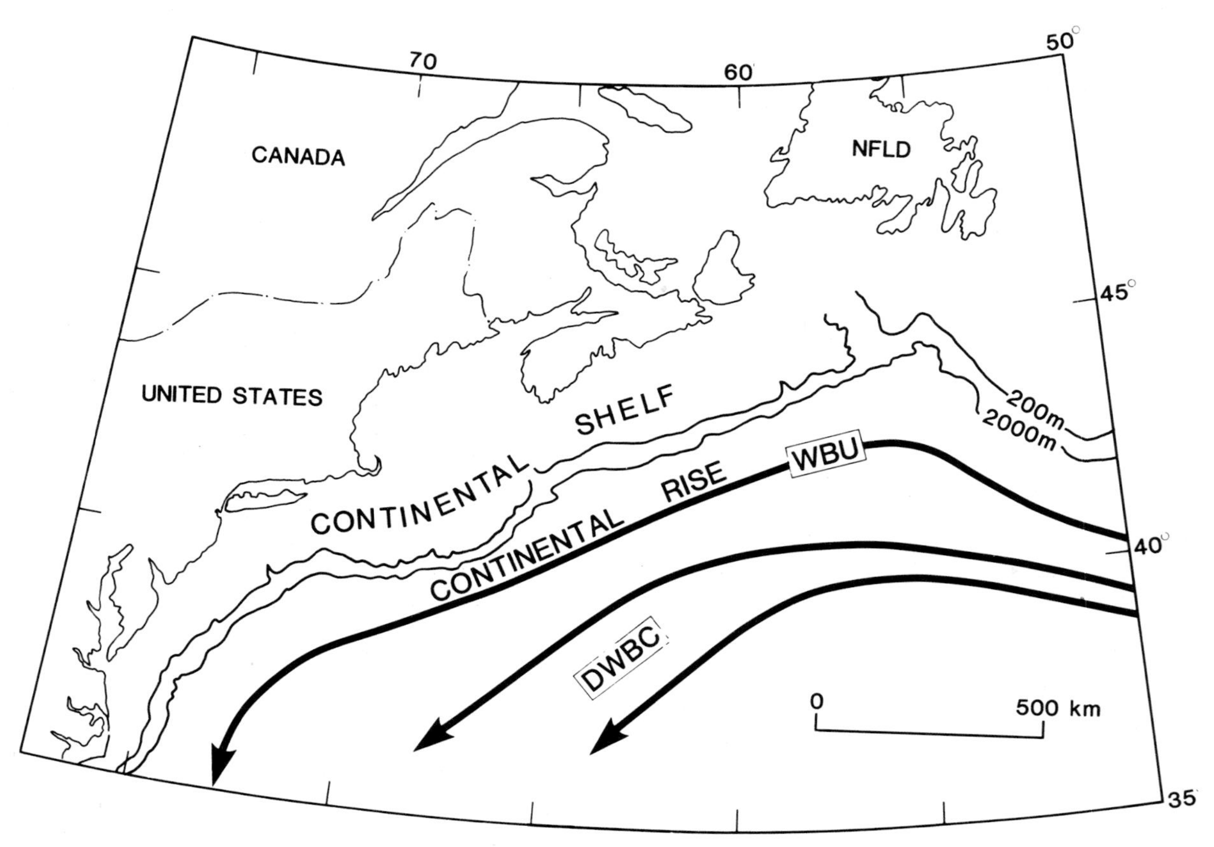

Figure 2.16 Approximate tracks of Western Boundary Undercurrent (WBU) and Deep Western Boundary Current (DWBC) relative to the eastern continental margin of North America. After Hollister & McCave (1984).

and the *Deep Western Boundary Current* (DWBC), which occupies the same region at depths of 4000–5000 m (Richardson *et al.* 1981). The WBU is derived from the Norwegian Sea, whereas the DWBC appears to be formed of Antarctic bottom water (Hogg 1983). These currents carry a dilute suspended load – generally <0.1–0.2 g/m^3 – that forms the thick bottom *nepheloid* layer (Ewing & Thorndike 1965, Biscaye & Eittreim 1977). Concentrations may briefly reach values much higher, up to at least 12 g/m^3 (Biscaye *et al.* 1980, Gardner *et al.* 1985). Most of the fine-grained suspended material is winnowed from the seafloor; the rest is probably added to the current by cascades of cold shelf water or lutite flows originating at the shelf-slope break (McCave 1972).

The WBU and DWBC are capable of long-distance transport of fine-grained sediments. According to Heezen & Hollister (1971), distinctive red mud derived from the weathering of Carboniferous and Triassic bedrock in the Gulf of St Lawrence area (45°N latitude) has been transported at least as far south as the Blake Plateau (30°N), a distance of about 2000 km.

On the Newfoundland Rise (Carter & Schafer 1983), the high-velocity core flow of the WBU ($U \leq 35$ cm/s) intersects the bottom at depths of 2600–2800 m, and is capable of transporting sediment grains of approximate diameter 0.1 mm 1–15% of the time. The sea bed beneath this core zone is sandy. Finer grains are effectively maintained in suspension as a nepheloid layer up to 800 m thick.

On the continental rise off Nova Scotia, the high-velocity core flow of the DWBC ($U \leq 70$ cm/s) is at depths of 4500–5000 m (Richardson *et al.* 1981, Bulfinch & Ledbetter 1984). Characteristics and effects of the DWBC have been studied in great detail during the multidisciplinary 'High Energy Benthic Boundary Layer Experiment' (HEBBLE; for an excellent summary of findings, see Nowell & Hollister 1985). The bottom beneath the DWBC consists of coarse silts moulded into longitudinal ripples (Bulfinch & Ledbetter 1984, Swift *et al.* 1985, Tucholke *et al.* 1985). The silt size interval most affected by the core flow is 5–8ϕ. Net accumulation rates are not high (5.5 cm/1000 years), but instantaneous rates may be, due to alternation of periods of rapid erosion and rapid deposition from a highly concentrated nepheloid layer (Hollister & McCave 1984). The bottom flow over time is very complicated, and may involve significant variations in flow speed and reversals of flow direction. Times of strongest and most variable flow are called 'deep-sea storms' by Hollister & McCave (1984). The 'storms' last from a few days to several weeks, are characterized by current speeds in excess of 20 cm/s, and result in high concentrations of suspended sediment. Based on a five year record in the HEBBLE area off the coast of Nova Scotia, about three such 'storms' occur each year, and occupy about 35% of the time (Hollister & McCave 1984).

Contour current deposits, or *contourites* (Hollister & Heezen 1972), may be treated as two end members: (a) muddy contourites, and (b) sandy contourites. Muddy contourites are fine grained; mainly homogeneous and structureless; thoroughly bioturbated; and only rarely show irregular layering, lamination and lensing. They are poorly sorted silt- and clay-size sediments with up to 15% sand. They range from finer-grained homogeneous mud to coarser-grained mottled silt and mud, and their composition is most commonly mixed biogenic and terrigenous grains. According to Hollister & McCave (1984), short-term depositional rates of mud can be extremely high, about 17 cm/year, followed by rapid biological reworking.

Sandy contourites comprise thin irregular layers (<5 cm) that are either structureless and thoroughly bioturbated, or may possess some primary parallel or cross lamination which may be accentuated by heavy mineral or foraminiferal concentrations (Bouma & Hollister 1973). Grading may be normal or inverse, and bed contacts may be sharp or gradational. Grain size ranges from coarse silt to, rarely, medium sand, with poor to moderate sorting. The sandy facies, which may be rich in biogenic sand grains, is produced by winnowing of fines by stronger flows (Driscoll *et al.* 1985).

Muddy and sandy contourites commonly occur together in vertical coarsening-upward to fining-upward sequences (Faugères *et al.* 1984, Stow & Piper 1984). A complete sequence shows inverse grading from a fine homogeneous mud, through a mottled silt and mud, to a fine-grained sandy contourite facies, and then normal grading back to a muddy contourite (Fig. 9.15b). The changes in grain size, sedimentary structures and composition probably are related to long-term (1000–30 000 year) fluctuations in the mean current velocity (Stow & Piper 1984).

In particular successions, it may be difficult to distinguish between mud turbidites and muddy contourites (Bouma 1972, Stow 1979). Also, the reworking of sand turbidites can result in bottom-current-modified turbidite sands, believed to be common on continental slopes and rises. In the central parts of ocean basins, bottom currents are known to construct large sediment drifts (Ch. 9) of almost pure biogenic material (McCave *et al.* 1980, Stow & Holbrook 1984). Such biogenic contourites may be indistinguishable from true pelagites.

High-level escape of mud from the shelf

Suspended sediment concentrations in shelf areas may be quite high due to input of mud-laden river water, or stirring of the bottom by strong waves, tidal currents, or internal waves at density interfaces (Cacchione & Southard 1974). This suspended sediment may be advected off the shelf by ambient currents, possibly wind-driven, or by transport in cascading cold water that may flow off the shelf in the winter months (McCave 1972, McGrail & Carnes 1983). Suspensions of fine-grained sediment may also leave the shelf as dilute turbidity currents (lutite flows), moving along the bottom onto the lower slope and rise, or along density interfaces in the ocean water (Fig. 2.17) (McCave 1972, Gorsline *et al.* 1984). These dilute suspensions may move down a smooth upper slope as unconfined sheet flows, or may be confined by gullies or canyons in which suspension is augmented by weak tidally-forced flows (Shepard *et al.* 1979, Gorsline *et al.* 1984). There is evidence that most mud transported off the shelf by dilute turbidity currents (lutite flows) bypasses the slope, leading to maximum rates of accumulation on the continental rise (Nelson & Stanley 1984).

On narrow shelves, plumes of suspended sediment from river deltas can extend beyond the shelf-slope break (Emery & Milliman 1978, Thornton 1981, 1984), directly contributing fine-grained sediments to slope and rise areas. In polar areas, sediment-laden spring melt water may actually flow from the land across the surface of floating sea ice, and deposit its load directly onto the continental slope (Reimnitz & Bruder 1972).

Mud that leaves the shelf either by 'high-level' escape in river plumes, by dilute turbidity-current flow, or by movement along density interfaces in the water column over the slope eventually settles to the sea floor to form the bulk of hemipelagic deposits. Deposition rates are on the order of 10–60 cm/1000 years (Krissek 1984, Nelson & Stanley 1984). As is true for strictly pelagic sediments, the finest particles in the high-level suspensions are probably carried to the bottom as aggregates in the form of faecal pellets (Calvert 1966, Schrader 1971, Dunbar & Berger 1981).

It is appropriate here to differentiate between the terms pelagic and hemipelagic as they apply to sedimentary deposits. In this book, pelagic grains are defined as those grains that initially enter the marine hydrosphere beyond the shelf-slope break, or are

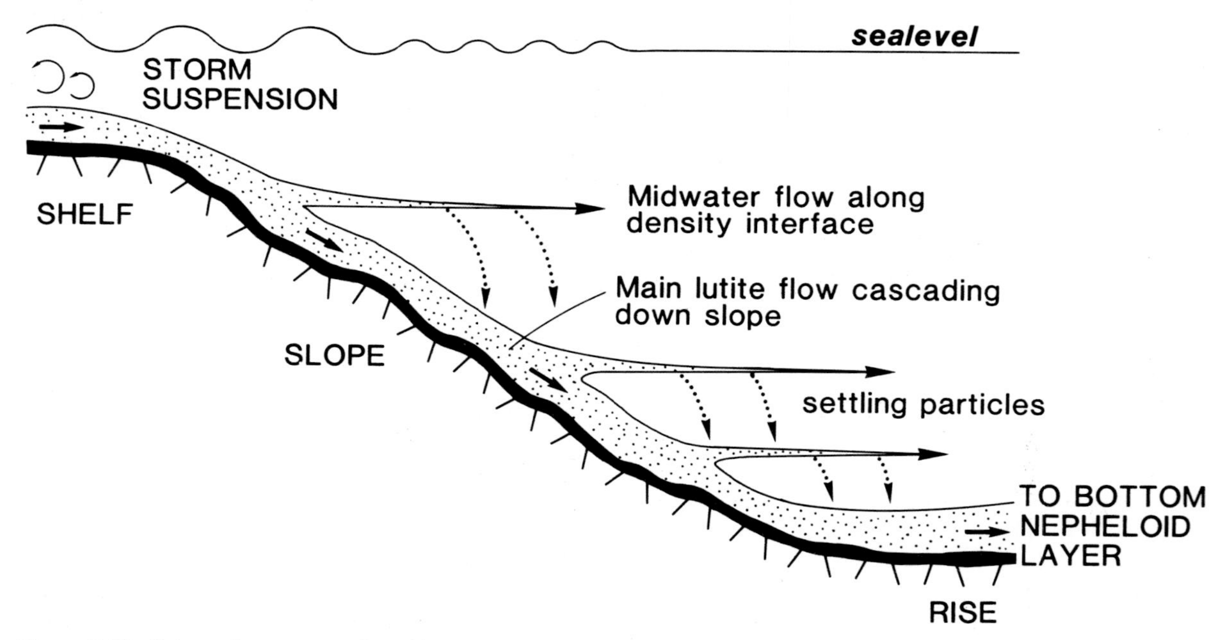

Figure 2.17 Schematic representation of lutite flows with cascading. Decrease in width of arrows indicates decrease in concentration. After McCave (1972).

created by organisms in this open-ocean region, and that subsequently settle to the sea floor. Pelagic particles, therefore, include biogenic siliceous skeletons (diatoms, radiolaria), calcareous skeletons (foraminifera, nannofossils), wind-blown dust or volcanic ejecta that lands on the sea surface of the open ocean, and debris liberated by melting icebergs. Hemipelagic sediments contain a pelagic component, generally 5–50% but locally as much as 75%, with the remainder consisting of terrigenous mud that initially enters the ocean at the coast, either by coastal erosion or through river systems (deltas, estuaries etc.). This terrigenous component moves across the shelf-slope break by processes outlined in this section.

Fine grained suspensions that move seaward across the shelf-slope break may vary seasonally in (a) grain size and (b) content of suspended organics. In anaerobic/dysaerobic basins, this fine-scale seasonal cyclicity can be preserved in the sediment record; on oxygenated basin slopes all such lamination would be destroyed by burrowers. Dimberline & Woodcock (1987) and Tyler & Woodcock (1987) convincingly argue that submillimetre-thick interlaminations of silt and organics in the Silurian Welsh Basin are a result of alternations of (a) spring algal blooms with (b) increased winter discharge of silt into the basin. The assumption of annual cyclicity leads to reasonable sedimentation rates of 60–150 cm/1000 years (Dimberline & Woodcock 1987). A general depositional model (Fig. 2.18) involves bottom-hugging and mid-water dilute flows advected off the shelf during fair-weather periods (hemipelagic laminated silts/organics, depending on season), and during storms (silty/sandy graded beds with irregular order of internal structures).

Sediment slides

The downslope component of gravity can cause sediment masses previously deposited on a slope to move by increments or during a single episode into deeper water. Slow downslope movement without slip along a single detachment surface, i.e. without *failure*, is referred to as *creep*. No structures in ancient successions have been unambiguously attributed to creep. More

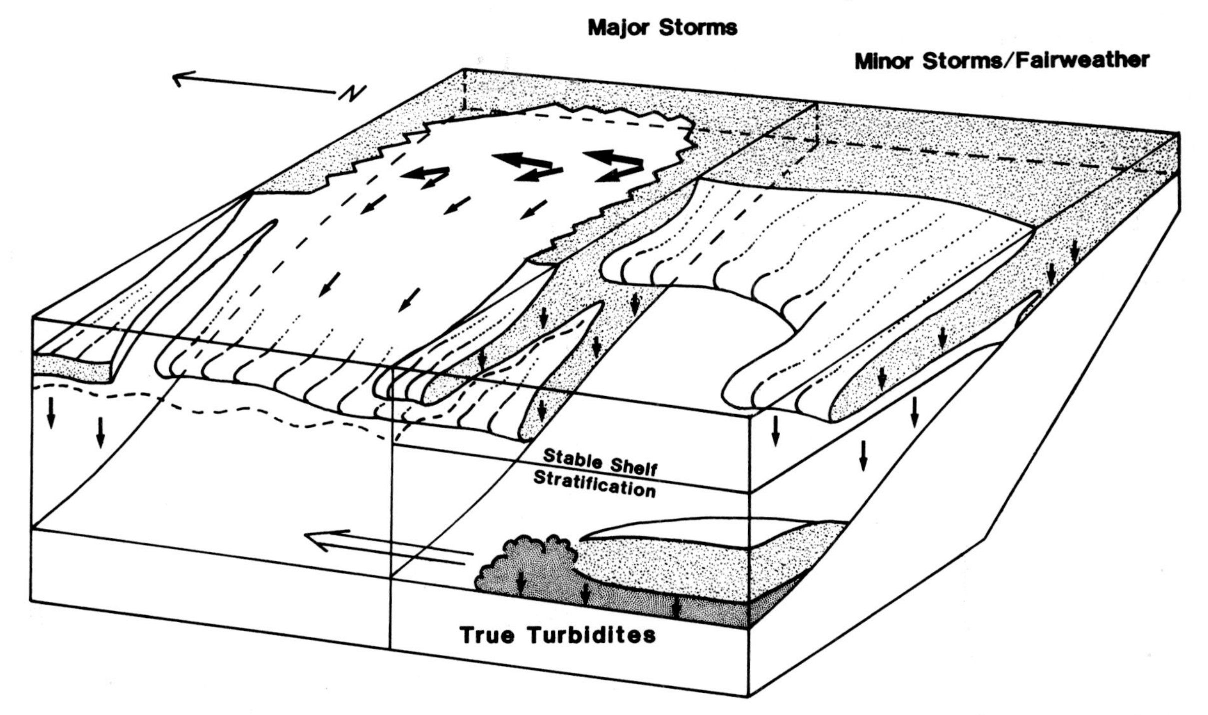

Figure 2.18 Model for hemipelagites (Bailey Hill Formation) and turbidite sands (Brimmon Wood Member) in the Welsh Basin. Silts, fine sands and organics were advected off the shelf by wave and current action, forming dilute bottom and mid-water flows. Annual seasonal layering in the hemipelagites was preserved under anaerobic/dysaerobic conditions. After Dimberline & Woodcock (1987).

rapid downslope movements subsequent to failure events generate sediment slides and debris flows. *Sediment slides* can result in little-deformed to intensely folded, faulted and brecciated masses that have translated downslope from the original site of deposition (coherent slides). If the primary bedding is entirely destroyed by internal mixing, with soft muds and/or water mixed into the slide, then it may transform into a debris flow. Slides and debris-flow deposits ('debrites') are common surficial sediments in the modern oceans. For example, Embley (1980) claims that 'at least 40% of the continental rise of eastern North America . . . is covered by a veneer of mass-flow deposits [slides] including debris flows [i.e. 'debrites']'. Debrites have already been described, and the internal characteristics of coherent slides are described in Chapter 3. This section will therefore be restricted to conditions for initiation of sliding. Whether the material subsequently remains coherent or mixes to produce a debris flow will depend primarily on the relationship of the water content to the liquid limit of the material, the internal strength of the material, and the duration of movement. On marine slopes, there is probably a complete gradation between coherent slides and thoroughly mixed debrites.

When sliding occurs, the movement occurs along a basal failure surface at some depth below the sea floor. Along this surface, the shear stress produced by the sum of gravitational acceleration and cyclic accelerations due to seismic shocks and passing surface or internal waves exceeds the internal shear strength of the sediment. The shear strength depends on a variety of sediment properties like water content, texture, pore pressures, and organic content.

Fine-grained sediment has a variety of geotechnical properties (Bennett & Nelson 1983) that are useful indicators of its physical state and that help determine under what conditions the sediment will fail and generate a slide. Conditions that favour initiation of slides in muddy terrigenous sediments are a function of (a) bottom slope (Moore 1961), (b) sedimentation rates (Hein & Gorsline 1981), and (c) the response of the sediment to cyclic stress produced by earthquake shaking (Morgenstern 1967). Sedimentation rates on basin-margin slopes vary widely, but Hein & Gorsline (1981) conclude that a rate of 30 mg/(cm^2 year) must be attained before slope failures become common. For example, in the Santa Barbara Basin, California Borderland, sedimentation rates exceed 50 mg/(cm^2 year), and debris flows are widespread on slopes of $<1°$ (Hein & Gorsline 1981).

The sedimentation rate effectively determines the water content and shear strength, S, of the sediment, although shear strength is also a function of other variables such as content of organic matter (Keller 1982) and generation of gas in the sediment by decay of organics or by clathrate decomposition (Carpenter 1981). According to Keller (1982), 'cohesive sediments with greater than about 4–5% organic carbon [have] . . . (1) unusually high water content, (2) very high liquid and plastic limits, (3) unusually low wet bulk density, (4) high undisturbed shear strength, (5) high sensitivity, (6) high degrees of apparent overconsolidation, and (7) high potential for failure [by liquefaction] in situations of excess pore pressure'.

The stability of sediments on a sloping bottom has traditionally been analysed using a static infinite slope model (Moore 1961, Morgenstern 1967), in which the lateral extent of the slope is much greater than the thickness dimension of the surficial sediment (Fig. 2.19). As pointed out by Booth *et al.* (1985), the model expresses the balance between resisting forces and shearing forces. Safety factor, SF, is defined as the ratio of the two forces, and is given by:

$$SF = [1 - \psi/(\gamma' Z \cos^2\alpha)] \tan\phi/\tan\alpha \quad (2.14)$$

where ψ = excess pore fluid pressure, Z = depth below the surface, $\gamma' = \rho_s \mathbf{g}'$, ρ_s = sediment density, reduced gravity $\mathbf{g}' = \mathbf{g}\int(\rho_s - \rho)/\rho_s$, ρ = density of seawater, $\mathbf{g}$ = gravitational acceleration, ϕ = angle of internal friction, and α = slope angle. Case studies (Athanasiou-Grivas 1978) show that the probability of failure is low for $SF > 1.3$, and that failure is a virtual certainty for $SF < 0.9$.

Booth *et al.* (1985) provide a nomogram that allows determination of SF under undrained conditions given sedimentation rate, coefficient of consolidation, sediment thickness, slope angle, and angle of internal friction. Excess pore pressure, u, is obtained from consolidation theory (Gibson 1958), under the assumption that these excess pressures are entirely the result of trapping of pore water in compacting, fine-grained sediment of low permeability.

The infinite slope model can be extended to the case of superimposed ground accelerations due to earthquakes (Morgenstern 1967, Hampton *et al.* 1978). In this case, an earthquake safety factor, ESF, can be expressed as (Booth *et al.* 1985):

$$ESF = (SF\,\gamma' \tan\alpha)/(\gamma\, a_x + \gamma' \tan\alpha) \quad (2.15)$$

where $\gamma = \rho\mathbf{g}$, and a_x = horizontal acceleration coefficient expressed in terms of gravity (e.g. 0.1 $\mathbf{g}$).

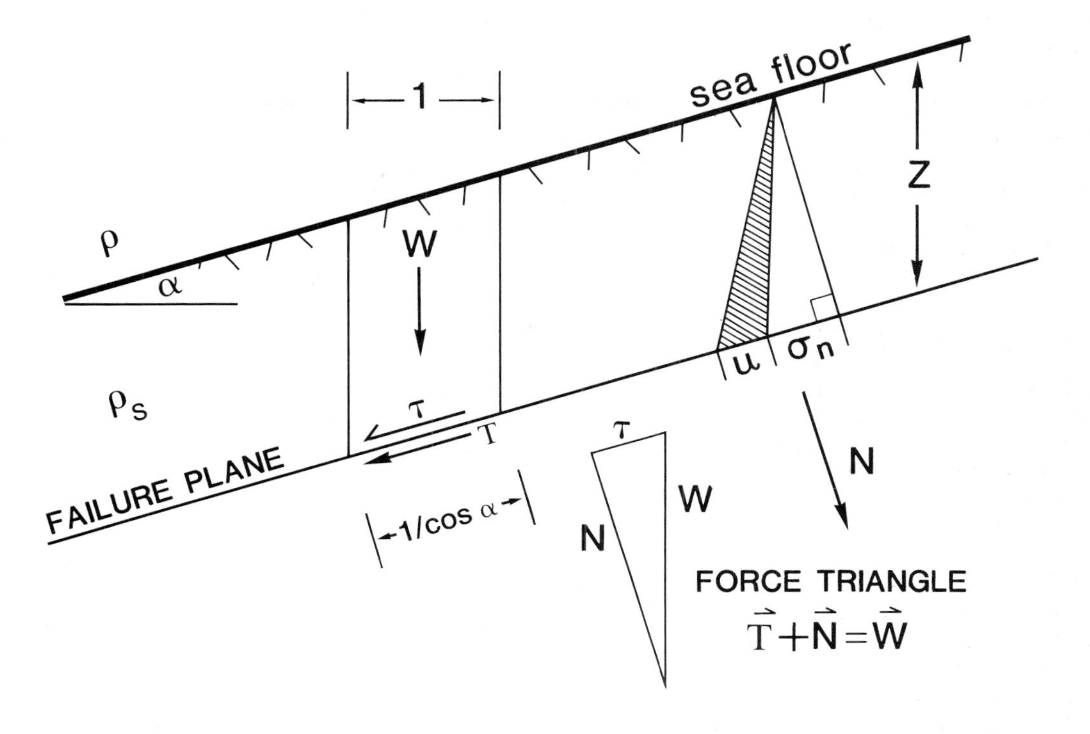

$$W = \gamma' Z \; ; \; \gamma' = \rho_s g'$$

$$N = W \cos \alpha \; ; \; \sigma_n = W \cos^2 \alpha - u$$

$$T = W \sin \alpha \; ; \; \tau = W \sin \alpha \cos \alpha$$

$$SF = \frac{\text{RESISTANCE FORCE}}{\text{SHEAR FORCE}} \cdot \frac{\text{UNIT AREA}}{\text{UNIT AREA}} = \frac{\text{STRENGTH}}{\text{SHEAR STRESS}}$$

$$= \frac{\sigma_n \tan \phi}{\gamma' Z \sin \alpha \cos \alpha} = \frac{(\gamma' Z \cos^2 \alpha - u) \tan \phi}{\frac{1}{2} \gamma' Z \sin 2\alpha} \; ; C = O$$

For slope $< 20°$, $\sin 2\alpha \simeq 2 \tan \alpha$; therefore

$$SF = [\cos^2 \alpha - u/(\gamma' Z)] \tan \phi / \tan \alpha$$

$$\text{i.e. } SF = [1 - u/(\gamma' Z \cos^2 \alpha)] \tan \phi / \tan \alpha \qquad (2.14)$$

Figure 2.19 Summary of variables affecting sediment stability on sloping bottoms. ρ, water density; ρ_s, sediment density; *N*, normal force due to sediment load on potential failure plane at depth *Z*; σ_n, effective normal pressure at depth *Z*; *u*, excess pore fluid pressure at depth *Z*; α, slope angle; *W*, sediment weight of a unit prism above the failure plane, where the area of the base of the prism is cos α; τ, shear stress at depth *Z*; T , shearing force along the failure plane. Other variables are explained in the text. Calculation of safety factor, *SF*, is based on strength of a material in which cohesion, *C*, is zero (compare with Eqn 2.6).

Figure 2.20 is taken from Booth *et al.* (1985), and allows estimation of the earthquake-induced horizontal ground acceleration required to reduce *ESF* to 1.0 for a wide range of slopes and safety factors, and a reasonable range of specific weights. The increase in excess pore pressures caused by ground shaking (Egan & Sangrey 1978) must be taken into account in estimating safety factor. Clearly, 'even small earthquake-induced accelerations are very detrimental to the stability of a submarine slope' (Morgenstern 1967).

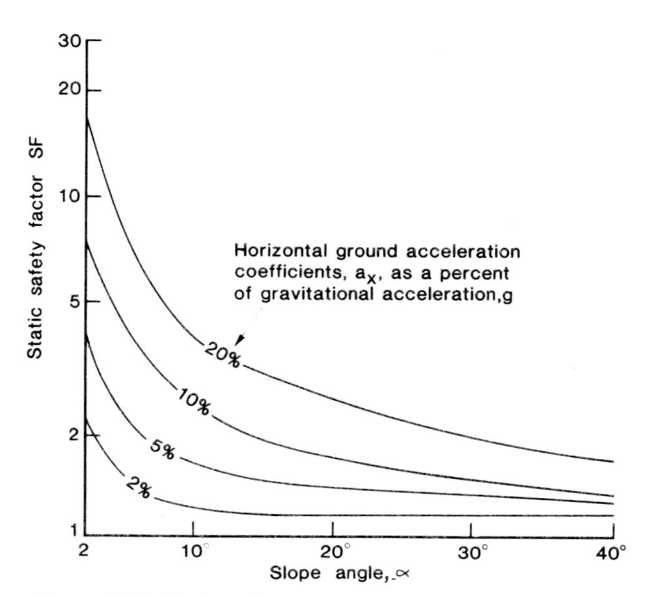

Figure 2.20 Horizontal ground accelerations, a_x, required to reduce the static safety factor to a value of 1 for given slope angles and sediment density in the range 1.5–2.0 g/cm^3 (simplified from Booth *et al.* 1985). For example, if $SF = 2$ and $\alpha = 10°$, ground accelerations of about 0.07 **g** or greater will reduce safety factor to 1 or less, and failure will be likely. On the same slope with $SF = 4$, accelerations of at least 0.2 **g** would be needed to cause failure. Note that *SF* is itself a function of bottom slope, as well as excess pore fluid pressures (Eqn 2.14).

Summary

The main mechanisms for long-distance lateral transport of sediment in the deep sea are: (a) debris flows, (b) high- and low-concentration turbidity currents, (c) deep, thermohaline, clear-water currents that commonly flow parallel to bathymetric contours, (d) movement of dilute mud suspensions in water masses that drift or cascade off the shelf, and (e) mass movement in the form of sediment slides. The deposits of debris flows – debrites – are poorly sorted, generally lack stratification, have a poorly developed clast fabric, and stand up above their surroundings as irregular mounds with a tapered marginal snout. Deposits of high-concentration turbidity currents commonly show evidence for dispersive pressure (traction carpets) and pore-fluid escape (dish structures or pillars), but these processes are restricted to the time of deposition and tell us little about the long-distance transport mechanism.

Fine-grained sediment may be deposited by low-concentration turbidity currents, by bottom currents (contour currents), or by settling from dilute suspensions at the top of, or within, the water column (hemipelagites). The deposits have some diagnostic characteristics, but are in many cases difficult to interpret because of a superposition of more than one process, or because of post-depositional bioturbation.

Sediment slides may involve all size grades of sediment, but are most common in poorly consolidated, water-rich mud that characterizes areas with high depositional rates. The susceptibility of sediment masses to sliding, even on very low slopes, is sharply increased by cyclic vibrations generated by earthquakes.

Fundamental contributions to the understanding of ancient deep-water facies and environments have been made by Professor Emiliano Mutti, Parma University (left); Professor Arnold Bouma, Institute of Coastal Studies, Louisiana State University (centre), and Dr. Franco Ricci Lucchi, Bologna University, (right). Photographed during the international COMFAN II meeting in Parma, Italy (Sept. 1988).

CHAPTER 3

Deep-water facies and depositional processes

Introduction

The most widely used facies classification for deep-water siliciclastic sediments is that of Emiliano Mutti and Franco Ricci Lucchi (Mutti & Ricci Lucchi 1972, 1974, 1975, Mutti 1977). Their classification (Fig. 3.1) has proven extremely useful as a tool for interpreting deep-water deposits. Recent research in modern and ancient deep-water environments, however, has led to the recognition of facies that do not easily fit into the Mutti and Ricci Lucchi scheme (e.g. contourite deposits, oozes). This chapter outlines a new classification scheme for deep-water facies that incorporates the latest research data. While we emphasize deep-water siliciclastic facies, we include the associated pelagic-hemipelagic facies that may be largely biogenic in composition. We do not deal with resedimented carbonates separately, as these are considered to be broadly similar to the siliciclastics, but with some notable differences (see Stow 1984; Colacicchi & Baldanza 1986). This scheme has recently been outlined in detail elsewhere, with a full set of representative field and core photographs of facies (Pickering, Stow, Watson & Hiscott 1986a). Brief summaries of the scheme, based on the preliminary manuscript of Pickering *et al.* (1986a), were published by Stow (1985, 1986). There are minor modifications to the earlier versions of the scheme in this chapter, particularly for Facies Classes A, F and G. In this chapter, photographs are used to illustrate most, but not all facies. Cross-references to facies photographs in other chapters are provided.

Classification scheme

For the purpose of our classification, we use the term 'facies' to mean a body of sedimentary rock/sediments with specific physical, chemical and biological characteristics. The chief attributes used to define the different facies are bedding style and thickness, sedimentary structures, composition and texture.

For the sake of brevity, we have mainly used the terminology for modern, unconsolidated sediments throughout this chapter. The terms gravel, sand, mud, silt and clay are, therefore, used to include the ancient

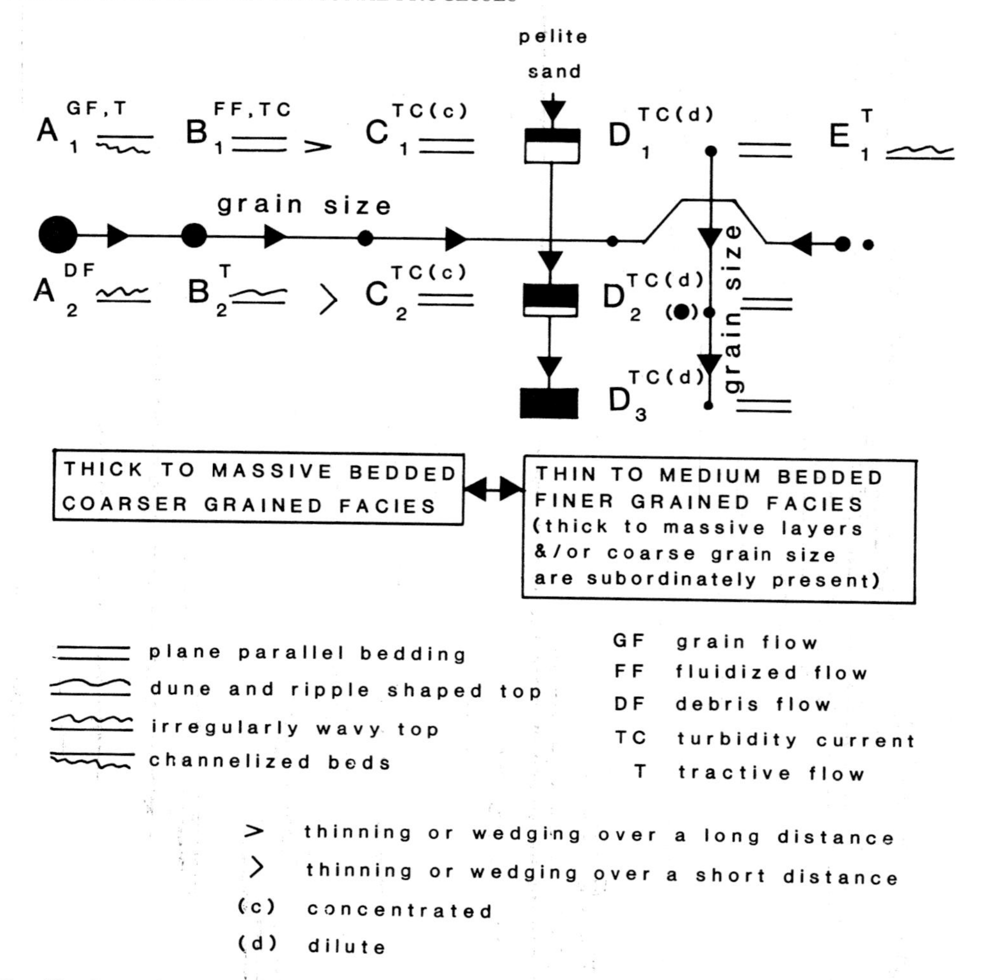

Figure 3.1 Classification of deep-water deposits redrawn from Mutti & Ricci Lucchi (1975).

lithified rock types conglomerate, sandstone, mudstone, siltstone and claystone. Bed thicknesses are defined according to Inman (1954): laminae, less than 1 cm; very thin beds, 1–3 cm; thin beds, 3–10 cm; medium beds, 10–30 cm; thick beds, 30–100 cm; and very thick beds, greater than 100 cm thick.

Our classification (Table 3.1, Fig. 3.2) is hierarchical. *Facies classes* are divided into two or more *facies groups*, which are each further subdivided into constituent *facies*. The seven facies classes are defined largely on (a) texture of the gravelly, sandy or silty divisions of the beds (Fig. 3.3); (b) relative thickness of mud interbeds or caps; and also on (c) internal organization for Facies Class F, and on composition for Facies Class G (Table 3.2). For the second-order classification, Facies Classes A–E can be divided into disorganized and organized facies groups (A1, A2, etc.), i.e. groups with beds that lack clear stratification or grading and groups with beds that show clearly defined sedimentary structures. Facies Class F is mainly disorganized and can be divided into two groups, characterized respectively by (1) exotic clasts and (2) contorted strata. A third group was included in this class by Stow (1985, 1986), but the scheme was revised to include these deposits in Facies Class A (Pickering *et al.* 1986a, this book). Facies Class G is divided into biogenic oozes, muddy oozes, biogenic muds, and chemogenic sediments.

For the purposes of large-scale mapping and reconnaissance fieldwork, subdivision of strata into facies classes or facies groups may be appropriate. For much more detailed analysis, recognition of the more specific individual facies will be necessary. In erecting models

Table 3.1 List of facies classes, groups and facies. After Pickering *et al.* (1986a).

Class / Group / Facies	See text for detailed descriptions
A	Gravels, muddy gravels, gravelly muds, pebbly sands, ≥5% gravel
A1	Disorganized gravels, muddy gravels, gravelly muds and pebbly sands
A1.1	Disorganized gravel
A1.2	Disorganized muddy gravel
A1.3	Disorganized gravelly mud
A1.4	Disorganized pebbly sand
A2	Organized gravels and pebbly sands
A2.1	Stratified gravel
A2.2	Inversely graded gravel
A2.3	Normally graded gravel
A2.4	Graded-stratified gravel
A2.5	Stratified pebbly sand
A2.6	Inversely graded pebbly sand
A2.7	Normally graded pebbly sand
A2.8	Graded-stratified pebbly sand
B	Sands, ≥80% sand grade, <5% pebble grade
B1	Disorganized sands
B1.1	Thick/medium-bedded, disorganized sands
B1.2	Thin-bedded, coarse grained sands
B2	Organized sands
B2.1	Parallel-stratified sands
B2.2	Cross-stratified sands
C	Sand-mud couplets and muddy sands, 20–80% sand grade, <80% mud grade (mostly silt)
C1	Disorganized muddy sands
C1.1	Poorly sorted muddy sands
C1.2	Mottled muddy sands
C2	Organized sand-mud couplets
C2.1	Very thick/thick-bedded sand-mud couplets
C2.2	Medium bedded sand-mud couplets
C2.3	Thin-bedded sand-mud couplets
C2.4	Very thick/thick-bedded, mud-dominated, sand-mud couplets
D	Silts, silty muds, and silt-mud couplets, >80% mud, ≥40% silt, 0–20% sand
D1	Disorganized silts and silty muds
D1.1	Structureless silts
D1.2	Muddy silts
D1.3	Mottled silt and mud
D2	Organized silts and muddy silts
D2.1	Graded-stratified silt
D2.2	Thick irregular silt and mud laminae
D2.3	Thin regular silt and mud laminae
E	≥95% Mud grade, <40% silt grade, <5% sand and coarser, ≤25% biogenics
E1	Disorganized muds and clays
E1.1	Structureless muds
E1.2	Varicoloured muds
E1.3	Mottled muds
E2	Organized muds
E2.1	Graded muds
E2.2	Laminated muds and clays
F	Chaotic deposits
F1	Exotic clasts
F1.1	Rubble
F1.2	Dropstones and isolated ejecta
F2	Contorted/disturbed strata
F2.1	Coherent folded and contorted strata
F2.2	Brecciated and balled strata
G	Biogenic oozes (>75% biogenics), muddy oozes (50–75% biogenics), biogenic mud (25–50% biogenics) and chemogenic sediments, <5% terrigenous sand and gravel
G1	Biogenic oozes and muddy oozes
G1.1	Biogenic ooze
G1.2	Muddy ooze
G2	Biogenic muds
G2.1	Biogenic mud
G3	Chemogenic sediments

to describe and define the various deep-water sedimentary environments, it is often useful to lump facies together, and it is at this level of description that our facies classes and groups become particularly useful.

We have retained the general outlines of the Mutti and Ricci Lucchi scheme. The main differences are (a) the abolition of their Facies E that is now distributed among the other facies, (b) the restriction of Facies Class D (their Facies D) to silt and silt–mud units, rather than including sands, (c) the addition of a much-needed new Facies Class E for muds, and (d) the definition of three tiers of classification rather than two, allowing for a greater number of facies within a scheme that is still manageable.

We do not use the term 'facies association' in our classification because facies associations represent the

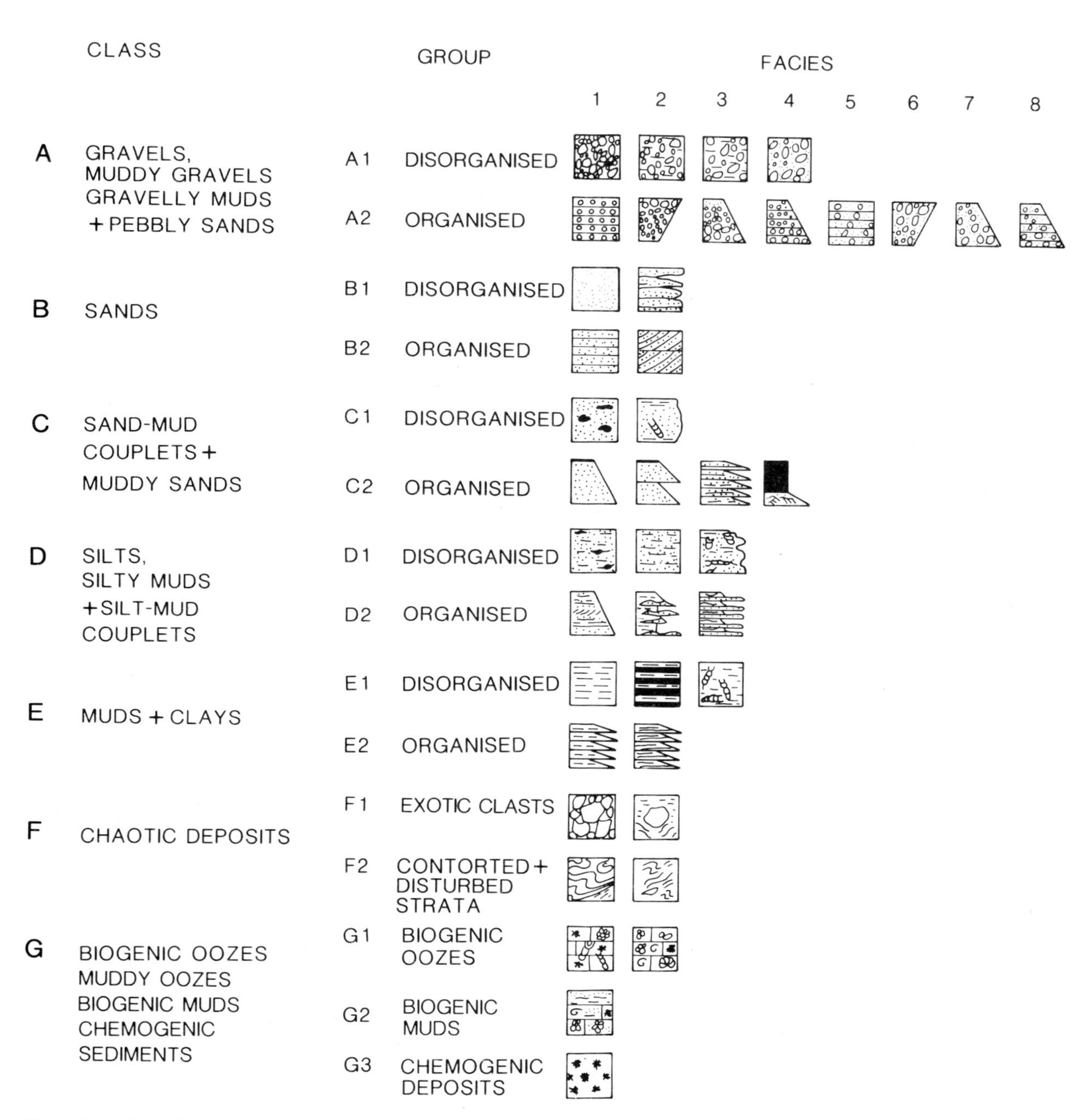

Figure 3.2 Classification scheme for deep-water sediments used in this book, from Pickering *et al.* (1986a). Facies classes are defined on the basis of grain size (Facies Classes A–E), internal organization (Facies Class F) and composition (Facies Class G). Facies groups are distinguished mainly on the basis of internal organization of structures and textures. Individual facies are based on internal structures, bed thicknesses and composition.

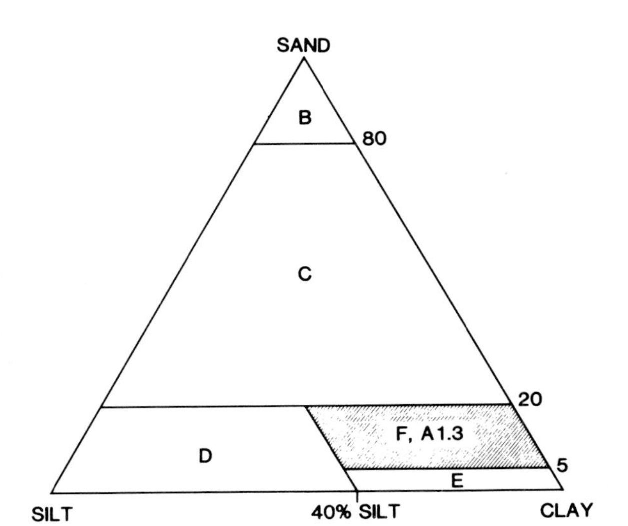

Figure 3.3 Sketch of textural limits of Classes B, C, D and E, neglecting gravel and biogenic clasts. For Class B, the mud component (silt + clay) forms an interstitial matrix, whereas for Classes C, D and E the mud is both matrix and interbedded, genetically related, mud. The ruled field is only occupied by chaotic or mudflow deposits, or by clayey muds with scattered ice-rafted sand and gravel.

Table 3.2 Criteria for recognition of facies classes.

Class	Texture of basal coarse divisions % gravel	% sand	% mud	Typical ratio of mud caps to basal divisions[a]
A	≥5	<95	<95	<1:10
B	<5	≥80	<20	<1:10
C	<5	20–80	<80*	<1:1, rarely to 4:1
D	~0	0–20	>80*	<1:1
E	~0	<5	≥95*	NA
F	----------- chaotic -----------			NA
G	Chemogenic; or <5% terrigenous sand and gravel, mixtures of hemipelagic mud & biogenics, biogenics 25–100%.			NA

NA = not applicable due to muddy nature of all deposits.
[a]Ratios are appropriate for DSDP-type cores at depths exceeding about 500 m. The ratios would need slight modification for rocks, and more substantial modification for piston cores (see compaction data in Hamilton 1976).
*For classes C, D and E, much of the mud component in basal 'cohesionless' divisions is silt, not clay.

temporal and spatial arrangement of facies in the sedimentary record. Our facies classes and groups, however, are composed of facies taken out of context from adjacent facies.

Facies Class A: gravels, muddy gravels, gravelly muds, pebbly sands, ≥5% gravel grade

Facies Class A consists of the coarsest grained members of deep-water clastic sediments, with greater than 5% pebble-grade or coarser material. This facies class includes clast-supported gravels, gravels with a supporting sand matrix, muddy gravels and gravelly muds. The latter two lithologies may contain more mud than gravel, but their transport process may be identical to that of deposits of this class with a smaller amount (<10%) of mud matrix; these mud-rich deposits were included in Facies Class F by Stow (1985, 1986).

Deep-water gravels and pebbly sands are commonly termed 'resedimented' to set them apart from the fluvial and shallow-marine deposits; they are believed to have accumulated first in shallow water and subsequently to have been transported into deeper water (Walker 1975). In many cases, features of resedimented gravels taken in isolation from the associated facies and other features of a succession are not sufficient evidence for deep-water deposition.

Walker (1975, 1976, 1977, 1978) established a descriptive model for resedimented conglomerates (Fig. 3.4) that incorporated the sedimentary structures into a sequence analogous to the Bouma sequence (Bouma 1962) for turbidites. Walker believed that the depositional process and the rate of deposition determined the degree of organization in the final conglomerate deposit. His fundamental dichotomy between 'organized' and 'disorganized' conglomerates forms the basis of most other classifications (e.g. Kelling & Holroyd 1978, Piper *et al.* 1978), including ours. Walker (1975) speculated that conglomerate facies are arranged spatially such that disorganized beds are most proximal, inverse-to-normally graded beds are intermediate in position, and graded and graded-stratified beds are most distal. Surlyk (1984) has carefully documented the spatial distribution of conglomerate facies for 15 km away from a steep, faulted basin margin in east Greenland, and finds no such spatial pattern. Instead, conglomerate facies appear to have no simple relationship with proximality.

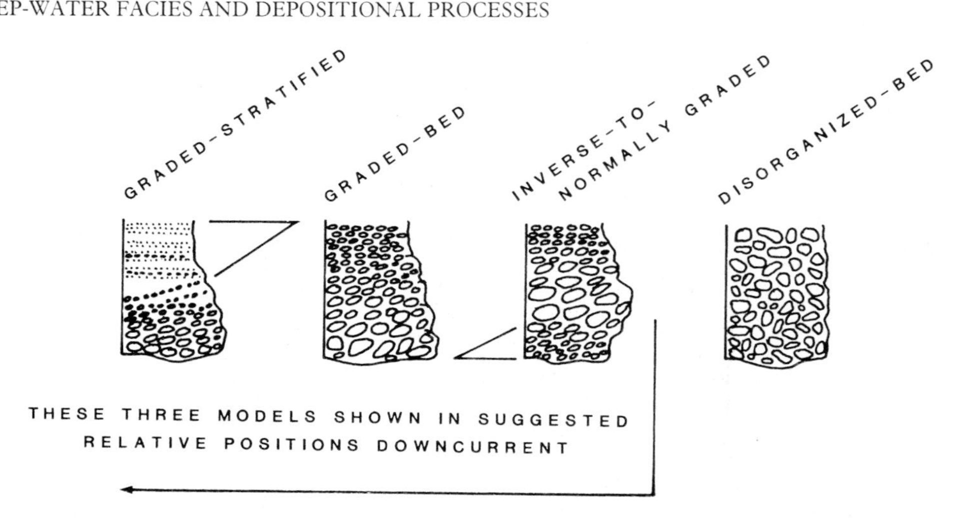

Figure 3.4 Four models for resedimented conglomerates redrawn from Walker (1978). Characteristics are: 1, 'graded-stratified' model – imbrication, stratification, no normal grading; 2, 'graded bed' model – imbrication, no inverse grading, no stratification; 3, 'inverse-to-normally graded' model – imbrication, no stratification; and 4, 'disorganized-bed' model – imbrication rare, no grading, no inverse grading, no stratification.

FACIES GROUP A1 – DISORGANIZED GRAVELS, MUDDY GRAVELS, GRAVELLY MUDS AND PEBBLY SANDS

Gravels in this group may be supported by clast contacts, by a sand matrix, or by a mud matrix. Bed thickness varies, although these deposits tend to occur in medium to thick and very thick beds. The shape of beds generally reflects the topography over which the sediment gravity flows travelled. Four facies are recognized: A1.1, disorganized gravels; A1.2, disorganized muddy gravels; A1.3, disorganized gravelly muds; and A1.4, disorganized pebbly sands.

FACIES A1.1 – DISORGANIZED GRAVEL

In general, this facies is more thickly bedded than other gravel facies. Exceptionally, single beds may be several tens of metres thick; more commonly, thicknesses are from 0.5–5 m. In some cases, beds or layers may be thin to very thin stringers of gravel as little as one pebble thick. Beds may be flat-based to deeply scoured. Upper surface geometry may be irregular, wavy, or with individual clasts projecting out of the bed. Laterally, there may be gradual dilution of the clasts by sand, such that ill-defined stringers of clasts give a crude stratification and the gravel body has a pod-like shape. Alternatively, primary flow margins may be abrupt snouts with a shape like the margin of a water droplet on a smooth horizontal surface.

Clast size ranges from fine pebble to boulder grade, and beds are characteristically poorly sorted (Figs 3.5a & 5.25). Clast shape is dependent upon composition and inherited shape; consequently, disorganized gravels have been described with both angular and well-rounded clasts. Clasts lack a well ordered fabric, although concentrated, elongate clasts may exhibit a poorly defined parallel alignment with bedding, or a slight imbrication (Hiscott & James 1985). Variously oriented, plastically deformed, mudstone intraclasts are common. Facies A1.1 may pass laterally and/or vertically into Facies A1.4.

Transport process: high concentration turbidity currents or debris flows.

Depositional process: 'freezing' on decreasing bottom slopes due to intergranular friction and cohesion.

Selected references: Hendry (1973), Marschalko (1975), Carter & Norris (1977), Long (1977), Stanley *et al.* (1978), Surlyk (1978, 1984), Winn & Dott (1978), Johnson & Walker (1979), Nemec *et al.* (1980), Hein & Walker (1982), Hiscott & James (1985).

FACIES A1.2 – DISORGANIZED MUDDY GRAVEL

Facies A1.2 is matrix supported, structureless muddy gravel with 10–50% mud- or clay-grade matrix (Fig. 5.25). Units are medium bedded to very thick bedded. Bed shape may appear tabular in small outcrops, but many beds taper to a blunt snout. The bases of beds generally show little erosion into underlying units, but the tops of beds are commonly irregular and hummocky. Large cobbles and boulders may be evenly dispersed throughout the bed, may project above the top of the unit, or may define a coarse-tail grading. Enormous blocks or olistoliths may be contained in 100–200 m-thick megabeds with divisions identical to this facies, e.g. a block 300 × 150 × 30 m in a 170 m-thick megabed in Yugoslavia (Marjanac 1985), and platform carbonate slabs 50 m thick and about 1 km long in 'megaturbidites' to about 200 m thick in Spain (Labaume *et al.* 1983).

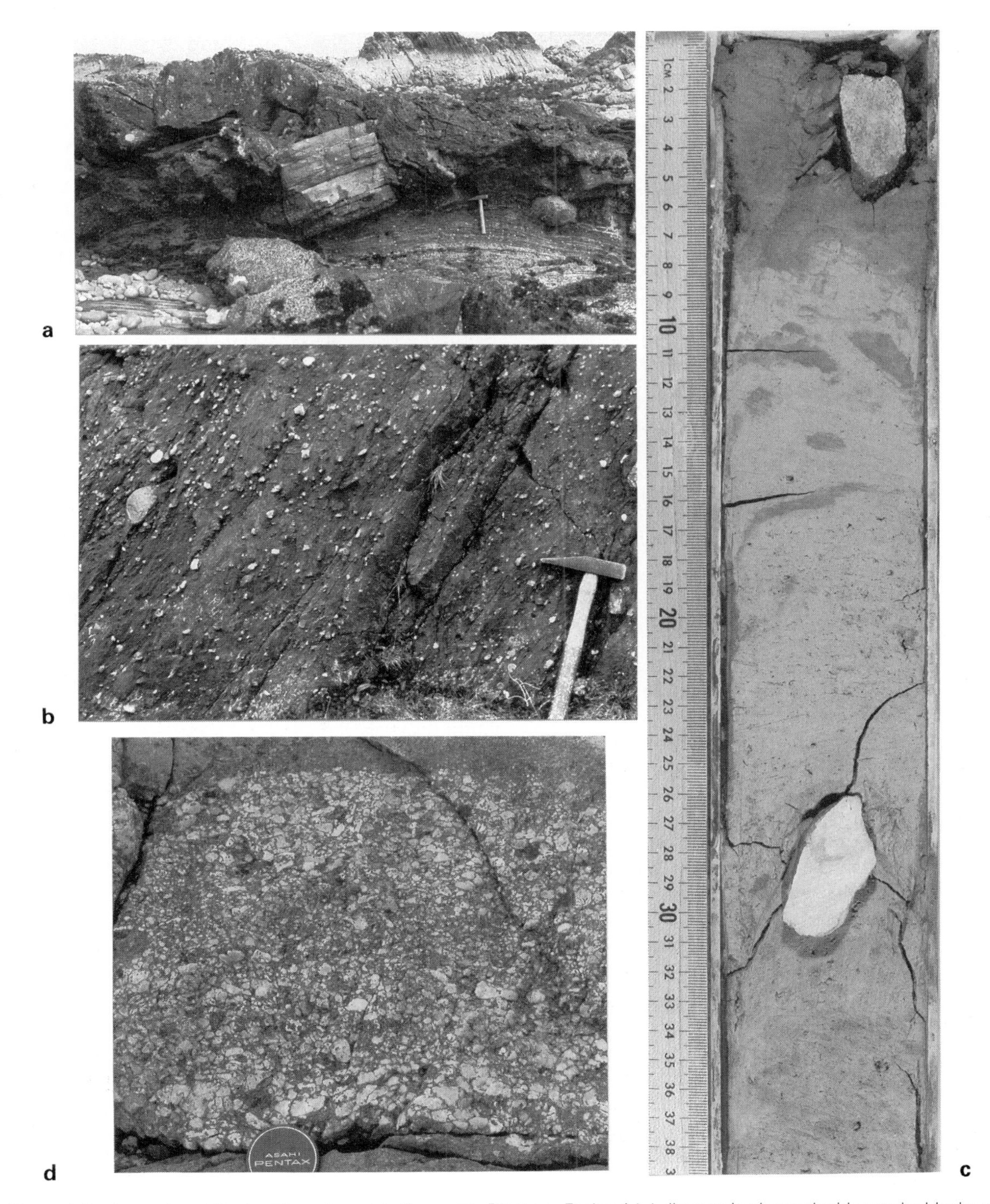

Figure 3.5 Examples of Facies Class A, mainly Group A1. **Photo a**, Facies A1.1 disorganized gravel with angular blocks of Devonian sandstone up to *c.* 1 m across, Upper Jurassic Boulder Beds, NE Scotland. **Photo b**, Facies A1.3 disorganized gravelly mud, upper Precambrian Kongsfjord Formation, N Norway (hammer length 35 cm). **Photo c**, upper part of a sharp-based, graded, ice-rafted unit cored at Site 645C, ODP Leg 105, Baffin Bay. Pale colour results from abundant detrital carbonate. Gravel content in lower part is 5–10%. Grading results from rapid initiation and slow decline of iceberg melting. The upper part of the unit is burrow mottled. Interval shown is 105–645C–3H–5, 0–39 cm. Water depth at the site is *c* 2020 m. **Photo d**, Facies A2.1 stratified gravel with abrupt top, upper Precambrian Kongsfjord Formation, N Norway (lens cap 5 cm across).

Clasts tend to show a polymodal grain size distribution, and have a poorly ordered fabric. If a fabric exists, it occurs as a poorly defined parallel to sub-parallel lamination and/or a crude alignment of platy or rod-shaped clasts in bedding. Clast composition may be igneous, sedimentary, metamorphic, lithified biogenic material, or unlithified sediment. This facies is commonly associated with other Class A deposits or with deposits of Class F.

Transport process: cohesive debris flows. Enormous slabs may slide into place on a cushion of overpressured or liquefied mud (Labaume *et al.* 1983).

Depositional process: 'freezing' on decreasing bottom slopes due to intergranular friction and cohesion.

Selected references: Jeffery (1922), Crowell (1957), Johnson (1970), Hampton (1972), Mutti & Ricci Lucchi (1972), Rodine & Johnson (1976), Embley (1976), Enos (1977), Kurtz & Anderson (1979), Winn & Dott (1979), Damuth & Embley (1981), Page & Suppe (1981), Lowe (1982), Naylor (1982), Labaume *et al.* (1983), Middleton & Southard (1984), Hiscott & James (1985), Marjanac (1985).

FACIES A1.3 – DISORGANIZED GRAVELLY MUD

Facies A1.3 includes (a) the pebbly mudstones and olistostromes described from many rock successions, particularly at ancient active continental margins, and (b) ice-rafted deposits with high concentrations of gravel. Characteristics are much like those of Facies A1.2, except that the deposits contain 50–95% mud- or clay-grade sediment (Figs 3.5b, 5.36 & 10.21d). Beds range from decimetres to metres in thickness, although individual beds may be tens of metres thick. Most beds are laterally discontinuous and highly irregular in shape, and display marked variations in the degree of internal organization, matrix content and bed shape over very short lateral distances. Grading is absent in pebbly mudstones and olistostromes, but may be present in ice-rafted units (Fig. 3.5c). Internally, however, these latter deposits are disorganized.

Clast compositions are like those for Facies A1.2, with the common addition of abundant silt-mud chips and slabs. In ancient examples, the ductility contrast between the matrix and clasts may result in a tectonically sheared matrix surrounding relatively undeformed clasts.

Transport process: cohesive mud flows (debris flows), and settling from melting icebergs.

Depositional process: for mud flows, 'freezing' on decreasing bottom slopes as shear stress at the base of the flow becomes less than the cohesive strength. For ice-rafted sediments, grain-by-grain deposition

Selected references: as for Facies A1.2.

FACIES A1.4 – DISORGANIZED PEBBLY SAND

Facies A1.4 is distinguished by the dispersion of larger clasts in a sand matrix. Mud-grade sediment can account for up to a few per cent. Bed shape and thickness are similar to Facies A1.1. Where clasts are widely dispersed, the definition of bedding surfaces is poor. Scouring, loading, and large sole marks are well documented from pebbly sands. Clasts of fine to coarse pebble grade appear to be most common, with dispersed cobbles and boulders being less common. Clast concentration is variable, with irregular patches and stringers of more concentrated clasts occurring down to one pebble in thickness; mudstone clasts are also common to Facies A1.4, and where concentrations are very high, and the clasts are angular, the beds are best termed mud-flake breccias.

Grading, stratification, and preferred orientation of the larger clasts are generally absent. Larger clasts may be concentrated toward the base of a bed and then pass abruptly up into dilute pebbly sand. Alternatively, there may be a gradual upward decrease in the size of the 'floating' clasts to give the appearance of a coarse-tail grading.

Transport process: high concentration turbidity currents.

Depositional process: rapid collective grain deposition of a pebble-sand mixture due to increased intergranular friction as the flows decelerate.

Selected references: Dzułynski *et al.* (1959), Bartow (1966), Ricci Lucchi (1969), Walker & Mutti (1973), Carter & Lindqvist (1975), Lowe (1976a), Surlyk (1978), Walker (1978), Winn & Dott (1978), Hein (1982), Hein & Walker (1982).

FACIES GROUP A2 – ORGANIZED GRAVELS AND PEBBLY SANDS

Organized gravels and pebbly sands are described with close reference to the evolutionary scheme of Lowe (1982). Our recognition of eight facies (Table 3.1) may appear cumbersome, but the facies are easy to recognize, and in predominantly conglomeratic successions these divisions should prove considerably more useful than any simpler classification scheme.

FACIES A2.1 – STRATIFIED GRAVEL

Stratification in resedimented gravels is most commonly reported in fine pebble gravels (Figs 3.5d, 3.6a) and pebbly sands; the best ancient examples of stratified, coarse grained, clast-supported gravels known to us have been described by Winn & Dott (1977, 1979) from southern Chile. These beds are lenticular and wedge-shaped with inclined strata up to 12 m thick and with dune-shaped bodies up to 4 m thick. Individual strata range from a single pebble thickness to over 1 m thick. Scours and erosional sole marks have been described from these ancient gravels.

Imbrication is well developed: Winn & Dott (1977, 1979) report *ab* planes of clasts dipping upstream with *a*-axes parallel to the flow direction. Large-scale stratification, especially where stratal boundaries are not sharp, may be difficult to distinguish from individual, stacked, graded and structureless gravel beds.

Modern gravel waves of 2–5 m height and 50–100 m wavelength have been described from water depths of 2000–4500 m on the Laurentian Fan (Piper *et al.* 1985). These are inferred to be

Figure 3.6 Examples of Facies Class A, Group A2. **Photo a**, Facies A2.1 stratified gravel showing cross stratification, Cambro–Ordovician Cap Enragé Formation, Quebec, Canada. 15 cm notebook for scale. **Photo b**, Facies A2.2 inversely graded gravel, Upper Ordovician – Lower Silurian Milliners Arm Formation, central Newfoundland, Canada. **Photo c**, Facies A2.3 normally graded gravel in shallow channel, Milliners Arm Formation, central Newfoundland, top to left, hammer length 35 cm. **Photo d**, Facies A2.4 graded-stratified gravel, Milliners Arm Formation, central Newfoundland, 15 cm scale. **Photo e**, Facies A2.5 stratified pebbly sand, Milliners Arm Formation, central Newfoundland, top to left, hammer length 35 cm. **Photo f**, Facies A2.8 graded-stratified pebbly sand, Milliners Arm Formation, central Newfoundland, 15 cm scale.

bedload deposits of the powerful 1929 Grand Banks turbidity current that travelled at speeds in excess of 18 m/s (Heezen & Ewing 1952, Piper *et al.* 1985). Although not sampled, internal sedimentary structures in these gravel waves probably include medium- to large-scale cross stratification.

Transport process: high concentration turbidity currents.

Depositional process: grain-by-grain deposition from suspension and then traction transport as bed-load.

Selected references: Winn & Dott (1977, 1979), Hein (1982), Piper *et al.* (1985).

FACIES A2.2 – INVERSELY GRADED GRAVEL

Inversely graded gravels make up a significant proportion of many resedimented, coarse grained successions. Beds are commonly lenticular, with basal erosion, lateral thinning, and variations in clast concentration causing complex bed shapes. Inversely graded beds reach a maximum thickness of several metres, but most commonly are 0.5–4 m thick. Poor sorting and large clast sizes are typical of this facies. Clast imbrication is well developed.

The entire bed may be inversely graded (Figs 3.6b, 5.34), or the gravel overlying the basal inversely graded part may be structureless or normally graded. In thicker inversely graded beds, the lowest 5–20% of the bed commonly contains finer clasts (usually fine to coarse pebbles) than the immediately overlying part. The transition between these two parts of the bed is commonly abrupt; the coarser clasts are concentrated at a certain distance above the base of the bed.

The tops of inversely graded gravels may be abrupt, showing a sharp break between gravel and sand, or there may be an upward increase in sand content such that the uppermost part of a bed has a bimodal size distribution.

Transport process: high concentration turbidity currents.

Depositional process: rapid deposition of a concentrated traction carpet/dispersion near the bed due to increased intergranular friction. The inverse grading and strong imbrication are both caused by intense grain interaction.

Selected references: Davies & Walker (1974), Surlyk (1978, 1984), Howell & Link (1979), Johnson & Walker (1979), Winn & Dott (1979), Nemec *et al.* (1980), Watson (1981), Hein (1982).

FACIES A2.3 – NORMALLY GRADED GRAVEL

Normally graded gravels tend to be finer grained than inversely graded or disorganized beds within any single succession. However, bed thicknesses are similar to those of the other gravel facies, ranging from about 0.5 to several metres (Figs 3.6c & 5.43b). Beds show marked thickness changes as a result of localized deep scours and more gradual, large-scale down-cutting into pebbly sands. Normally graded gravel appears to be one of the most abundant of the clast-supported facies.

Normal grading occurs in several modes. Abruptly graded beds mainly show a coarse-tail grading where the coarsest clasts are only present in the lowest part of the bed and rapidly give way upward to fine pebbles. Gradual distribution grading from cobbles to granule sand is less common, but can occur in very thick beds. Imbrication appears to be less well developed in this facies than in the inversely graded gravel facies (Facies A2.2).

Transport process: high concentration turbidity currents.

Depositional process: grain-by-grain deposition from suspension. The clasts undergo little or no traction transport after reaching the bed, probably because of relatively rapid deposition.

Selected references: Marschalko (1964), Hendry (1972, 1978), Mutti & Ricci Lucchi (1972), Davies & Walker (1974), Walker (1977), Winn & Dott (1978), Nemec *et al.* (1980), Hein (1982), Hein & Walker (1982), Surlyk (1984).

FACIES A2.4 – GRADED-STRATIFIED GRAVEL

Walker (1975, 1976) established a 'graded-stratified bed model' in which pebbly sands with parallel, inclined and cross stratification overlie graded gravel (Fig. 3.6d). Beds of graded-stratified gravel are generally thinner bedded and finer grained than other clast-supported gravels. Bed shape is less variable, with sharp planar bases, although some scouring characterized by trough-shaped scour-and-fill stratification is common.

Clast size may decrease progressively toward the top of the bed. Alternatively, the lower clast-supported part of the bed may show 'delayed grading'. Graded-stratified gravels are considered to be transitional between clast-supported cobble/pebble gravels and graded-stratified pebbly sands (Facies A2.8).

Transport process: high concentration turbidity currents.

Depositional process: grain-by-grain deposition from suspension followed by traction during deposition of the upper part of the bed only.

Selected references: Hubert *et al.* (1970), Hendry (1972), Mutti & Ricci Lucchi (1972), Davies & Walker (1974), Rocheleau & Lajoie (1974), Walker (1975, 1976, 1977, 1978), Aalto (1976), Surlyk (1978, 1984), Johnson & Walker (1979), Hein (1982), Hein & Walker (1982).

FACIES A2.5 – STRATIFIED PEBBLY SAND

Beds of stratified pebbly sands show an alternation of pebble- and sand-rich layers (Fig. 3.6e). Commonly, individual strata have gradational contacts with both normal and inverse grading. Strata may pinch and swell and split into irregular stringers and lenses. Stratification may also occur on a finer scale with alternations of (a) granule sand with a few pebbles and (b) coarse to medium grained sand.

Bed shape is extremely variable, with typical bed thicknesses from 0.5–3 m. Defining individual flow deposits may be difficult because a pebbly sandstone unit may be composite. Generally, entire beds are poorly sorted; coarse pebbles and, rarely, cobbles, may be present as irregular stringers and scour fills.

Transport process: high concentration turbidity currents.

Depositional process: grain-by-grain deposition from suspension followed by traction transport as bed load.

Selected references: Hendry (1973, 1978), Hein (1982), Hein & Walker (1982), Lowe (1982), Surlyk (1984).

FACIES A2.6 – INVERSELY GRADED PEBBLY SAND

Pebbly sands with well developed inverse grading throughout are analogous to the inversely graded gravels of Facies A2.2, and appear to be relatively rare. It is more usual to find a pebble-free zone several centimetres thick that passes rapidly up into structureless or normally graded pebbly sand. Alternatively, there may be alternations at a scale of 5–10 cm of (a) pebble-poor zones and (b) pebbly sand, giving an indistinct stratification. In such cases, it is very difficult to define individual sedimentation units.

Facies A2.6 tends to occur in planar-based beds that are generally thinner than beds of other pebbly-sand facies. Well developed, multiple, inversely graded layers suggest a transition into thicker bedded granule sands with 'near-horizontal stratification' (Facies B2.1).

Transport process: high concentration turbidity currents.

Depositional process: rapid deposition by frictional 'freezing' of a traction carpet driven along by shear at the base of the flow.

Selected reference: Lowe (1982).

FACIES A2.7 – NORMALLY GRADED PEBBLY SAND

Normally graded pebbly sands are very common in deep-water clastic successions. Generally, this facies is thicker bedded than the stratified and inversely graded pebbly sands. Common scour structures tend to give most beds an irregular appearance. Bed contacts may be diffuse where amalgamation occurs and clast concentrations are low.

Facies A2.7 typically displays well defined normal grading that is most commonly coarse-tail grading, although distribution grading also occurs. There are many reported examples of beds 2–3 m thick showing normal grading from base to top.

Transport process: high concentration turbidity currents.

Depositional process: grain-by-grain deposition from suspension, with rapid burial and no significant traction transport on the bed.

Selected references: Hubert *et al.* (1970), Aalto (1976), Long (1977), Walker (1977, 1978), Stanley *et al.* (1978), Watson (1981), Hein (1982).

FACIES A2.8 – GRADED-STRATIFIED PEBBLY SAND

Facies A2.8 is similar to Facies A2.7 in terms of bed thickness, bed shape and clast size. Lateral transitions between these two facies are common. Basal scouring is common, and upper bed contacts are sometimes poorly defined where stringers of pebbles occur high in the bed.

Beds of this facies show an overall grading from base to top, although layers of coarser clasts are repeated upward throughout beds (Fig. 3.6f). However, clasts coarser than very fine pebbles and granules appear to be confined to the lower graded portions of beds. Stratification may be parallel, oblique, multiple sets, draping scours, or megaripple cross-bedding. Facies A2.8 is considered to be transitional between graded-stratified gravels of Facies A2.4 and the sands of Facies Class B.

Transport process: high concentration turbidity currents, becoming more dilute with time at a single locality.

Depositional process: grain-by-grain deposition from suspension. Initially, deposition is so rapid that no subsequent traction transport takes place. At higher levels in the deposit, grains are transported as bedload to form stratification before being buried.

Selected references: Hubert *et al.* (1970), Rocheleau & Lajoie (1974), Aalto (1976), Walker (1978), Hein (1982), Hein & Walker (1982), Surlyk (1984).

Facies Class B: sands, ≥80% sand grade, <5% pebble grade

This class comprises sand beds with <20% mud and silt matrix, and <5% pebble-grade material. Facies Class B is divided into an organized and a disorganized facies group, based on the presence or absence of well defined sedimentary structures. Bed thickness and shape are highly variable. Most beds in this facies class cannot be described successfully using the Bouma (1962) scheme for classic turbidites.

Deposits with characteristics common to our Facies Class B are well documented; for example, some of the 'arenaceous facies' of Mutti & Ricci Lucchi (1972), the structureless beds of Sanders (1965), some 'fluxoturbidites' of Stanley & Unrug (1972), Dzułynski *et al.* (1959) and Kuenen (1964).

FACIES GROUP B1 – DISORGANIZED SANDS

Disorganized or structureless sands, comparable to Walker & Mutti's (1973) Facies B1 and B2 massive sands with or without dish structure, and Mutti & Ricci Lucchi's (1972) Facies B sands, are recorded from many flysch successions (Stauffer 1967, Carter & Lindqvist 1975, Keith & Friedman 1977, Piper *et al.* 1978, Hiscott & Middleton 1979, Cas 1979, Hiscott 1980, Lowe 1982). Facies Group B1 consists of two facies.

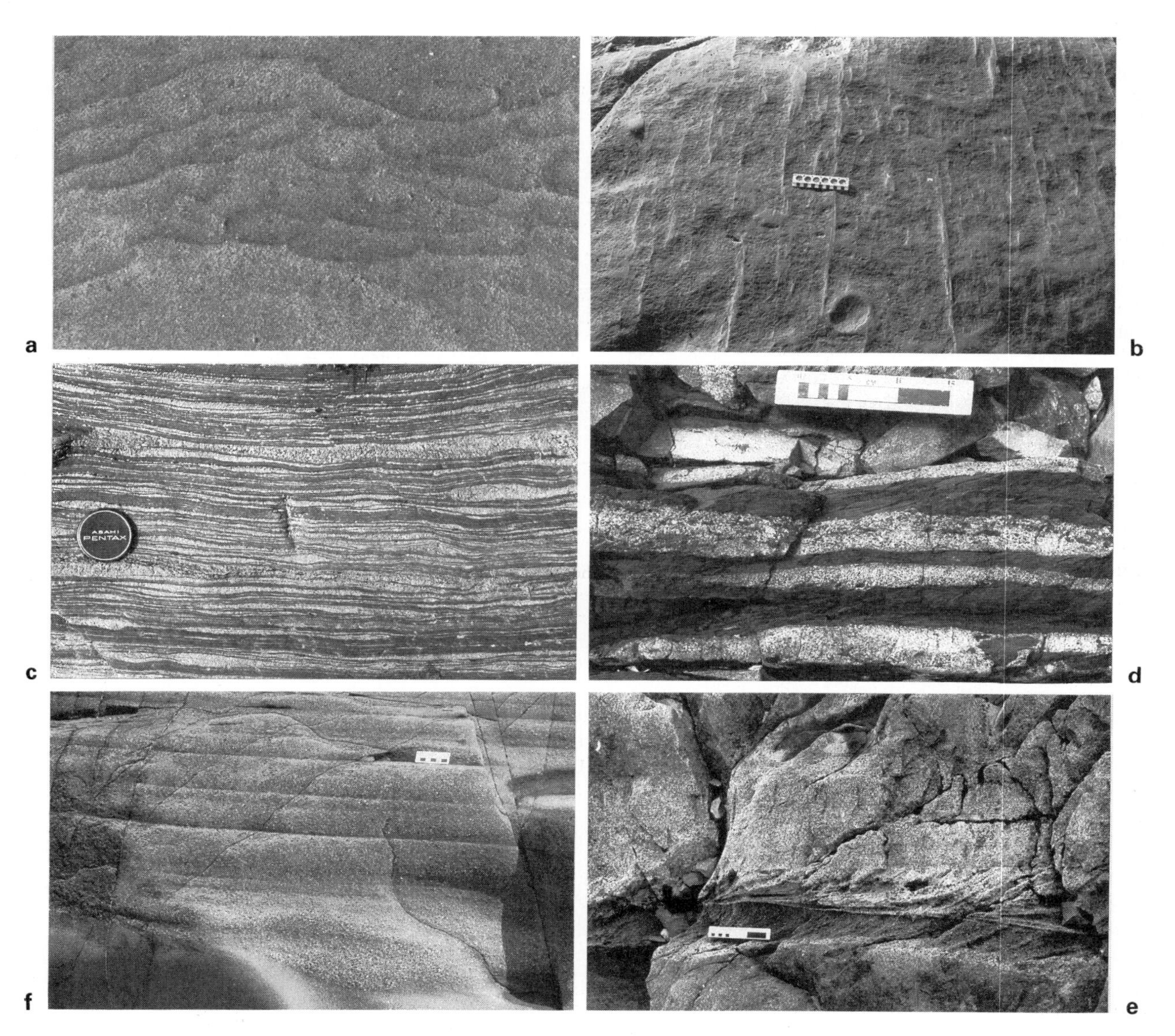

Figure 3.7 Examples of Facies Class B. **Photo a**, Facies B1.1 disorganized sand with dish structures in slightly oblique cross-section, Cambro–Ordovician Cap Enragé Formation, Quebec, Canada (view 40 cm wide). **Photo b**, Facies B1.1 disorganized sand with vertical fluid-escape pillars, Ordovician Tourelle Formation, Quebec, Canada, 15 cm scale. **Photo c**, Facies B1.2 thin-bedded, coarse grained sands as lenticular beds 1–3 cm thick interbedded with Facies D2.2, Upper Jurassic Boulder Beds, NE Scotland (lens cap 5 cm across). **Photo d**, Facies B1.2 thin-bedded, coarse grained sands (some at granule grade), Upper Ordovician – Lower Silurian Milliners Arm Formation, central Newfoundland, Canada, 15 cm scale. **Photo e**, Facies B2.1 parallel-stratified sand, Ordovician Cloridorme Formation, Quebec, Canada, 5 cm scale. Individual stratification bands are inversely graded, and are an internal sedimentary structure within a thicker graded bed. **Photo f**, Facies B2.2 cross-stratified sand, upper Precambrian Kongsfjord Formation, N Norway, 15 cm scale. The overlying sand belongs to Facies B2.1.

FACIES B1.1 – THICK/MEDIUM-BEDDED, DISORGANIZED SANDS

Facies B1.1 consists of laterally continuous, parallel-sided to highly irregular, medium to thick beds (Fig. 7.27). Sole marks tend to be rare. Grading is absent, or poorly developed as a coarse-tail grading with small pebbles and granules concentrated in a thin basal layer. The most obvious internal sedimentary feature may be fluid-escape structures. These tend to occur in the upper half of beds and include subvertical sheet structures (Laird 1970), dish structures (Fig. 3.7a), and fluidization pipes and pillars (Fig. 3.7b) (Wentworth 1967, Lowe & LoPicollo 1974, Lowe 1975). Dish structures seem to be characteristic of the better sorted sands of this facies, in which upward percolation of escaping pore water is able to form relatively impermeable 'consolidation laminations' (Lowe & LoPicollo 1974) that, when breached, develop a characteristic dish shape. In more poorly sorted sands, pore-fluid escape is not general, but is localized along pillars and sheets (e.g. Hiscott & Middleton 1979).

Transport process: high concentration turbidity currents.

Depositional process: rapid mass deposition due to intergranular friction in a concentrated dispersion near the bed. The resultant open grain packing may collapse during or after deposition of the entire bed, resulting in escape of surplus pore fluids and formation of fluid-escape structures.

Selected references: Stauffer (1967), Middleton (1970), Carter & Lindqvist (1975), Aalto (1976), Lowe (1976b), Keith & Friedman (1977), Piper *et al* (1978), Cas (1979) (a2 division), Hiscott & Middleton (1979), Jordan (1981), Hein (1982), Lowe (1982) (S3 division), Surlyk (1984).

FACIES B1.2 – THIN-BEDDED, COARSE GRAINED SANDS

Facies B1.2 is distinguished from the other facies in the class by its thin-bedded nature and very coarse grain size (Figs 3.7c & d, 5.33E). Pebbles are rare. Angular silt and mud clasts may occur. The beds are internally structureless.

Beds are typically irregular with common wedge-shaped or pinch-and-swell geometry. Tops are sharp. This facies is associated with Facies B2.2 in many cases, and in some respects resembles Facies E of Mutti & Ricci Lucchi (1972, 1975), but is generally coarser grained and lacks internal structures. There is no grading, and rarely small pebbles may occur within beds as stringers. In some beds, pebbles protrude above the bed top into overlying facies.

Transport process: bed-load transport beneath turbidity currents or strong bottom currents. Transport may be short, and the major process may be winnowing out of finer grain sizes.

Depositional process: grain-by-grain deposition of bed load.

Selected references: Mutti & Ricci Lucchi (1972, 1975), Mutti (1977).

FACIES GROUP B2 – ORGANIZED SANDS

This group includes any sands showing clearly defined sedimentary structures that are clearly not part of the Bouma sequence for sand-mud turbidites. Various fluid-escape structures may be present, but do not destroy substantial amounts of the original structures. Our Facies Group B2 contains deposits that show many of the features of Mutti & Ricci Lucchi's (1972) subfacies B1 and B2. We recognize two facies in this group.

FACIES B2.1 – PARALLEL-STRATIFIED SANDS

Hiscott & Middleton (1979, 1980) were the first to document this facies in detail. Much of the stratification in this facies is defined by bands several centimetres to 10 cm thick, each showing an inverse distribution grading (Fig. 3.7e) although, overall, there is a normal grading within the bed. The base of stratification bands may be erosive over lateral distances of several metres. From the base to the top of each stratification band, Hiscott & Middleton (1979) recognized the following sequence of structures and textures: (a) a basal horizontal or near-horizontal erosional surface; (b) a subdivision of inversely graded sand, typically grading from 2–3 ϕ to approximately 1 ϕ, and (c) a subdivision of structureless or massive -1 to 1 ϕ sand, commonly with well developed grain imbrication. Sand between divisions of near-horizontal stratification is structureless and may contain fluid-escape structures like Facies B1.1. Bed tops grade into silt and may have an upper division of ripple lamination.

Transport process: high concentration turbidity currents.

Depositional process: 'freezing' of successively generated (and successively thinner) traction carpets at the base of the flow. Intense grain interaction in this layer produces both the strong imbrication and the inverse grading. Structureless divisions record rapid grain-by-grain fall-out from suspension or 'freezing' of a thicker, unsorted layer (like Facies B1.1).

Selected references: Chipping (1972), Mutti & Ricci Lucchi (1975) (subfacies B1), Hendry (1972), Hiscott & Middleton (1979, 1980).

FACIES B2.2 – CROSS-STRATIFIED SANDS

The granule grade to coarse grained sands of Facies B2.2 are much better sorted than sands of other facies. The thin-bedded varieties of this facies are distinctive in having a very coarse grain size for their thickness.

The beds consist of cross-stratification in sets that are typically from 10–25 cm thick (Fig. 3.7f). They may occur as solitary sets or cosets that are tabular to trough shaped. Internally, the lamination may have a low dip, or may rest at or near the angle of repose. Beds are commonly irregular, with lensing, splitting and amalgamation; basal contacts may be erosive and the tops are sharply defined. Foresets are defined by alternations of coarser and finer grained layers with concentrations of coarser grained material toward the toes of the foresets. Oversteepened and recumbently folded cross-stratification (similar to Type 1 of Allen & Banks 1972) may occur in Facies B2.2. Beds rarely contain abundant mud chips.

Transport process: bed-load transport beneath dilute turbidity currents or strong bottom currents in confined channels.

Depositional process: avalanching (grain flow) or intermittent suspension transport of grain dispersions over the crests of medium- to large-scale bedforms, or into scours.

Selected references: Hubert (1966b), Scott (1966), Piper (1970), Mutti & Ricci Lucchi (1972), Mutti (1977), Hiscott & Middleton (1979), Hein (1982), Hein & Walker (1982), Pickering (1982a), Lowe (1982). NB lenticular complexes with relatively thin-bedded 'dune' forms were defined as Facies E by Mutti & Ricci Lucchi (1972) and Mutti (1977); we assign such beds to this facies.

Facies Class C: sand–mud couplets and muddy sands, 20–80% sand grade, <80% mud grade (mostly silt)

Most beds of Facies Class C can be described using the Bouma (1962) sequence for classic turbidites. Bed shape is variable and cannot be used to differentiate the constituent facies. In general, however, beds are sheet like. In many cases, the deposits of single sediment gravity flows are graded, with most of the mud present as a cap on the top of the bed. In these cases, the lower sandy divisions may not be particularly muddy, even though the entire deposit is best called a muddy sand.

Deposits of this class characterize flysch successions (e.g. Kuenen & Migliorini 1950, Dzułynski *et al.* 1959, Bouma 1962, 1964, Bouma & Brouwer 1964, Dzułynski & Walton 1965). The depositional processes for sand beds in this class are well understood because of considerable experimental and theoretical work (Kuenen 1951, Middleton 1966a & b, 1967, Pantin 1979, Lüthi 1981, Southard & Mackintosh 1981, Parker 1982, Middleton & Southard 1984).

This facies class is similar to, although not strictly analogous to, Facies C1 and C2 of Mutti & Ricci Lucchi (1972, 1975) and Walker & Mutti (1973). Facies Class C beds are the 'classic' or 'classical' turbidites of Walker (1976, 1978, 1984). The terms 'proximal' and 'distal' as qualifying terms for the beds of this class (Walker 1967, 1970) are not employed in our classification to distinguish thicker from thinner beds, because it is now recognized that thin beds need not be distal relative to the point of sediment supply (Nelson *et al.* 1975).

Facies Class C is divided into an organized and a disorganized group. Facies distinctions are made on the basis of textural homogeneity for the disorganized group, and bed thickness for the organized group. Bed thickness is a useful environmental indicator and permits easy recognition in the field. Bed thickness is broadly related to grain size (Sadler 1982) and to the sequence of internal sedimentary structures, such that the facies represent a gradational spectrum from the coarsest grained and thickest to the finest grained and thinnest beds.

FACIES GROUP C1 – DISORGANIZED MUDDY SANDS

FACIES C1.1 – POORLY SORTED MUDDY SANDS

Facies C1.1 is characterized by a high content of silt- and clay-grade sediment (up to 80%) in poorly sorted beds (Figs 3.8a, 5.33F & H). Maximum grain sizes are typically in the range of coarse to fine grained sand. Normal distribution grading may occur and, in some cases, a coarse-tail grading may be present in very coarse to coarse grained sand in the lower part of the bed. The uppermost part of the bed is silty mud and the lower part is muddy sand. Bounding surfaces are generally clearly defined, with the bases showing a range of sole markings. The tops are planar or gradational.

Internal sedimentary structures are mainly absent, but indistinct parallel lamination may occur in the lowest few centimetres, as well as convolute lamination associated with pseudonodules of silty mud. These liquefaction structures can give the beds a swirled appearance, and are partly responsible for the fact that they have been termed 'slurry beds' by some workers (Morris 1971, Hiscott & Middleton 1979, Strong & Walker 1981).

Silty mud clasts or 'chips' occur in varying proportions (Fig. 3.8a). In some beds, large rafts up to several metres long are found 'suspended' within the deposit. These rafts may be much longer than the beds are thick (Hiscott 1980).

Transport process: mud-rich, very high density turbidity currents or fluid sand-mud debris flows.

Depositional process: rapid mass deposition due to increased intergranular friction or cohesion. The deposit remains sufficiently plastic for gravitationally-induced loading to occur.

Selected references: Wood & Smith (1959), Burne (1970), Morris (1971), Skipper & Middleton (1975), Mutti *et al.* (1978), Hiscott & Middleton (1979), Hiscott (1980), Pickering (1981b), Strong & Walker (1981), Pickering & Hiscott (1985).

FACIES C1.2 – MOTTLED MUDDY SANDS

Mottled muddy sands occur mostly as very thin to medium beds (<1 cm to about 20 cm), with an irregular to sheet-like shape. Both tops and bases of beds may be sharp or gradational and are variable from one part to another of the same bed. Cross lamination and parallel lamination are rare (Fig. 9.3a), but irregular

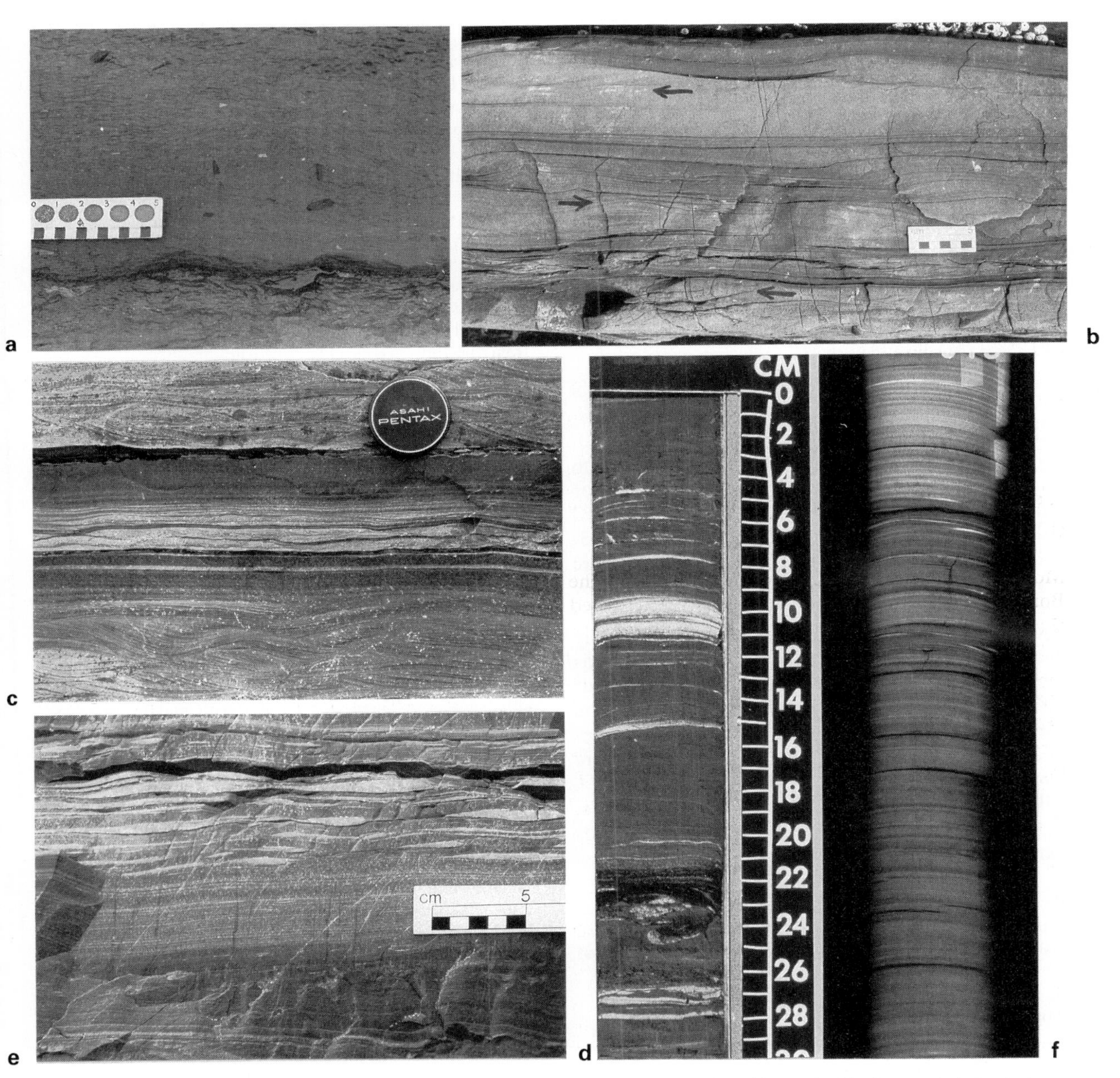

Figure 3.8 Examples of Facies Classes C and D. **Photo a**, Facies C1.1 poorly sorted muddy sand with dispersed shale clasts, Ordovician Tourelle Formation, Quebec, Canada, scale in centimetres. **Photo b**, Facies C2.4 thick to very thick bedded, mud-dominated sand-mud couplet (sand division only; mudstone cap = 100 cm), Ordovician Cloridorme Formation, Quebec, Canada. Flow, based on ripple and climbing-ripple lamination, was initially from right to left, then from left to right, and finally again from right to left (3 arrows). Scale 5 cm. **Photo c**, Facies D2.1 graded-stratified silt, upper Precambrian Kongsfjord Formation, N Norway (lens cap 5 cm across). **Photo d**, Facies D2.1 graded-stratified silt, Mississippi Fan interchannel deposits from DSDP Leg 96 drilling. Interval shown is 096–615–19X–01, 20–50 cm. **Photo e**, Facies D2.2 thick irregular silt and mud laminae, upper Precambrian Kongsfjord Formation, N Norway. **Photo f**, X-radiograph of Facies D2.3 thin regular silt and mud laminae, Mississippi Fan interchannel deposits from DSDP Leg 96 drilling. Interval shown is 096–615–11H–03, 95–120 cm.

concentrations (layers or lenses) of coarser material are common. An indistinct normal or inverse grading can be present.

Bioturbation is often pervasive and may mask any primary physical structures. The grain size is mostly fine sand (gradational to the mottled silts of Facies D1.3), and beds are poorly to moderately well sorted. The grains are often of both terrigenous and biogenic origin, depending on the original sediment source.

Transport process: winnowing of fines, and short, repeated bed-load transport of sands by strong bottom currents.

Depositional process: grain-by-grain deposition of coarse load, and thorough post-depositional mixing of sand and mud by burrowers.

Selected references: McCave *et al.* (1980), Stow (1982), Stow & Piper (1984), Gonthier *et al.* (1984).

FACIES GROUP C2 – ORGANIZED SAND–MUD COUPLETS

Facies Group C2 consists of moderately well sorted to poorly sorted sand–mud couplets showing partial or complete Bouma sequences (Fig. 2.5). Beds tend to show marked normal grading (Kuenen 1953, Ksiazkiewicz 1954). Both tool and scour marks are common sole markings. The bases of beds may show deep scour structures, load structures, or may be smooth and planar. The tops of beds are generally smooth to planar if the upper part of the bed contains substantial amounts of mud. Bioturbation may occur within or throughout the bed, but is more common toward the tops of beds. Liquefaction structures, including convolution and fluid-escape structures, are common.

The Bouma (1962) sequence of internal structures, T_{abcde}, may be complete, but more commonly beds show base-, top- or middle-absent sequences.

We recognize four facies in this group, the first three based entirely on bed thickness (Table 3.3): Facies C2.1, very thick and thick-bedded sand–mud couplets; Facies C2.2, medium bedded sand–mud couplets; and Facies C2.3, thin-bedded sand–mud couplets.

Beds in Facies C2.1 generally begin with Bouma division a, those in Facies C2.2 with division b, and those in Facies C2.3 with division c. A distinctive form of Facies C2.3 is very thin beds, usually <3 cm thick, with sand:mud > 1.0, in which there are low amplitude ripples (<2 cm high) with relatively long wavelengths up to decimetres. These ripples may occur with stoss-side erosion and only lee-side preservation, followed abruptly by a silt/mud drape giving the beds a 'form surface'. Some of Mutti's (1977) Facies E examples are included in this category.

All beds of Facies C2.1 and C2.2, and many examples of Facies C2.3, are the deposits of turbidity currents. Facies C2.1 is deposited from high concentration flows, whereas Facies C2.3 is deposited from relatively dilute currents. Facies C2.2 is deposited from turbidity currents of intermediate character.

The predominant depositional process is grain-by-grain deposition from suspension, followed either by burial (Bouma division a) or by tractional transport as bed load (Bouma divisions b, c). The muddy upper divisions are deposited in the same manner as mud turbidites of Facies Class D.

Unlike the thicker bedded facies, Facies C2.3 may also be deposited from contour currents of variable strength (Ch. 9, Fig. 9.3b & c), although the precise range of contourite characteristics is still poorly understood. This alternate process interpretation for Facies C2.3 was not considered by Pickering *et al.* (1986a).

Facies C2.4 comprises thick to very thick bedded, mud-dominated sand-mud couplets, generally 1–15 m thick, that show internal evidence of flow reversals during deposition (Fig. 3.8b), and that commonly have a mud cap that accounts for about 80% of the thickness of the deposit. The sandy basal divisions are graded, although the grading may be step-wise with mud breaks between divisions with opposed senses of flow (Hiscott & Pickering 1984, Pickering & Hiscott 1985). Internal structures include mega-ripple form sets, ripple and climbing-ripple lamination, wavy and parallel lamination, and pseudonodules. Pickering & Hiscott (1985) interpret these unusual turbidites as the deposits of large-volume, high concentration turbidity currents confined within small basins, such that multiple deflections and reflections of the initial current occur during deposition of the sand-silt load. The thick silty mud caps are deposited by rapid settling of flocs formed in the highly concentrated mud cloud that becomes ponded above the basin floor after cessation of flow. In order to emphasize the

Table 3.3 Classic turbidites of Facies Group C2.

Facies	Bed thickness	Common Bouma divisions	Typical basal grain size	Amalgamation	Packeting	Typical sand:mud
C2.1	>30 cm	ab(c) (d)	vc–m	c	c	4:1
C2.2	10–30 cm	(a)bcd	c–vf	p	p	1:1
C2.3	<10 cm	(b)cd	f–vf	r	p	<1:1

c = common, p = present, r = rare, vc = very coarse, c = coarse, m = medium, f = fine, vf = very fine.

importance of reflections and ponding in deposition of Facies C2.4, Pickering & Hiscott (1985) called the emplacing flows *contained turbidity currents*.

Selected references: literature on Facies C2.1, C2.2 and C2.3 is abundant. Instead of providing a long list, we refer the reader to references in the books edited by Doyle & Pilkey (1979), Siemers *et al*. (1981) and Tillman & Ali (1982). Examples of Facies C2.4 are described by van Andel & Komar (1969), Ricci Lucchi & Valmori (1980), Ricci Lucchi (1981), Hiscott & Pickering (1984) and Pickering & Hiscott (1985).

Facies Class D: silts, silty muds, and silt–mud couplets, >80% mud, ≥ 40% silt, 0–20% sand

Facies Class D contains those sediments that are dominantly silt and clay grade. Key synthesis papers include those by Mutti (1977), Nelson *et al*. (1978), Piper (1978), Stow (1979, 1981), Stow & Lovell (1979), Lundegard *et al*. (1980), Kelts & Arthur (1981), Stanley & Maldonado (1981), Gorsline (1984) and Stow & Piper (1984). The deposits range from sediments with over 90% medium to coarse grained silt, to those with about 40% silt, much of which may be very fine grained. The coarser silts are commonly in distinct beds or laminae interstratified with mud and clay. This class encompasses a wide range of sedimentary characteristics: (a) beds over 1 m thick to laminae less than 1 mm thick; (b) parallel-sided, lenticular or highly irregular layers; (c) structureless or thinly laminated sediment with a variety of other small-scale current-generated structures; (d) poorly developed to absent normal or inverse grading, and graded-laminated units, and (e) layers of coarse grained silt with a relatively high proportion of fine grained sand and sections with only 10% fine silt laminae in mud.

Sediments in this class include those transported by most of the main processes outlined in Chapter 2. In particular, they may form from the tail of high concentration turbidity currents, the body of low concentration turbidity currents, or may be deposited from suspension in deep-water bottom currents. Silts and clays can also be transported by surface currents and winds, to settle through the water column and contribute to the hemipelagic deposits of Facies Class G. Facies Class D deposits are commonly the materials that form coherent and incoherent slides, giving rise to some of the facies of Class F.

Within Class D, we recognize two main facies groups, disorganized (D1) and organized (D2), both of which contain several facies. As with Facies Group C2, the subdivision of Facies Group D2 is based on layer thickness.

FACIES GROUP D1 – DISORGANIZED SILTS AND SILTY MUDS

Facies Group D1 contains all those silts, muddy silts and irregularly interlayered silts and muds that show little regular or consistent organization. They may, however, show poor indistinct grading and irregular layering and lensing.

FACIES D1.1 – STRUCTURELESS SILTS

Facies D1.1 commonly occurs as medium- to thick-bedded, parallel-sided, essentially structureless silts. There may be a poorly defined normal and, rarely, an inverse grading at the base of the bed. Commonly, the sediments are fine to coarse silt size, and are often sandy with floating mud clasts. The silts may range from poorly sorted to well sorted.

Transport process: high concentration, silt-dominated turbidity currents, or highly fluid, silty debris flows.

Depositional process: rapid mass deposition from a concentrated dispersion, due to a combination of increased cohesion and intergranular friction.

Selected references: Piper (1973, 1978), Jipa & Kidd (1974), Stanley & Maldonado (1981), Stow (1984).

FACIES D1.2 – MUDDY SILTS

Facies D1.2 occurs as thin- to thick-bedded, poorly sorted, essentially structureless muddy silts. Grading is absent unless as an ill-defined normal grading. Typically, the base of the bed is sharp, possibly resting on a scoured surface, whereas the upper surface is gradational into finer grained facies. Bioturbation is commonly localized in the upper part of Facies D1.2 beds.

Transport process: high concentration, mud-dominated turbidity currents. Some sediment creep or sliding may contribute to transport.

Depositional process: rapid deposition of silt grains and mud flocs from suspension with no size sorting either in the viscous sublayer or on the bed.

Selected references: Piper (1978), Chough & Hesse (1980), Stow (1984), Wetzel (1984).

FACIES D1.3 – MOTTLED SILT AND MUD

Typically, Facies D1.3 consists of very thin beds, laminae, lenses and mottles of silt in mud. Bed shape is characteristically irregular and both bases and tops of the beds vary from sharp to gradational. Normal and inverse grading may occur on the scale of individual laminae and over intervals up to several tens of centimetres thick, although the grading is mostly irregular in nature. Sorting is poor to moderate. Bioturbation is extensive. With increasing grain size, this facies grades into Facies C1.2 (mottled muddy sands); fine grained examples grade into Facies E1.3 (mottled muds).

Transport process: long-lived bottom currents.

Depositional process: grain-by-grain deposition from suspension, with subsequent pervasive bioturbation destroying most of the original physical sedimentary structures.

Selected references: Piper & Brisco (1975), Stow (1982), Faugères *et al.* (1984), Gonthier *et al.* (1984), Stow & Piper (1984).

FACIES GROUP D2 – ORGANIZED SILTS AND MUDDY SILTS

The facies of Group D2 consist of silts and silty muds either as discrete beds or as interlaminated units of mud and silt. This group also includes fissile, organic-rich muds with silt lenses or laminae. Stratification, grading and a predictable sequence of sedimentary structures are common attributes of Facies Group D2.

FACIES D2.1 – GRADED-STRATIFIED SILT

Facies D2.1 is thin to medium bedded (<30 cm), rarely thick bedded. Soles are often sharp and scoured, whereas bed tops tend to be gradational. Normal distribution grading prevails. Internal sedimentary structures can be described using the Bouma (1962) turbidite model (Figs 3.8c & d, 10.21a, b, c & f). In many cases, Facies D2.1 occurs as beds that are thoroughly laminated. Deposits are of silt grade, grading upward into clay-grade sediment. This facies, to a certain extent, overlaps in character with sandy Facies C2.2 and C2.3.

Transport process: low concentration turbidity currents.

Depositional process: grain-by-grain deposition from suspension, followed by traction transport along the bottom to produce lamination. Clay-grade tops are deposited from suspension as flocs, with no subsequent traction transport.

Selected references: Piper (1973, 1978), Jipa & Kidd (1974), Pickering (1982a, 1984a), Stow & Piper (1984).

FACIES D2.2 – THICK IRREGULAR SILT AND MUD LAMINAE

The sediments of Facies D2.2 are typified by medium to thick lenticular silt laminae in mud (Figs 3.8e; 10.21d & f), and/or thin, irregular, convolute and sub-horizontal silt laminae. In some cases, extreme loading of the silt laminae into the underlying muds produces deep, irregular load structures with intervening mud flame structures protruding upward into the silt layers, or detached load balls (pseudonodules). Typically, silt : mud ratios exceed 2:1. Facies D2.2 often contains thick silt laminae with sharp, commonly rippled tops and scoured, sharp bases. An internal micro-lamination and slight normal grading may be present through individual laminae. Groups of laminae may be arranged in normally graded, laminated units showing partial structural sequences. The sediment grade is typically medium to coarse silt interlaminated with fine silt and mud. Facies D2.2 is gradational with Facies D2.1 and D2.3. Facies D2.2 is equivalent to the Piper (1978) E1 and Stow & Shanmugam (1980) T0–2 divisions (Fig. 2.9).

Transport process: low concentration turbidity currents, or relatively weak bottom currents.

Depositional process: fairly rapid grain-by-grain deposition from suspension (turbidites only), followed by traction transport of the silt load.

Selected references: Piper (1972a, 1972b, 1973, 1978), Stow (1976, 1979, 1981, 1984), Nelson (1976), Nelson *et al.* (1978), Lundegard *et al.* (1980), Stow & Shanmugam (1980), Kelts & Arthur (1981), Pickering (1982b), Stow & Piper (1984), Gardiner & Hiscott (1988).

FACIES D2.3 – THIN REGULAR SILT AND MUD LAMINAE

Facies D2.3 occurs as thin to medium beds containing horizontal silt laminae in mud, with some slightly lenticular, indistinct and wispy silt laminae (Fig. 3.8f). These laminae are commonly grouped into graded-laminated units in which successive silt laminae become finer grained upward over intervals of about 2–10 cm. These units show regular sequences or partial sequences of structures. Silt : mud ratios range from about 2:1 to about 1:2. The silt laminae show sharp to gradational tops and bases. Grain sizes are mostly medium to fine silt and clay. Facies D2.3 is gradational with Facies D2.2 and E2.1. This facies is equivalent to the Piper (1978) E1, E2 and the Stow & Shanmugam (1980) T2–6 divisions (Fig. 2.9).

Transport process: mainly low concentration turbidity currents, but possibly weak bottom currents.

Depositional process: for turbidites, slow uniform deposition from suspension, with shear sorting of silt grains and clay flocs in the viscous sublayer of the turbidity current as described by Stow & Bowen (1980). Examples deposited by bottom currents would indicate an environment hostile to burrowers.

Selected references: as for Facies D2.2, and Stow & Bowen (1980).

Facies Class E: ≥95% mud grade, <40% silt grade, <5% sand and coarser grade, ≤25% biogenics

Facies Class E includes some of the finest deep-sea sediments, both silty clays and clays. Layers vary from very thick beds (several metres thick) and essentially unbedded thick sections, to very thin beds and laminae. Layers are mostly parallel-sided, although broad basal scours may occur. Contacts may be sites of bioturbation. Individual beds can be structureless, graded, irregularly or finely laminated, and can show varying degrees of bioturbation.

Modern representatives of this facies class are widespread, and include the pelagic (red) clays of abyssal oceanic depths, thick proximal mud turbidites, thin basinal turbidites, muddy contourites, and much of the hemipelagic cover of continental slopes and rises. These deposits may undergo later creep, or may contribute to submarine slides or debris flows.

There have been few syntheses of data from deep-sea muds, so that little is known in detail of environments and processes of deposition. Efforts in this direction have been made by Piper (1978), Stow & Lovell (1979), Stanley & Maldonado (1981), Stanley (1981), Gorsline (1980, 1984), Hoffert (1980), and Stow & Piper (1984).

As with other facies classes, there appears to be a natural distinction between a disorganized (E1) and an organized (E2) facies group. Many of the attributes of the fine grained facies in this class are best studied with the aid of X-radiographs of thin slabs (rocks or cores), SEM or with the aid of peels (rocks).

FACIES GROUP E1 – DISORGANIZED MUDS AND CLAYS

Facies Group E1 comprises muds, silty muds, and clays, often in thick, uniform, essentially structureless sections. Sedimentary features are generally subtle. Origin is in many cases enigmatic. Three facies are recognized.

FACIES E1.1 – STRUCTURELESS MUDS

Structureless muds commonly occur in thick sections (one to tens of metres thick); bedding is poorly defined or absent. This facies appears to be common in both modern and ancient successions. While there is a notable absence of structures, both primary and secondary, an indistinct textural, compositional, or colour banding or lamination, and zones of burrow mottling, may be locally developed. The muds are either clay grade, or mixed silt and clay grades; sand-sized material is less common. Composition may be remarkably uniform, being dominantly terrigenous. Facies E1.1 tends to occur in association with biogenic muds (G2), mud turbidites (E2.1), slides (F2) and muddy debris flow deposits (A1.3).

Transport process: largely unknown, but is likely to include thick, mud-rich turbidity currents and lateral transfer of hemipelagic material by deep-ocean currents or by sliding.

Depositional process: probably rapid deposition of flocs, based partly on lack of significant planktonic biogenics. Rapid deposition may result from ponding of mud-rich turbidity currents in confined basins (Pickering & Hiscott 1985).

Selected references: Piper (1978), Stanley & Maldonado (1981), Stanley (1981), Kelts & Arthur (1981), Stow (1984), Pickering & Hiscott (1985).

FACIES E1.2 – VARICOLOURED MUDS

Many mud-dominated successions contain not only the classic 'pelagic red clay', but also interbedded muds of various colours (red, green, brown, grey, etc.), mostly lacking sedimentary structures; these are assigned to Facies E1.2. Such successions may be up to tens of metres thick. Individual beds or layers are defined on the basis of colour changes. Typically, layers are parallel-sided with either abrupt or gradational bases and tops. Primary physical structures are mostly absent, whereas secondary structures such as bioturbation, mottling, and burrowing tend to be common. Grain sizes are mainly silt grade to clay grade. Facies E1.2 is predominantly terrigenous in composition ($\leq$25% biogenic material), often with a significant volcanigenic fraction and authigenic minerals. Ferromanganese nodules and crusts are common and enrichment in trace metals may occur. Subtle differences in chemical composition, oxidation states and the amount of organic carbon control the colour differences. High organic carbon content can give a black mud (Facies E2.2). Facies E1.2 is commonly associated with biogenic oozes and muddy oozes (G1) and fine grained turbidites (D2 and E2).

Transport process: lateral transport of hemipelagic material by ocean currents and/or aeolian action; *in situ* biogenic production.

Depositional process: settling of individual particles or particle aggregates (faecal pellets).

Selected references: Arthur & Natland (1979), Hoffert (1980), Arthur *et al.* (1984), Jenkyns (1986).

FACIES E1.3 – MOTTLED MUDS

Facies E1.3 consists of relatively uniform, thin to thick intervals of mud that are poorly bedded. Relict primary sedimentary structures include wavy, indistinct, or fine parallel layering (Fig. 5.16). Bioturbation (mottling) and indistinct burrows are ubiquitous. Clay-grade material is dominant, but relatively silty intervals (silt mottles, pockets and blebs) may be common. Where silt content becomes substantial, Facies E1.3 grades into Facies D1.3. Compositionally, this facies shows considerable variation, but typically consists of mixed terrigenous and biogenic components, with or without volcanigenic debris. The facies is often associated with probable contourite deposits (D1.3 and C1.2), biogenic muds (G2) and fine grained turbidites (D2, E2).

Transport process: transport as suspended load in bottom currents, commonly contour currents. Silt grains may have moved as bed load beneath the same currents.

Depositional process: settling of particles or particle aggregates from suspension. Extensive post-depositional bioturbation.

Selected references: Stow (1982), Faugères *et al.* (1984), Gonthier *et al.* (1984).

FACIES GROUP E2 – ORGANIZED MUDS

Muds and clays that show some internal organization are assigned to Facies Group E2. There is a complete gradation with the facies of Group E1. Facies include muds, silty muds and clays in thin isolated beds or thicker layers. There are two distinctly different facies: graded muds, and finely layered or laminated muds.

FACIES E2.1 – GRADED MUDS

Facies E2.1 is widespread, especially in deep basinal settings, although it can occur more proximally. It can occur as single, isolated, graded beds >1 m thick, or as thick, repetitive successions of graded muds of variable thickness. Individual beds may show broad scoured bases; thicker beds can be lenticular. Most beds show normal grading. Bioturbation becomes increasingly more important toward the top of many graded mud beds.

Colour or compositional grading is generally more marked than grain-size grading, although beds of dominantly terrigenous composition tend to be more silty toward the base, with increasing clay content upward. These beds may have thin silt laminae at the base. Wherever biogenic material is present, the overall texture tends to be slightly coarser. Facies E2.1 is gradational with silt–mud turbidites (D2.3 and D2.1), and commonly occurs in association with Facies Classes F and G.

Transport process: low and high concentration turbidity currents.

Depositional process: grain-by-grain or floc settling during flow deceleration. No significant traction precedes burial. Post-depositional bioturbation.

Selected references: Piper (1978), Kelts & Arthur (1981), Stow (1984).

FACIES E2.2 – LAMINATED MUDS AND CLAYS

Facies E2.2 occurs in thin to thick mud-dominated sections (up to tens of metres thick) and typically constitutes from 10–60% of sections in which it occurs. Individual beds or intervals range from 1 cm to decimetres in thickness, commonly with fine parallel lamination or distinct 'varves' (Figs 5.16, 33G & 35). The laminae or 'varves' show slight colour, compositional and/or textural variations. Bioturbation is generally absent, although small-scale local burrows may be evident. Clay-grade and fine silt-grade sediments dominate this facies and, in some cases, there are thin silt laminae.

Composition is mixed terrigenous and biogenic, with from 1–10% organic carbon (more rarely >20%) and common but minor iron sulphides. Facies E2.2 includes, in modern examples, the precursors of the marine 'black shales' of the ancient geological record. This facies is commonly associated with oozes and biogenic muds (Facies Class G) and with fine grained turbidites (E2.1 and D2).

Transport process: *in situ* production of organic matter in the overlying water column, transport of hemipelagic material by ocean currents, periodic introduction of mud as suspended load in low concentration turbidity currents.

Depositional process: grain-by-grain or aggregate settling from the transporting currents. Little bed-load traction of the silt fraction. 'Varves' are related to periodic fluctuations of terrigenous input. Anoxic bottom waters favour the preservation of organic matter.

Selected references: Arthur *et al.* (1984), Stow & Dean (1984), Dimberline & Woodcock (1987), Tyler & Woodcock (1987).

Facies Class F: chaotic deposits

Facies Class F consists of more or less chaotic mixtures of sediments, including other deep-water facies emplaced by large-scale downslope mass movements or by ice rafting. We exclude deposits resulting from *in situ* liquefaction. The thickness and shape of the deposits in this facies class ranges from single, isolated clasts (pebbles, cobbles, boulders) to whole sections of continental slope up to hundreds of metres thick. There are two facies groups: F1, exotic clasts; and F2, contorted/disturbed strata.

Lateral and vertical transitions between facies can occur over very short distances, e.g. from the central to marginal parts of submarine slides. Where vertical/lateral facies changes are abrupt and abundant, a researcher may choose to describe a succession only in terms of facies groups.

FACIES GROUP F1 – EXOTIC CLASTS

Facies Group F1 comprises blocks or clasts, either in complete isolation from, or associated with, other similar clasts. Generally, any interstitial 'matrix' is very poorly sorted, and shows a bimodal to polymodal grain-size distribution, although there may be overall systematic textural and compositional variations laterally. The 'matrix', or sediments enveloping the exotic (F1) clasts, is commonly much finer grained than the clasts. There is never any conclusive evidence that the 'matrix' is contemporaneous with clast deposition. Rather, textural relationships suggest that the 'matrix' percolated into the interstices or draped the clasts after deposition.

In many ancient examples, Group F1 deposits have been affected by subsequent tectonism. In some cases, shaly 'matrix' may have acquired a fissility due to compaction alone, but in other examples it may be pervasively sheared during tectonic deformation, even though the exotic clasts are relatively undeformed (Abbate *et al.* 1970).

Within this group, two facies are recognized: F1.1, rubble; and F1.2, dropstones and isolated ejecta.

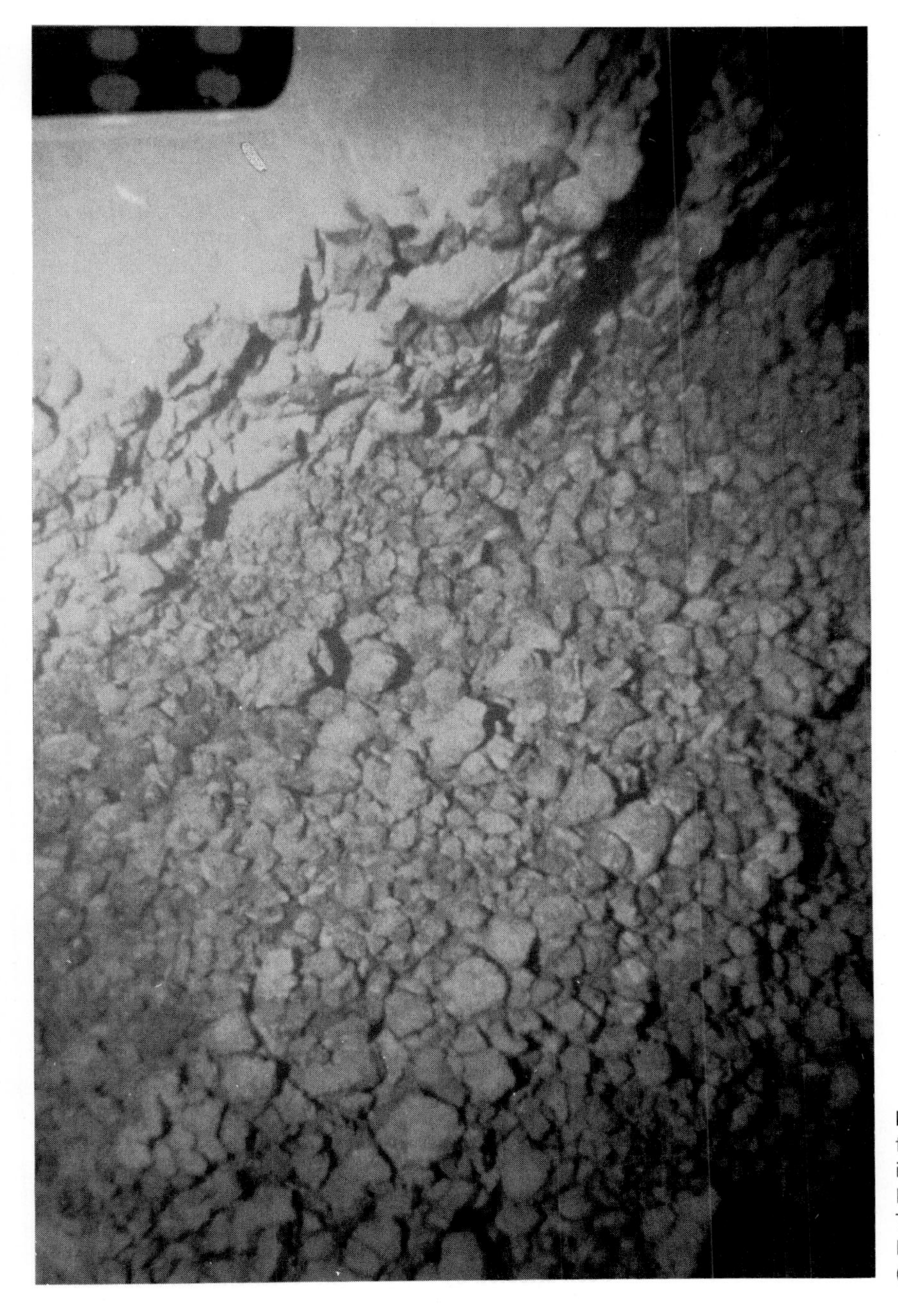

Figure 3.9 Facies F1.1 rubble from 3800 m water depth near the intersection of the Mid-Atlantic Ridge and Oceanographer Transform Fault (Lat. 35° 05′ N, Long. 34° 55′ W). Fault scarp (upper left) has throw of *c.* 1–2 m.

FACIES F1.1 – RUBBLE

Facies F1.1 consists of a chaotic jumble of mainly angular to sub-angular cobbles and boulders, usually as a talus or scree fringing relatively steep submarine cliffs and slopes (Fig. 3.9, 9.29). In ancient examples, the matrix tends to resemble a rock flour devoid of any current-related sedimentary structures; infilling of residual hollows between the larger clasts is indicated by sediment drape. In many modern examples, the rubble is either devoid of interstitial sediments or contains limited and patchy matrix.

The clasts produce mainly compressional deformation and rupturing of underlying sediments in the form of depressed bedding, small-scale syn-sedimentary faults, buckling, and attenuation of beds surrounding the clasts. Wherever large blocks occur, detailed sedimentary logging and mapping may be required to reveal sediment drape and the 'exotic' nature of the blocks; faunal data may be necessary to establish age relationships of the rubble constituents in this facies.

Clast compositions include igneous, metamorphic and sedimentary rock types. The degree of lithification of the rubble clasts varies considerably. Also, clasts may have experienced internal wet-sediment deformation. Isolated, displaced clasts, or exotic blocks, are well documented from ancient successions, e.g. limestone blocks up to 250 m in maximum dimension from the Mesozoic Laganegro Basin southern Italian Apennines (Wood 1981), isolated blocks to 50 × 50 × 100 m in the Longobucco Group of southern Italy (Teale 1985), and the so-called 'fallen stack' of Devonian Caithness Flagstone (45 × 27 × ? m) in the Upper Jurassic of northeast Scotland (Bailey & Weir 1932, Pickering 1984b). Facies F1.1 tends to be associated in particular with deposits of Facies Groups A1 and F2.

Transport process: submarine rockfalls, avalanching and sliding along overpressured glide planes, ?debris flow.

Depositional process: cessation of movement because of basal friction. Some large blocks travelling in debris flows may have become 'grounded', even though the rest of the flow continued into the basin, leaving little or no depositional record in the vicinity of the block.

Selected references: Bailey & Weir (1932), Flores (1955), Abbate *et al.* (1970). Hsü (1974), Surlyk (1978), Pickering (1984b), Teale (1985).

FACIES F1.2 – DROPSTONES AND ISOLATED EJECTA

Dropstones and isolated ejecta of Facies F1.2 are generally substantially larger in size than their 'matrix', or host sediment, occurring either in isolation (Figs 3.5c top & 5.33G), as groups, or in lumps of till. In high latitudes, however, large quantities of gravel-, sand-, silt-, and even clay-size material can be derived from melting icebergs (Srivastava, Arthur *et al.* 1987), and a true gravel facies (Facies A1.3) is developed. The composition, shape and other attributes of the clasts are mainly a function of the source area. For example, dropstones may show glacial striations, polishing, or faceted faces. Such features are supportive of a glaciomarine interpretation, but are not diagnostic of that environment as glacial clasts may be reworked in other environments.

Transport process: encased within, or carried on top of floating ice. Very rarely, the rafting agent may be seaweed or floating trees. Volcanic bombs are ejected during eruption, and follow ballistic trajectories to the sea surface.

Depositional process: grain-by-grain settling. For ice-rafted material, the release is triggered by (a) melting, or (b) the sudden overturning of sediment-laden icebergs.

Selected references: Boltunov (1970), Ovenshine (1970), Anderson *et al.* (1979), Gravenor *et al.* (1984), Edwards (1986), Srivastava, Arthur *et al.* (1987).

FACIES GROUP F2 – CONTORTED/DISTURBED STRATA

Penecontemporaneously deformed packets of beds and layers resulting from lateral translation of parts or the entire sediment packet along discrete shear or glide surfaces are assigned to Facies Group F2. The thickness of the packets is typically on the order of metres, although the range is from centimetres to hundreds of metres. Bounding surfaces range from smooth, planar and parallel-sided, to highly irregular with deep erosional scours at the base of the packets. Internally, bedding varies from virtually undisturbed to highly disturbed, and it is this criterion that separates Facies F2.1, coherent folded and contorted strata, from Facies F2.2, dislocated and brecciated strata.

Modern and ancient examples of this facies group are described in texts edited by Doyle & Pilkey (1979), Watkins *et al.* (1979), Saxov & Nieuwenhuis (1982) and Jones & Preston (1987).

FACIES F2.1 – COHERENT FOLDED AND CONTORTED STRATA

Facies F2.1 comprises folded and contorted, essentially coherent to semi-coherent strata in irregularly-shaped layers or horizons from centimetres to tens of metres thick (Figs 3.10a & b, 5.44 & 10.21g). Discrete internal glide or shear surfaces may be visible and define the bounding surfaces of the deposit (Figs 3.10c & 5.42a). Upper and lower bounding surfaces vary from smooth and planar to highly irregular; there are commonly dramatic changes in layer thickness along strike.

Internal structure, bed thickness and grain sizes of this facies are highly variable. There may be a consistent sense of overturning in any folded interval, and locally such folds may be of relatively constant wavelength and amplitude. Typically, it is the coarser grained beds and laminae that have preserved their lateral continuity in this facies; mud-rich sediments tend to deform plastically and may inject microfaults (Pickering 1983b, 1987a). Facies F2.1 often passes over short lateral and vertical distances into Facies F2.2.

In ancient rocks, the scale of many Facies F2.1 layers often precludes an appreciation of their vertical and especially lateral dimensions. In modern continental-margin and other deep marine environments, this facies is recognized on the basis of deep and shallow seismic profiles, long-range sidescan sonar, and deep-sea drilling. In seismic profiles, Facies F2.1 horizons typically appear as acoustically unstratified sediments and/or chaotic acoustic horizons. Sidescan images show irregular hummocky topography. In cores, this facies is suggested by frequent reversals of bedding dip, and by stretched, attenuated,

Figure 3.10 Examples of Facies Class F. **Photo a**, Facies F2.1 coherent folded and contorted strata in a sediment slide, 'Black Flysch' (Deva Formation), Arminza, northern Spain, 1 m scale. **Photo b**, X-radiograph of Facies F2.1 from the main Mississippi Fan channel from DSDP Leg 96 drilling. Interval shown is 096–621–32H–01, 40–90 cm. **Photo c**, draped slide scars upslope from Class F sediment slides, Eocene Ainsa Sequence, Hecho Group, northern Spain. Trees give scale. The scene has been tilted so that the lowest strata have a gentle dip comparable to that expected on basin slopes, or major levees of submarine channels.

sheared, faulted (usually microfaulted), folded and overturned strata.

Transport process: slides and rotational slumps, either due to depositional overloading of weak sediments, or to cyclic or single shocks (earthquakes, tsunami).

Depositional process: cessation of movement on decreasing bottom slopes because gravity forces no longer exceed or balance basal and internal friction.

Selected references: Moore (1961), Laird (1968), Lewis (1971), Ricci Lucchi (1975a), Roberts *et al*. (1976), Woodcock (1976a, b, 1979a & b), Clari & Ghibaudo (1979), Doyle & Pilkey (1979), Watkins *et al*. (1979), Pickering (1982b, 1984a & 1987a), Saxov & Nieuwenhuis (1982).

FACIES F2.2 – BRECCIATED AND BALLED STRATA

Facies F2.2 is gradational from F2.1, and is typified by highly brecciated and balled strata in a chaotic jumble of fragments, in layers of varying thickness but generally thinner than those of Facies F2.1. Layer shape is highly variable from parallel-sided to very irregular and lens-shaped. Typically, the layers consist of a relatively fine grained 'matrix' with fragments of original bedding and laminae in angular to well rounded pieces (Fig. 5.42b, c & d). Facies F2.2 is an intraformational deposit with relatively uniform composition for any given layer. A crude ghost stratigraphy may be present and commonly many of the fragments are folded/contorted as isolated prolapse structures and single folds on a scale of decimetres.

The layers of balled strata, so named because of the roundness of the fragments, are the most common variant of Facies F2.2. Typically, bounding surfaces are smooth, although layer thickness is variable along strike. Relict bedding is visible as plastically deformed bundles of laminae/beds, or as 'wisps' within an almost 'homogenized' layer. Layers of balled strata may pass over short lateral distances through a zone of pervasively microfaulted sediment into Facies F2.1.

The brecciated variants of Facies F2.2 are characterized by abundant, poorly imbricated, unordered, elongate, angular to subangular intraformational fragments, usually of fine grained sands, silts, muds and clays, in layers on the order of decimetres thick. Upper and lower surfaces tend to be irregular. Undercutting of partially eroded sediments may occur. Upper surfaces may be draped by overlying sediments. Commonly, brecciated strata are associated with fluid-escape structures.

Transport process: gravity-induced sliding, during which internal deformation and brecciation occurs.

Depositional process: as for Facies F2.1.

Selected references: as for Facies F2.1.

Facies Class G: biogenic oozes (>75% biogenics), muddy oozes (50–75% biogenics), biogenic muds (25–50% biogenics) and chemogenic sediments, <5% terrigenous sand and gravel

Biogenic and chemogenic sediments are ubiquitous throughout the world's oceans, and are widely recognized in ancient deposits. Most facies of Class G result either from slow settling of material through the water column in the absence of substantial bottom currents, or from direct chemical precipitation. However, many of the processes that are proposed for the accumulation of some of these sediments require advective currents to furnish (a) suspended sediments, (b) nutrients and (c) other chemicals to the water column. Three distinctive facies groups can be distinguished: G1, biogenic oozes and muddy oozes; G2, biogenic muds; and G3, chemogenic sediments.

The chief distinguishing features of Groups G1 and G2 include: (a) evidence for low to very low rates of sediment accumulation and continuous bioturbation (except in anoxic basins, Byers 1977); (b) an absence of primary sedimentary structures or other evidence of sustained bottom currents; (c) an essentially uniform composition within any given succession except for regular cyclicity related to climatic or other controls; (d) a variable biogenic component, mainly of planktonic tests; (e) a very fine grained, often far-transported, terrigenous component; and (f) commonly a significant authigenic fraction.

The true biopelagic sediments (Facies Group G1) accumulate in the open ocean, and primarily consist of whole or disarticulated skeletons of planktonic organisms, together with some minor amounts of very fine silt and clay, much of which reaches the open ocean by aeolian transport. The actual proportion of terrigenous material may be increased by preferential dissolution of the biogenic fractions. Total dissolution results in the varicoloured pelagic clays of Facies E1.2 (≤25% biogenics; often <1%).

Biogenic muds (Facies Group G2) accumulate on continental margins and in other settings near terrigenous sediment sources. They consist of indigenous biogenic material (>25–50%) and silt- and clay-grade terrigenous detritus. Lower biogenic contents result in transitions with Facies E1.1 and E1.3. Both these latter facies and Group G2 deposits can be broadly referred to as hemipelagites.

Chemogenic sediments (Facies Group G3) are composed almost entirely of authigenic minerals such as ferromanganese nodules and phosphorites. These deposits are very complex, and a satisfactory discussion is far beyond the scope of this book. The interested reader is referred to comprehensive books on these intriguing deposits by Horn (1972), Glasby (1977), Bentor (1980) and Baturin (1982).

FACIES GROUP G1 – BIOGENIC OOZES AND MUDDY OOZES

FACIES G1.1 – BIOGENIC OOZES

Biogenic oozes are most typical of the open ocean basins far from terrigenous sources. They have been the subject of much study over the past 125 years; hence, their sedimentary characteristics are well known. Some important syntheses are to be found in Arrhenius (1963), Hsü & Jenkyns (1974), Cook & Enos (1977) and Jenkyns (1986).

Oozes are composed predominantly of the tests of planktonic organisms (>75%), either calcareous (coccoliths, foraminifera, pteropods and nannoplankton) or siliceous (radiolaria, silicoflagellates), or a mixture of both (Berger 1974). These major components are soluble to a significant degree in sea water. The other components (Lisitzin 1972) may include very fine grained terrigenous material (principally quartz, feldspars and clay minerals), volcanigenic minerals (e.g. palagonite and derived clay minerals), authigenic minerals (e.g. phosphates, barite, zeolites, ferromanganese nodules and coatings), and rare extraterrestrial material (tektites). Under normal oxic conditions, the organic carbon content is extremely low, but under anoxic conditions pelagic black shales can contain >20% organic carbon (Isaacs 1981, Arthur *et al.* 1984).

Rates of accumulation are low, commonly from <1 mm/1000 years to 10 mm/1000 years, although this can be an order of magnitude higher under zones of upwelling. The sediments are usually thoroughly homogenized by bioturbation, and devoid of any primary current-formed structures. A variety of burrow types may be preserved (Fig. 3.11), dependent on varying environmental factors such as water depth, grain size, rate of sediment accumulation and redox conditions (Seilacher 1967, Werner & Wetzel 1982). Some of the main diagnostic trace fossils are *Zoophycus, Chondrites, Planolites, Scolicia, Trichichnus, Teichichnus* and *Lophoctenium*.

The grain size of biogenic oozes is largely dependent on the composition of the biogenic components. Coccolith plates are very small (clay size), whereas some foraminifera-rich or diatom-rich oozes may have a mean grain size that is in the silt range. The terrigenous fraction is largely clay grade. Full grain-size analyses of pelagic oozes are rarely conducted because data on the hydraulic equivalence of biogenic particles (Berthois & le Calves 1960, Maiklem 1968, Berger & Piper 1972, Braithwaite 1973) are too meagre to allow useful interpretation.

The characteristics, composition and distribution of the different types of biogenic oozes are dependent on (a) the water depth and the location of the carbonate compensation depth (CCD); (b) the source and type of terrigenous/volcanic material, together with the processes that supply this material; (c) surface water productivity and the type of biogenic materials produced; (d) surface currents and bottom circulation patterns; (e) climate and basin physiography; and (f) the physiochemical conditions (Lisitzin 1972, Berger 1974).

Figure 3.11 Example of Facies G1.1: a burrowed radiolarian chert of Caradoc age from north central Newfoundland, New Bay area, Canada. Width of slab is 6 cm.

Biogenic components settle relatively slowly through the water column, are buried very slowly, and therefore are exposed for relatively long time periods, either in the water column or on the sea floor. The actual processes of settling as single grains or as larger flocs and pellets are discussed by Gorsline (1984) and McCave (1984).

Depositional process: grain-by-grain or aggregate settling through the water column.

Selected references: Arrhenius (1963), Lisitzin (1972), Berger (1974), Broecker (1974), Hsü & Jenkyns (1974), Jenkyns (1986), Werner & Wetzel (1982), Arthur *et al.* (1984), Gorsline (1984), Isaacs (1984), McCave (1984).

FACIES G1.2 – MUDDY OOZE

There is a continuum of facies from biogenic ooze with >75% biogenic material to pelagic clay with <25% biogenics (Facies E1.2). Muddy biogenic oozes are a relatively common intermediate sediment type, with attributes that are transitional between an ooze and a clay. They differ from true hemipelagic sediments in possessing little terrigenous silt, and in being an open-ocean rather than continental-margin facies. Layers of this facies are devoid of physical sedimentary structures, but are commonly pervasively bioturbated.

Transport process: introduction of clays by aeolian action and/or as suspended load of very dilute ocean currents.

Depositional process: grain-by-grain or aggregate settling through the water column.

Selected references: as for Facies G1.1.

FACIES GROUP G2 – BIOGENIC MUDS

FACIES G2.1 – BIOGENIC MUD

Both this facies and its facies group were named 'hemipelagite' in an earlier published version of our facies scheme (Pickering *et al.* 1986a), but inspection of the attributes and interpretations of Facies Class E indicates that many Class E muds resemble hemipelagic deposits described in the literature. For this reason, the genetic term 'hemipelagite' does not appear here as a facies name, and Facies G2.1 is restricted to 25–50% biogenic content from its previous 5–75% range (Pickering *et al.* 1986a). No requirement is placed on silt content, although most examples of this facies contain mixtures of silt and clay in the terrigenous fraction. Sand content may range up to 15%, but is mainly of biogenic origin.

These sediments are poorly sorted and show no systematic grading. In published descriptions of sedimentary successions, they are frequently referred to with some Facies Class E deposits as 'background sedimentation', 'normal', 'ubiquitous' or 'interbedded' facies. At high latitudes (e.g. in the Arctic Ocean and Baffin Bay), ice rafting may be a major contribution to fine grained muds and biogenic muds.

Hemipelagites (Facies G2.1, E1.1, E1.3 ± other Class E deposits) are often not described in detail although they constitute the greater part of many successions. Particularly useful detailed descriptions of modern hemipelagites include those of Stanley *et al.* (1972), Moore (1974), Rupke (1975), Kolla *et al.* (1980), Stanley & Maldonado (1981), Hill (1984b), Isaacs (1984) and Thornton (1984). A number of papers describe possible hemipelagites from ancient slope and basinal environments (Hesse 1975, Piper *et al.* 1976, Ingersoll 1978b, Hicks 1981, Pickering 1982b). Such facies are commonly homogeneous and structureless, with poorly defined bedding or no bedding. There are no primary current-formed structures such as lamination or erosional contacts, although a depositional lamination may be preserved under anoxic conditions (e.g. Isaacs 1984, Thornton 1984) or in Precambrian strata deposited before the advent of vigorous burrowers.

Under normal oxic conditions, bioturbation is ubiquitous and pervasive, commonly resulting in a completely homogenized sediment with a mottled appearance. Burrow traces may be preserved, with the same major trace fossil assemblages as for Facies G1.1. Iron sulphide filaments (mycelia) and mottles are also common features of hemipelagic deposits.

The biogenic input to muds and biogenic muds may be calcareous or siliceous. It is commonly of mixed planktonic and *in situ* bathyal benthic species. The terrigenous components are mostly uniform in any given area of an ocean and display very gradual compositional trends toward the source area(s). The nature of hemipelagites varies considerably throughout the geologic record and in modern oceans, depending on the tectonic, climatic, source area and oceanic physiochemical conditions. There may be far-transported, wind-blown and ice-rafted terrigenous detritus, constituting substantial proportions of the total sediment volume in some cases.

Transport process: introduction of terrigenous materials by aeolian action, ice rafting, suspension transport in mid- and bottom-water currents from river deltas and other coastal areas.

Depositional process: grain-by-grain or aggregate settling through the water column.

Selected references: Stanley *et al.* (1972), Hesse (1975), Rupke (1975), Piper *et al.* (1976), Stanley & Maldonado (1981), Pickering (1982b), Gorsline (1984), Gorsline *et al.* (1984), Isaacs (1984), Thornton (1984).

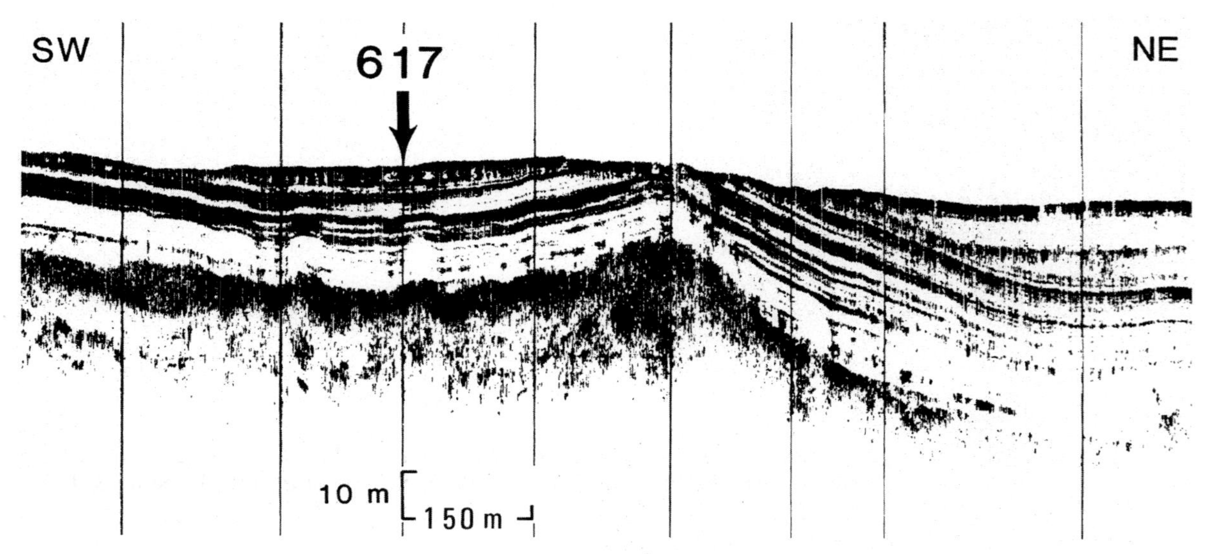

4.5 KHz high-resolution EDO deep-towed seismic reflection profile over channel margin-levee-overbank from middle fan region of the Mississippi Fan, Gulf of Mexico, in the vicinity of DSDP Leg 96 Site 617 in a water depth of 2478.5 m. Note the erosional surface and the onlap by muddy intrachannel sediments that plugged most of the channel during the rise in sea level from the Pleistocene to Holocene and Recent (cf. Pickering *et al.* 1986b).

CHAPTER 4

Controls on sedimentation and sequences

Introduction

There are many variables that control the sedimentation patterns in deep-marine clastic systems, as indeed in other sedimentary environments (cf. Blatt *et al.* 1980, Ch. 1). Invariably, these controls are interactive, and in any clastic system it may be possible to identify one or several primary and several secondary variables. The main controls are: (a) sediment type and grain size; (b) eustatic and local sea-level changes; (c) tectonics; (d) rates of sediment supply and accumulation; (e) geometry and size of receiving basin, and (f) ocean current circulation patterns, governed in part by the Coriolis effect.

These six controls or variables are governed by even larger scale processes such as the relative rates of generation and destruction of oceanic crust, the disposition of the continents, global climatic changes and possibly Milankovitch cyclicity. The maximum depth for accumulation of deep-marine carbonates is dependent upon the position of the carbonate compensation depth (CCD) and aragonite compensation depth (ACD), in turn influenced by ocean circulation, latitude and sea-water chemistry. The net result of the spatial and temporal variation of these controls is the generation of vertically and laterally changing facies, facies associations and sequences. Such changes result in geographically and stratigraphically distinct units or 'packets' of sediments, each with its own environmental interpretation. Preferred vertical sequences of facies and bed thickness are considered at the end of this chapter in order to provide a framework for Parts II and III of this book.

Other general reviews of the controls on deep-marine sedimentation may be found in Pickering (1982c), Shanmugam & Moiola (1985), Shanmugam *et al.* (1985), Stow (1985, 1986), Stow *et al.* (1985), Jenkyns (1986), and books written or edited by Hsü & Jenkyns (1974), Talwani *et al.* (1979), Warme *et al.* (1981), Kennett (1982), Scholle *et al.* (1983), Stow & Piper (1984a) and Bouma *et al.* (1985a).

Sediment type and grain size

Figure 4.1 shows the main distribution of sediment types throughout the world oceans. In deep water, the

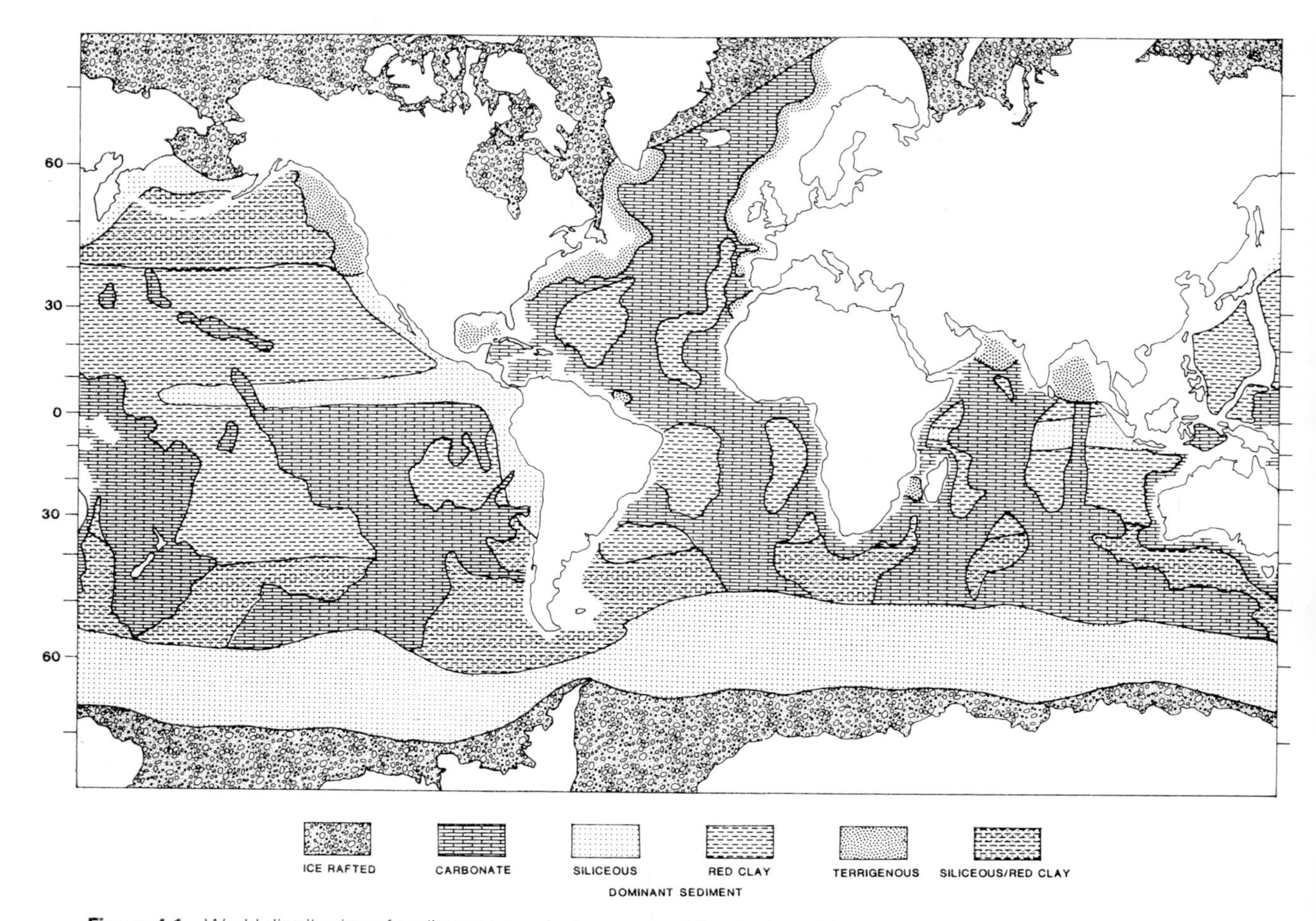

Figure 4.1 World distribution of sediment types in the oceans. After Jenkyns (1986).

principal redeposited sediments are terrigenous gravel, sands, silts, muds and clays. In low latitudes, biogenic carbonate debris, derived from carbonate platforms, can furnish substantial volumes of resedimented material that may form a predominantly carbonate deep-marine clastic system. In forearc and marginal basins at convergent margins, particularly around intra-oceanic arcs, volcaniclastic materials form the main source of sediments (Fisher & Schmincke 1985, Cas & Wright 1987).

Locally fast rates of pelagic/hemipelagic sediment accumulation may form thick successions of calcareous and siliceous oozes and muds (Facies Class G). Black shales rich in organic carbon (Facies E2.2) may accumulate under favourable conditions of productivity and anoxia or oxygen minima in the bottom waters. Evaporites also provide a source of redeposited sediments, as in the deeper parts of the Red Sea. Black shales and evaporites, however, are rarely volumetrically significant in deep-marine clastic systems.

The nature of deep-marine pelagites (Facies E1.2, G1.1, G1.2) and hemipelagites (Facies G2.1, E1.1, E1.3 and +/− other Class E deposits) depends on many factors, particularly the position of the CCD, below which siliceous oozes and red clays tend to accumulate (Figs 4.2a & b). However, near continental margins, especially those associated with deltaic systems, siliciclastic turbidite systems tend to dilute carbonate detritus (Figs 4.2a & b).

Grain size is an important variable. Coarse-grained systems tend to have large volumes of Facies Classes A, B and C, commonly supplied from relatively infrequent high velocity and/or high concentration flows, and generally fed from coarse-grained coastal systems on narrow shelves where longshore-drift currents are intercepted by canyons. The coastal plains associated

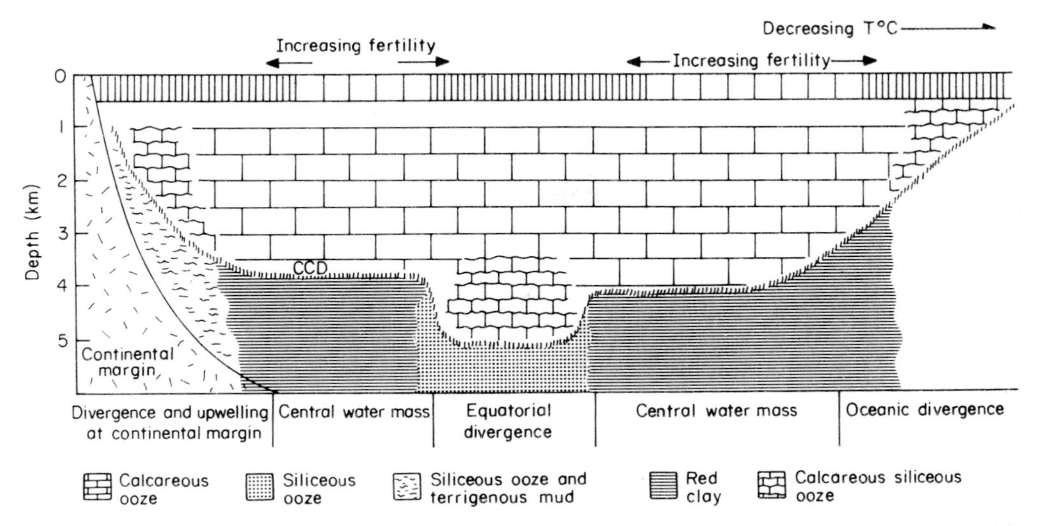

Figure 4.2a Schematic diagram, based on Pacific Ocean, to show main location of sediment types relative to the CCD and near-surface organic productivity. After Ramsay (1977).

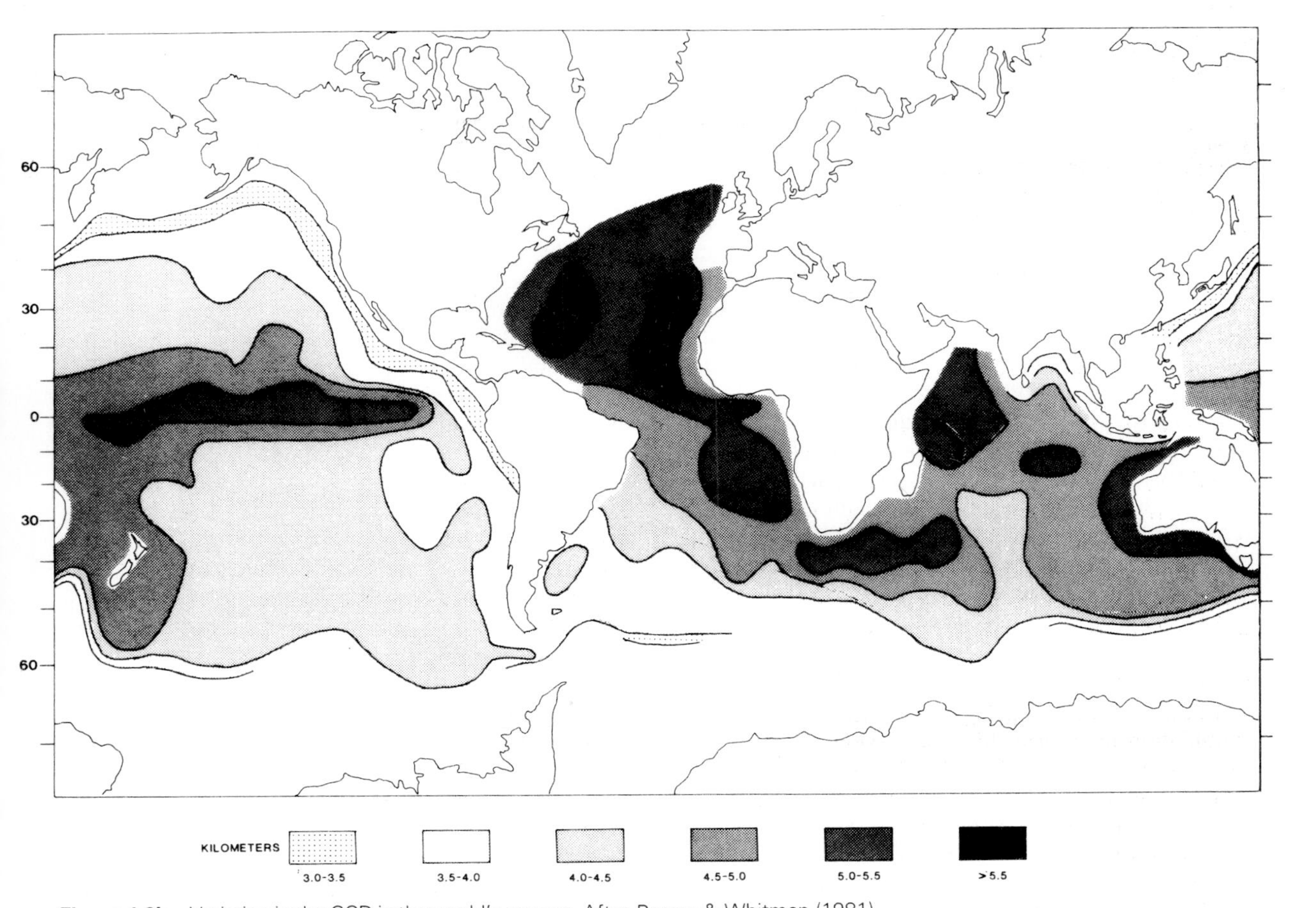

Figure 4.2b Variation in the CCD in the world's oceans. After Barron & Whitman (1981).

with these systems tend to be relatively steep with alluvial systems dominated by bed-load transport (e.g. Yallahs Fan Delta–Canyon–Fan, Wescott & Ethridge 1982). Many of the small-radius submarine fans and gullied slopes in the offshore California Borderland, and in deep-water fan-deltas of fjords, are of this type. These coarse-grained systems contain the greatest proportion of deep-water cross-stratification, probably concentrated within channels as side-bars and mid-channel bars (Hein & Walker 1982).

Steep submarine slopes, such as those adjacent to seamounts and fault scarps, have the most coarse-grained sediments of all the clastic systems, with abundant Facies Classes A and F as rockfalls (Fig. 3.9), debrites and slides: seamount talus slopes and fracture zones are examples.

Fine-grained deep-marine systems tend to be volumetrically much larger than coarse grained systems, not only because muds and clays are about an order of magnitude more abundant globally than sands and gravels, but also because, generally, they are supplied by a wider range of more frequent processes, such as plumes from rivers with large suspended-sediment loads, storms and tidal currents. In fine-grained clastic systems, Facies Classes C, D and E predominate; examples include the largest submarine fans, associated with the world's largest rivers, such as the Bengal, Indus, Amazon, Niger, Congo, Mississippi and Laurentian fans. Systems in which Facies Classes E and G predominate include ponded basins adjacent to mid-ocean ridges, some open-ocean abyssal plains, sediment-starved trenches such as the Tonga-Kermadec Trench, and sediment drapes on submerged seamounts and other aseismic ridges and fracture zones.

Sediment type is partly dependent upon latitude. In high latitudes, glaciers, icebergs and ice shelves may contribute substantial volumes of terrigenous sediments to the shelf and basin; shelf accumulations can be redeposited into deeper water. Dropstones may occur in any water depth. In low to intermediate latitudes, aeolian processes may provide an important contribution to the pelagic component of deep-marine sediments, particularly the abyssal red clays containing 'loess' (Lisitzin 1972).

Eustatic and local changes in sea level

The increased understanding of deep-marine clastic systems, both siliciclastic and carbonate, has led to an appreciation of just how important global (eustatic) and regional changes in sea level are in controlling sedimentation. Indeed, arguably, this variable exerts the most dramatic and predictable control on ocean-basin sedimentation and palaeoecology.

In this book, we use the sea-level curves of Haq *et al.* (1987) (Fig. 4.3), updates of earlier sea–level curves (Vail *et al.* 1977, Vail & Mitchum 1979, Vail & Todd 1981). The true significance and global applicability of these sea-level curves is controversial (Cloetingh 1986, Miall 1986). Nevertheless, they have had a profound effect on our approach to deep-marine systems. The sea-level curves may be resolved into: (a) first-order, 300–200 m.y. cycles; (b) second-order, 80–10 m.y. cycles, and (c) third-order, 10–1 m.y. cycles (Fig. 4.3); Haq *et al.* (1987) further refine their curves into sequence chronozones or cycles, designated mega-cycles, supercycles and cycles. The first-order, and many of the second-order, cycles are explained by changes in mid-ocean ridge volumes (Pitman 1979, Sheridan 1986), whereas the third-order, and some second-order, cycles are well explained by changes in global ice volume (Donovan & Jones 1979). Schlanger *et al.* (1981) postulate variations in the amount of mid-plate volcanism as a major factor in producing fluctuating sea levels.

Pitman (1979) recognized that the global volume of ocean basins and therefore sea level is controlled by variations in four principal parameters, with associated estimated maximum rates of sea-level change of: (a) 6.7 mm/1000 years due to changes in the volume of mid-ocean ridges; (b) 2 mm/1000 years due to changes in the net volume of sediment delivered to ocean basins; (c) 1.6 mm/1000 years due to the collision of the Indian and Karakoram–Asian plates, i.e. variations in the aerial extent of oceanic versus continental crust because of orogenesis, and (d) 0.2 mm/1000 years due to variations in the volume of seamounts. The only viable explanation of the most rapid inferred fluctuations in sea level, of the order of 10 m/1000 years, must be glacio–eustatic effects (Pitman 1979, Pitman & Golovochenko 1983). This also applies to most third-order cycles with rates of change in sea level of at least 10 cm/1000 years (Fig. 4.3). Global climatic changes, therefore, apparently exert a major control on the nature of deep-marine systems, through lowered sea level during glaciations when the volume of the polar ice-caps expands, and raised sea level during warmer periods when large volumes of water stored at the poles are released by melting.

Highstands of sea level are associated with an amelioration in climate, increased biological diversification, reduced mid-water oxygen concentration, an open-ocean shoaling of the CCD (Fig. 4.4), a prolifer-

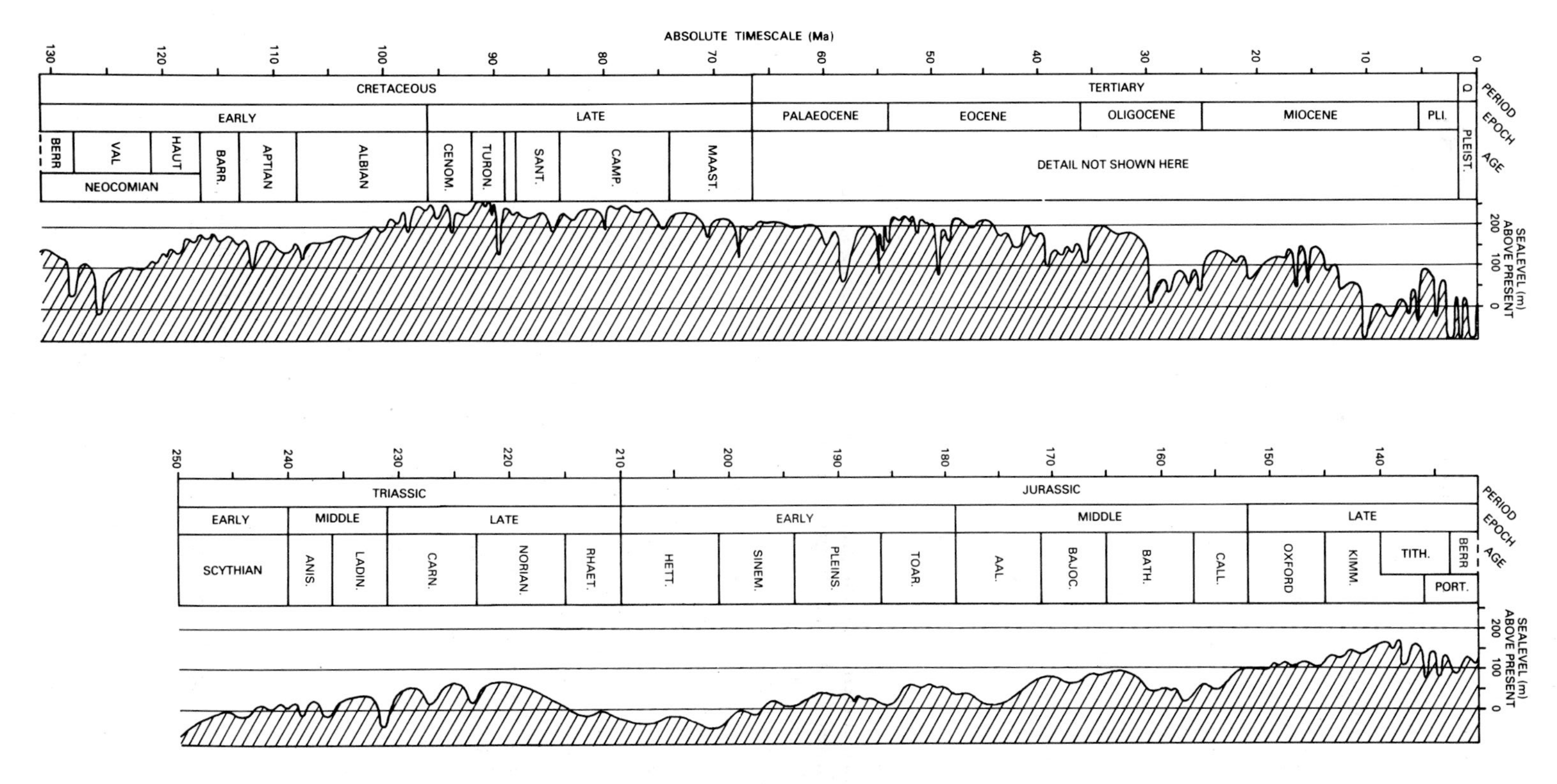

Figure 4.3 Global sea-level curve since 250 Ma (as proposed by Haq *et al*. 1987).

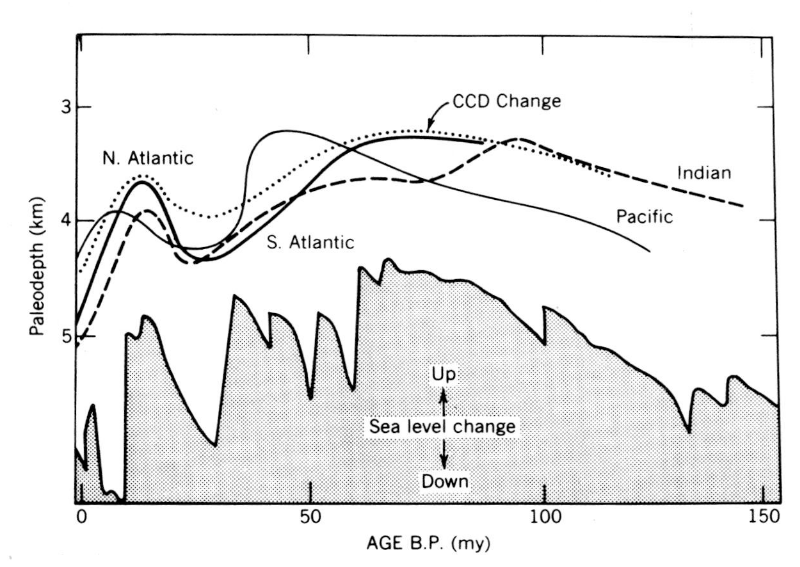

Figure 4.4 Fluctuations in the depth of the CCD since 150 Ma in the Atlantic, Pacific and Indian oceans and the variation in global sea level. After Kennett (1982).

ation of deep-marine 'condensed successions' of pelagic sediment, and the accumulation of muddy abandonment facies over many deep-marine clastic systems. At the present time, during the Holocene raised sea level, most of the world's deep-marine clastic systems are essentially dormant and being mantled by pelagic/hemipelagic sediments. The surfaces of many submarine fans are veneered by about 1 m of light-brown, pelagic, foraminiferal ooze or marl, representing approximately the last 11 000 years; for example on the Amazon Fan (Damuth & Kumar 1975, Damuth & Flood 1985), and the Mississippi Fan (Bouma, Coleman *et al.* 1986).

Lowstands of sea level generally are associated with global climatic cooling (glaciations), prominent and widespread unconformities, or hiatuses, on the continental slope and shelf, and enhanced rates of sediment accumulation in deep-marine clastic systems, for example lowstand submarine fans (see Ch. 7 for explanation and discussion).

Tectonics

The importance of tectonics in controlling deep-marine sedimentation has long been appreciated, but the theory of plate tectonics resulted in more sophisticated models for tectonically-controlled systems. An historical perspective of the development of ideas on sedimentation and tectonics is presented by Mitchell & Reading (1986). Tectonic activity during sedimentation is considered in detail in Part III of this book, in relation to: (a) evolving and mature passive margins; (b) active convergent margins, and (c) oblique-slip continental margins (Ch. 10, 11 & 12 respectively).

Plate-scale tectonic processes may also be responsible for the widespread relative rises and falls of sea level that influence deep-marine sedimentation. Recall (a) that Haq *et al.* (1987) infer that third-order cycles of apparent sea-level variation (Fig. 4.3) are of global extent, and (b) that the rates of sea-level rise and fall that characterize these cycles (>10 cm/1000 years), if indeed eustatic, can only be explained by growth and decay of continental ice sheets (Donovan & Jones 1979, Pitman 1979). These third-order cycles are as common in the Cretaceous as at other times (Fig. 4.3), yet there is strong evidence that the Cretaceous was free of any significant continental ice (Barron 1983), leaving no plausible driving mechanism for the hypothetical rapid changes in sea level. Watts *et al.* (1982), Cloetingh (1986) and Miall (1986) question the evidence for the global extent of many sea-level excursions and Schwan (1980) and Bally (1980, 1982) suggest that many of the prominent unconformities assigned to sea-level low stands by the EXXON group (Vail *et al.* 1977, Vail & Mitchum 1979, Vail & Todd 1981) instead correlate closely with orogenic episodes and times of major reorganization of plate motions. The implication is that (a) third-order cycles of Haq *et al.* (1987) may not be of global extent and therefore may not have a eustatic origin, and (b) rapid apparent rises and falls of sea level, although of regional importance (e.g. North Atlantic area), may have a tectonic origin linked to broad subsidence or uplift. Uchupi & Aubrey (1988) document considerable variations in sea level related to differing, consanguinious, relative uplift and subsidence along the western edge of the North American Plate unrelated to the Quaternary glaciations/deglaciations: they interpret these local sea-level changes as due to terrane accretion events.

What is a plausible tectonic explanation for apparent sea-level cyclicity with third-order character? Cloetingh (1986) points out that modern lithospheric plates are characterized by large-scale intraplate stress fields, either tensional or compressive. Plates under tension tend to subside, whereas plates under compression tend to be elevated, with associated rates of the order of 1–10 cm/1000 years. Continental margins situated on plates that experience fluctuations in the magnitude and orientation of intraplate stress will preserve a stratigraphic record indicating fluctuations in sea level; magnitudes of intraplate stress and relative height of sea level are coupled in this model. Stress provinces (i.e. regions over which intraplate stresses have a common orientation and sense) may have dimensions approximately the same as the dimensions of the lithospheric plates, so that global sea-level changes cannot be unequivocally demonstrated from only a few passive margins. Cloetingh (1986) cautions that the EXXON sea-level curves (Haq *et al.* 1987) are based almost entirely on data from North America (including the Gulf Coast), northwest Africa, and the North Sea, so that the claim that widespread transgressions and regressions are the result of important variations in global sea level is not compelling. Fluctuations in intraplate stresses could produce the same effect.

Irrespective of whether the more rapid relative sea-level fluctuations are of global (Haq *et al.* 1987) or only regional extent (Cloetingh 1986), they have a profound influence on local deep-marine sedimentation, and must be considered in the interpretation of ancient deep-marine successions. We only wish to caution here that regional tectonics and intraplate stresses may be responsible for many apparent changes in sea level, and that the global extent of third-order cycles of sea-level change remains unvalidated.

Local tectonic processes, such as fault activity, may exert a dramatic short-term influence on sedimentation (e.g. Kimmeridge–Volgian times in the Northern North Sea). During earthquake activity lasting minutes to weeks, submarine faulting can alter relief by tens of metres or more, generating various sediment gravity flows and mass flows. Halokinesis (cf. Talbot & Jackson 1987) may be very important in creating intraslope basins by forming ridges on the sea floor behind which sediments may become ponded (e.g. as documented for the northern slopes of the Gulf of Mexico (Bouma 1983). Thus, of all the variables controlling sedimentation, tectonic processes provide one of the few mechanisms operating on a large and effectively instantaneous scale. The process–response models for tectonic control mainly vary with: (a) plate tectonic setting; (b) rates of tectonic processes (Fig. 4.5), both short-term and long-term (cumulative) effects, and (c) grain size and sediment type.

The magnitude, frequency and nature of faulting, whether normal, reverse or strike-slip, are important variables in determining the processes of sediment redeposition, together with the sediment volumes involved. Strong earthquakes are most likely to generate large-volume slope failure and result in thick-bedded or very thick-bedded deposits, for example Facies C2.4 (Ch. 3). Active continental margins are most susceptible to frequent strong seismicity, whereas mature passive margins experience considerably fewer earthquakes of similar magnitude, although they do occur periodically, e.g. the Grand Banks earthquake off southern Newfoundland in 1929 (Heezen & Ewing 1952, Heezen & Drake 1964, Shepard & Dill 1966, Piper *et al.* 1985).

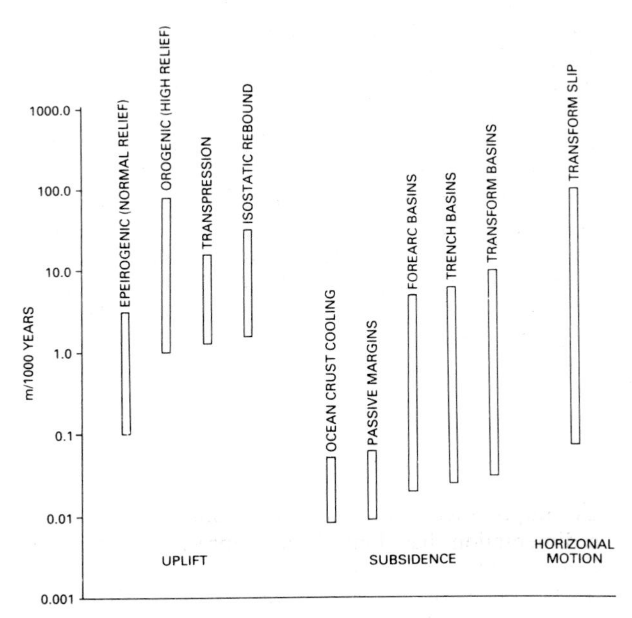

Figure 4.5 Log-plot to compare time intervals for tectonic uplift, subsidence and horizontal plate motions, based on data from Howell & von Huene (1980) and Blatt *et al.* (1980). After Stow *et al.* (1985).

The longevity of depositional systems is controlled primarily by tectonics. Long-lived, relatively stable basins typify mature passive continental margins, whereas active convergent margins are characterized by many variably-sized, commonly young, basins. Relative plate motions at strike-slip or oblique-slip margins may result in ephemeral deep-marine clastic systems, as sediment entry points are separated from the associated clastic system, for example the Magdalena Fan off southern California. The sustained major collision and uplift between the Indian and Asian plates since 50 Ma, manifest in the Himalayan and related mountain belts, has supplied enormous volumes of sediment to the southwest into the Indus Fan, and to the southeast into the Bengal Fan. However, the eastern part of the Bengal Fan is being deformed and accreted to the Sunda Arc in the eastward-dipping Java Trench (Curray & Moore 1974).

Rates of sediment supply and accumulation

The rates at which sediments are supplied to the receiving basin, and the rate at which they accumulate (Fig. 4.6), are important variables that depend largely upon other factors such as: (a) rates of uplift and erosion in the source area; (b) residence times for sediments in intervening temporary storage, e.g. fluvio-lacustrine, coastal and shallow-marine shelf environments; (c) sediment transport processes from shallow to deep water (Ch. 2); (d) grain size and sediment type, a function of the original source rock composition and climate/vegetation-controlled weathering and erosion processes; (e) steepness of the gradient between the terrestrial hinterland and the deep-marine basin, and (f) marine factors, including wave climate, tidal currents, mid- and bottom-water currents, water temperature and chemistry (e.g. salinity), Coriolis effect, upwelling and organic productivity.

Large rivers, such as the Mississippi, Amazon, Indus, Ganges and Niger rivers, have a fast delivery rate of fine-grained suspended sediments to the shelf, and downslope onto associated submarine fans. As a first approximation, the rate of supply to deep basins, given a fast delivery rate to the shelf, will be a function of: (a) shelf width; (b) currents redistributing sediment along or over the shelf-break, e.g. tidal, longshore-drift and storm cells, (c) the location of submarine canyons intercepting shelf-sediment transport paths, and (d) sediment stability, related to many variables including tectonics.

Sediment accumulation rates show a dramatic variation from hundreds of metres being deposited essentially instantaneously as large-scale, continental-margin slides, to slow pelagic deposition, typically 1–60 mm/1000 years (Fig. 4.7). However, pelagic sediment accumulation rates may reach about 100

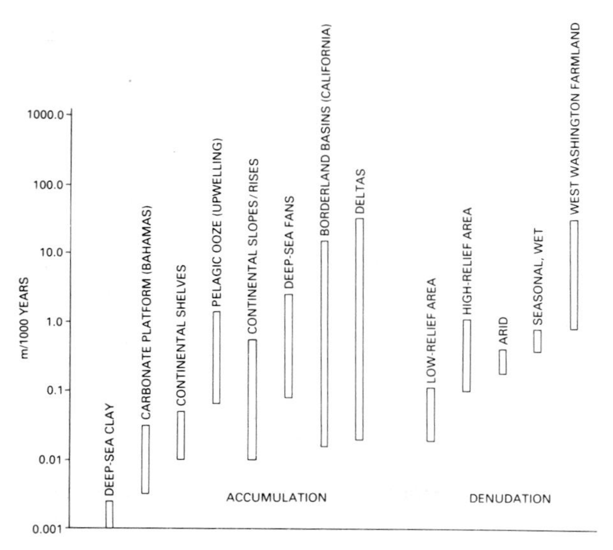

Figure 4.6 Log-plot to compare time intervals for the accumulation of various sediment types with terrestrial rates of denudation in different climates, based on data from Howell & von Huene (1980) and Blatt *et al.* (1980). After Stow *et al.* (1985).

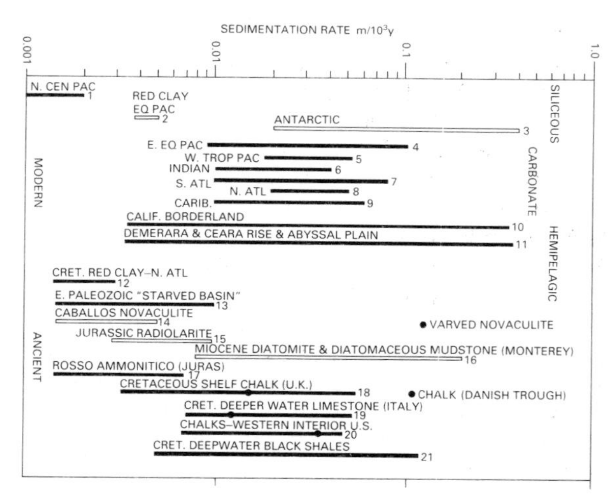

Figure 4.7 Rates of sediment accumulation of some modern and ancient pelagic and hemipelagic sediments. Note that estimated rates for ancient sediments are approximately 60–70% of modern rates. (After Scholle & Ekdale (1983).

mm/1000 years on outer shelves and upper slopes in areas of upwelling. Hemipelagites commonly accumulate at 5–300 mm/1000 years (Fig. 4.7). Turbidite systems, such as submarine fans, vertically aggrade at approximately 100–200 mm/1000 years, although rates up to about 11 000 mm/1000 years are recorded from the Mississippi Fan (Bouma, Coleman *et al.* 1986). The estimated frequency of turbidity-current events typically ranges from about one hundred to several thousand years between flows, and about one order of magnitude less frequent for redeposited carbonates (Howell & Normark 1982, Stow *et al.* 1985, Dimberline & Woodcock 1987).

Geometry and size of basin

The geometry and size of the basin receiving sediments is an important variable because it may severely constrain the shape of the depositional system (Pickering 1982c) and, therefore, the three-dimensional distribution of facies, facies associations and sequences. Whether a basin is over- or under-supplied, with respect to sediment input, will depend upon the rate at which sediments are delivered; tectonic activity may alter the basin geometry during sedimentation.

Small slope basins in areas of fast sediment accumulation will tend toward over-supply, and subsequent spill-over into neighbouring basins. Such a process is occurring in the basins of the offshore California Borderland, where the filling of nearshore basins leads to a progressive oceanward migration of depocentres (e.g. Douglas & Heitman 1979). Small slope basins on accretionary prisms also are susceptible to over-supply. An important aspect of over-supplied basins is the dominant influence of extrabasinal factors in controlling sedimentation. Increased rates of sediment accumulation, and tectonism reducing the basin dimensions, may exert profound effects on such basins.

Under-supplied basins are the most common type, although with time many tend toward over-supply. The extensive abyssal plains tend to remain under-supplied, as do large tracts of continental slopes and rises. In these under-supplied basins, the shapes of depositional systems are most likely to develop an internal equilibrium with their environment. Morphologically well developed, large, radial to elongate fans typify under-supplied basins. In contrast, small, fault-controlled, over-supplied basins, as in the offshore California Borderland, tend to contain morphologically poorly developed fans, with shapes governed by the basin margins.

Linear or ribbon-like deep-water clastic systems will form in a variety of environments and scales, e.g. in submarine trenches, submarine canyons, gullies and valleys, and some linear slope basins (Ch. 6). In these systems, facies distribution may show axial proximal to distal changes, but the detailed clastic architecture will be governed by many factors, including rate of sediment supply, location of sediment entry points, tectonics, basin topography, grain size and sediment type.

The location and number of sediment entry points exerts an important influence on the distribution of sediments within the basin. Closely spaced slope gullies and canyons favour overlapping depositional systems, as in a slope apron (Ch. 5). Examples include steep submarine slopes, such as along active fault margins. Fewer major entry points to a large ocean basin, with fast rates of sediment accumulation, tend to be associated with large, morphologically well developed submarine fans like the Indus and Bengal fans of the Indian Ocean.

Ocean current circulation and Coriolis effect

Surface ocean currents, and in turn deeper marine circulation (Figs 4.8a & b), are influenced mainly by atmospheric circulation (winds), latitudinal temperature gradients, the Earth's rotation, ocean-basin size, depth, physiography, inter-connectedness to other oceans, and basin orientation. The deep currents exert a major control on the resuspension and deposition particularly of finer grained sediments, e.g. contourites, hemipelagites and pelagites. Figure 4.9 shows the present deep-water circulation in the Atlantic Ocean and the location of large sediment drifts that are preferentially developed where flow patterns favour the long-term accumulation of fine-grained sediments. Unfortunately, detailed palaeoceanographic reconstructions that include ocean circulation patterns are generally at best speculative for many ancient deposits. However, in order to emphasize the significance of such current patterns, and because they are not considered in depth elsewhere in this book, this section is relatively detailed.

In this book, the main focus is on deep circulation driven by temperature and salinity differences, i.e. thermohaline circulation. However, surface and internal waves, and shallower wind-driven or tidal currents and cells may be important in suspending shelf and canyon-head sediments that are subsequently redeposited in deeper water. Seasonal variations in the position of mid- and bottom-water sediment plumes

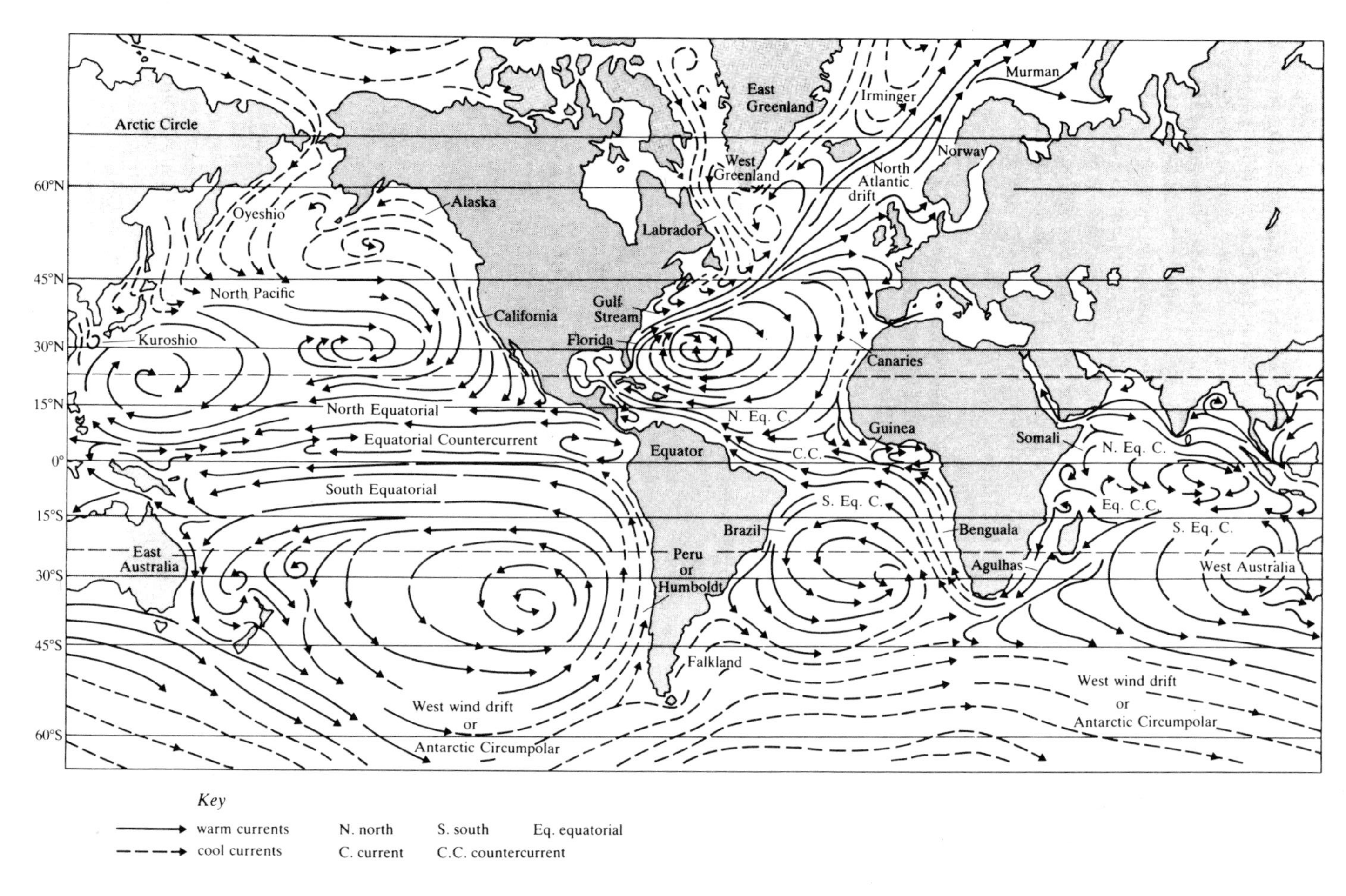

Figure 4.8a World ocean current circulation in surface waters.

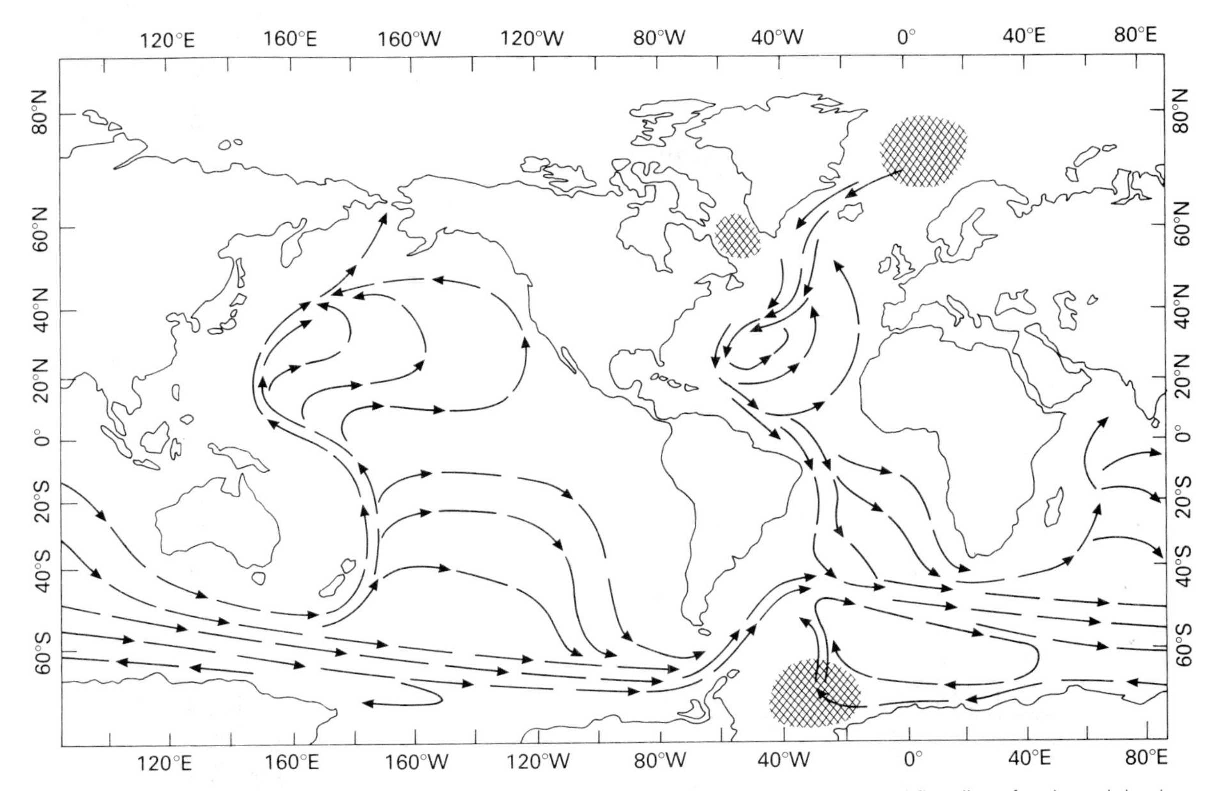

Figure 4.8b The cross-hatched areas are regions of production of bottom waters. Schematized flow lines for abyssal depths.

will affect the amount of shelf spill-over of suspended fine-grained matter (see also Ch. 2). For example, off southern California, Karl (1976) demonstrated the changes in suspended sediment concentrations between winter and summer (Fig. 4.10). In winter, compared to the summer months, storms produce higher concentrations of suspended sediment nearer the shelf-break where turbulence diffuses the plumes, and breaking internal waves are capable of suspending sediments farther offshore. Thus, offshore from California, the main contribution to hemipelagic sedimentation may spill over the shelf-break into deeper water during the winter months.

Ocean currents control the mixing of cold and warm water masses, biological productivity, sites of upwelling, and the distribution of various chemogenic sediments (Facies Group G3) such as phosphorites. Oceanic circulation, interacting with water depth and basin physiography, oxygen minima or anoxia, and biological productivity, govern the potential distribution of, for example, organic carbon-rich black shales (Facies E2.2).

Although modern ocean circulation patterns are well known (Fig. 4.8) and despite their profound impact on the distribution of fine-grained facies, there is little data on ancient ocean current patterns, except constraints provided by DSDP and ODP drilling. For example, a comparison of Palaeocene and Oligocene world oceans, based partly on DSDP data (Fig. 4.11) shows the important effect that the distribution of continents plays in controlling ocean circulation. In the Palaeocene, prior to the collision of India with the Asian Plate, there was an equatorial circum-global Pacific–Tethys current, and clockwise subpolar gyres inferred for the Southern Pacific and Atlantic. By the Oligocene, the remnant of Tethys was a relatively small fragmented ocean basin, and much of the present-day ocean circulation was established, including the Circum-Antarctic Current.

Results from various DSDP sites around the Central Atlantic show that even in the Early Cretaceous, a proto-Gulf Stream was established (Fig. 4.12), with pelagites and hemipelagites comprising varve-type laminations, graded claystones and limestone-shale couplets (Robertson 1984, Sheridan, Gradstein *et al.* 1983). The fine-grained varve-type lamination, formed

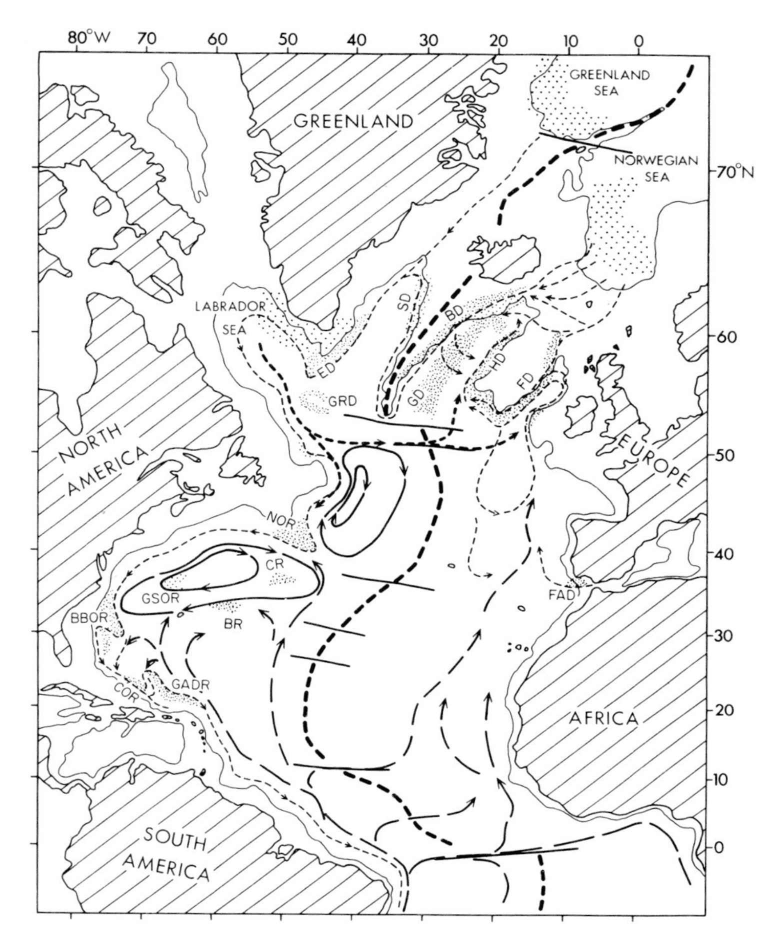

Figure 4.9 North Atlantic present-day deep-water circulation and major sediment drifts (close stipple). FAD, Faro Drift; FD, Feni Drift; HD, Hatton Drift; GD, Gadar Drift; BD, Bjorn Drift; SD, Snorri Drift; ED, Eirik Drift; GRD, Gloria Drift; NOR, Newfoundland Outer Ridge; CR, Corner Rise; BR, Bermuda Rise; GSOR, Gulf Stream Outer Ridge; BBOR, Blake Bahama Outer Ridge; COR, Caicos Outer Ridge, and areas of bottom water production shown in wide stipple. After Stow & Holbrook (1984).

by fluctuating proportions of terrigenous plant material, marine plankton and clastics, may reflect short-periodicity (tens to hundreds of years) climatic changes. The graded claystones and black shales represent fine-grained turbidite redeposition from within or near the oxygen minimum zone on the upper continental slope. The limestone–shale couplets suggest climatic variation on time scales of 20 000–60 000 years, with the organic-rich shales formed during wetter periods when abundant plant material entered the sea (Robertson 1984). The abundance of radiolaria in the pelagic chalks suggests upwelling to produce fertile surface waters, possibly due to the inflow of nutrient-rich waters from western Tethys (Fig. 4.12). This is just one example of the importance of palaeoceanography in interpreting ancient deep-marine mixed siliciclastic and carbonate sediments, including organic-rich shales

The Earth's rotation causes an apparent deflection of oceanic currents to the right in the northern hemisphere, and to the left in the southern hemisphere. This deflection relative to the Earth's surface is called the Coriolis effect, and accounts for the deflection of equatorward-flowing bottom currents to the western side of the ocean basins, and differences in the heights of right- and left-hand levees of submarine channels (Komar 1969).

Sequences

Sequences, rhythms or cycles, define an essentially progressive change in bed thickness and/or grain size, both for vertically-stacked beds or packets of beds. For consistency, the term sequence is adopted in this book.

Sequences were first recognized in flysch successions by Vassoyevich (1948), although vertical trends were recorded as early as 1897 by Bertrand. Research from 1960–80 emphasized the importance of asymmetric

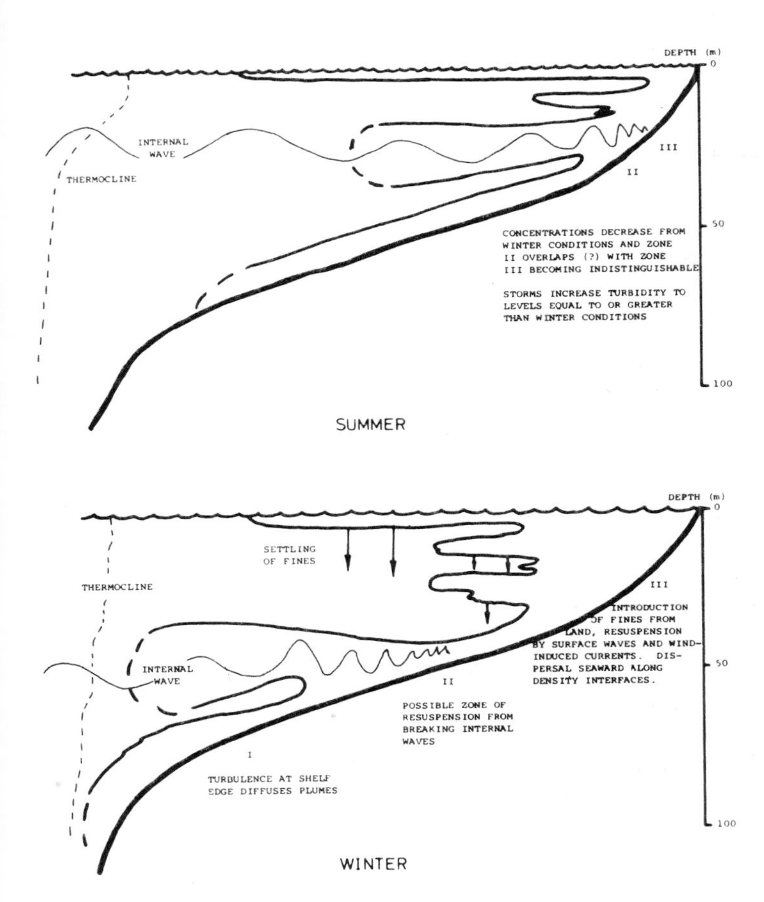

Figure 4.10 Model for turbid layer development in typical winter and summer conditions for the southern California shelf. After Karl (1976).

thinning- or thickening-upward sequences in deep-water successions (e.g. Nederlof 1959, Ksiazkiewicz 1960, Kelling 1961, Dzułynski & Walton 1965, Kimura 1966, Sestini 1970, Sagri 1972, Mutti & Ricci Lucchi 1972, 1975, Mutti 1974, 1977, Ricci Lucchi 1975b, Shanmugam 1980, Ghibaudo 1980). The objective recognition of sequences at a variety of scales is important in understanding intrabasinal and extra-basinal controls on sedimentation.

By convention, thickening- and/or coarsening-upward sequences are designated as 'negative', and thinning- and/or fining-upward sequences as 'positive'. This terminology is based, to a large extent, upon the sequence classification of Ricci Lucchi (1975b).

As suggested by Ricci Lucchi (1975b) and Shanmugam (1980), the largest scale sequences are defined as first-order sequences, and generally involve hundreds to thousands of metres of succession, often trans-formational, in which there is an upward change in grain size, bed thickness and sediment composition. Overall progradation or retreat of major related depositional environments may generate first-order sequences. Examples include: (a) the approximately 600 m-thick Namurian Edale Shales, Mam Tor Sandstones, Shale Grit, Grindslow Shales, Kinderscout Grit thickening- and coarsening-upward sequence, northern England, interpreted as an upward-shallowing from deep-water basinal shales to delta-top sandstones (Allen 1960, Walker 1966, Collinson 1969, 1970, McCabe 1978); (b) the progradational suites of basin plain, outer fan, middle fan, inner fan turbidites and hemipelagites in the northern Apennines, Italy (Ricci Lucchi 1975); (c) the pelagic starved-basin to outer fan deposits of the Middle Ordovician Blockhouse Formation, Tennessee (Shanmugam 1980); (d) the overall 3200 m-thick, thinning- and fining-upward sequence of the inner, middle, outer fan turbidite system of the upper Precambrian Kongsfjord Formation, northern Norway (Pickering 1981a, 1985), and (e) the Carboniferous (Mississippian) progradational sequences described from the Ouachita Mountains, Oklahoma and Arkansas (Niem 1976).

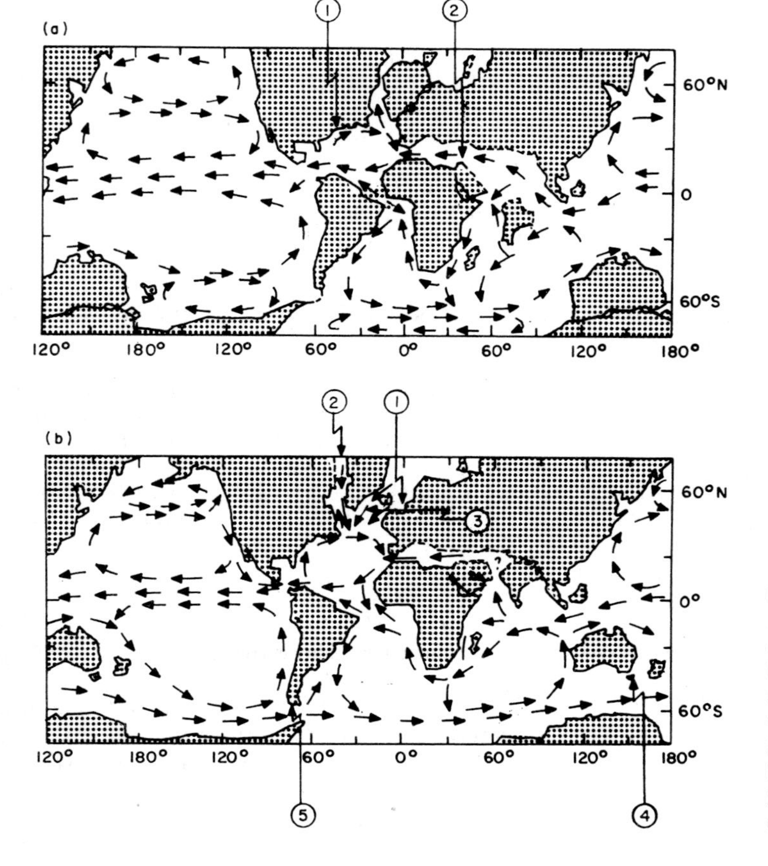

Figure 4.11 Schematic reconstruction of the Palaeocene and Oligocene distribution of continents and ocean surface-circulation patterns. Numbers refer to following: (a) Palaeocene; 1, Proto Gulf Stream; 2, Tethys Current; (b) Oligocene; 1, Norwegian–Greenland Sea; 2, North Labrador Passage; 3, Greenland–Iceland–Faroe Ridge; 4, South Tasman Rise; 5, Drake Passage. From Leggett (1985) after Haq (1981).

First-order sequences reflect major changes in sediment input to a basin, probably governed by long-term changes in sea level, overall subsidence or uplift of a basin, major progradation of linked clastic systems (e.g. delta–slope–fan systems), large-scale strike-slip displacements, and gradual peneplanation or rejuvenation of a source area. First-order sequences can describe the overall history of related depositional systems and, with good chrono- and litho-stratigraphic control, it may be possible to elucidate the specific factors generating such sequences.

Second-order sequences are defined on a scale of tens to hundreds of metres. They are identified as progressive changes in the thickness of beds or packets of beds, all within the same depositional system. Examples of second-order sequences are documented from the upper Precambrian Kongsfjord Formation, northern Norway, where hundreds of metres of fan-fringe sediments show increased or decreased frequency of outer-fan lobe deposits (Pickering 1981b). Second-order sequences may develop because of extrabasinal or intrabasinal processes governing the distribution of sediments within a clastic system.

Second-order sequences showing a thickening- and coarsening-upward trend develop on oceanic plates as they approach subduction zones and receive increasing amounts of terrigenous sediments. This vertical change is from open-ocean pelagites, to hemipelagites and thin bedded, fine grained turbidites, to thicker bedded and coarser grained trench/slope turbidites (e.g. Schweller & Kulm 1978).

Third-order sequences occur on a scale of metres to tens of metres, and rarely at slightly larger or smaller scales. Third-order sequences are the most commonly reported, but most subjective and controversial. Sequences on this scale are believed to reflect intrabasinal controls. Examples include claims of thinning- and fining-upward submarine channel fills (e.g. Mutti & Ricci Lucchi 1972, 1975, Ricci Lucchi 1975a & b, Mutti 1977, 1979, Walker 1978, Normark 1978,

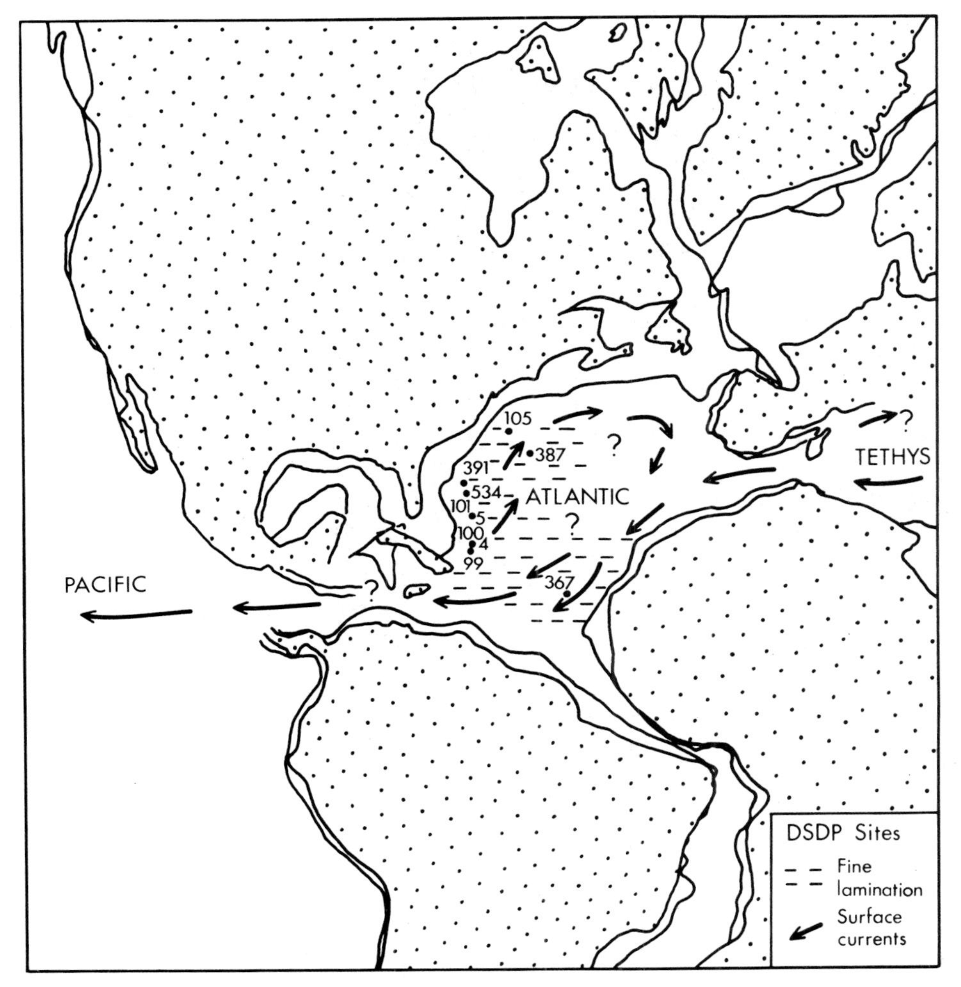

Figure 4.12 Reconstruction of the Central Atlantic Ocean in Early Cretaceous time, showing location of some DSDP sites, the inferred surface circulation and probable distribution of varve-type laminated sediments. After Robertson (1984).

Hendry 1978, Pickering 1982a), and thickening- and coarsening-upward fan-fringe and lobe deposits (e.g. Mutti & Ricci Lucchi 1972, 1975, Ricci Lucchi 1975a & b, Mutti 1977, Normark 1978, Walker 1978, Ghibaudo 1980; Pickering 1981b).

The basis for recognition of third-order sequences is almost always entirely subjective, and in many cases random alternations of bed thickness (and/or grain size) can generate asymmetric cycles that are as convincing as those that have been subjectively identified in field studies (Hiscott 1981). This is particularly true for inferred submarine-fan deposits, and claims that third-order thinning- or thickening-upward sequences are useful tools for environmental assessment (e.g. Ricci Lucchi 1975b) should be viewed with caution.

Statistical tests for the presence of asymmetric sequences can be designed to evaluate (a) whether trends of either decreasing or increasing bed thickness (or grain size) predominate in a selected section, and (b) whether a small segment of such an interval that appears to exhibit thickening or thinning upward in the field does in fact possess a statistically verifiable trend. Query (b) is not so interesting as query (a), because even undisputed asymmetric cycles can be produced with low probability by random variations in bed thickness in long stratigraphic sections.

A non-parametric test for supporting or rejecting the presence of an asymmetric cycle in a pre-selected interval (Charles W. Harper, pers. comm. 1987) involves the calculation of a coefficient of rank correlation called Kendall's τ (Kendall 1969, pp. 3–7, 1976, pp. 26–8) that compares the rank (in thickness or grain size) of each bed with all other beds above and below. τ varies from -1 to $+1$ (0 = no asymmetric trend), with thinning-upward sequences having $\tau < 0$ and thickening-upward sequences $\tau > 0$ (for assignment of highest ranks to the thickest beds). Variation of τ depends on the number of beds in the sequence, and the null hypothesis of no asymmetric trend can be evaluated at any appropriate significance level.

Waldron (1987) outlines how a signs-of-differences test proposed by Moore & Wallis (1943) can be used to address query (a), i.e. whether a long stratigraphic section has sequences with consistent asymmetry. The test compares the total number of decreases in bed thickness observed in the section, m (i.e. the number of negative differences in thickness from each bed to the next overlying bed) to the mean number of decreases, μ_m, expected in a section with N beds, where:

$$\mu_m = (N - 1)/2 \qquad \text{for } N > 12 \qquad (4.1)$$

standard deviation of m, σ_m, is given by:

$$\sigma_m = \sqrt{(N+1)/12} \qquad (4.2)$$

The test statistic, z, is given by:

$$z = (m' - \mu_m)/\sigma_m \qquad (4.3)$$

where $m' = (m - 0.5)$ if $m > \mu_m$, and $m' = (m + 0.5)$ if $m < \mu_m$. In a random succession, z is approximately normally distributed with mean of zero and standard deviation of 1.0. The null hypothesis that sequences with consistent asymmetry are not present can be assessed at various significance levels. If the null hypothesis is rejected, and $z > 0$, the sequences thin upward; if $z < 0$, the sequences thicken upward. Waldron (1987) also indicates how data can be smoothed to filter out the effect of random fluctuations superimposed on truly cyclic, asymmetric sequences. Note that a significant result using this signs-of-difference test might result from the sections showing a steady trend of bed thickness or grain size with no cyclicity. Conversely, a mixture of positive and negative asymmetric sequences could cancel one another out giving a non-significant value of z.

There is an important conceptual difference between asymmetric sequences that show a gradual and progressive trend in bed thickness (or grain size) and those that show step-like trends. Consider two thinning-upward sequences of ten beds above a fine grained interval, one progressive and the other involving three steps (Fig. 4.13). Excluding the possibility of ties, for the gradual sequence (Fig. 4.13a), the probability, P_G, of observing such an ordered sequence is given by:

$$P_G = \sum_{i=1}^{i=9} P(t_{i+1} < t_i) = 1/9! = 2.7 \times 10^{-6} \qquad (4.4)$$

where t_i is the thickness of the ith bed. The calculated value of P_G was obtained by considering the probability that ten bed thicknesses taken from the populations of beds would occur in only this one regular arrangement of consistently decreasing thickness from base to top.

If (a) the sequence shows a step-wise decrease in bed thickness above a thin-bedded interval, (b) beds tend to occur in 'packets' of either thick, medium, or thin beds (Fig. 4.13b), and (c) the three types of packet are equally abundant in a section, then the probability, P_{SW}, that a step-like sequence will occur above a thin-bedded interval is given by:

$$P_{SW} = (1/2)^3 = 0.125 \qquad (4.5)$$

Clearly, $P_{SW} >> P_G$ in this case, and the step-like sequence is far less likely to represent a statistically significant trend.

In Figure 4.13b, the packets of quite different bed thickness really represent three different 'states' of the deep-water system, each perhaps corresponding to a different environment (e.g. lobe axis versus lobe fringe), and each with its own distribution of bed thickness. In our experience, many deep-water fan deposits are characterized by step-like sequences (cf. Ricci Lucchi 1975b), and alternations of 'packets' with substantially different bed thicknesses (e.g. Fig. 7.9). Within packets, the thickness of a bed is not entirely independent of the thickness of associated beds, and the tests of Kendall (1969, 1976) and Waldron (1987) are inappropriate.

Instead, we suggest subdivision of successions with step-like changes in bed thickness or grain size into 'states' (facies) based on these characteristics, and then application of Markov Chain analysis to identify preferred sequences of these states. Procedures are given by Powers & Easterling (1982) and Harper (1984). This approach was used successfully by Hiscott (1980) to

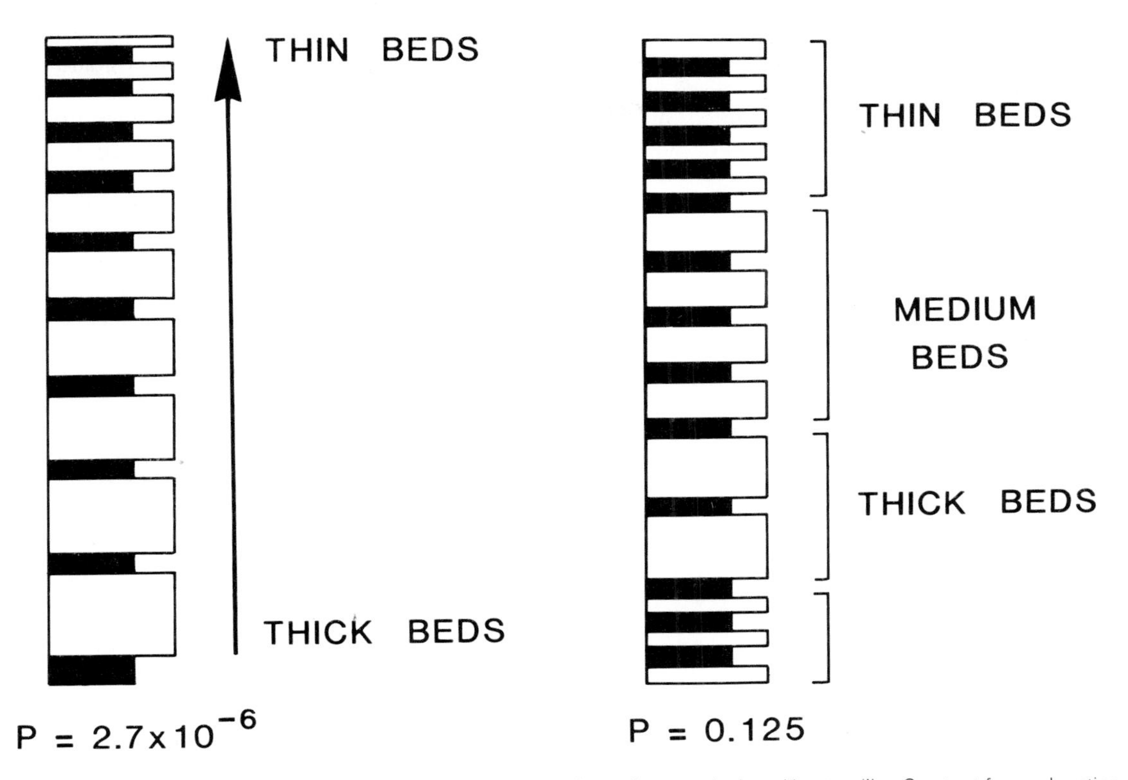

Figure 4.13 Idealized types of sequence (shown as thinning-upward trend): a, gradual, and b, step-like. See text for explanation.

identify a tendency for thinning- and fining-upward in part of the Ordovician Tourelle Formation, Quebec, Canada, but unsuccessfully by Hein (1979, to identify asymmetric trends in the less organized submarine channel fills of the Cambro-Ordovician Cap Enragé Formation, Quebec.

In spite of our view that third-order asymmetric sequences require careful verification by statistical tests, we encourage continuing efforts to objectively document such sequences, because third-order sequences may provide the most sensitive geological indicators of short-term fluctuations in sea level (Mutti 1985), processes of channel and lobe switching (Ch. 7), climatic changes in a source area, pulses of tectonic activity, and retrogressive slope failure (Pickering 1979).

Thinning- and fining-upward sequences appear to be more plentiful than positive sequences at the second- and third-order levels. If sea-level fluctuations are partly responsible for many sequences (Mutti 1985), then the difference in abundance of positive and negative sequences may indicate that falls are more rapid than rises in sea level. An alternative, perhaps more plausible explanation is that falls in sea level lead to an increased frequency of erosive events, whereas rises, on average, are associated with less erosive sediment flows. Thus, sequences associated with a rise in sea level have the greatest preservation potential. A good example of the latter case is an approximately 200 m-thick thinning- and fining-upward sequence (Fig. 7.3), generated by a rise in sea level from the Pleistocene to Holocene in a middle-upper fan channel of the Mississippi Fan (Pickering *et al.* 1986b).

Milankovitch cyclicity

Since the publication of the Milankovitch Theory in 1924, there has been considerable controversy regarding the recognition and significance of Milankovitch cycles (cf. Imbrie & Imbrie 1979). Essentially, Milankovitch Theory postulates that secular variations in the orbital parameters (eccentricity, obliquity, precession) cause long-term changes in the solar radiation flux to the Earth's surface. The fluctuating radiation levels appear to have caused climatic variations which led to the succession of glacial and interglacial phases during the Pleistocene (Fig. 4.14). Imbrie *et al.* (1984) have recently confirmed that there is a close connection between global ice volumes monitored by oxygen isotopes and varying insolation. It remains unclear how very small changes in the radiation budget led to large climatic variations.

Pelagic and hemipelagic sequences frequently exhibit decimetre to metre-scale cyclicity. It has been argued that this cyclicity reflects orbital-climatic control of sedimentation (cf. Fischer 1986 for review). Some of the best examples of such cyclicity come from the Mesozoic where major changes in ice volume appear not to have been a factor. Barron *et al.* (1985) have demonstrated that limestone/laminated shale alternations from the middle Cretaceous of the USA could have resulted from variations in monsoon intensity that were controlled by the orbital cycles. Spectral analysis has recently been applied to ancient successions as an objective method for detecting regular sedimentary cycles. For example, Weedon (1986, 1989, Fig. 4.15) has inferred that three regular cycles detected with power spectra for the Lias (Jurassic) of Switzerland can be attributed to the orbital cycles.

Global changes in climate exert a primary control on most sedimentary systems. Ice maxima and the associated lowstands in sea level cause an increased rate of clastic influx, with correspondingly larger grain-size populations reaching deep-marine environments. Although it is possible to relate variations in clastic influx to sea-level and climatic changes, it is considerably more difficult to tie such changes to fluctuating radiation levels to the Earth, especially over considerable geological time periods. We are currently unable to assess the utility of Milankovitch Theory in interpreting deep-marine systems, but determining the significance of such extraterrestrial controls poses a challenge for future research to address. If the theory is to prove useful, ancient successions must be dated on the same time scales as the Milankovitch cycles, i.e. 10,000–50,000 years. It is not yet clear in what range of deep-marine facies Milankovitch cyclicity might be detected.

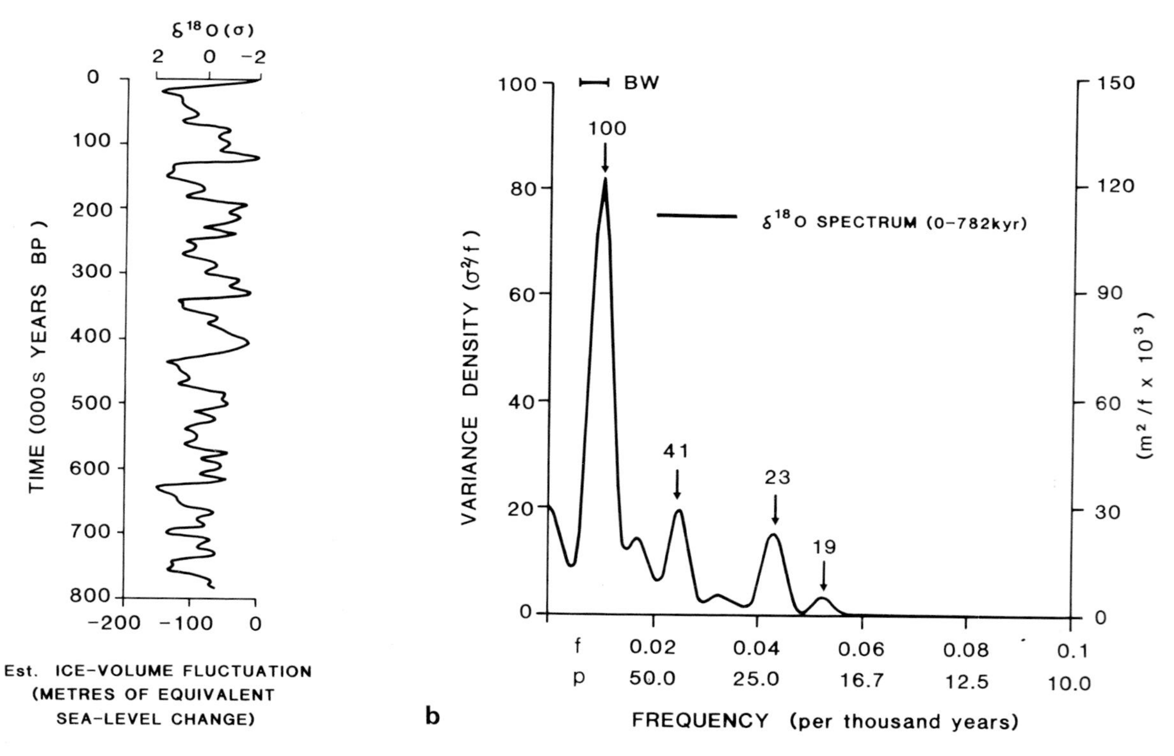

Figure 4.14: (a) Variations in ice volume over the last 780 000 yr as recorded by sea water $\delta^{18}O$ values (modified after Imbrie *et al.* 1984). (b) Power spectrum of the isotope curve revealing regular cycles of 100 000 yr (eccentricity), 41 000 yr (obliquity = tilt), 23 000 yr and 19 000 yr (precession) (modified after Imbrie 1985). The resolution of these spectra (0.0072c per 1000 years) is shown by the bandwidth (BW).

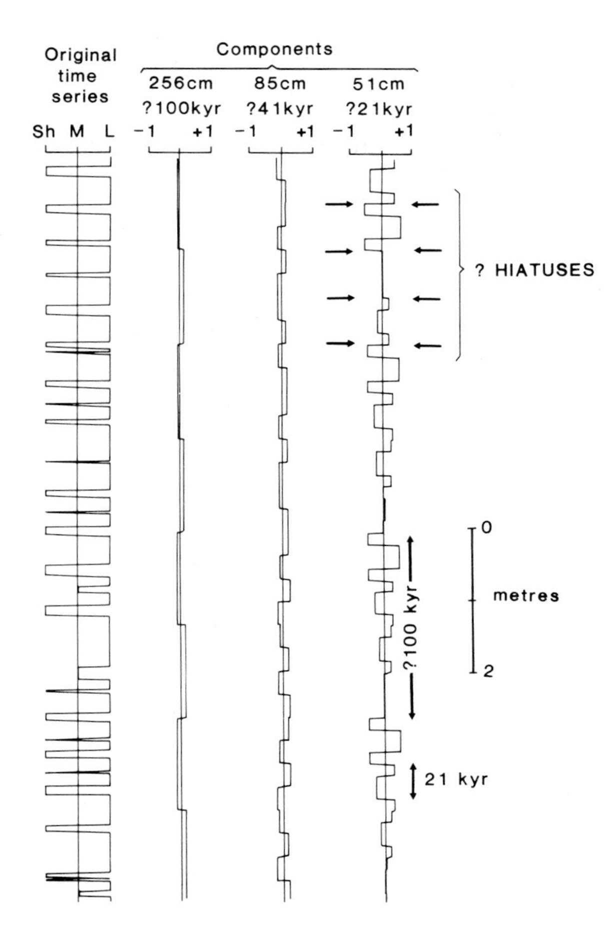

Figure 4.15: Measured section data from the early Pliensbachian (Jurassic), Breggia Gorge, Switzerland (left column) – Sh = shale; M = marl; L = limestone. Spectral analysis of data shows three regular sedimentary cycles of 256, 85 and 51 cm wavelength, illustrated here after filtering, thought to correspond to the 100 000, 41 000 and 21 000 yr orbital cycles (left to right column, respectively). The 51 cm (? 21 000 yr) cycle shows amplitude variations perhaps recording modulation of the 21 000 yr precession cycle by the 100 000 yr eccentricity cycle. Arrows denote inferred hiatuses based on apparent disruption of the pattern of amplitude variations at top of section. After Weedon (1989).

PART 2

Deep-water basin elements

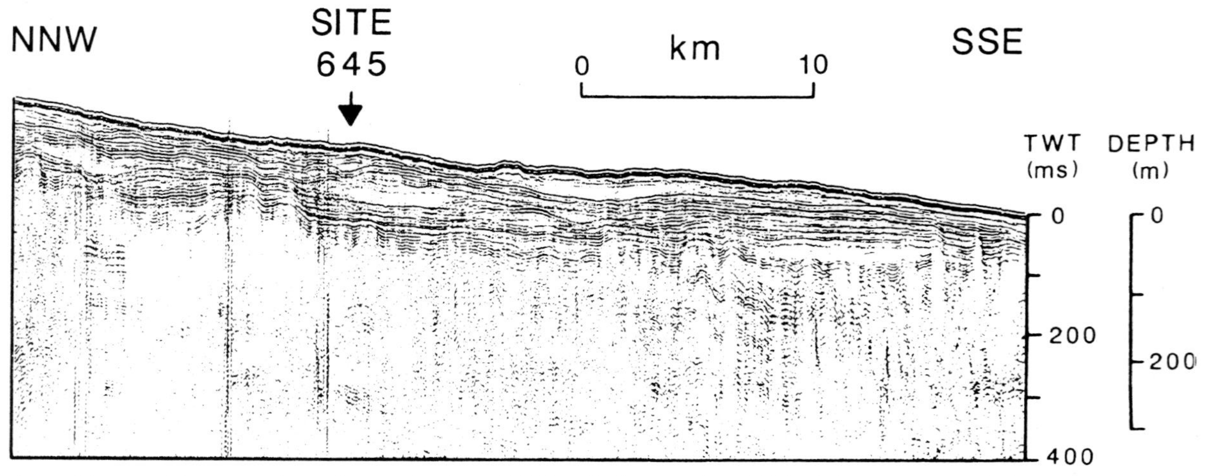

Air-gun high-resolution seismic profile passing from N-NW to S-SE through ODP Site 645, west-central Baffin Bay, showing several acoustically transparent debris flow lenses in the upper 170 m of sediment. This profile runs along the slope at a water depth of about 2000 m, and was obtained on CSS *Hudson* cruise 87–033.

CHAPTER 5

Slope aprons and slope basins

Introduction

Wagner (1900) originally defined the *continental slope* as that area on the continental margin between the shelf break and the abyssal floor. The current definition of the continental slope was proposed by Heezen *et al.* (1959, pp. 18–19) for 'that relatively steep (3–6°) portion of the sea floor which lies at the seaward border of the continental shelf.' Slope-aprons comprise steep slopes and rises that border shelf basins, large ocean basins, flanks of oceanic ridges, isolated seamounts and plateaus. The distinction, in part, between the continental slope and a slope-apron is that the continental slope marks the transition zone between continental and oceanic crust, whereas slope-aprons define local areas of high bathymetric gradient, on continental, oceanic or transitional crust.

Both the continental slope and slope-aprons are important sites for the transfer of clastic material to deep-marine basins by various resedimentation processes. At some sites, there is significant accumulation of economic hydrocarbon and offshore mineral deposits (e.g. Bouma 1979, Winn *et al.* 1981, Galloway & Hobday 1983). Due to the widespread economic, political and geologic interest there has been extensive work done on modern continental slopes and slope-aprons from seismic, geological and marine geotechnical engineering perspectives (Bouma 1979). A complete literature review is beyond the scope of this book; the reader is referred to the following publications which present much of the available references in this field (Bouma 1979, Cook & Enos 1977, Doyle & Pilkey 1979, Watkins *et al.* 1979, Cook *et al.* 1982, Saxov & Nieuwenhuis 1982, Galloway & Hobday 1983, Stanley & Moore 1983, Embley & Jacobi 1986, Stow 1986).

A variety of different morphologies and sedimentary facies occurs on continental slopes and slope-aprons depending on the tectonic, erosional and depositional history of the area. The modern continental slope begins beyond the shelf break at 45–300 m below sea level. Slope physiography reflects progradational and/or erosional history, faulting and the type of sediment supply (Fig. 5.1). Prograding slopes vary in width from <1 km to >200 km, commonly have gentle gradients (2–7°), and a complete range of biogenic and clastic facies. At sites of active synsedimentary tectonism,

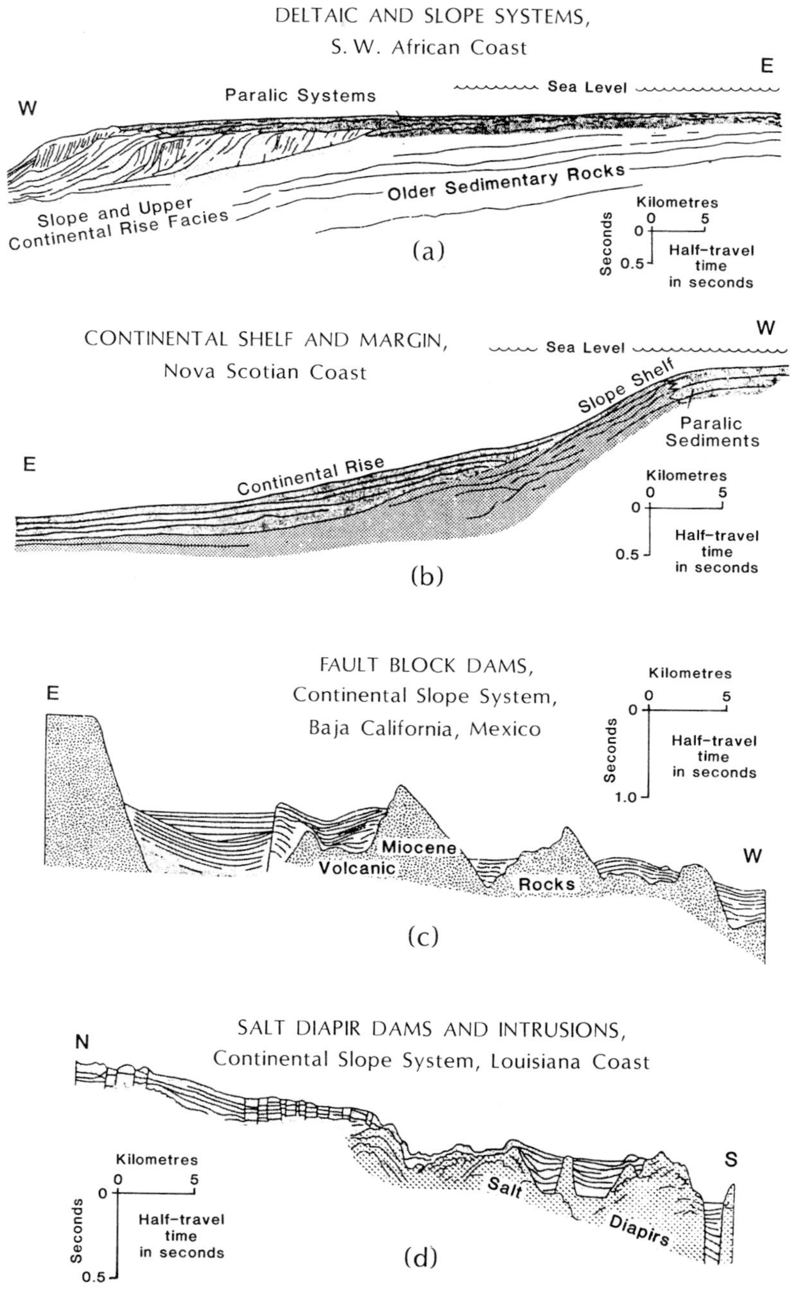

Figure 5.1 Seismic reflections that characterize several styles of deposition along some Holocene continental margins. a, complex patterns (including rotated and resedimented clinoforms) in deltaic and slope systems along southwestern African coast; b, continental rise onlap along Nova Scotia coast, c, fault basins containing superimposed (uplap) slope deposits dammed behind fault blocks, Baja California, Mexico; d, slope deposits trapped in salt basins and behind salt ridges along Louisiana coast, Gulf of Mexico.

other irregularities develop on the slope, including small intraslope-basins behind salt or mud diapirs, roll-over and thrust-faulted anticlines, slide blocks, small horst-and-graben or pull-apart structures and fault-scarps (Fig. 5.1). Some slopes are aggradational scarps (i.e. many carbonate slopes) or are erosional, with exposed bedrock (Fig. 5.1c); these types are mainly sites of sediment bypassing and nondeposition.

The upper slope on many modern continental margins is typically a zone of sediment bypassing, with local sediment ponding (Fig. 5.1c & d), progradation, and gully or channel fills. Most modern slopes, however, may not be truly representative of preserved ancient slope systems. The present day systems were affected by the rapid postglacial rise in sea level with, in most cases, the result being an erosional regime.

Stow (1986) identifies differences in facies associations between upper and lower slope-apron muddy facies. The upper slope-apron is marked by more abundant 'higher energy' features, including more re-sedimented facies, slide scars, erosional gullies and channels; in contrast, the lower slope-apron is characterized by more 'lower energy' features, with greater frequency of fine-grained turbidites, isolated submarine channel fills, debris flows, slide and slump scars, and local interfingering with contourite drifts and/or submarine-fan deposits. The distinction between upper slope-apron and lower slope-apron is not precise.

No comprehensive 'facies model' exists for deep-water slope sediments. This is mainly due to the paucity of data, partly a function of the difficulty in studying modern continental slopes (which are only recently accessible through modern techniques of sampling and surveying) and due to the fact that many ancient clastic slope successions tend to be preserved only in special tectonic settings (i.e. commonly in very rapidly subsiding basins – such as syn-breakup, post-breakup phases of passive margins, active fault-bounded basins, trench-slope settings or wrench-fault basins associated with transform fault terrains Fig. 5.2. In general, there seems to be better preservation of ancient carbon-

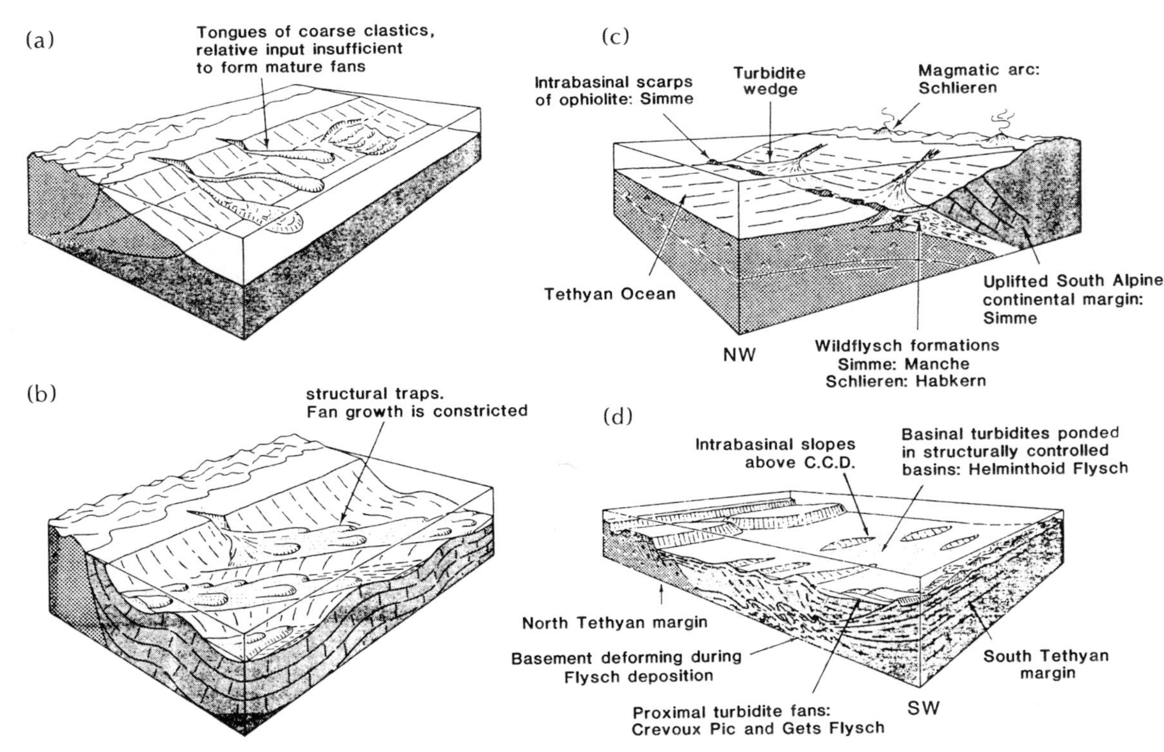

Figure 5.2 Immature ('non-fan') deep-marine systems as envisaged for some of the flysch successions in the western Alps. a, coarse gully and fan systems on a passive margin; b, ponded basins due to structural traps; c, turbidite wedges in trenches and intrabasinal scarps of subduction zones, and d, proximal turbidites on progressively deformed mobile basement and intrabasinal slopes. After Homewood & Carron (1983).

ate shelf-to-slope successions. This may reflect the greater compactness and, hence, ease of recognition of carbonate system shelf-breaks compared to those of siliciclastic systems. Some similarities in facies patterns exist between the ancient clastic slope and carbonate slope settings – but there are also some fundamental differences (McIlreath & James 1984).

A particular problem with the development of a 'slope' facies model is the basic tenet of facies modelling, that there is 'non-randomness' in the stacking of facies and that there is a *predictable* pattern in the horizontal arrangement of facies, which is reflected in vertical profiles. Slope-apron deposits tend to be poorly organized, displaying at most thin isolated thinning- and fining-upward sequences due to infill of a single channel or gully system (Fig. 5.3). Thus, facies are relatively isolated, with either poorly-ordered vertical stacking or chaotic facies distributions. This apparent 'chaos' in some natural systems and the inapplicability of the facies-model approach to these systems has been noted previously (e.g. Anderton 1987).

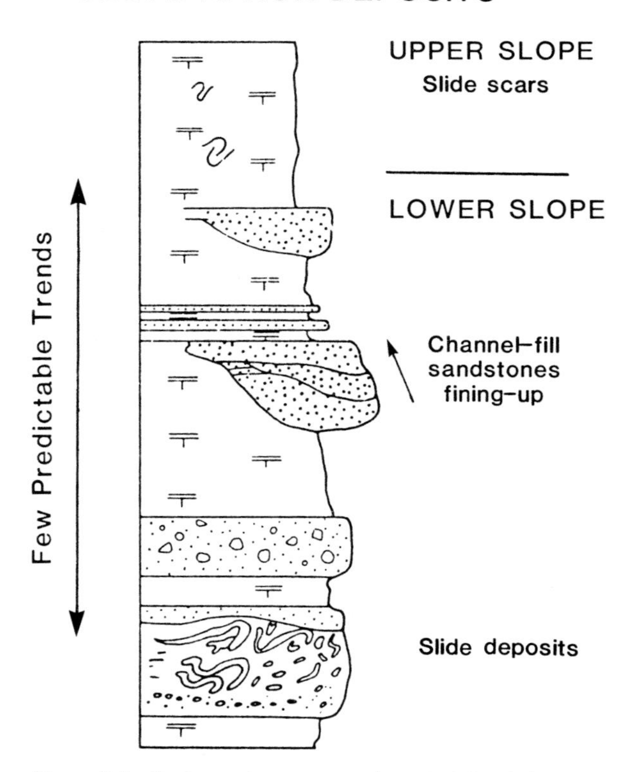

Figure 5.3 Facies and sequences characteristic of slope-apron environments. After Nilsen (1978) modified from Mutti & Ricci Lucchi (1972).

As emphasized for shelf sandstones by Cant & Hein (1986), a fundamental approach in the development of facies models is to consider those factors that influence and cause variations in sedimentation. One can assess the major controls on the distribution of facies and erect models to account for this variability rather than trying to draw 'ideal' facies models for each type of sedimentary environment – this is the so-called 'second generation' of facies modelling (Walker 1987). Thus, in this chapter we outline the variables that are important in developing slope-apron morphologies and deposits, and attempt to show how they can be expressed, using modern and ancient case studies.

Despite our impression that no comprehensive facies models can be defined for slope-sedimentation, the reader should be aware of the views of Stow (1986), who has summarized three composite slope-apron models – normal (clastic), faulted, and carbonate (Table 5.1); he also proposes at least ten different types of slope-aprons depending on their primary morphotectonic setting. The main subsea landforms recognized by Stow (1986) include: an abrupt shelf-break, fault-scarp and reef-talus wedges, slump and slide scars, irregular slump and debris flow masses, small channels and gullies, larger more complex dendritic canyons, isolated lobes, mounds and drifts, and broad smooth or current-moulded surfaces.

Seismic data have provided considerable information on continental slopes. Various patterns of reflectors, designated as seismic facies, have been recognized (e.g. Sangree *et al.* 1978). These include (Fig. 5.4): (a) mounded chaotic facies, the result of slumps, slides and other mass movements; (b) onlapping fill facies, due to infilling of slope irregularities by turbidity currents; (c) divergent and parallel-layered facies, a result of deposition from low-energy turbidity currents and hemipelagic sedimentation; and (d) sheet drape facies, resulting from blanketing of slope irregularities by hemipelagic sediments. Other more complex patterns of seismic reflections, including oblique and sigmoidal progradational seismic facies, are due mainly to interactions between shelf-margin and slope sedimentation processes (Sangree *et al* 1978).

In this chapter, the normal (clastic) and faulted types of slope-aprons are discussed, with emphasis on the morphology, and the seismic and sedimentary facies of the smaller physiographic features. The larger submarine canyons, gullies and valleys are discussed in Chapter 6, submarine fans in Chapter 7 and contourite drifts in Chapter 9. Some of the most spectacularly preserved ancient successions showing shelf-to-slope transitions occur in carbonate terrains, so a brief consideration of such margins is included in this chapter.

Major external controls on slope development

The major external factors controlling transport and sedimentation of material on deep-water slopes are similar to the controls on the distribution of almost all other clastic sedimentary facies. These factors are all somewhat interdependent, and in many cases only the relative influence of the various factors is discernible. The following factors will be discussed with particular reference to slopes: (a) sea-level variation; (b) sediment input; (c) subsidence; and (d) oxygen-content of bottom waters. Oxygenation of the bottom water is important, in that the location of the oxygen minimum and upwelling along continental slopes influences the organic content of the sediments, which in turn controls the distribution of epi- and infaunal populations and various pelagic and fine grained deep-sea sediments (e.g. black organic-rich shales, phosphorites and diatomites). Oxygen content is controlled by many processes; among the most important are ocean circulation, climate and sediment input. Chapter 4 presents a more general discussion of the variables controlling deep-water sedimentation.

Table 5.1 Main characteristics of modern and ancient slope-aprons (modified from Stow, 1986).

	Slope-aprons		
	Normal (clastic)	Faulted	Carbonate
Occurrence	Between shelf and basin floor, margins surrounding ocean basins and shelf, slope, or marginal sea basins		
	Widespread	Tectonically active regions	Low latitudes and ocean-ridge flanks
Shape	Narrow, rectilinear, elongate parallel to margin		
Dimensions	Width variable 5–300 km	Width variable 5–50 km	Width variable 5–300 km
Gradient	From 10° (upper) to 1° (lower)	Can be stepped with steeper parts >45°	Steep off reefs very gentle off oceanic ridges
Chief morphological elements	Shelf break, smooth open slope, gullies, channels, canyons, slide scars, slide and debris flow masses, base-of-slope sediment drifts and isolated lobes, transverse fracture zones across ocean-ridge flanks		
Processes and dispersal pattern	Linear distribution of sediment input points; resedimentation processes (all types) mainly downslope; bottom currents (all types) up-and-down channels and along-slope; pelagic settling widespread		
Facies associations	Fine-grained turbidites and hemipelagites dominant; slides and debrites common; minor coarse-grained turbidite channel fill facies; interbedded contourites and isolated contourite drifts.	Coarse, medium and fine-grained turbidites dominant; slides and debrites common; hemipelagites and pelagites common to minor; contourites minor.	Coarse, medium and fine-grained calciturbidites, pelagites and hemipelagites common; slides and debrites common;
Horizontal facies distributions	Fairly irregular	May have slope-parallel distribution of coarse to fine facies	Fairly irregular
Vertical facies sequences	Coarsening-upward and fining-upward sequences may occur related to slope progradation, sea-level fluctuation or tectonic activity; similar sedimentation-related sequences can occur in channels and lobes		

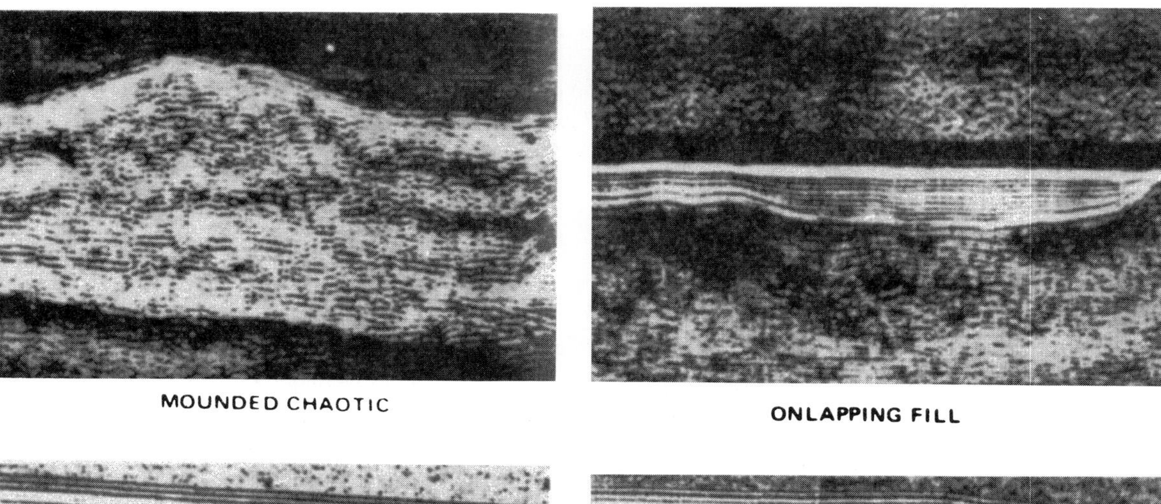

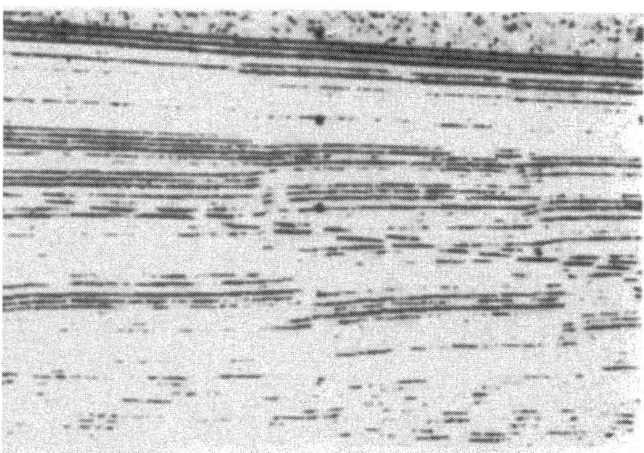

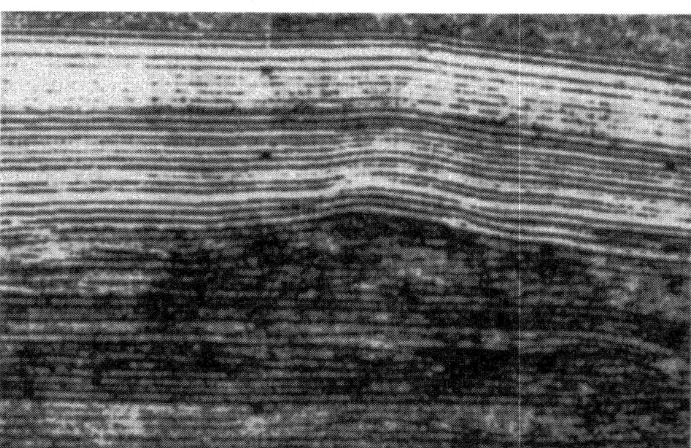

Figure 5.4 Interpretations of seismic facies: mounded chaotic – mass-transported facies; onlapping fill – turbidites; divergent and parallel layered – turbidites and hemipelagic sediments; sheet drape – hemipelagic sediments. Scales vary, vertical exaggeration ~ 2.5 : 1. From Sangree *et al.* (1978).

SEA-LEVEL VARIATION

In deep-sea sedimentary environments, sea-level variation mainly affects the sediment supply to the shelf-break and upper-slope areas. Also, sediment type (e.g. black shales, phosphatic sediments, diatomites, etc.) is among other factors partly a function of sea-level and climate (see Ch. 4; and later in this section). In an idealized siliciclastic model (Figs 5.5 & 6), highstands are marked by trapping of coarse-grained sediment in nearshore and shelf areas, with dominantly fine-grained hemipelagic sedimentation occurring beyond the shelf break. Coarse-grained sediment gravity flows are confined to submarine canyons which transfer material from nearshore sites to the offshore base-of-slope. During lowstands, there may be exposure of the shelf, with fluvial downcutting, nearshore sediment bypassing, and high sediment accumulation rates in offshore sites, much of which occurs as progradation of slope-aprons (Figs 5.5 & 6). Depending on the width of the continental shelf, deltas or littoral drift cells may input their load directly at the shelf-break, either into confined submarine canyons or onto unconfined slope aprons (Fig. 5.6) (Stanley *et al.* 1972b).

SEDIMENT INPUT

The grain size of material being transported to the slope-apron and the dominant mode of transportation affects the general shape of the continental slope and

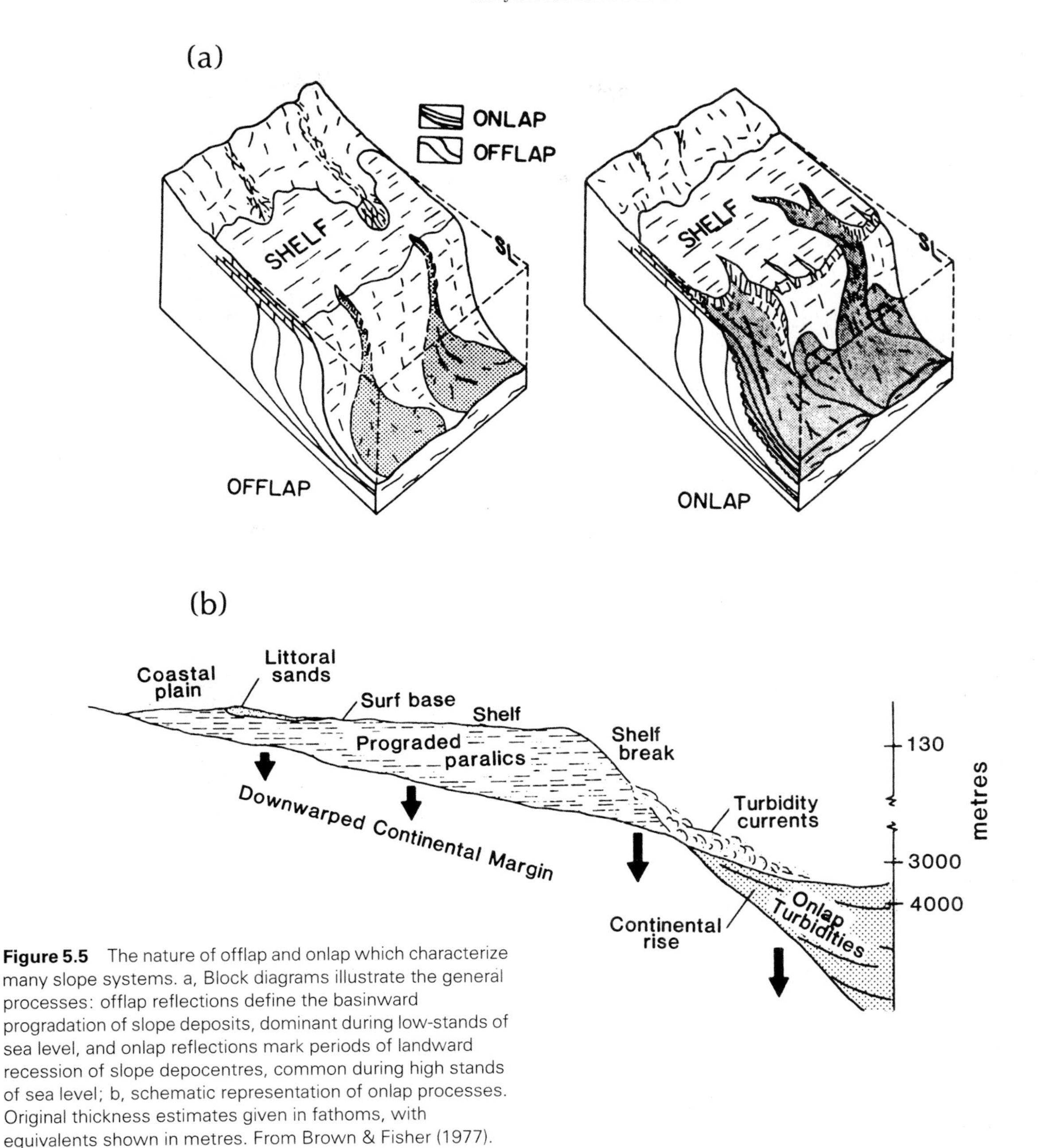

Figure 5.5 The nature of offlap and onlap which characterize many slope systems. a, Block diagrams illustrate the general processes: offlap reflections define the basinward progradation of slope deposits, dominant during low-stands of sea level, and onlap reflections mark periods of landward recession of slope depocentres, common during high stands of sea level; b, schematic representation of onlap processes. Original thickness estimates given in fathoms, with equivalents shown in metres. From Brown & Fisher (1977).

slope-apron. In regions of mainly fine-grained, hemipelagic sedimentation with littoral or contour-current reworking, slope-aprons tend to have a relatively gentle gradient (although local gradients may be higher), with thick accumulations of fine-grained material, most of which may move downslope by nearly continuous creep, sliding and faulting. Examples of this type of system include the slope-margins of Santa Barbara Basin (Thornton 1984) (Figs 5.7 & 8, Table 5.2) and portions of the continental margin off Nova Scotia (Fig. 5.1b) (Hill 1984b, Mosher 1987).

By contrast, where bedload is transported to the

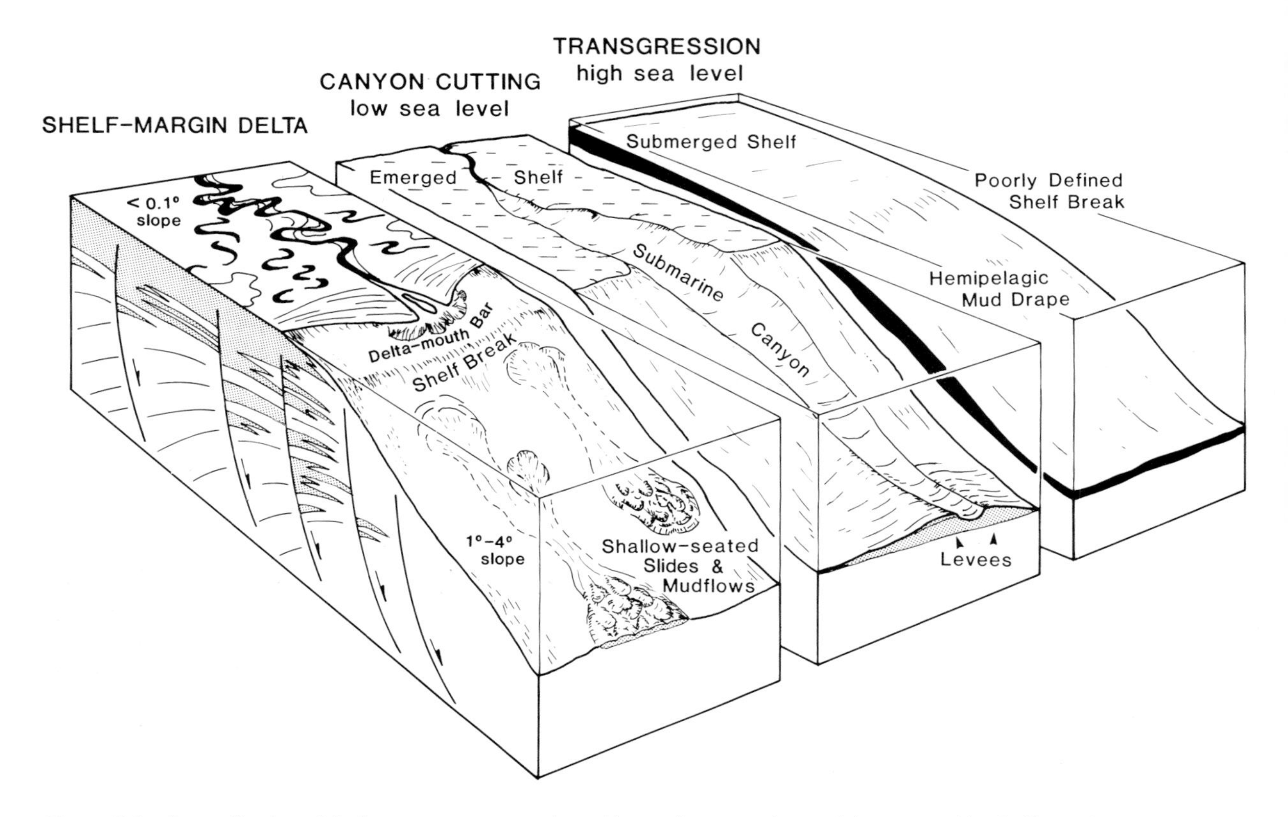

Figure 5.6 Generalized model of contemporaneous deposition and structural growth in an unstable shelf margin system. Structural styles, depositional environments and delta types are highly variable. Shelf-margin deltas occur during low stands of sea level. Submerged shelf margins, with possible trapping of material in cross-shelf submarine canyons occur during high stands of sea level. Modified from Winker & Edwards (1983).

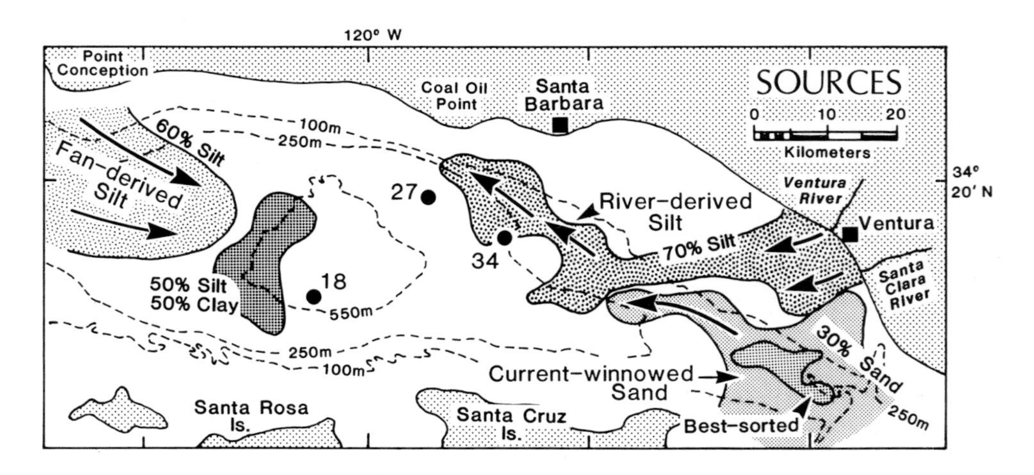

Figure 5.7 Sediment sources, transport paths and transport processes in Santa Barbara Basin, California Continental Borderland. X-radiographic prints of surficial cores (black dots: 18, 27, 34) are shown in Fig. 5.16. From Thornton (1984).

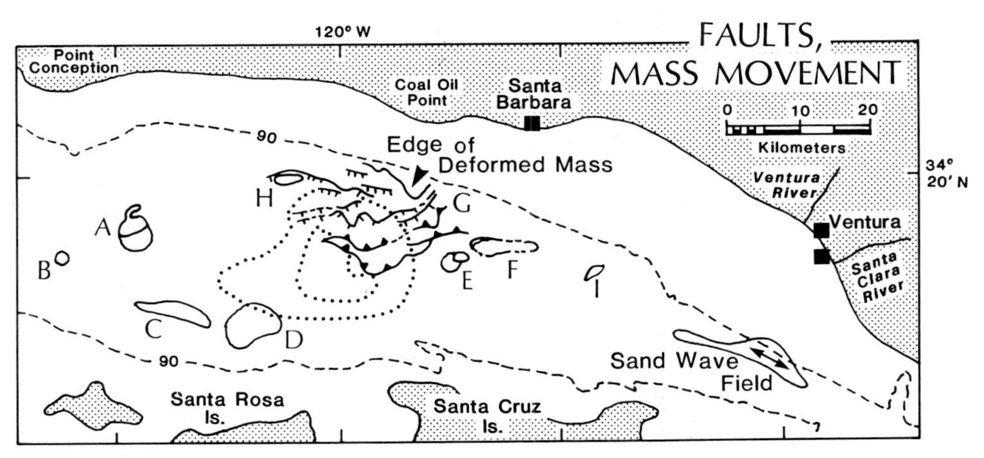

Figure 5.8 Mass-movement features, faults and sand wave field, Santa Barbara Basin, California Continental Borderland. Arrow on sand wave field denotes predominant direction of crests. From Thornton (1984).

Table 5.2 Characteristics of mass-movement zones and Slope Area G, Santa Barbara Basin (refer to Figs 5.7 & 8 for location of mass-flow zones) from Thornton 1984, in Stow & Piper 1984. X-radiographic prints of surficial cores from flows G and F are shown in Fig. 5.16. For further description of surficial facies and geotechnical properties see Hein (1985).

Flow	Area (km^2)	Gradient	Average deposit thickness	Type
G	200	highly variable 0.06–1.8° 0.13–4.0°	20 m	sediment creep, gravity faulting, reverse faulting
D	26	2–4.4°	5 m ?	mud flow
A	18	0.56–1.2°	5.8 m	mud flow
C	15	2.34–5.2°	3 m ?	mud flow
F	11	1.5–3.3°	5 m ?	mud flow
H	10	1.45–3.2°	19 m	slide zone
E	5	1.44–3.2°	3 m	slump and mud flow
I	4	1.10–2.4°	1 m	slump and mud flow
B	3	1.2–2.6°	2 m ?	mud flow ?

shelf-break by deltas or as littoral sand sheets, the upper slope and slope-apron have a much steeper gradient. If there is little subsequent resedimentation of material, the continental margin shows steeply dipping, prograding clinoforms with some vertical upward growth to balance compactional subsidence (Fig. 5.1a). Some clinoform units may be separated by unconformities. Subsequent failure of the clinoforms and resedimentation results in the generation of a complex stratigraphy downslope. An example of this type of margin is the delta and slope system off West Africa (Fig. 5.9). Commonly, rapid progradation of delta or littoral sheet sand onto a steeply dipping upper slope-apron results in the generation of turbidity currents and sandy debris flows which move coarse material farther downslope. Thus, shelf-break and upper slope-aprons in such a setting are characterized by sediment bypassing, with most of the material accumulating as lower slope-aprons or submarine-fan complexes at the base-of-slope (Ch. 7).

On a much smaller scale, Syvitski *et al.* (1988) have modelled the two-dimensional accumulation of material on slope-aprons fed by deltas along graben margins. The individual basin-infill succession is controlled by: (a) basement geometry; (b) type and amount of sediment (i.e. discharge of bedload, Q_b, and discharge of suspended load, Q_s); and (c) oceanic setting of the basin (Fig. 5.10). If a basin receives most of its input as suspended load (model 1: Fig. 5.10), the dominant input is via spreading turbid plumes. Sediment rains out continuously from the plume to give an exponential basinward decrease in rates of sediment accumulation. Prodelta muds prograde over basinal muds; eventually a critical slope angle is reached and mudslides are generated. These mudslides tend to be large-scale and infrequent (Syvitski & Farrow 1989) (Fig. 5.11). Initially, the slides commence with creep, which is succeeded by the development of a shear surface or glide path, along which listric faults develop. As the main slide block moves downslope, a tensional depression forms downslope of the scar (caudal pull-apart, Fig. 5.11), and, in compressive zones at the toe of the slide, folding and thrust faulting occur (Fig. 5.11).

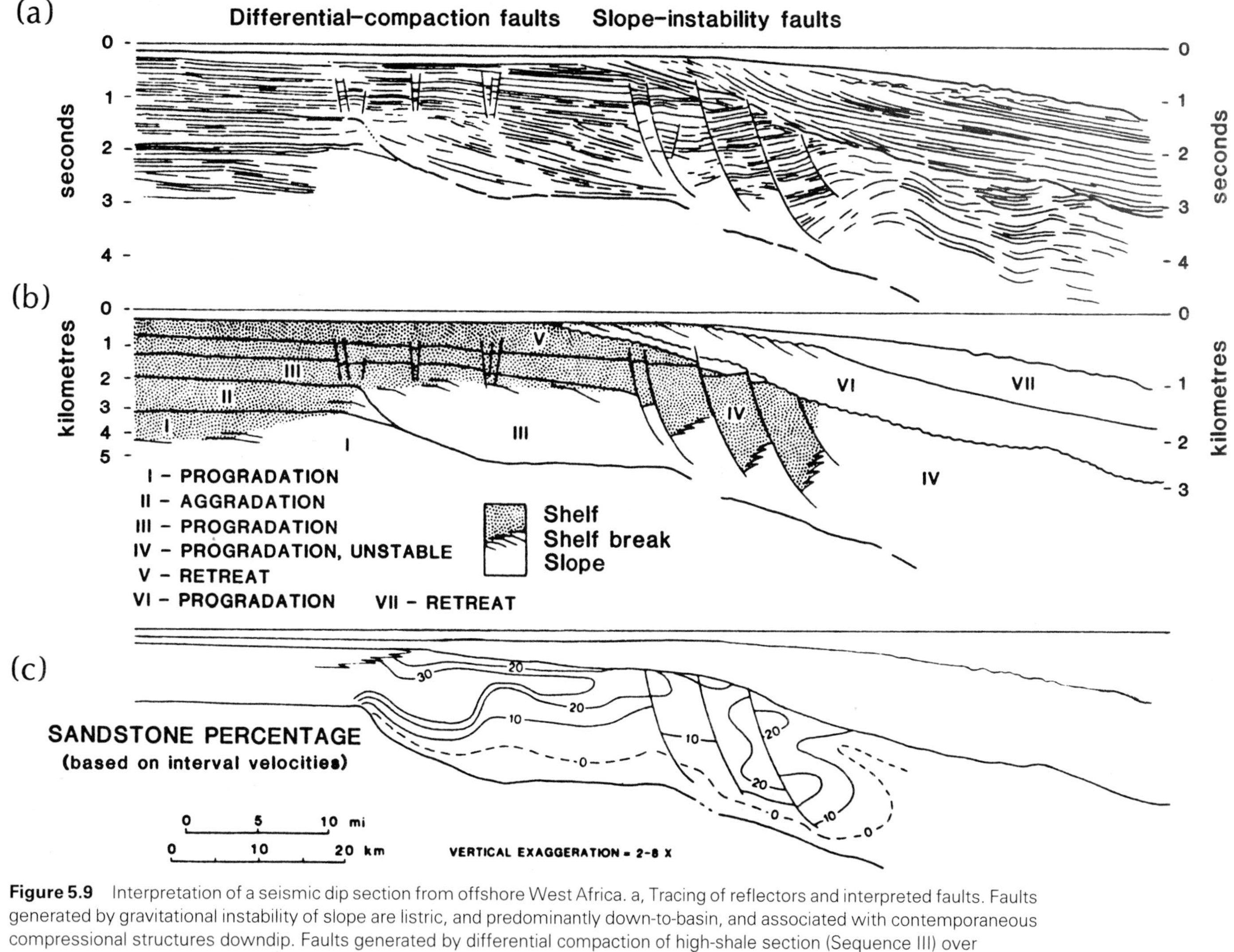

Figure 5.9 Interpretation of a seismic dip section from offshore West Africa. a, Tracing of reflectors and interpreted faults. Faults generated by gravitational instability of slope are listric, and predominantly down-to-basin, and associated with contemporaneous compressional structures downdip. Faults generated by differential compaction of high-shale section (Sequence III) over carbonate bank and slope (Sequence II) are higher-angle, more symmetrically distributed, and not listric: displacement decreases to zero at depth; b, interpretation of shelf and slope facies and classification of seismic sequences according to migration of the shelfedge; c, sandstone percentage reflects high subsidence rate and rapid accumulation of shallow-water sediments deposited near the shelf margin. From Winker & Edwards (1983).

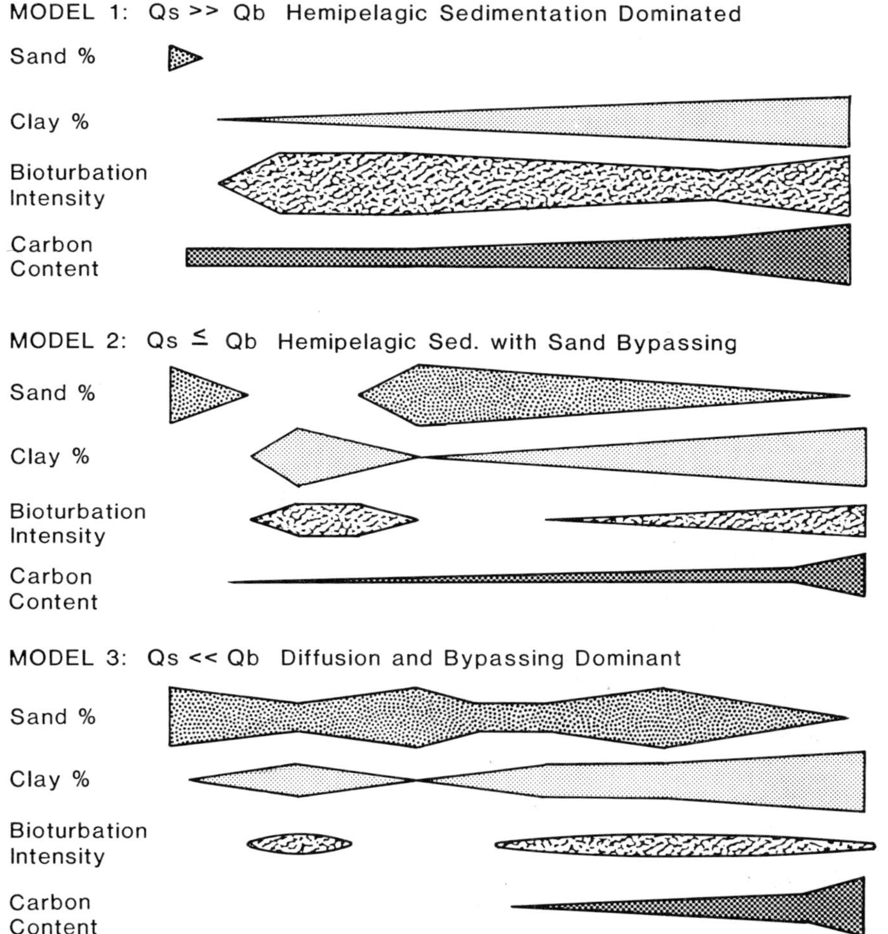

Figure 5.10 Schematic representation of proximal to distal (left to right) trends of basin infilling due to the growth of prograding deltas. Q_s: discharge of suspended sediment into the basin; Q_b: discharge of bed load sediment into the basin. Hemipelagic sedimentation is under seaward flowing river plumes. Sand bypassing is by turbidity currents. Downslope diffusion is by creep and small slides. For a further discussion of the model see the text and Syvitski *et al.* (1988). From Syvitski & Farrow (1989) in Whateley & Pickering (1989).

At the other end of the spectrum, where the marine basin receives most of its sediment as bedload (model 3: Fig. 5.10), prodelta slope sediments fail continuously as coarse-grained sediment gravity flows. The rapid deposition of coarse-grained sediments on steep (15–25°) delta foresets commonly leads to liquefaction and retrogressive slide failure. Numerous chutes are cut into the delta fronts (Fig. 5.12). Channels are separated by interchannel highs, which receive mainly fine-grained hemipelagic fallout from river plumes. On very gullied prodelta slopes, there is little net sediment accumulation and most of the sediment accumulates downslope as submarine fans. High-resolution seismic profiles of many bedload-dominated prodelta slopes show deep-seated growth faults, with smaller listric riders (Syvitski & Farrow 1989) (Fig. 5.11). Activation of smaller listric faults generates finer grained sediment gravity flows. Thus, the basin receives both fine- and coarse-grained sediment gravity flow deposits.

It is difficult to extend this modelling to a three-dimensional, unconfined system such as a continental margin. However, similar variations occur in sedimentation processes, geometry of the subsea landforms and the resultant stratigraphy of slope-aprons (compare Figs 5.13, 9 & 11) suggesting that similar processes are responsible. Thus, in terms of mechanisms for sediment transfer to the deep-sea, the major controls on the development of slope-apron systems include: (a) importance of hemipelagic sedimentation from fine-grained plumes; (b) importance of bedload deposition

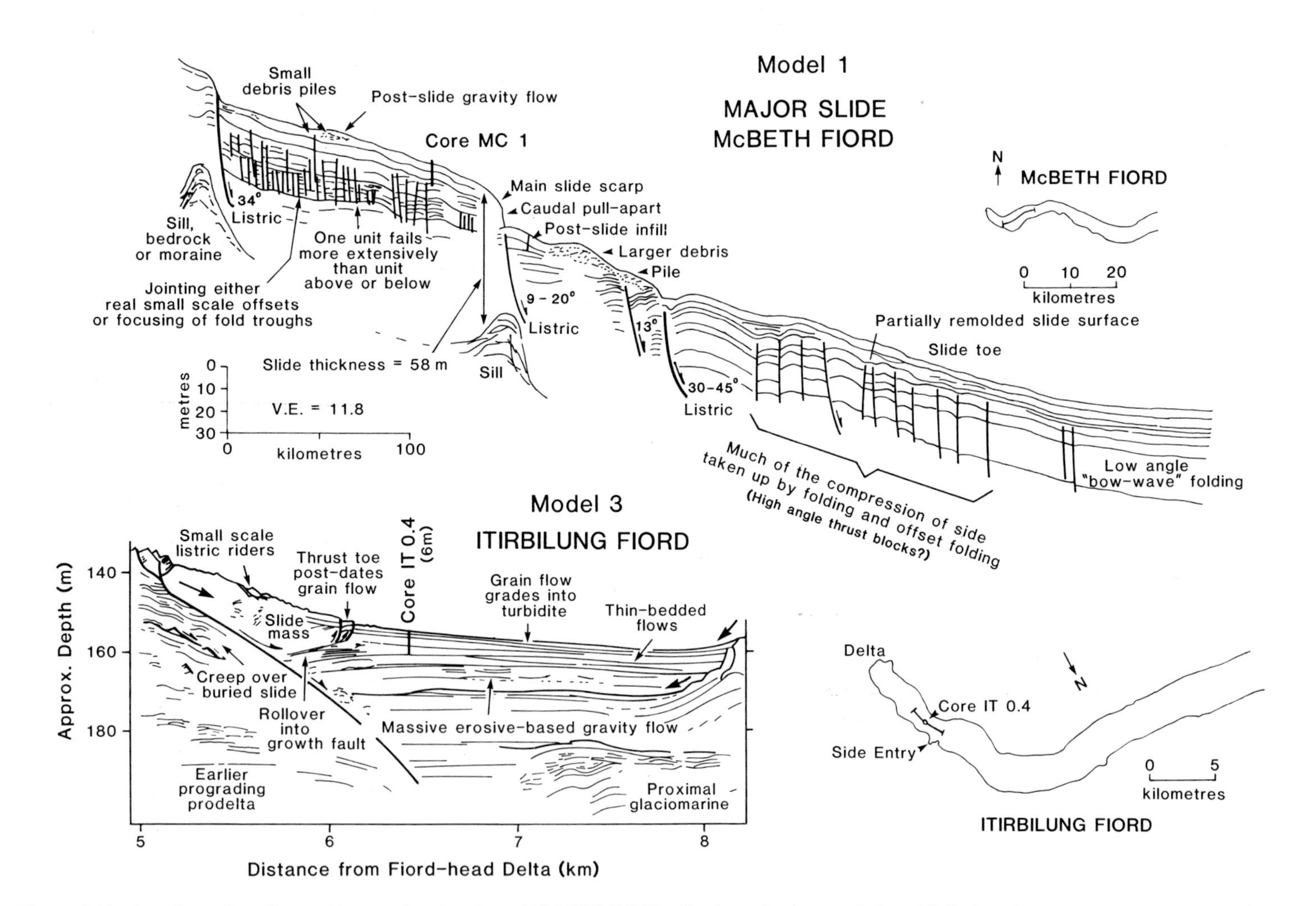

Figure 5.11 Locations of profiles and interpretive sketches of HUNTEC@ DTS reflection seismic records from McBeth and Itirbilung fjords, Baffin Island. For descriptions of sedimentary facies in cores see Hein & Syvitski (1987). See Fig. 5.10 and text for description of the models. After Syvitski & Farrow (1989) in Whateley & Pickering (1989).

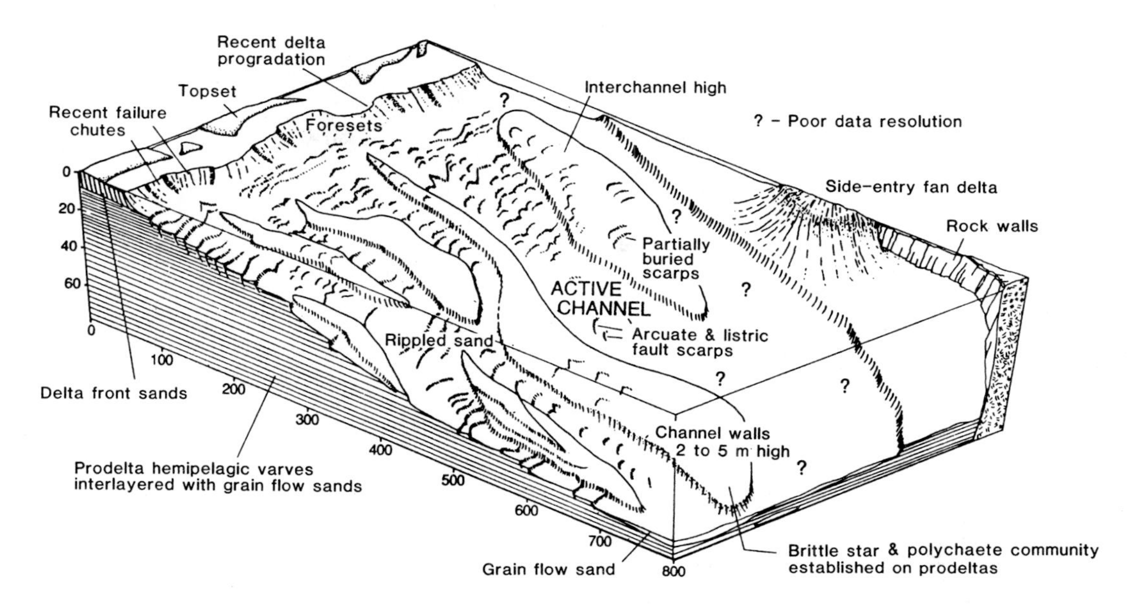

Figure 5.12 Interpretive sketch of side-scan sonar profiles and submersible dive observations, prodelta slope of Itirbilung fjord, Baffin Island. The concave arcs within the channels are arcuate and listric fault scarps *not bedforms* (identified by submersible observations). The one ripple sand zone is indicated in the front-centre part of the diagram. Vertical and horizontal scales in metres. From Syvitski & Farrow (1989) in Whateley & Pickering (1989).

at shelf-break fronts and the progradation of slope-aprons; (c) proximal upper slope bypassing, with offshore base-of-slope sedimentation from turbidity currents, and sandy debris flows; and (d) downslope 'diffusive' processes, mainly creep and sliding.

As well as the quantity and grain size of sediment, another important aspect affecting sedimentation on slope-aprons is the type of sediment source – whether a point source (e.g. a delta distributary), a line source (e.g. a fan-delta complex or littoral reworking of shallow-marine sand sheets along fault-scarps) or a sheet source (e.g. shallow marine sand sheets or the continual rainout of coarse detritus from a disintegrating ice shelf in glacial-marine settings). Each of these different types of sediment source will affect the geometry of the slope-apron package. In addition to the style of sediment input at the shelf-break and upper-slope, the physiography of the shelf-break will also affect how the sediment is transferred to the upper slope (Vanney & Stanley 1983). In gullied shelf-break and upper slope settings, coarse clastic sediment is funnelled down narrow conduits, mainly as sediment gravity flows (Belderson & Stride 1969, McGregor *et al.* 1982, May *et al.* 1983). In contrast, on smooth shelf-breaks and upper slopes, sediment may cascade or disperse downslope as unconfined sheets (Stanley 1974).

SUBSIDENCE

In sedimentary basins, subsidence results from a combination of tectonics, sediment and water loading, and compaction of older sediments. The amount of sediment accumulating in an area depends upon the rate of sediment input and also the rate of relative rise in base level (usually sea level). On unstable shelf margins and slope-aprons, subsidence associated with sedimentation includes: (a) depression of the crust by sedimentary loading; (b) extensional thinning of the sedimentary section by listric normal faulting; (c) salt withdrawal; and (d) sedimentary compaction (Fig. 5.13) (Winker & Edwards 1983). These factors combine with the tectonically-induced subsidence, for example lithospheric thinning, to produce the clinoforms seen in seismic sections of continental margins with varying internal characteristics and overall geometries. Synsedimentary tectonism is important in the development of offshore topographies, such as slope-basins (Fig. 5.1c & d), or offshore troughs, which trend parallel or oblique to the strike of the continental margin (Fig. 5.2b & d).

Different subsidence patterns, whether tectonically and/or sedimentologically-induced, produce different patterns of slope deposition. If subsidence rates (or

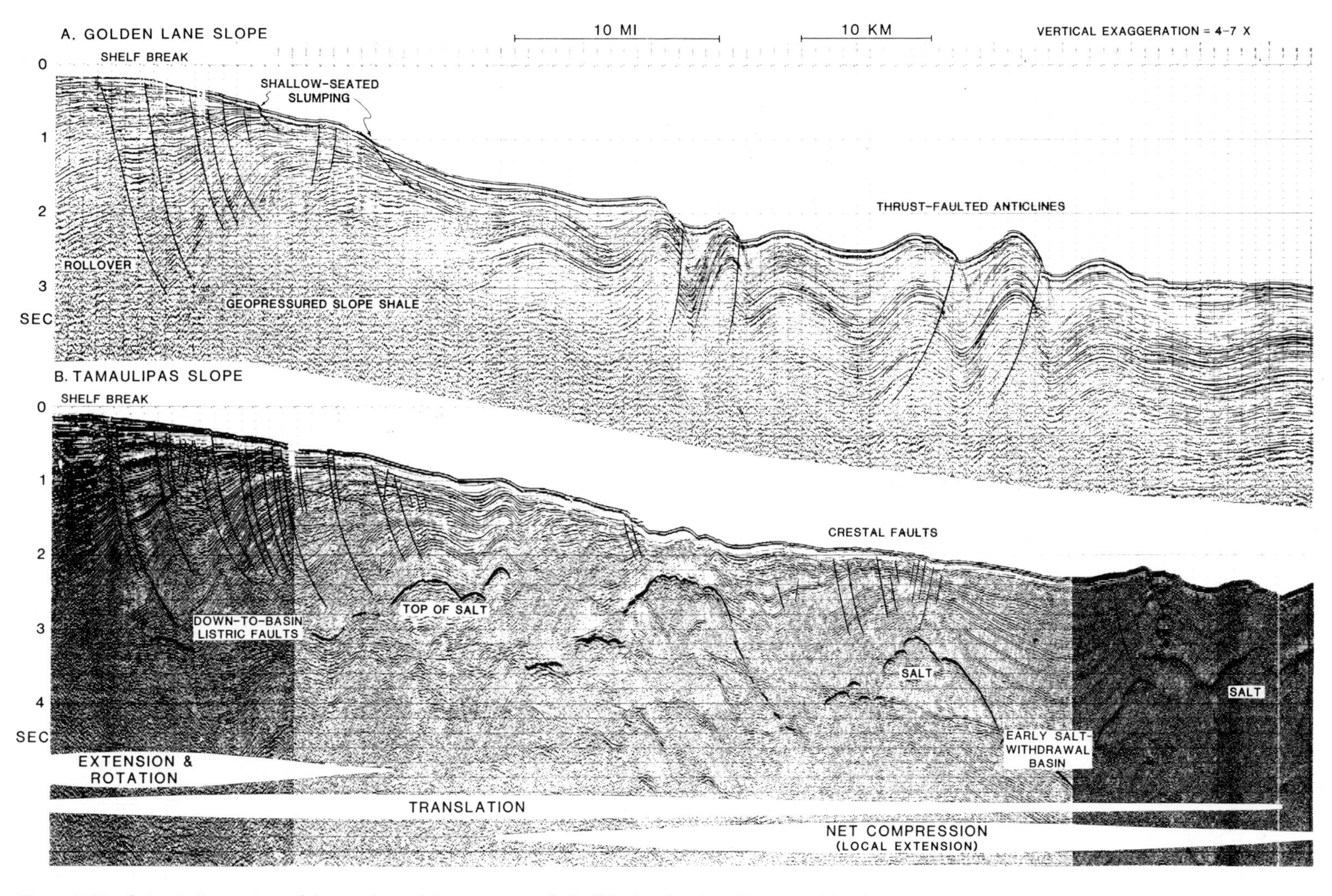

Figure 5.13 Seismic dip sections of the continental slope, western Gulf of Mexico. Sections illustrate origin of contemporaneous structures through deep-seated sliding of the continental slope. Growth faults originate in tensional regime along top of slope: decollement is in undercompacted shale (A) or salt (B). Ridges cored by shale (A) or salt (B) originate in compressional regime along lower slope. These ridges may evolve into diapirs as shelf margin progrades over them. Most of the compression is probably accommodated by thrust faulting rather than folding. Approximate depth conversion: 1 s = 2500 to 3500 ft (760 to 1070 m). From Winker & Edwards (1983).

uplift) exceed sediment accumulation rates, then the observed sedimentary patterns may be tectonically controlled; conversely, if sediment accumulation rates exceed subsidence rates (or uplift), then the observed sedimentary patterns may be primarily sedimentologically controlled. The various styles of uplap, offlap and onlap on slope-aprons are also influenced by variable tectonics and sediment supply (Fig. 5.14, cf. Brown & Fisher 1977).

OXYGEN CONTENT OF BOTTOM WATERS

Usually most organic matter, whether terrestrial or marine in origin, is chemically degraded as it settles

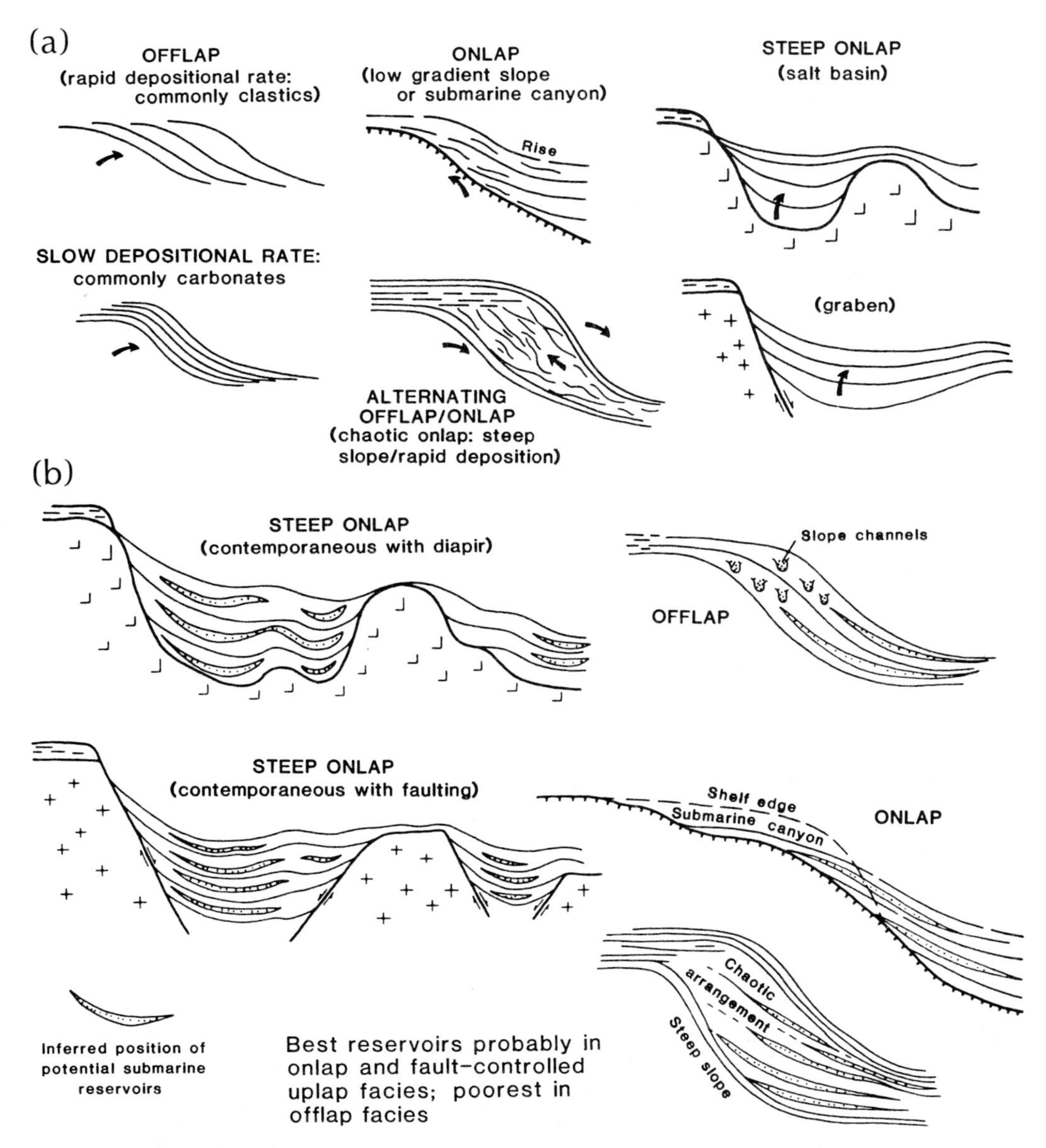

Figure 5.14 Seismic–stratigraphic slope facies patterns and inferred distribution of submarine fan reservoirs. a, Schematic representation of reflection patterns that characterize offlap, onlap and uplap slope facies; b, inferred distribution of submarine fan sandstone facies in three principal types of slope systems.

through an oxidizing water column, or is biologically degraded on the seafloor due to benthic reworking by epi- or infaunal populations. Under special oceanographic conditions of high supply of carbon-rich sediment and/or low dissolved oxygen content in the water column, carbon-rich sediment can accumulate. In the modern, well-aerated oceans this condition is uncommon, occurring on shelf-break and upper-slope regions under waters of exceptionally high organic productivity (cf. upwelling). This situation occurs in coastal zones of upwelling, where cold currents rise up the slope (e.g. off Peru–Chile, California, southwestern Africa, among others). Here, offshore winds drive the surficial waters out to sea. Because of the surface drift away from shore, there is a compensatory return flow of deep, oxygen- and nutrient-rich waters to the surface, creating a zone of very high bioproductivity (Tolmazin 1985). Oxygen consumption of the increasing amounts of organic matter sinking through the water column causes an oxygen minimum layer to occur beneath the zone of high organic productivity (Leggett 1985, Tolmazin 1985) (Fig. 5.15). For example, along the eastern Pacific continental margin, an oxygen minimum layer (< 1 ml/l of dissolved oxygen) occurs in intermediate water masses, commonly at depths between 250 and 1500 m (Ingle 1981). The extent of the oxygen-minimum zone depends on the rate of upwelling, rate of production of organic matter, circulation of the water masses, and rate of uptake of dissolved oxygen (Ingle 1981). Silled basins, such as those in the California Borderland or modern fjord environments, can locally develop anoxic or euxinic bottom waters.

The occurrence of the oxygen-minimum layer allows the accumulation of organic-rich sediment under reducing conditions. Because of the locally anaerobic to dysaerobic conditions, such sediments are commonly well-laminated, and generally lack trace fossils (Fig. 5.16, cores 18 & 27). In zones of higher oxygen levels, the sediments become increasingly reworked, with destruction of the primary laminae (Fig. 5.16, core 34) and a reduction in the organic-carbon content. Laminated muds can accumulate in different zones of deep-water slopes, depending on the slope physiography, the

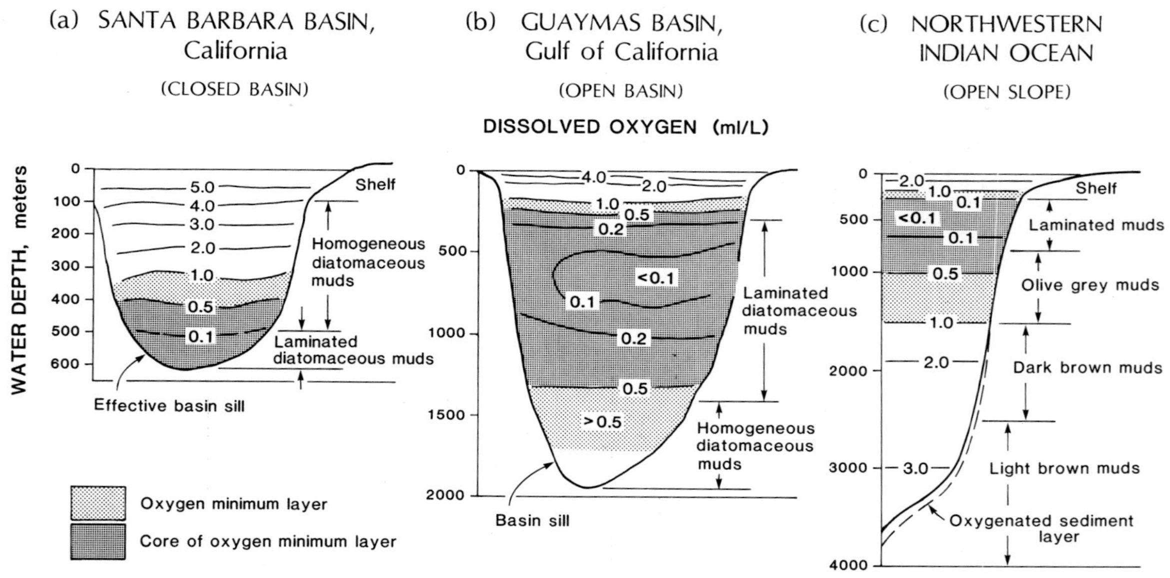

Figure 5.15 Relationship of mid-water oxygen minima to deposition and preservation of laminated sediments in three Recent continental margin settings. a, a closed basin (Santa Barbara Basin, California see Figs 5.7 & 5.8), with a basin sill within the core of the oxygen minimum layer and consequent preservation of laminated diatomaceous muds on the basin floor; b, an open basin (Guaymas Basin, Gulf of California) with a sill below the core of the oxygen minimum layer and consequent restriction of laminated muds to basin slopes, and c, an open continental slope (northwestern Indian Ocean) with an oxygen-rich surface layer, well developed oxygen minimum layer, and increasingly aerobic deep-water masses against the slope. Note that the laminated muds are confined to areas underlying those portions of oxygen minima marked by dissolved oxygen values less than 0.5–0.2 ml/l leading to the exclusion of burrowing organisms and consequent absence of bioturbation. From Ingle (1981).

depth of impingement of the oxygen-minimum core, and the depth of effective basin sills (Ingle 1981) (Fig. 5.15).

A variety of models for slope sedimentation can be established for the occurrence of laminated, organic-rich, muds depending upon: (a) the basin configuration (i.e. closed basin, open basin or open slope); (b) stratification of the water masses; (c) lateral and vertical mixing of the water masses (i.e. vigorous versus sluggish); (d) input rate of terrestrial and marine organic matter; (e) carbonate compensation depth; (f) sea-level variation, and (g) climate. Such models have been used to describe the distribution of deep-marine pelagic sediments, including diatomites (McKenzie *et al.* 1979, Govean & Garrison 1981, Ingle 1981, Soutar *et al.* 1981, Jenkyns 1986), black shales (i.e. marly or siliceous sediment with organic content >1%, local values up to 30%; Weissert 1981, Leggett 1985); radio-

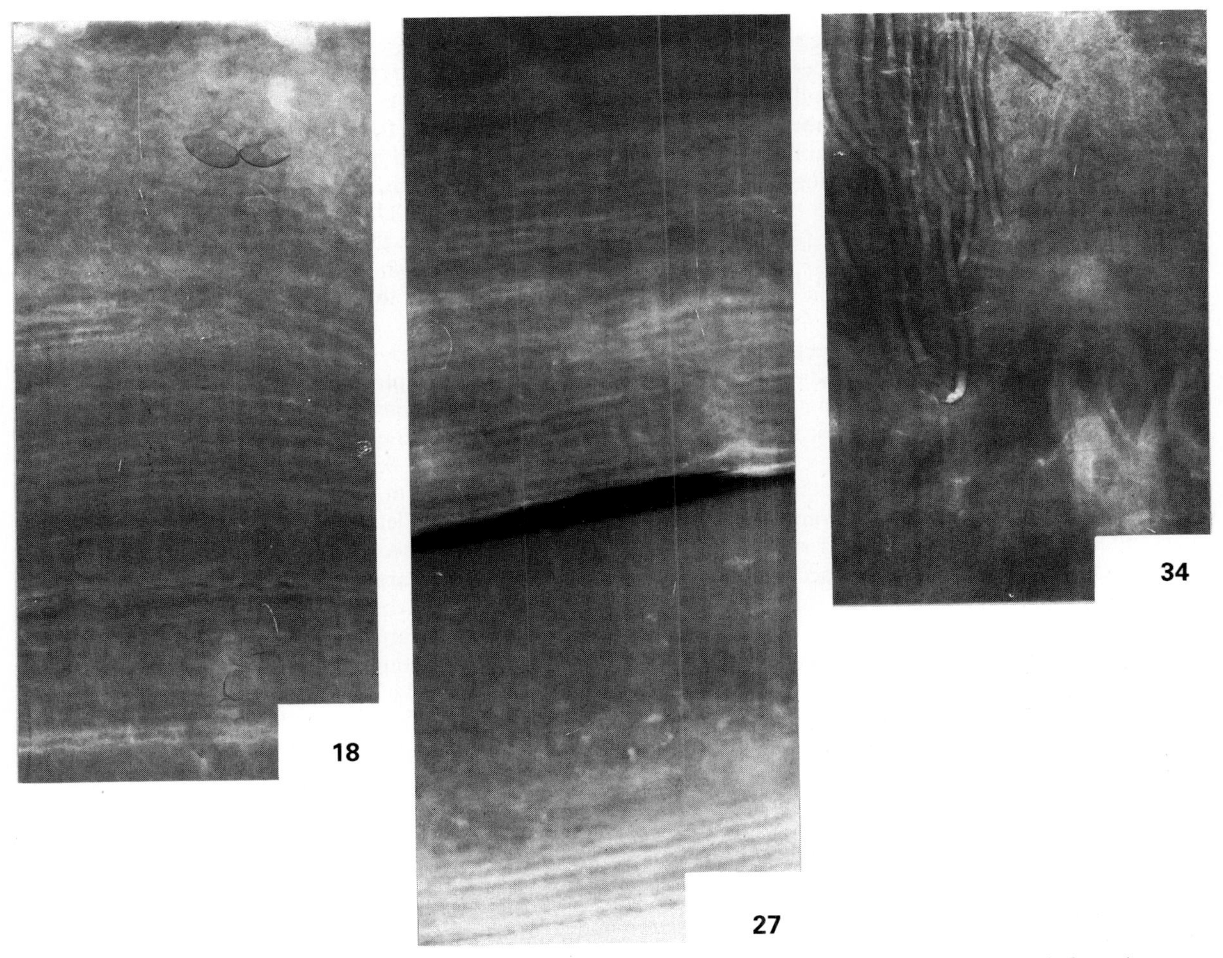

Figure 5.16 X-radiographic prints of cores from Santa Barbara Basin. For core locations see Fig. 5.7. Core 18 is from the hemipelagic core of the basin underlying the oxygen-minimum layer. Note the preservation of the laminae (Facies E2.2) with bioturbation (Facies E1.3) in the upper few centimetres. This suggests that there may be local turnover or intrusion of oxygenated bottom waters at the present time into the core of the basin. Core 27 is from the surface of flow 6 on the north slope of the basin (Figs 5.7 & 5.8), at water depths near the upper limit of the oxygen minimum layer (Fig. 5.15). The upper surficial sediment consists of alternating laminated (Facies E2.2) and more bioturbated (Facies E1.3) units, suggesting that there may be periodic variations in the rate of sediment supply as well as the intrusion of oxygenated bottom waters to this part of the slope. Core 34 is from the surface of flow F, above the oxygen minimum layer (Fig. 5.15). This core is completely bioturbated (Facies E1.3). See Hein (1985) for further discussion of facies characteristics.

larites, phosphorites, and graptolitic shales (Jenkyns 1986).

Styles of slope deposits

Case studies are selected from a variety of tectonic and climatic regimes, including: (a) fault-controlled scarps of graben-fills along extensional parts of a transform margin; (b) fault-controlled slope-basins on slope and deep shelf settings; (c) glacial-marine successions on a passive margin; (d) resedimented deep-marine carbonates on a passive margin; and (e) slopes associated with subduction zones. The case studies discussed in this section are not meant to be 'end-members' – rather they are selected to show the range of variability of facies and their associations on deep-marine slopes. Other types of slope deposits certainly exist.

MODERN OCCURRENCES: SLOPE BASINS, ACTIVE AND PASSIVE MARGINS

Slope basins are common to many deep-marine slopes. The occurrence of such basins depends on the existence of a downslope-deflecting or trapping ridge to pond sediment on the slope. Such ridges are tectonic and/or sedimentary in origin, and include (a) bounding-fault blocks of tilted half-grabens (Fig. 5.1c & 14); (b) detached slide blocks downslope from slide or slump scars (Fig. 5.39); (c) mudlump and salt diapirs (Fig. 5.13); and (d) tectono–sedimentary accretionary ridges at the base of slopes on the arc-ward side of subduction zones.

On passive margins, slope-basin sediments may be preserved in successions which have not undergone deep erosion. On active convergent margins, slope basins are best developed in prisms in which there is abundant sediment supply from both the overriding and subducting plates, resulting in the development of thick accretionary wedges. Internally, accretionary wedges generally consist of imbricate masses of folded, flattened and sheared sediments, including mélange, separated by slope-margin and slope-basin sediments. Such slope-basins are younger and less deformed than the surrounding accretionary-prism sediments (Moore & Karig 1976). Ancient successions of slope-basin deposits from lower slope and trench settings include recycled Franciscan conglomerate within older Franciscan mélange in central California (Cowan & Page 1975) and Palaeogene 'Great Valley outliers' which rest on Franciscan mélange in the Coast Ranges of northern California (Berkland 1972). The preservation of such slope-basin deposits in forearcs depends, amongst other factors, upon the relative ratios of sedimentation rate from both the overriding and subducting plates, and the rate and style of deformation (Ch. 11). As discussed by Moore & Karig (1976) the accretion of material at the lower trench slope causes uplift and rotation of the thrust slices, and of overlying slope and slope-basin sediments. With progressive deformation, motion along some of the thrust faults diminishes, and the inactive thrust-fault ridges become buried by slope deposits.

CASE STUDY: SERAM TROUGH, IRIAN JAYA SLOPE, JAVA TRENCH, INDONESIA

In 1984–5 the Indonesian–Dutch Snellius-II Expedition obtained a series of high-resolution shallow penetration seismic reflection profiles of the slope along the Java Trench. The following description of an intra-slope basin along this margin is from John Woodside (pers. comm. 1988) and is part of a more comprehensive paper on the Seram Trough (Jongsma *et al.* in press).

Line 23 (Fig. 5.17) is an excellent example of the seismic stratigraphy and structural relationships of slope-basins within the Indonesian convergent margin. The slope on the overriding plate (to the right, Fig. 5.17, 0600–0320) consists of a Late Mesozoic carbonate platform succession, which is gently folded. Overlying Plio–Pleistocene sediments are bounded below by a major decollement surface, above which the slope sediments are being displaced downward along listric faults, which sole in the major decollement surface. Sediments are moving as slide blocks along the listric faults, although there is some overturning of the tops of strata within the slide blocks (Fig. 5.17, 0430–0400) due to landward (arc-ward) tilting of the margin by underthrusting from below (to the left of Fig. 5.17). Near the base of slope, beneath the slope-basin (Fig. 5.17, 0300–0270) there are two primary detachment surfaces, which are stacked 'piggyback' upon one another, and can be interpreted as buried duplex thrusts. This duplex structure is further evidence of arc-ward tilting of the margin due to subduction–accretion processes.

The decollement surface can be traced into the trough. Reflectors from the displaced slope material can be traced into the slope-basin. The slope-basin sediments are Plio–Pleistocene in age, with seven stratigraphic units emplaced in 5.5 Ma (Jongsma, pers. comm. 1988). Reflectors within the slope-basin show onlap relationships with both the arc-bounding slope (to the right, Fig. 5.17, 0230–0330), and with the

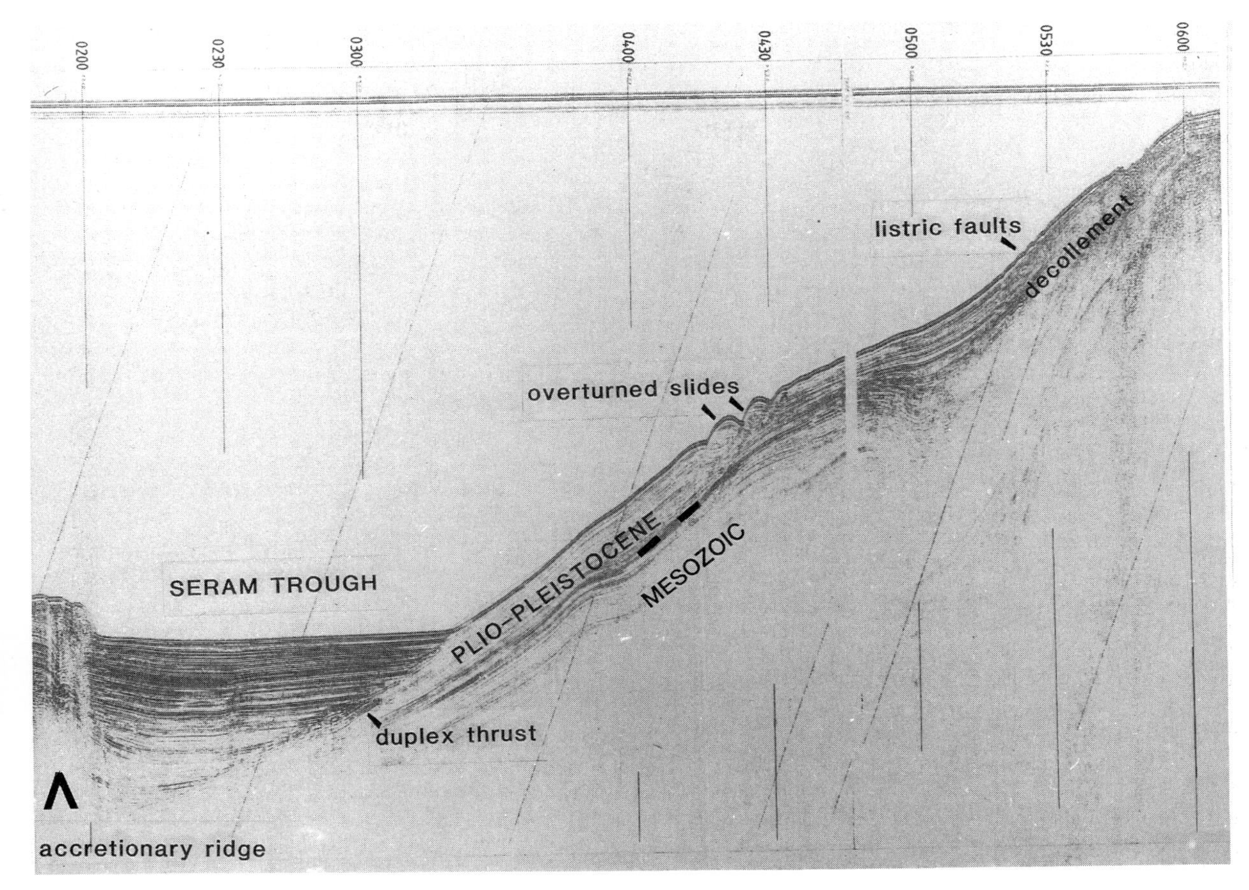

Figure 5.17 High-resolution shallow penetration seismic reflection profile of the Seram Trough, Irian Jaya slope, Java Trench, Indonesia. Obtained by the 1984–5 Indonesian–Dutch Snellius-II Expedition. Line 23 from Jongsma *et al.* (in press). Four-second sweep of two-way travel time is shown on the vertical scale.

accretionary-ridge slope (i.e. thrust-block ridge) (to the left, Fig. 5.17, 0200). Compared to other profiles of this margin, the trough has become wider on Line 23, with almost 1.0 second (two-way-travel time) thickness of strata onlapping a section of flat-lying trough fill. The onlapping relationships between the slope deposits and the slope-basin fill suggest that slides, slumps and sediment gravity flows (most likely turbidity currents for the more even, prominent reflectors) contribute directly to infill of the slope-basin.

As with other accretionary prisms within Indonesian-type convergent margins, this example shows abundant sediment supply to the lower trench slope, with input from both the forearc and oceanic plate. The amount of sediment supply from the accretionary ridge depends upon the relative ratio of sediment supply to the slope-basin and rate of fault movement along the thrusts. If there is a relatively lower rate of sediment supply to the basin (compared with the rate of uplift along the thrust faults), the accretionary ridges will become exposed. Subsequent resedimentation of material along the flanks of the accretionary ridges as slides, slumps and debris flows will contribute to infill of the slope-basin (as in the case of Line 23 of the Seram Trough). If the sediment accumulation rate exceeds the rate of motion along the basin-bounding faults, then the slope sediments will mask the fault trace, the accretionary material will be buried, and there will be minimal sediment supply from the offshore accretionary ridge. Slope sediments then prograde from the landward (arc-ward) margin to the lower trench slope and there is no development of slope-basins. This latter type of margin with dominant sediment supply from the over-thrusting plate is characteristic of the eastern Apenninic Calabrian margin, which Mascle & Mascle (1983/1984) called the Apenninic type, interpreted as a 'mature stage' of tectono-sedimentary accretionary prisms at convergent margins.

CASE STUDY: ORCA BASIN, TEXAS–LOUISIANA SLOPE, GULF OF MEXICO

The irregular topography of the Texas–Louisiana continental slope is a result of active diapirism and associated faulting and slumping (Bouma, Coleman *et al.* 1986). Many of the diapirs are salt injections (Lehner 1969). This diapirism leads to the formation of intraslope basins, of which two types are recognized: (a) those due to blockage of submarine canyons or channels (i.e. Pigmy Basin, Bouma, Coleman *et al.* 1986); and (b) those interdomal basins isolated from the surrounding seafloor by diapirs that effectively block the entry of bottom-hugging sediment gravity flows and density currents on the slope (i.e. Orca Basin, Bouma, Coleman *et al.* 1986, Leventer *et al.* 1983).

The following case study of Orca Basin, in the northwest Gulf of Mexico (Fig. 5.18), illustrates an example of the interplay of the bottom water characteristics in determining the nature of the sedimentary facies within an intraslope basin (Fig. 5.19). Orca Basin is structurally produced, hypersaline and anoxic. The basin covers an area of 400 km^2 and reaches a maximum water depth of 2400 m. This interdomal basin is unique in that its floor is covered by a 170m–thick anoxic brine (Fig. 5.18).

The interface between the basin floor and the overlying brine occurs at 2230 m and is sharp, evident by a strong reflector on seismic profiles (Fig. 5.20). The walls of Orca Basin are quite steep (5–19°), and there is extensive sliding and slumping of material into the basin (Fig. 5.20). Seismic records show a dominant hyperbolic pattern, related to the pervasive instability of material along the basin margins, mainly related to diapirism. Thus the detailed morphology along the basin margin is quite irregular with numerous slides, slumps and faults occurring on all scales (Feeley 1982, Bouma, Coleman *et al.* 1986).

Orca Basin was cored to a total depth of 92.5 m subbottom. Sediments are mainly interbedded mass-movement and pelagic/hemipelagic deposits, comprising grey clay and silty clay with black, anoxic mud and silty/clay intercalations. Much of the sediment is structureless with the exception of the well-laminated units. These thick structureless units may be a result of uniform settling of material from nepheloid flows or

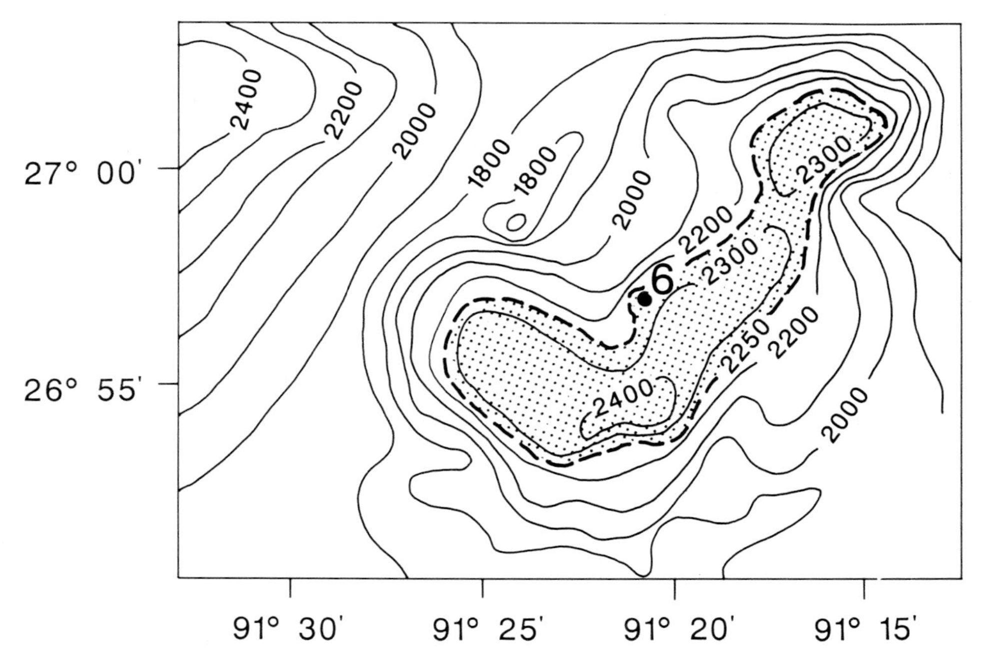

Figure 5.18 Bathymetric map of the Orca Basin in the northwestern Gulf of Mexico showing the location of the piston core EN32–PC6 within the basin. The stippled region represents the modern extent of the hypersaline, anoxic brine.

low concentration turbidity currents which travelled along the pycnocline between the hypersaline brines in the centre of the basin and the ambient sea water (Pickering, unpublished data 1988).

Detailed isotopic and microfossil analysis from piston cores shows relationships between anoxia, glacial meltwater and microfossil preservation (Leventer *et al.* 1983). The faunal composition of the sediments for the last 29 ka BP depends on both preservation and climatic changes. Isotopically depleted glacial meltwater was introduced into the Gulf at approximately 16 ka BP. Coinciding with this glacial meltwater influence, anoxic conditions prevailed within the intraslope basin, as shown by the preservation of laminated sediments, and increased preservation of pteropods, planktonic foraminifera and radiolaria (Fig. 5.19).

Water masses were stratified within the slope basin, and the increased input of terrigenous organic matter may have partially induced the anoxia (Leventer *et al.* 1983). At about 13 ka BP, oxygenated conditions resumed, with a depletion in the microfossil abundance and lack of preservation of laminated sediments. At this time there was a reduction of the meltwater into the Gulf. At 8.5 ka BP Orca Basin again became anoxic due to the development of the hypersaline brine. Coincident with this brine formation there was enhanced microfossil preservation. The very fossiliferous layers at 5.8, 4.4 and 3.7 ka BP are thought to be due to a decreased terrigenous dilution. Thus, the presence of anoxic bottom waters and the development of a stable stratified water mass within the intraslope basin was the dominant controlling factor in the preservation of microfossils and laminated sediments.

ANCIENT SUCCESSIONS

Most ancient deposits interpreted as slope and slope-apron successions are 'noncyclic' accumulations. This apparent lack of cyclicity within deep-marine slope and slope-apron deposits may indeed reflect the natural 'near randomness' of sedimentation in these settings. Two major styles of noncyclic accumulations will be discussed: (a) sediment gravity flow deposits, occurring as individual bed accumulations, sheets, gully-fill and lobe deposits; and (b) synsedimentary slope failures, showing slide masses, intraformational truncation zones and synsedimentary shear zones.

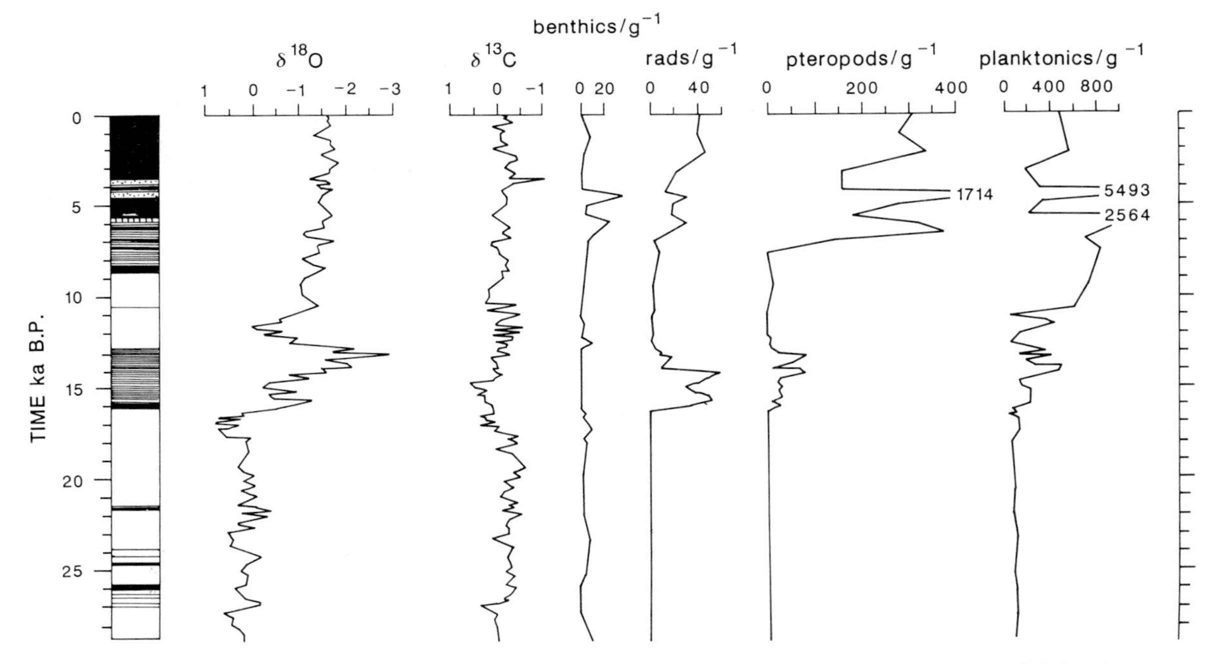

Figure 5.19 Plot of lithologic column, δ $^{18}O_{PDB}$ (‰), δ $^{13}C_{PDB}$ (‰), no. benthic foraminifera/gram, no. radiolarians/gram, no. pteropods/gram and planktonic foraminifera/gram versus time for piston core EN32–PC6. Only microfossils > 150 μm were counted. Isotopic analyses were performed on *Globirgerinoides ruber*, white variety, 180–250 μm.

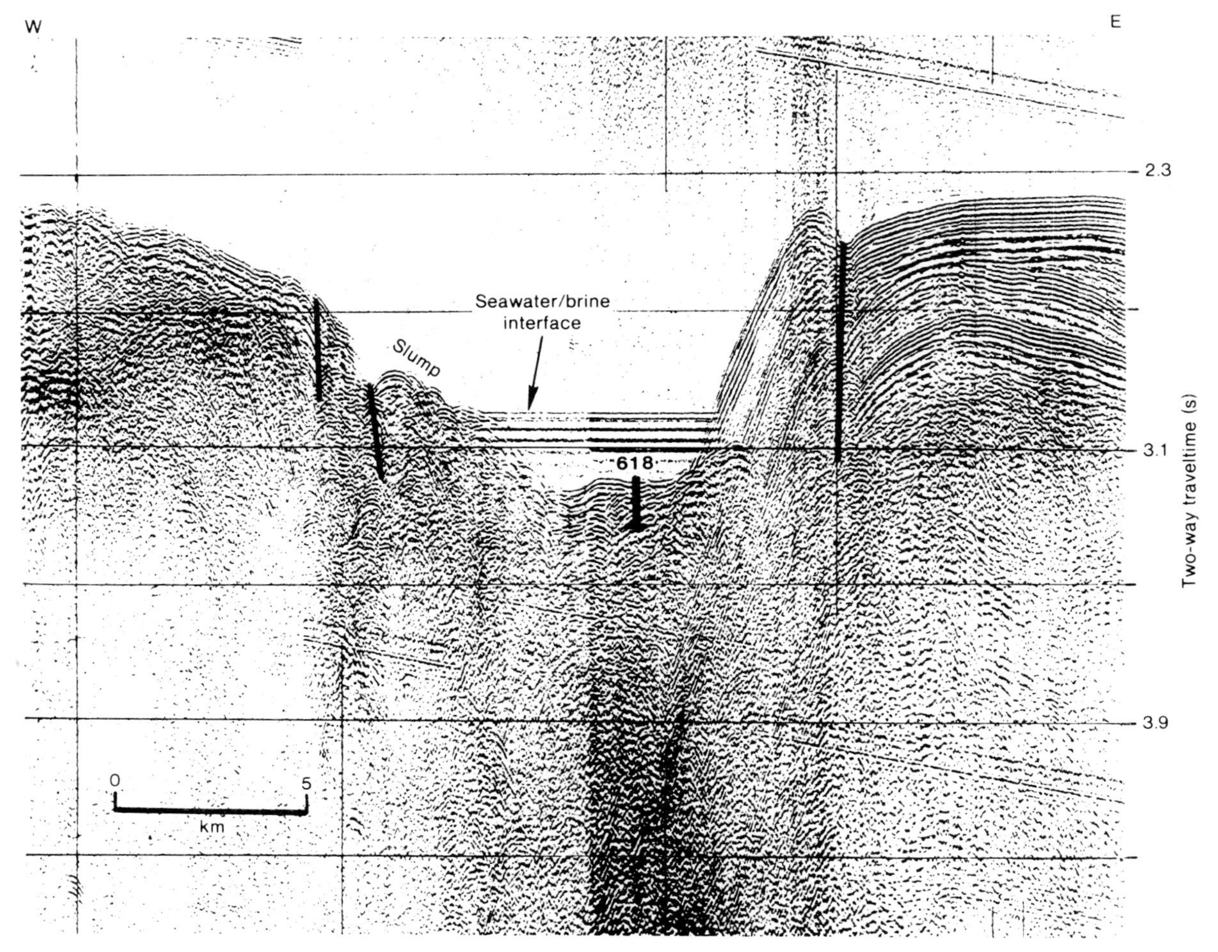

Figure 5.20 Air gun (80 in^3) record (USGS Line 69) over Orca Basin and location of DSDP Site 618. Vertical scale in seconds (two-way travel time). Borehole length for DSDP 618 approx. 92.5 m.

SEDIMENT GRAVITY FLOW DEPOSITS

Case study: submarine fault scarp of extensional origin, Jason Deposit, Yukon Territory, Canada The Jason stratiform sulphide deposit, Yukon Territory, Canada (Fig. 5.21) shows a succession of mudflow, turbidite, slide and debris flow deposits emplaced along the margins of a graben (Winn *et al.* 1981; Winn & Bailes 1987), probably a base-of-scarp in an intrabasin site as opposed to a base-of-continental-slope setting. The graben formed during the Late Devonian due to extension related to transform motion between the North American Plate and an oceanic plate to the west.

Stratiform sulphides (Fig. 5.22), including galena, sphalerite, and barite, were precipitated from hydrothermal solutions, expelled along synsedimentary fault zones. The mineralization ranges from massive and thickly bedded (Fig. 5.22A), to bedded sulphide and sulphate alternating with argillaceous material (Fig. 5.22B). Underlying these are localized zones of silicified, carbonatized and brecciated rock (Fig. 5.22C), interpreted as stockwork mineralization. The mineralization is interstratified with, and also cross-cuts, deep-marine coarse clastic rocks of the Lower Earn Group (Winn & Bailes 1987).

Synsedimentary faults formed the graben margin, controlled sedimentation of deep-marine clastics of the Lower Earn Group, and focused the discharge of the metalliferous fluids. Mineralization and sulphate precipitation resulted from the interaction of hydrothermal fluids with basin-floor sediments. Associated shales are laminated and lack bioturbation. These

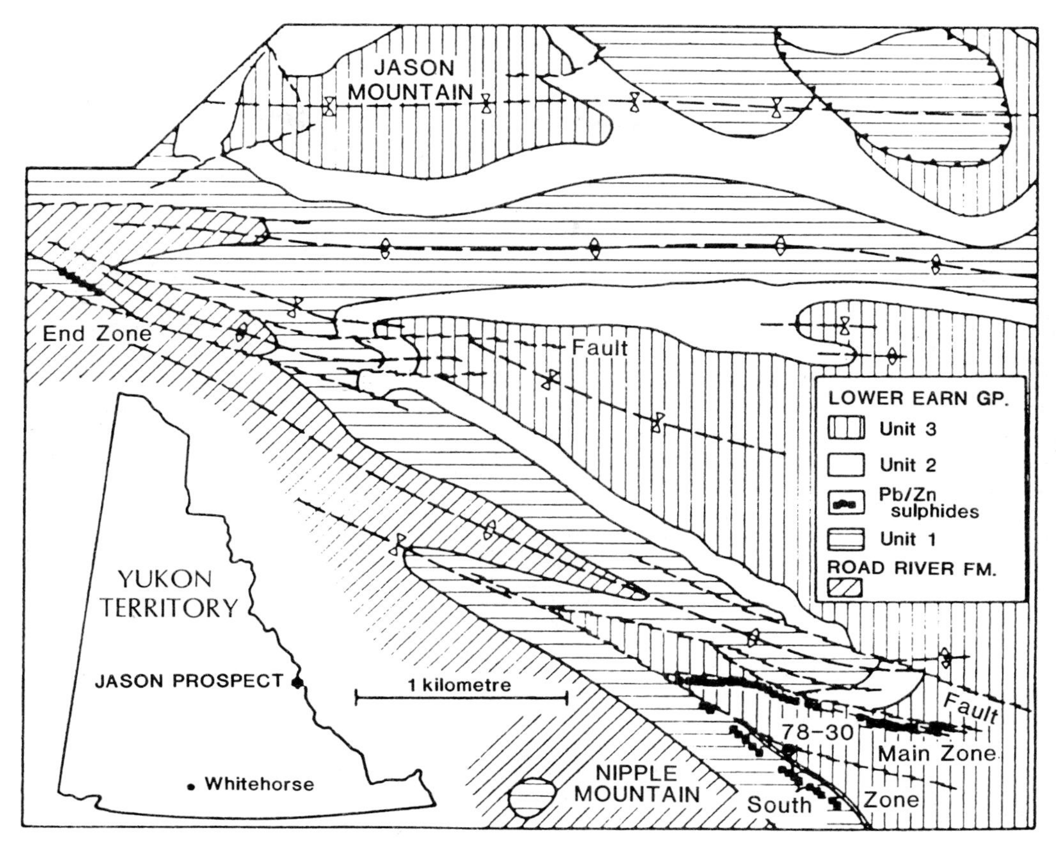

Figure 5.21 Location and geological map of the Jason lead–zinc deposit, Yukon, Canada. Units 1 and 3 are shale, sandstone and minor conglomerate; Unit 2 is the conglomeratic member. Surface traces of stratiform mineralization are shown (South, Main and End zones). Note location of core 78–30 through the South zone (see Fig. 5.24). From Winn & Bailes (1987).

features, along with the preservation of the lead-zinc sulphides, suggest anoxia (or low oxygen levels) during deposition. The inferred protected or 'silled' graben is analogous to modern fjord environments (Syvitski *et al.* 1987).

The inferred paleogeography of the Lower Earn Group in the vicinity of the Jason sulphide deposit is shown in Figure 5.23. Faulting is synsedimentary, shown by abrupt thinning of slope-apron deposits, and the abundance of laterally discontinuous, locally-derived mass failure deposits. Most of the Lower Earn Group consists of laminated argillites (Facies E2.2), with less common sandstones (Facies Class B), conglomerates (Facies Class A), sedimentary breccias (Facies Group F2) and minor black limestones (Facies Group G1). In the centre of the graben, the Lower Earn Group is 1500 m thick but thins to less than 400 m at the graben margin. Two dominant styles of coarse clastic sedimentation occur: (a) confined deposition within channels, trending along the centre of the graben, and (b) unconfined mass-flow deposits along the graben margin (Fig. 5.23).

The mid-graben channels range up to 20 m deep, possibly form a braided pattern, and control distribution of the coarsest material in the centre of the graben. Many turbidity currents were confined within the channels but larger flows overspilled the channels and deposited sediment along the flanks of the channels. Extremely large-volume events blanketed most of the basin-floor with mud, and in some cases, sand. Channel-fills have thinning- and fining-upward sequences, due to progressive channel abandonment. Sedimentary facies of the coarsest channel fills are mainly clast-supported, graded and imbricated conglomerate (Facies A2.3). Some parallel stratified (Facies A2.1, A2.4), inversely-graded (Facies A2.2)

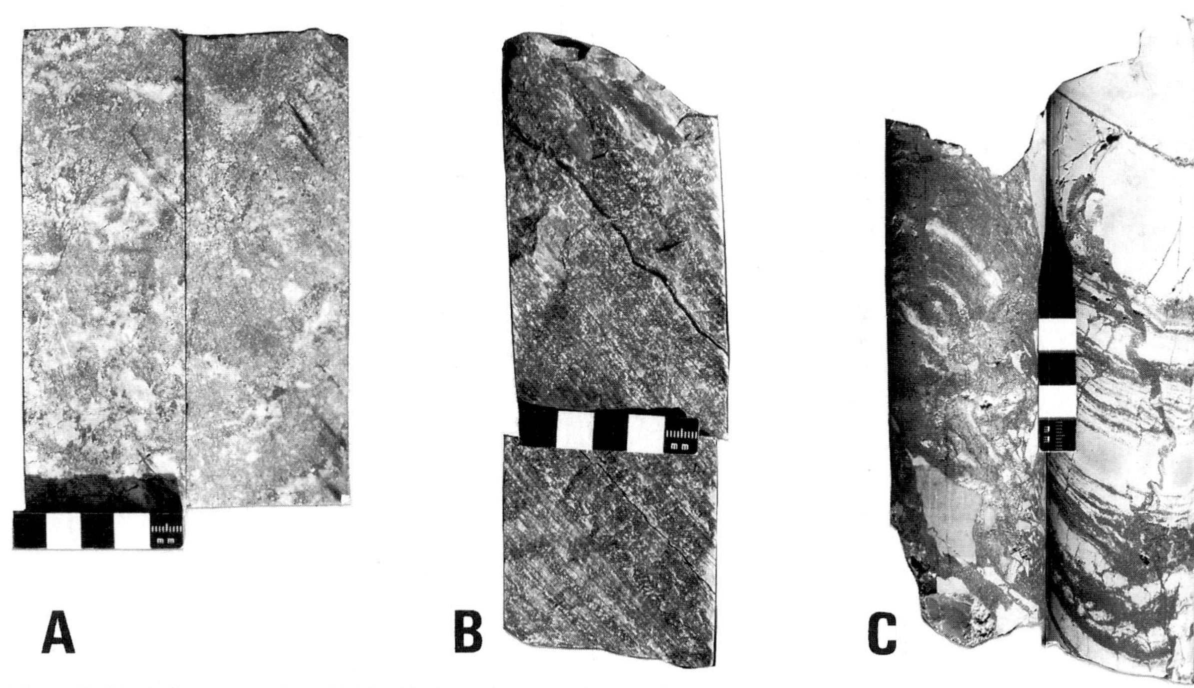

Figure 5.22 A, Representative, thick-bedded, proximal sulphide bed from South Zone, Jason deposit. Rock consists of a mixture of galena, sphalerite, quartz, siderite, pyrite and pyrrhotite; B, bedded sulphide and sulphate from the Main Pb–Zn zone, Jason deposit. Rock consists largely of laminated barite, sphalerite, and galena (mostly in light laminae) and argillaceous material (dark laminae); C, core section through hydrothermal feeder conduit beneath South Zone showing silicified shales cut by veins containing siderite, chert, pyrite, pyrrhotite, galena and sphalerite. Scales are in centimetres. From Winn & Bailes (1987).

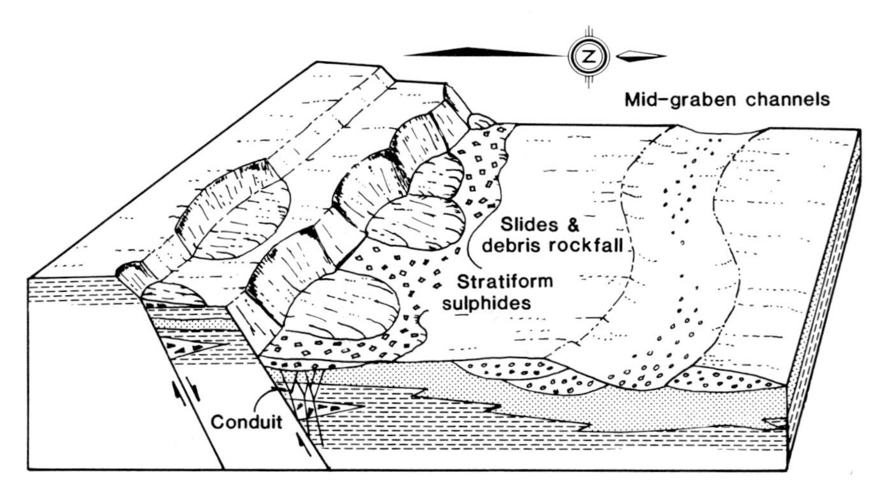

Figure 5.23 Schematic diagram of depositional environments of the Lower Earn Group in the vicinity of the Jason lead–zinc sulphides. The MacPass graben, one side of which is depicted, appears to have been a small intra-basinal margin, within a larger deep-marine basin. Clastics have two sources: locally from the faulted graben margin and externally from a transform plate margin to the west. The latter clastics were funneled through the graben largely by a system of braided channels tens of metres deep and hundreds of metres wide. Note the position of the hydrothermal conduit beneath the stratiform deposits. From Winn & Bailes (1987).

and disorganized (Facies A1.1) conglomerate also occurs. Laterally equivalent facies, along the margins of the channels, are much more continuous (traceable for tens of metres along strike), finer grained (mainly sandstones and shale, Facies Classes B & C) and thinner bedded (maximum 1 m thick). Sandstones show flutes and grooves on the base of beds, are normally graded, and show Bouma (1962) sequences (Facies Group C2, commonly T_{b-e}, T_{a-e} and T_{abe}).

In contrast, the graben-margin sediments are much finer grained and thinner bedded, and comprise laminated shale (Facies E2.2), sedimentary breccia (Facies F2.2) and pebbly mudstone (Facies A1.3) (Figs 5.24 & 25). The coarse-grained component of the pebbly mudstone is chert pebbles. Laminated-shale intraclasts (petrographically identical to the laminated Lower Earn Group shale), are up to 1 m in length, and were presumably derived by sliding of cohesive, unlithified, slope material. Two vertical facies trends were noted: (a) gradations from contorted and microfaulted units (Facies F2.1) to homogenized, sandy/pebbly mudstone (Facies A1.3), and (b) gradations from mudclast-supported beds (Facies Class F) to homogenized, sandy/pebbly mudstone (Facies A1.3). These trends are interpreted as representing a process continuum from slide/glide → debris flows → less viscous submarine mudflows.

Individual slide units and sedimentary breccias reach a maximum of tens of metres thick, and are traceable for distances up to 1 km. Locally, recurrent movement and uplift of the graben floor occurred, as suggested by the resedimentation of basin-floor sands, laminated shales and gravels into slide deposits (Fig. 5.25D). In some cases, mineralized beds show sediment slide structures. Periodic uplift along the faults is shown by some interbedding of the intraclast

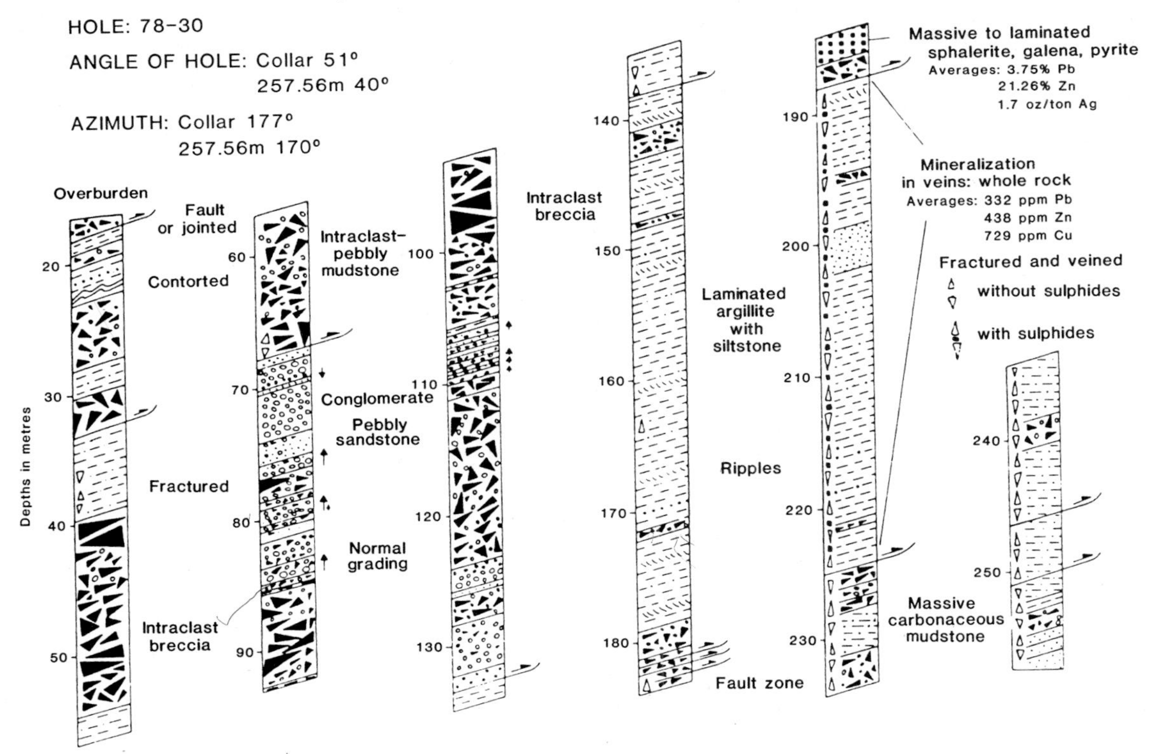

Figure 5.24 Schematic diagram of core 78–30 through the South Zone (see Fig. 5.21). Core is representative of lithologies along the margins of the MacPass graben. Core was drilled at an angle into steeply dipping unit 1 rocks; actual bed thickness is about 75% of what is shown. None of the faults shown repeat the section. Fault symbols do not indicate the direction of fault movement. Note the abundance of laminated shale (Facies E2.2) and mudclast breccia (Facies Class F, Facies A1.3). From Winn & Bailes (1987).

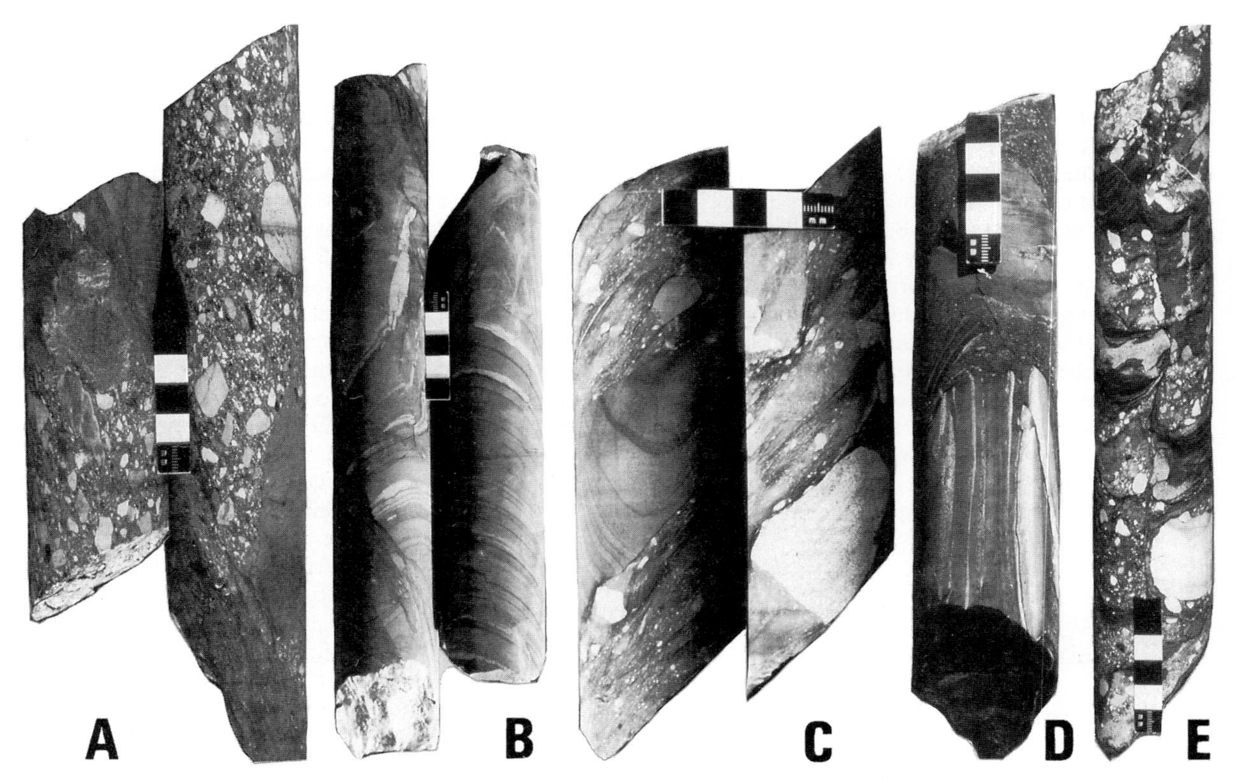

Figure 5.25 Sedimentary fabrics of mudclast breccias (Facies Class F, Facies A1.3) and conglomerate (Facies Class A) from the inferred graben margin, Jason deposit, Yukon, Canada; A, poorly sorted conglomerate from unit 1 above the South Zone. Note mudclast in left core piece is mineralized; B, sedimentary breccia from unit 1 made up of mudclasts with only small amounts of mud between the clasts. Note soft-sediment folding and faulting within clasts; C, mudflow breccia from unit 1 made up of chert pebbles and mudclasts. Clasts are suspended in a heterogeneous matrix of mud and sand; D, sedimentary breccia clasts in a heterogeneous sand-mud matrix, E, sedimentary breccia consisting of pebbles, pebbly sandstone clasts, mudclasts and sedimentary breccia fragments in a mudstone matrix. Scale in cm. From Winn & Bailes (1987).

breccias (Facies Class F) with better sorted conglomerates, pebbly sandstones (Facies Group A2), and argillites (Facies Group E2) (Fig. 5.25).

In summary, the graben margin is characterized by a thin-bedded succession of shale, mudflow, slide and slump deposits, with less common occurrences of coarse clastics and thick-bedded units rich in lead–zinc sulphides, some of which have wet-sediment deformation structures. Interbedded shales are laminated and lack bioturbation (Facies E2.2). Few overall trends occur in the succession. Most of the mass flows are inferred to have been generated from local fault scarps because of their localized extent, the dramatic thickness variations, and close association with the stratiform lead–zinc deposits. It is clear that, in this type of setting, submarine mass wastage events may generate a number of spatially separate slope failures that affect separate lithologies on varied slopes. Thus, a variety of resedimented deposits may be separate yet contemporaneous. In addition, some of the failed masses may subsequently fail again, thus producing a complex mix of mass-flow deposits.

Case study: slope and deep shelf gully sandstones on a fault-controlled submarine ramp, Jameson Land, East Greenland

The Mesozoic sections of East Greenland show several examples of deep-marine fan and 'nonfan' coarse clastic sediments deposited on a faulted deep-marine shelf margin (Surlyk 1973, 1975, 1978, 1984, 1987; Surlyk *et al.* 1973). Some of these successions are discussed in Ch. 10 (Fig. 10.27).

The Upper Jurassic Hareelv Formation consists of 200–500 m of black shale with thick-bedded sandstones as the dominant lithology (Surlyk 1987). The sandstones are interpreted as deposits of a submarine ramp (Ch. 7) flanking a fault-bounded deep-marine shelf basin (Fig. 5.26). Sediment was supplied along a line source from a sand-dominated shelf. Seaward of the shelf was a narrow, fault-controlled slope, against which slope-apron sediments of the clastic submarine ramp onlapped. Most sandstones were deposited at the base-of-slope and on the basin floor, although some of the outcrop belt may have been deposited directly on the slope.

Three main facies associations occur: a black mudstone 'background' facies (Facies Class E), thin turbidite sandstones (Facies Group C2), and structureless sandstones (Facies B1.1) deposited from sediment gravity flows. Within the black mudstone facies (Facies E2.2), laminations average 1 mm thick and are generally not bioturbated. At some levels, wormlike burrows and meandering trails occur. The organic content is high, with total organic carbon values from 6 to 12%. This facies forms about 50% of the Hareelv Formation. Very thin bedded (millimetre to centimetre thick) sandstones are interbedded with the black mudstone facies. These very thin bedded sandstones have sharp bases and tops and are mainly structureless, although some faint parallel lamination and diffuse normal grading occur. These sandstones are interpreted as thin turbidites (Facies Group C2), with Bouma (1962) T_a, T_{ab} and T_b sequences.

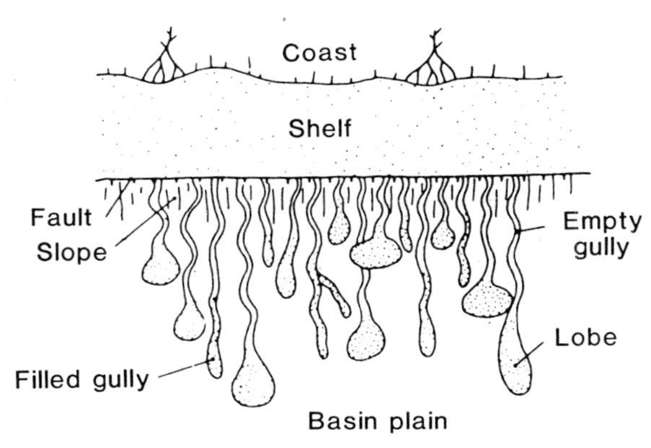

Figure 5.26 Depositional model for a fault-controlled submarine clastic ramp, based on the Hareelv Formation, Jameson Land, East Greenland. From Surlyk (1987).

Thick-bedded and very thick-bedded (0.5 to 50+ m) sandstones are ungraded and virtually devoid of primary sedimentary structures (Facies B1.1), with the exception of faint parallel stratification in some of the very thick beds. These occur as two major types: (a) laterally discontinuous fills of steep-sided (to overhanging) gullies or scours, and (b) parallel-sided beds which are laterally continuous for hundreds of metres. The gully-fill sandstones are common within the outcrops, comprising up to 50% of the formation volume, and alternate randomly with the laterally-extensive sandstones (Fig. 5.27). Most of the gullies have a sinuous to meandering form. There is a virtual absence of overbank deposits along the margins of the gully-fills, and infilling of individual gullies and scours appears to have been accomplished by a few flows.

Figure 5.27 Closely spaced, structureless massive gully sandstones (Facies B1.1) in partly scree-covered dark mudstone (Facies Class E). Note lack of organization of the system. From Surlyk (1987).

In addition to the three major facies associations, wet-sediment deformation, including liquefaction structures, clastic dykes and sills, are common within the Hareelv Formation. At the contact between the structureless thick-bedded sandstones and the encasing 'background' black mudstone, the margins of the thick-bedded sandstones are commonly modified by loading and liquefaction (Fig. 5.28). Features along the margins include subhorizontal convex ridges (up to several tens of metres long); large, diffuse dish structures (length up to 1.5 m and amplitudes of 0.3 m); large concentric curved joints that parallel the loaded ridges of the gully margins; and sandstone dykes (1 mm–20 cm thick) and sills (1 mm–several m thick). Dykes may crosscut bedding for hundreds of metres of vertical section; more commonly dykes only cross a few metres of section. Dykes often connect with sills.

There are no consistent vertical trends within the measured sections. Sandstone bodies appear and terminate randomly, both vertically and laterally. Within the sandstones there are no trends of grain size, bed thickness, sedimentary structures, or facies patterns. On a regional scale, for approximately 75 km in a north–south direction or 60 km in an east–west direction, there are some variations in the sandstone: mudstone ratio but there are no consistent patterns in this variation and no clear facies trends. The complete lack of apparent organization and the extremely irregular vertical stacking of sandstone bodies suggests that these deposits did not form on a submarine fan. The random juxtaposition of gully and lobe deposits shows that the depositional system was disorganized – with individual flow events randomly initiated along the fault-scarp line-source and travelling independently for variable distances downslope. High-density turbidity currents, perhaps triggered by reactivation of basin-margin faults, are thought to have emplaced the sands. The general characteristics of this system are summarized in Figures 5.29 and 30.

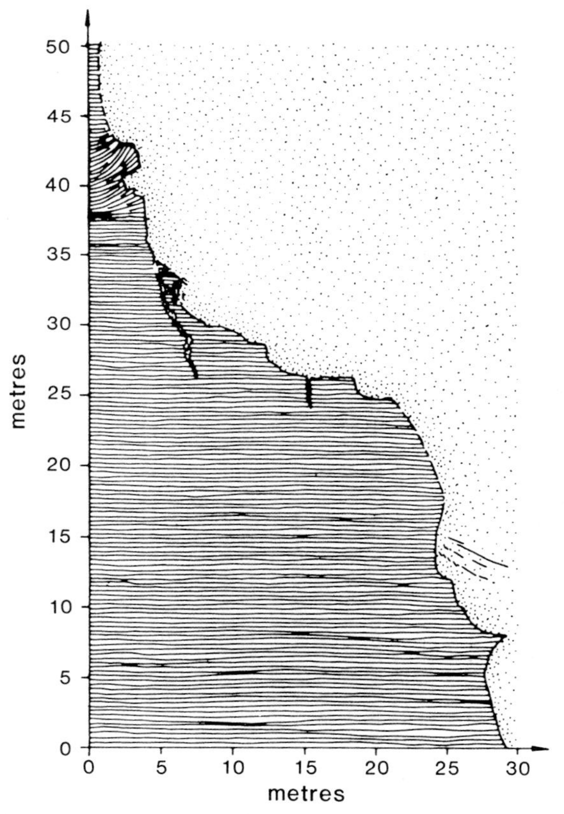

Figure 5.28 Vertical measured section of deep gully filled with structureless sandstone (Facies B1.1). Note subvertical load-deformed gully margin and ptygmatically folded sandstone dykes. From Surlyk (1987).

The Hareelv Formation is thus interpreted as the product of essentially continuous hemipelagic sedimentation of black shales under anoxic or poorly oxygenated conditions in relatively deep water. This hemipelagic sedimentation was periodically interrupted by high-density turbidity currents which transported large volumes of sand into gullies and scours. Sandstone dykes, sills and liquefaction features were associated with dewatering of the gully-fill sand upon deposition and compaction. Source areas for the sand were shallow shelf areas located to the northwest and northeast of the basin (Fig. 5.29). Flows were triggered along the entire slope (Figs 5.26 & 29) by movements along the basin-margin faults. The generation of the flows from a line source resulted in a random distribution of sandstone bodies (Fig. 5.30) and the preclusion of the development of a submarine-fan system.

The thick and very thick sandstone bodies are potential stratigraphic traps of hydrocarbons, because of their isolated occurrence within thick, organic-rich oil-prone source rock. This formation may serve as a model for an unusual type of reservoir with analogues in the North Sea (Surlyk 1987). There is a remarkable similarity in geometry between the gully and channel features described here, and features observed at the fronts of sandy deltas undergoing synsedimentary faulting (compare Fig. 5.29 with Fig. 5.12).

Case study: Huronian glaciomarine sedimentation, Early Proterozoic continental margin, Gowganda Formation, Elliott Lake, Ontario, Canada The Gowganda Formation (Huronian), near Elliott Lake in northern

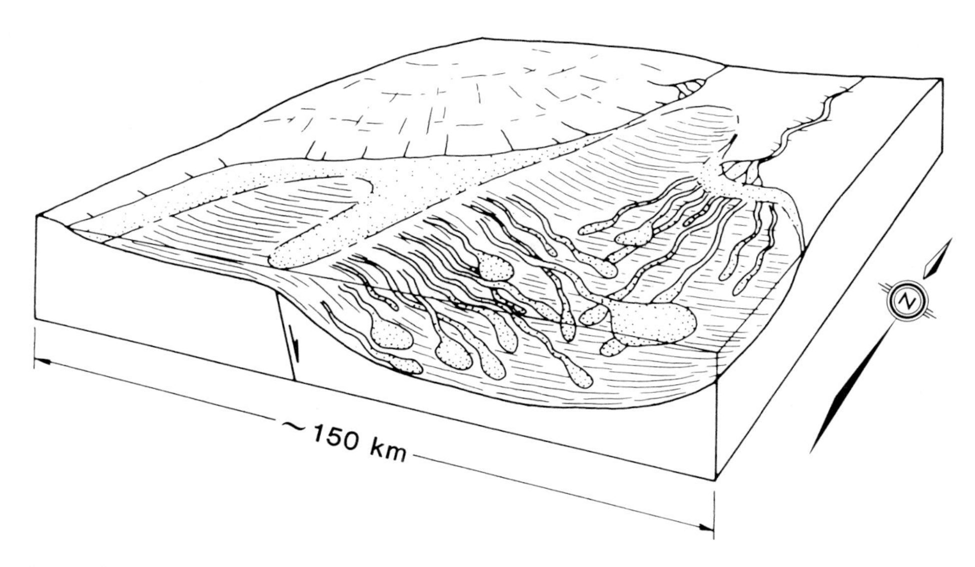

Figure 5.29 Palaeogeographic reconstruction of Jameson Land area, East Greenland, showing deposition of basinal Hareelv Formation during late Oxfordian. Size of sandstone bodies greatly exaggerated. From Surlyk (1987).

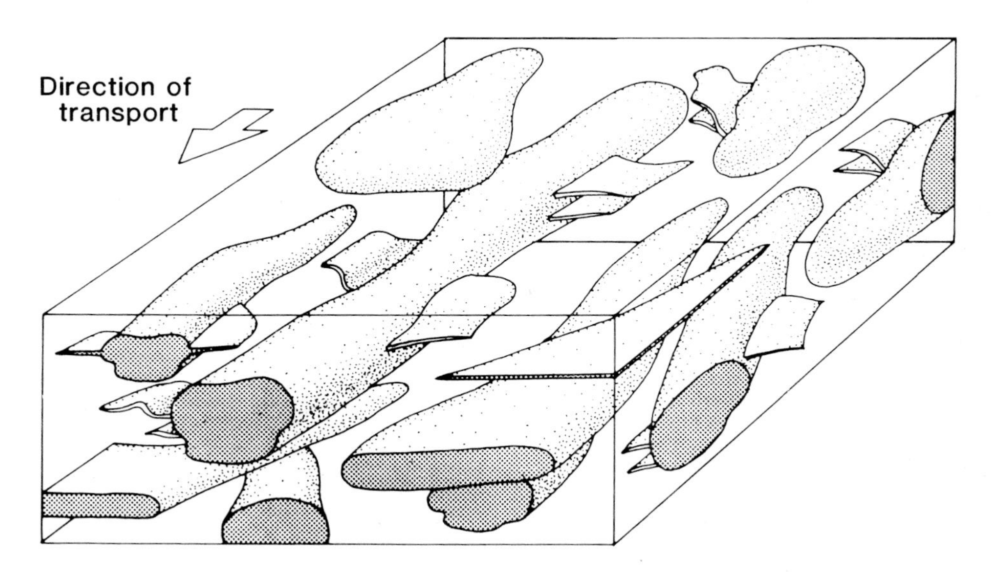

Figure 5.30 Schematic three-dimensional geometry and lateral relationships of reservoir sandstone bodies within submarine gully and lobe system of Hareelv Formation. Winglike extensions of sandstone bodies represent thick sandstone sills and dykes. Block width and height are approximately 310 and 125 m, respectively. Arrow indicates transport direction. From Surlyk (1987).

Ontario (Fig. 5.31), is world famous as an example of an ancient glacial deposit. It is part of the Huronian Supergroup (2.2–2.7 Ga), occurring as a southward-thickening wedge, up to 10 km thick, deposited on a mature (?) passive continental margin (Fralick & Miall 1981). Earlier work on the Gowganda Formation interpreted the unit as a succession of continental tillites and varved argillites (Lindsey 1969, 1971). More recently Miall (1983, 1985) has interpreted most of the formation as glacial-marine slope deposits. There is an absence of obvious traction current features, except near the base of the formation, and no evidence of wave

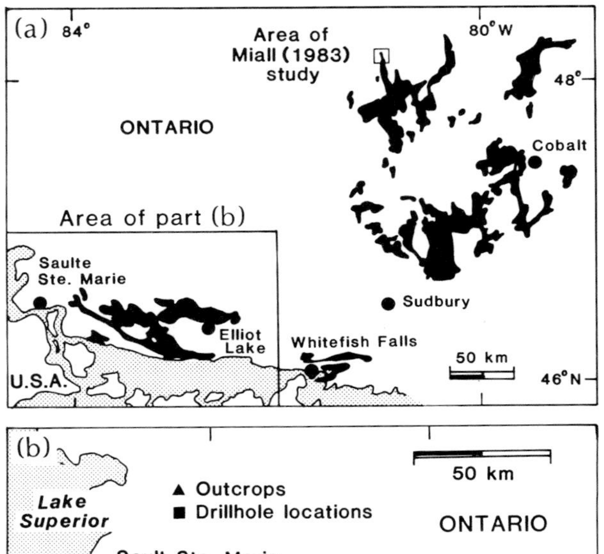

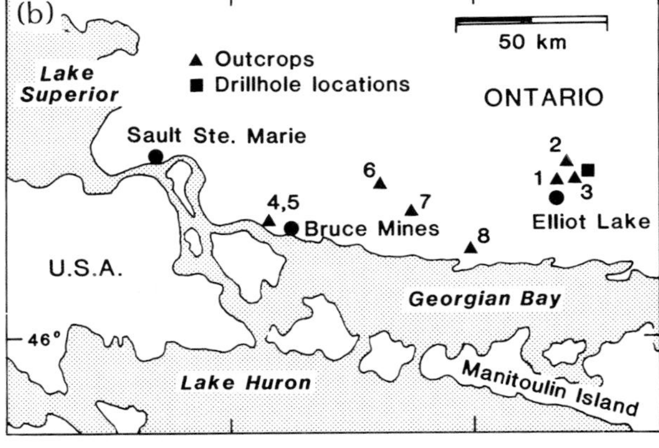

Figure 5.31 a, Outcrop distribution of the Gowganda Formation, north of Sudbury, Ontario; b, location of drillholes and outcrops studied by Miall. From Miall (1985).

activity. The Gowganda is underlain by a very thick shallow marine to non-marine Huronian succession, and was deposited on continental crust. The setting is inferred to be deep shelf to slope setting rather than rise (Miall, pers. comm. 1987). Continental glaciers provided the coarse sediment to the margin by shedding debris-laden icebergs, which then rafted material across the shelf to the slope (Miall 1985).

A generalized stratigraphic section of the Gowganda Formation in the subsurface is given in Figure 5.32. The lowermost 50 m of the Gowganda Formation is interpreted as a continental slope deposit (Miall 1985), and mainly consists of gravelly and sandy sediment gravity-flow deposits (Facies Classes A & B) (Figs 5.33 & 34) and mudstones (Facies Class E) (Fig. 5.35) containing dropstones, interbedded with minor 'rain-out units' of chaotic gravelly-sandy-mudstone (Facies F1.2) (Fig. 5.36). Overlying the slope deposits is a submarine channel system, 10–50 m thick, infilled with hemipelagic mudstones (Facies Class E) containing dropstones (Facies F1.2) and showing synsedimentary deformation structures. These mudstones are interbedded with channel-fill deposits, some of which are submarine point-bar sequences, up to 4.5 m thick. The channel-fill is interpreted to have formed during a high sea level stand and reduced sediment supply, but it is unclear how this relates to the glacial advance/retreat cycles. Capping the section is a blanket of structureless 'rain-out' gravelly-sandy-mud (Fig. 5.36), up to 55 m thick, which is succeeded by a younger succession of sediment gravity flow deposits (Facies Groups A & B).

Much of the mudstone is structureless; other parts contain thin siltstone or fine sandstone laminae, with rhythmic bedding (Fig. 5.35). Many of the rhythms show normal grading on a microscopic scale, but lack the cyclicity characteristic of lacustrine varves (Jackson 1965). Within the rhythmic mudstones there are common dropstones, and rare sandstones with synsedimentary deformation structures. Dropstones were rafted by floating ice; the deformed sandstones are interpreted as slurried beds (Miall 1985). Superficially the fine-grained facies in this association resemble the hemipelagic deposits and fine-grained turbidites seen on the modern Scotian slope (Hill 1984a & b). Other distinctive features attesting to the glacial influence on sedimentation of the Gowganda Formation include the occurrence of faceted clasts, and groove casts that are interpreted as iceberg scours (Miall 1985).

The first stage of deposition consisted of the rapid advance of glacial ice across a continental shelf (Fig. 5.37a). The beginning of Gowganda sedimentation was marked by a rapid transgression. Floating icebergs rained out an irregular blanket of gravelly sandy mud interspersed with hemipelagic mud. Local debris flows and slides resulted in an interfingering with the underlying Serpent Formation. Following this, an initial phase of continental-slope progradation occurred (Fig. 5.37b) as debris flows from multiple sediment sources blanketed the slope (cf. Surlyk 1984). Drifting icebergs shed coarse debris, which formed thick 'rain-out' blankets. Many of these deposits subsequently slid downslope to be redeposited along the slope-apron as sediment gravity flow deposits. A few kilometres downslope, turbidity currents were generated from the least viscous debris flows and sediment gravity flows. These turbidity currents capped individual debris flow deposits or travelled further downslope and were deposited as superimposed graded

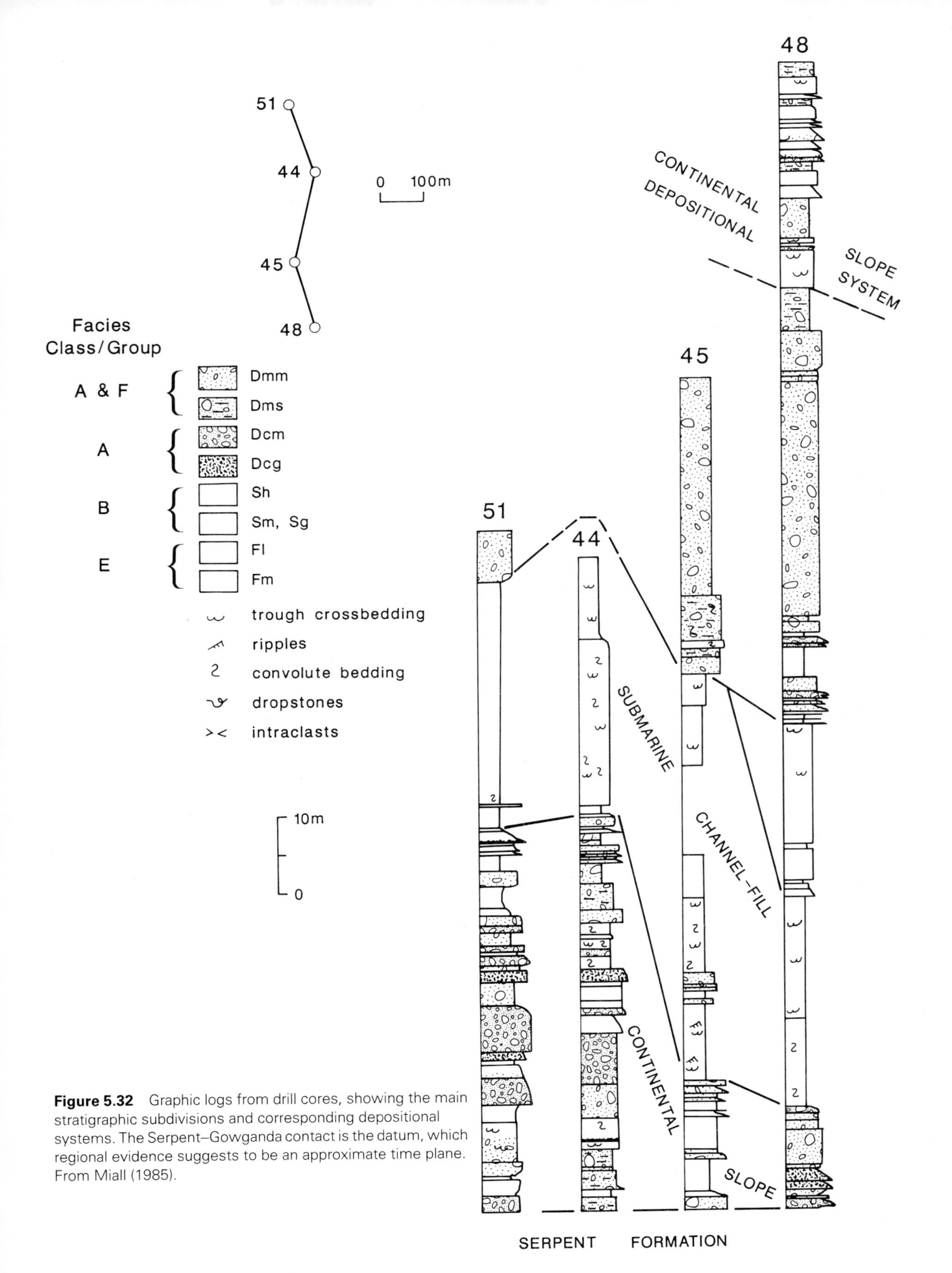

Figure 5.32 Graphic logs from drill cores, showing the main stratigraphic subdivisions and corresponding depositional systems. The Serpent–Gowganda contact is the datum, which regional evidence suggests to be an approximate time plane. From Miall (1985).

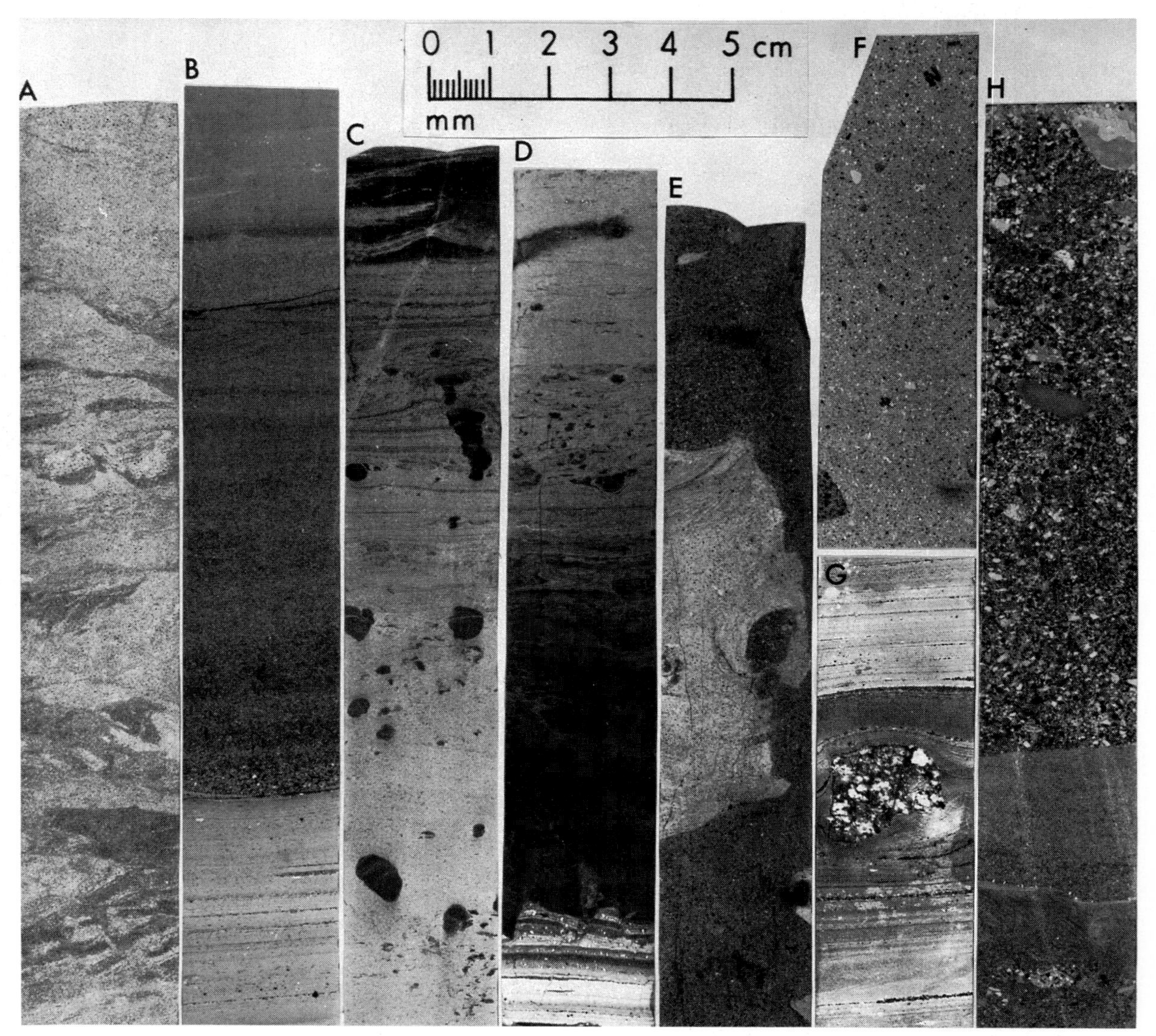

Figure 5.33 Sedimentary facies of sandstone and congomerate from the Gowganda Formation, near Elliott Lake, Ontario. For drill hole locations see Fig. 5.31. A, structureless standstone, with wet-sediment deformation structures (Facies B1.1), Hole 3. B, graded, sandstone to laminated mudstone (Facies Group C2), with Bouma sequences T_{abcde}, Hole 2. C, laminated fine-grained sandstone (dark) and silty mudstone (light) (Facies B2.1). Note occurrence of pseudonodules, Hole 3. D, interbedded sandstone and mudstone (grey shades) with wet-sediment deformation structures and synsedimentary faults (Facies Class C and Facies Group F), Hole 2. E, thin sandstone bed (Facies B1.2). Note deformed light grey siltstone intraclast, Hole 3. F, Poorly sorted pebbly sandy mudstone (Facies C1.1) with scattered dropstones, Hole 3. G, granite pebble dropstone (Facies F1.2) in laminated mudstones (Facies E2.2), Hole 2. H, poorly sorted pebbly sandy mudstone (Facies C1.1) resting with sharp contact on a sandstone showing ripple-drift cross-lamination (Facies Group C2), Hole 3. From Miall (1983).

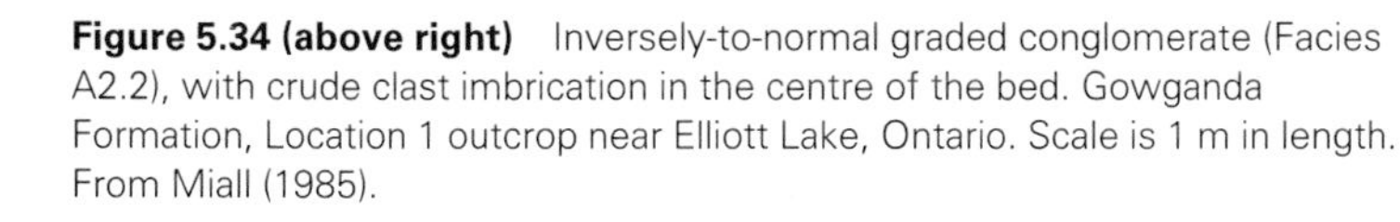

Figure 5.34 (above right) Inversely-to-normal graded conglomerate (Facies A2.2), with crude clast imbrication in the centre of the bed. Gowganda Formation, Location 1 outcrop near Elliott Lake, Ontario. Scale is 1 m in length. From Miall (1985).

Figure 5.35 (lower right) Rhythmic-bedded mudstones (Facies E2.2). Outcrop location 6 (Fig. 5.31). This was one of two outcrops studied by Jackson (1965). Note flow roll structure in sandstone bed to right of lens cap. From Miall (1985).

Figure 5.36 (above left) Typical outcrop of poorly sorted pebbly sandy mudstone (Facies A1.3), with giant granite clast. Scale is approximately 1.5 m long. Outcrop location 1 (Fig. 5.31). From Miall (1985).

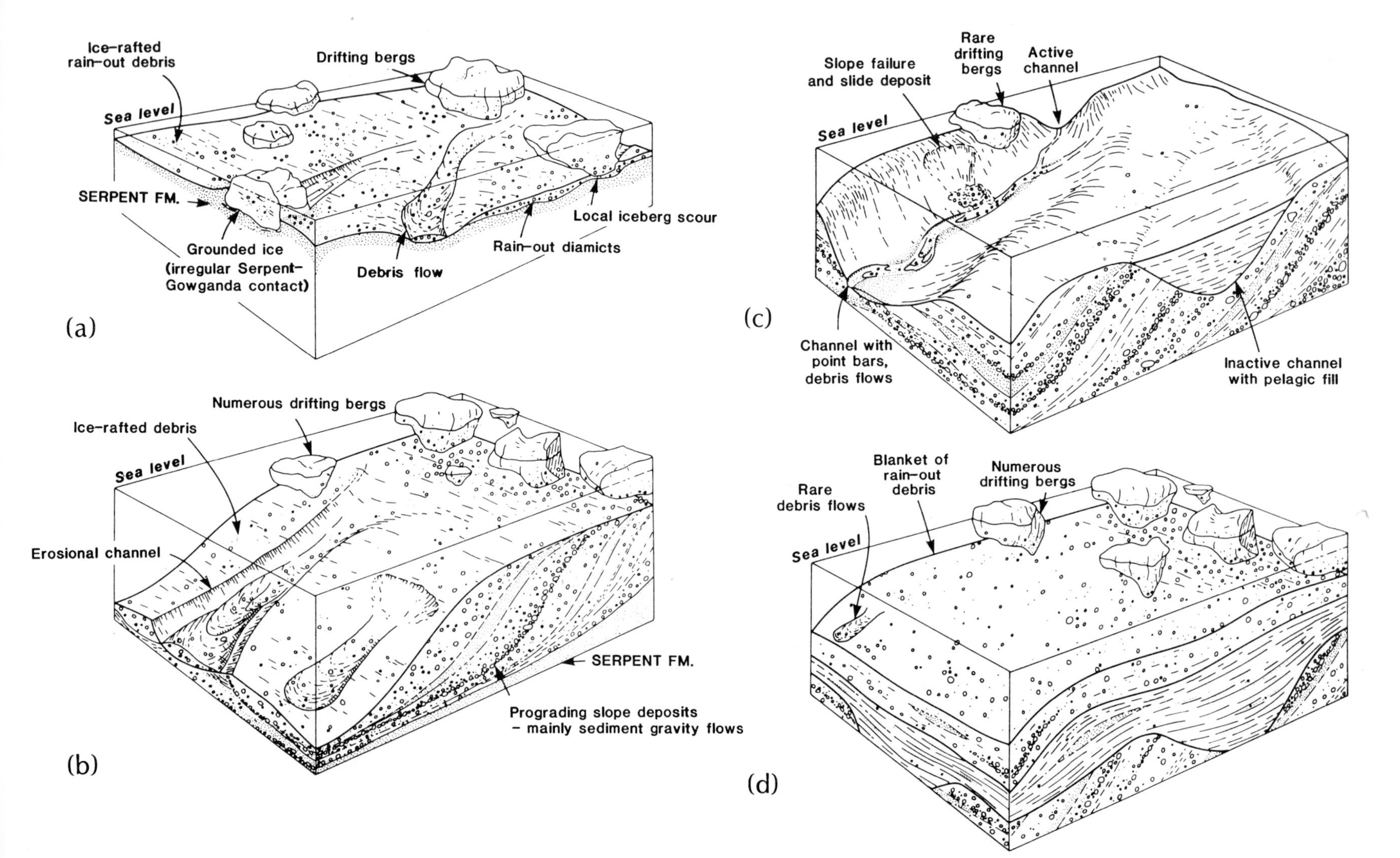

Figure 5.37 Evolution of the continental margin inferred from the glaciomarine Gowganda Formation in the drillhole area. Area shown is approximately 600 m wide, parallel to the depositional strike (which is oriented east–west). Schematic environmental models for four stages. a, Initial rapid advance of the ice across the continental shelf; b, first phase of continental slope progradation; c, channel cut and fill during reduction of sediment supply; d, pebbly-sandy-mudstone blanket formed by rain out during ice readvance, at beginning of new phase of slope progradation. From Miall (1985).

sandstone units. Between periods of active sedimentation, hemipelagic muds with dropstones accumulated. The third stage was marked by a reduction in sediment supply, cut and fill of submarine channels, and the draping of irregular slope topography by pelagic sediments (Fig. 5.37c). Locally, slope failures were generated along the margins of the active channels, and rare drifting icebergs provided dropstones to the system. The final stage was a structureless sediment blanket formed by 'rain out' of debris during an ice readvance (Fig. 5.37d). During this stage there was renewed slope progradation. Sediment was probably supplied to the slope-apron by floating icebergs. At first a structureless blanket of gravelly sandy mud formed, and was succeeded by sediment gravity flow deposits. The resultant complex stratigraphy and facies patterns of the overlying material make it difficult to reliably reconstruct further evolution of the continental margin.

Clearly, slope-apron sedimentation under glacial-marine conditions is complex. Similar patterns of sedimentation on glacio-marine continental slopes have been documented by Visser (1983) and Visser & Loock (1987) for the Dwyka Tillite in South Africa, by Socci & Smith (1987a & b) for Precambrian deposits in the Boston Basin, and by Eyles (1987) for the Yakataga conglomerate, Middleton Island, Alaska.

SYNSEDIMENTARY SLOPE FAILURES

Synsedimentary slope failures, involving the resedimentation of slope material as slide/glide and slump deposits, have been most studied in ancient resedimented deep-marine carbonates (Davies 1977, Hubert *et al.* 1977, Cook 1979, Coniglio 1986, James & Stevens 1986), although these features have also been reported from ancient clastic slopes and slope-aprons (Williams *et al.* 1969, Woodcock 1976a, 1979a, Belt & Bussieres 1981, Pickering 1982b, 1984a & b, 1987a, Farrell 1984, Jacobi 1984, Gawthorpe & Clemmey 1985).

Case study: Lower Palaeozoic deep-marine slope failures, Cow Head Group, Western Newfoundland, Canada

The Cow Head Group is a well exposed deep-marine carbonate–shale succession in western Newfoundland (Fig. 5.38). This carbonate succession contains intraformational truncation surfaces, slide masses and synsedimentary shear zones (Facies Group F2) (Fig. 5.39). Comprehensive work by James & Stevens (1986) and Coniglio (1986) shows that the succession represents a southeast-dipping base-of-slope carbonate apron. Abundant carbonate conglomerates and 'megaconglomerates' are interpreted as debris-flow deposits (Hiscott & James 1985).

A review of the terminology and framework of synsedimentary slope failures is given by Coniglio (1986) and is summarized as follows. In the description of the sedimentary slope failures, Coniglio (1986) has used the following definitions of Nardin *et al.* (1979): a *slide* is the movement of a rigid, internally deformed mass along a discrete shear surface. Slides are further subdivided into *slumps*, in which there is rotation on a curved shear surface, and *glides*, in which the mass is translated along a plane or surface. Scale problems at outcrop make it difficult to distinguish glides and slides, hence the general term slide was used in the description of the slope failures within the Cow Head Group. Sliding results in a *slide mass*, which is laterally displaced relative to the underlying sediments (Fig. 5.39). This lateral displacement results in local repetition of stratigraphy, and the formation of an angular disconformity called an *intraformational truncation surface*. Similar dislocations and truncation surfaces are seen on modern continental slopes, where they are referred to as *slide scars* or *slump scars* (e.g. Enos & Moore 1983, Jacobi 1984). The surface along which the translation has occurred is a shear plane or a more diffuse shear zone. Shear zones form immediately below a truncation surface, or below sediment masses which have crept downslope without detachment (Fig. 5.39). Internal deformation in slide masses is common, often concentrated at the leading edge, the base, and in some cases throughout the entire slide mass (cf. Nardin *et al.* 1979, Cook & Mullins 1983, Gawthorpe & Clemmey 1985).

The identification of intraformational truncation surfaces and slide masses commonly requires extensive lateral outcrop of exposure. Small-scale deformation structures are commonly associated with the shear zone deformation, and in areas of limited outcrop these can be used to help identify sediments which underwent periodic slope failure. Coniglio (1986) has shown a catalogue of many of these small-scale deformation structures associated with synsedimentary shear. Generally, these small-scale structures are from shear zones which are a metre or less in thickness, and include those shear zones associated with slope creep, below truncation surfaces, and/or those in basal zones of slide masses. The geometry of the deformation includes intrafolial folding, brecciation, rotation of slabs, 'fitted lenticular bedding,' and small-scale homogenized or micro-faulted/folded zones (Fig. 5.40).

Several of the Cow Head outcrops show the range of variation in the small-scale deformation structures. In

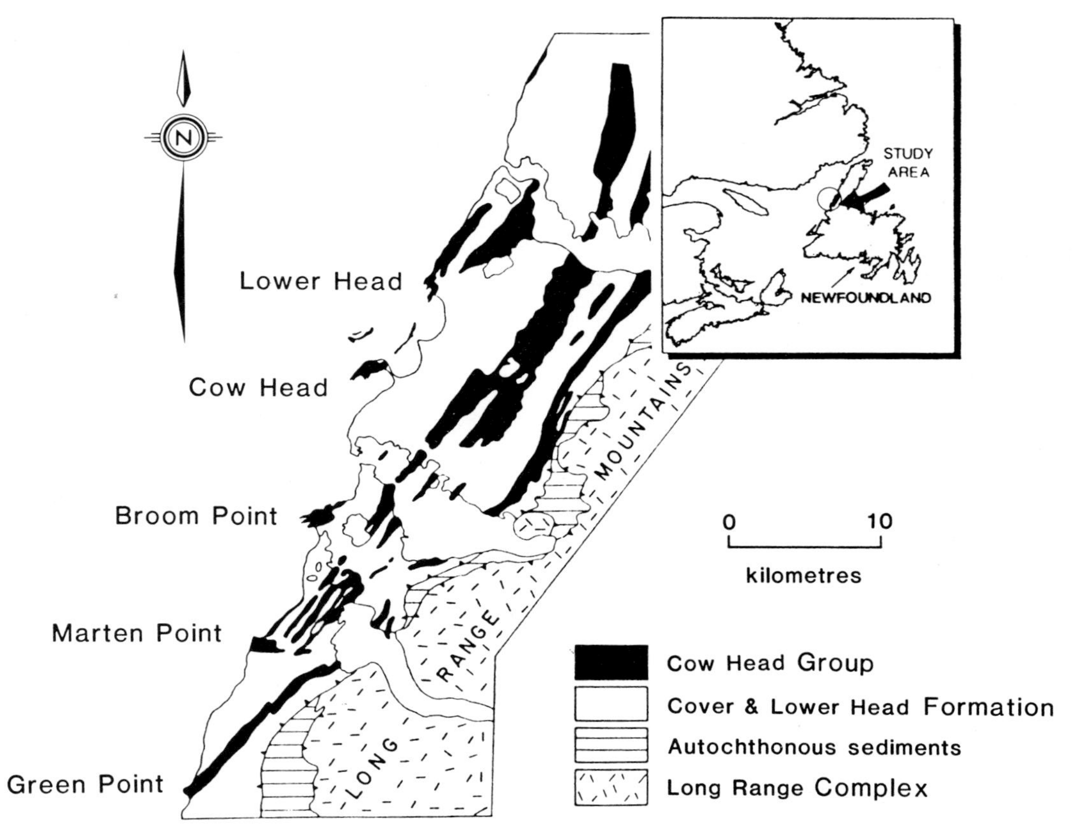

Figure 5.38 Outcrop locations and geologic setting of the Cow Head Group, western Newfoundland, Canada. From Coniglio (1986).

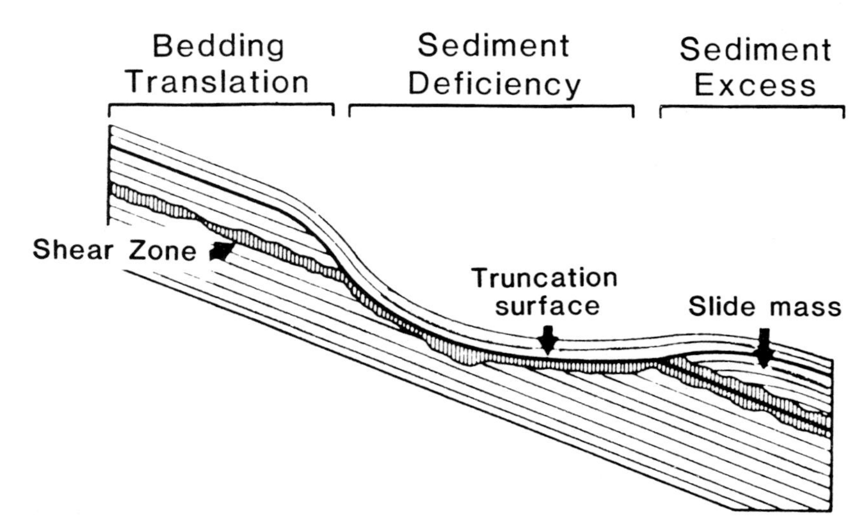

Figure 5.39 Simplified schematic diagram of submarine failure, illustrating synsedimentary shear zones. From Coniglio (1986).

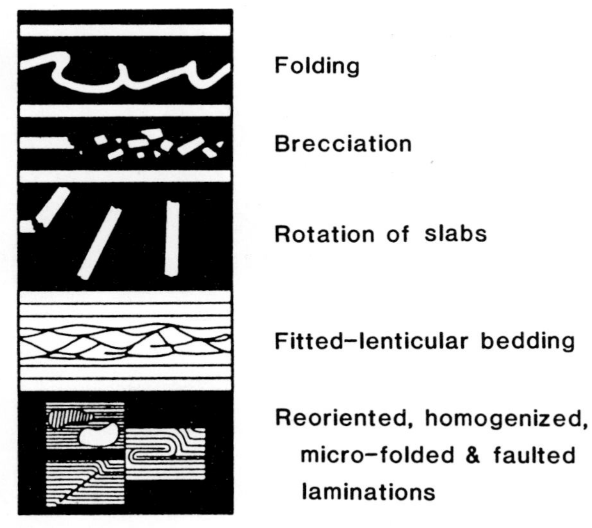

Figure 5.40 Schematic summary diagram of small-scale deformation (millimetre to decimetre scale) structures resulting from synsedimentary shear in the Cow Head Group. From Coniglio (1986).

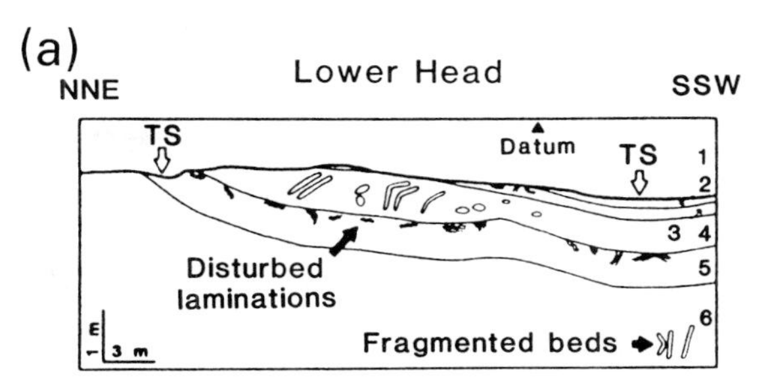

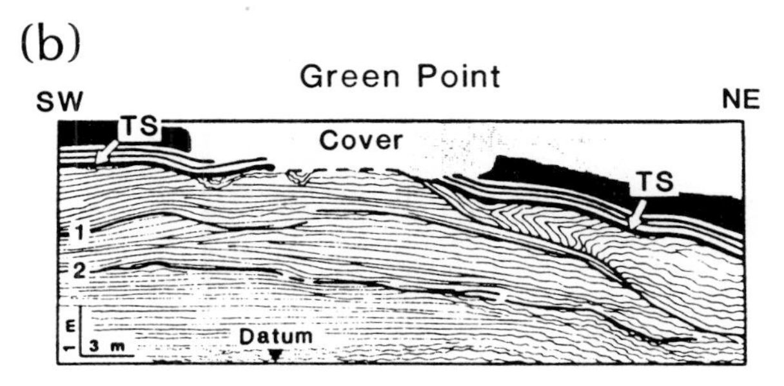

Figure 5.41 Intraformational truncation surfaces, Cow Head Group, Newfoundland. a, Measured outcrop of Lower Head truncation surface. Rotated mudstone slabs and disturbed laminations in marls are shown schematically in their approximate positions. Bed numbers as indicated; b, measured outcrop of Green Point truncation surface. '1' and '2' are small displacement synsedimentary faults that are traceable over most of the outcrop. Wavy lines indicate the distribution of disturbed bedding as illustrated in Figure 5.44b. Drag folds are shown in exact positions immediately underlying truncation surface. From Coniglio (1986).

the first example from Lower Head (Figs 5.38 & 41a) there is a prominent truncation surface, in which 2.5 m of truncation occurs in the measured section, and an additional 3–4 m of truncation further along strike (TS in Figs 5.41a, 42a & b). The depression created by the truncation was infilled with thinly-bedded parted grainstones and mudstones. Sediments beneath the truncation surface consist of black shales (Facies E1.1), finely laminated marls (Facies Group G1) and ribbon limestones (Facies Groups F & G). They show various types of small-scale deformation structures, including: fragmented and rotated limestones (Facies F2.2) (Figs 5.41a, 42b, c & d). Some fragments show random orientations; others have subparallel orientations at high angles to bedding (Figs 5.41a & 42d). In the Cow Head Group, the maximum amount of removed section is 7 m. Other ancient studies document 15–25 m-deep cuts into slope sediments of the Piedmont Basin in Northern Italy (Clari & Ghibaudo 1979) and up to 100–150 m of truncation of Palaeozoic slope carbonates in the Sverdrup Basin (Davies 1977).

The second example at Green Point (Figs 5.38 & 41b) shows a distinct slide mass, which reaches a thickness of 4 m and is laterally traceable along strike for 180 m. The only exposed lateral margin (Fig. 5.43a) shows approximately 1 m of basal erosion. The transported sediment comprises mainly thin-bedded, burrowed and fissile mudstones (Facies E1.3). Near the margin, disruption of bedding and homogenization of internal sedimentary structures (Facies F2.2, F2.1) occurs; elsewhere, bedding appears relatively undeformed. Throughout most of the outcrop, aside from the lateral margin, there is no evidence of a basal shear zone, and the contact between the slide mass and underlying sediment shows no erosion or deformation.

The third example is from the Middle Ordovician at Cow Head South (Fig. 5.38), where a slide mass disconformably overlies thin-bedded, fissile Lower Ordovician grainstones. The slide (Facies F2.1) is approximately 1 m thick and is overlain by a 1.5 m thick normally graded conglomerate (Facies A2.3) (Fig. 5.43b). Near the base of the slide mass, an anastomosing web of millimetre-thick, argillaceous partings occur, interpreted as fitted-lenticular bedding (Facies F2.2) (Fig. 5.43c & d). This type of deformation has also been reported in Cambro–Ordovician carbonate slides in Nevada (Cook 1979, Cook & Mullins 1983).

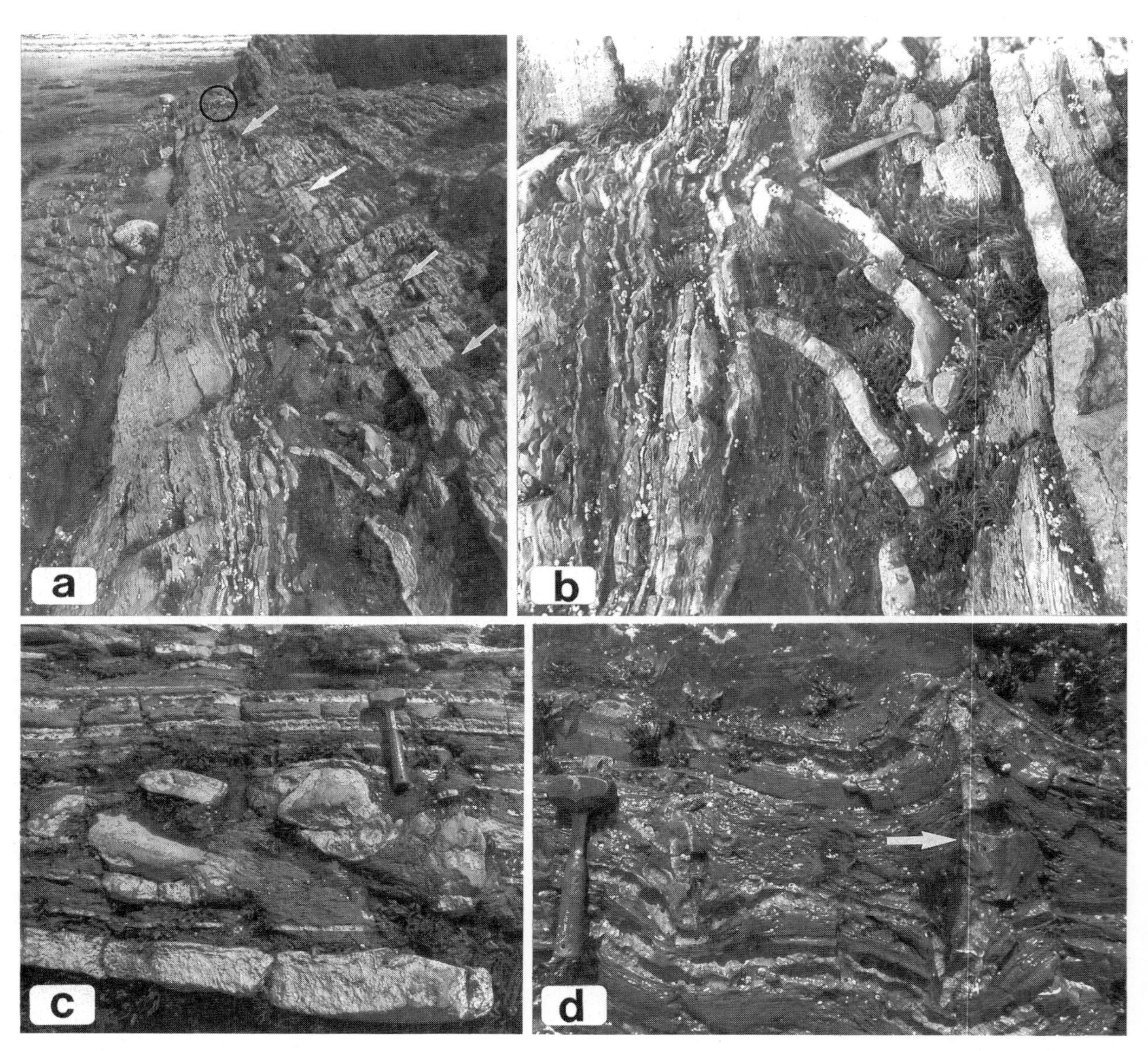

Figure 5.42 Intraformational truncation surface at Lower Head, Cow Head Group, Newfoundland. a, Arrows point to truncation surface. A close-up of the rotated mudstone slab in centre of photograph is shown in (b). Knapsack for scale is circled. Stratigraphic top is to upper right; b, fragmented, subparallel aligned mudstone slabs in ribbon limestone succession (bed 4 in Fig. 5.41 a) Stratigraphic top is to right; c, a cluster of fragmented mudstone beds in ribbon limestone succession (bed 6 in Fig. 5.41 a) Stratigraphic top is up; d, vertically oriented mudstone slab (arrow) in a ribbon limestone succession (bed 6 in Fig. 5.41a) with a drape of finely laminated marls and mudstone beds. This illustrates that bed fragmentation and rotation of the slabs occurred in uncompacted sediment, a synsedimentary disruption. (Facies F2.2, Facies Groups F and G). Stratigraphic top is up. From Coniglio (1986).

Figure 5.43 Slide masses and related deformation, Cow Head Group, Newfoundland. a, slide mass of burrowed mudstone containing domains of homogenized and chaotic bedding (Facies Group F2). Basal scour of underlying argillaceous sediments and lateral pinchout demonstrate this to be a slide mass. Stratigraphic top to left, Green Point; b, slide mass (Facies Group F2) (bracketed) overlain by a graded conglomerate (Facies A2.3). Knapsack for scale near top slide mass. Stratigraphic top is up, Cow Head south; c, detail of basal irregular partings passing upward into more regularly bedded sediment. Arrows point to truncation surface. In the overlying slide mass, fitted lenticular bedding defined by irregular, anastomosing web of partings. Stratigraphic top is up; d, mudstones within this outcrop show partings similar to the fitted lenticular bedding of (c). Stratigraphic top is up. From Coniglio (1986).

The geometry resembles the small 'bowl-shaped' dislocations reported by Hein (1985) in surficial cores from deformed clayey silt on deep-marine slopes of the California Borderland. In these modern sediments, some of the bowl-shaped dislocations are associated with microfaults with up to 3 cm displacement. They are interpreted as internal shear planes within small surficial slides or slumps (Hein, 1985). On a decimetre to metre scale, Guiraud & Seguret (1987) describe a similar geometry for dislocations, interpreted as hydroplastic microfaults within clayey siltstones.

The final example of small-scale deformation structures in the Cow Head Group is from shear zones related to slope creep. The shear zones are parallel to bedding and synsedimentary in origin. At Green Point (Fig. 5.38), there is a 2 m-thick interval of parted limestones which shows several laterally extensive zones of internally deformed sediment (Facies Group F2). The disrupted zones are 20–30 cm thick, consisting of folded (Facies F2.1) and fragmented (Facies F2.2) mudstones set in a marl matrix (Facies G1.2) (Fig. 5.44a). In most cases, the deformed zones do not extend more than 200 m along strike, and attain thicknesses up to 1 m. At north Cow Head (Fig. 5.38), thicker bands of folded (Facies F2.1) and fragmented (Facies F2.2) beds occur as isolated units or clusters within a marl matrix; elsewhere intrafolial drag folds (Facies F2.1) are developed within otherwise undisturbed sediment (Fig. 5.44b). Similar shear zones have been reported by Hill *et al.* (1982) in slope sediments of the Canadian Beaufort Sea that have undergone creep.

The excellent exposures of the Cow Head Group clearly document the relationship between small-scale deformation features, larger scale truncation surfaces, synsedimentary shear zones and slide masses. In study areas of limited lateral outcrop or in subsurface studies, recognition of possible synsedimentary slope failure depends upon the identification of associated small-scale deformation structures. The identification of synsedimentary slope failure is extremely important from both a stratigraphic as well as economic point of view. Repeated or missing section may result from such failure, and this model of slope development may be helpful in identifying missing litho- and biostratigraphic section in ancient slope deposits.

Figure 5.44 Possible synsedimentary creep shear zones, Cow Head Group Newfoundland. a, Thirty centimetre thick shear zone contains fragmented and folded mudstones (light grey) (Facies Group F2) and a marl matrix (dark grey) (Classes F and G). Note planar bedded, undisturbed mudstones occur on both sides of the shear zone. Stratigraphic top is up, Green Point; b, recumbent drag fold in a shear zone suggests transport of overlying sediment mass to the northeast towards left in this two dimensional outcrop. This illustrates clearly that deformation may be extremely localized and in much of the shear zone it is not possible to tell whether relative movement of sediment packages has occurred. Hammer butt rests on top of shear zone. Stratigraphic top is up, Cow Head north. From Coniglio (1986).

6 m-deep conglomerate-filled U-shaped channel in Upper Cretaceous Rosario Formation 400 km south of the US-Mexican border, Baja California. These volcaniclastic intrachannel sediments are nested within a larger 1 km-wide channel near the base of a 7 km-wide submarine canyon. Photograph courtesy of Dr. Cathy J. Busby-Spera, University of California, Santa Barbara, CA (cf. Morris & Busby-Spera 1988, Bulletin American Association Petroleum Geologists **72**, 717–37).

CHAPTER 6

Submarine canyons, gullies and other sea valleys

Introduction

Submarine canyons are common features along continental margins, for example, along the east coast of the USA (Fig. 6.1), and may cut into crystalline rock (around Sri Lanka and the tip of Baja California), consolidated and under-consolidated sediments (the eastern continental margin of the USA), and even evaporites (Congo Canyon). Canyons may be tens of metres to hundreds of metres deep from rim to floor, have widths varying from tens of metres to tens of kilometres, lengths from kilometres to hundreds of kilometres, and show considerable variability in cross-sectional profile (Figs 6.2 & 3). The smallest linear conduits sometimes associated with canyons, but often occurring independently along continental slopes, and generally an order of magnitude smaller, are referred to as submarine gullies. Since there is a complete size gradation from canyons to gullies, both are included in this chapter. Submarine channels and some 'troughs' that do not appear to form an integral part of submarine fans are also dealt with in this chapter since they display many features in common with canyons. In ancient successions, the distinction between a submarine canyon and other deep marine channels or troughs may be impossible.

The study of submarine canyons began in earnest about fifty years ago, accelerated by technological advances in deep-marine surveying apparatus and techniques, for example seismic profiling and side-scan sonar. Much of this research has been concerned with the detailed recording of bathymetric charts and explanations as to the origin of canyons. To date, the main thrust of our understanding of submarine canyons has been inspired by the work of F. P. Shepard and co-workers (e.g. Shepard 1951, 1955, 1963, 1966, 1975, 1976, 1977 & 1981, Shepard & Marshall 1969, 1973a & b, 1978, Shepard *et al.* 1979). Amongst the many explanations of canyon origin, the most frequently cited are: (a) subaerial river-cut valleys now submerged by raised sea level (Spencer 1903, Bourcart 1938); (b) glaciated valleys now below sea level (Shepard 1933); (c) erosion by turbidity currents (Daly 1936); (d) underground water circulation dissolving rocks at continental margins (Johnson 1939, 1967); (e) tsunami-cut canyons (Bucher 1940); and (f) structural control

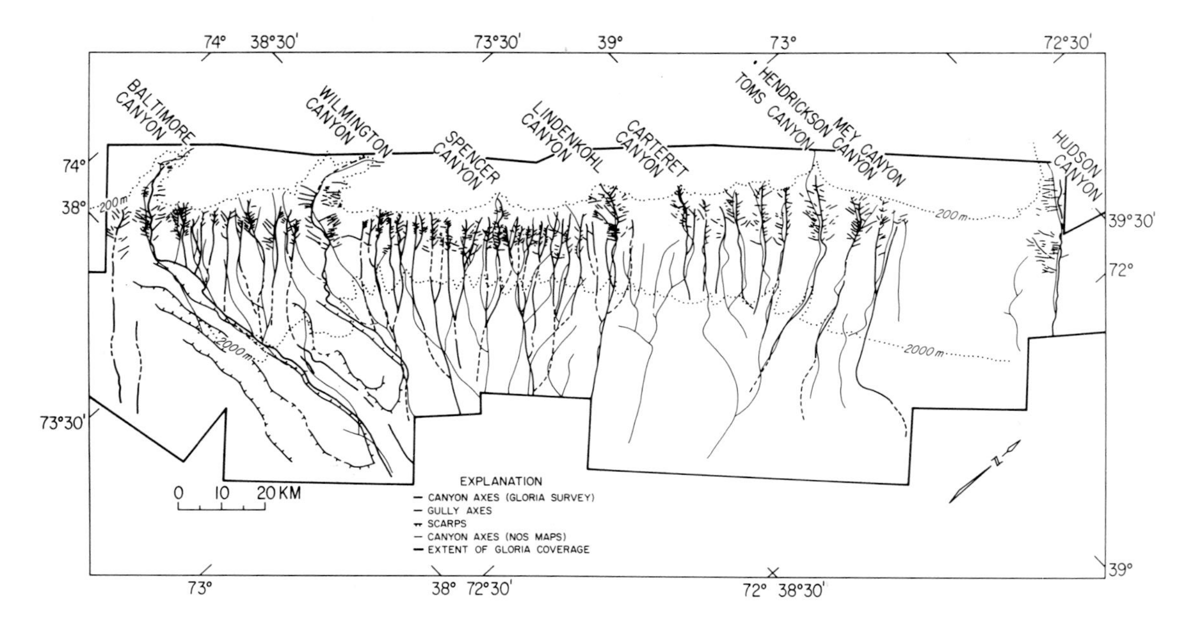

Figure 6.1 Location of submarine canyons and gullies along the eastern US continental margin derived from sonographs (after Twichell & Roberts 1982). Hatchured lines define possible slide blocks southeast of Wilmington and Baltimore Canyons. Boxed area is continental margin studied using GLORIA side-scan sonar.

enhanced by erosive processes (Kenyon *et al.* 1978, Picha 1979, Berryhill 1981). Present-day studies tend to emphasize the complex history of many modern canyons and gullies, with a tendency to define the main factors governing the development of a particular conduit rather than suggesting a single controlling process. Naturally, the interpretive problems are magnified when studying ancient canyons, gullies and other sea valleys. However, the advantage of studying ancient canyons is that they record a complete cut and fill history that is incomplete in modern canyons. In this chapter, most attention is focused on modern canyons and canyon processes, with shorter sections on ancient canyons, gullies, and other sea valleys and troughs.

The work of Daly (1936), and subsequent researchers, has shown that various sediment gravity flows, particularly turbidity currents, have been important in the development of many modern and ancient submarine canyons. In a study of many canyons and basins off southern California, Moore (1965) showed that turbidity currents have been the main sediment transport process during the last several million years. He recognized the importance of turbidity current erosion, particularly with respect to high density, high velocity currents generated in the region of canyon heads associated with sand transport by longshore drift.

Dill (1964) believed that one of the major factors governing the deepening of canyons was sand creep processes occurring over long time periods. This hypothesis was supported in an experiment where stakes were placed in a line across canyon floors and the central or 'axial' stakes were observed to have moved farther downslope than the 'lateral' stakes. Although sand creep may be important, more catastrophic gravity flows are suggested by the disappearance of some stakes, large concrete blocks and a car body (Dill 1964) – the latter, no doubt, having produced an 'autobreccia'!

Ancient canyons are filled mainly with turbidity current deposits (Almgren 1978, Picha 1979), and modern canyons may contain coarse-grained sediments tens to hundreds of kilometres from the inferred source, deposited by turbidity currents and other sediment gravity flows. Low density, low velocity turbidity currents are more widely reported from modern canyons, for example off California (Shepard & Marshall 1978), and appear to have been important in the deposition of thin-bedded, fine-grained ancient canyon sediments (Nelson *et al.* 1978).

Table 6.1 Summary characteristics of deep-marine channels based on published data, after Carter (1988). See pp. 155–9.

	Channel Length (km)	Width floor (km)	Width shoulder (km)	Depth (m)	Levees+ high (m)	Levees+ wide (km)	Axial slope	Pelagic drape	Heads (water depths in m)	Ends (water depths in m)	Age (Ma)
NAMOC	3800	1.5–7.5	6–16	100–200	RB 18–87	≤30	1:1500 to 1:2100	Yes	*Continental slope off Greenland	Sohm Abyssal Plain (5000)	?50
Cascadia	2200	0.65–4.5	4–7	40–320	RB 30+	≤7.5	1:625 to 1:4000	No†	Western US margin	Tufts Abyssal Plain (3500)	5
Bounty	1000+	1–2	5–7	150–650	LB 32–84	10–15	1:400 to 1:3300	Yes	Otago continental margin	SW Pacific Abyssal Plain (4500+)	50+
Vidal	800+	0.5–1.5	2–4	100–220	No	—	1:1100	Yes	Demerara Abyssal Plain	Barracuda Abyssal Plain (5400)	?
Maury	3500	3–10	5–15	*c.* 100–300	No	—	1:500 to 1:1000+	?No	Upper Rockall Basin	Iberian Abyssal Plain (5300)	≥13 (?60)
Surveyor	700	0.5–3	5–8	100–450	RB 40	13	1:500 to 1:1000	?No	*Alaskan Slope	Aleutian Trench (5200)	≤5
Equatorial Atlantic	1275	2–4	4–8	50–200	LB ≤25	1–7	1:1040	Yes (≤60m)	Brazil continental rise (4300 m)	Fernando de Noronha fracture zone (5450)	15–8
Valencia	200	0.5–4	5–10	200–350	RB ≤40	1–2	1:150 to 1:600	?No	Ebro Fan	Valencia Fan (Balearic Abyssal Plain) (2700)	5 (?20)

+ RB/LB Right or left bank, viewed down-channel
* not contiguous with fan/canyon system
† Holocene silt turbidites

Characteristics of modern canyons

Most submarine canyons show the following features (Shepard 1977): (a) sinuous courses, partly true meanders; (b) canyon floors deepen quite consistently seaward; (c) canyons lose their deep V-shaped profiles at about the same distance from the shore as the adjacent continental slope shows a marked decrease in slope gradient, becoming a part of the continental rise; (d) canyons are rarely found where the continental slopes are gentle; (e) canyons are cut into rocks of all degrees of hardness as well as into unconsolidated sediments, and (f) they have tributaries, better developed at the canyon heads than in their lower reaches. The physiography of modern submarine canyons shows considerable variability (Fig. 6.2), and in general most canyons are cut into the upper part of basin slopes or the continental slope.

Most of the world's large submarine canyons are associated with major rivers, whereas many of the smaller canyons, such as along the continental shelf and slope off California and Baja California, are not necessarily associated with rivers. Redondo Canyon (Fig. 6.2e) typifies these smaller canyons: the canyon heads in about 15 m of water approximately 300 m from the shoreline, extending seaward for about 15 km to the head of a submarine fan. The canyon is up to 1.6 km wide with a maximum recorded depth of 395 m, and shows a longitudinal decrease of gradient from about 15° at the head to less than 2° at the mouth. The southern wall is steeper, straighter and higher than the north wall. While Redondo Canyon is not related to a specific river system today, Yerkes *et al.* (1967) described a buried channel, filled by the Gardena Sand, that was formed by a stream that terminated at the head of the modern canyon. The position of Redondo Canyon is structurally controlled: the southern wall of the canyon is the surface expression (sedimentologically modified) of the Redondo Canyon Fault. It has been suggested that the ancient 'Gardena' river excavated a course along the fault trace (Yerkes *et al.* 1967).

Berryhill (1981) describes a large, buried, near-linear submarine canyon or trough crossing the seaward margin of the continental shelf off Louisiana, infilled by Pleistocene–Holocene sediments (Fig. 6.4a & b). The original length was about 90 km, maximum width at the shelf edge approximately 16 km, and the maximum eroded depth is estimated to have been at least 305 m (Berryhill 1981). High-resolution seismic-reflection profiles A, B and C (Fig. 6.4b), located in Figure 6.4a, reveal that: the headward region (Profile A) was covered by about 85 m of post canyon-fill sediments; at least three separate phases of canyon excavation occurred (Profiles B & C), with the oldest being numbered '1' in Figure 6.4b, and the margins of the canyon are, at least in part, fault controlled.

Canyons along mature passive continental margins tend to develop longitudinal and cross-sectional profiles that change relatively consistently downslope, perpendicular to the slope contours. Canyons along active, convergent, destructive or strike-slip margins are more variable. Underwood & Bachman (1982) describe canyons from the Middle America Trench off Mexico, where the largest canyons trend downslope into the trench, but smaller canyons die out or are confined to 'perched basins' in the trench slope (Fig. 6.5). Thus, the smaller canyons are tectonically dammed by either the trench-slope break or, if the forearc basin is filled, by a tectonically formed ridge on the lower slope.

The distribution of canyons along continental margins, in some cases, has been shown to be related to slope gradient; for example, Twichell & Roberts (1982) studied the continental slope between Hudson and Baltimore Canyons (Fig. 6.1) and found that: (a) canyons are absent where the gradient is less than 3°; (b) canyons are spaced 2–10 km apart where gradients range from 3–5°, and (c) where gradients are greater than 6°, canyons occur 1.5–4 km apart. Two of the canyons, Wilmington and South Wilmington (Fig. 6.2f) have been studied in detail between 2300–2400 m water depth (Stubblefield *et al.* 1981, 1982, McGregor *et al.* 1982), and serve as interesting examples of canyons with different physiography.

Wilmington Canyon extends down the continental

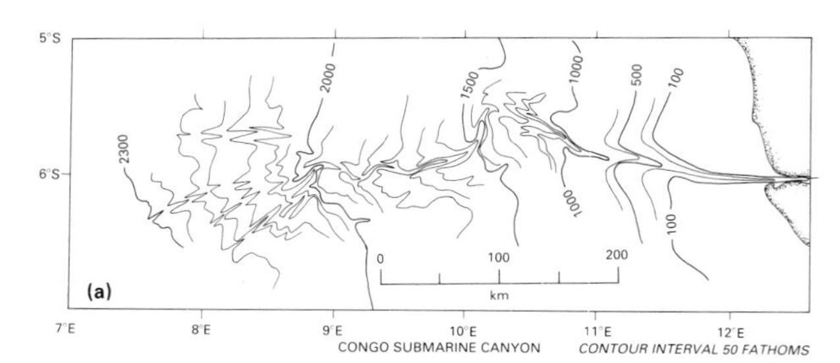

Figure 6.2 Examples of submarine canyons to show their morphology: a, Congo Canyon, off central West Africa; b, Monterey and Carmel Canyons; c, Scripps Canyon, offshore California Borderland; d, Newport Canyon, offshore California Borderland; e, Redondo Canyon, offshore California Borderland; f, Wilmington and South Wilmington Canyons, eastern US margin.

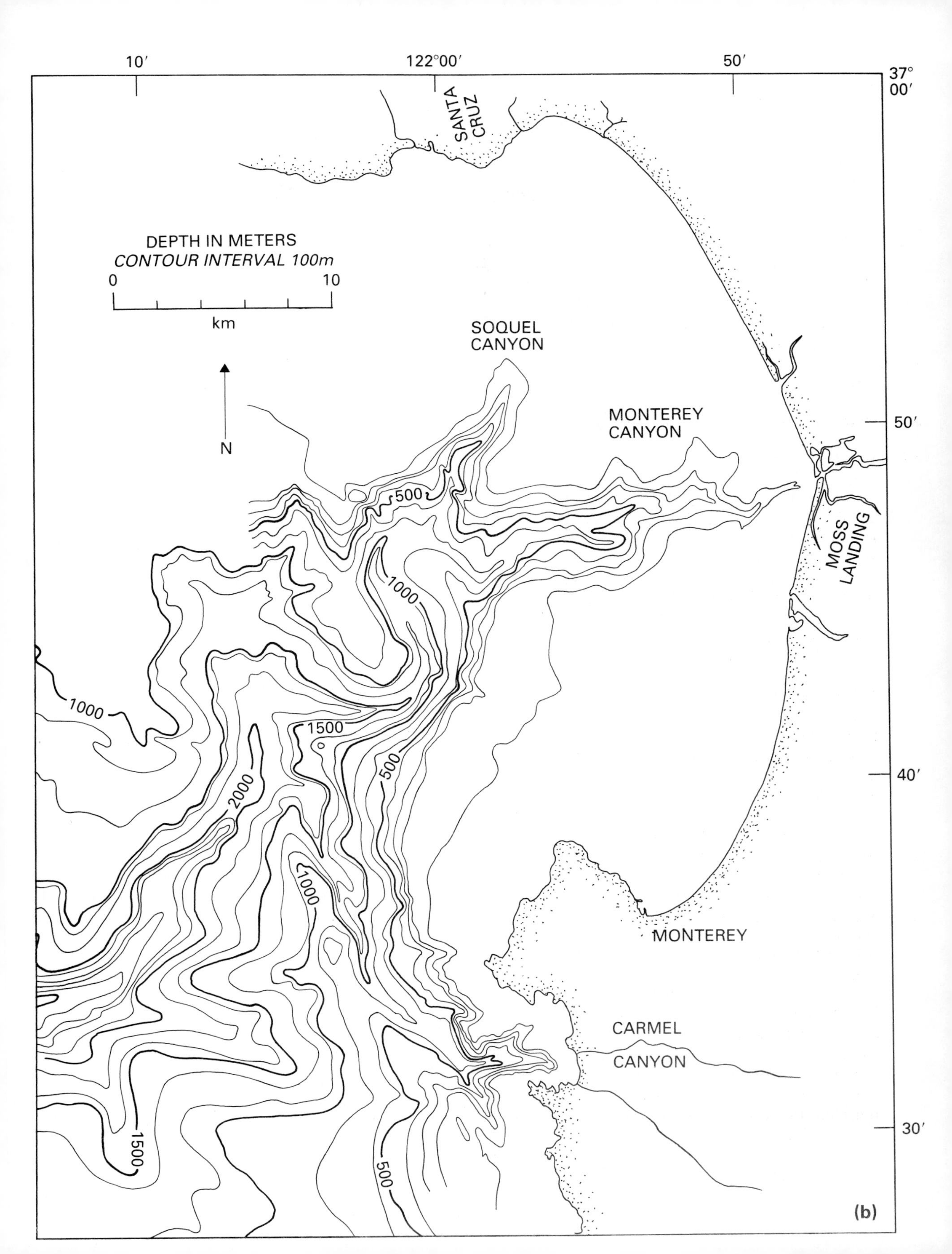
10′
122°00′
50′
37° 00′
SANTA CRUZ
DEPTH IN METERS
CONTOUR INTERVAL 100m
0
10
km
N
SOQUEL CANYON
MONTEREY CANYON
MOSS LANDING
500
1000
1000
1500
500
2000
1000
1500
500
50′
40′
30′
MONTEREY
CARMEL
CANYON
(b)

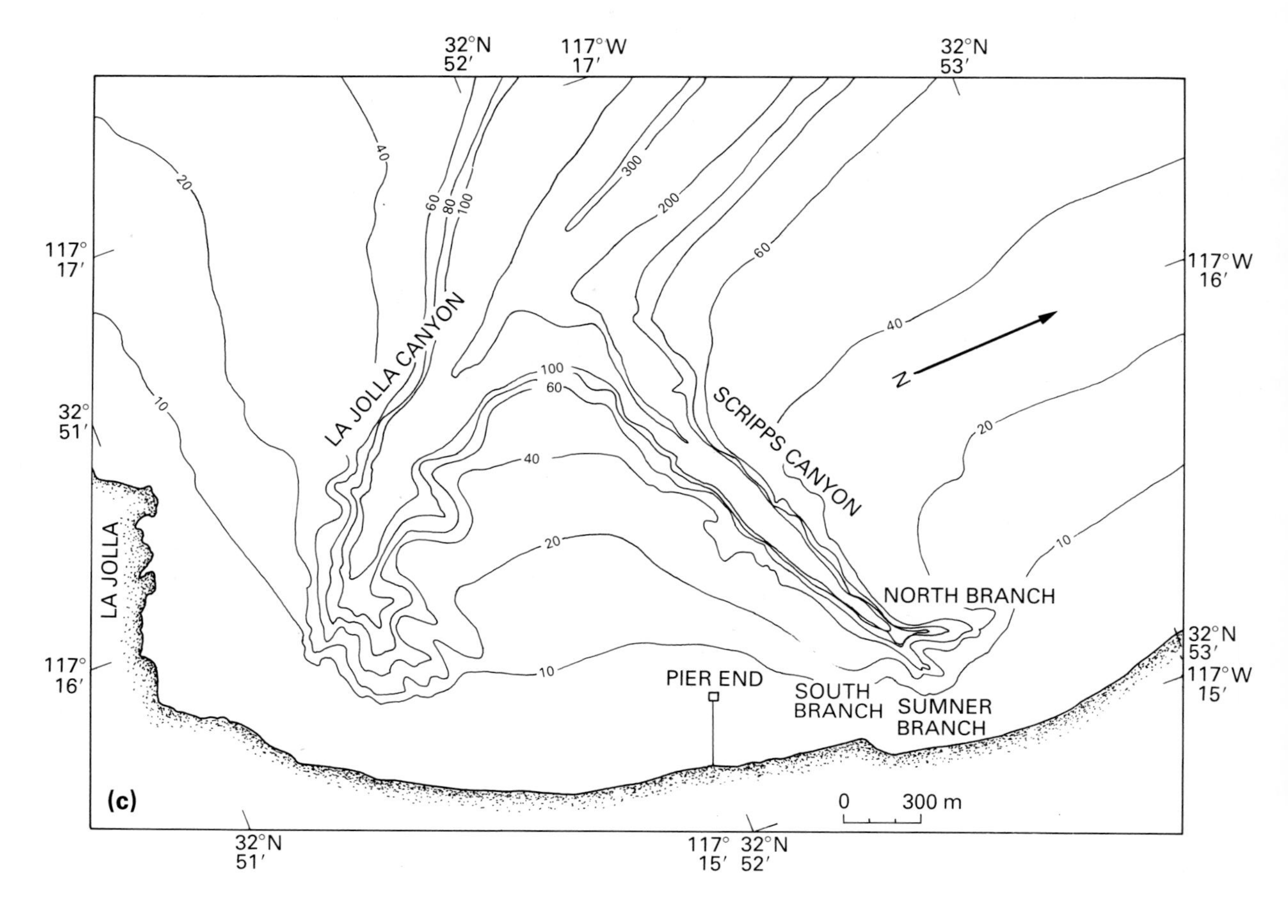

slope and across the continental rise, and has a meandering channel with steep undercut outer banks and gently sloping inner banks. Localized sediment failure is inferred from the many step-like depressions on the steep outer banks of the meanders. On the slope, the axial gradient is 2.5°. The meanders have wavelengths of 500–2000 m. In contrast, South Wilmington Canyon has no meanders and the channel floor features include deformed and displaced sediment, upturned clay beds, disaggregated gravels, loosely bound gravel conglomerates in a reddish-brown matrix, and even a tabular structure resembling a tree-cast, within thin-bedded reddish-brown sandstone. Cobbles and rock fragments occur, with evidence of current scour around the large cobbles. This record suggests that the gravels and sandstones were deposited in shallow water, possibly near the top of the slope during lowered sea level, and subsequently transported downslope *en masse* as a slide. The axial gradient of the canyon on the slope is 7°, which is about three times that for Wilmington Canyon.

Both Wilmington and South Wilmington Canyons contain an associated dendritic drainage pattern of gullies. The canyons are 8 km apart, and are incised into a slope with an average gradient of 8°. It seems likely, therefore, that the initial sea-floor gradient, sediment type and sediment sources were the same during the development of Wilmington and South Wilmington Canyons. The fact that Wilmington Canyon has eroded back into the shelf, whereas South Wilmington Canyon begins on the upper continental slope, suggests that Wilmington Canyon is older. If this is the case, and if these canyons typify many other canyons, we may postulate that canyons tend to begin as straight conduits that show a temporal change toward lower gradients and eventually develop a meander system.

Research on the intensively dissected USA Middle Atlantic continental slope has led Farre *et al.* (1983) to propose a dynamic working hypothesis to explain the evolutionary phases in canyon development (Fig. 6.6). The starting point is localized slope failure and the

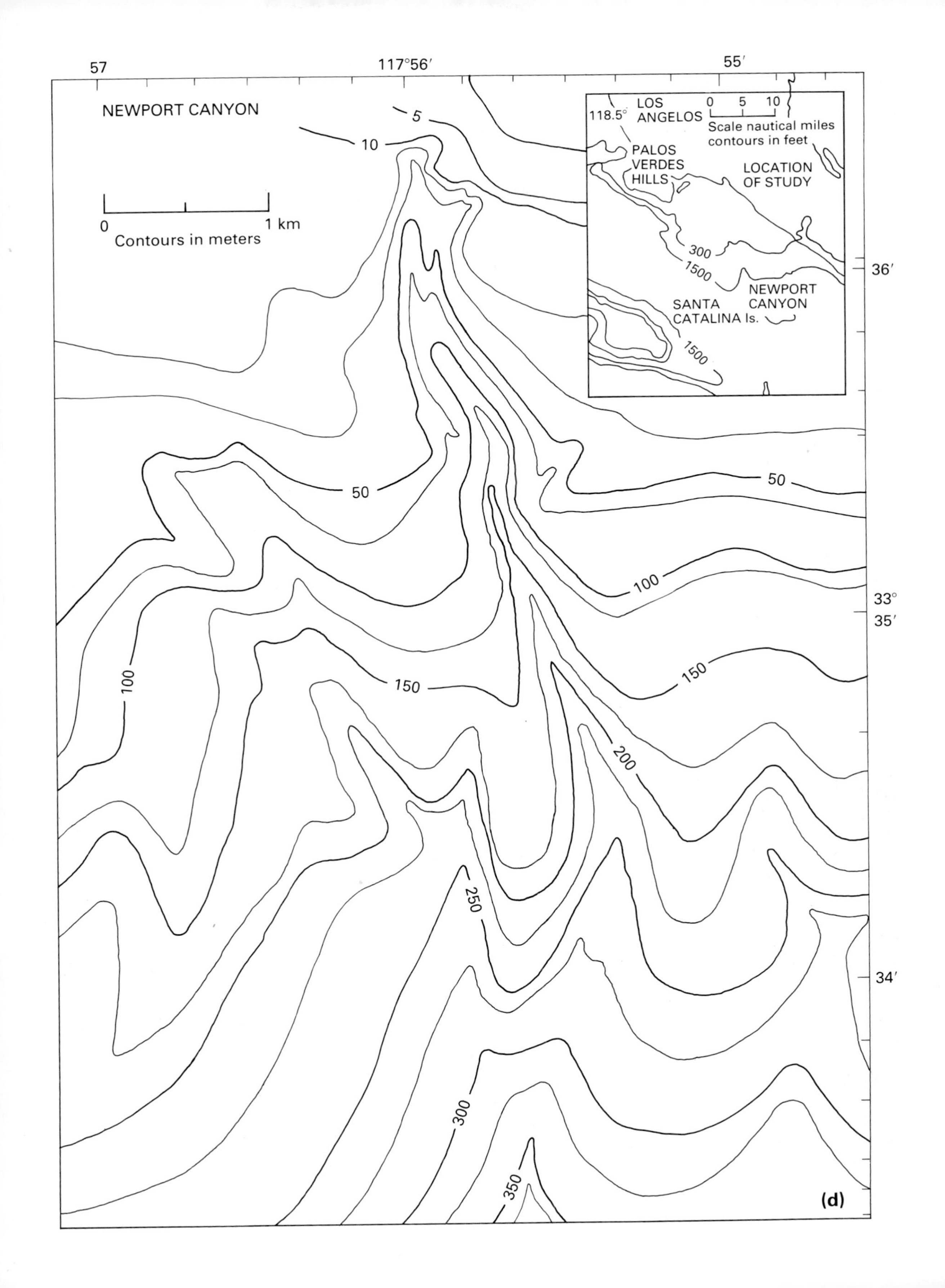

57
117°56′
55′
NEWPORT CANYON
0
1 km
Contours in meters
5
10
50
100
150
200
250
300
350
36′
33°
35′
34′
118.5°
LOS ANGELOS
0 5 10
Scale nautical miles
contours in feet
PALOS VERDES HILLS
LOCATION OF STUDY
300
1500
NEWPORT CANYON
SANTA CATALINA Is.
1500
(d)

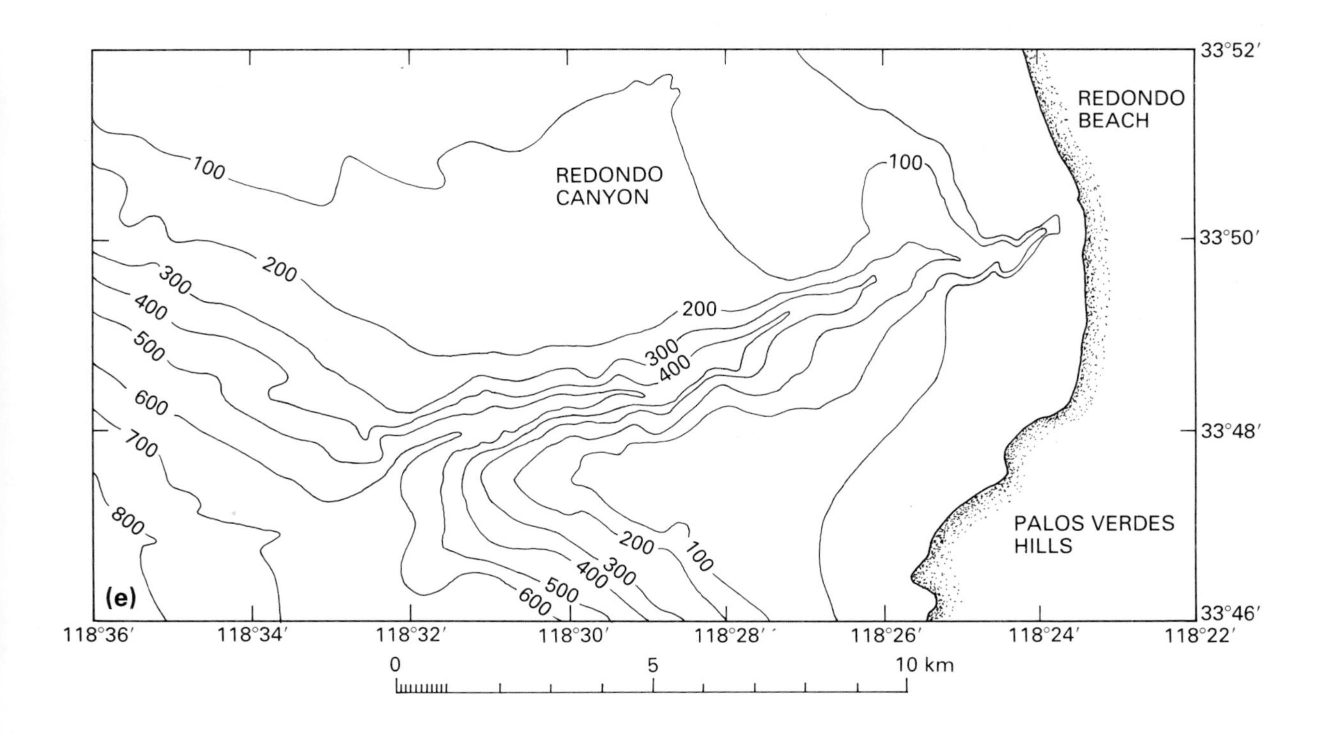

REDONDO CANYON
REDONDO BEACH
PALOS VERDES HILLS
100
200
300
400
500
600
700
800
(e)
33°52′
33°50′
33°48′
33°46′
118°36′
118°34′
118°32′
118°30′
118°28′
118°26′
118°24′
118°22′
0
5
10 km

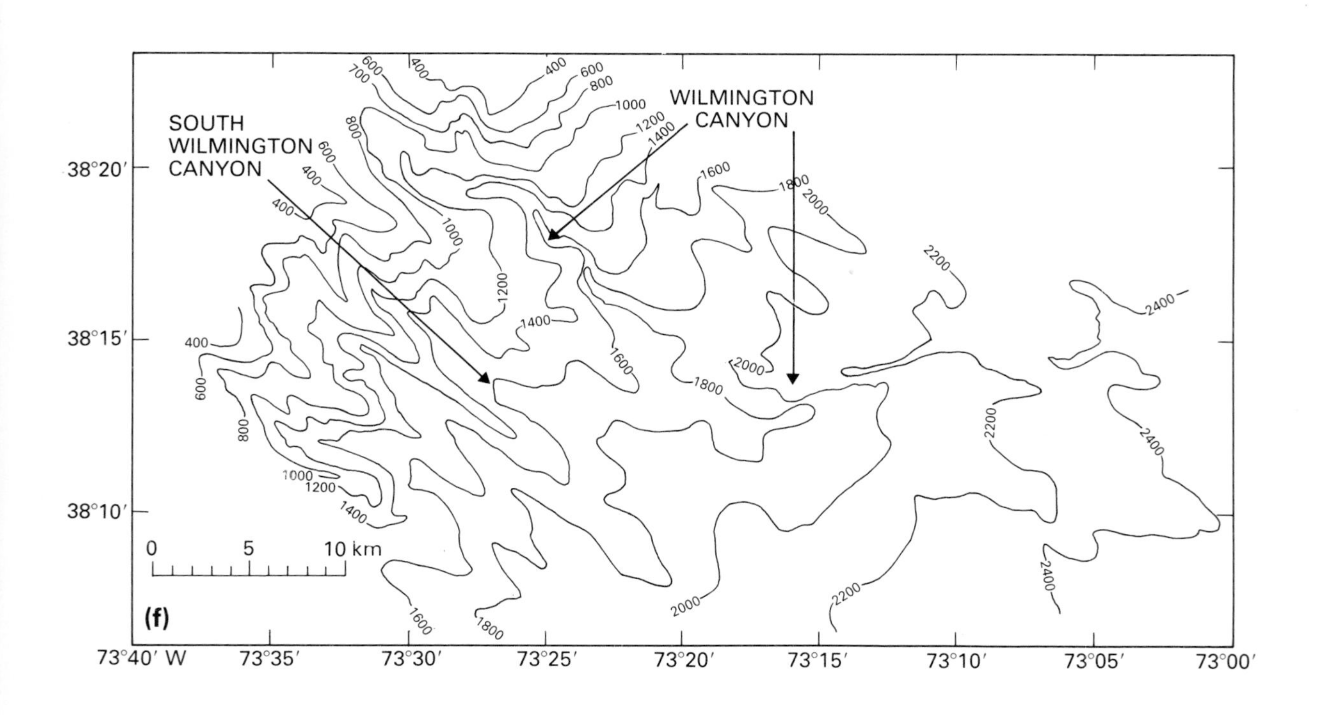

SOUTH WILMINGTON CANYON
WILMINGTON CANYON
400
600
700
800
1000
1200
1400
1600
1800
2000
2200
2400
(f)
38°20′
38°15′
38°10′
73°40′ W
73°35′
73°30′
73°25′
73°20′
73°15′
73°10′
73°05′
73°00′
0
5
10 km

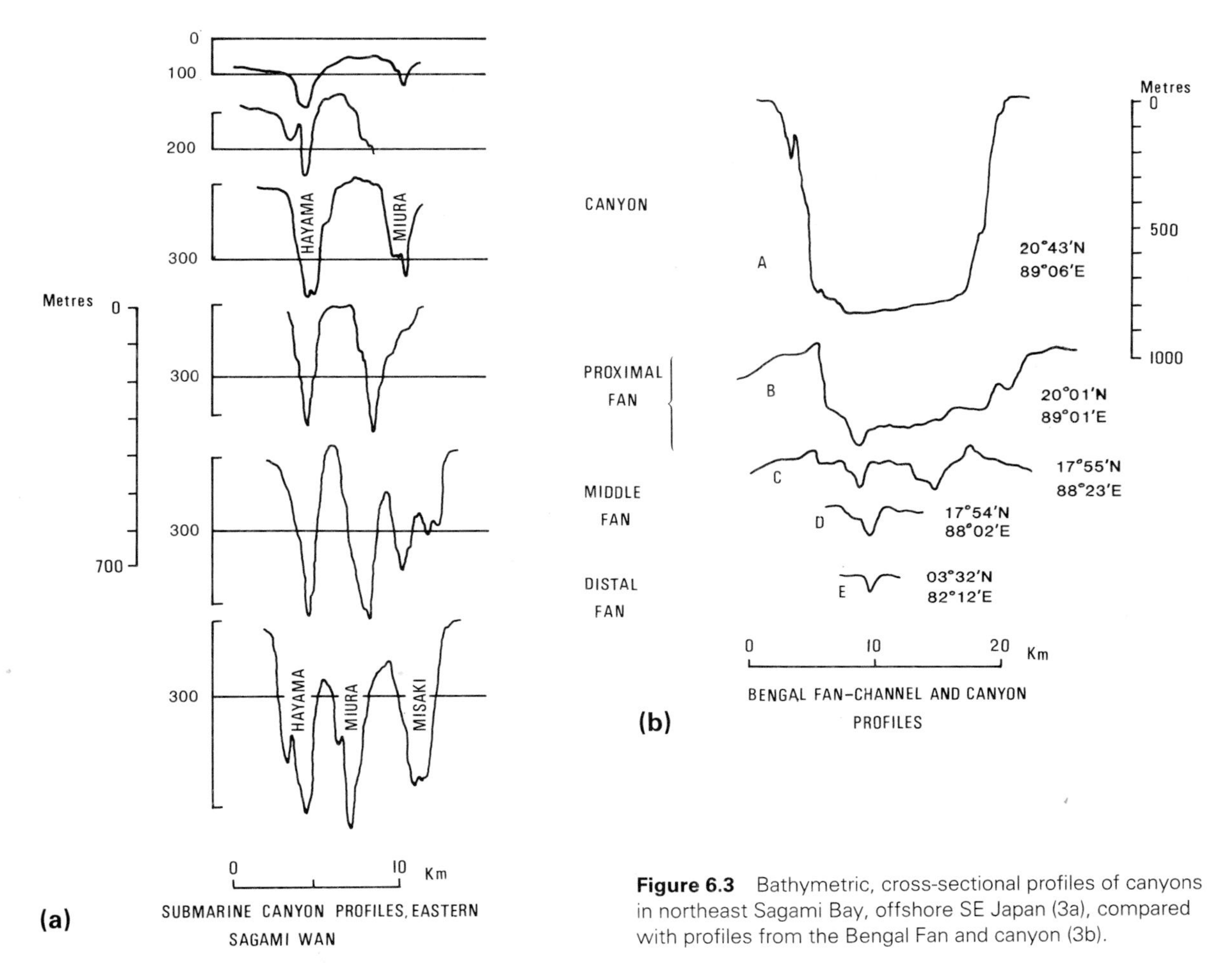

Figure 6.3 Bathymetric, cross-sectional profiles of canyons in northeast Sagami Bay, offshore SE Japan (3a), compared with profiles from the Bengal Fan and canyon (3b).

headward growth of erosional scours and scars by a variety of mechanisms. Changing physical and/or, environmental conditions may modify or halt the initial phase of potential canyon development. Once the valley head breaches the shelf-break, then slope-bypassing of shelf sediments to the deeper water environments may occur, and canyon erosion may be accelerated because of the increased frequency of sediment transport processes within the canyon. Naturally, changing conditions may halt the erosion process at any time.

The most commonly applied methods for studying modern canyons involve shallow seismic and side-scan sonar surveys, together with dredging or box-coring. An example where dredging and seismic profiling have determined the stratigraphy of basin sediments is for Ascension Canyon, offshore Californian margin (Nagel *et al.* 1986) (Fig. 6.7). Ascension Canyon comprises eight heads that notch the continental shelf at the 150 m isobath, coalescing to form a single canyon at the 2700 m isobath (Fig. 6.7). The two NW heads are V-shaped and discrete features, whereas the six SE heads are more diffuse.

The late Pleistocene and Holocene evolution of the six SE canyon heads was controlled largely by tectonic uplift of the Ascension–Monterey basement high at the SE terminus of Outer Santa Cruz Basin. This uplift created slope instability and enhanced canyon erosion. Dredges of Ascension Canyon suggest that the oldest wall rocks, in the deep western part where Ascension Canyon dissects the Santa Cruz High, are Jurassic/Cretaceous to Pliocene serpentinites and mafic volcanics of the Franciscan Complex. In the SE part of the canyon, laminated diatomaceous opaline mudrocks of the upper Middle Miocene were recovered, and correlated with the Monterey Formation (>6.6 Ma) cropping

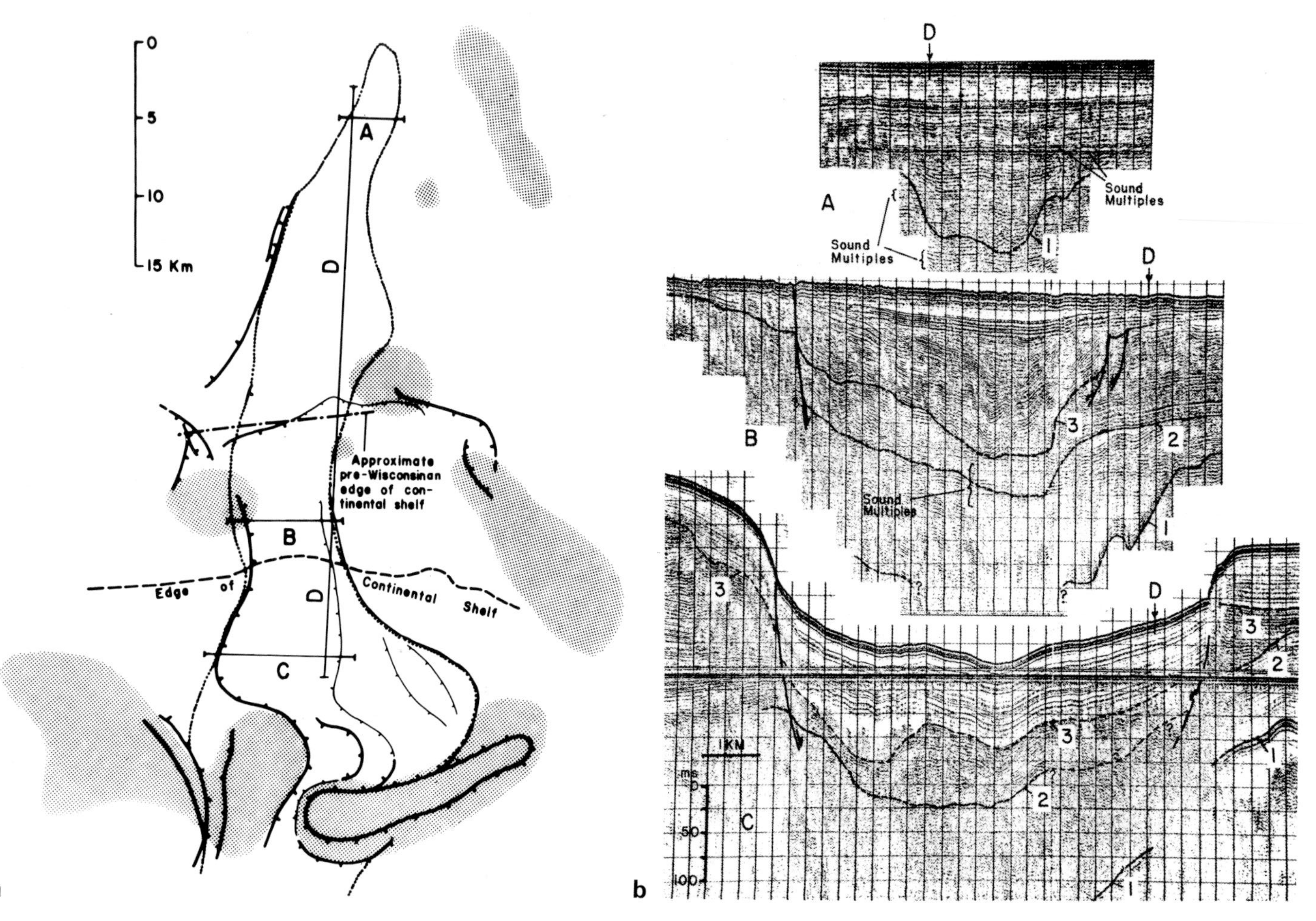

Figure 6.4 **a**, Structural setting of a buried submarine canyon about 100 km south of Atchafalaya Bay, Mississippi Delta, on the edge of the continental shelf. Dots show margins of canyon; stippling indicates diapiric structures; thick unbroken lines are faults that either offset the sea floor or are at depths of less than 5 milliseconds; hachures indicate direction of downthrow on faults. A, B and C indicate location of east–west profiles shown in Fig. 6.4b; **b**, seismic profiles A, B and C located in Fig. 6.4a. Note the fault control on the margins of the submarine canyon. After Berryhill (1981).

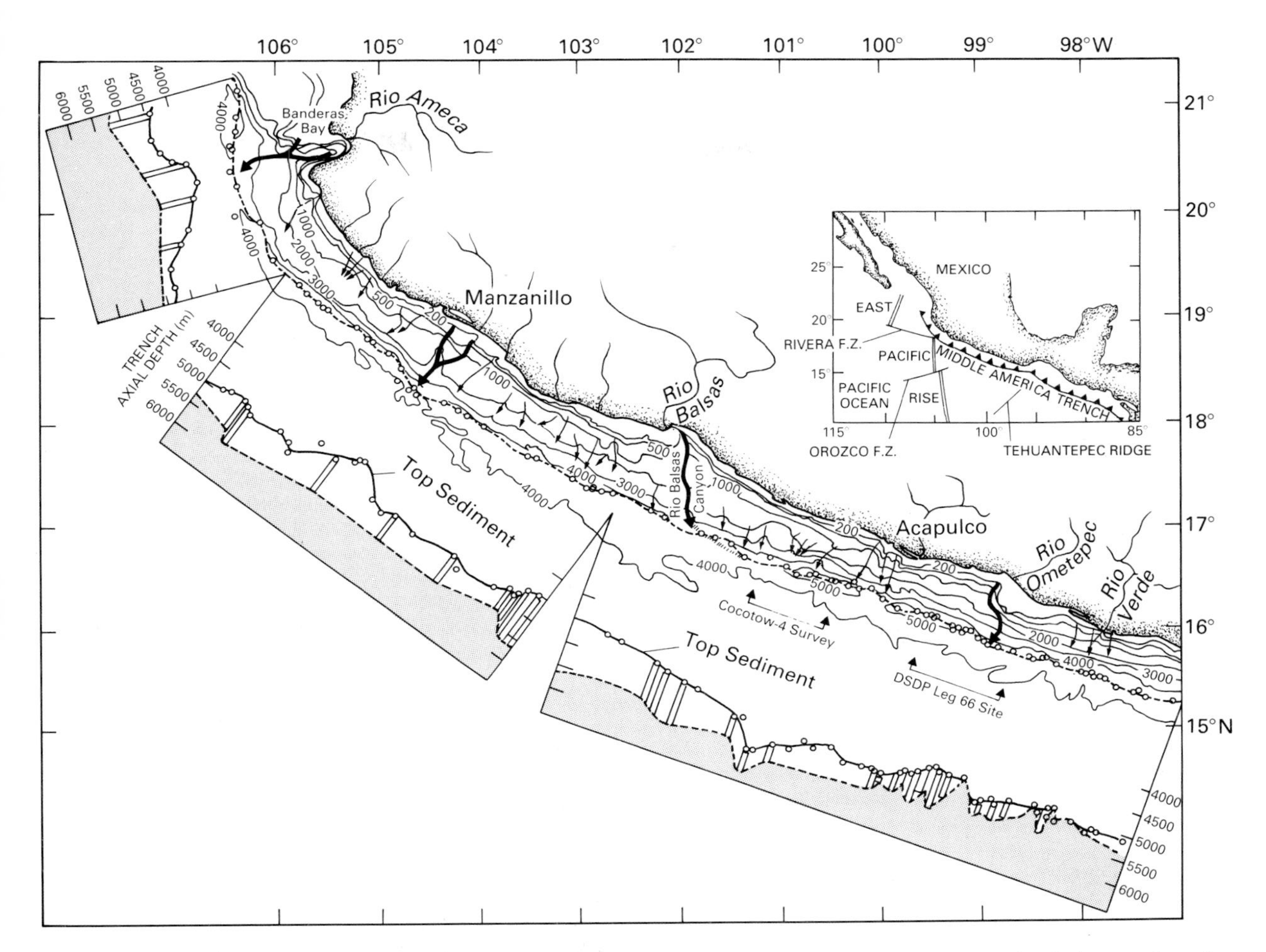

Figure 6.5 Simplified bathymetric map of the Middle America Trench to show the location and spacing of main large and small canyons. Axial depths are in corrected metres, with sediment thicknesses calculated assuming $v = 2.0$ km/s such that a 0.5 s penetration represents 500 m of sediment. After Underwood & Bachman (1982).

out in sea cliffs southwest of the San Gregorio fault zone at Point Ano Nuevo. The most commonly dredged rock types were Pliocene tuffaceous sandstones, siltstones and mudstones, correlated to the onshore Purisima Formation (6.6–2.8 Ma) that unconformably overlies the Monterey Formation. The only other rock type is oil-saturated structureless arkosic sandstones, lacking diagnostic faunas and of uncertain stratigraphic affinity. In addition, unconsolidated Quaternary sediments were recovered (<2.8 Ma).

CURRENTS IN SUBMARINE CANYONS

Current-meter data have been collected in submarine canyons to depths of over 4000 m (Shepard *et al.* 1979). Generally, the currents alternate directions, flowing up and down canyon with periodicities from 15 min to 24 h. The longest recorded unidirectional flows are five days down-canyon, off the Var River, France (Gennesseaux *et al.* 1971), and three days down-canyon in the Hudson Canyon off New York (Cacchione *et al.* 1978), in both cases with variable speeds.

Progressive vector plots of measured current data from many canyons tend to show a net down-canyon flow, although the results from the Eastern Seaboard off the United States show approximately equal amounts of up- and down-canyon flows (Fig. 6.8). The time periods vary over which current speeds change considerably. The periodicity of most currents approximates to semi-diurnal tidal cycles at depths usually

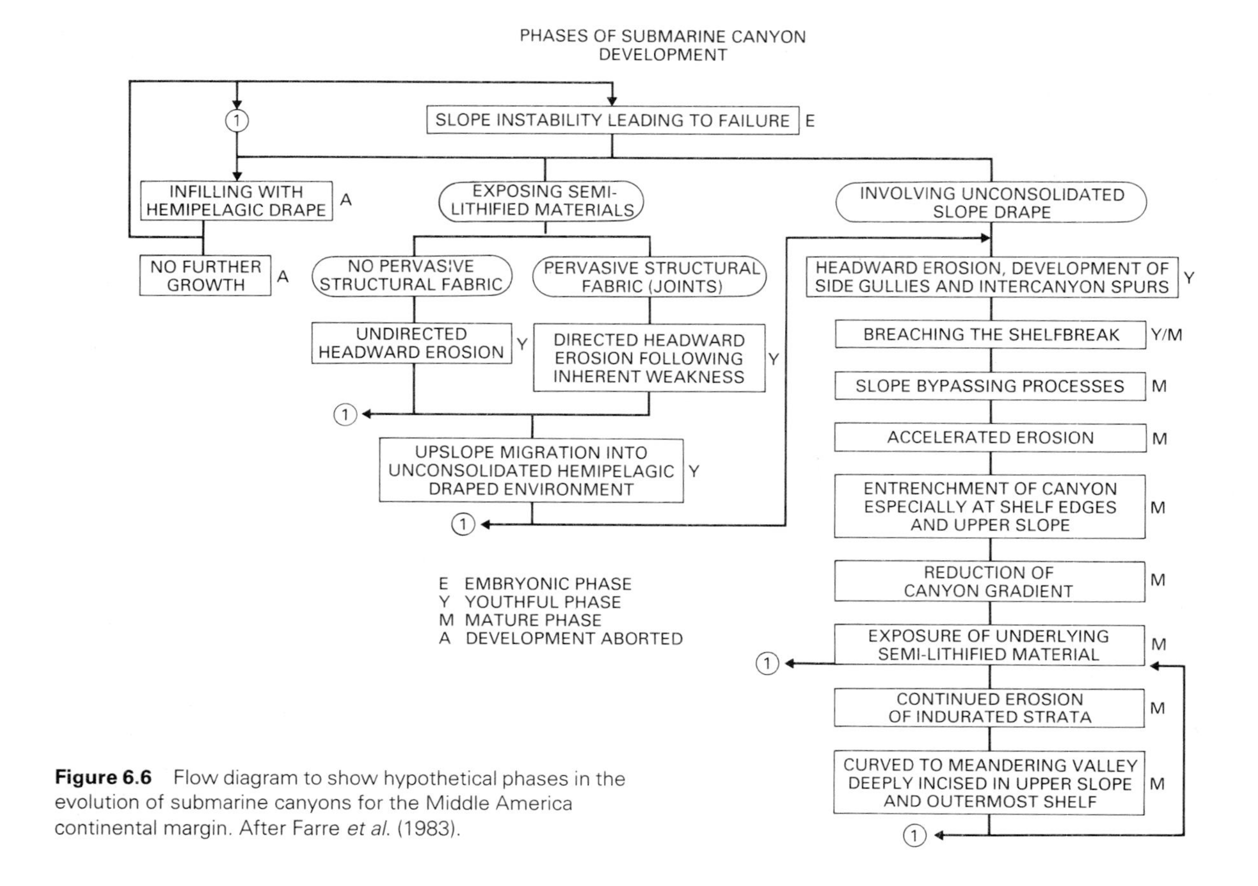

Figure 6.6 Flow diagram to show hypothetical phases in the evolution of submarine canyons for the Middle America continental margin. After Farre *et al.* (1983).

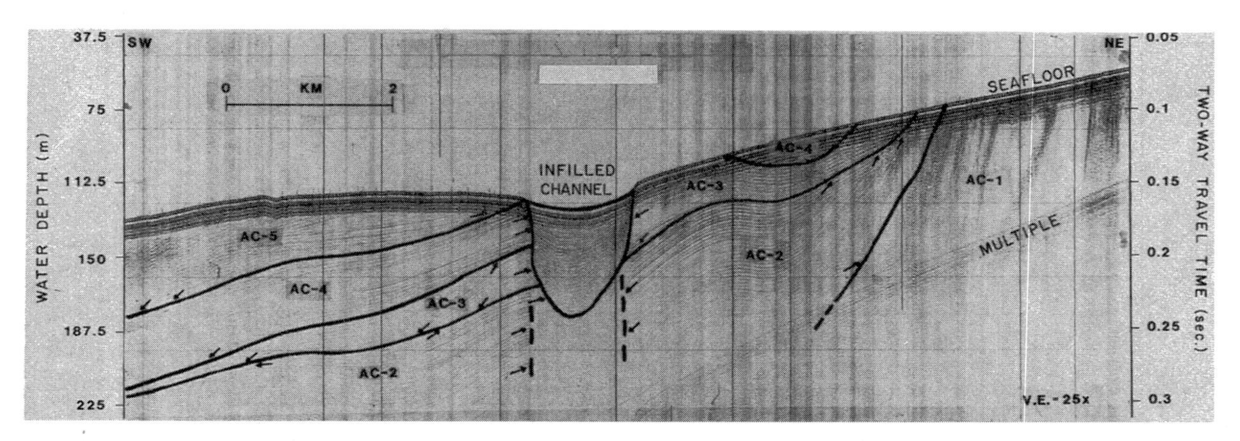

Figure 6.7 1 kJ sparker profile of buried submarine canyon, west of the San Gregoria fault zone and north of Monterey Bay, offshore California Borderland. The table shows the main aspects of the five depositional units, AC–1 to AC–5. AC–5, unconsolidated sediment, Late Pleistocene; AC–4, unconsolidated sediment, Middle Pleistocene; AC–3, unconsolidated sediment, early Pleistocene; AC–2, diatomaceous, tuffaceous siltstones (Purisima Fm.), Late Miocene to Pliocene; AC–1, diatomaceous, opaline mudstones (Monterey Fm.), Middle Miocene to early Late Miocene. After Nagel *et al.* (1986).

greater than about 200 m (e.g. Shepard *et al.* 1974); however, in canyons associated with small tidal ranges, such as off the west coast of Mexico, the length of a canyon–current cycle only approaches the tidal frequency at much greater depths. Figure 6.9, taken from Shepard *et al.* (1979), shows the relationship between the average tidal cycle periods of the up- and down-canyon alternating flows compared with the depth of the canyon axis and also the tidal range. In general, we may conclude that small tidal ranges and shallow depths tend to be associated with short average cycles, whereas large tidal ranges and/or deep water tend to be associated with long average cycles.

Although most currents flow up or down canyon, variations occur (Fig. 6.10) and, in some cases such as in Hueneme Canyon off the Santa Clara Delta, California, there is a considerable spread of flow directions (Fig. 6.10b). Hudson Canyon current data (Fig. 6.10c) apparently bear little or no relationship to the canyon orientation compared to the good agreement shown for Carmel Canyon off California (Fig. 6.10a). Currents that flow at an angle to the 'normal' up- or down-canyon direction are referred to as 'crosscanyon flows' (Shepard & Marshall 1978). Most crosscanyon flows are common to wide-valleyed canyons, for example in the Kaulakahi Channel of northwest Kauai; however, Hudson Canyon data reveal little correlation between the canyon axis and current flow directions (Fig. 6.10c) despite the relatively narrow canyon profile. Strong crosscanyon flows tend to occur at low tide, possibly related to strong wind-driven currents becoming effective at slack low tide. In the Santa Barbara Channel, west of Santa Cruz Island, California, the crosscanyon bottom flows are mainly eastwards, similar to the direction of the surface currents in this area (Shepard & Marshall 1978). The origin of crosscanyon flows is poorly understood at the present time, and Shepard *et al.* (1979) suggest that where canyons have wide floors, canyon currents meander along the canyon axis in a similar manner to the way in which a small subaerial stream meanders in a wide valley.

Current data from some canyons, for example Carmel Canyon, show a temporal effect that is considered to reflect the up-canyon advance of internal waves. Time-velocity patterns of up- and down-canyon flows in Carmel Canyon appear similar when shifted time-wise. This time shift indicates the up-canyon advance of internal waves at shallow depths in canyons such as Carmel, Monterey, Hydrographer and Santa Cruz canyons, and at greater depths in Carmel, San Clemente Rift Valley and Kaulakahi Canyons (Fig. 6.11). In other cases, for example Fraser Sea Valley, Redondo and Santa Cruz Canyons at shallow depths, and Monterey Canyon at greater depths, the internal waves are recorded as advancing down-canyon (Fig. 6.11). Shepard *et al.* (1979) ascribe the down-canyon advance of internal waves in Santa Cruz, Santa Barbara and Rio Balsas Canyons to the introduction of a moving water mass into the head of the canyons.

Where relatively shallow current-meter stations are sited, the evidence suggests a correlation between wind speed and the magnitude and direction of currents within canyons (Fig. 6.12). Pressure waves, preceding a storm, may be responsible for at least some, or part of these current patterns. In other cases, however, there appears to be no correlation; for example, during a storm in La Jolla Canyon with 65 km/h onshore winds, maximum current speeds increased as wind speeds rose, although the up- and down-canyon periodicity did not vary until finally a large down-canyon surge up to 50 cm/s was recorded (Shepard & Marshall 1973a & b). Unfortunately, the current meter was damaged during this surge and, therefore, any additional increase in speed that may have occurred went unrecorded: the meters were retrieved 0.5 km down-canyon partially buried by sediments and kelp. Also, during this storm, and probably contemporaneous with the current surge, a trough with walls 0.5 m high was excavated into the silty sand of the canyon floor. Shepard & Marshall (1973a & b) ascribed the current surge with associated erosional features to the passage of a storm-generated turbidity current flowing down-canyon. Similar down-canyon currents, with velocities up to 190 cm/s, have been reported from the head of Scripps Canyon during an onshore storm (Inman *et al.* 1976), and from other canyons (Gennesseaux *et al.* 1971, Reimnitz 1971, Shepard *et al.* 1975).

The measured current velocities are, at times, sufficient to transport finer grained sediment. If larger grain sizes, such as pebbles, cobbles and boulders, are to be transported within canyons, something that appears to have been the case from the 'lags' in many canyons (Felix & Gorsline 1971, Herzer & Lewis 1979), then catastrophic processes like turbidity currents and debris flows must be an important factor in supplying sediments to canyon systems. The 'ambient' or 'normal' contemporary sedimentation within canyons appears to be mainly the deposition of finer grained suspended matter presumably entrained by the periodic up- and down-canyon currents (Drake *et al.* 1978). While sediment transport by high density, high velocity turbidity currents appears to be unimportant in canyons at the present time, the ancient record suggests that this was not always the case (e.g. Picha 1979). Indeed, the presence of symmetrical, sharp-crested, ripples in muddy sediments at approximately 3540 m water depth in Great Abaco Canyon, Blake Plateau, suggests considerably stronger currents than the maximum 10 cm/s recorded by current meters (Mullins *et al.* 1982). There is, therefore, abundant evidence from modern and ancient, fine to coarse, canyon-floor sediments to show that the present inactivity of submarine canyons is probably unrepresentative of many important sediment-transport mechanisms that operate infrequently (catastrophically) in canyons at the present, or at times of lowered sea level in the past.

SUBMARINE GULLIES

Many submarine canyons show an extensive dendritic drainage pattern especially near the canyon heads. McGregor *et al.* (1982) describe gullies arranged in a pinate pattern around the axes of South Wilmington and North Heyes Canyons, occurring where steep slopes show seaward gradients of up to about 20°. These gullies typically have widths of 75–250 m, depths of 10–20 m and lengths of the order of 1 km. Kenyon *et al.* (1978) describe a deeply canyoned, gullied and wet-sediment deformed section of the continental slope between southwestern Ireland and Spain. They record numerous slide folds with wavelengths of 1.5–2.5 km developed on slopes of 3–5°, bordering the Landes Marginal Plateau, and numerous gullies where the overall gradient is 5–9° west of the Celtic Sea and Armorican Shelf. Thus, it appears that gullies are associated with relatively steep submarine slopes, whereas more gently-inclined slopes appear to be characterized by sediment slides and fewer gullies.

Gullies are not only associated with submarine canyons but also are common features in many delta fronts and prodeltas. Prior *et al.* (1981) have shown that the central and upper parts of a fan-delta in Howe Sound, British Columbia, are characterized by gullies or chutes incised perpendicular to the fan delta: the gullies tend to have widths of 10–30 m (with gentle side slopes) and depths of 3–5 m. One of the most interesting features of the Howe Sound fan-delta is the association of gullies with predominantly coarse-grained deposits, and gradients up to 27°. Prior *et al.* (1981) also described gullies or chutes from Adventfjord, off Spitsbergen, and demonstrated that these gullies act as sediment transfer systems for sediments from the upper delta slopes into deeper water. The source regions appear to be slide scars and not the continuation of seaward-dipping large channels.

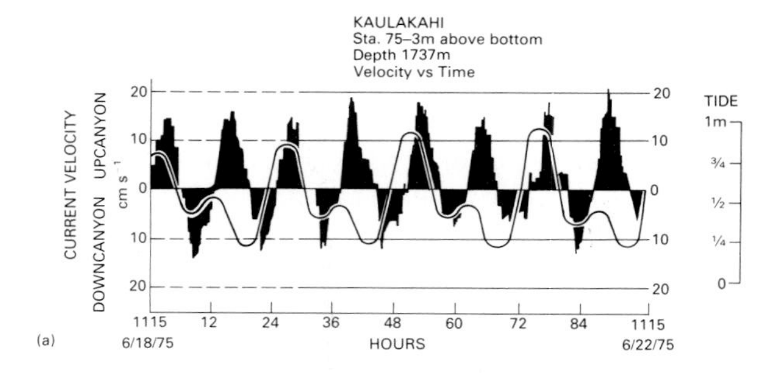

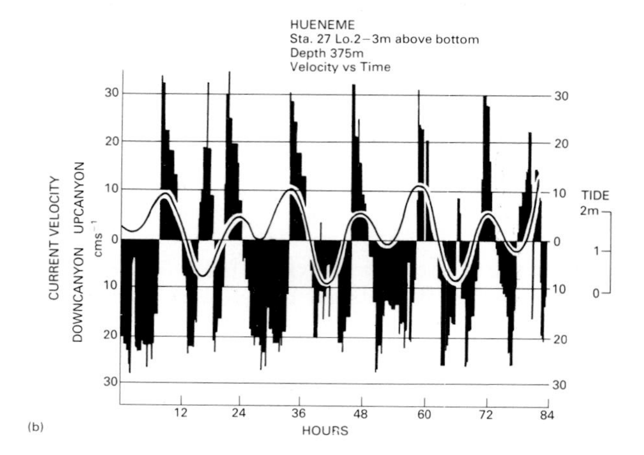

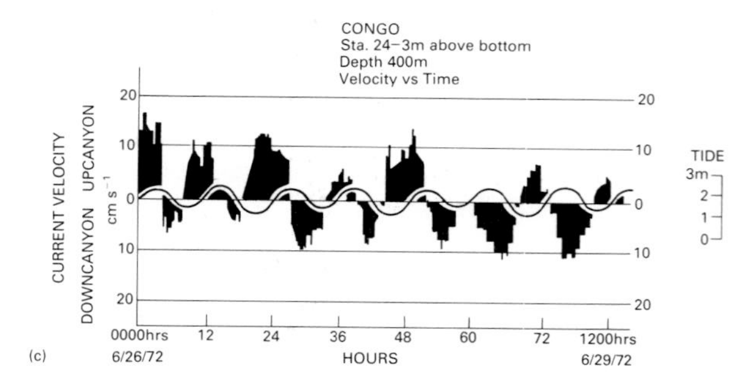

Figure 6.8 Diagrams to show the periodicity of oscillating up- and down-canyon currents. Velocities are corrected for divergence from canyon axis by cosine of angle between flow direction and axis. Tide relationship obtained from the predicted tide at the nearest land station. a, Kaulakahi Canyon, between Kauai and Niihau Islands, Hawaii; b, Hueneme Canyon, offshore CA Borderland; c, Congo Canyon, West Africa; d, Hydrographer Canyon, US continental margin off Massachusetts. Compilation from various sources, after Shepard & Marshall (1978).

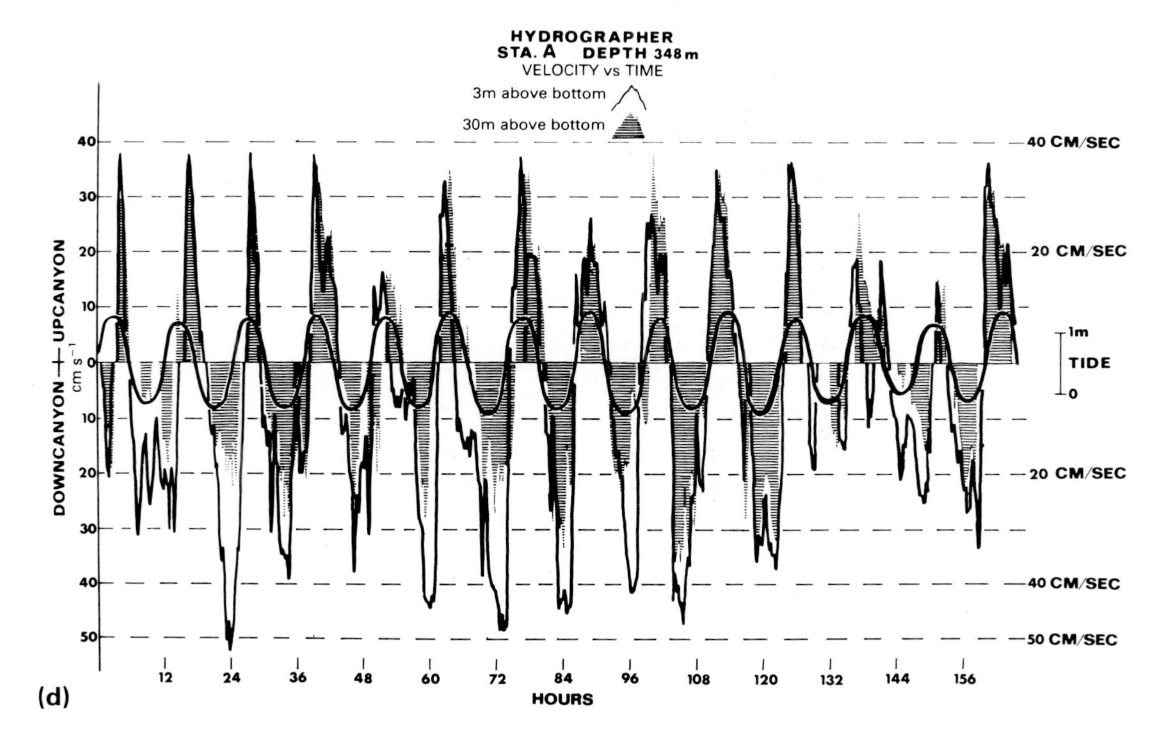

STRUCTURAL/TECTONIC CONTROL

It has been appreciated for some time that the geometry of submarine canyons is strongly controlled by faults and intersecting joint patterns in basement rocks (Prior & Doyle 1985). Submarine canyons ascribed to fault control include Redondo Canyon, offshore California Borderland (Yerkes *et al.* 1967), and many of the canyons along the continental slope between southwest Eire and Spain (Kenyon *et al.* 1978). The NW–SE oriented continental slope on the northern side of the Bay of Biscay tends to display downslope, WSW, or SSE directed canyon axes (Fig. 6.13), with few canyons actually trending directly downslope. Original Mesozoic basement relief, prior to the opening of the Bay of Biscay, is thought to have been responsible for the grouping of these canyon-axis trends. The canyon-axis trends along the north Spanish margin are also believed to be a result of the re-activation of Hercynian NE–SW and NW–SE faults (Fig. 6.13).

In the US Atlantic continental margin, at water depths up to 2450 m in the Carteret Canyon area off New Jersey, Prior & Doyle (1985) describe the geometry of canyons and inter-canyon ridges that are controlled primarily by fracture and joint patterns in the middle Eocene rocks of the lower continental slope. The joints in the Eocene basement rocks consist of two dominant orientations that have fashioned a rectilinear, rectangular, fabric to the entire lower slope. Rockfall has modified the morphology of the canyon walls, leaving the canyon floors strewn with Eocene angular chalk blocks (Facies F1.1). The phase of enhanced rockfall correlates well with the period of maximum upslope extent of the Western Boundary Undercurrent (*c.* 10 000–6000 years BP), when carbonate dissolution in bottom sediments was particularly intense (Prior & Doyle 1985).

Great Abaco Canyon, incised into Blake Plateau, has a relatively straight profile devoid of well developed tributaries and has been ascribed to structural control on its location by the Great Abaco Fracture Zone during pre-Santonian times (Mullins *et al.* 1982). The walls of the canyon are distinctly terraced with vertical to overhanging rock cliffs (mainly of Cretaceous limestones and dolostones) that separate intervening, sediment-covered, slopes. The maximum relief of Great Abaco Canyon is 3385 m, and it extends from about 1300 m depth on Blake Plateau to 4800 m in the Blake–Bahama Basin, over a horizontal distance of 130 km. Mullins *et al* (1982) interpret the canyon history as: (a) faulting; (b) subsidence; (c) scouring by mass wasting, and (d) erosional down-cutting.

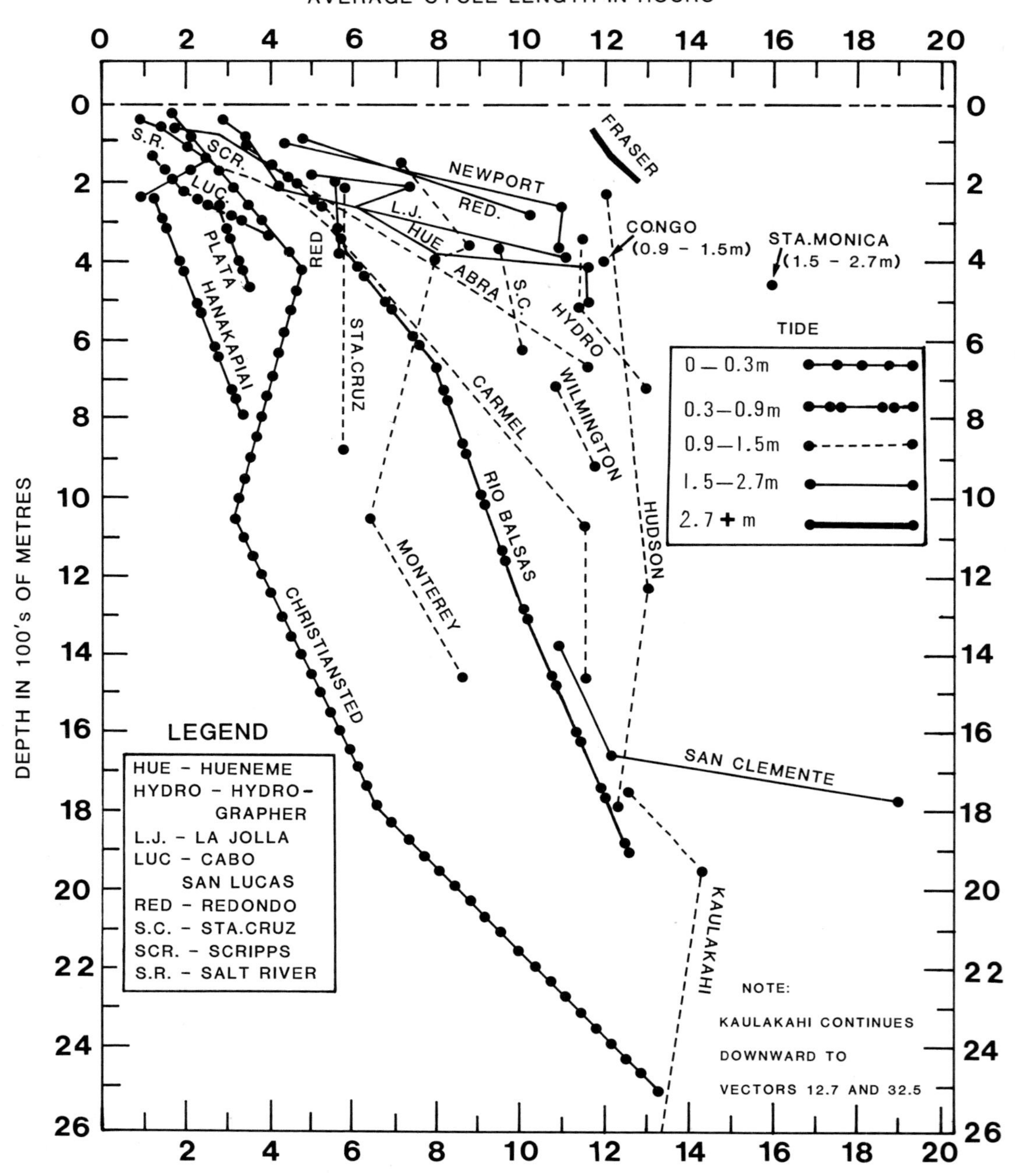

Figure 6.9 Graph to show the relationship between average cycle of the up- and down-canyon currents against the depth of the valley axis. The associated tidal range is shown by the different line legends. Small tidal ranges and relatively shallow depths are generally associated with short average cycles. Large tidal ranges and/or deep water generally correlate with long average cycles of canyon currents. After Shepard *et al.* (1979).

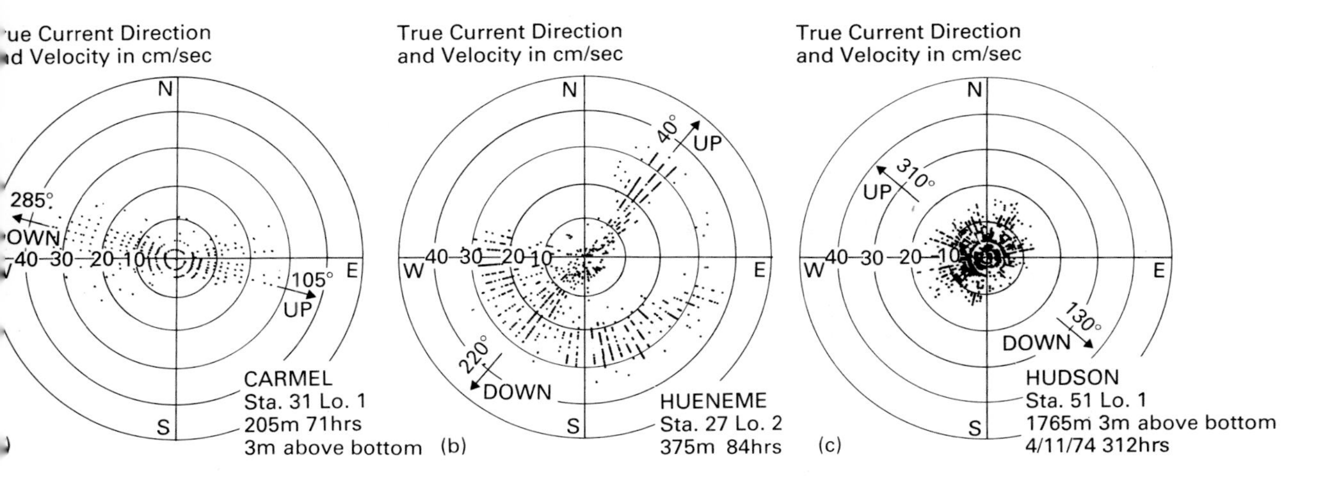

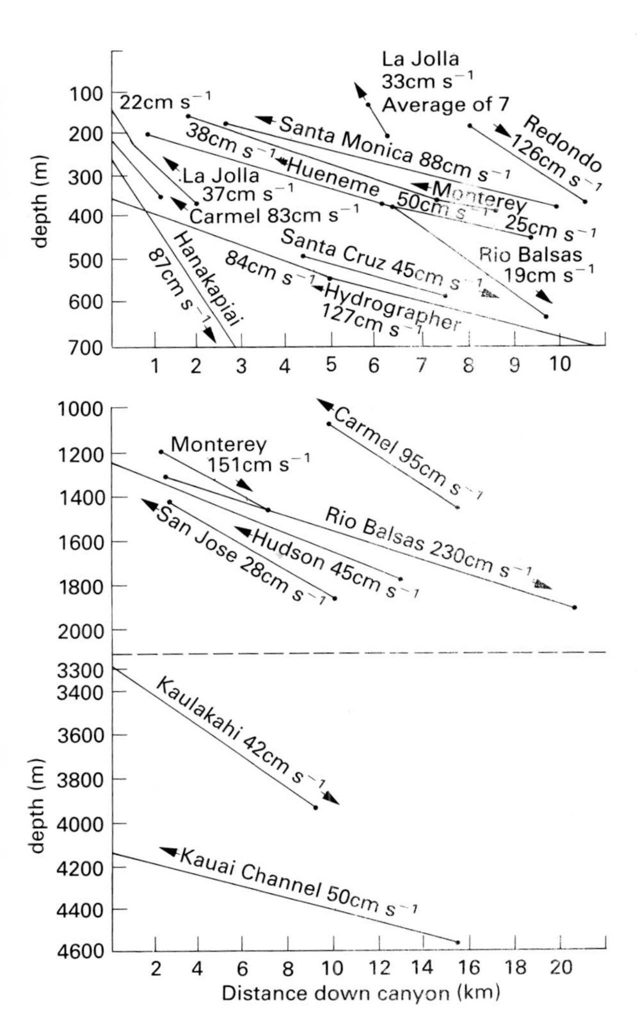

Figure 6.10 Polar plots of canyon currents (after Shepard & Marshall 1978). a, Carmel Canyon, offshore CA Borderland, is an example of currents with flow directions close to that of the canyon axis. There appears to be no simple relationship between the current directions and canyon axis in Hueneme Canyon, offshore CA Borderland b, and Hudson Canyon, off New York, C.

Figure 6.11 Graphs to show the most likely direction and approximate speed of advance of internal waves up and (less frequently) down the axes of various submarine canyons. The speed of advance is approximate due to matching errors between current stations, particularly in cases where the current appears to be up-canyon. After Shepard & Marshall (1978).

The characteristics of fault-bounded and fault-controlled canyon topography are regarded as the presence of undrained depressions on the sea floor, trough-shaped floors and straight, steep walls, and the possible absence of tributary valleys on the sides (Shepard 1981). Presumed fault canyons, displaying many of these features, are well documented from the world's oceans, for example, off the California Borderland and Baja California, off the 'South-western Approaches' to the English Channel, off northern Spain and south of the eastern Aleutian Islands. In the Aleutian examples, however, Gates & Gibson (1956) report some V-shaped canyon cross-sectional profiles, in common with some of the Californian canyons. It seems, therefore, that the shape of transverse canyon profiles may prove to be unreliable in recognizing fault-influenced canyons.

Canyons in oblique-slip mobile zones may be tectonically truncated and offset along major fault

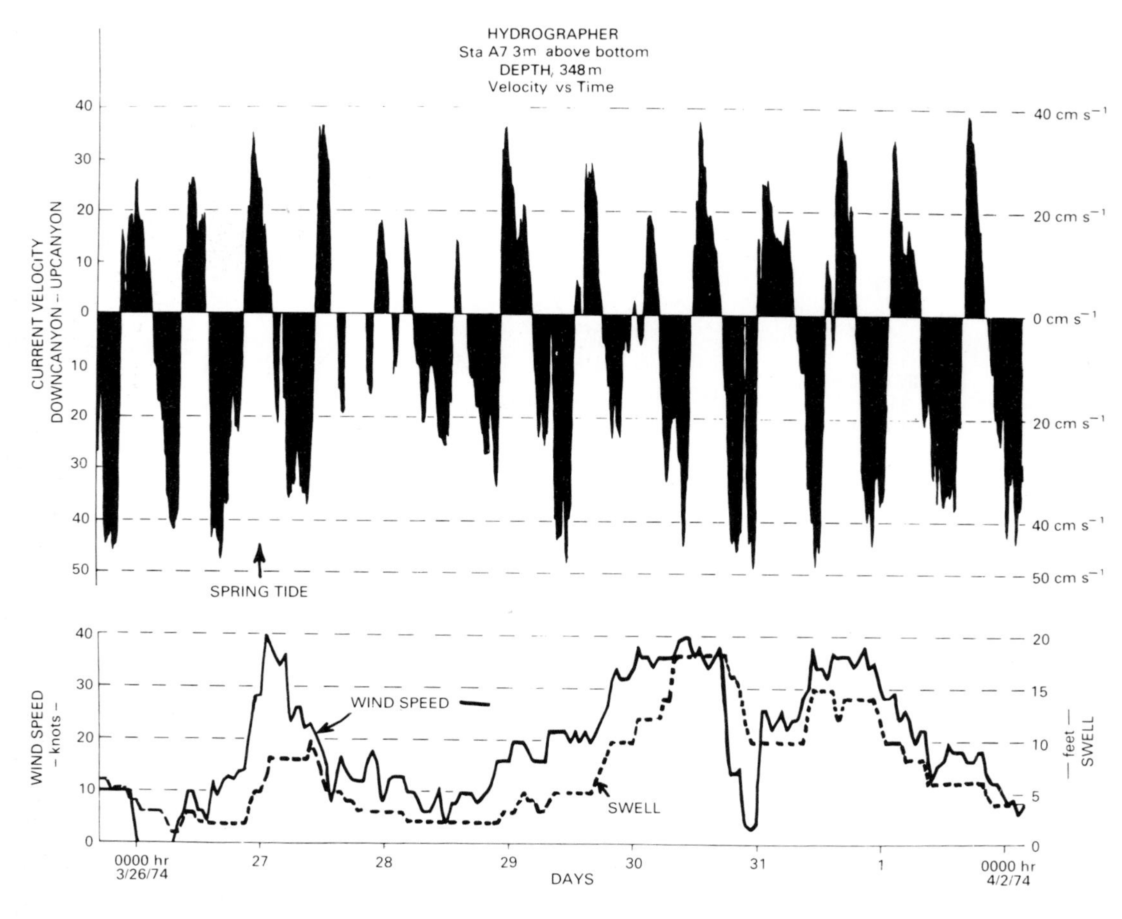

Figure 6.12 Relation of wind speed and height of swell to the magnitude of up- and down-canyon currents during a storm period in Hydrographer Canyon, US continental margin off Massachusetts. The slowest currents occurred during periods of reduced wind speeds and reduced swell. After Shepard & Marshall (1978).

zones. As an example, Nagel *et al.* (1986) document the structural (see Ch. 12) and sedimentological evolution of Ascension Canyon, San Clemente Basin, offshore California Borderland.

While it is difficult to demonstrate that diapirism can be important in determining the location and development of canyons, there is some evidence to suggest that this is indeed the case. The Congo Canyon is associated with salt diapirism (Shepard & Emery 1973); and Figure 6.14 shows that on some of the seismic profiles faulting appears to be related to diapir intrusion, and that the canyon walls are fault-bounded in places. The Congo Canyon, therefore, is an example of canyon development and location that may be, at least in part, related to halokinesis.

SHIFTING LOCUS OF COARSE-GRAINED CLASTIC INPUT

An important factor in canyon development is the change in sedimentation patterns with time. Felix & Gorsline (1971) have shown just how important this can be in the case of Newport Canyon, located off Newport, California.

The origin of Newport Canyon has been ascribed to lateral shifts in: (a) the position of the major clastic source, the Santa Ana River, during the Pleistocene and Holocene, and (b) the point of intersection of longshore drift currents on a narrow shelf. Sediment gravity flows, generated from the spill-over of shelf sands onto the continental slope, are believed to have

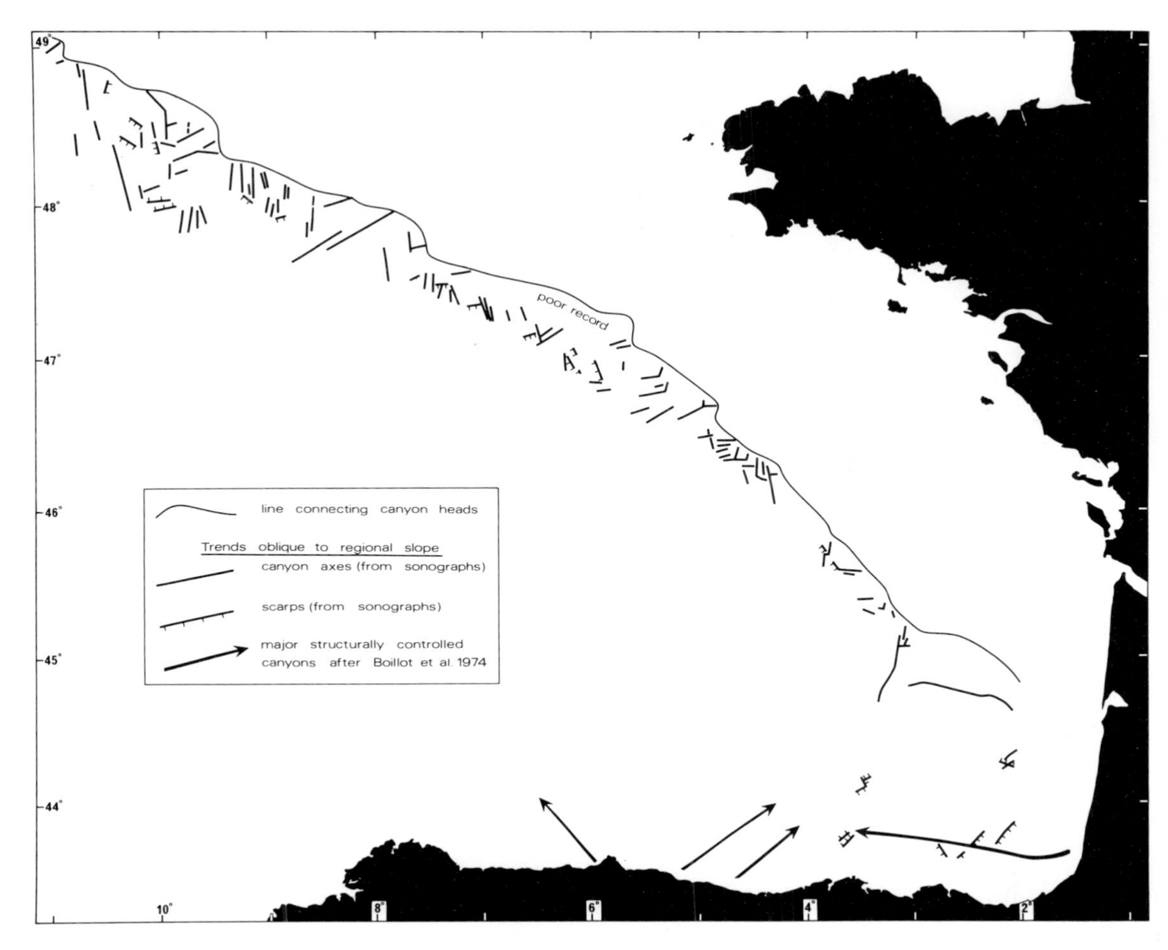

Figure 6.13 Map of the continental slope in the Bay of Biscay to show the preferred alignment of the axis of the main submarine canyons, but excluding the down-slope oriented axes since such directions may not indicate a structural/basement control. The contour along the landward extremity of the canyon heads may indicate offsets in the early continental shelf. After Kenyon *et al.* (1978).

excavated channels that developed into the present day Newport Canyon. Changes in the location of the river mouth have shifted the point of longshore drift convergence and sand concentration about 1 km northwest of the present canyon head, such that only fine-grained, organic-rich sediments are being deposited in the head. Sand that would have entered the canyon, prior to the relocation of the river-input point, is now accumulating as a lobate shelf deposit north of Newport Canyon, and is beginning to spill over the shelf-break, possibly to initiate the excavation of a new canyon.

BIOLOGIC EROSION

Boring, scavenging and browsing organisms are potentially important in the development of some submarine canyons because their activity may lead to sediment instability and failure, with significant erosion of canyon walls. Bio-erosion, for example, is reported from Barrow Canyon in the Beaufort Sea, Arctic Ocean (Eittreim *et al.* 1982). The current circulation patterns on the shelf and at the shelf-break are believed to be important in the development of Barrow

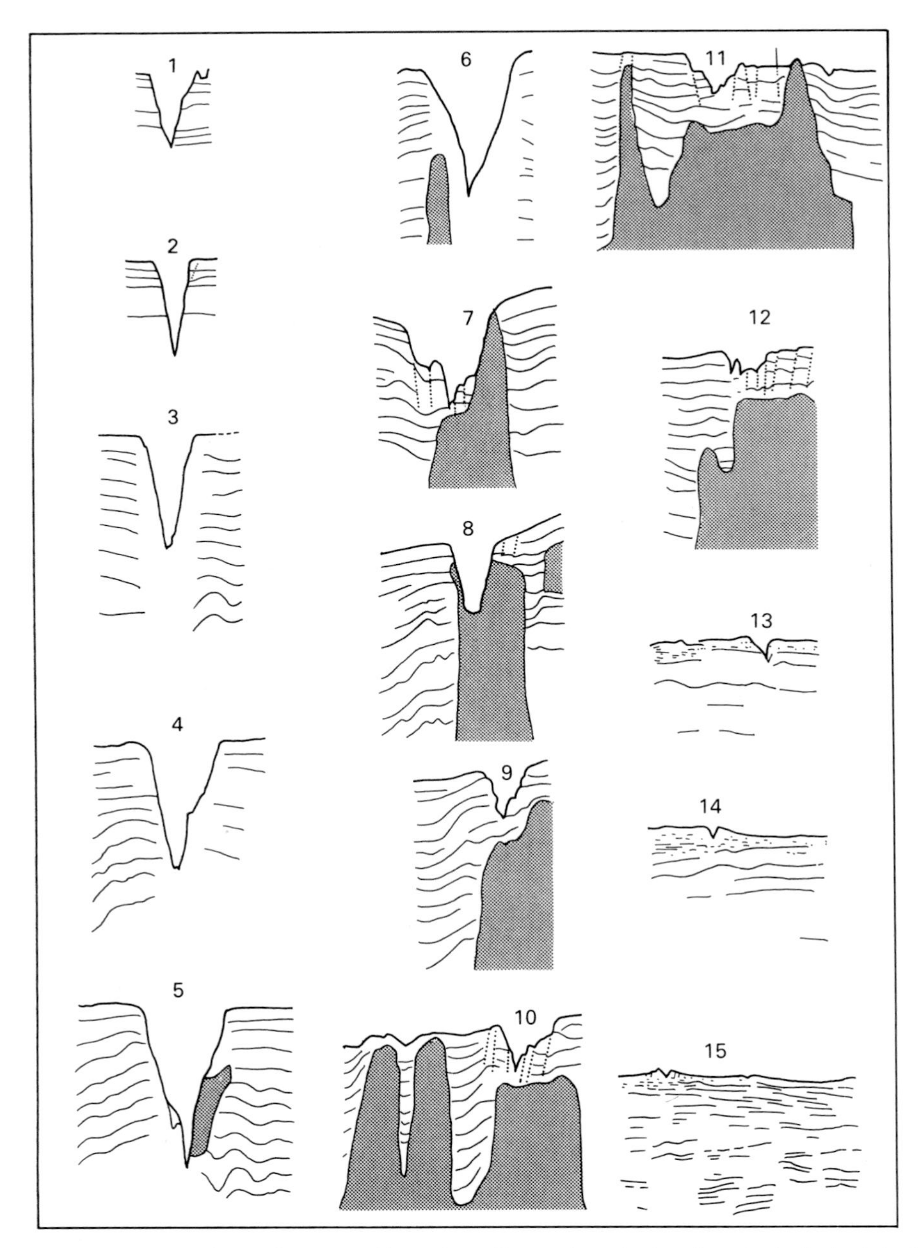

Figure 6.14 Line drawings across Congo Canyon and inner fan valley to show the location of salt diapirs (stipple). After Shepard & Emery (1973).

Canyon, being associated with occasional upwelling events that bring nutrient-rich water up the canyon along the western wall. The upwelling has led to a large population of burrowing organisms with corresponding fast rates of bio-erosion.

CASE STUDY: MODERN SEDIMENTATION, QUINAULT CANYON

The Holocene rise in sea level profoundly altered the nature of submarine canyon sedimentation throughout the world's continental margins such that fine-grained sediments are now preferentially accumulating in canyons. Quinault Canyon, off the coast of Washington State, USA, provides a case study of contemporary sea-level controlled fine-grained canyon sedimentation.

Quinault Canyon (Fig. 6.15a) intercepts Columbia River sediments on the continental shelf of Washington State. The canyon incises the continental shelf and slope approximately 140 km NNW of the mouth of the Columbia River, and extends northwest from the shelf-break in about 200 m water depth to 1800 m where it swings southwest across the wide lower continental slope to end in Cascadia Basin (Fig. 6.15a). Quinault Canyon, with its numerous tributaries, is divided into an upper (200–1600 m depth) and lower canyon (>1600 m) that eventually joins the lower reaches of

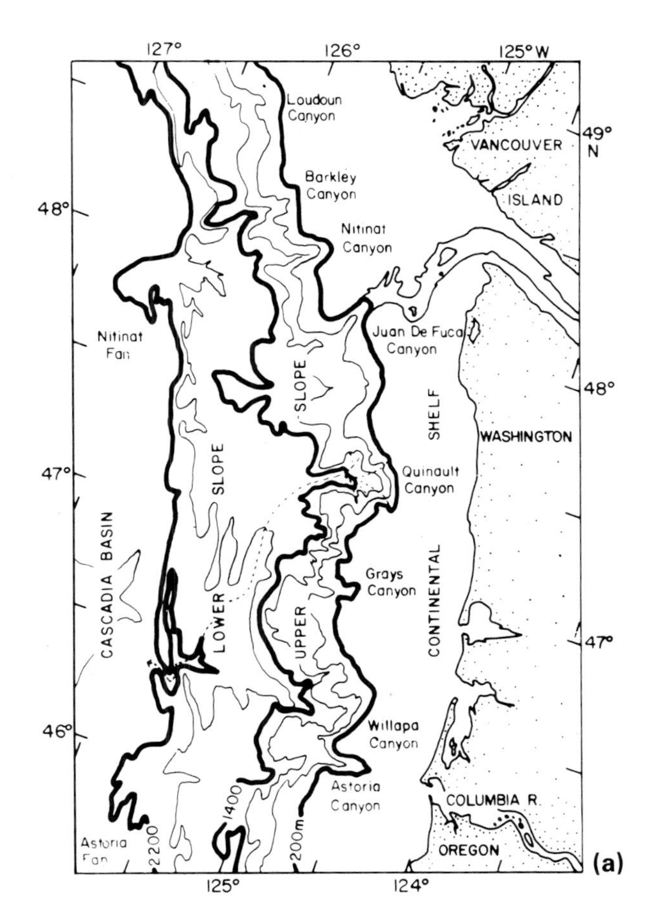

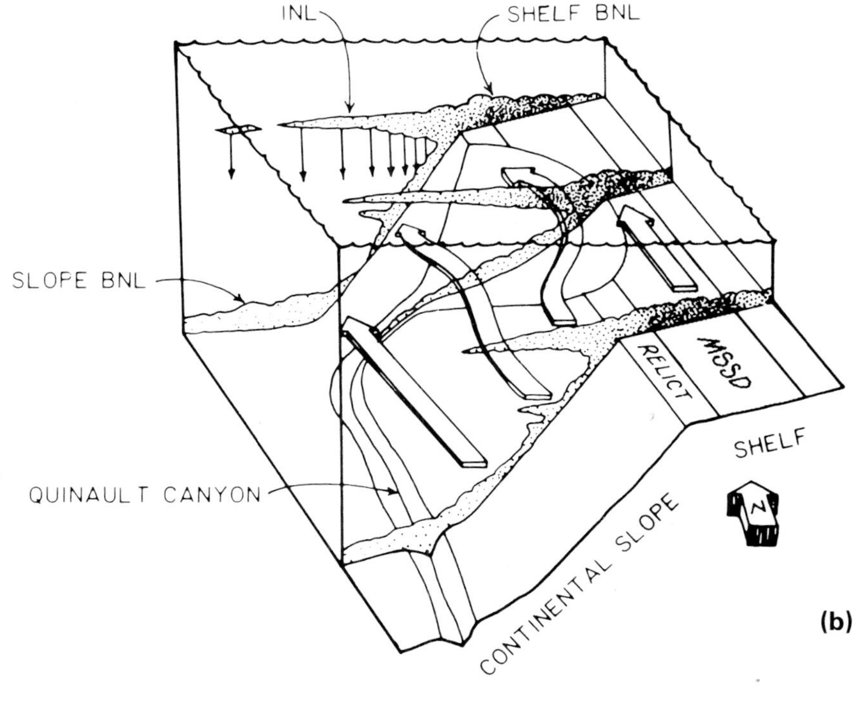

Figure 6.15 a, Location of Quinault Canyon, Washington continental margin (after Thorbjarnarson *et al.* 1986), and b, schematic representation of distribution of suspended particulate matter with regional mid- and bottom-water flow cells (after Carson *et al.* 1986). Shelf sediments intersecting the canyon head consist of relict deposits near the shelf edge, and a zone of modern sandy silts (MSSD) to the east. Sediment deposited in the canyon is derived mainly from the INL (Intermediate Nepheloid Layer), which in turn originates from the shelf BNL (Bottom Nepheloid Layer). Advective transport of suspensates is largely parallel to the shelf edge. Over the canyon head, near-bottom flow is deflected to follow the bottom contours. The deeper, and relatively narrow, segments of the canyon exert a negligible effect on the path of regional currents, with flow being predominantly across the canyon.

Grays and Willapa Canyons (Thorbjarnarson *et al.* 1986). The upper canyon walls are steep, from 6–30°, and they narrow to 3 km apart at about 1600 m water depth, with the axial gradient decreasing to 0.3° at 1800 m depth approximately 30 km seaward of the shelf-break (Carson *et al.* 1986).

The Columbia River is the main source of sediments for the Washington Shelf, with predominant NNW-directed transport paths during the winter storms. Sandy silts are the main sediments. Fine silts (5–9 μm) dominate the canyon deposits, whereas coarse silts (20–40 μm) typify the shelf sediments at the canyon head (Snyder & Carson 1986). Cutshall *et al.* (1986), using ^{60}Co and ^{152}Eu radionucleides, have demonstrated that sediments from the Columbia River are transported into Quinault Canyon, to depths of 1500–2000 m, over periods less than a few years!

Carson *et al.* (1986) have derived a model for contemporary sedimentation in Quinault Canyon (Fig. 6.15b), in which they identify: (a) northward-advected material, resuspended mainly during winter storms, as a bottom nepheloid layer (BNL) that spills over the shelf, and (b) cross-canyon advection cells to produce an intermediate nepheloid layer (INL) that spills over the upper slope and canyon, with rapid settling of particle aggregates. Baker & Hickey (1986) estimate that the settling velocities of these amorphous aggregates in the INL, including faecal pellets, range from 100–200 m/day. In the last hundred years, sediment accumulation rates appear to have been up to four times greater in the canyon compared to the adjacent slope (Carpenter *et al.* 1982). Sediment accumulation rates are greater in the upper canyon because it underlies the INL. Upper canyon sediment accumulation rates, calculated from profiles of excess ^{210}Pb and ^{137}Cs (Thorbjarnarson *et al.* 1986) are about 14×10^{-2} g/cm^2/year, and in the lower canyon about 3×10^{-2} g/cm^2/year: sediment 'focusing' in topographic depressions gives an average canyon-floor value of 12×10^{-2} g/cm^2/year, compared to 5×10^{-2} g/cm^2/year for the canyon walls.

Within Quinault Canyon, the cut-off depth for storm-related resuspension is generally less than 250 m. Changes in particle concentration, within 50 m of the canyon floor, are coupled to shelf-edge processes (Hickey *et al.* 1986). Maximum deep canyon flow is up to 15 cm/s, surging first up- and then down-canyon during storms (Carson *et al.* 1986). Today, there is negligible downslope transport along the canyon axis, local resuspension of canyon sediments, or deposition from particles in the ambient BNL (Baker & Hickey 1986). However, vertical cells (± 100 m) are recorded due to isopycnal adjustments to the quasi-geostrophic along-isobath flow, and as a result of semidiurnal tidal oscillations (Hickey *et al.* 1986).

Ancient submarine canyons

Several well-documented ancient submarine canyons have been reported (for a review, see Whitaker 1974, Stanley & Kelling 1978). Recognition relies upon good lateral and vertical continuity of exposure, together with a clear stratigraphy. The importance of studying ancient submarine canyons is that they can reveal a complete cut and fill history unseen in modern examples, and some canyon fills provide excellent hydrocarbon traps as in the Tertiary of northern California Sacramento Valley (Almgren 1978).

Early well-documented examples of ancient submarine canyons include the six Silurian (Ludlow) canyons of the English Welsh Borderland in which inferred syn-sedimentary faults controlled the location of some canyons. Whitaker's (1962) model (Fig. 6.16) contains many features subsequently identified in modern canyons, and was a very perceptive study that remains applicable to many present studies.

The age of the sediments that the Ludlow canyons were eroded into, when compared to the age of the canyon-infill, suggests that these canyons were excavated and filled over periods up to a few million years. Whitaker (1962) inferred a fault control on the location of at least some of the canyons since they appear to follow the NE–SW Caledonian trend in the British Isles. An interesting aspect of the Ludlow canyons is that the indigenous fauna differs considerably from the adjacent shelf sediments, for example unusual faunas including starfish appear to have thrived within the canyons (Whitaker 1962).

One of the best described ancient canyon systems occurs in the Tertiary of the southern Sacramento Valley, California (Almgren 1978). Three canyons and their infills are recognized: (a) Martinez Canyon, cut in the late early to middle Palaeocene and filled in the early late Palaeocene; (b) Meganos Canyon, cut and filled in latest Palaeocene, and (c) Markley Canyon, cut in the early Oligocene and filled in the late Oligocene to early Miocene (Fig. 6.17). The canyons record three marine cycles of similar history. Each cycle began with a marginal-marine to marine transgressive sand unit, followed by subsidence to bathyal depths, then upward shoaling periods of varying degrees. A fourth and final cycle began with shallow-marine, followed by non-marine, sedimentation. The depositional cycles each

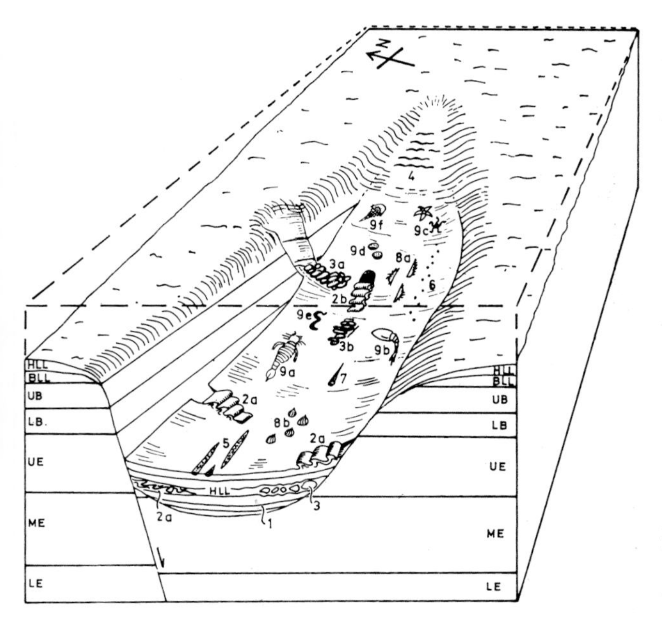

Figure 6.16 Schematic reconstruction of a submarine canyon head in the Silurian (Ludlow) of the Welsh Borderland, UK. Data compiled from six ancient canyons. LE, ME, UE are Lower, Middle and Upper Elton Beds, respectively; LB, UB are Lower and Upper Bringewood Beds, respectively; BLL, HLL are basal and upper Leintwardine Beds, respectively. 1, concave-up bedding or channel fill; 2, sediment slides; 3, boulder beds; 4, ripples with alignment of crests; 5, grooves parallel to axis and filled with disarticulated shells; 6, skip casts parallel to axis; 7, prod casts parallel to axis; 8, oriented fossils (a = *Saetograptus leintwardinensis* parallel with axis; b = *Sphaerirhynchia wilsoni* and *Camarotoechia nucula* with umbones up-channel); 9, unusual fauna concentrated within channel (i = Eurypterida; ii = Phyllocarida; iii, Asterozoa; iv, Echinoidea; v, Annelida, vi, Xiphosura). After Whitaker (1962, 1974).

preceded periods of tectonism and erosion in which the canyons were excavated. The development of these canyons occurred approximately during the period in which Nilsen & Link (1975) postulate no significant right lateral (dextral) displacement along the San Andreas Fault in central and northern California. The main sediment infill of the canyons are shales, clays and fine-grained sandstones of Facies Classes D, E and G.

In contrast to the geometrically well preserved canyons mentioned above, Berryhill (1981) reports a north–south trending Tertiary canyon, crossing the seaward margin of the continental shelf off south-western Louisiana, in which the estimated original width of up to 16 km is now only 7 km because of deformation by salt diapirism. The original canyon length was 90 km, and the maximum eroded depth may have been 305 m. The sedimentary infill is mainly transgressive and regressive deltaic deposits, together with sediment slides. Longitudinal, axial, progradation of sediments predominated along the headward reaches of the canyon, but nearer the shelf edge, filling from the sides became volumetrically more significant.

Other interesting ancient canyons include the Nesvačilka and Vranovice Canyons of the ancient Tethyan margin in Czechoslovakia (Picha 1979). Both canyons have been traced in the subsurface for more than 30 km, with NW–SE trending axes, and towards the southeast they merge into a single large canyon (Fig. 6.18). In their upper reaches, the canyons have widths of 1–3 km with lateral slopes of 30–35°, but farther downslope the widths increase up to about 10 km in Nesvačilka Canyon. The sedimentary fill comprises mainly sandstones and conglomerates of Facies Classes A, B and C in the lower parts and silty mudstones (Facies Classes D and E) in the upper parts. The margins show pebbly mudstones, sediment slides (Facies Class F) and other sediment gravity flow deposits. The mudstones contain from 1–9% organic matter. In Nesvačilka Canyon, the sedimentary fill increases in thickness from 800 m to 1060 m over a canyon axis-parallel distance of 5.5 km. The time taken to infill the canyons is estimated at 10–12 m.y. This represents the final depositional phase in the history of the canyons, and there is little record of the erosional phase. Finally, three distinct foraminiferal faunas are recognized within the canyon deposits, indicating the complex nature of canyon sedimentation, and probably typical of many other canyon fills: (a) indigenous canyon-dwellers; (b) syn-sedimentary faunas flushed from estuaries and the open shelf, and (c) faunas reworked from older sediments.

The nature of canyon sedimentary facies, perhaps, is amongst the most varied of any deep-marine depositional system, emphasizing the need to identify canyons on stratigraphical, structural and palaeontological criteria, with sedimentary characteristics as supportive or simply additional descriptive elements.

Other modern and ancient deep-marine valleys

There are many deep-marine troughs and valleys, or channels, that because of their size or location are not defined as submarine canyons. However, many of their physiographic features are common to documented canyons, such as a linear trend, steep-sided walls and a thalweg channel with essentially axial funnelling of

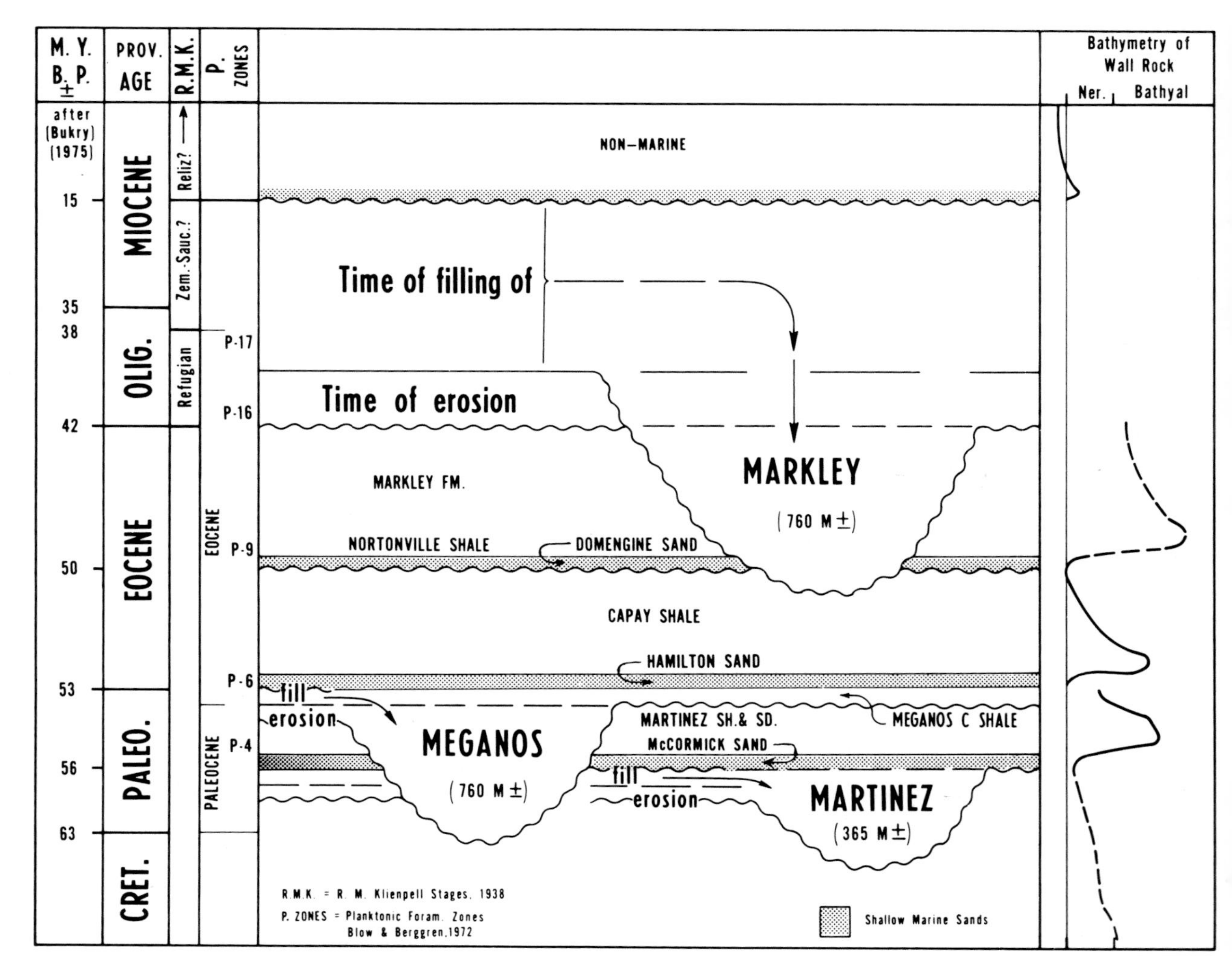

Figure 6.17 Schematic diagram of age relationships for Tertiary canyons in northern California. Periods of erosion, infill and estimated water depths are shown. Thickness of each canyon fill is given. After Almgren (1978).

sediment gravity flows. The larger of these sedimentary systems could be considered as linear basins, but because of the characteristics listed above, they are grouped with submarine canyons.

Carter (1988) has documented the characteristics of the better-studied deep-marine valleys or channels (Table 6.1 & Fig. 6.19), and believes that in many cases their origin can be traced back to the rifting phase in the evolution of passive margins; thus, they may act as sediment conduits for tens of millions of years. Naturally, deep-marine valleys can only remain active and open if the rate of excavation, ultimately governed by the rate of subsidence of the terminus of the channel (its base-line), exceeds the rate of sediment infilling (Carter 1988). Thus, we might predict that in the early stages in the evolution of a passive continental margin, when subsidence greatly exceeds deep-marine sediment accumulation rates, there are enhanced rates of valley development. However, as the subsidence rates exponentially decrease, eventually pelagic and hemipelagic sediment accumulation rates, at 1–10 cm/1000 years, become sufficient to bury the valleys.

Deep-marine valleys, or channels, vary from hundreds of kilometres to approximately 4000 km in length, excavated into continental slope and rise sediments or abyssal plains. Axial gradients, typically less than 1:1000, generally mirror those of the adjacent sea floor. Levees, up to about 80 m high and 5–15 km in

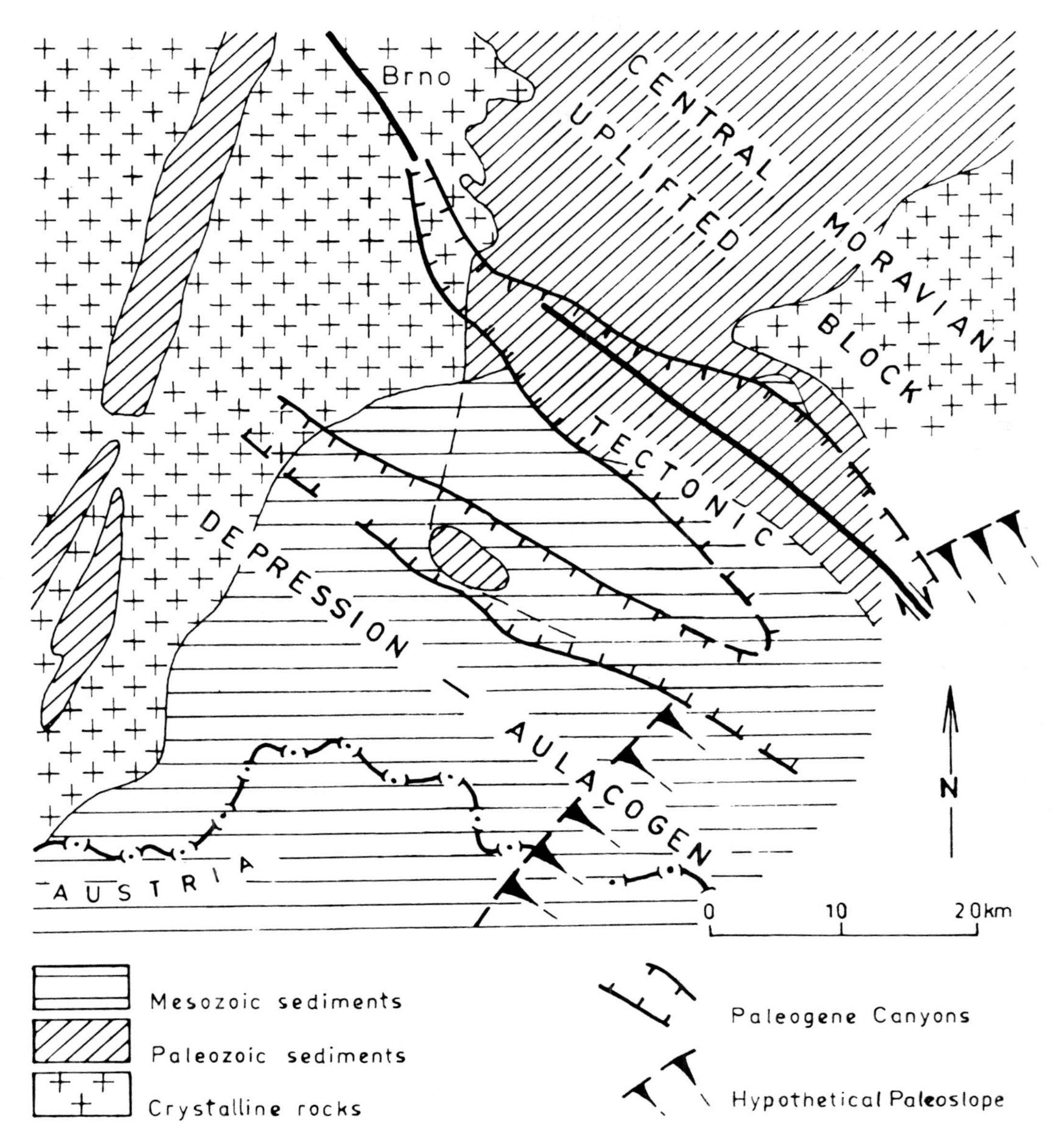

Figure 6.18 Location map of buried Palaeogene submarine valleys below the Neogene of the Carpathian flysch. After Picha (1979).

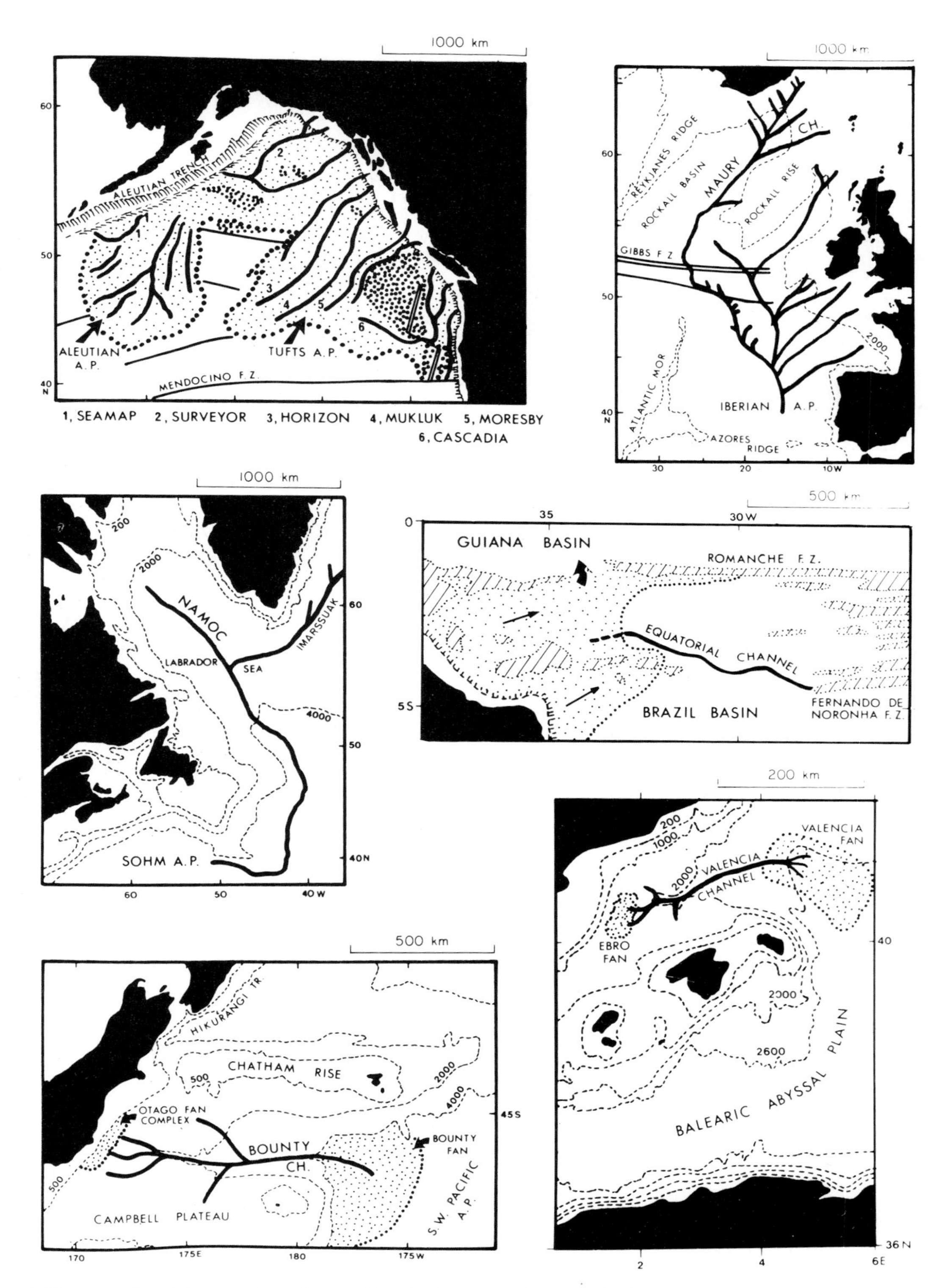

Figure 6.19 Examples of deep-marine valleys and 'non fan' channels. See Table 6.1 for summary of these, and other, valleys. See text for explanation. After Carter (1988).

width, may be asymmetric in scale due to the Coriolis effect; however, in equatorial valleys where the Coriolis effect is minimal, such as in Vidal channel (Embley *et al.*1970), this effect is not observed.

Deep-marine valleys may follow the base of the continental rise, such as Valencia Channel (Fig. 6.19) or, more commonly, develop at large angles to continental margins and even cut through fracture zones. For example, Cascadia Channel cuts through the Blanco Fracture Zone (Wilson *et al.* 1984) and it appears that the fracture zone bypassing occurred approximately 4.5 Ma associated with a major reorientation of the spreading direction in the Juan de Fuca-Blanco-Gorda area (Riddihough 1984, Wilson *et al.* 1984, Wilson 1986).

The Northwest Atlantic Mid-Ocean Channel (NAMOC) was the first deep-sea channel to be described in detail (Ewing *et al.* 1953, Hèezen *et al.* 1969, Egloff & Johnson 1975, Chough & Hesse 1976, Kidd & Roberts 1982), and fed by numerous tributaries, extends from the northern end of the Labrador Sea to the Sohm Abyssal Plain (Fig. 6.19).

The Northwest Atlantic Mid-Ocean Channel (NAMOC), one of the largest deep-marine valleys with a length of 3800 km, is a depositional–erosional feature with extensive natural levees (Chough & Hesse 1976, Hesse & Chough 1980). These levees border the NAMOC throughout its course and pass laterally into flood-plain-like areas either side of the channel (Hesse *et al.* 1987). The levees average 60 m above the adjacent sea floor and extend up to 100 km away from the NAMOC. The right-hand levee is both higher and wider than the left-hand levee (up to 90 m), with a width difference of 5–10 km as a consequence of the Coriolis effect. The channel width is 2–5 km at the floor and 6–19 km at the levee crests. A study by Hesse *et al.* (1987) of the NAMOC has shown that the meandering thalweg channel is associated with submarine 'point bars' of thick-bedded, coarse-grained, turbidites and gravel layers (? Facies Classes A, B, & C). In contrast to these 'channel-fill facies', the 'spill-over facies' comprise thin-bedded, fine-grained, turbidites (Facies Classes D, E). Turbidity current activity in the NAMOC has been strongly controlled by fluctuations in sea level, with a marked decrease about 7000 years BP associated with the rapid deglaciation (Hesse *et al.* 1987).

The Equatorial Mid-Ocean Canyon, located within the Fernando de Noronha Basin off Brazil, is a relict deep ocean channel that was active during the Miocene (Fig. 6.19). This channel is representative of other 'mid-ocean channels' (see also Heezen *et al.* 1959, 1960, 1969, Heezen & Menard 1963, Heezen & Hollister 1971), and in common with others, it is mainly associated with an abyssal province and trends sub-parallel to the adjacent continental margin. The Equatorial Mid-Ocean Canyon is continuous for at least 1275 km, is 5–8 km wide and up to 200 m deep with an axial gradient of 1:1000. Subsurface mapping of the canyon shows a 75–150 km buried portion extending onto the continental rise. The canyon has natural levees, with the north or left levee being up to 25 m higher than the southern levee, this being attributed to the Coriolis effect.

Piston core data suggest that the canyon has been inactive for at least the late Quaternary and that the system is now being buried (Damuth & Gorini 1976). Damuth & Gorini (1976) show that the Equatorial Mid-Ocean Canyon has a trend perpendicular to the continental margin in its buried sourceward reaches, and that, contrary to earlier interpretations (Heezen *et al.* 1960), its origin probably was not tectonically controlled. While the canyon path is structurally constrained, the canyon was eroded by sediment gravity flows from the Brazilian continental shelf. It was abandoned and filled when the main direction of sediment dispersal shifted from southeastward along the Fernando de Noronha Basin to northeastward into the adjacent Guiana Basin, thereby altering the principal sediment-transport paths in the region.

Channel margin deposits from the late Precambrian Kongsfjord Formation submarine fan, Finnmark, North Norway. Cliff section about 15 m high, SE Kongsfjord east of Veines. Note normal bedding to right and lateral migration of channel suggested by progressive cut-and-fills to left or north (cf. Pickering 1982a).

CHAPTER 7

Submarine fans

Introduction

Submarine fans have kindled more interest than any other deep-marine environment because the sands and gravels deposited on fans have the potential to act as reservoirs for hydrocarbons (Walker 1978, Stow 1985). Modern fans were evident on early bathymetric maps of the ocean-basin margins (Menard 1955, Gorsline & Emery 1959, Shepard & Einsele 1962, Hand & Emery 1964). Most early studies dealt mainly with gross morphology, but in some cases closely-spaced bathymetric and high-resolution seismic profiles allowed recognition of distinctive morphologic features, including channel patterns. An example is the pioneering study of Redondo Fan, California, by Haner (1971), who recognized sinuous and braided channel patterns, provided data on channel dimensions for channels as shallow as a few metres, and illustrated down-fan changes in Bouma divisions of turbidites using box-cores.

Sophisticated studies of modern fans using deep-tow instruments, including side-scan sonar, began with the work of Normark (1970), with similar techniques subsequently applied to numerous modern fans worldwide (Barnes & Normark 1984). These fans can, in most cases, be divided into (a) an *upper fan* just downslope from the feeder channel, or canyon, and characterized by a single deep channel with levees, (b) a *middle fan* characterized by multiple distributary channels (most inactive) and relatively smooth depositional lobes with shallow channels and relatively high sand content, and (c) a *lower fan* with few if any channels, and a smooth surface with low gradient (Fig. 7.1). Studies of ancient fans commonly use the subdivisions *inner*, *middle* and *outer* fan, rather than upper, middle and lower fan. Care must be exercised in equating these terms, because some ancient fan models (e.g. Mutti & Ricci Lucchi 1972) apply the terms middle and outer fan to regions that all fall within the middle fan as defined by Normark (1970). The lower fan of Normark (1970) would be classified as part of the basin plain in the scheme of Mutti & Ricci Lucchi (1972).

The earliest recognition of ancient submarine fans was based on facies associations, presence of channels, and palaeocurrent maps (Sullwold 1960, Jacka *et al.*

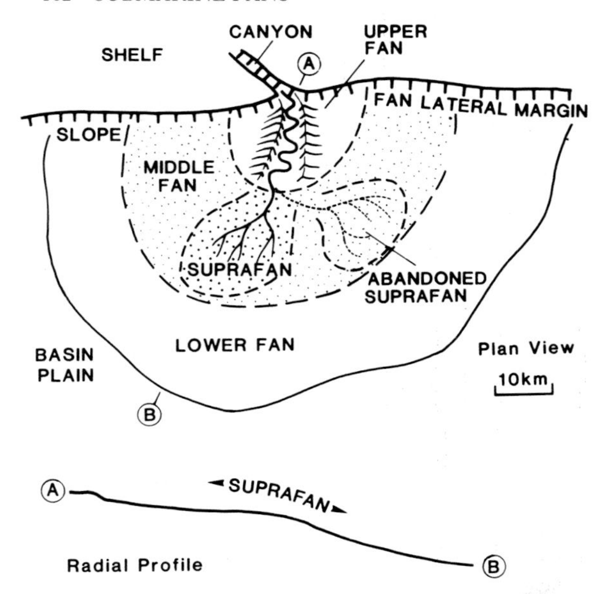

Figure 7.1 Submarine fan model of Normark (1970, 1978) emphasizing growth by successive addition of suprafan lobes on the middle fan. The middle fan region is stippled. The fan lateral margin was not specified by Normark.

1968). In general, however, ancient submarine fans cannot be recognized on the basis of radial palaeocurrents and shape, but instead must have their origin deduced from facies associations and vertical sequences of facies, as first proposed by Mutti & Ghibaudo (1972) and Mutti & Ricci Lucchi (1972). According to the latter authors, submarine fans consist of (a) an *inner fan* characterized by conglomerate and coarse sandstone facies (our Facies Classes A and B) in large channels cut into fine-grained deposits (our Facies Class E); (b) a *middle fan* consisting of packets of sandstone and minor amounts of conglomerate (Facies Classes A and B) in thinning- and fining-upward sequences, alternating with packets dominated by Facies Classes C, D and E, and (c) an *outer fan* with few or no channels, and parallel-sided turbidites arranged in thickening- and coarsening-upward sequences (Fig. 7.2).

The models proposed for modern and ancient fans in the 1970s are currently in a state of flux. Modern fans show a high degree of variability in morphology and scale (Pickering 1982c, Barnes & Normark 1984, Stow *et al.* 1984), and there seems to be little justification for erecting a universal model. Our understanding of the processes responsible for deposition on modern fans are continually being modified as a result of (a) side-scan sonar imagery, particularly imagery of meandering and braided channels (Garrison *et al.* 1982, Damuth *et al.* 1983a, Belderson *et al.* 1984, Kastens & Shor 1985), and (b) deep drilling, to date concentrated on the Mississippi Fan (Bouma, Coleman *et al.* 1986). As an example, the uppermost 100–150 m of channel deposits drilled on the Mississippi Fan (Pickering *et al.* 1986b) consist predominantly of silt-, mud- and clay-rich, very thin/thin-bedded sediments (Figs 7.3 & 10.22), remarkably like facies that have been interpreted as interchannel or levee deposits in the ancient record.

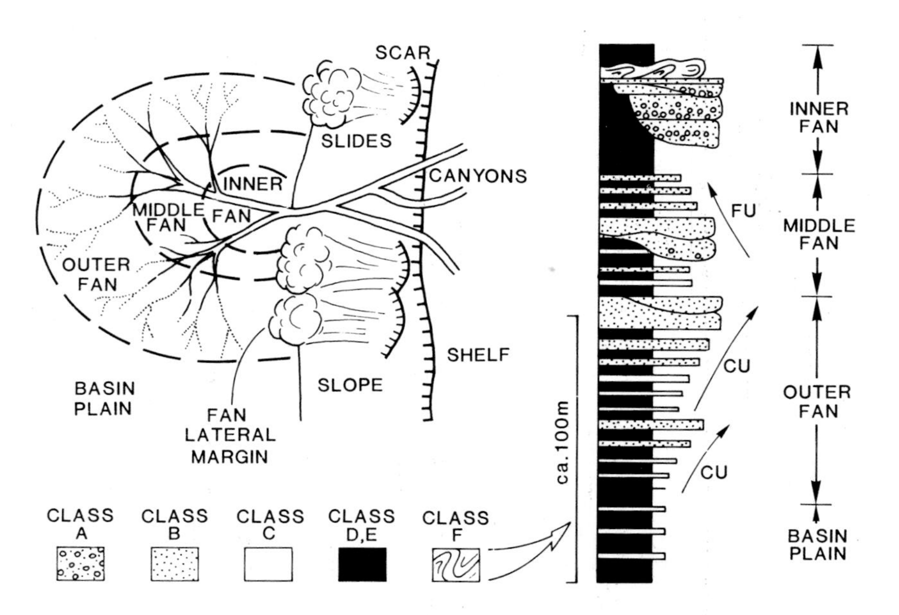

Figure 7.2 Submarine fan model of Mutti & Ricci Lucchi (1972), based on relationships observed in ancient deposits. Facies classes are those defined in Chapter 3, not those of Mutti & Ricci Lucchi (1972). The fan lateral margin was not specified by Mutti and Ricci Lucchi.

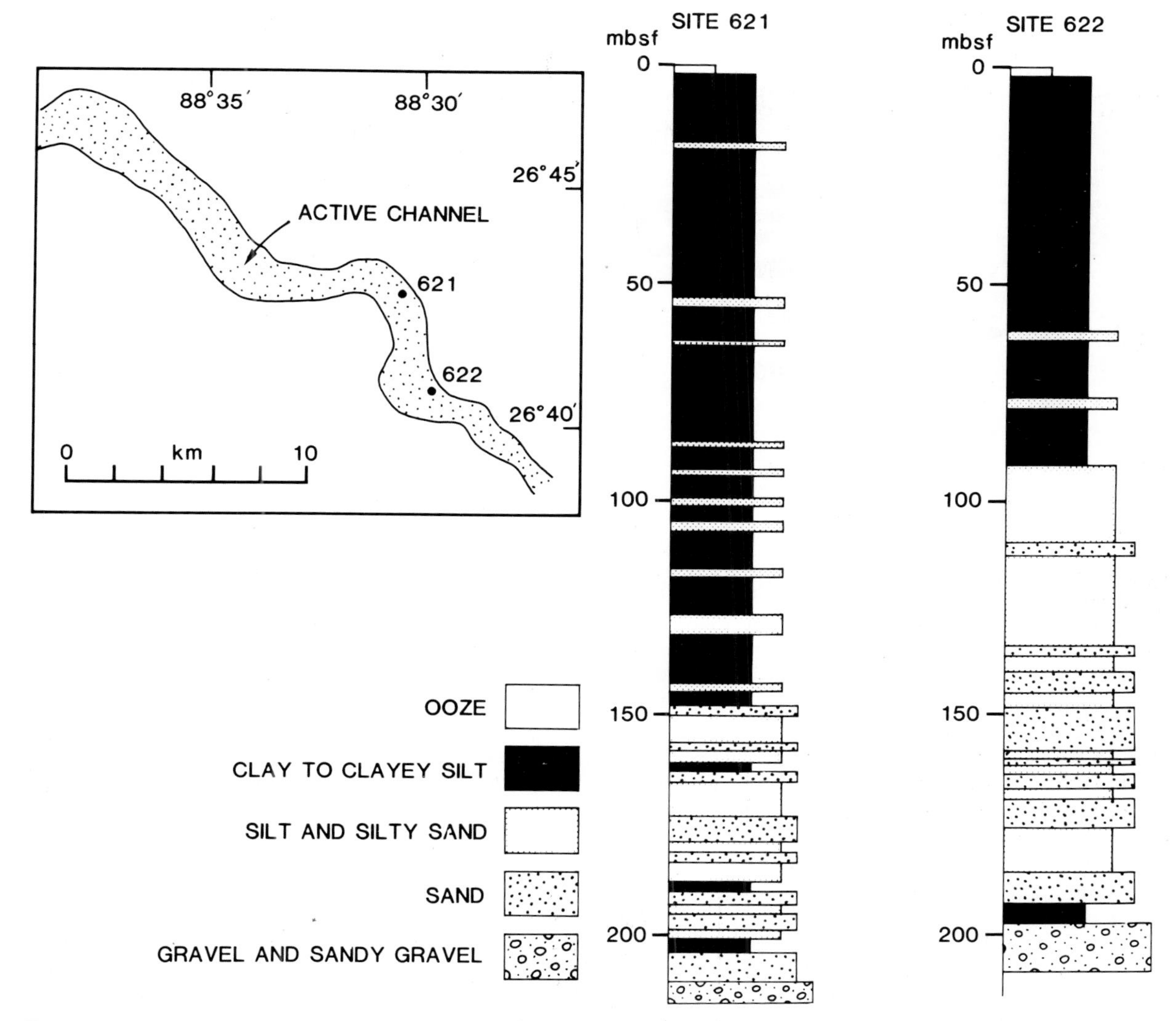

Figure 7.3 Interpretive lithostratigraphy of DSDP Leg 96 sites 621 and 622 from the meandering channel of a channel–levee–overbank system on the Mississippi Fan. Inset shows location of the sites. mbsf = metres below sea floor.

Ancient fans are also seen to be more complex than suggested by models of Mutti & Ricci Lucchi (1972), Ricci Lucchi (1975b), and Walker (1978). Facies sequences reflect a complex interaction between sediment texture, rate of sediment supply, tectonic regime, sea-level rises and falls, and fan processes like channel switching, mass wasting, etc. Many ancient fan deposits do not show the abundance of thinning- and thickening-upward sequences implicit in published models (Hiscott 1981, Chan & Dott 1983, McLean & Howell 1984), and there is often no statistical support given in journal publications for asymmetric sequences (see Ch. 4). There is now some consensus that those thickening-upward sequences that do occur form not by lobe progradation but instead by subtle lateral shifts in the site of turbidity–current deposition to form compensation cycles (Fig. 7.4) (Mutti & Sonnino 1981, Mutti 1984, Ricci Lucchi 1984). When channels are exposed at outcrop, the fill may consist of very fine-grained facies (e.g. Garcia-Mondejar *et al.* 1985, p. 321) unlike those predicted by extant fan models, although seismic data from some modern fans (e.g.

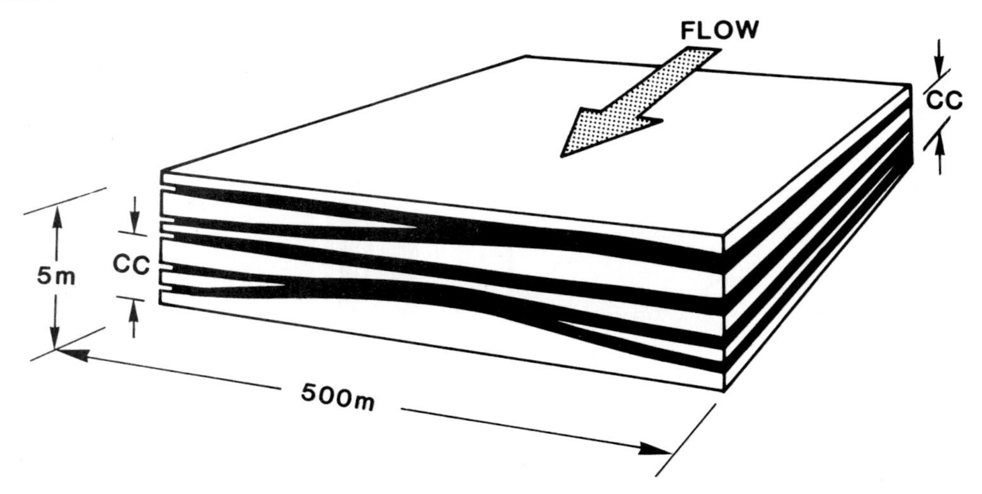

Figure 7.4 Block diagram illustrating the origin of compensation cycles (CC) by lateral shifts in the thickest parts of successive turbidites, resulting in a smoothing of bottom topography and in formation of thickening-upward cycles.

Indus Fan, Kolla & Coumes 1987) does suggest common fining-upward sequences as channel fills of the upper and middle fan segments.

The major difficulty in correlating attributes of ancient deep-marine fans with processes observed on modern fans is the difference in scale of observation (Normark *et al.* 1979, Shanmugam *et al.* 1985), along with the fact that most modern fans have been essentially inactive since the last rise in sea level, and are blanketed by hemipelagic mud (Facies Class E). Regarding scale, features observed in the best outcrops are in all but exceptional circumstances smaller than the best resolution that can be attained with available ship-board deep-sea imagery (Fig. 7.5). Even deep-tow instrument packages only have a maximum resolution of 2–3 m vertically and 10–20 m horizontally (Normark *et al.* 1979). For this reason, scientists working on ancient deposits often think on an entirely different scale than ship-based marine geologists. As an example, thickening-upward sequences that were traditionally interpreted as the products of lobe progradation are generally on the order of 10–20 m thick, or even thinner, consisting of less than ten beds (Ricci Lucchi 1975b). In contrast, modern suprafan lobes are on the order of 100–600 m thick (Emmel & Curray 1981, Garrison *et al.* 1982, Bouma *et al.* 1984, Nelson *et al.* 1984), strongly suggesting that the thin sequences recognized in ancient deposits have some other origin than lobe progradation (or aggradation) followed by abandonment.

We will begin this chapter with a brief summary of what we believe are the major controls on fan shape and facies distribution, based on many recent publications (e.g. Stow *et al.* 1984, Bouma *et al.* 1985b, Mutti 1985 and Mutti & Normark 1987) but of necessity only a summary of their main points. This will be followed by case studies of selected modern and ancient fans, chosen in order to illustrate the range of scales and facies that typify submarine fans. In all examples, facies are classified according to the scheme presented in Chapter 3.

Major controls on submarine fans

SEDIMENT TYPE

The two major end-member grain-size populations in fan deposits are either sand-rich or mud-rich systems. Sand-rich populations are characteristic of small, radial fans with an immediate source from sands and silts transported along the outer edge of the shelf and subsequently trapped in the heads of submarine canyons. The fans of the California Continental Borderland, from which the Normark (1970) fan model was derived, are of this type, as are the ancient fans that form the basis of the Walker (1978) fan model. The hydrodynamic consequence of a paucity of mud is that turbidity currents lack a quasi-continuous fluid phase that is significantly more dense than seawater, and therefore decelerate more rapidly as sand load is lost through deposition. The turbidity currents are, therefore, less efficient in transporting sand and silt load away from the canyon than they would be if they

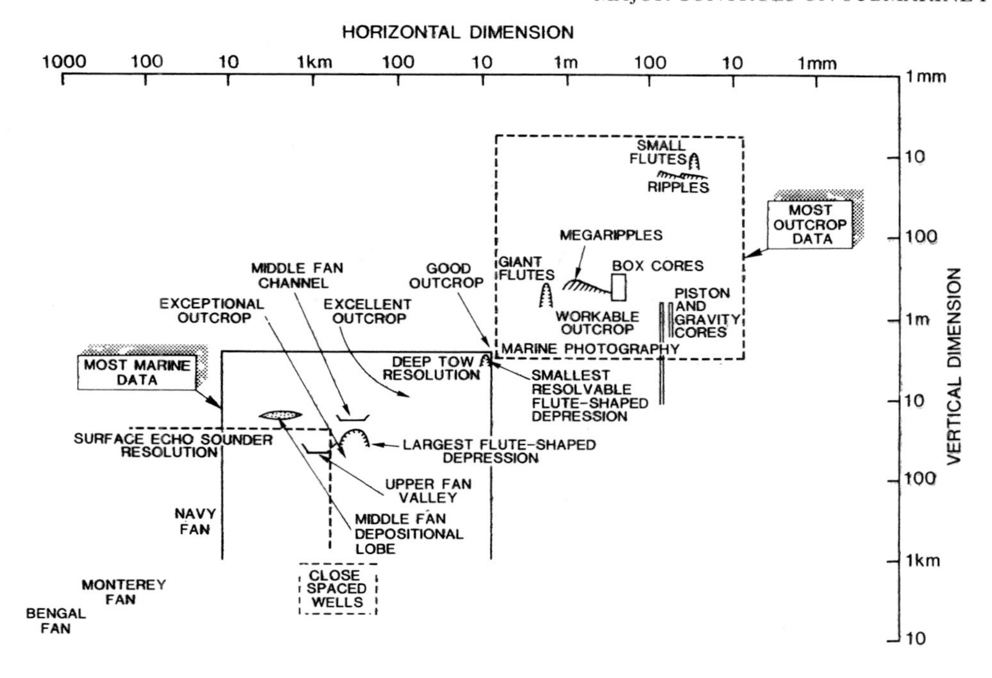

Figure 7.5 Size of submarine fans and component morphological and sedimentological features compared with the limits of resolution of various marine survey techniques.

contained more suspended mud (see Ch. 2). For this reason, sand-rich fans have been called *low-efficiency fans* by Mutti (1979).

Mud- or silt-rich grain populations are characteristic of large, elongate fans fed by major rivers draining areas of high sediment yield, particularly areas with a humid climate or active continental glaciation, both conducive to the production of large quantities of mud. Modern examples include the Astoria, Bengal, Indus, Mississippi, Amazon, Laurentian and Rhône Fans. These are the *high-efficiency fans* of Mutti (1979). Such fans commonly cannot be easily divided into upper, middle and lower fan segments (Pickering 1982c), and do not possess the distinct suprafan lobes of the Normark (1970) fan model.

TECTONIC SETTING AND ACTIVITY

The direct effect of tectonic setting is in the determination of basin size and shape, bottom gradients, and type and rate of sediment supply. On mature passive continental margins, the main tectonic process is slow thermal subsidence, which has little effect on the development of large, mature fans with low surface gradients. At active margins, which include convergent, oblique-slip and young rifted margins with rapidly tilting fault blocks (e.g. Surlyk 1978, Stow *et al.* 1982), depositional basins may be small and irregular in shape, so that fans, particularly if they are of the delta-fed high-efficiency type, will be constrained by the basin geometry, and will not be able to develop the predicted elongate shape (cf. Pickering 1982c). Fans at active margins are also often transient, because active vertical and horizontal movements of the crust may cut off sediment supply. The fan sediments may also be deformed and uplifted even as fan growth continues (e.g. Nicobar Fan, Bay of Bengal, Bowles *et al.* 1978). Unlike most submarine fans, the Mozambique Fan has been structurally constrained to develop *parallel* with the African coastline (Mozambique) and the Madagascar Ridge/Madagascar (cf. Kolla *et al.* 1980a & b, Droz & Mougenot 1987).

Along the steep marine/non-marine transitions of some oceanic islands or active continental margins, fan-deltas may pass seaward into small, coarse-grained slope aprons (Wescott & Ethridge 1982, Postma 1984) that share facies types with the inner parts of small, coarse-grained, fully marine fans. This similarity of facies may lead to confusion in the interpretation of ancient deposits (e.g. alternative interpretations of the

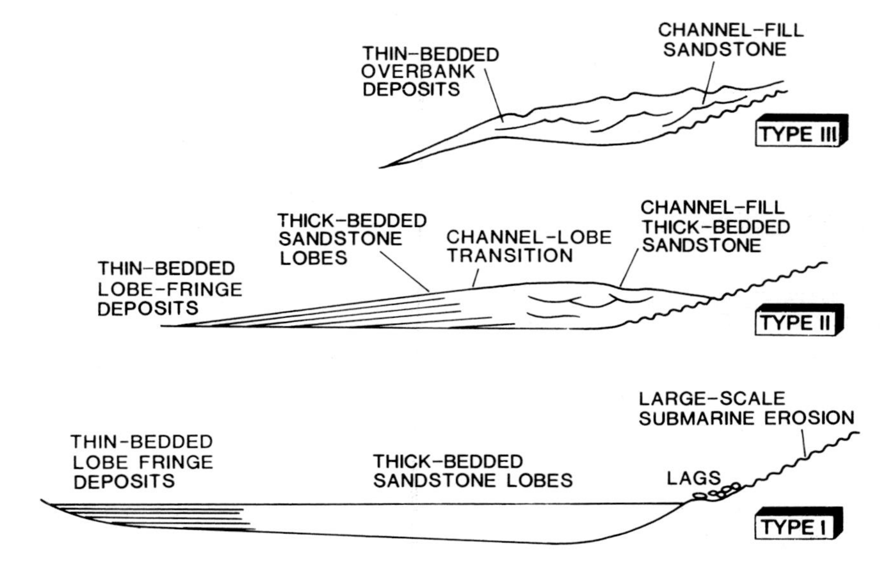

Figure 7.6 Three main types of turbidite systems recognized by Mutti (1985). The systems differ from one another mainly in terms of where sand is concentrated.

Brae Oilfield coarse clastics by Harms *et al.* 1981, Stow *et al.* 1982 and Turner *et al.* 1987).

SEA-LEVEL FLUCTUATIONS

Sea-level variations may be (a) eustatic (global), or (b) regional or relatively local events caused by tectonic processes (see Ch. 4). Regardless of cause, relative high stands of sea level result in reduced supply of terrigenous sediment to fans due to preferential deposition in coastal systems (e.g. estuarine embayments) and on a relatively broad shelf. Major fan processes then become: (a) slow deposition of hemipelagites (Facies Class E); (b) reworking of surface sediments by thermohaline bottom currents, and (c) mass wasting, producing chaotic deposits of Facies Class F (e.g. Stow 1981).

Mutti (1985) has developed a remarkably perceptive model for the evolution of fan systems during rises and falls of sea level that incorporates the features of high- and low-efficiency fans into a single scheme. Mutti recognizes three types of turbidite system (Fig. 7.6), differing in the distribution of sand deposits and in relative scale. The three types may develop independently of one another, or may develop in sequence in large delta-fed systems as sea level varies. The discussion that follows pertains to large delta-fed systems, and is therefore most applicable to fans in orogenic settings with fast rates of sediment supply. When sea level falls, large volumes of delta-front and delta-slope sediment become unstable and generate a series of large-volume, high-momentum turbidity currents through retrogressive slope failure. These turbidity currents do not deposit in proximal erosional channels, but rather carry their load far out into the basin to form extensive sheet-like sand bodies, or packets, 3–15 m thick, that grade distally into finer grained lobe-fringe deposits. Individual sand beds have flat bases. In vertical sections, the sand bodies alternate with lobe-fringe deposits to form units several hundreds of metres thick that are classified as Type I systems by Mutti (1985), and that correspond to the high-efficiency fans of Mutti (1979). As the volume of individual turbidity currents decreases, either due to slowly rising sea level or to smaller slope failures on a slope of reduced gradient, sands are not carried as effectively into the basin, and instead accumulate at the distal ends of distributary channels and on small lobes immediately downfan from the channels. The result is a low-efficiency fan of the type described by Normark (1970), with sand-rich suprafan lobes. These are classified as Type II systems by Mutti (1985). Continued sea-level rise results in mud deposition on the fan surface, but in the vicinity of a large river delta, eventual progradation of the delta to

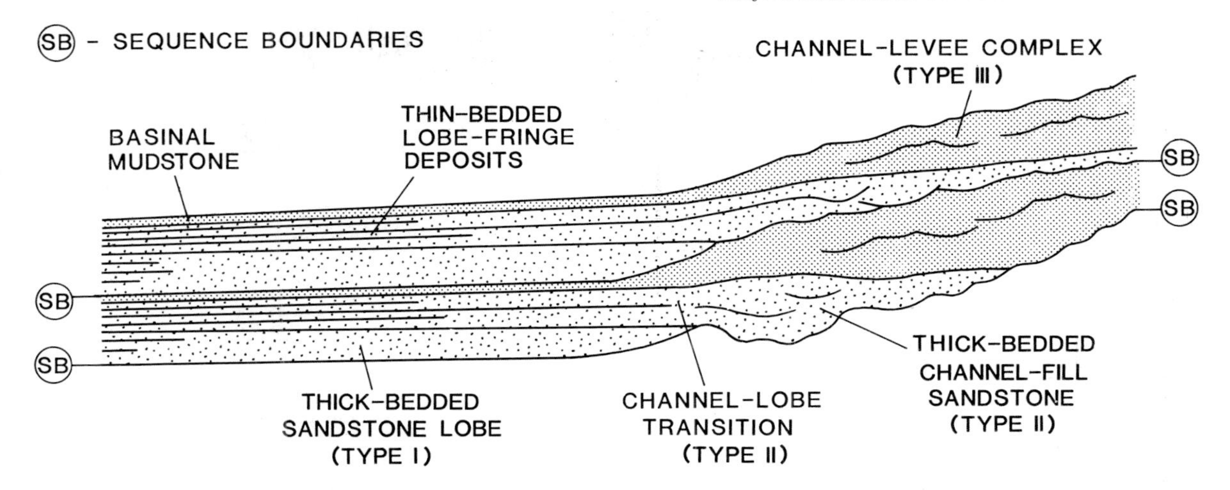

Figure 7.7 Changing character of sequence boundaries of Mutti (1985) from the basin margin into the deeper parts of the basin. In this hypothetical example, the sequences are complete and show an evolution from Type I to Type II to Type III systems.

the shelf edge during the high stand leads to deposition on the fan surface of mud-rich channel-levee complexes in which sands are restricted to the proximal parts of small channels. The bulk of the sediment is mud that represents overbank deposition from the channels. These muds undergo slumping and sliding from levee crests, producing characteristic multiple angular unconformities. These mud-rich deposits are assigned to Type III systems by Mutti (1985).

The full evolutionary sequence from Type I to Type III is only developed for delta-fed fans with fast rates of sediment accmulation, such that large volumes of unconsolidated sediment can accumulate on the shelf during high stands of sea level to guarantee a source for subsequent large low-stand failures. Where shelf sedimentation rates are slower, low-stand Type II systems alternate vertically with mud blankets of intervening high stands.

A fundamental conceptual breakthrough in the Mutti (1985) model is the suggestion that sand packets in a vertical succession of ancient fan deposits may not represent deposits of suprafan lobes like those described by Normark (1970), but instead may represent extensive sand sheets produced during a low stand of sea level, perhaps covering all of what might previously have been considered as middle fan and lower fan. In this case, mudstone units between sandstone packets were never lateral equivalents of sandy facies, but instead were deposited uniformly over the fan surface during a time of higher sea level. Likewise, upper-fan channel-levee complexes are younger than, not equivalent to, middle- and lower-fan deposits (see also Mitchum 1985).

According to Mutti (1985), channel-levee complexes and upslope deltaic sediments of the high stand provide the bulk of material that fails during the next drop in sea level, producing Type I deposits of a younger sequence. The result is that sequence boundaries in the basin are marked by the sudden introduction of sheet-like sand bodies, whereas sequence boundaries on the slope and base-of-slope are marked by erosional unconformities (Fig. 7.7). On a smaller scale, Mutti (1985) proposes that minor short-term fluctuations of sea level are responsible for 3–15 m-thick alternations of thick-bedded (Facies C2.1 and C2.2) and thin-bedded (Facies C2.3 and D2.1) lobe and lobe-fringe deposits so commonly seen in ancient fan deposits (Figs 7.8 & 9). The wire-line log response that characterizes such packets of thinner and thicker bedded strata is like that illustrated by Hsü (1977) for the Repetto Formation of the Ventura basin (Fig. 7.10).

The general model of Mutti (1985) is a useful guide to facies development in delta-fed systems of tectonically active areas (e.g. Morgan & Campion 1987), but its applicability to large passive-margin fans like the Indus Fan has been questioned by Kolla & Coumes (1987). On the Indus Fan, low-stand deposits are of Type III (i.e. channel-levee complexes), not Type I, and the deposition of channel-levee complexes continues from the time of a low sea-level stand throughout the subse-

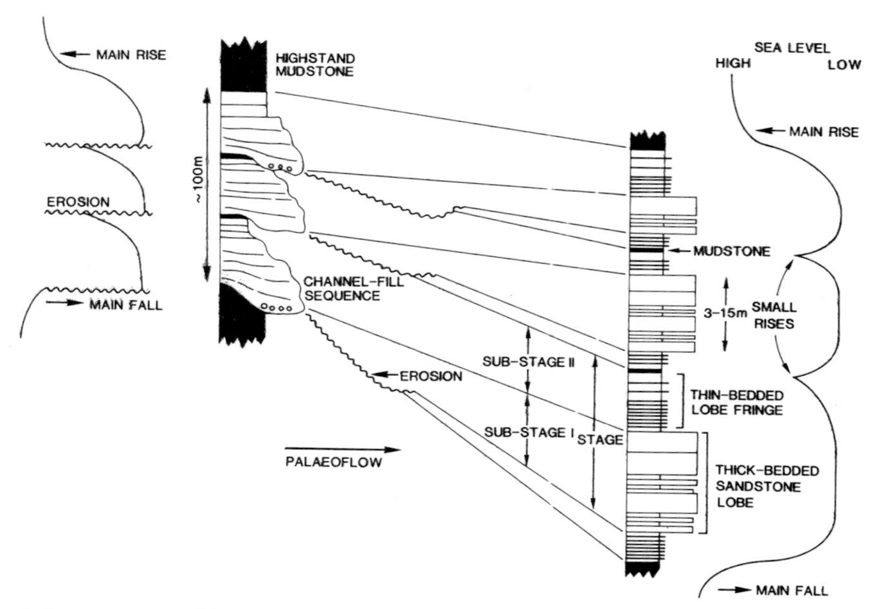

Figure 7.8 Sea-level changes and resulting facies cyclicity in a fan system comprising of channel-fill and lobe sediments. Sub-stages result from minor oscillations in sea level, superimposed on a major fall/rise event. Minor falls in sea level result in proximal unconformities and distal sandstone packets (Facies Class B & C) 3–15 m thick. Minor rises result in proximal backfilling of channels and distal deposition of thin-bedded lobe-fringe deposits (Facies Class D).

Figure 7.9 Alternating sandstone packets (numbered) and shaly lobe-fringe and interchannel deposits in the Ordovician Tourelle Formation of Quebec, Canada. Note that packet 5 has an erosional base. The depth of this channel is 9 m. INJ = interval of clastic injection described by Hiscott (1979).

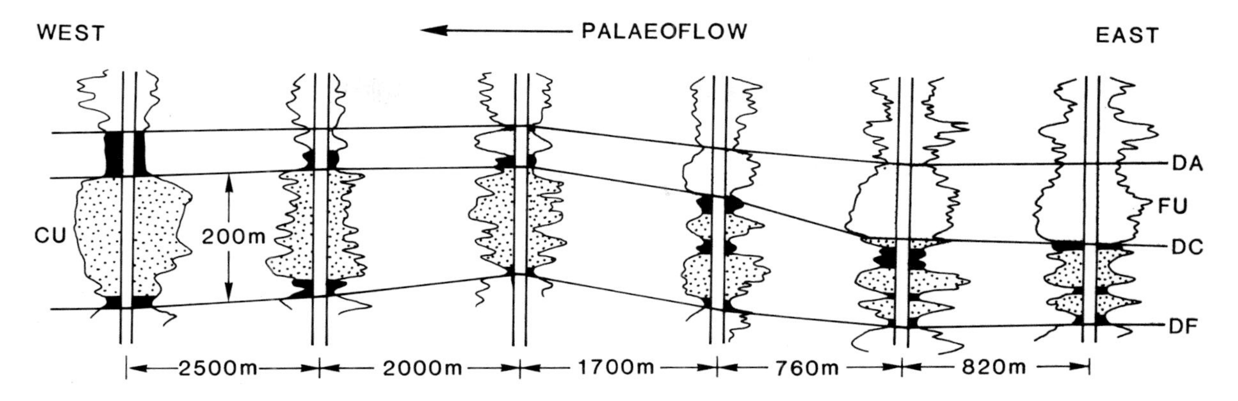

Figure 7.10 Self-potential (left) and resistivity (right) logs through sandstone packets of the Repetto Formation, Ventura Basin, California (Hsü 1977). Note the eastward (upcurrent) thinning and increase in shale content of the packet between markers DC and DF, and the westward (downcurrent) pinch out of the packet between markers DA and DC.

quent period of rising sea level. Kolla & Coumes (1987) attribute the apparent lack of any Type I deposits in modern, large, passive-margin fans to generally fine sediment size and great distances to the uplifted source areas.

Modern fans

An excellent compilation of the attributes of 21 modern fans is presented by Barnes & Normark (1984). Table 7.1 is a summary of some of this data. Our treatment of modern fans will be confined to three case studies of well documented fans, these being the Navy, Rhône, and Amazon Fans (Table 7.1). We will make no attempt to formulate a general model for modern fans, but instead will take an entirely descriptive approach.

CASE STUDY: NAVY FAN

Navy Fan is a small, upper Pleistocene, sand-rich fan (now inactive) located in the California Continental Borderland (Fig. 7.11). The fan has been studied in great detail by W.R. Normark and D.J.W. Piper and co-workers (Normark & Piper 1972, 1984, Normark *et al.* 1979, Piper & Normark 1983, Bowen *et al.* 1984), using single-channel and high-resolution seismic profilers, deep-tow instruments with side-scan sonar, and colour photography. More than 100 piston and gravity cores have been taken to complement the geophysical data. The fan has an irregular shape due to confinement in a tectonically active basin, but can still be clearly divided into upper, middle, and lower fan areas. The fan deposits have a maximum thickness of 900 m (Normark & Piper 1972).

The upper fan has a single, relatively straight, leveed valley about 400 m across that decreases in depth from 50 m to 15 m over a distance of 8 km. The entire valley–levee complex is 4–5 km across. The levee surface is characterized in places by numerous, irregular, commonly steep-sided depressions 4–8 m deep and 50–100 m wide. These may be slide scars or large erosional scours formed when turbidity currents spill out of the valley. The middle fan has been examined in most detail, and is marked by three distributary channels up to 500 m wide and 5–15 m deep, with poorly developed levees along their upper reaches, passing downfan into three smooth depositional lobes 2–3 km in length. The channels and lobes comprise <20% of the area of the middle fan. Only one of the channel–lobe systems was active immediately prior to the Holocene sea-level rise; the other lobes are separated from the main distributary channel by low levee deposits. Acoustic mapping and coring indicates that the most sand-rich deposits occur within the depositional lobes. Muddy facies characterize interchannel areas, lobe-fringe areas, and the lower fan (Normark *et al.* 1979).

Navy Fan is important because of the information it has provided about middle fan processes and lobe switching (Normark *et al.* 1979, Piper & Normark 1983, Bowen *et al.* 1984). Sharp channel bends at the apices of abandoned lobes indicate that upward growth (aggradation) of a depositional lobe eventually forces the channel to switch to a new course along one of the marginal depressions at the edge of the lobe (Fig. 7.12);

Table 7.1 Dimensions, setting and texture of 'modern' fans.

Name	Region	Margin type[1]	Dimensions					Shape	Grain size	
			L (km)	W (km)	A (km^2)	T_{max} (m)	V (km^3)		max.	mean
Amazon Fan	Brazil	P	700	250–700	3.3×10^5	4200	$>7 \times 10^5$	Fig. 7.21	pebbles	clay
Bengal Fan	Indian Ocean	P	2800	1100	3×10^6	>5000	4×10^6	elongate	medium sand	mud
Crati Fan	S. Italy	A	16	5	60	30	0.9	elongate	medium sand	mud
Ebro Fan	E. Spain	P	100	50	5×10^3	370	1.7×10^3	oval	medium sand	fine sand
Indus Fan	Indian Ocean	P	1500	960	1.1×10^6	>3000	10^6	fan	sand	mud
La Jolla Fan	California	T	40	50	1200?	1600	1175?	pear	gravel	fine sand
Laurentian	E. Canada	P	1500	400	4×10^5	2000	10^5	elongate	gravel	very fine sand
Mississippi	Gulf Mexico	P	540	570	$>3 \times 10^5$	4000	3×10^5	conical	gravel	silty mud
Navy Fan	California	T	25	25	560	900	75	Fig. 7.14	gravel	sandy silt
Nile Fan	Mediterranean	P	280	500	7×10^5	>3000	$>1.4 \times 10^5$	fan	sand	silty mud
Rhône Fan	S. France	T	440	210	7×10^4	1500	1.2×10^4	Fig. 7.16	fine sand	clay silt

[1] Margin types: A = accretionary, P = passive, T = transform.

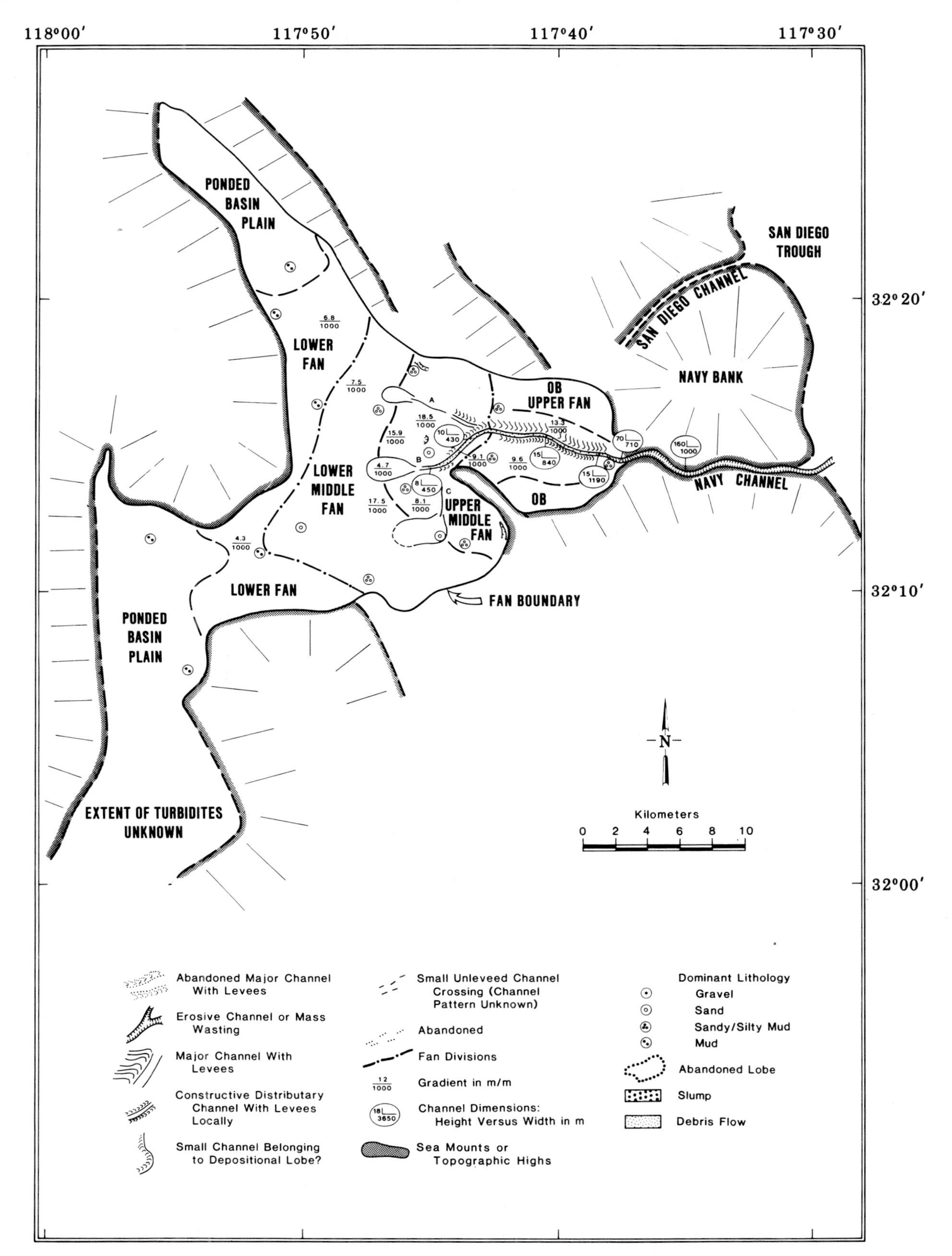

Figure 7.11 Schematic representation of Navy Fan, showing morphological divisions, channel pattern, channel dimensions, fan-surface gradients, dominant lithologies, and basin configuration.

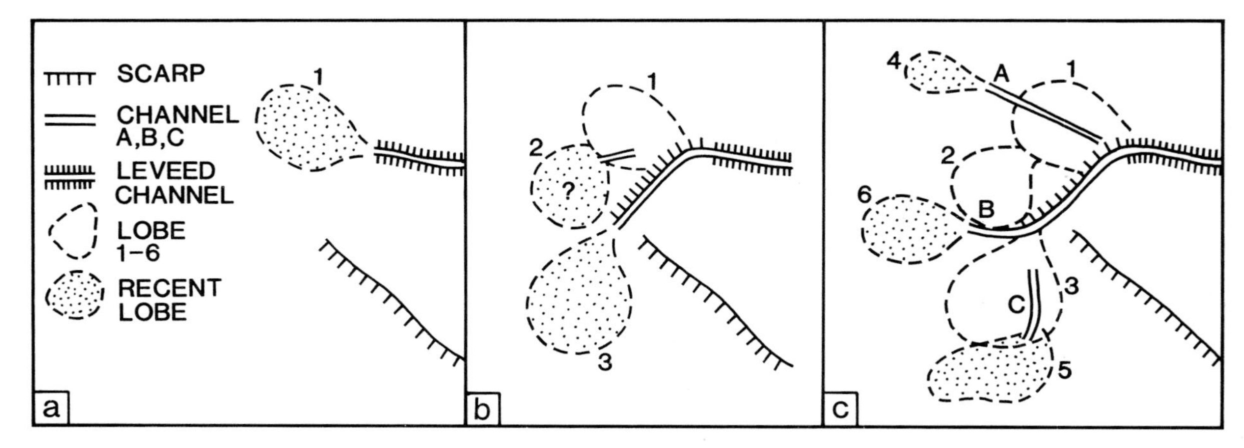

Figure 7.12 Development of depositional lobe pattern on Navy Fan. Channels A, B and C are shown on Figure 7.11.

a new lobe then begins to form at the mouth of the new channel. Turbidity currents that subsequently travel down the channel system are forced to negotiate a series of sharp bends. Thin sand-rich flows are able to remain in the channel, but thicker mud-rich turbidity currents are not (Bowen *et al.* 1984). Instead, the upper part of the turbidity current may flow along a straight path over the levee crest at the outside of the bend, resulting in: (a) formation of large erosional flute-like scours; with dimensions of 50 to >500 m in width at the levee crest, (b) deposition of a mud turbidite of Facies Class D on top of the abandoned lobe, and (c) crippling of the lower part of the turbidity current that remains in the channel, because of the momentum lost when the upper part of the flow was stripped away. Piper & Normark (1983) call this process *flow stripping*. The lower part of the flow that remains in the channel is enriched in sand by this process, but because of diminished flow power some of the sand load is dropped just beyond the bend and does not reach the new depositional lobe (Fig. 7.13).

The suprafan (middle fan) channels are partly erosional, as shown by abrupt steps and terraces, with relief of at most a few metres, that cross the channel floors. These are interpreted by Normark *et al.* (1979) to result from differential erosion of sandy and muddy beds in the walls of the channels, but some of the terraces may be depositional in origin, like the terraces described from the ancient by Hein & Walker (1982). In spite of evidence for some erosion into older deposits, the smooth surfaces of abandoned lobes on Navy Fan indicate that channels do not gradually prograde across their depositional lobes (Normark *et al.* 1979). Instead, lobes appear to be aggradational features that are periodically abandoned in favour of a nearby depression on the middle fan surface.

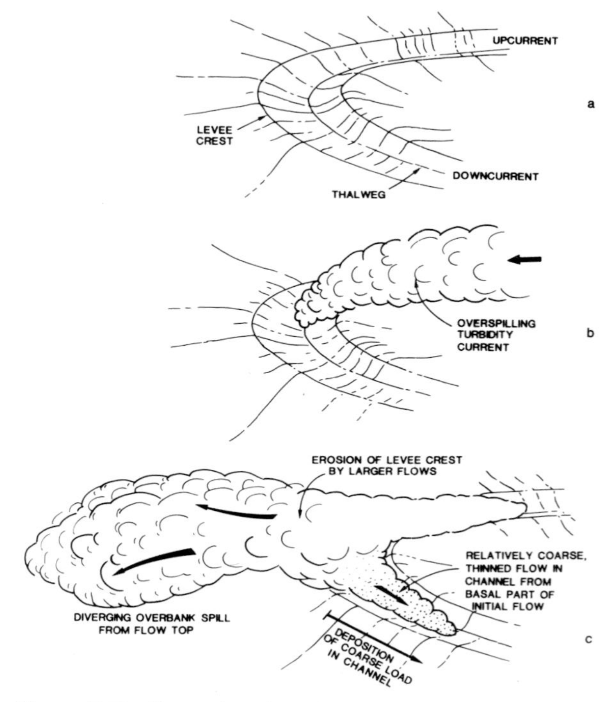

Figure 7.13 Illustration of the flow stripping hypothesis of Piper & Normark (1983). Channel curvature (a) causes eventual splitting of the initial flow (b) into two parts (c). Loss of momentum by overbank spill (c) results in deposition of sand just beyond the channel bend.

Side-scan sonar maps of the surface of Navy Fan reveal a wide range of *mesotopography*, defined as relief

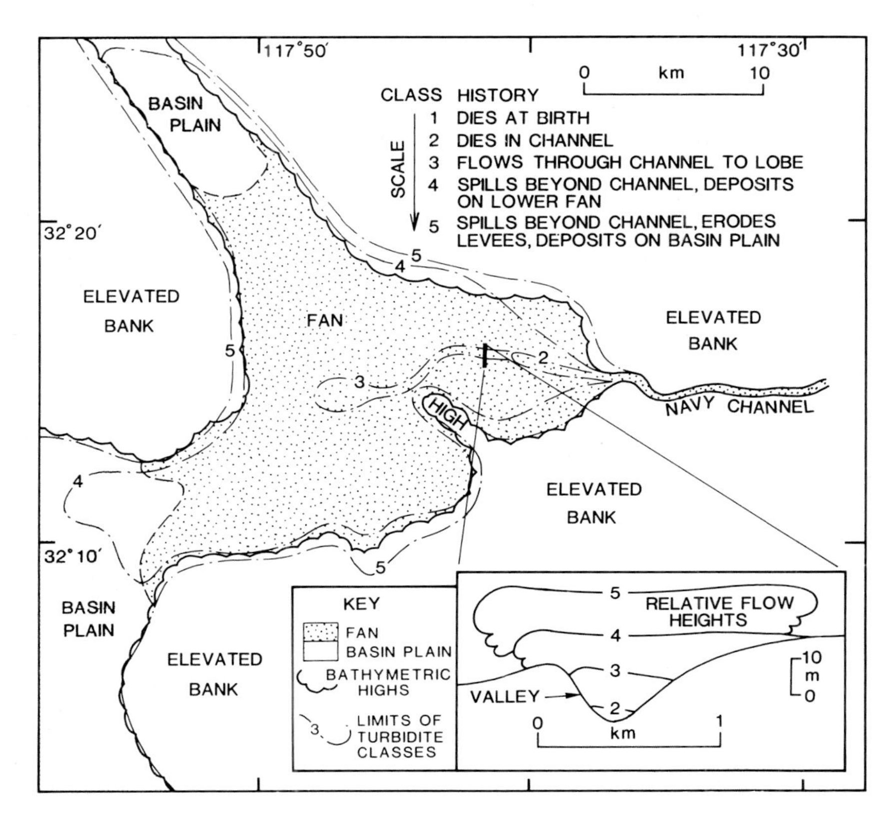

Figure 7.14 Schematic map of Navy Fan and cross-section of upper fan valley, showing thickness and extent of turbidites of size classes 2–5.

features too large to be observed on sea-floor photographs, but too small to be adequately defined from the sea surface using convential acoustic profiling systems. The flute-like scours and channel terraces mentioned above fall into the category of mesotopography. Interchannel areas on the middle fan are characterized by an abundance of mesotopographic features, mainly scours and hummocky topography. In contrast, the depositional lobes and the lower fan are free of a significant amount of mesotopographic relief.

Evidence from Navy Fan suggests that the style of fan growth depends strongly on the size of turbidity currents (cf. Mutti 1985). Flows of small volume are entirely confined to the channel system, whereas the largest of flows escape the influence of the channels, cause extensive erosion of levees and the proximal parts of the fan surface, and deposit over large areas of the lower fan and basin plain. Piper & Normark (1983) recognize five arbitrary sizes of flow (Fig. 7.14) that reach successively larger portions of the fan surface, and that are inferred to have recurrence intervals of from one year to several thousands of years (Fig. 7.15).

CASE STUDY: RHÔNE FAN

The area of Rhône Fan is more than 100 times that of Navy Fan (Table 7.1), although its maximum thickness is only slightly greater (i.e. 1500 m). Rhône Fan has been studied with echo sounders; a SEABEAM multiple-narrow-beam depth recorder; single channel, multichannel and high-resolution seismic profilers; and piston and gravity cores. The fan is now inactive, except for post-glacial sliding of slope sediments onto the fan surface and sliding of levee deposits (Normark *et al.* 1984), but has a history of active sedimentation from early Pliocene to late Wisconsinan (Droz & Bellaiche 1985). The fan has an elongate shape (Fig. 7.16), partly dictated by the pre-fan bathymetry of the basin.

There are at least four canyons that incise the slope above the Rhône Fan, but only one of these, the Petit-Rhône Canyon, is connected to the channel system on the fan surface. This canyon is characterized by large meanders, and terraced flanks due to erosion along the cut banks of tight meanders of a thalweg channel that

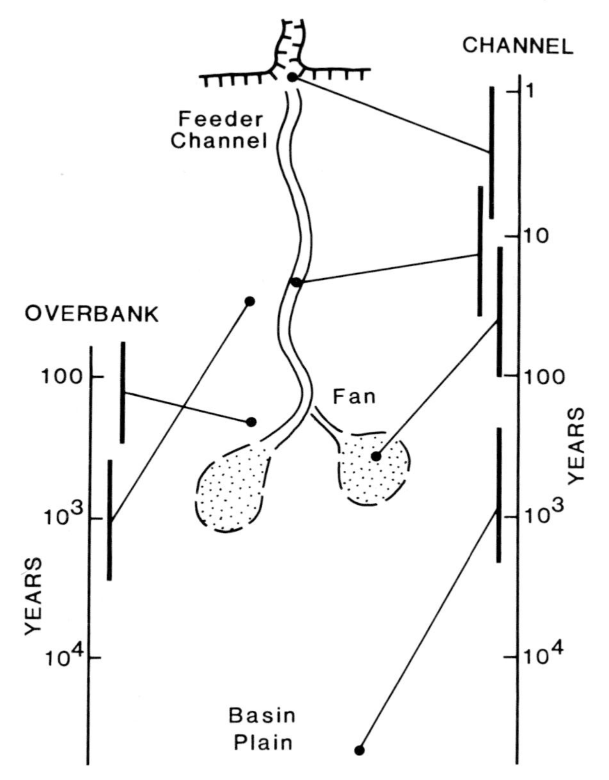

Figure 7.15 Turbidity current frequency of recurrence in different environments of active fans (Piper & Normark 1983).

defines the canyon axis. The slopes surrounding the canyons display large slide scars; the slide material came to rest on the fan surface (Fig. 7.16) to either side of the upper fan channel. A major control of slope instability has been the deep movement of Messinian salt, that has a surface expression in growth faults both on the slope and in the upper fan area. Halokinesis also affects the middle and lower fan areas, producing surface domes and depressions that can easily be confused with levees and channels respectively.

The upper fan has a single meandering channel, 2–5 km wide, flanked by levees 50–75 m high (Droz & Bellaiche 1985). The channel has a thalweg 40–50 m deep and 500–600 m wide. In seismic profiles of what Droz and Bellaiche define as the 'upper series' (those deposits associated with the present fan position), the upper fan is characterized by a stack of eight lenticular acoustic units (Fig. 7.17), each about 100 m thick and about 70 km wide, and consisting of central chaotic facies grading laterally into converging-bedded or transparent facies. These are interpreted as channel–levee complexes, each deposited during a low stand of sea level. The middle fan is also characterized by lenticular acoustic units, but in this case they are not stacked vertically, but instead are displaced laterally from one another through time (Fig. 7.18). Each unit is somewhat thicker than beneath the upper fan (maximum 150 m), but lateral shingling leads to an aggregate thickness of only 500 m. The lower fan is also formed of shingled and stacked lenticular acoustic units <70 km wide and <80 m thick, but these cannot be correlated with those higher on the fan because of disruption of seismic markers by salt domes and associated faults. In the upper fan, lenticular acoustic units beneath the middle and lower parts of the fan are each interpreted as the depositional record of a low stand of sea level (Droz & Bellaiche 1985). The attributes of these acoustic units on the Rhône Fan are summarized in Figure 7.19.

Deep seismic data indicate a complex early history for the Rhône Fan, that is divided into two intervals ('basal series' and 'lower series') by Droz & Bellaiche (1985). The 'basal series' was deposited by turbidity currents emanating from all four slope canyons. The 'lower series' was deposited on a fan supplied by the Petit-Rhône Canyon, but with its upper fan channel about 20 km to the east of the present channel (Fig. 7.20). This channel was then blocked by three large slides, forcing a new channel to form at the present position, and initiating deposition of the 'upper series'. Just before the last rise in sea level, a slide blocked the present upper fan channel about 50 km from the canyon, causing deposition to shift to a new fan lobe (Fig. 7.16) that is anomalous in that it rests on what is otherwise clearly the upper fan: lobes generally are restricted to the middle fan.

CASE STUDY: AMAZON FAN

The Amazon Fan has an area of 3.3×10^5 km^2, of the same order of size as the Laurentian, Mississippi, and Nile Fans (Table 7.1), and about five times as large as the Rhône Fan. The fan began to form in the Miocene. The principal work on this fan has been done by J. E. Damuth and coworkers (Damuth & Kumar 1975, Damuth & Embley 1981, Damuth *et al.* 1983a & b, Damuth *et al.* 1983b, Damuth & Flood 1984, Damuth *et al.* 1988, Manley & Flood 1988). Techniques have included echo sounding, GLORIA long-range side-scan sonar, single-channel and high-resolution seismic profiling, seismic refraction, piston and gravity coring (44 cores), and bottom photography.

Like the Rhône Fan, the Amazon Fan is elongate, and has large sediment slides and debrites covering a

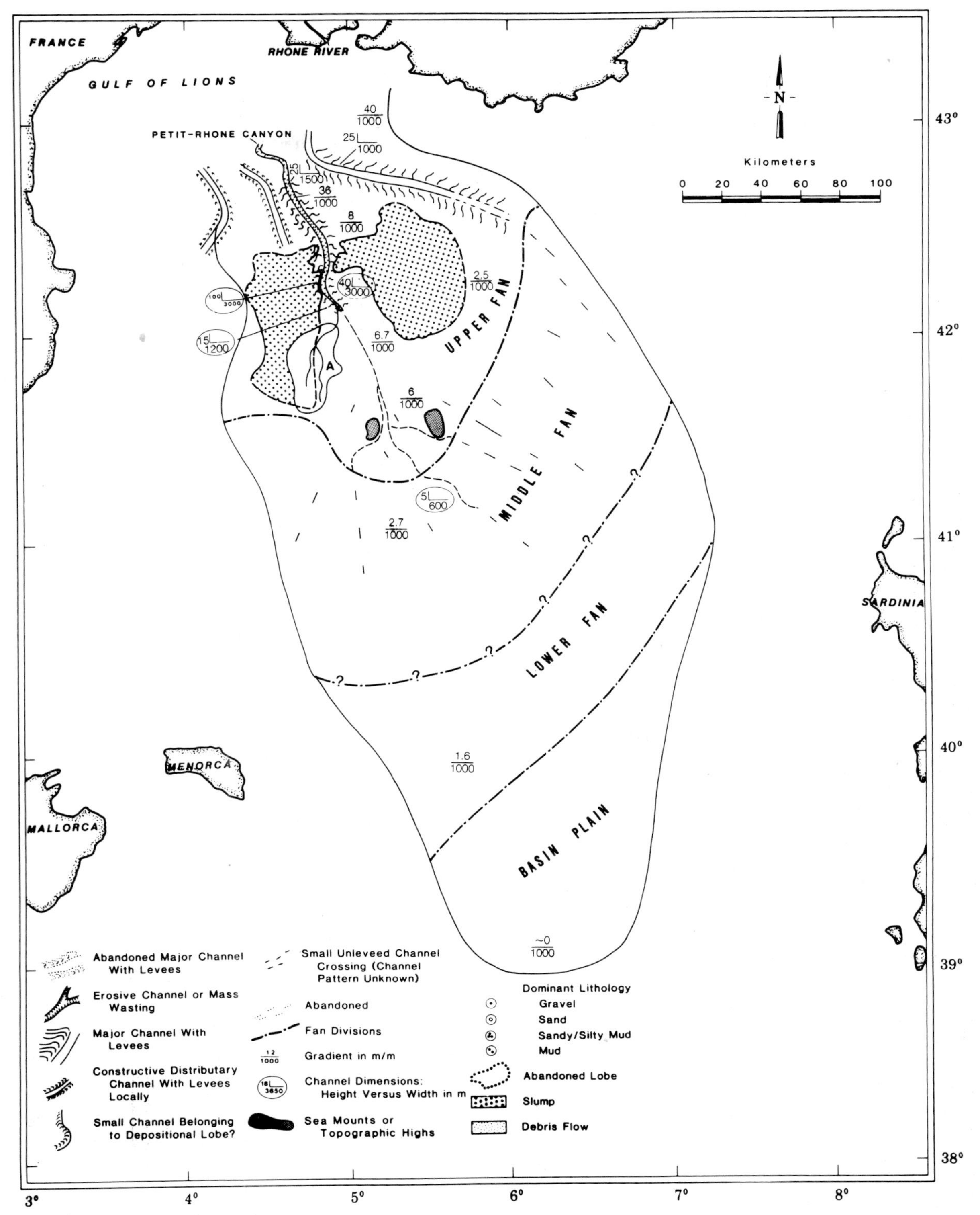

Figure 7.16 Schematic representation of Rhône Fan showing morphologic divisions, channel pattern, channel dimensions, fan-surface gradients, dominant lithologies, and basin configuration. Salt diapirs are shown with the pattern for seamounts. Letter A denotes the youngest depositional lobe.

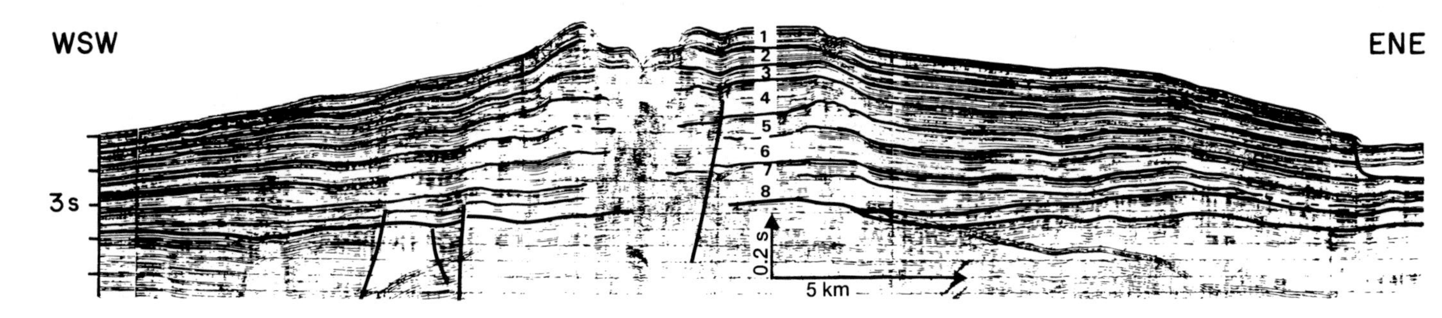

Figure 7.17 Air-gun seismic profile of upper Rhône Fan with eight lenticular acoustic units highlighted.

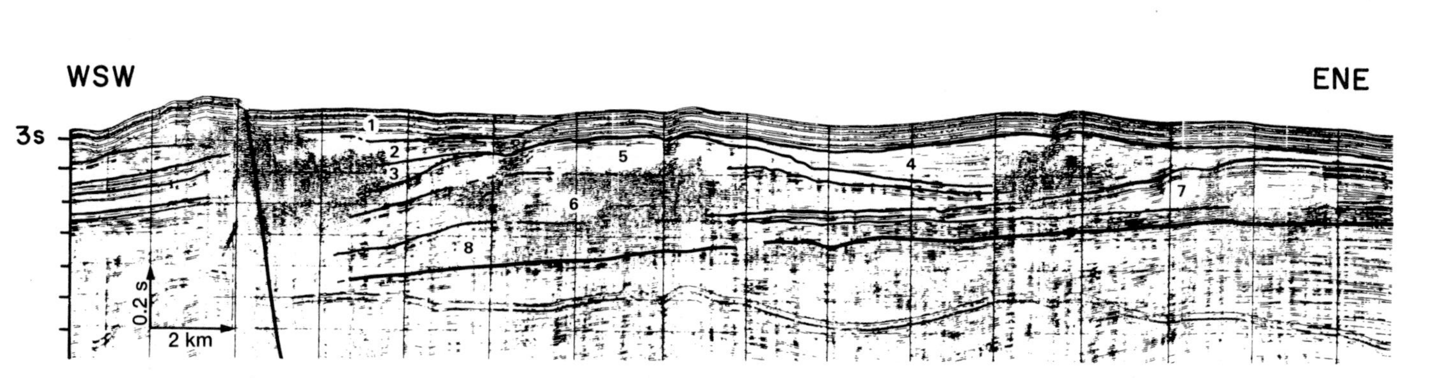

Figure 7.18 Air-gun seismic profile of middle Rhône Fan with the same eight lenticular acoustic units highlighted as in Figure 7.17.

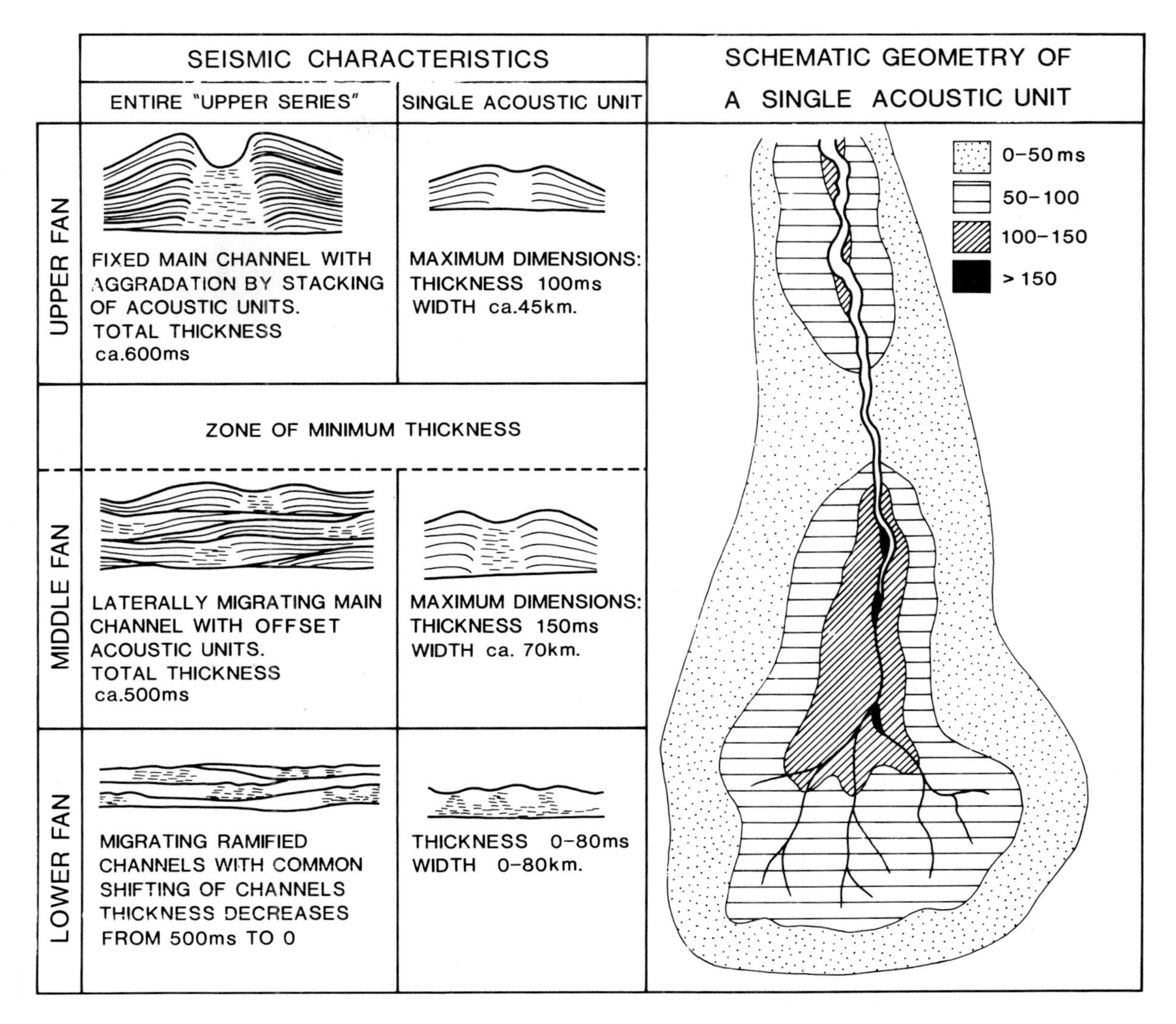

Figure 7.19 Schematic diagram showing relative sizes, shapes, and genesis of lenticular acoustic units (channel–levee systems) on Rhône Fan. Thicknesses are in milliseconds (ms) of two-way travel, where 1 ms corresponds to about 0.75 m of sediment.

significant part (14%) of the fan surface (Fig. 7.21). The slides/debrites (Facies Class F) average 10–50 m thick, and are of Wisconsinan age (Damuth & Embley 1981). The maximum thickness of the fan is 4–5 km. As defined by Damuth and coworkers, the upper fan consists of the Amazon Submarine Canyon, and a central distributary channel that divides into four prominent leveed distributaries. The central distributary channel is up to 2500 m wide and 250 m deep. It is perched on top of a levee system 50 km wide and 1 km thick, that rises up to 300 m above the adjacent fan surface. The four distributaries of this main channel also rest on thick levee complexes. The middle fan is characterized by numerous large, meandering, leveed distributary channels that result from bifurcation of the four upper-fan channels. The levee systems of these middle-fan channels overlap one another and coalesce to form two large channel–levee complexes, called the Western and Eastern Complexes (Fig. 7.21). The downslope terminations of these channel–levee complexes define the boundary between the middle fan and the lower fan. Interpretation of seismic profiles

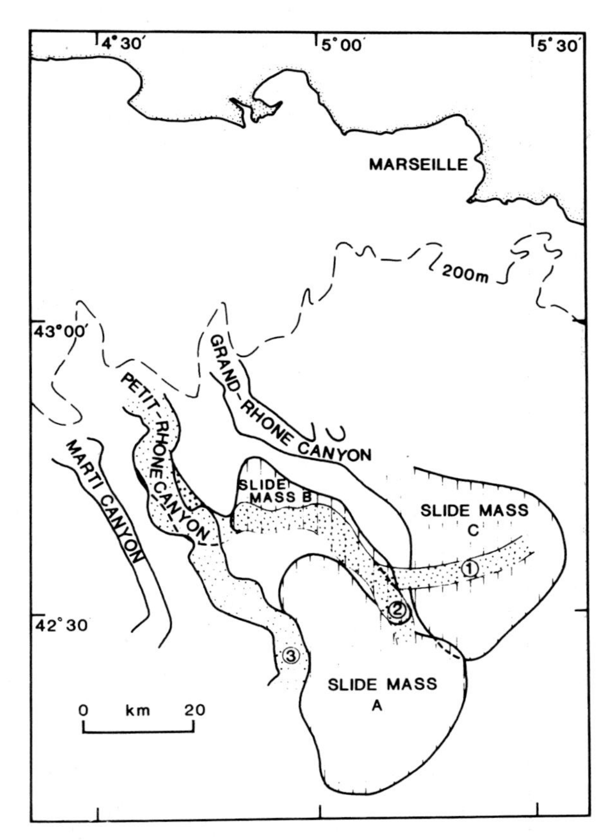

Figure 7.20 Successive positions of the upper fan channel of Rhône Fan, with positions 1 and 2 corresponding to the time of deposition of the 'lower series'. These channels were then blocked by slide mass A, B and C, and the upper fan channel shifted to its present position 3.

(Damuth *et al.* 1983b) indicates that only one channel system is active at any one time on each channel–levee complex, with the margins of its levees overlapping the levees of older systems (Fig. 7.22). The lower fan is relatively smooth, but, unlike the lower fans of Normark (1970), is covered with numerous small, unleveed distributary channels that are characterized by tight meander loops (Fig. 7.23), cut-offs, and abandoned meander loops (Damuth *et al.* 1983a, 1988). Many of these channels extend for considerable distances down the fan surface without bifurcating, but when bifurcation occurs, it appears to have been the result of crevassing through channel walls and levees at the outside bends of meander loops (cf. flow stripping of Piper & Normark 1983).

According to Damuth *et al.* (1983b, 1988), the growth pattern of the Amazon Fan is controlled by the growth and abandonment, through avulsion, of channel–levee complexes like the Western and Eastern Complexes. A single active upper-fan/mid-fan channel deposits levee sediments that eventually partly bury adjacent, older levee systems (Fig. 7.22). The channel then shifts to a margin of the former levee, and deposits another levee sequence. These repeated shifts result in accumulation of a large channel-levee complex on the order of 1 km thick. Major avulsion high on the fan can lead to complete abandonment of the entire channel–levee complex, and initiation of deposition on a previously inactive part of the fan. The channel–levee complexes are characteristic of the upper and middle fan, and are formed predominantly of muddy overbank turbidites (Facies Group D2). On the Amazon Fan, turbidity currents with sandier loads effectively bypass the channel–levee complexes and deposit sandy turbidites of Facies Group C2 on the lower fan and Demerara Abyssal Plain. This effective transfer of sands to the lower fan is characteristic of the high-efficiency fans of Mutti (1979).

The Amazon Fan currently receives no sediment through Amazon Submarine Canyon, and probably has never done so at times of high sea level. During low stands of sea level, however, the continental shelf off the Amazon River was emergent (Damuth & Kumar 1975), and the river discharged its load directly into the canyon. It is at these times that the channel–levee complexes were active. Piston core data indicates that during low stands, the *entire* fan was active, and receiving sediment at rates >25 cm/1000 years. Flow of turbidity currents in the channel systems may have been so common at these times as to approximate to a geologically continuous process. Damuth *et al.* (1983a, 1988) believe that quasi-continuous flow is necessary to explain the highly meandering channel patterns of the middle and lower fan areas, patterns that are identical to those found in some rivers.

Ancient fans

Barnes & Normark (1984) summarize the attributes of ten ancient fans, but many more have been described in the literature. We will describe two ancient fan deposits: (a) the Tyee Formation of Oregon, and (b) the Milliners Arm Formation of Newfoundland.

Another well studied ancient fan inferred to have been deposited on a late Precambrian passive margin is the Kongsfjord Formation of northern Norway (Pickering 1981a & b, 1982a & b, 1983a, 1985), but space does not permit its inclusion as a case study in this

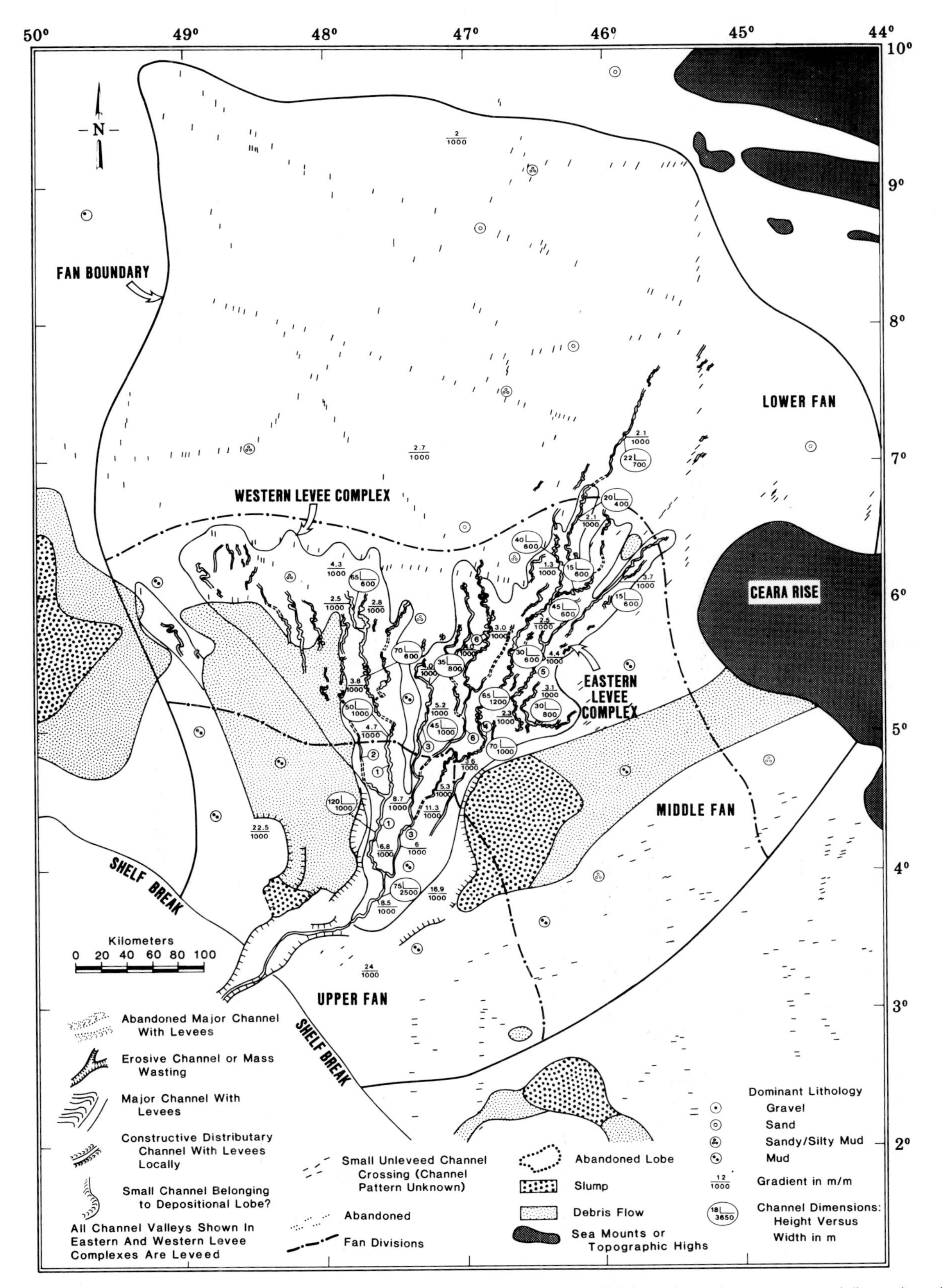

Figure 7.21 Schematic representation of Amazon Fan showing morphologic divisions, channel pattern, channel dimensions, fan-surface gradients, dominant lithologies, and basin configuration. Numbers 1 through 6 beside major distributary channels indicate relative ages of channels in order of increasing age.

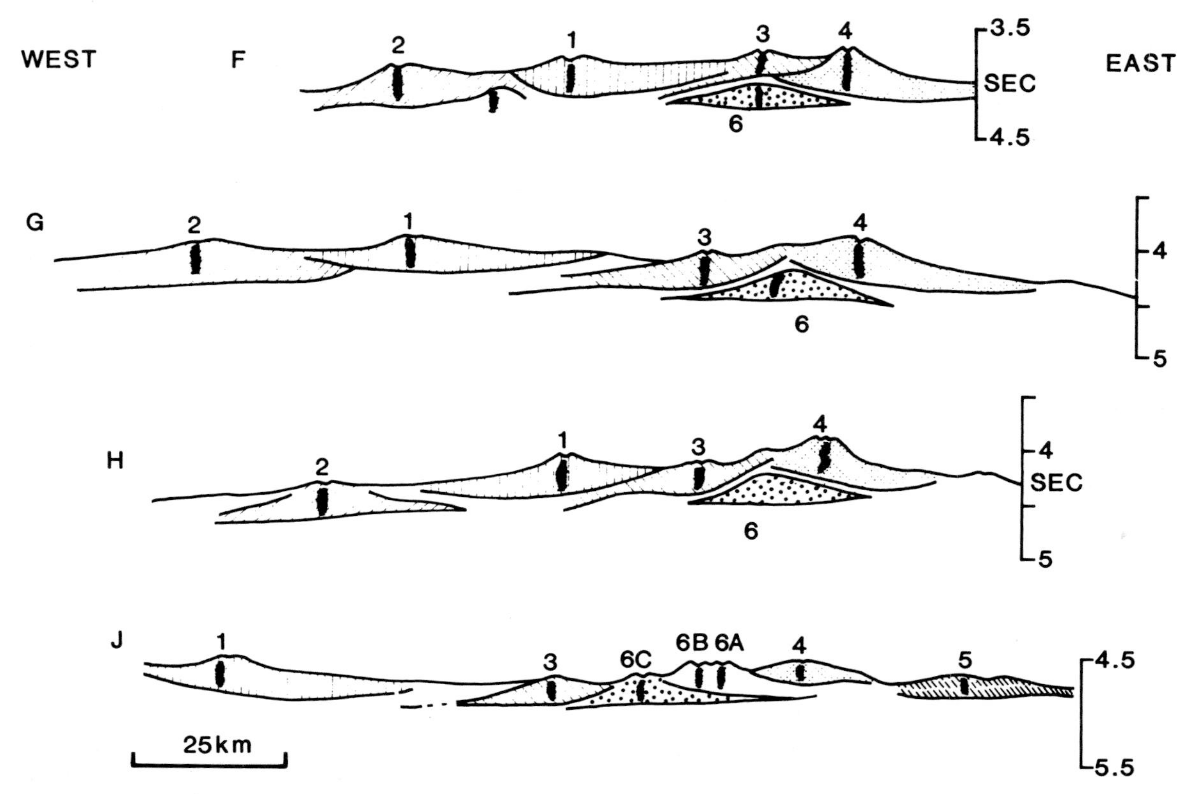

Figure 7.22 Line drawings traced from original seismic profiles showing stratigraphic relationships between major channel–levee systems on the upper part of the middle Amazon Fan. Profile F is the most proximal and profile J the most distal. The black area beneath each channel axis represents high-amplitude reflectors that may be generated by coarse sediments. Channel–levee complex 1 is the youngest, and 6C the oldest. One second of two-way travel is about 0.75 km of sediment.

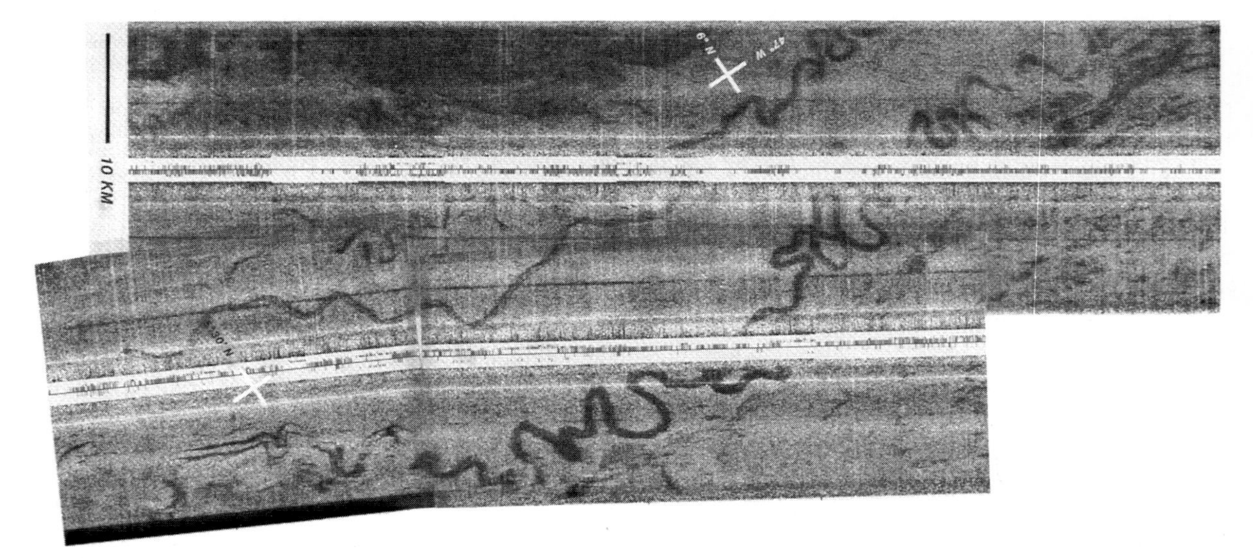

Figure 7.23 GLORIA long-range side-scan sonar mosaic showing meandering distributary channels on the middle Amazon Fan. The image is corrected for slant range, ship speed and beam pattern. Ship tracks are along the white band down the centre of each sonograph.

chapter. An interesting facies association of the Kongsfjord Formation that deserves special mention, however, is one that formed at the lateral margin of the submarine fan. The sediments are called *fan lateral margin deposits* (Macpherson 1978, Pickering 1983a), and are characterized by (a) a relatively high proportion of fine-grained sandstone and siltstone turbidites (Facies C2.3, D2.1), (b) relatively small channels developed at various angles to the regional basin slope direction, (c) lobes associated with the channels, and (d) abundant clastic dykes and other wet-sediment deformation. The dykes originated in channel sandstones, and are preferentially located near the margins of overlying channel sandstones (cf. Hiscott 1979).

The two ancient fans we describe below serve as an introduction to the complexity of ancient fans. It should be stressed that most well-documented ancient fans appear to have been deposited at active continental margins, including foreland basins. Except for the Kongsfjord Formation and, perhaps, Lower Palaeozoic fans of North Greenland (Surlyk & Hurst 1984), the literature is devoid of well exposed, large-scale, passive-margin fans, probably because such fans (e.g. Mississippi, Amazon, Laurentian Fans) can only become subaerially exposed through orogenic processes, leading to severe dismemberment as the sediments are thrust away from the oceanic or transitional crust on which they originally accumulated. Remnants of such passive-margin fans in orogenic belts cannot normally be satisfactorally re-assembled to permit description of the entire fan.

CASE STUDY: TYEE FORMATION

The Tyee Formation is a 3 km-thick Eocene deposit in Oregon, USA, that consists, in vertical section, of a shallowing succession from a deep turbidite basin, through fan-like and slope deposits, to a mixed delta and shelf succession (Fig. 7.24). The turbidite deposits are sand-rich (Lovell 1969), and were fed from multiple feeder channels that crossed the basin slope from a source area to the south in the Klamath Mountains (Fig. 7.25). Depositional rates were fast, about 70 cm/1000 years (Chan & Dott 1983) as the delta prograded rapidly from the south (progradation rate of 12.5–25 m/1000 years; Heller & Dickinson 1985).

The Tyee Formation was deposited in a small, tectonically active forearc basin (Chan & Dott 1983) at a time of globally-low sea level (Heller & Dickinson 1985). Multiple shallow feeder channels generated an unconfined, sheet-like deposit that shows little of the facies organization implicit in the Normark (1970) and

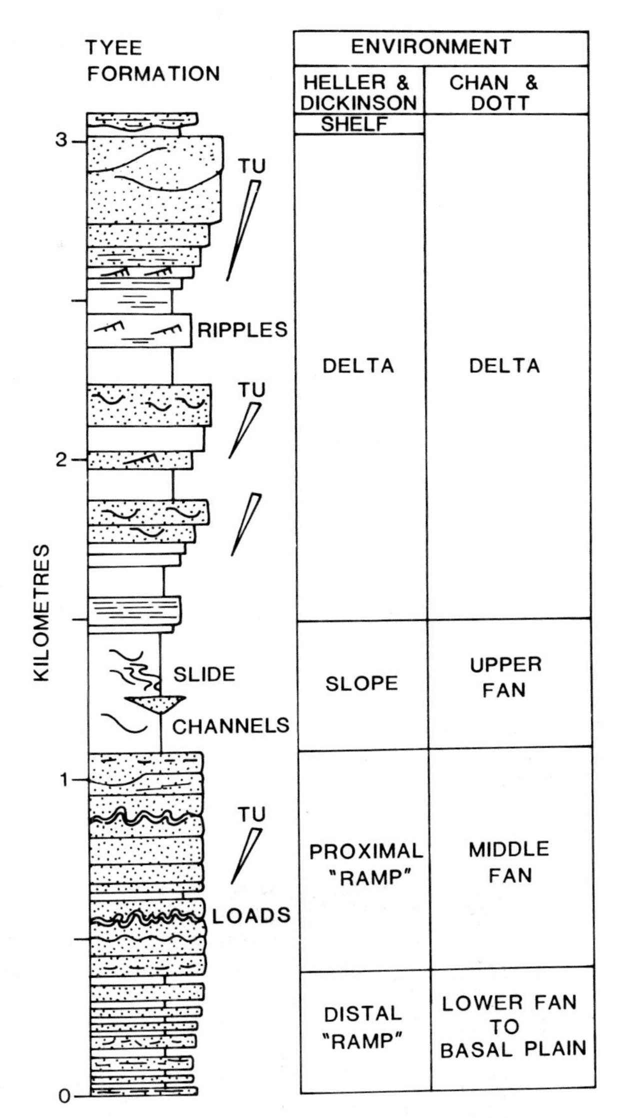

Figure 7.24 Simplified stratigraphic section through the Tyee Formation on and near the west side of Tyee Mountain, with interpretations of Heller & Dickinson (1985) and Chan & Dott (1983).

Mutti & Ricci Lucchi (1972) fan models. Instead, thick, sheet-like, commonly amalgamated sandstones of Facies Group B1 (sand:shale > 9:1) pass basinward into thinner bedded sandstones of Facies Group C2 with a gradual increase in shale content (down to sand:shale = 2:1). There is no differentiation into channel and interchannel deposits, precluding assignment of even the more proximal deposits to an upper- or middle-fan setting (Chan & Dott 1983, Heller &

Dickinson 1985). This was clearly an oversupplied sandy system, and conforms closely with the Type I systems of Mutti (1985). Lack of a single feeder channel induced Heller & Dickinson (1985) to propose a new name, *submarine ramp*, for the depositional setting of the Tyee Formation, but we will persist in describing these rocks as *submarine fan* deposits, recognizing that deposit shape is in most cases not that of a simple fan radiating from a point source. In terms of geometry alone, the Tyee 'Fan' is an elongate system.

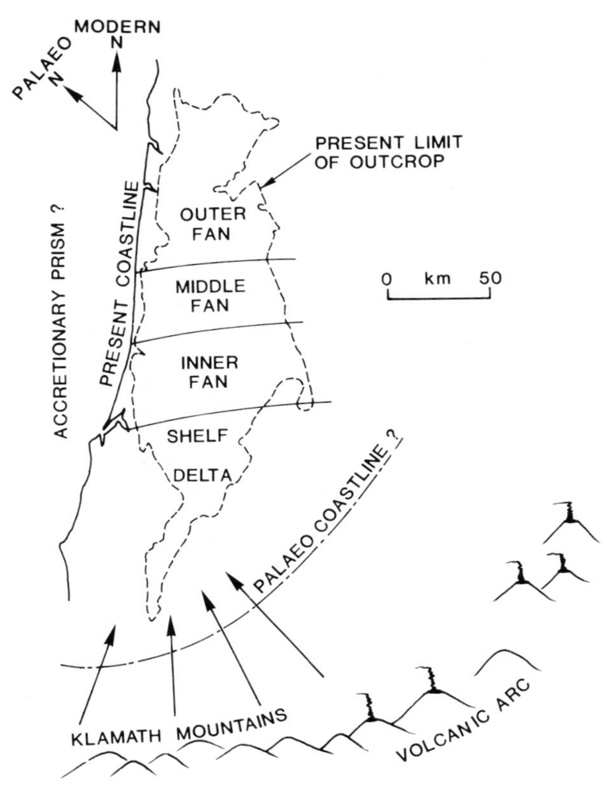

Figure 7.25 Palaeogeographic sketch map for Tyee Formation deposition. Modern north is at the top of the map. Palaeoflow in the Tyee Formation is from modern south to north.

The upper-fan region of Chan & Dott (1983), described as basin slope by Heller & Dickinson (1985), is a muddy succession characterized by a series of channels, up to 350 m wide and 40 m deep, filled with either (a) essentially 100% Facies Class B (mostly Group B1) sandstones (Fig. 7.26), or (b) thin-bedded fine-grained sandstone, siltstone and mudstone (Facies Groups C2, D2 and E2) identical to the facies into which the channels are cut. According to Heller & Dickinson (1985), the mud-filled channels tend to occur stratigraphically above the sand-filled channels. They attribute the sands to backfilling of the channels at the base of the slope: channels higher on the slope received no sand fill. It is also possible that some of the mud-filled 'channels' are really draped slide scars or megascours (cf. Normark *et al.* 1979).

Because of the lack of clear distinguishing features of middle- and lower-fan environments, we will describe the Tyee 'Fan' in terms of proximal and distal segments corresponding to the *proximal ramp facies* and *distal ramp facies* of Heller & Dickinson (1985), and to the mid-fan and lower-fan/basin-plain facies of Chan & Dott (1983). The proximal deposits have sand:shale from 2:1 to 9:1. Most sandstone beds are sheet-like, 1–3 m thick, poorly graded, amalgamated, and belong to Facies Class B (Fig. 7.27). Some beds are separated by thin mudstone layers. Only a few of the thickest beds have Bouma T_b, T_c, and T_d divisions at the very top. Asymmetrical cycles cannot be recognized in the field (Chan & Dott 1983) or through application of statistical techniques (Heller & Dickinson 1985). Only at the inferred base of the slope can a few broad, flat-

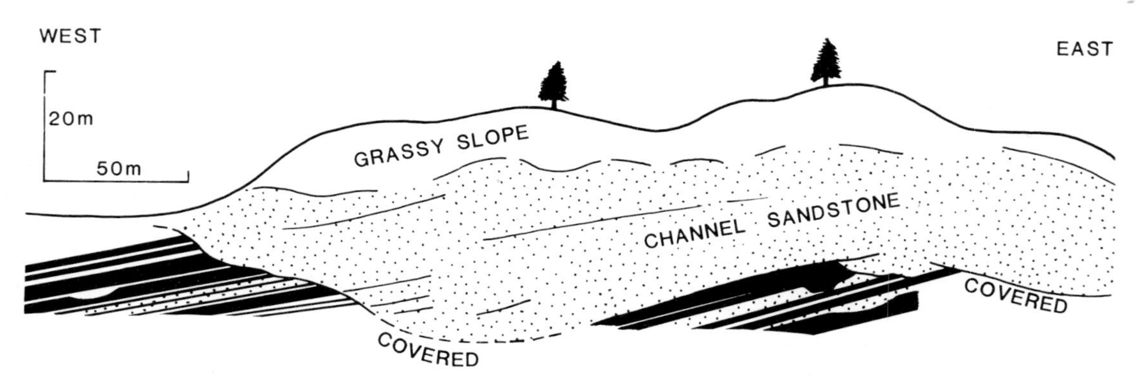

Figure 7.26 Sketch of channel cut into thin-bedded facies and filled with structureless Facies Class B sandstones.

Figure 7.27 Proximal sheet-like sandstones of Facies Class B (plus minor Class C), Tyee Formation, 10.5 km east of Highway 101, Smith River Road 48. Thickness shown is *c.* 10 m.

based channels be recognized. Elsewhere, all bedding contacts are essentially flat.

The distal fan deposits consist of sandstone beds 0.1–1.0 m thick, interbedded with mudstone layers up to 1 m thick. Sand:shale ranges from about 7:1 to about 2:1, a range not unlike that for the proximal deposits, although the distal deposits are on average more mud-rich. The sandstones belong to Facies Group C2, and contain various combinations of Bouma divisions, while the thicker beds contain all the divisions (Chan & Dott 1983).

The fan deposits of the Tyee Formation are about 1 km thick, while the overlying slope sequence is only 500 m thick. Heller & Dickinson (1985) believe that unconfined, sand-rich fans with multiple feeder channels appear to be associated with relatively thin slope deposits because of the rapid rate of fan (ramp) aggradation in the basin. There is ultimately little bathymetric relief between the edge of the shelf (delta front) and the basin.

CASE STUDY: MILLINERS ARM FORMATION

The Milliners Arm Formation (MAF) is a 4.5 km-thick, progradational fan succession of Late Ordovician to Early Silurian age in the Dunnage Zone of the Appalachian Orogen in Newfoundland (Watson 1981, Arnott *et al.* 1985). The sedimentological data and interpretations that follow are entirely from Watson (1981).

The MAF can be conveniently divided into lower and upper parts. The lower part (3.1 km) is characterized by sandstone packets, each about 10–20 m thick, separated by units of thin-bedded turbidites and shales from about 10 m thick up to about 250 m thick in those parts of the section that lack sandstone units. The upper part of the formation (1.4 km) is characterized by channelized conglomerate and sandstone bodies. Average palaeoflow is from north to south, but with considerable dispersion about the mean, particularly in facies associations that are non-channelized. The best sections through the lower part of the MAF are narrow coastal exposures that generally do not allow tracing of beds for more than a few tens of metres along strike. In exceptional circumstances, however, lateral tracing of packets for up to 600 m is possible. The upper part of the formation outcrops in long strike sections that allow detailed bed-by-bed and packet-by-packet correlations up to 3 km.

The lower part of the MAF can be divided into five facies associations: three non-channelized associations and two channelized associations (Table 7.2). The non-channelized and channelized associations alternate throughout the lower part of the formation.

In non-channelized intervals, the preferred sequence of facies associations is NCF1 → NCF2 → NCF3. NCF1 constitutes 72% of all non-channelized deposits in the lower part of the formation, and consists predominantly of thin and very thin-bedded siltstone turbidites, thin-bedded sandstone turbidites, and structureless mudstone. Individual beds are laterally continuous for >250 m without changes in either bed thickness or internal sedimentary structures. Where asymmetric cycles were noted, they most commonly show a thickening and coarsening upward. NCF2 forms only 11% of the thickness of the lower part of the MAF. Maximum thickness of the association is 10 m, with most occurrences being 3–4 m thick. This association is more sand rich and thicker bedded than NCF1, and is heterogeneous, with interbedded thin and medium beds. Packets of NCF2 are continuous for hundreds of metres along strike, although individual sandstone beds may be lenticular or wedge-shaped. NCF3 forms 17% of non-channelized deposits, and is the thickest bedded and coarsest grained of the three non-channelized associations. Sandstone beds are amalgamated, and sand:shale approaches 100%. Sandstone packets vary in thickness from 2–25 m, with the thickest packets having the most coarse-grained facies. The base of packets are relatively sharp, but there is no evidence of significant erosion. Asymmetric cycles are uncommon. Instead, packets are in many cases coarsest in the centre, and finer both at the top and the base. The coarse central parts (Fig. 7.28) consist of lenticular bodies of pebbly sandstone that contrast in their shape with the essentially sheet-like nature of the packets themselves. In one example, the coarsest parts of successive beds are offset from one another, producing a shingling effect.

The NCF3 sandstone packets are interpreted as non-channelized depositional lobes that grade distally into lobe-fringe (NCF2) and fan-fringe (NCF1) deposits. The interlayering of packets of the three facies associations, with the sequence NCF1 → NCF2 → NCF3 indicates that the dominant growth pattern of the fan at some times was repeated progradation of sandy lobes. Parts of the section that are characterized by alternations of NCF1 and NCF2, or NCF2 and NCF3, probably developed by rapid lateral switching of lobes, with no requirement for progradation.

The channelized facies associations are more-or-less equally abundant (CF1:CF2 = 55:45). CF1 consists of strongly lenticular, amalgamated structureless and graded sandstones and pebbly sandstones with parallel and cross stratification. Pebbly sandstones may form 50% of CF1 units. The base of the units are sharp and erosive, being characterized by either abundant rip-up clasts forming a basal lag, or step-wise downcutting to minimum depths of 2–3 m. The top of CF1 packets tend to show either a step-like thinning upward into overlying fine-grained deposits of CF2, or are abrupt. This upper transition zone is characterized, in about 1/3 of all packets, by Facies B2.2 reworked, cross-stratified, lenticular units to 4 m thick. This facies is almost exclusively confined to the top of CF1 units. CF2 consists of thin and very thin-bedded siltstone turbidites, isolated siltstone ripple trains, and minor

Table 7.2 Composition of Non-Channelized Fan (NCF) and Channelized Fan (CF) facies associations, Lower Milliners Arm Formation (from Watson 1981).

Association	Per cent by facies													
	A1.4	A2.7	A2.8	B1.1	B2.1	B2.2	C1.1	C2.1	C2.2	C2.3	D2.1	D2.2	D2.3	E1.1
NCF1	—	—	—	—	—	—	3	—	8	21	40	—	25	1
NCF2	—	—	—	—	—	—	—	11	24	32	28	—	5	—
NCF3	—	10	4	23	—	—	—	45	15	3	—	—	—	—
CF1	3	22	10	27	1	3	—	28	4	2	—	—	—	—
CF2	—	—	—	—	—	—	2	4	—	9	30	13	40	2

thin-bedded sandstone turbidites, in units 2–35 m thick. These facies are organized into well defined thinning-upward sequences several metres thick. Palaeoflow data are more diverse than for CF1.

Watson (1981) interprets associations CF1 and CF2 as channel and interchannel deposits, respectively. The thinning- and fining-upward sequences that characterize the top of CF1 units are believed to indicate progressive channel abandonment (Mutti & Ghibaudo 1972) and/or shingling of the thickest parts of beds in a wide channel (Martini & Sagri 1977). Facies B2.2 reworked deposits that occur at the top of sandstone packets are interpreted as channel-margin sediments, formed beneath non-depositing flows that spilled over the channel levees. The specific process involved may have been flow stripping (Piper & Normark 1983). Channel migration led to the superposition of channel-margin deposits on top of channel deposits.

Channelized and non-channelized facies associations alternate throughout the lower part of the MAF, and are interpreted as lower and middle fan deposits. The poor development of asymmetric cycles results in uncertainty as to whether many of the sandstone packets represent channel fills or lobes.

The upper part of the MAF is about 1.4 km thick, and consists predominantly of thick units of conglomerate and pebbly sandstone of Facies Class A. Several field photographs from this interval appear in Chapter 3 (Figs 3.6b, c, d, e, f & 7d). Five facies associations are recognized (Table 7.3), the first two being mainly coarse grained and the last three being finer grained. Abundant channeling and the coarse nature of the deposits led Watson (1981) to interpret all these sediments as upper fan deposits.

Facies association IF1 constitutes 12–70% of measured sections, and is dominated by Facies Class A clast-supported conglomerates. Amalgamated units average 20 m thick, and are interlayered with packets of

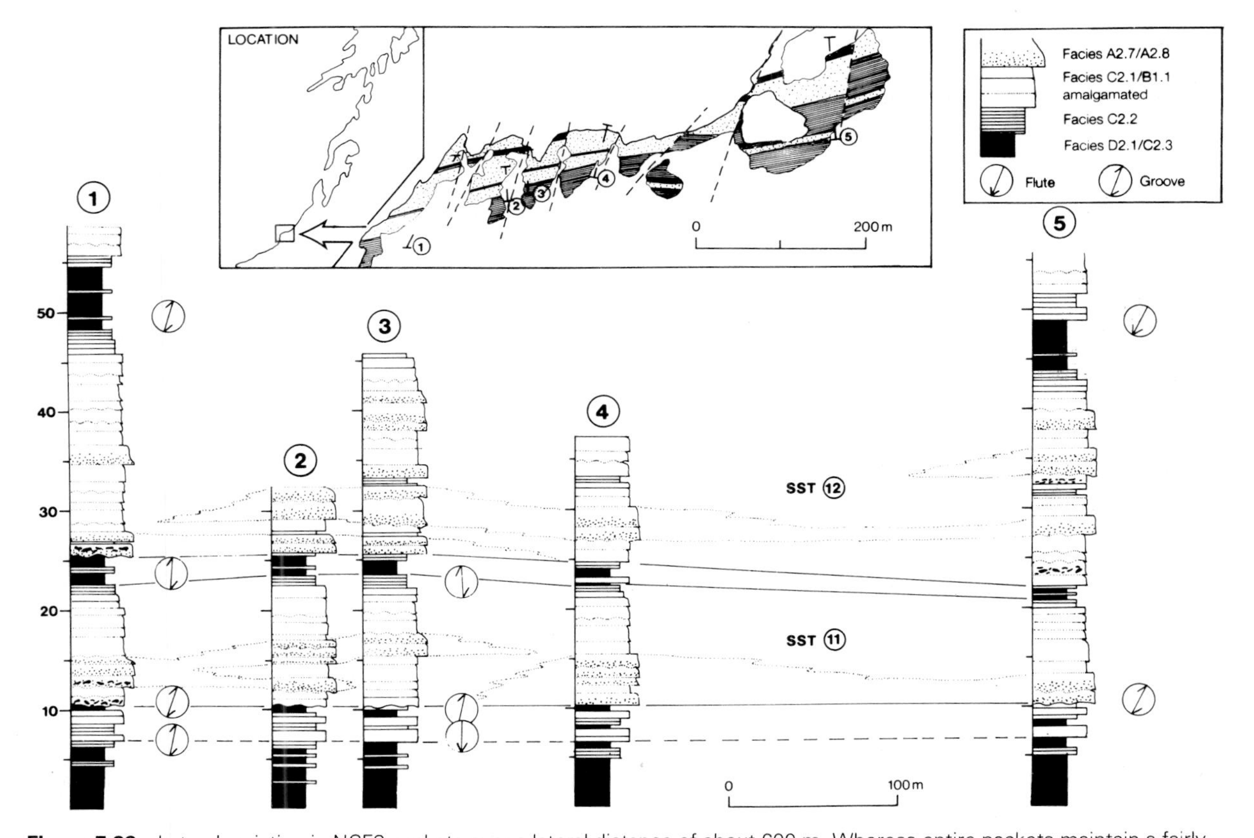

Figure 7.28 Lateral variation in NCF3 packets over a lateral distance of about 600 m. Whereas entire packets maintain a fairly constant thickness, pebbly sandstone facies are lenticular. Palaeocurrent measurements are plotted with north at the top of the diagram.

Table 7.3 Composition of Inner Fan (IF) facies associations, upper Milliners Arm Formation (from Watson 1981).

Association	Per cent by facies																
	A1.1	A1.4	A2.2	A2.3	A2.4	A2.7	A2.8	B1.1	B1.2	B2.1	B2.2	C2.1	C2.2	C2.3	D2.1	D2.2	D2.3
IF1	13	11	23	27	4	12	8	2	—	—	—	—	—	—	—	—	—
IF2	1	5	—	4	3	23	15	18	—	3	—	25	3	—	—	—	—
IF3	—	—	—	—	—	—	—	—	17	—	65	—	—	5	8	3	2
IF4	—	—	—	—	—	—	—	4	—	1	—	22	30	26	15	—	2
IF5	—	—	—	—	—	—	—	—	—	—	—	—	1	12	61	6	19

IF2 to form multiple stacked units up to 130 m thick. Association IF2 forms 20–40% of measured sections, and is dominated by graded and stratified pebbly and non-pebbly sandstones in units 2–18 m thick. The composite IF1/IF2 units have erosive bases, with demonstrable erosion of up to 18 m into underlying thin-bedded units. The erosion is accomplished by a series of steep steps separated by flat terraces. Internal erosion is also characteristic of the IF1/IF2 units.

Association IF3 is volumetrically minor, and consists of strongly lenticular units, <7.5 m thick, of relatively well sorted, cross-bedded sandstones (Facies B2.2) and gravel lags (Facies B1.2). These deposits almost always overlie composite IF1/IF2 units and underlie IF4/IF5 units. Cross-bed azimuths differ from sole markings in both underlying and overlying units by 30–60°.

Association IF4 forms 3–33% of measured sections, in units 1–20 m thick, and consists predominantly of classic turbidites of Facies Groups C2 and D2. Where units can be traced along strike for distances of several tens of metres, they are seen to have a wedge shape that results from gradual thinning (or thickening) of all sandstone beds in the unit. Both grain size and the proportion of beds starting with Bouma T_a and T_b divisions decrease in the direction of thinning. Palaeoflow for IF4 units differs by up to 50° from the mean direction in IF1/IF2 units.

Facies association IF5 accounts for 30% of sections through the western part of the upper MAF, but forms only 2% of more easterly sections (Fig. 7.29). Siltstone turbidites of Facies Group D2 form 85% of this association (Table 7.3). Units are 4–40 m thick, and commonly gradationally overlie units of IF4. Contorted bedding and truncation surfaces attest to gravitationally-induced sliding. Palaeoflow is broadly the same as for IF1/IF2 units.

In vertical sections through the upper part of the MAF, there is a clear tendency for facies associations to succeed one another in the order IF1 → IF2 → IF3 → IF4 → IF5 (Fig. 7.30). Partial sequences occur as a result of erosion beneath IF1/IF2 conglomerate and pebbly sandstone units. Fully developed sequences from IF1 to IF5 are most easily explained by channel-migration processes, with IF1 deposits corresponding to the channel axis and IF5 deposits to levee and overbank sediments. The reworked deposits of IF3 are assigned to the channel margin, and are believed to form during spillover of non-depositing flows, a suggestion supported by the palaeoflow divergence between axial deposits and cross-beds of IF3. Association IF2 was deposited in the channel system, and is intimately associated with conglomerates of IF1. This lateral segregation of conglomerates and pebbly sandstones in the channel suggests a system of thalweg channels (IF1) flanked and separated by terraces (IF2) (Fig. 7.31), much like the depositional model proposed by Hein & Walker (1982) for similar deposits in Quebec, Canada.

The entire Milliners Arm Formation is the deposit of a prograding submarine fan system. Oscillations between channelized and non-channelized lobe and lobe-fringe deposits in the lower part of the formation, however, indicate that fan growth was irregular, with frequent lobe abandonment. The upper fan valley was of the order of 5–10 km across, a size that corresponds to valleys of modern fans with a radius of about 50–100 km (Normark 1978, Fig. 6).

Summary

Case studies show that submarine fans, both modern and ancient, show a high degree of variability in scale, mean grain size, and facies distribution. In general,

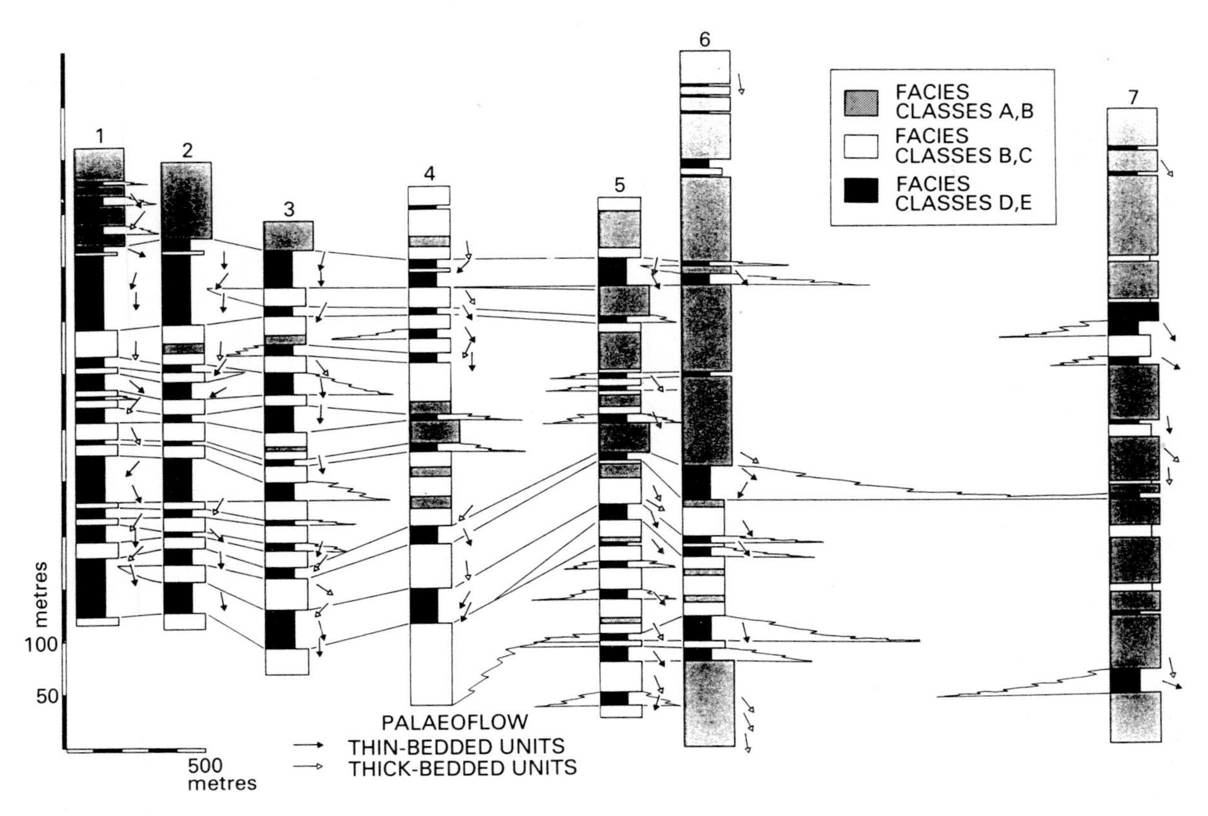

Figure 7.29 Simplified correlated sections in the upper part of the MAF, showing general eastward decrease in finer grained facies, and the importance of channels.

the world's largest fans are delta fed and mud rich, but have a low probability of being preserved in the ancient record, as a result of severe dismemberment during subsequent orogenesis. Sand-rich fans tend to be smaller, and are characterized by well developed channels and lobes as predicted by the Normark (1970, 1978) and Walker (1978) fan models. There is a growing awareness that sea-level fluctuations exert a major control on (a) the delivery of sands to submarine fans, and (b) the growth pattern and sand distribution on fans (Mutti 1985). In particular, the attributes that were originally assigned to deposits of the middle and lower fan may not reflect proximality, but rather the relative height of sea level. The importance of asymmetric cycles in defining fan environments has probably been overestimated, and in any case the explanations for convincing thinning-upward and thickening-upward cycles are not yet entirely clear. It appears that most thickening-upward cycles are compensation cycles (Mutti & Sonnino 1981), and not progradational cycles. An important process in the plugging of channels and the initiation of the abandonment that generates thinning-upward cycles may be flow stripping (Piper & Normark 1983).

As concluded by Shanmugam *et al.* (1985), the fundamental weakness in our current understanding of submarine fans stems from the lack of continuous long cores from modern fans that could allow meaningful correlations to be made between fan environments and facies, particularly for features that can be observed at the scale of the outcrop. An exception is the Mississippi Fan, drilled on Leg 96 of the Deep Sea Drilling Project (Bouma *et al.* 1985, Bouma, Coleman *et al.* 1985b). Until such cores become available for several modern fans, the distinction between high- and low-efficiency fans, the assessment of the influence of sea-level variations on fans of different sizes, and a precise explanation for cyclicity, will remain speculative.

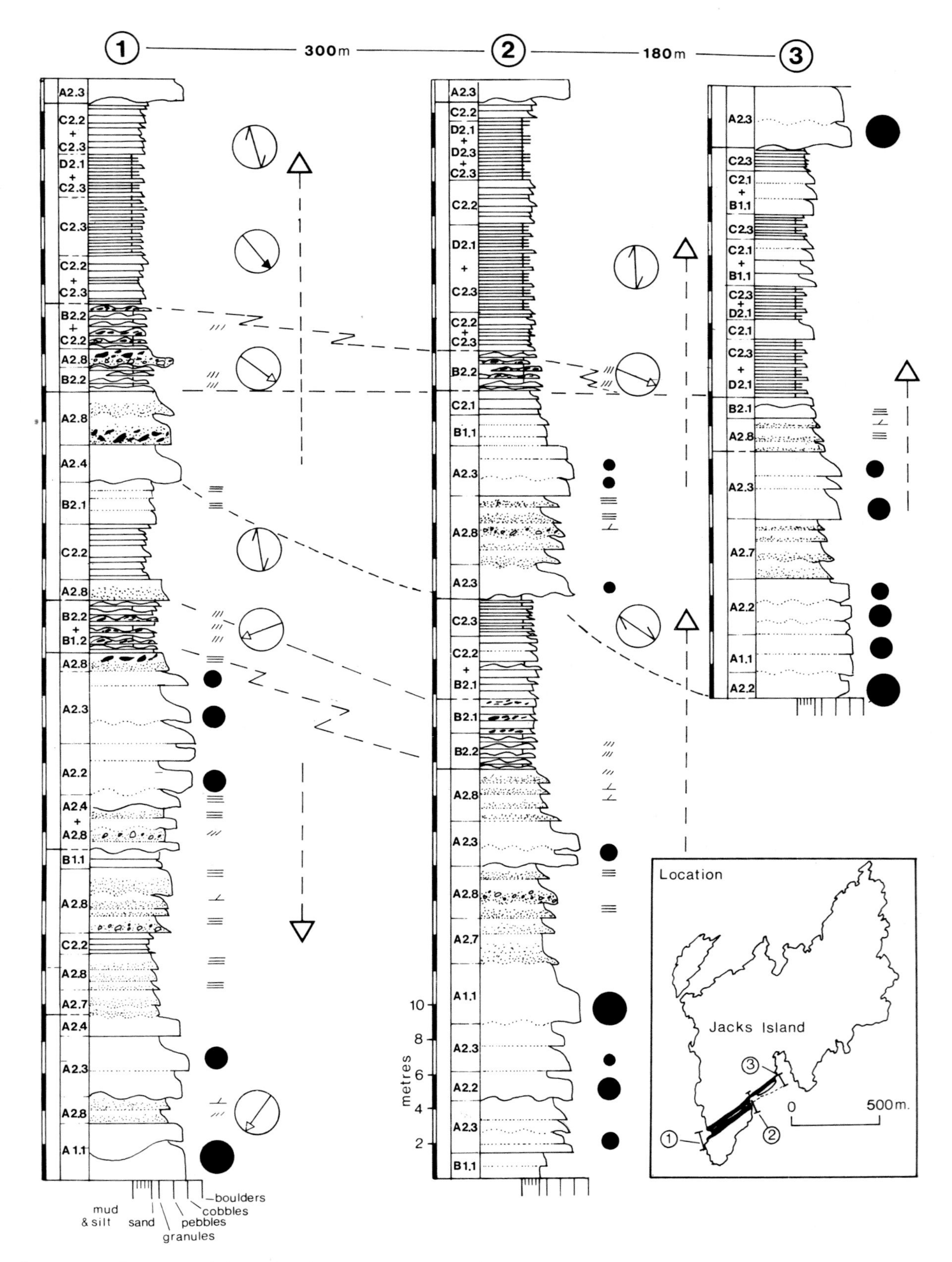

Figure 7.30 Correlation of detailed sections in the upper part of the MAF, showing tendency for fining-upward sequences with partial or complete cycles of IF1 → IF2 → IF3 → IF4 → IF5. Coarse-grained packets tend to be lenticular.

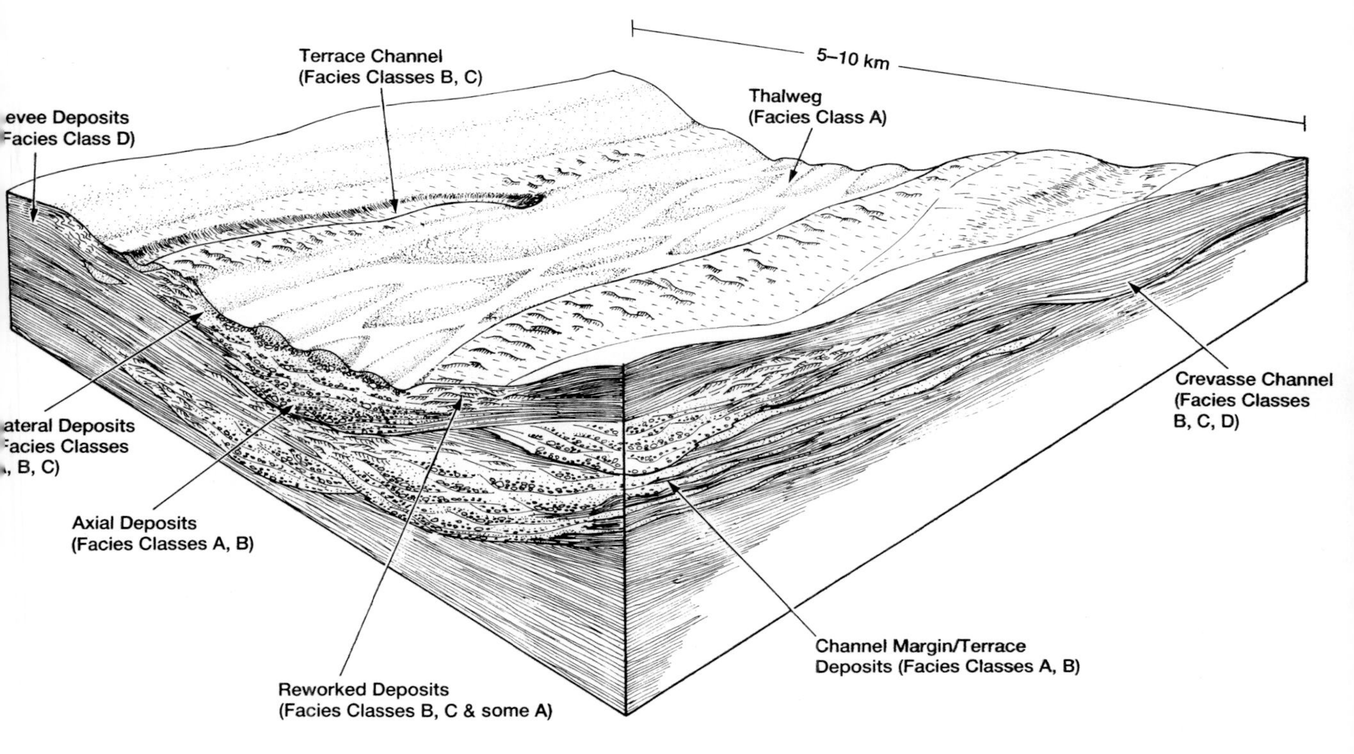

Figure 7.31 Summary block diagram showing the interpretation proposed by Watson (1981) for inferred upper fan channel deposits of the upper part of the Milliners Arm Formation.

Example of sheet turbidite system from an ancient foreland basin, Miocene Marnoso-arenacea Formation, Coniale, Santerno Valley, northern Apennines, Italy. Vertical section through about 80 m of stratigraphy. Thick bed at top of section is the famous Contessa megaturbidite (cf. Ricci Lucchi & Valmori 1980).

CHAPTER 8

Sheet systems

Introduction

This chapter, in three parts, is concerned with sheet depositional systems that cover essentially horizontal, relatively large areas of the sea floor either at abyssal depths (often greater than 4000 m) or in shallower basins below storm wave-base. Part I deals mainly with open ocean sheet systems such as modern abyssal plains with some case studies in order to emphasize the features of such systems and, indeed, to show that while generalizations can be made, each system has unique features. Part II concerns elongate sheet systems, particularly modern trench floors, arc-related and foreland-basin floors, again with case studies. Finally, Part III includes a discussion of ancient deep-marine plains and summarizes the main attributes of such sheet systems. The small ponded basins associated with slopes are described in Chapter 5.

While we recognize a clear distinction between modern open ocean abyssal systems (Part I) and trench floors (Part II), often such differentiation is impossible in ancient systems. Furthermore, plate movements mean that many open ocean plains become incorporated into part of active trenches prior to severe deformation during subduction–accretion processes (Ch. 11). Thus, in any dynamic reconstruction for ancient deep-marine plains, account must be taken of the fact that the plate tectonic setting has changed through time. While some readers may find this chapter contains 'strange bedfellows', we would emphasize that both open ocean plains, trenches, arc-related and foreland basins may contain similar sheet-like sedimentation.

In modern environments, we can precisely characterize and describe abyssal plains and other deep-marine plains in terms of shape, size, physiography, sediment types and distribution, and sediment source. However, in ancient successions, many of these features are inferred from limited data. This is especially true in attempting to define water depths; therefore, the less precise terms 'basin plain', 'basin floor', 'deep-marine' or 'deep-sea plain' are preferred to abyssal plain. The main attribute of clastic systems discussed in this chapter is that sedimentation is in the form of extensive sheet deposits known or believed to have been laid down in the relatively deep parts of the basin or as part of a much larger ocean floor. There are no constraints on bed thickness, grain size, composition or plate tectonic setting. In many cases, limited shallow channels/valleys may also form part of such sheet systems.

It is important to stress that while pelagic sediments commonly occur in many sheet systems, they are rarely the dominant lithology. Turbidites tend to infill depressions rapidly and mantle uneven topography, thereby accounting for the bulk of sediments in most sheet systems, including abyssal plains.

I: Modern abyssal plains

Abyssal plains were discovered in 1947 when M. Ewing and J. L. Worzel modified the echo sounder aboard *R/V 'Atlantis'* so that it could record depths to 4000 fathoms (approx. 8000 m). Modern abyssal plains cover approximately 4.4% of the Earth's surface (Fig. 8.1, areas from Weaver *et al.* 1987) as essentially horizontal areas of the ocean floors where gradients typically are less than 1:1000 (Heezen & Laughton 1963). There are few recent general publications on abyssal plains, except the book edited by Weaver & Thomson (1987).

Abyssal plains may be thousands of kilometres in length and up to hundreds of kilometres wide, and range from elongate to highly irregular in shape. Sediment thicknesses generally are less than a few hundred metres, although local maxima of 1.5–2 km have been recorded (e.g. Collette *et al.* 1969, Ewing & Hollister 1972, Wang *et al.* 1982). The larger abyssal plains tend to be seaward of major continental drainage areas, with single or multiple sediment supply points such as submarine canyons and deep-marine valleys (see Ch. 6). About 75 abyssal plains are recognized in the world's oceans (Fig. 8.1), with a rather uneven distribution between the oceans. The Atlantic Ocean contains 29, the Arctic Ocean ten, and the Pacific Ocean has seven (Weaver *et al.* 1987). The largest is the Enderby Abyssal Plain in the Antarctic Ocean at about 3.7×10^6 km^2.

Seismic profiling shows many of the abyssal/deep-sea plains to be the result of sediment draping, infilling

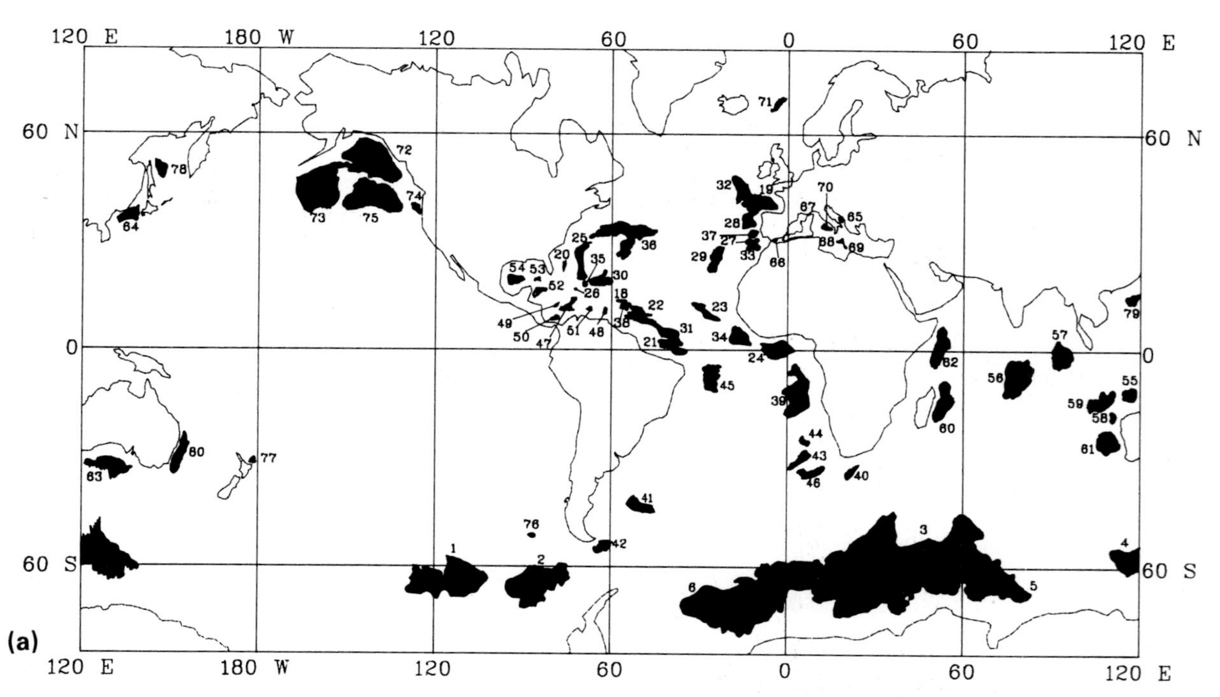

Figure 8.1a Distribution of abyssal plains in the world's oceans. See also Figure 8.1b. *Antarctic Ocean*: 1, Amundsen; 2, Bellinghausen; 3, Enderby; 4, South Indian; 5, Valdivia; 6, Weddell. *Arctic Ocean*: 7, Barents; 8, Boreas; 9, Canada; 10, Chukchi; 11, Dumshaf; 12, Fletcher; 13, Greenland; 14, Mendeleyev; 15, Northwind; 16, Pole; 17, Wrangel. *Atlantic Ocean (N)*: 18, Barracuda; 19, Biscay; 20, Blake Bahama; 21, Ceara; 22, Demerara; 23, Gambia; 24, Guinea; 25, Hatteras; 26, Hispaniola; 27, Horseshoe; 28, Iberian; 29, Madeira; 30, Nares; 31, Para; 32, Porcupine; 33, Seine; 34, Sierra Leone; 35, Silver; 36, Sohm; 37, Tagus; 38, Vidal. *Atlantic Ocean (S)*: 39, Angola; 40, Aghulas; 41, Argentine; 42, Burdwood; 43, Cape; 44, Namibia;

and smoothing the uneven basement topography by horizontally stratified sediments. Seamounts, ocean ridges, abyssal hills and fracture systems disrupt the otherwise planar topography. The distribution of facies is mainly influenced by the shape of the basin, the location of sediment sources, temporal variations in the relative importance of various sediment sources, sea-level fluctuations and the tectonic history of the plain.

Turbidites in abyssal plains commonly onlap or offlap abyssal hills. The hills locally deflect turbidity currents on the plains. Pelagic sediments commonly mantle these hills. Weeks & Lattimore (1971) document abyssal hills between the Pioneer Ridge and the Murray Fracture Zone, 400–500 km from the base of the continental slope, with relief up to 600 m, but seldom exceeding a few hundred metres, and with widths up to 2.5 km.

Abyssal plains may be the final destination of the finest grained and farthest transported terrigenous sediments; also, it is here that large amounts of truly

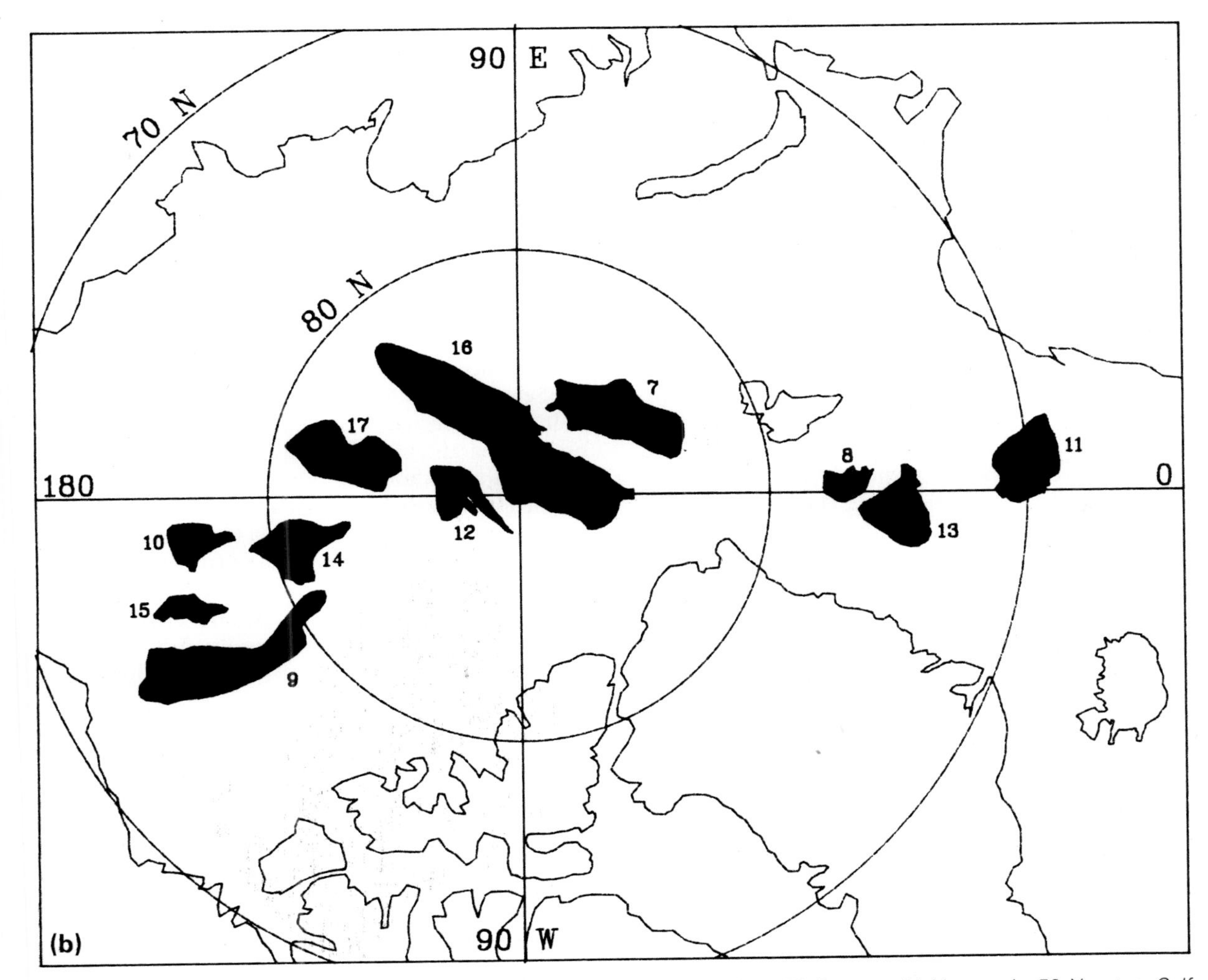

45, Pernambuco; 46, Town. *Caribbean*: 47, Columbian; 48, Grenada; 49, Jamaican; 50, Panama; 51, Venezuela; 52, Yucatan. *Gulf of Mexico*: 53, Florida; 54, Sigsbee. *Indian Ocean*: 55, N. Australian; 56, Mid Indian; 57, Cocos; 58, Cuvier; 59, Gascoyne; 60, Mascarene; 61, Perth; 62, Somali; 63, S. Australian. *Sea of Japan*: 64, Japan. *Mediterranean*: 65, Adriatic; 66, Alboran; 67, Balearic; 68, Sicilia; 69, Sidra; 70, Tyrrhenan. *North Sea*: 71, Norway. *N. Pacific*: 72, Alaska; 73, Aleutian; 74, Cascadia; 75, Tufts. *S. Pacific*: 76, Mornington; 77, Raukumara. *Sea of Okhotsk*: 78, Okhotsk. *South China Sea*: 79, South China Sea. *Tasman Sea*: 80, Tasman. After Weaver & Thomson (1987).

pelagic sediment occur as pteropod, foraminiferal, diatomaceous, nannofossil and radiolarian oozes (Facies Group G1), although such pelagites occur elsewhere in ocean basins and upon topographic highs such as abyssal hills far from areas of substantial terrigenous clastic influx. In the deepest parts of the oceans and seas, such biogenic oozes may be replaced by the 'red clays' (Facies E1.2) formed both from the disintegration of *in situ* minerals and sediments/rocks, and accumulation of aeolian dust. In these plains, the major controls on pelagic sedimentation are the relative position of the calcite compensation depth (CCD) and the organic productivity of the surface waters.

Although abyssal plains are defined as sheet systems, this should not necessarily imply that they comprise only parallel, horizontally stratified sediments at depth. For example, in the north Iberia Abyssal Plain, where there are five NNE–SSW trending, en echelon, turbidite-filled basins deeper than 5500 m (from 4–10 km wide and 19–65 km long), the sediment fills, dating from at least 8 Ma, are tilted eastwards due to more rapid tectonic subsidence of the eastern margins (Addy & Kagami 1979). Thus, while the basin deposits are classified as sheet systems, their overall seismic geometry is that of a wedge because of active tectonics during sedimentation.

In recent years, one of the most interesting and possibly useful features to emerge from many modern and ancient deep-marine plains is the occurrence of megabeds or ponded turbidites. Examples include: the correlation of six individual turbidites that flowed for over 300 km in the 15 000 km^2 Horseshoe Abyssal Plain, with three beds having volumes of 5.7, 8.0 and 12.1 km^3 (Hoyt & Fox 1977); the correlation of three turbidite layers for about 32 km in the 37 km by 15 km Western Alboran Abyssal Plain (Bartolini *et al.* 1972); the basin-wide extent of three turbidites that travelled about 100 km across the Hispaniola–Caicos Abyssal Plain, with maximum thicknesses of 2 m, volumes from 0.89 to 3.08 km^3, and covering areas ranging from 3500–5200 km^2 (Bennetts & Pilkey 1976); and in the Puerto Rico Trench with up to 1.7 km of sediment fill, individual graded beds up to 6.5 m thick that can be traced for up to 200 km, with the largest bed averaging 3 m thick, occurring over 10 000 km^2, and having a volume of 0.03 km^3 (Conolly & Ewing 1967). These thick beds were deposited from ponded (contained) turbidity currents. They provide not only useful environmental indicators but also important chronostratigraphic horizons for correlation purposes.

Deep-marine plains are often associated with 'sediment waves' (Ch. 9), for example in a 300 m-thick zone 550 m beneath the 3260 m-deep Mendeleyev Abyssal Plain – this zone in turn overlies about 1500 m of horizontally stratified sediments (e.g. Weeks & Lattimore 1971, Mudie *et al.* 1972).

Before looking at some specific case studies of abyssal plains, brief summaries of some of the more interesting aspects of abyssal-plain sedimentation from various parts of the world's oceans are described in a 'regional' context.

NORTH ATLANTIC

A considerable part of our present knowledge about the North Atlantic abyssal plains stems from work in the 1950s such as that by Ericson *et al.* (1951, 1952), Heezen *et al.* (1951, 1954), and Luskin *et al.* (1954); more recent work is summarized in Weaver & Thomson (1987). The best known plains in the west include the Sohm, Hatteras (see case studies below) and Nares Abyssal Plains, the latter being the deepest plain of the North Atlantic with an area of about 95 000 km^2. In the east, the larger plains include Biscay, Tagus, Iberia and Horseshoe Abyssal Plains, with smaller examples such as Cape Verde, Madeira and Seine Abyssal Plains. For the eastern North Atlantic plains, the principal sediment sources are the Anglo–French margin, the Iberian Peninsula and North Africa. Abyssal hills and other submarine highs have resulted in a complex, irregular, abyssal-plain system with relatively sinuous links between some of the basins, for example the Theta Gap that joins the Biscay and Iberia Plains. In all the eastern North Atlantic plains, there appears to be a westward increase in the amount of pelagic sediment, paralleled by a general reduction in the grain size of turbidites (Horn *et al.* 1972).

In the Madeira Abyssal Plain, Weaver & Rothwell (1987) have mapped out twelve individual turbidites and intervening pelagic sediments on the basis of colour, organic-carbon and calcium carbonate content, and a fine-scale stratigraphy based partly on micropalaeontology. Figure 8.2 shows the characteristics of the correlated turbidites. Weaver & Rothwell (1987) plotted isopach maps for these deposits in which fine-grained T_{de} (E1–E3 of Piper 1978) turbidites exceed 5 m thickness – in the case of the 'F Turbidite' (Fig. 8.3), deposit volume is estimated at 12.6 km^3. The turbidites thicken westwards into the basin, although the silty bases coarsen eastwards towards the source on the NW African margin.

Weaver & Rothwell (1987) show that the pelagic

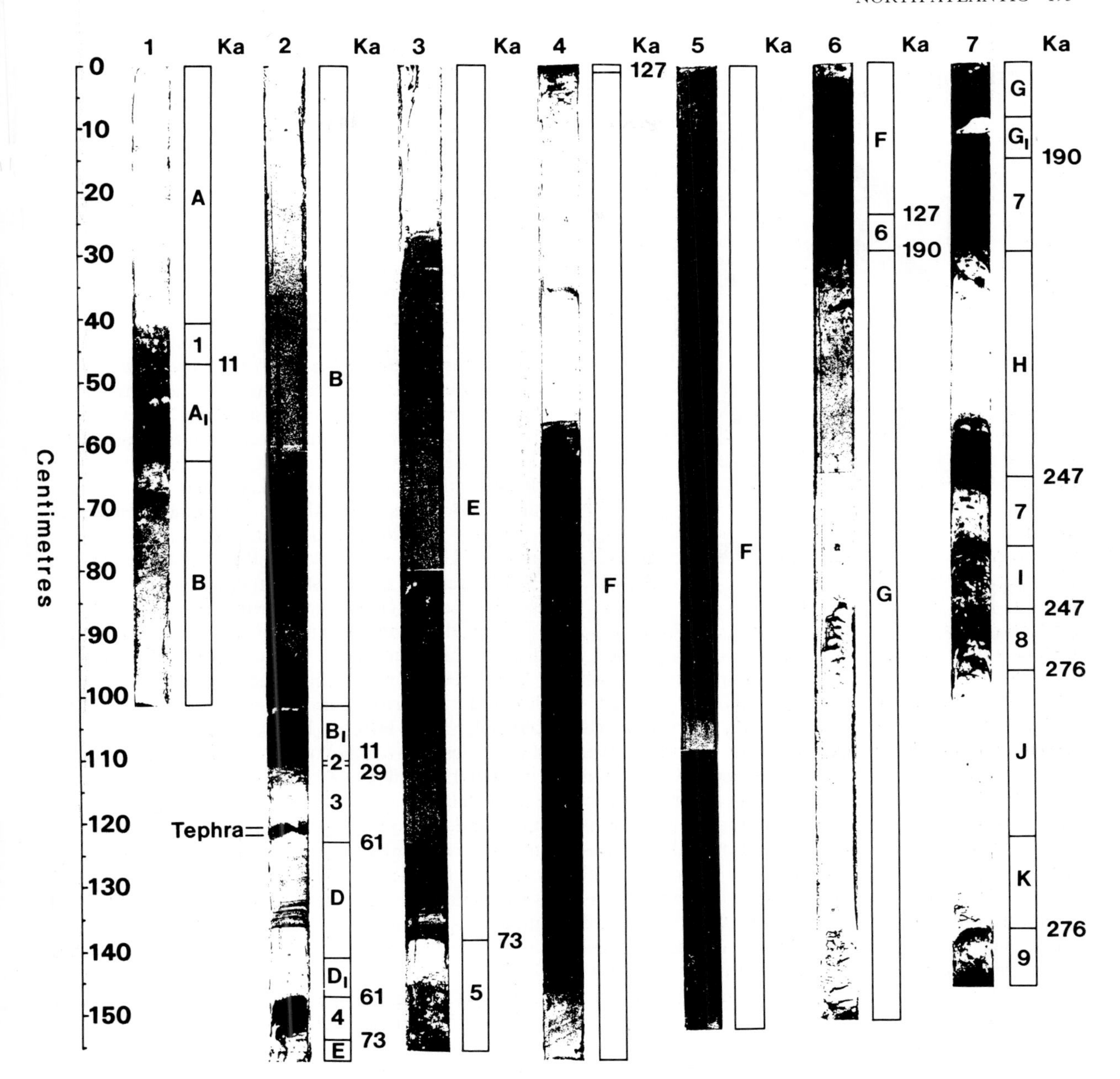

Figure 8.2 Piston cores from Madeira Abyssal Plain (D10688). Lettered units are interpreted as turbidites and numbered units refer to intervening pelagic layers. Numbers to right of columns are ages based on oxygen–isotope stratigraphy. Note: bioturbation of pelagic units and upper part of turbidites; the coarse-grained, laminated, base of turbidites B, B1 and G, and to a lesser extent in turbidites E and F; essentially structureless nature of thick turbidites; distinct colour changes in organic-rich turbidites A1, E, F and H representing relict oxidation fronts. From Weaver & Rothwell (1987).

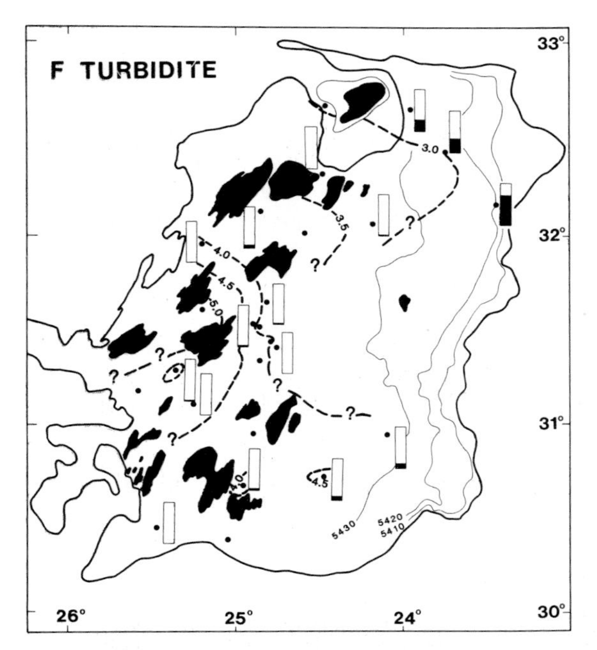

Figure 8.3 Isopach map in metres of Turbidite 'F' from Madeira Abyssal Plain, N. Atlantic. Shaded areas are abyssal hills; 5400 m isobath is approximate boundary of plain; columns represent basal metre of turbidite in which shaded part is T_d, or coarse-grained T_{bc} division. After Weaver & Rothwell (1987).

sediment composition has varied from calcareous marls and oozes during periods of continental glaciation, with sedimentation above the CCD, to pelagic clay deposited below the CCD during interglacials. Furthermore, Weaver & Kuijpers (1983) demonstrate that the emplacement of the thick turbidites occurred during sea-level changes at the beginnings and ends of glacial periods, not only during low-stands of sea level as predicted for sandy turbidites by Vail *et al.* (1977), and Shanmugam & Moiola (1982). In a study of the southern Brazil Basin, Johnson & Rasmussen (1984) also document the emplacement of turbidites during glacial-interglacial transitions.

MEDITERRANEAN SEA

In the Mediterranean, the largest plain is the Balearic Abyssal Plain (Hersey 1965a & b), defined in the Mediterranean below the 2850 m contour, and containing an estimated 1200 m of mainly turbidite fill (Menard *et al.* 1965). Core data indicate an overall southward decrease in bed thickness and grain size of the turbidites away from the Rhône Fan. The sea floor also slopes gently towards the southeast. However, some turbidites are derived from the North African continental margin. A particularly interesting feature of this plain is that underlying Messinian salt deposits have formed domes that in 70% of the examples recorded by Stanley *et al.* (1974) have penetrated the surface with relief of at least 10 m. In some cases, domes have acted as dams behind which sediment has been ponded. In contrast to most abyssal plains shaped by depositional processes, the distribution of sediments within the Balearic Abyssal Plain is controlled by salt diapirism causing folding and faulting (Fig. 8.4). In fact, locally, deformation rates exceed rates of sedimentation, resulting in a lack of correlation of individual turbidites in this plain (Stanley *et al.* 1974). Other Mediterranean plains include the Tyrrhenian, Messina, and Herodotus Abyssal Plains, together with numerous small basin plains off the west coast of the Peloponnesus and along the flanks of the Mediterranean Ridge in the Ionian Sea (Fig. 8.5) – the latter examples are not at abyssal depths but are characterized by relatively flat floors and horizontally stratified sediments. Herodotus Abyssal Plain, in the eastern Mediterranean, is interesting in that it lies between the passive and aseismic continental margin of North Africa and the Mediterranean Ridge, which is an active feature subject to strong neotectonic deformation.

'Megabeds' or very thick-bedded turbidites (Facies C2.4) are reported from some of the Mediterranean abyssal plains. Hiecke (1984), in the Ionian Abyssal Plain, describes a Holocene homogenite or unifite (Stanley 1981), with a 3 m-thick graded sandy base and structureless mud cap. This bed is greater than 12 m thick and, based on the assumption that the minimum thickness is 10 m over an area of at least 1100 km^2, it has a volume of at least 11 km^3. The bed contains benthic foraminifera that suggest water depths greater than 200 m, and its origin is ascribed to submarine slope failure to the north during the collapse of Santorini Caldera 3500 years BP. Also, upper Pleistocene megabeds in excess of 10 m thick are reported from Herodotus Abyssal Plain (Cita *et al.* 1984). These beds are pale brown to whitish, coarse-grained, carbonate-rich bioclastic sediments containing mainly displaced shallow-water faunas. Individual beds are believed to involve sediment volumes of about 10 km^3. Cita *et al.* (1984) divide the turbidites of Herodotus Abyssal Plain into those that are thin-bedded (type A) and megabeds (type B) that are up to an order of magnitude thicker. The megabeds are believed to have resulted from the sudden collapse of bioclastic sediments at one of two canyon heads on the North African continental margin during lowered sea-level stands, while the thin-bedded

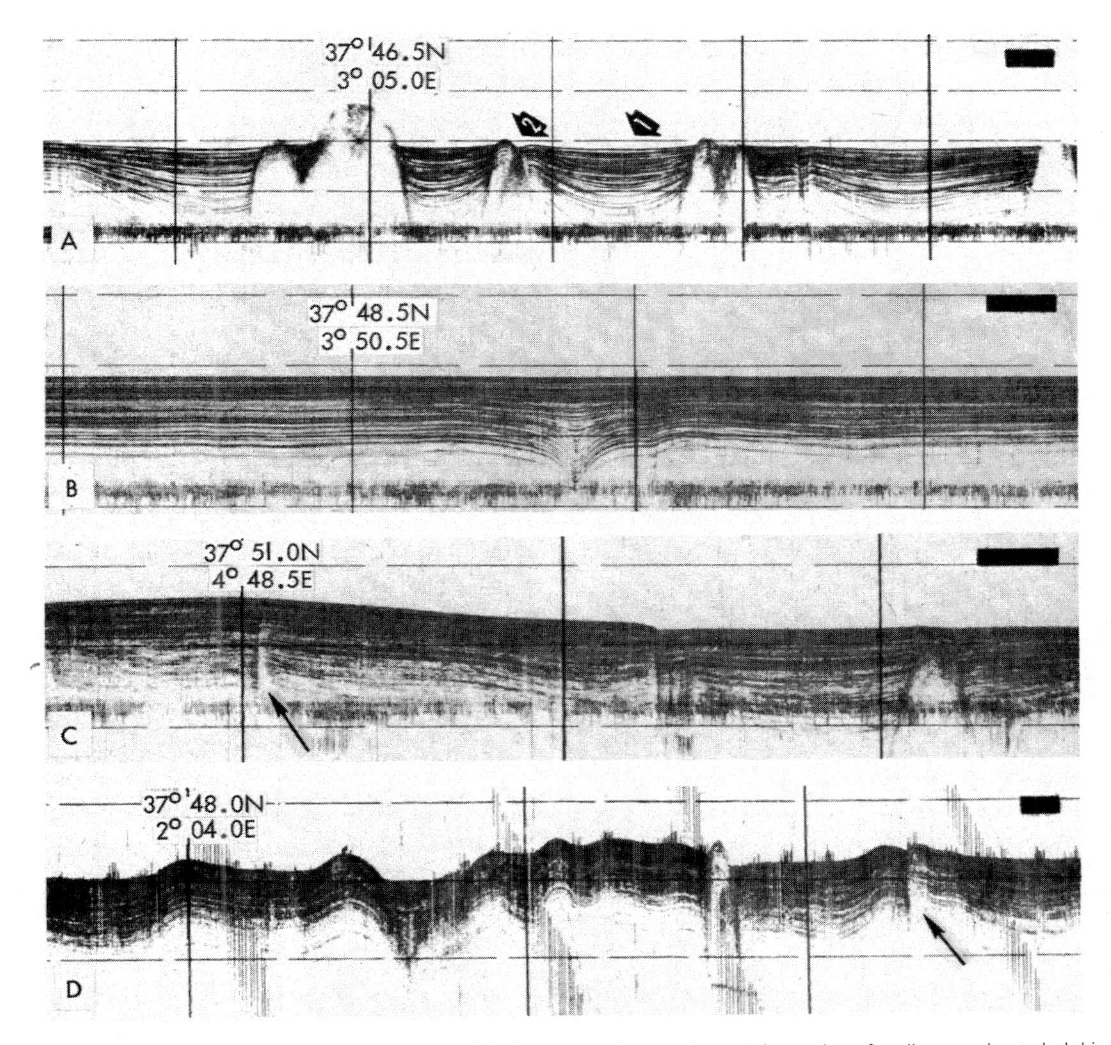

Figure 8.4 3.5 kHz record from Balearic Abyssal Plain, Mediterranean Sea, to show deformation of sediments due to halokinesis. Arrows are faults; arrow '2' shows salt-dome piercement, and arrow '1' shows bathymetric depression. After Stanley *et al.* (1974). Thick horizontal bars are 1 km; thin horizontal bars are 36.6 m apart.

turbidites came from the Nile Cone. Flow paths are complex around abyssal highs, with sediment by-pass in some areas.

NORTH PACIFIC

In the North Pacific, the abyssal plains are well protected from any substantial terrigenous influx in the north, west and part of the east. This is due to small basins, island arcs, trenches and troughs acting as effective barriers and traps to turbidity currents, unlike in the Mediterranean plains. However, in the NE Pacific, turbidites extend for over 1700 km seaward off the coast of Oregon, having come mainly from the Columbia River, via Willapa Canyon, across Cascadia Abyssal Plain and through the Blanco Fracture Zone onto Tufts Abyssal Plain. While numerous conduits for coarse clastic detritus existed towards the close of the Pleistocene, most coarse material only reached as far north as the Gulf of Alaska Abyssal Hills (Horn *et al.* 1972).

The Aleutian Abyssal Plain (see case studies) predates the development of the Aleutian Arc and Trench, after which a thick pelagic cover (up to 96 m in the west

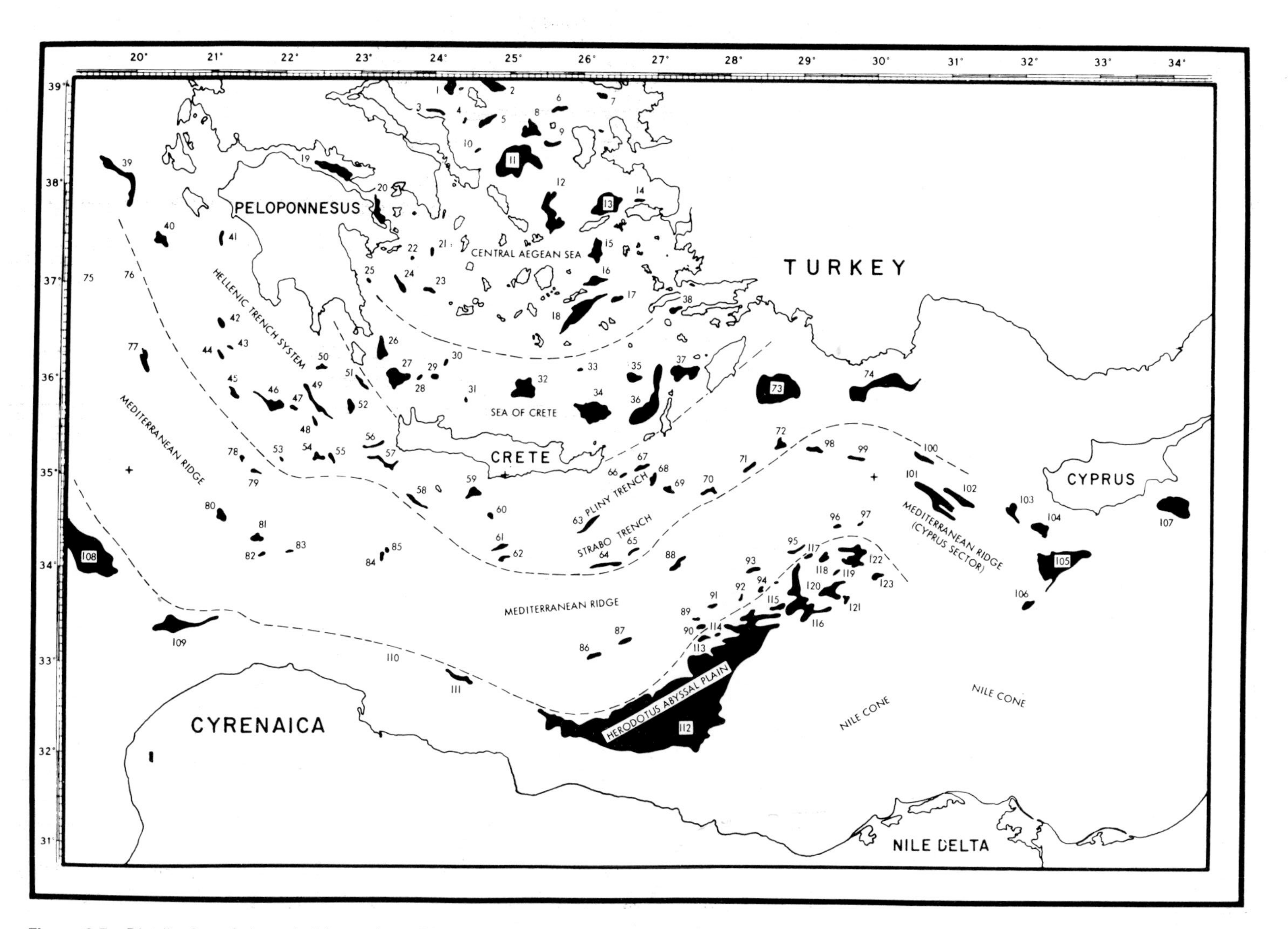

Figure 8.5 Distribution of abyssal plains and small-basin floors in the eastern Mediterranean. After Stanley (1973).

central part) blanketed the older turbidite fill of the abyssal plain (Hamilton 1967). Bering Abyssal Plain now traps most turbidites, and Ewing *et al.* (1965) demonstrated, using continuous sub-bottom reflection profiles, that the southeastern part of the plain is underlain by 1000 m of turbidites compared to only 470 m in the central part (Scholl *et al.* 1968). In common with abyssal plains the world over, the Holocene transgression has led to extensive pelagite/hemipelagite blankets over turbidite-dominated successions in these abyssal plains (see also Griggs *et al.* 1969, Cascadia Abyssal Plain; Carson & Arcaro 1983, Juan de Fuca Abyssal Plain).

A feature of some abyssal plains is the occurrence of channels and valleys incised into otherwise flat surfaces. For example, the eastern Central Pacific Abyssal Plain contains channels filled with calcareous turbidites cutting the siliceous, pelagic, sediments – the turbidity currents originated on topographic highs of the Central Basin Rise, flowing to abyssal depths through three graben-formed valleys (Orwig 1981). Reflection profiles in this area show common transparent, highly stratified, and opaque layers. DSDP Leg 66 (Watkins, Moore *et al.* 1982) cored these transparent layers, showing them to be Quaternary to Oligocene cherts and oozes, with no carbonate in the siliceous layers because this area has been below the CCD since the beginning of sedimentation in the Cretaceous. There are considerable variations in thickness of the flat-lying, often ponded, stratified turbidites, ranging from 300–500 m in the deepest parts of the abyssal plain, compared to 200–300 m for the radiolarian oozes. The turbidites extend up to, overlap, and onlap seamounts and abyssal hills, with turbidites distributed in structurally controlled basement channels (Orwig 1981).

EQUATORIAL ATLANTIC

Oceanward of the Amazon Fan, there are a number of relatively small abyssal plains such as Ceara, Barracuda and Demerara Plains. In a 3.5 and 12 kHz seismic study of this region, Damuth (1975) showed that these abyssal plains have different acoustic properties; he therefore inferred differences in their sedimentation patterns. Ceara and Barracuda Abyssal Plains are characterized by continuous, distinct, sharp echos with continuous sub-bottom reflectors over tens to hundreds of kilometres. In contrast, the proximal parts of Demerara Abyssal Plain near the Amazon Fan are characterized by very prolonged echos with alternating zones of parallel sub-bottom reflectors and zones of intermittent 'mushy' sub-bottom reflectors. From this data, Damuth (1975) concluded that turbidites from the Amazon Fan are ponded against the North Brazilian Ridge; some turbidity currents probably also flowed around the eastern tip of the ridge to the Ceara Abyssal Plain.

ARCTIC OCEAN

During the most active phases of sedimentation on abyssal plains, turbidites are preferentially deposited and, therefore, volumetrically more significant than pelagites and hemipelagites. For example, Campbell & Clark (1977) in a study of more than 100 cores from the Canada Abyssal Plain demonstrated the abundance of turbidites, including T_a and T_{ab} beds. The highest concentration of turbidites occurs in the northeast and southeast of the plain (27%), compared to 2.5% in cores from the north central part of the plain, suggesting that the former areas are nearer to the sediment sources. The mineralogy appears to indicate derivation from the Lower Palaeozoic metasedimentary rocks of the Canadian Archipelago and also, possibly, from more localized slope failure during the Pliocene–Pleistocene lowered sea-level stands. Pipette analysis of the sediments (Campbell & Clark 1977) indicates that the average size fraction of these silty turbidites lies in the range medium to very fine silt. Sedimentation rates based on ^{14}C dates are in the range 4–463 mm/1000 years.

CASE STUDY: SOHM AND HATTERAS ABYSSAL PLAINS, WESTERN CENTRAL ATLANTIC OCEAN

Sohm Abyssal Plain (Heezen *et al.* 1959) lies south of Nova Scotia and the Grand Banks, and has an unusual T-shape (Fig. 8.1). It is approximately 340 km wide, ranges in depth from 5400–6000 m, with gradients from 1:1000 in the north to 1:5000 in the south. The basin floor in the eastern and western arms slopes towards the 'T-junction' and then southwards between the Bermuda Rise and the western flank of the Mid-Atlantic ridge. The plain receives sediment from three principal sources: Gulf of Maine to the west; Laurentian Channel to the north, and the Mid-Ocean Canyon to the east. Minor and local sources include abyssal hills, knolls and seamounts within and adjacent to Sohm Plain. The plain was formed in the Late Triassic to Early Jurassic (Sheridan, Gradstein *et al.* 1983, DSDP Leg 76).

The thickest and coarsest sediments appear to be in the N–S branch with typical sands being very fine-

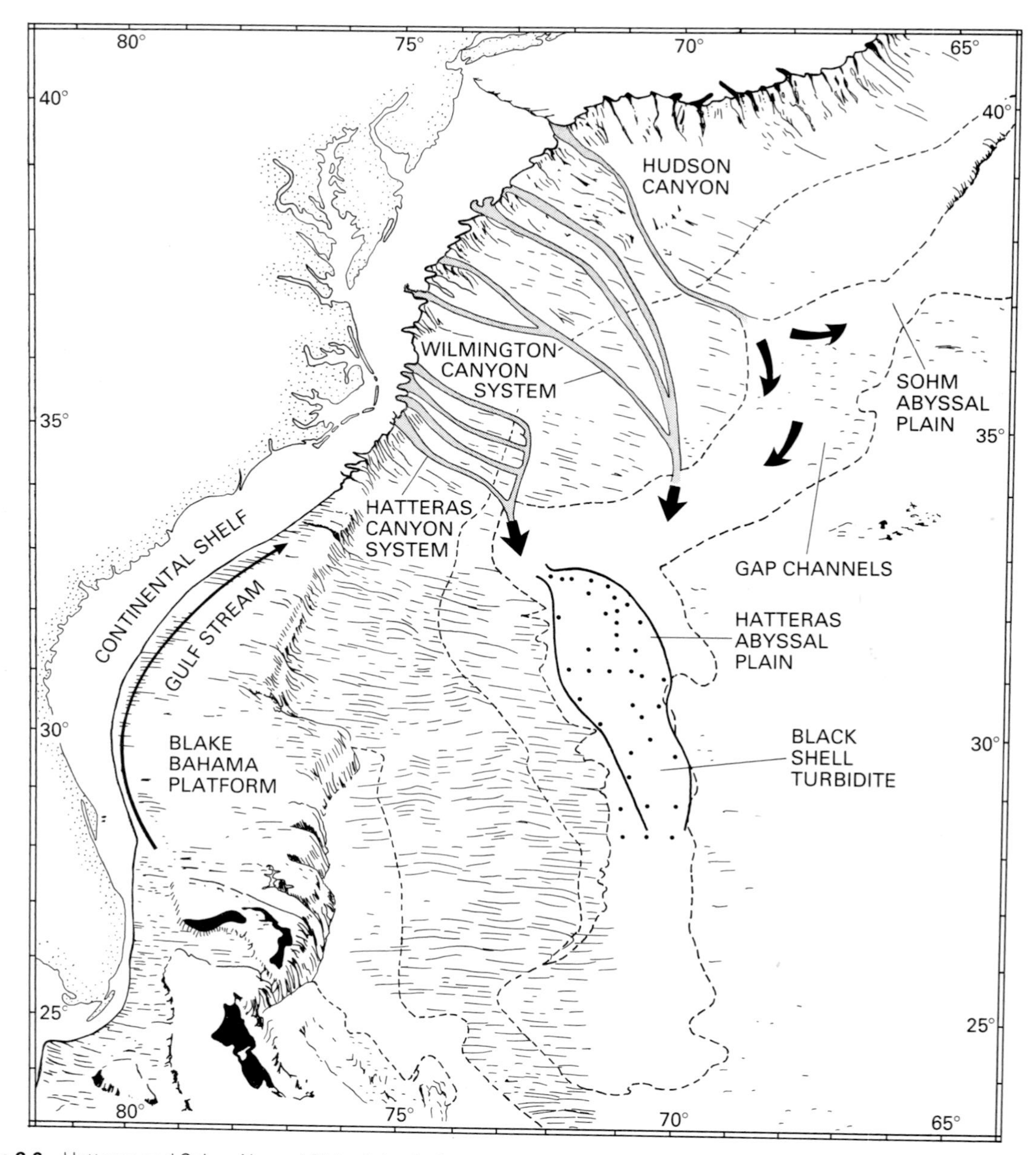

Figure 8.6 Hatteras and Sohm Abyssal Plain, Atlantic Ocean, with the main sediment entry points and paths shown by arrows. The distribution of the Black Shell turbidite is shown by close stipple. From Elmore *et al.* (1979).

grained, poorly sorted, fine-skewed and ranging from mesokurtic to extremely leptokurtic (Horn *et al.* 1971). Sediment dispersal within the plain is complex because of the multiple sediment sources, the relatively large volumes of flows, and the many obstacles around which currents must flow on the plain. The Grand Banks earthquake of 1929 on the St Pierre slope (Piper & Normark 1982, Wang *et al.* 1982) generated a turbidity current that travelled 1700 km, including 1300 km over the Sohm Abyssal Plain, transporting 160 km^3 of sediment and depositing a turbidite over an area of about 470 000 km^2 (Fruth 1965).

Hatteras Abyssal Plain, at the base of the continental rise off the east coast of the USA in water depths of 5100–5450 m, is a N–S elongate plain 1185 km long by 150–300 km wide, with bottom gradients of 1:3000 to the south (Fig. 8.1). It is bordered to the northwest by a series of lower continental-rise hills with relief of 100 m or less. The rest of the western margin is the Blake–Bermuda Outer Ridge, while abyssal hills form an eastern boundary along the entire length of Hatteras Abyssal Plain. The southern boundary is the Nares Abyssal Plain and the edge of the Greater Antilles Outer Ridge. As with the Sohm Abyssal Plain, there are multiple sediment sources, namely Hatteras, Wilmington and Hudson Canyons. However, in the past, currents flowed north from an unspecified western/southwestern source to fill a depression lying between Hudson Canyon and the Kelvin Seamounts (Horn *et al.* 1972).

Upper Pleistocene and Holocene sediments of Hatteras Abyssal Plain are mainly turbidites. Pelagites are minor, and contour currents appear to be 'an inconsequential sediment source' (Cleary *et al.* 1977). Many of the turbidite beds are medium-grained sands, becoming fine/very fine-grained towards the south. Very fine-grained carbonate turbidites that do occur in the south may have come from Silver Abyssal Plain to the south or the Blake–Bahama Plain to the west (Horn *et al.*1972). Individual sand layers often do not correlate from core to core, suggesting an irregular sediment distribution pattern on the plain (Ericson *et al.* 1961). Horn *et al.* (1971) recognize a progressive decrease of mean and maximum grain size at the base of turbidites both laterally and down-current from the main avenues of flow.

Perhaps one of the most outstanding discoveries of recent years in Hatteras Abyssal Plain is the identification of an upper Pleistocene megabed, the 'Black Shell turbidite' (Elmore *et al.* 1979). The turbidite contains 2–50% blackened mollusc shell fragments, has an estimated volume of 100 km^3, and can be traced in 35 piston cores over 44 000 km^2 (Fig. 8.7). The Black Shell turbidity current flowed uninterrupted for at least 500 km from north to south to form a tongue-shaped turbidite 100–140 km wide and as much as 4 m in thickness. The maximum thickness of the sand part of the bed occurs in the centre of the plain, whereas the maximum thickness of the mud cap is displaced eastward of the axis of the basin. Elmore *et al.* (1979) believe that this impressive turbidite originated about 16 000 years BP as a massive shelf-edge slide – the sand coming from the shelf and most of the mud from the upper continental slope. The Black Shell turbidite

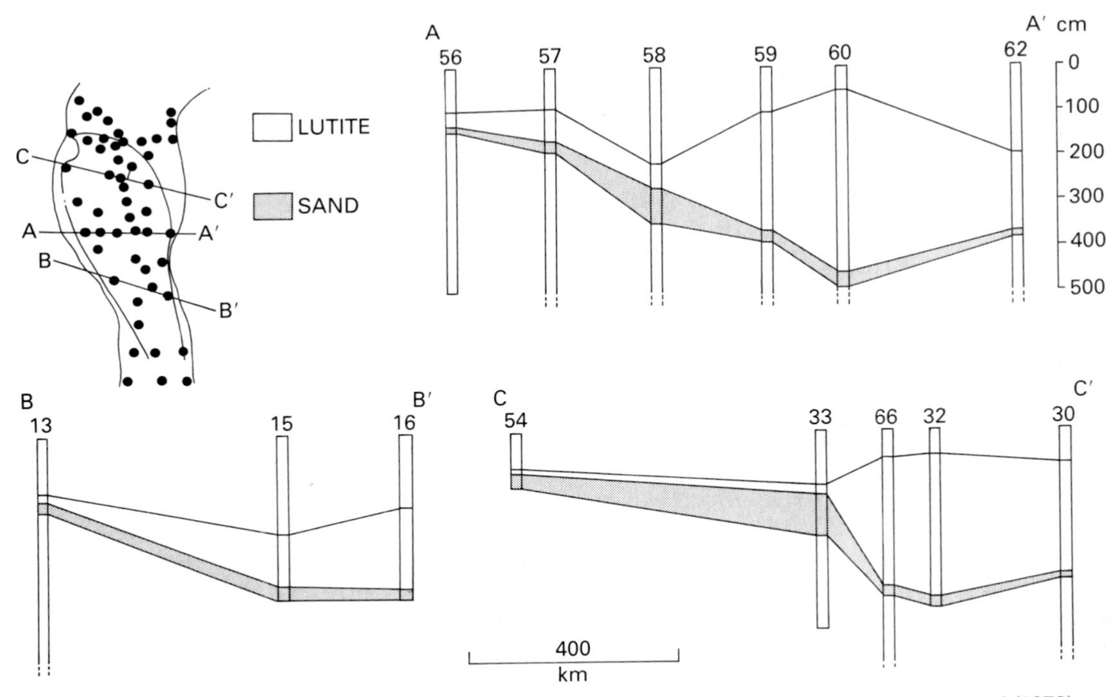

Figure 8.7 Cross-section of the Black Shell turbidite to show sand and mud-cap thickness. After Elmore *et al.* (1979).

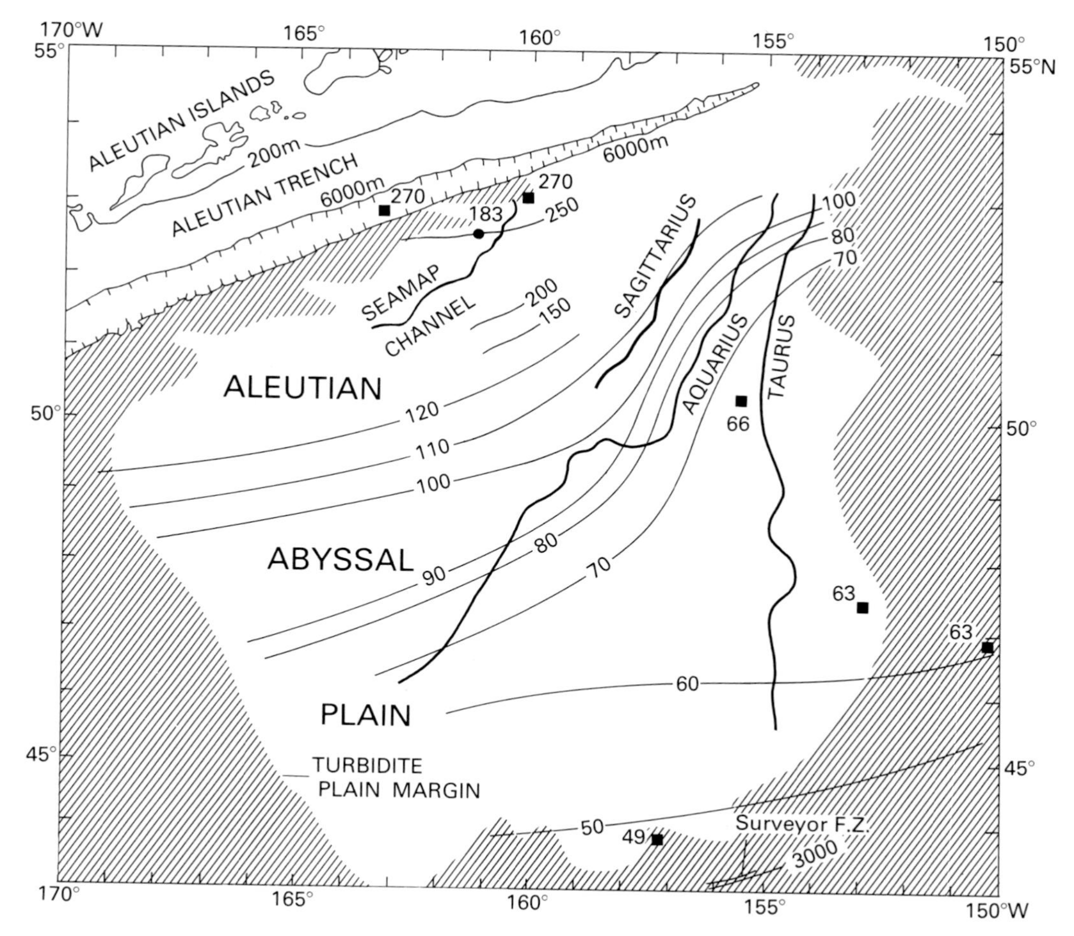

Figure 8.8 Aleutian Abyssal Plain showing isopachs (contoured in metres) of pelagic sediment thicknesses overlying the turbidite systems. Deep marine channels shown. Isolated numbers indicate measured thickness of pelagic sediments. After Hamilton (1973).

appears to be the largest individual turbidite yet traced continuously over an abyssal plain; such events were probably much more frequent during the Pleistocene low sea-level stands.

CASE STUDY: ALEUTIAN ABYSSAL PLAIN

The Aleutian Abyssal Plain lies south of the Aleutian Trench, north of the Surveyor Fracture Zone, and is bounded to the east and west by abyssal hills and seamounts (Fig. 8.8). The Alaska and Tufts Abyssal Plains lie farther east. Turbidite deposition began in the Aleutian Abyssal Plain in the middle Eocene and effectively ended in the Oligocene (Scholl & Creager 1971), the plain now being covered by pelagic sediments. During the last 4.5 m.y., pelagic sedimentation rates have been approximately three times greater in the north (12.8 m/m.y.) than in the south (4.2 m/m.y.), with corresponding differences in thickness (Opdyke & Foster 1970). The source area for much of the older terrigenous sediments appears to have been the continental terrace of the Alaskan Peninsula that lies to the north.

Turbidites were supplied to the plain from a number of leveed channels that traverse the plain, especially Sagittarius, Aquarius and Taurus Channels (Mammerickx 1970), and Seamap Channel (Grim & Naugler 1969). A number of additional buried major channels have been surveyed (Hamilton 1973). These channels (Table 8.1) are depositional features and lie

Table 8.1 Dimensions and sediment thicknesses of deep-marine valleys and channels in the North Pacific (after Hamilton 1973).

Channel[1]	Levee height above channel (m)		Channel width (m)	Sediment thicknesses[2] (m)		
	western	eastern		pelagic	turbidite	total
Seamap (N)	64	53	1112	246	462	708
Seamap (S)	55	55	2780	223	282	505
Sagittarius (N)	20	18	5189	106	239	345
Aquarius (N)	60	48	1853	72	480	552
Taurus (N)	33	15	741	66	804	870
Taurus (S)	31	9	7042	—	—	518

[1]1970 expedition; (N) is north, and (S) is south crossing; [2]maximum under western (right-hand) levee.

along low ridges, with the western levees being higher and broader than those on the east due to the Coriolis effect (cf. Komar 1969); also, the general southward decrease in levee heights and thickness of turbidites suggests that most flows moved from the northeast to the southwest (Hamilton 1973).

The Aleutian Abyssal Plain has been and, indeed, still is affected by major tectonic activity associated with ocean fracture zones such as the Surveyor Fracture Zone. For example, elastic flexure of oceanic crust seaward of the Aleutian Trench has formed a rise or ridge (Walcott 1970) and this is associated with normal faulting (Hamilton 1973). Also, the region of abyssal hills to the northeast of the Aleutian Abyssal Plain, between Murray, Chirikof and Patton Seamounts, is rising relative to the Alaskan Abyssal Plain and undoubtedly this is affecting sediment distribution patterns on the plain.

The sedimentary record from the Aleutian Abyssal Plain has important implications for plate tectonics in this area. If, as seems likely, the turbidite source was from Alaska to the north and northeast, then either no trench, or a filled trench, must have separated Alaska from the plain from middle Eocene to middle Oligocene times. Because a Cretaceous trench and associated continental terrace existed in the present Kodiak–Shumagin–Sanak Islands area (Moore 1972), lack of an equivalent trench to the east seems unlikely, and the development of the plain is best explained by discontinuous plate motion to the middle Oligocene, and then the formation of a more active trench until the Pliocene to isolate the continental source from the Aleutian Abyssal Plain (Hamilton 1973). Thus, in contrast to the Sohm and Hatteras Abyssal Plains associated with a passive continental margin, the Aleutian Abyssal Plain has a more complex history because of its genesis near an active plate margin.

CASE STUDY: BLACK SEA EUXINE ABYSSAL PLAIN

Age estimates for the Black Sea vary from Precambrian to early Quaternary, with a Jurassic–Cretaceous age being favoured (Degens & Ross 1974). The origin of the Black Sea remains uncertain, although there is a growing body of evidence to favour genesis as a remnant backarc or marginal basin that formed behind an island arc during the destruction of Tethys (Biju-Duval *et al.* 1979, Zonenshain & Le Pichon 1986). There has been little tectonic activity during the last 5 m.y. The Black Sea is an oval basin with a pronounced E–W alignment. The crust beneath the sea is anomalous and probably oceanic, but considerably thicker (18–24 km) than more typical ocean crust (5–8 km): the sediment cover is 8–12 km thick, which is considerably greater than beneath abyssal plains of major oceans.

The Euxine Abyssal Plain (Fig. 8.9a) was formed during faulting along the southern, eastern and northern basin margins of the Black Sea, while the western part remained tectonically quiescent (Ross *et al.* 1974). Waves have no effect at depths of greater than 50 m. Shelf widths vary considerably from a few hundred metres to about 200 km. Sediment slides are common at the base of the basin slope and apron.

Much of the abyssal plain is underlain by horizontally stratified layers that are traceable for hundreds of kilometres. Sediments are mainly fine-grained and river-derived, although wind-transported material from the 'black storms' is present, especially in the north (Shimkus & Trimonis 1974). Rates of sediment accumulation range from 0–10 cm/1000 years in the west to 30 cm/1000 years in the east where turbidites are more common.

Ross *et al.* (1974) made the following conclusions about sediments reaching the Euxine Abyssal Plain: (a)

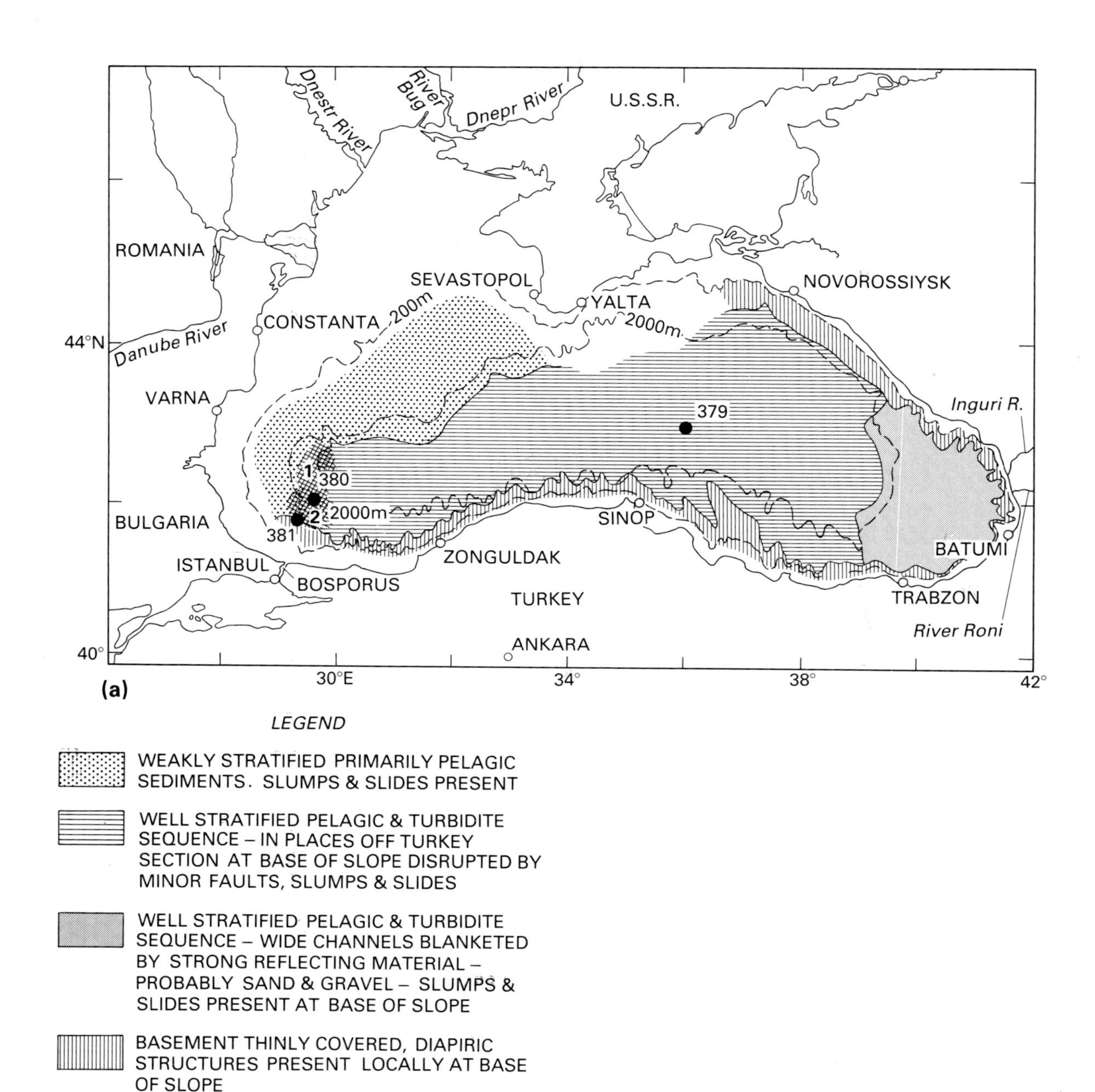

Figure 8.9(a) Physiographic map of the Black Sea (after Ross 1974), also showing DSDP Leg 42B Sites 379, 380 & 381. **(b)** Stratigraphy and correlation of DSDP Leg 42B Sites 379, 380 & 381. After Ross, Neprochnov *et al.* (1978).

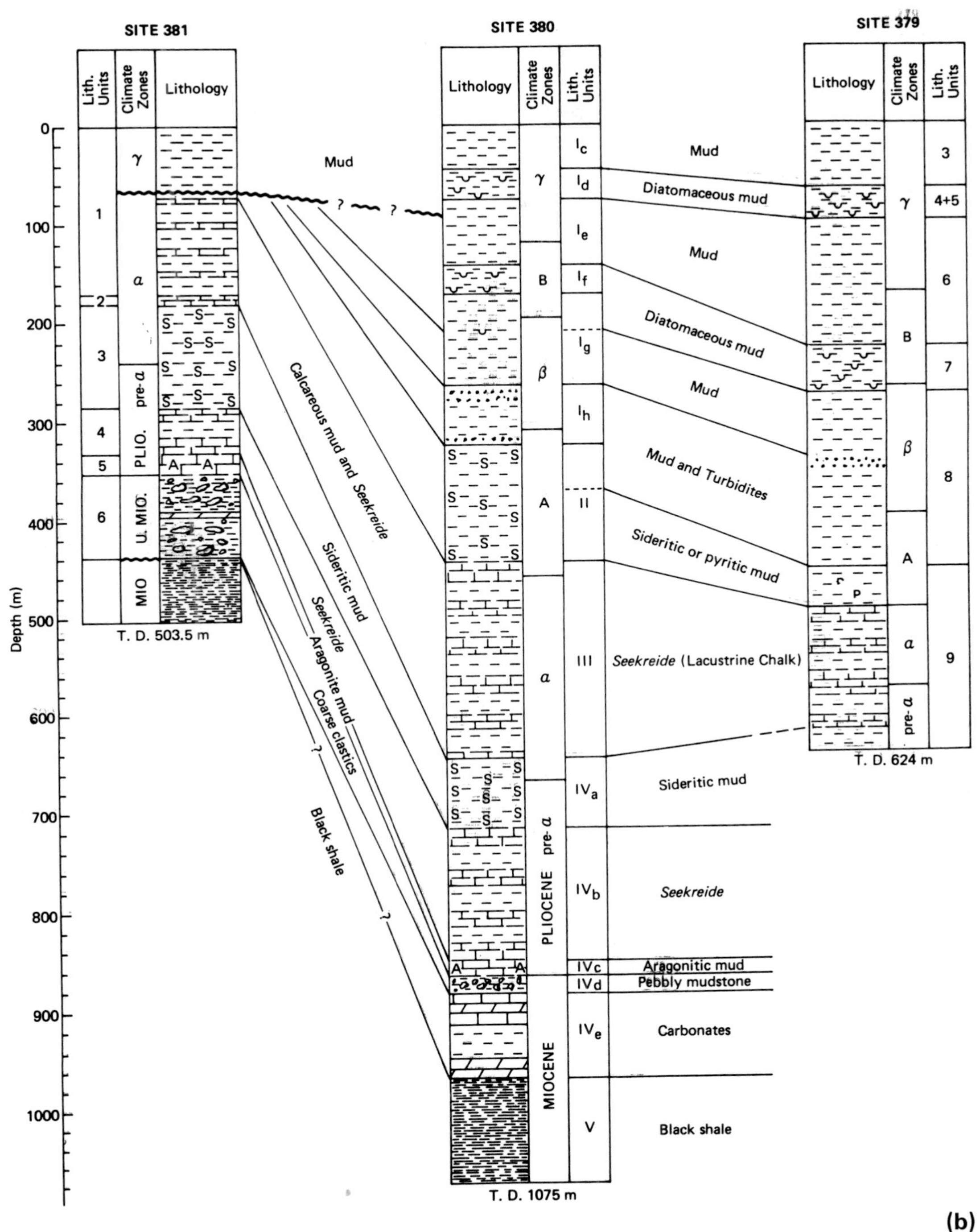
SITE 381
SITE 380
SITE 379
Lith. Units
Climate Zones
Lithology
Depth (m)
T. D. 503.5 m
T. D. 1075 m
T. D. 624 m
Mud
Diatomaceous mud
Mud
Diatomaceous mud
Mud
Mud and Turbidites
Sideritic or pyritic mud
Seekreide (Lacustrine Chalk)
Sideritic mud
Seekreide
Aragonitic mud
Pebbly mudstone
Carbonates
Black shale
Calcareous mud and Seekreide
Sideritic mud
Seekreide
Aragonite mud
Coarse clastics
Black shale
PLIO.
U. MIO.
MIO
PLIOCENE
MIOCENE

(b)

plumes of suspended sediment from rivers contribute most sediments as turbidites to deep water, with considerable seasonal fluctuations, e.g. seasonal percentages of annual suspended load discharge from the River Don are 1.6, 91.3, 5.6 and 1.5% in winter, spring, summer and autumn, respectively; (b) phytoplankton are the main suppliers of indigenous organic matter and production is more than ten times that brought in by rivers and about 50 times the production of macrophytes; (c) production of biogenic grains is five times the annual influx of terrigenous detritus, although 80% of the biogenic skeletons do not reach the sea floor; (d) carbonate grains are principally coccoliths (30% of all carbonate), and (e) Black Sea waters are supersaturated with respect to calcium carbonate. From their study, Ross *et al.* (1974) found that the abyssal plain contains widespread, carbonate-rich, mixed terrigenous and organic coccolith muds that are distinctly laminated, with greater than 3% organic matter – the concentration of organic matter being greatest in the central areas of the Black Sea.

DSDP Sites 379, 380 and 381 (Fig. 8.9a) were drilled in the Black Sea on Leg 42B (Ross, Neprochnov *et al.* 1978) and revealed mainly fine-grained deep-water muds and clays (Fig. 8.9b). These sections show an almost continuous sedimentary record from the Upper Miocene to the present. The major Pleistocene European glaciations are associated with thick terrigenous sections up to 450 m thick, while the interglacials show short-lived marine incursions. The lower Upper Miocene to early Pleistocene (Cromerian) is characterized by 'chemical' sediments that formed in a shallow restricted marine environment from Upper Miocene to Pliocene times: later sedimentation occurred in deep, fluctuating saline to freshwater, environments.

Thus, the Euxine Abyssal Plain is characterized by basinal turbidite sedimentation in a large enclosed sea, with very high organic productivity. Periods of anoxia are indicated by sapropel (black shale) and chalky ooze horizons, and laminated coccolith muds with high organic content.

II: Modern trench and arc-related basin floors

Modern deep-marine trenches, many forearc and backarc or marginal basins, and juvenile rift systems in evolving passive margins, are elongate, extremely complex depositional systems due to the relatively rapid tectonic changes altering basin shape, sediment input points and the morphology of the basin floors. Although it is outside the scope of this book to document the detailed geology of active margins (but see Burk & Drake 1974, Watkins *et al.* 1979, Leggett 1982, Watkins & Drake 1983, Howell 1985, Allen & Homewood 1986; Coward & Ries 1986), the aim of this section is to provide some insight into the sediment distribution and controls influencing sheet depositional systems in these environments. Although trenches and arc-related basins also contain submarine fans, canyons, slope-drapes, small intraslope basins etc. (Ch. 5, 11), these elements are dealt with elsewhere and the emphasis here is placed on sheet systems.

Modern trenches often have depths twice that of adjacent ocean floors and abyssal plains. Part of the difference in depth is the result of the arching of oceanic crust prior to downbending towards the trench floor, producing the prominent trench arches that are up to hundreds of metres high and typically 200–400 km wide. Subduction processes transport open-ocean abyssal plains into the trench realm, thereby dramatically changing the depositional environment, and nature and rates of sediment accumulation. Relatively unconfined, extensive sheet systems thus become highly confined linear basins in which sheet sedimentation may still predominate. For example, the eastward subduction of the Nazca Plate and the northeastward subduction of the Cocos Plate in the Pacific (Ch. 11) has transported open ocean-pelagic facies (Pliocene to upper Miocene brown clay with siliceous microfaunas, authigenic phillipsite, a dearth of terrigenous silt, and basal metal enrichment typical of abyssal pelagites) into the Peru–Chile–Ecuador, and Middle America Trenches, respectively (Schweller & Kulm 1978, Moore *et al.* 1982, von Huene *et al.* 1982). Schweller & Kulm (1978) recognize four facies associations in the eastern Pacific: (a) pelagic plate; (b) terrigenous plate; (c) trench wedge and (d) fan deposits (Fig. 8.10). The pelagic plate shows typical sediment accumulation rates of 2–5 mm/1000 years for regions below the CCD, whereas for the terrigenous plate rates are up to 175 mm/1000 years and 300–3000 mm/1000 years for the trench wedge and fan deposits (Kulm *et al.* 1973). Thus, while sheet deposition may continue as the plate moves from the open ocean to the trench realm, the nature of that sedimentation and the rate of accumulation tend to change, typically resulting in an overall coarsening- and thickening-upward sequence over hundreds of metres (Fig. 8.10). Eventually, the sheet sedimentation may give way to other sedimentary systems such as submarine fans.

While some trench floors are essentially flat and perhaps tens of kilometres wide, the Middle America

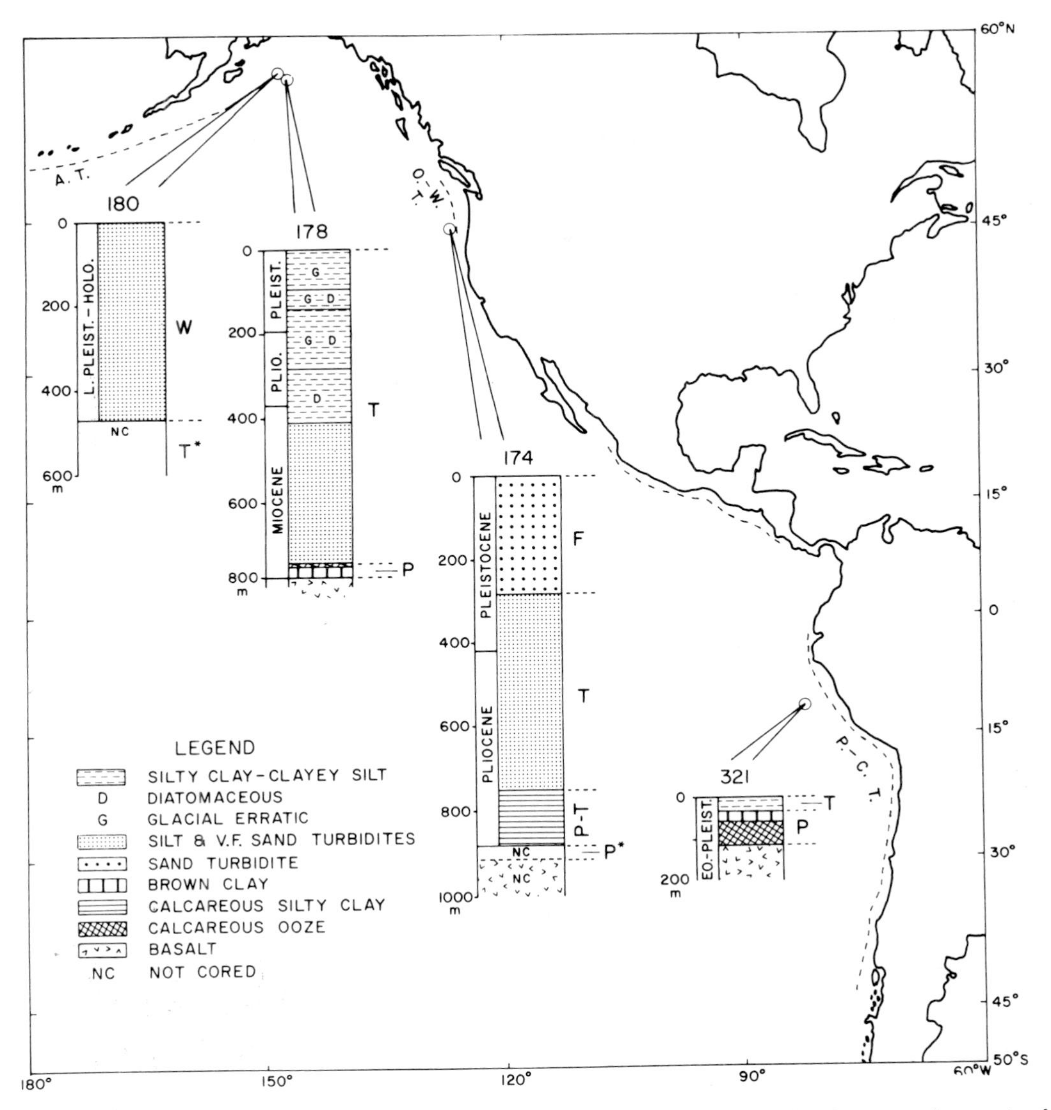

Figure 8.10 Eastern Pacific to show selected DSDP sites and the typical stratigraphy that develops on various parts of oceanic crust. P, pelagic sediments; T, terrigenous sediments; W, trench sediments; F, submarine fan sediments. A. T., Aleutian Trench; O.W.T., Oregon – Washington Trench; P.C.T., Peru – Chile Trench. See text for explanation. After Schweller & Kulm (1978).

Trench tends to be narrow with ridges separating areas of flat sea floor underlain by horizontal seismic reflectors, presumably representing ponded sheet sediments (von Huene *et al.* 1982). Based on the present convergence rates of the Cocos Plate with North America, the deepest upper Miocene beds drilled at Site 495, DSDP Leg 67, were deposited about 900 km seaward of the trench. In contrast, the ponded trench turbidites are less than 400 000 years old, showing the relative youth of the trench-floor sediments (von Huene *et al.* 1982). Sediment thicknesses vary considerably from the 'pelagic plate' to the trench wedge; a uniform thin sediment blanket of pelagites/hemipelagites 200–300 m thick, above Miocene oceanic crust, of the Cocos Plate is locally overlain by up to 625 m of turbidites in discontinuous sediment ponds (Moore *et al.* 1982; von Huene *et al.* 1982).

Trench widths vary from tens of kilometres to a few

kilometres, and individual trench floors also show considerable variations, for example the width of the Peru–Chile Trench varies from 2–5 km south of latitude 7° S to 8–14 km in the north, with sediment thicknesses from the seaward to the landward edges of the trench, respectively, of 170–250 m to 680–850 m in the north, and 160–425 m to 590–930 m in the south (Prince & Kulm 1975). The floor of the Peru–Chile Trench slopes locally from 5800 m to 6200–6300 m over less than 50 km resulting in southward transport of sediments from the northern Peruvian margin in recent times (Prince & Kulm 1975). In places, such as the Ecuador Trench, the turbidite plain is locally only 1 km wide, with a turbidite wedge up to 75 m thick (Lonsdale 1978): in this latter case, the depositional system probably is more similar to structurally confined deep-marine valleys than to other sheet systems. Indeed, McMillen *et al.* (1982) document several small channels on the floor of the Middle America Trench with a good correlation between sediment thicknesses and the position of feeder canyons, thereby showing that trench floors are by no means simple sheet systems but complex mixed depositional systems (also see Moore *et al.* 1982, Fig. 9).

In trenches, sediment thicknesses vary enormously. Fast rates of sediment accumulation together with slow rates of convergence perpendicular to the forearc region allow the development of substantial trench-fill sediments, given favourable sediment influx. The Oregon–Washington or 'Cascadia' Trench is an example of a bathymetrically masked trench because of the rapid rates of sediment accumulation (Nelson & Kulm 1973), whereas about 55% of the total length of Pacific trenches (about 30 000 km) have less than 400 m of sediment infill (Dickinson & Seely 1979). The Tonga–Kermadec Trenches, stretching for an aggregate 2700 km towards the NNE before turning sharply westward at about 16° S, have negligible sediment cover – the main sedimentation being Tertiary to Recent abyssal clays (Katz 1974).

Sheet systems not only occur in abyssal plains and trenches, but in other deep-marine plains, for example arc-related basin floors, such as forearc and backarc basins. Many of these basins are considerably larger, at least in width, than trench floors, for example the Grenada Basin (backarc) and Tobago Basin (forearc) associated with the Lesser Antilles or Barbados accretionary complex (Ch. 11). These basins are up to 100 km wide, the latter containing up to 3 km of undeformed sediments between Barbados Ridge and the volcanic island arc (Westbrook 1982).

Amongst the larger Pacific forearc basins are those off Java and Sumatra, with widths from 50–100 km and lengths of 150–500 km in 3500–4000 m water depths. These basins are separated by structural highs and contain up to 6 km of sediments, mainly as laterally extensive turbidite sheet systems with some small fans (Moore *et al.* 1982). Turbidites from the central Sumatra area are predominantly quartzose with andesitic detritus in varying amounts. A further example of Pacific, arc-related, sheet depositional systems occurs in the complex pattern of dislocated Tertiary active margins in the Sulu Sea. The Sulu Sea, covering an area of about 260 000 km^2 is divided into two basins by the Cagayan Ridge (Fig. 8.11a): (a) Outer Sulu Sea in 1500–2000 m of water, having formed inside an old island arc (Palawan Arc), and (b) Inner Sulu Sea, with its abyssal plain in 3800–5000 m of water, forming a marginal basin with oceanic crust (Mascle & Biscarrat 1979). The Inner Sulu Sea is partially filled mainly with horizontally layered turbidites, usually less than 2000 m in aggregate thickness. The southern and eastern margins of the Inner Sulu Sea correspond to a Neogene active plate margin; to the north, the South China Sea Abyssal Plain has been subducted westwards below Luzon at the Manila Trench since the middle or late Miocene (Mascle & Biscarrat 1979).

In a study of trace-fossil assemblages from the Sulu Sea (Fig. 8.11b), Wetzel (1983) concluded that the main controls on the ichnofacies were fast sediment accumulation rates (100–200 cm/1000 years) and low oxygen contents in the bottom waters. Wetzel (1983) was able to relate the trace-fossil assemblages to water depth, biogenic sedimentary structures and diversity (Fig. 8.11c). In slope and rise environments, trace-fossil assemblages are dominated by burrows with open tubes for optimum circulation of respiration water.

CASE STUDY: NORTH PACIFIC TRENCHES

The North Pacific is dominated by a number of major trench systems, namely the Oregon–Washington, Aleutian, West Japan and East Japan Trench systems, from east to west respectively. The West Japan System includes the Nankai Trough and Ryuku Trench along the northwest margin of the Philippine Sea, and the East Japan System includes, from the north, the Kuril–Kamchatka, Japan, Izu–Bonin (Ogasawara), Mariana, Yap and Palau Trenches (Fig. 11.24) (Sugimura & Uyeda 1973, Uyeda 1974). Trench widths vary considerably, for example 20 km in the 2000 km-long Kuril–Kamchatka Trench compared to 100 km in the 900 km-long Japan Trench. These trenches are amongst the deepest parts of the world's oceans with maximum water depths of 9550 m in the Kuril–Kamchatka Trench, 8412 m in the Japan Trench, and 11 034 m in the Mariana Trench.

Sedimentation patterns differ markedly between trenches, with 300–400 m of Upper Cretaceous–Holocene pelagites/hemipelagites and virtually no terrigenous turbidites in the Kuril–Kamchatka Trench, to 1–2 km of turbidites and hemipelagites in the Washington–Oregon Trench, both of which show landward-dipping layers because of tectonic tilting of horizontally deposited sediments. In the latter case, the lower part of this mainly middle/upper Pleistocene 1–2 km fill includes 630 m-thick of upper Miocene – middle Pleistocene hemipelagic/turbiditic sediments overlying 3–50 m of pelagites/hemipelagites above basaltic oceanic crust (Scholl 1974). Using an imbricate thrust model for the Oregon continental margin (Fig. 11.1), Kulm & Fowler (1974) argued on the basis of foraminiferal evidence and present water depths that abyssal–plain sediments were incorporated into the lower and middle continental slope at uplift rates from 12.5–250 cm/1000 years.

In the 4000 km-long Aleutian Trench, sediment thicknesses range from 1–2 km of turbidites above 200–400 m of Cretaceous and younger pelagites in the west-central parts, to a western end devoid of substantial turbidite thicknesses compared to an eastern, thick, lower Miocene–lower Pleistocene landward-

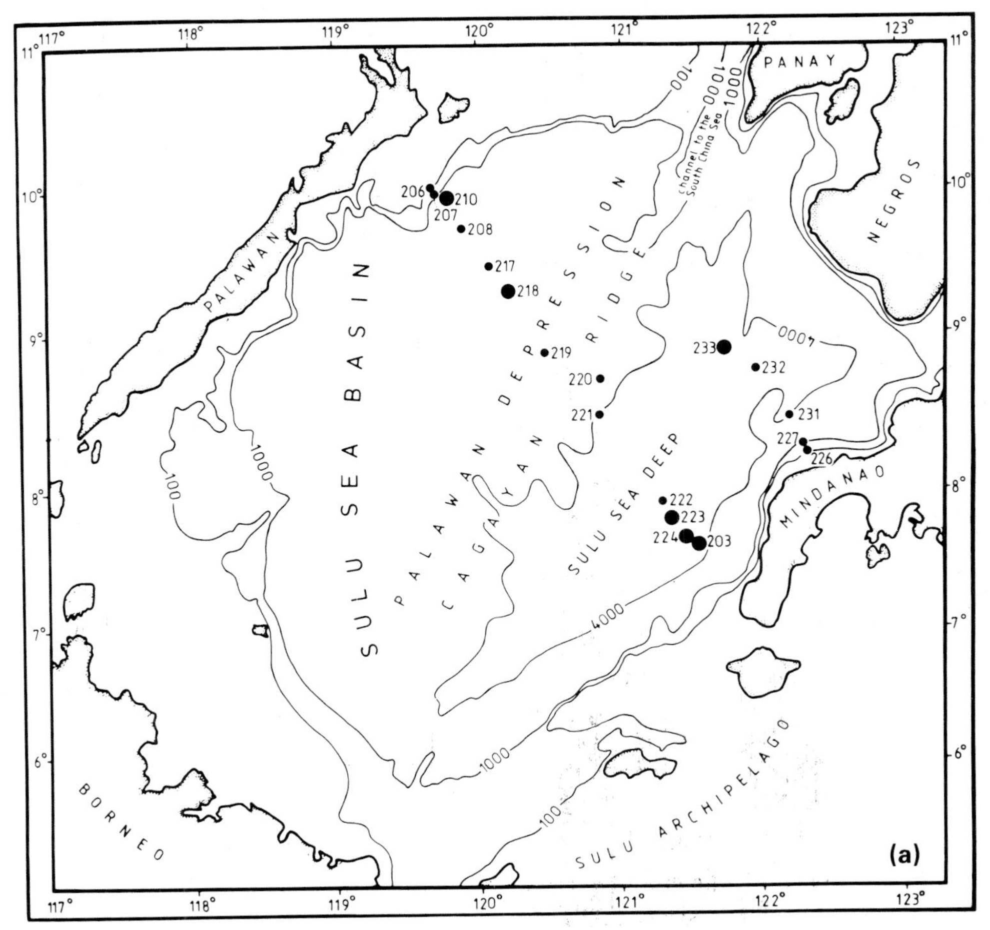

Figure 8.11 a, Map of the Sulu Sea Basin, W. Pacific to show location of short (small black dots) and long (large black dots) core sites; b, (overleaf) schematic drawings of trace fossils observed in the Sulu Sea cores, and c, diagram to show the relationship between biogenic sedimentary structures, degree of bioturbation, and trace-fossil diversity to water depth–diversity calculated for intervals of 500 m water depth from the upper part of diagram. After Wetzel (1983).

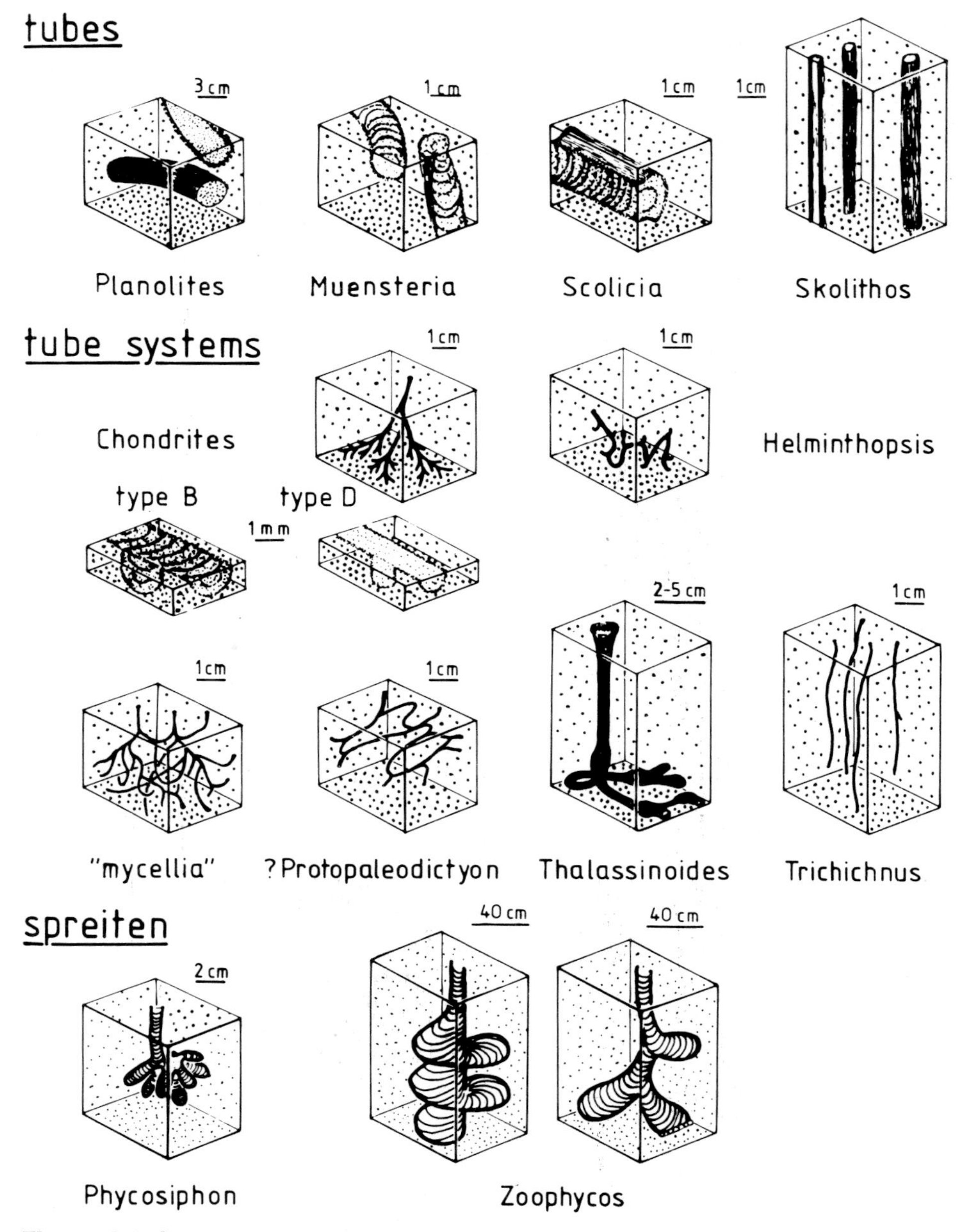

Figure 8.11b

dipping turbidite fill overlain by younger flat-lying turbidites. The basal pelagites/hemipelagites thicken westward to about 3 km of sediments at the western end of the trench (Scholl, 1974). Based on the Aleutian Trench, Scholl (1974) concluded: (a) that Pacific trenches of Mesozoic and Tertiary age probably lacked thick, laterally continuous turbidite fills, and (b) where trenches flank island arcs, pelagites/hemipelagites are dominant and occur in relatively thin successions compared to turbidite fills associated with trenches bordering continental margins.

Where present, turbidites within trenches tend to be somewhat coarser grained than in the surrounding trench-slope successions because of slope by-pass of the

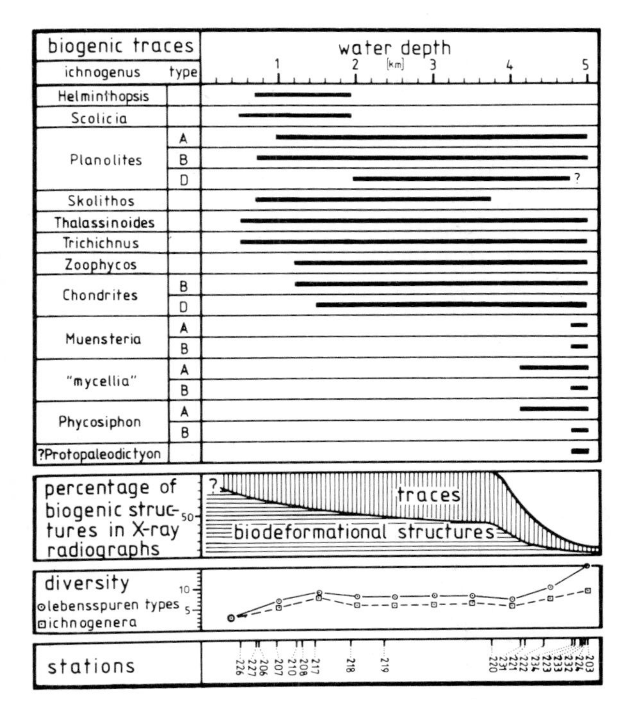

Figure 8.11c

coarse fractions through channels, canyons and other deep-marine valleys (e.g. von Huene & Arthur 1982).

Associated with the North Pacific, there are 'intra-slope terraces' making forearc slope basins up to 130 km in length and up to 75 km in width (Uyeda 1974), although many are considerably smaller and therefore are dealt with in the chapter on intraslope basins (Ch. 5). The larger slope basins have horizontal basin floors comparable in dimensions to many small, open-ocean abyssal plains and with widths far greater than most trench floors. Like abyssal plains, these contain, mainly, horizontally stratified sediments as sheet systems.

III: Ancient sheet systems

Ancient examples of abyssal plains and other deep-marine plains are difficult to identify, principally because of the lack of extensive outcrop over which bed-by-bed correlations can be made to establish the probable geometry of a basin and its floor. The difficulties are augmented by inadequacies in deducing palaeodepths even from diverse faunal assemblages: for pre-Mesozoic rocks this is impossible. Given the problems in interpretation, the term 'abyssal plain' is less useful for ancient deposits than 'deep-sea plain' or 'basin floor'.

Basin plains that developed on truly oceanic crust have a low preservation potential in the geological record because of subduction processes. Even where accreted as part of an accretionary complex, associated deformation and metamorphism generally are so intense that reconstructing the original depositional environment using palaeocurrent and bed-shape criteria have to be treated with extreme caution. Not surprisingly, some of the best documented deep-marine sheet systems appear to have formed in foreland basins (e.g. Ricci Lucchi 1975a, 1978, Ricci Lucchi & Valmori 1980, Hiscott *et al.* 1986, Pickering 1987b). These foreland-basin plains commonly escape the deformation and metamorphism of forearc trench fills, but are not good ancient analogues for the large, modern abyssal plains. Nevertheless, there are many similarities that justify some comparison.

In this section, a Cretaceous (Carpathians and East Alps), and a Miocene basin plain (Marnoso-arenacea Formation, Italy) are described as examples of ancient deep-marine plains. Other ancient examples include those described by Scholle (1971), Sagri (1972, 1974), Hesse (1974, 1975), Parea (1975), Hesse & Butt (1976), Robertson (1976), Bouma & Nilsen (1978), Ingersoll (1978a & b), Mutti & Johns (1979), Hiscott *et al.* (1986), Pickering (1987b) and numerous others.

CASE STUDY: MIOCENE MARNOSO–ARENACEA FORMATION (ITALY)

The Miocene Marnoso-arenacea Formation (Italian Apennines), up to 3500 m thick, outcrops for about 200 km parallel to the tectonic strike as an elongate NW–SE turbidite succession. It is interpreted as an example of deep-sea plain sedimentation (Ricci Lucchi 1975a, & b, 1978, Ricci Lucchi & Valmori 1980). The succession has an estimated volume of 28 000 km^3, and in 18 logged sections covering an area about 123 × 27 km, Ricci Lucchi & Valmori (1980) found that turbidites form 80–90% of the volume with hemipelagites accounting for the remainder. From about 14 000 bed-thickness measurements, they divided the beds into thick-bedded Facies Class B and C turbidites >40 cm thick, and thin-bedded Facies Class C and D turbidites <40 cm thick. Almost 40% of the thick-bedded turbidites could be traced over the whole study area, i.e. a maximum distance of 123 km.

The thin-bedded turbidites tend to show T_{cde} divisions, while the thick-bedded turbidites typically include the T_b division, up to 5 m thick in the exceptional Contessa Bed (see below). These beds may show base- and top-absent Bouma sequences. Most turbidites are fine-grained and display sheet-like geometry. One of the outstanding features of the Marnoso-arenacea Formation is the occurrence of very thick-bedded turbidites, or Contessa-like beds (named after the thickest, or Contessa Bed of Ricci Lucchi & Pialli 1973). These 'megabeds' are characterized by: (a) tabular or sheet-like geometry; (b) sand:mud ratios less than 1; (c) beds up to several metres thick; (d) subtle lateral changes in texture and sedimentary structures; (e) generally basin-wide aerial extent, and (f) up-current and basin-margin thinning (Ricci Lucchi & Valmori 1980). The Contessa Bed (Fig. 8.12a & b), up to 16 m thick, contains a very thick mudstone cap, as do other Contessa-like beds, and appears to have been deposited from a large turbidity current ponded within a confined basin.

The depositional basin for the Marnoso-arenacea was 'over-supplied' in that both normal and exceptional turbidite events had a large size compared to the area of the receiving basin. Flows were deflected along the basin axis at their entry points – the main sediment source being in the Southern Alps and shifting eastwards with time. While depth estimates are necessarily crude, Ricci Lucchi (1978) suggests the range 1000–3000 m from geometrical relationships such as: (a) the inferred distance from the source area versus minimum gradients for gravity transport; (b) ichno-

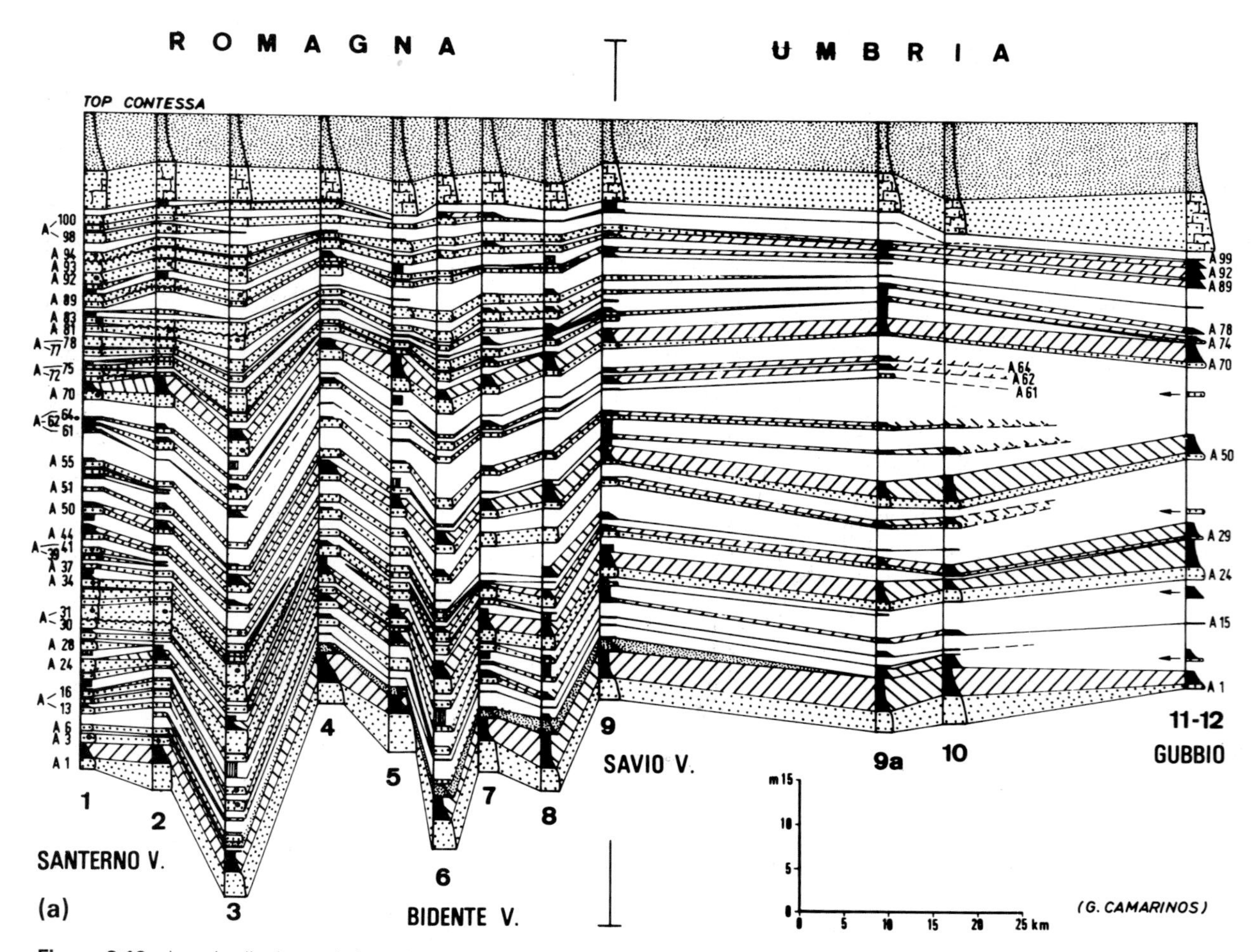

Figure 8.12 Longitudinal correlation of individual very thick-bedded turbidites in the pre-Contessa (a) and post-Contessa (b) intervals of the Miocene Marnoso–arenacea Formation, northern Italian Apennines. Black parts of columns represent mud-dominated parts with alternating turbidites and hemipelagites. After Ricci Lucchi & Valmori (1980).

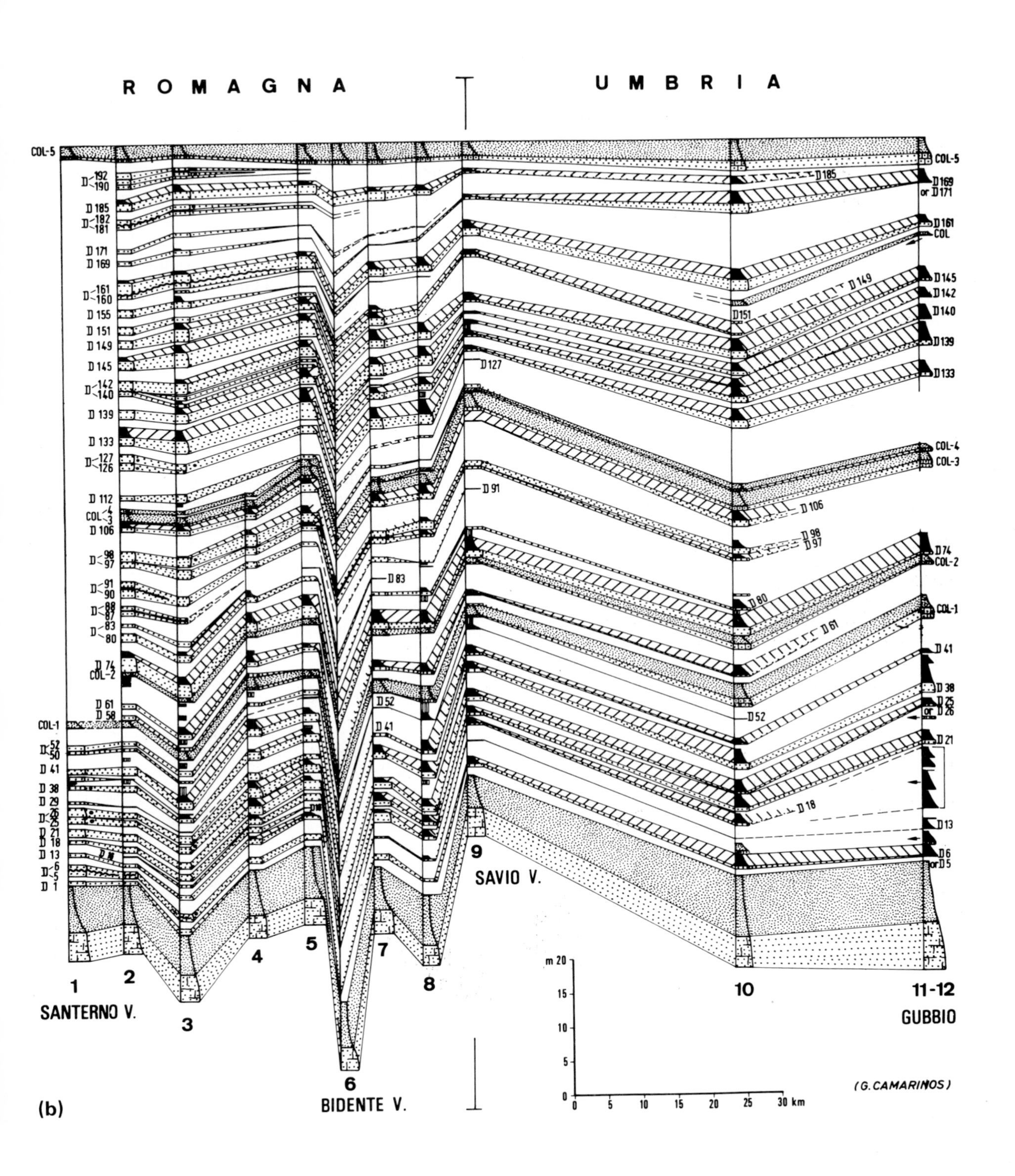

ROMAGNA
UMBRIA
SANTERNO V.
BIDENTE V.
SAVIO V.
GUBBIO
(G. CAMARINOS)

(b)

facies, and (c) the paucity of skeletal benthic remains. Sediment accumulation rates averaged 15–45 cm/1000 years, increasing to 75 cm/1000 years in the Tortonian.

CASE STUDY: CRETACEOUS FLYSCH ZONE (EAST ALPS)

This case study is based mainly on the work of Hesse (1974, 1982). The Flysch Zone of the East Alps consists of Cretaceous–Palaeogene trench and arc-related deposits that accumulated over 70 m.y. when the subduction zone associated with the trench was dormant (Fig. 8.13a). Some of the formations are interpreted as sheet depositional systems and the evidence for trench sedimentation is shown in Figure 8.13b and Table 8.2. This 500 km-long and 10 km-wide (possibly 80–100 km wide) trench was connected to the 1000 km-long Carpathian Trench, the western part of which is not demonstrably associated with subduction until the Oligocene.

In Aptian–Albian times, the trench contained a basin plain at least 100 km (possibly 200–300 km) in length and below the CCD. The existence of a basin plain is suggested by the long-distance correlation of individual beds for 50 km in the Campanian, and 115 km in the Aptian–Albian (Hesse 1974). The lack of: (a) syn-depositional deformation; (b) volcanism, and (c) volcaniclastic detritus, suggests that sediments accumulated in a trench that was not associated with active subduction. Major palaeocurrent reversals occur throughout the Cretaceous Flysch Zone (Table 8.2); changes in palaeoflow are associated with petrographic changes, i.e. different source areas. Also, the palaeocurrents show a remarkable consistency for any single time and suggest very confined axial turbidity currents.

Hesse (1974), in a study of the lower part of the 1500 m-thick Cretaceous–Palaeogene succession in Bavaria, near the tectonic base of the flysch belt, made a detailed study of the Aptian–Albian Gault Formation (Fig. 8.14). Within the approximately 200 m-thick Gault Formation, typical greywacke bed thicknesses are about 1 m, with some beds up to 4 m thick. Many beds show a basal Bouma T_b division. Sole marks consistently show palaeoflow from west to east throughout Aptian–Albian times. The intercalated packets of claystones average 75 cm with individual green, black and grey layers less than 20 cm thick. The green claystones are carbonate-free and represent hemipelagites, and the organic-rich, sometimes carbonate-rich, black claystones may represent hemipelagic/pelagic deposition during periods of basin anoxia. A packet of 55 glauconitic greywacke beds has been traced for up to 115 km along strike. Estimated minimum volumes of entrained sediment to form an 'average' 1 m-thick bed, assuming a basin width of 10 km, are of the order of 1 km^3, with possible maximum volumes of about 25 km^3. Sediment

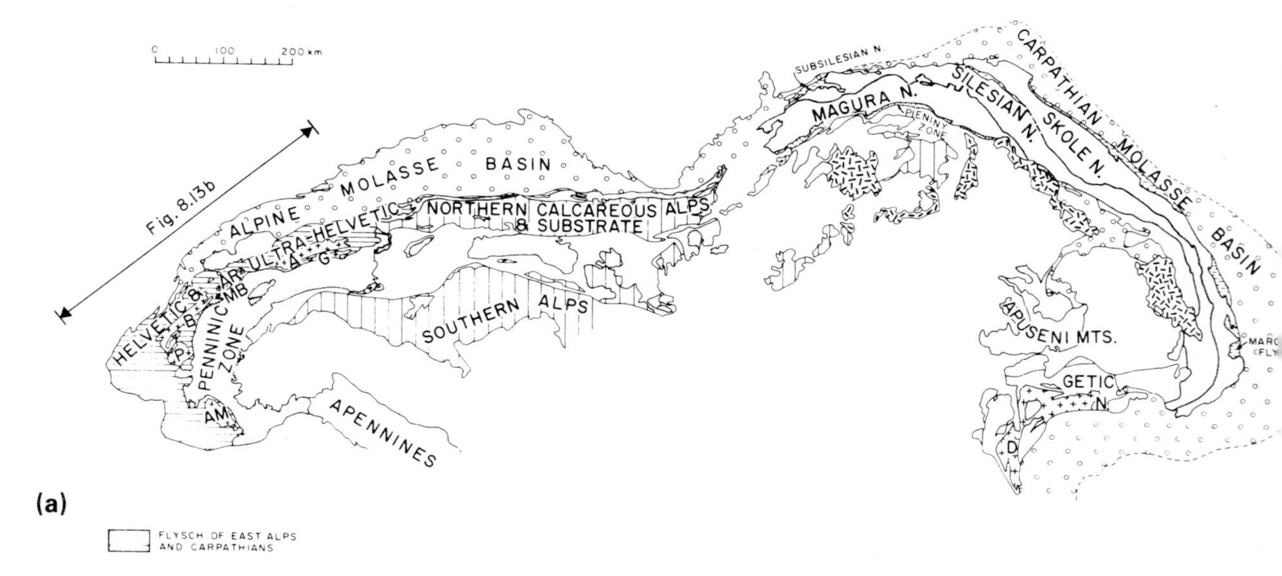

Figure 8.13 a, Flysch zones of the East Alps and Carpathians, and b, palaeocurrent pattern in the Flysch Zone for the Lower, Middle and Upper Cretaceous. Note the reversal of axial palaeocurrents, with length of arrows proportional to the number of measurements. See text for explanation. After Hesse (1982).

supply was axial from west to east with no evidence of lateral supply. The geometry of the beds suggests a sea floor with negligible relief.

Hesse (1974, 1982) envisages an Early Cretaceous abyssal plain or trench floor (Gault Formation) and island arc system associated with the closure of Tethys. Tectonic tilting of the trench axis produced systematic reversals of the dominant palaeocurrent direction. Subsequent sedimentation includes submarine fans and other depositional systems.

Summary

Many modern and ancient deep-marine plains (both open-ocean abyssal plains, trenches and arc-related basin floors) tend to have a number of features in common regardless of the specific details of the system: (a) aerially and/or axially extensive, sheet-like turbidites; (b) a high proportion, although variable, of hemipelagites and pelagites compared to other deep-water systems – however trench fills tend to be sandier; (c)

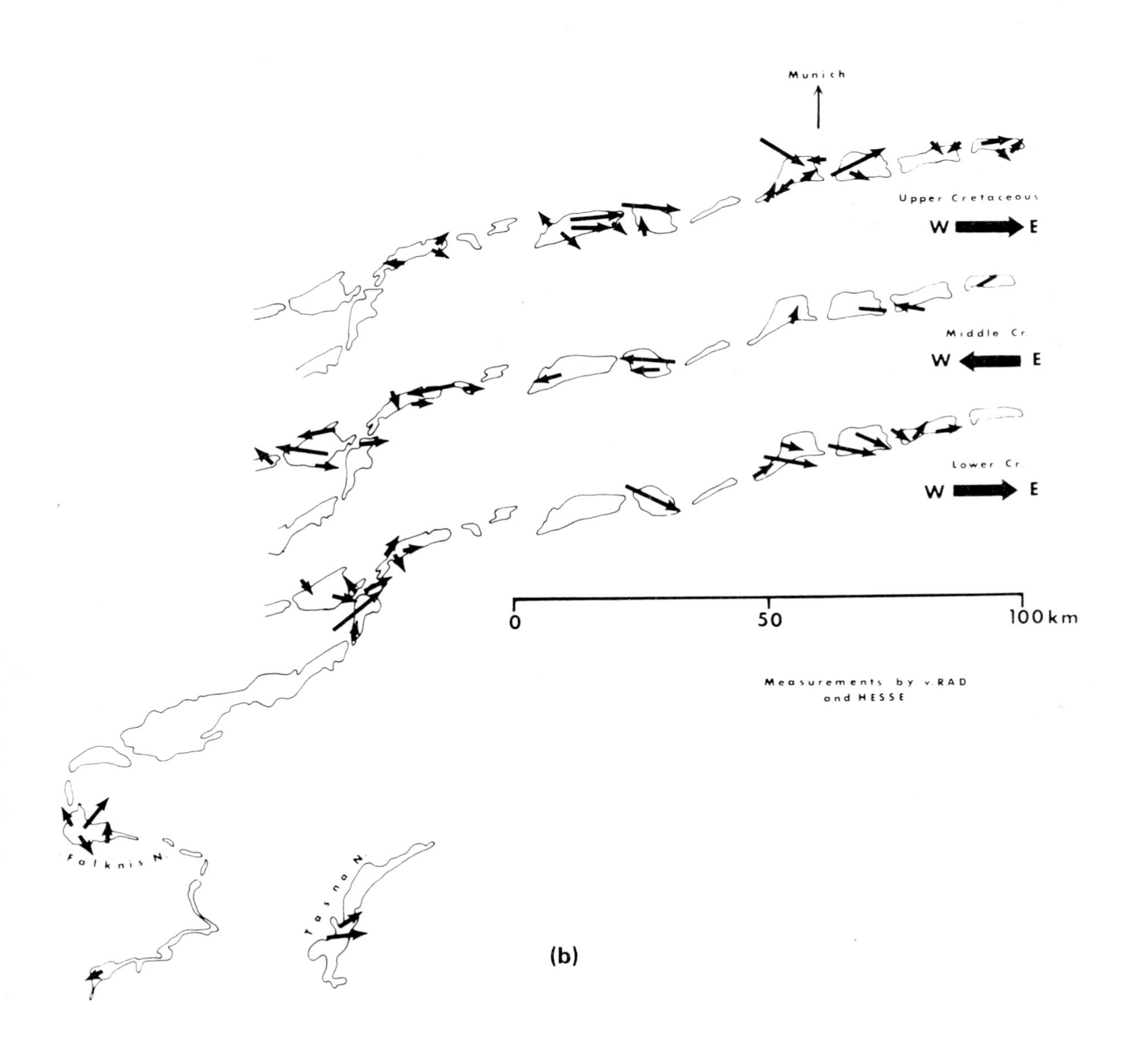

(b)

Table 8.2 Summary features of the Flysch Zone of the East Alps, Cretaceous–Palaeocene trench fill. After Hesse (1982).

Evidence for trench:

Dimensions: Length: 500 km (without Carpathians)
Width: 10–50 km (widening toward E: possibly > 100 km)
Depth: below palaeo-calcite compensation level.
Red claystone as equivalent of brown abyssal clay

Depositional environments:

(a) Elongate, nearly horizontal basin plain:
Palaeocurrent directions predominantly parallel to strike (basin axis)
Repeated reversal of current directions (change in azimuth by 180° in successive turbidites)
Long-distance continuity of individual turbidites
Gault Formation: 115 km
Zementmergel Formation: 50 km
Low gradients of downcurrent change in bed thickness, grain-size, etc. along individual beds (high gradients perpendicular to flow)

(b) Small to intermediate size deep-sea fan (Falknis-Tasna Nappes)

Limited number of lateral sediment sources suggests existence of slope basins that intercept access routes of turbidity currents

Lack of contemporaneous subduction:

Lack of volcanism and volcaniclastic detritus
Continuous sedimentation without major diastrophism for 70 Ma
Occurrence of slivers of basic and ultrabasic rocks (ophiolite suite) N and S of Flysch Zone

relatively rare wet-sediment sliding; (d) typical bathyal and abyssal faunas; (e) possible association with metaliferous, other chemogenic and fine-grained biogenic material, and (f) locally common megabeds (ponded turbidites) attributable to major basin-slope failure. These features, together with the overall stratigraphic framework of a succession may provide the best evidence for the identification of ancient sheet depositional systems. Furthermore, we would expect that the main controls on cyclicity in sheet systems are: (a) extra-basinal, such as climatically-influenced changes in sediment influx, or those that are tectonically-controlled, and (b) compensation cycles.

Perhaps the most likely confusion in the interpretation of ancient deep-marine plains is the distinction from fan fringe sedimentation, although clearly the two environments merge with one another. Ricci Lucchi (1978) has suggested the following distinctions: (a) hemipelagites/pelagites are thicker and more regularly interbedded with basin plain than with outer-fan turbidites – this means longer pauses elapsed between successive flows in a plain and only the larger turbidity currents reached this area; (b) a higher degree of correlation for basinal beds because of their greater volumes and momentum on reaching plains; (c) higher mud ratios than for the fan fringe, and (d) the occurrence of ponded flows as megabeds in plains and their absence from fan-fringe deposits.

Finally, many studies of ancient arc-related deep-marine plains (e.g. Ingersoll 1978a & b, Leggett 1980, Nilsen & Zuffa 1982, van der Lingen 1982, Chan & Dott 1983) have shown the complexity of such depositional systems in which preserved vestiges of sheet systems often form only a relatively small part. While this chapter has emphasized the obvious open-ocean abyssal plains, trench floors and arc-related basin floors, it must be stressed that the distinction between sheet and non-sheet systems is not always as sharp as the modern and ancient examples might suggest (e.g. trench systems, Ch. 11). However, whilst accepting that such deep-marine or deep-sea plains are not perfectly smooth horizontal to near-horizontal surfaces, to a first approximation the geometry both of most individual beds and the overall system is sheet-like.

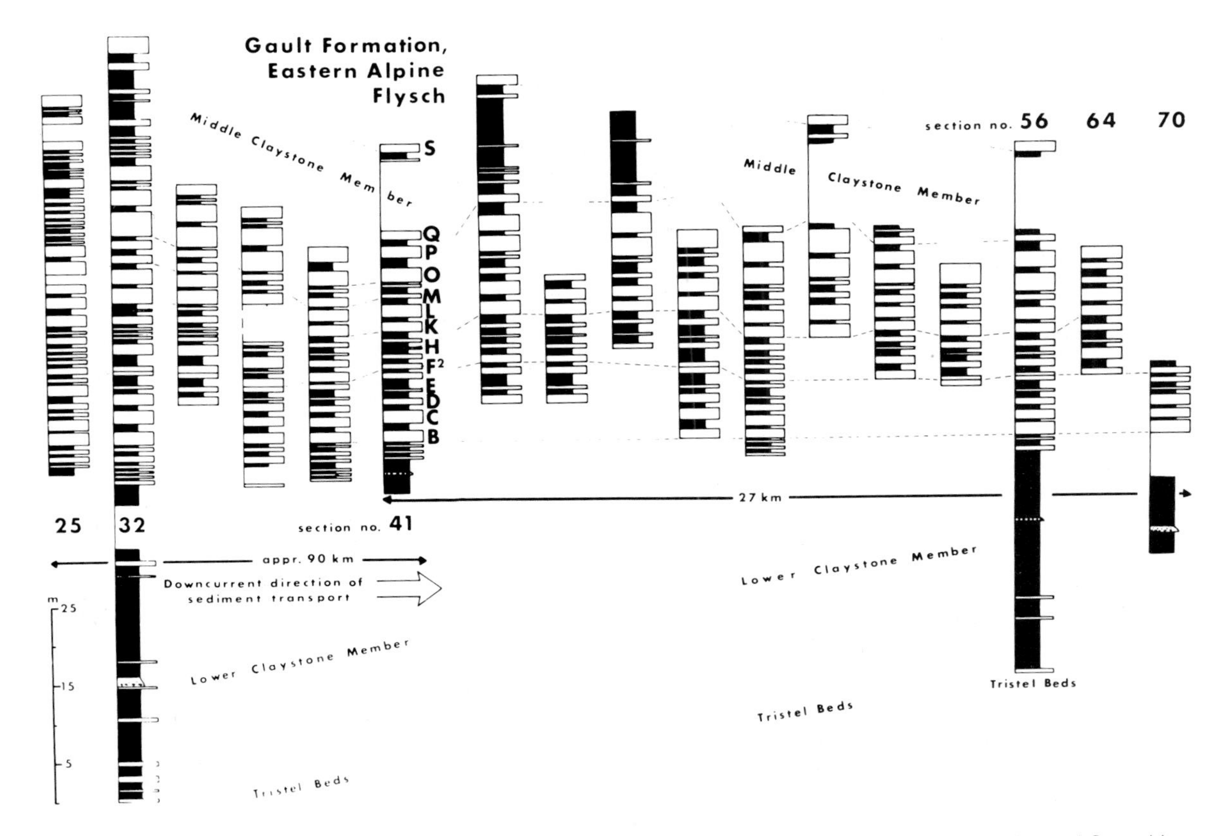

Figure 8.14 Example of correlation of turbidite sections from the Aptian – Albian Gault Formation, East Alps and Carpathians. Only the lower part of the formation is shown (i.e. Lower Claystone, Lower Greywacke and Middle Claystone Members). Bed-by-bed correlation is nearly perfect except for the most proximal section 25 which contains additional thin beds. Bed F2 is a relatively feldspar-rich marker bed. Letters refer to marker beds. After Hesse (1982).

Dr. Bill Normark, US Geological Survey, Menlo Park, California, has made major contributions to our understanding of modern submarine environments, particularly fans. Photograph taken at COMFAN II meeting in Parma, Italy (Sept. 1988).

CHAPTER 9

Contourite drifts

Introduction

A *contourite* was originally defined as the sediment deposited from thermohaline-driven bottom currents flowing parallel to bathymetric contours (Fig. 9.1) (Heezen *et al.* 1966). Recent definitions of a *contourite* are as follows: (a) 'a bed deposited or significantly reworked by a current that is persistent in time and space and flows along the slope in relatively deep water (certainly below wave base). The water may be fresh or saline; the cause of the current is not necessarily critical to the application of the term' (Lovell & Stow 1981); or (b) *current-drift* (contourite) as 'sediment reworked or redeposited by bottom currents, which are commonly geostrophic and contour-flowing, but which may be either fast or sluggish' (Robertson & Ogg 1986). *Contourite drifts* are large sedimentary bodies deposited beneath persistent deep-water currents. These sediment drifts have lengths up to hundreds of kilometres, widths of tens of kilometres, and relief of 200–2000 m (Johnson & Schneider 1969). Contourites have been identified from all the world's oceans (Heezen 1977), with the greatest concentration in the western North Atlantic Ocean, where more than a dozen of these sediment drifts have been identified (McCave & Tucholke 1986, Stow & Holbrook 1984). The extent and bathymetric distribution of drifts in the North Atlantic is shown in Figure 4.9.

During the 1970s, a major controversy arose concerning the relative importance of persistent bottom currents (or contour currents) versus episodic turbidity currents and related mass-flows in shaping continental margins (see Emery & Uchupi 1972, Gorsline 1978, 1980, Heezen & Hollister 1971, Hollister & Heezen 1972, Stow & Lovell 1979). Volumetrically, on most margins, episodic mass-flow processes tend to dominate. Sediment supplied to the margins by downslope movement can then be reworked or redistributed by bottom currents. In the western North Atlantic, with its well known sediment drifts, seismo-stratigraphic data show that contourites account for at most 10% of the stratigraphic section and, except for features like the Hamilton Spur and Blake Bahama Outer Ridge (Fig. 4.9), are nowhere a predominant facies (Piper, D.J.W. & Schafer, C.T., pers. comm. 1988).

No comprehensive facies model exists for contourite-drift sediments. This is partly a function of the difficulty in studying modern contourite drifts and the discrepancy in scale between modern sediment drifts and the scale of ancient outcrops. There are few well documented ancient contourite deposits – many of the reported possible cases are within a mixed turbidite-dominated/contourite (or bottom current) assemblage. In general, many land geologists associate deep-marine sediments mainly with various types of sediment gravity flow deposits (see Ch. 2 & 5). Paradoxically, many oceanographers studying contemporary ocean basins emphasize the importance of permanent/semi-permanent currents, including geostrophic and thermohaline flows.

The apparent discrepancy between land-based and marine-based studies is due to several factors. In modern studies, because of the short time scale of many oceanographic observations, there is an observational bias toward 'near-continual' processes such as the sweeping of continental rises by contour currents and other bottom currents (e.g. Hollister *et al.* 1978, McCave & Tucholke 1986) and the rain-out of material from nepheloid flows (e.g. Armi 1978, Armi & d'Asaro 1980, McCave 1983). The modern continental margins, characterized by high sea levels, may be inappropriate analogues for ancient successions (see Ch. 4 & 5). Direct observations are rare for the more sporadic mass-wasting phenomena, including slides, slumps, debris flows and turbidity currents. These are usually observed indirectly through observations of sediment core stratigraphy and structure. Finally, the preservation potential of thin, fine-grained bottom-current deposits is much less than that of thicker, coarse-grained sediment gravity flow deposits (Fig. 9.2). This is due to the greater potential for the thin, fine-grained deposits to be reworked by bioturbation or winnowing. In contrast, despite the low frequency of turbidity current and debris flow events, deposits resulting from these mechanisms have a greater preservation potential because of the greater volume, and in particular thickness, of deposition per event.

The strongest bottom currents may not deposit any sediment and can create unconformities in deep-ocean successions. In these cases, the only record of strong bottom-current activity may be condensed sections, lags, disconformities or unconformities. Kennett *et al.* (1975) showed that major erosion cycles exposed older pelagic sediments as bottom currents begin to operate in new parts of ocean basins, due to shifting plate positions. Thus the apparent diachronous nature of contourites is related to the broader aspects of palaeoceanography, including continental drift events, oceanic circulation patterns and climatic changes through time.

The final problem with the development of a contourite facies model lies in the basic philosophy of facies modelling. The relationship between depositional environments and the resulting vertical stratigraphic sequences were first emphasized by Johannes Walther, in his Law of Correlation of Facies (Walther, 1894, p. 979 – see Middleton 1973). In this Law, 'it is a basic statement . . . that only those facies and facies-areas can be superimposed primarily which can be observed beside each other at the present time' (as translated in Middleton 1973). Most facies models are successful because there are major facies differences between subenvironments – e.g. submarine channel → levee → interchannel tract. The facies sequences are then constructed by superimposing deposits of adjacent environments. Contourite-drift systems typically develop

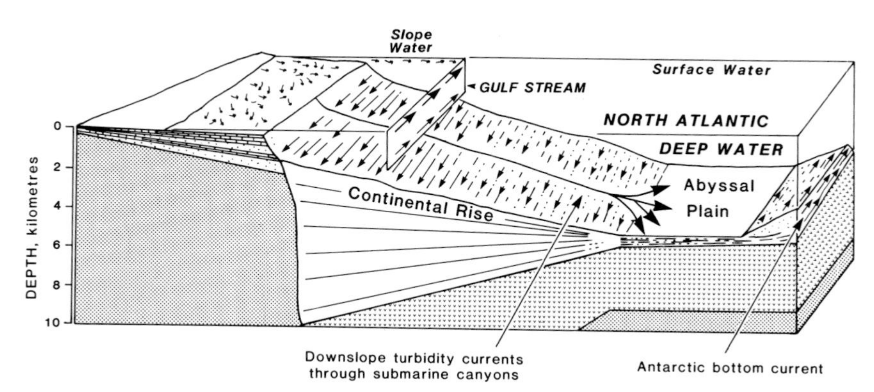

Figure 9.1 Schematic diagram showing deep-contour following currents shaping the continental rise off the eastern margin of North America. Continental rise sediments are shown as wedges.

in rather uniform bathyal/abyssal environments. Thus, for contourite drifts, one has to question whether there is sufficient variation in conditions, and in resulting facies, within the drift system to warrant a facies model.

In this chapter, factors controlling the distribution of modern sediment drifts are discussed. Some of the best examples are from the North Atlantic; hence, the modern case studies emphasize this region. A literature search reveals about 30 ancient successions in which there is a partial and/or probable contour current influence, compared to the hundreds of documented modern examples (see Stow & Lovell 1979). One ancient case study is described in detail.

CONTOURITE VERSUS TURBIDITE

A number of workers have formulated criteria for the distinction of turbidites from contourites (Hollister 1967; Bouma 1972, 1973; Hollister & Heezen 1972; Bouma & Hollister 1973; Stow 1979; Stow & Lovell 1979; Lovell & Stow 1981). The most recent syntheses of the characteristics of ancient sandy and muddy contourites are those of Lovell & Stow (1981) and Stow & Lovell (1979) (Tables 9.1 & 2 & Fig. 9.3). In many cases, the distinction between the two types of deposit is quite subtle, because contourites can result from (a) the reworking of older turbidites; (b) the 'piracy' of active downslope-moving fine-grained turbidity currents by along-slope bottom currents; and (c) fluctuating bottom-currents which, when depositing sediment, may mimic the grading patterns seen in turbidites. There are few regions where current meter data are available to substantiate the influence of along-slope currents (e.g. Carter *et al.* 1979, Ledbetter & Ellwood 1980, Hill & Bowen 1983, Hollister & McCave 1984). Consequently, many of the supposed diagnostic criteria for contourites may have been derived from examination of fine-grained turbidites or normal hemipelagic sediments.

Some of the more diagnostic features of contourites include (Lovell & Stow 1981) (Tables 9.1 & 2): the coarse lag concentrations, especially those with a largely biogenic component (commonly planktonic foraminiferal tests) (Carter *et al.* 1979); the inverse grading near the top of beds and a sharp upper contact (associated with the development of lags) (Carter *et al.* 1979, Lovell & Stow 1981); and grain fabric patterns, particularly those showing along-slope trends (Ledbetter

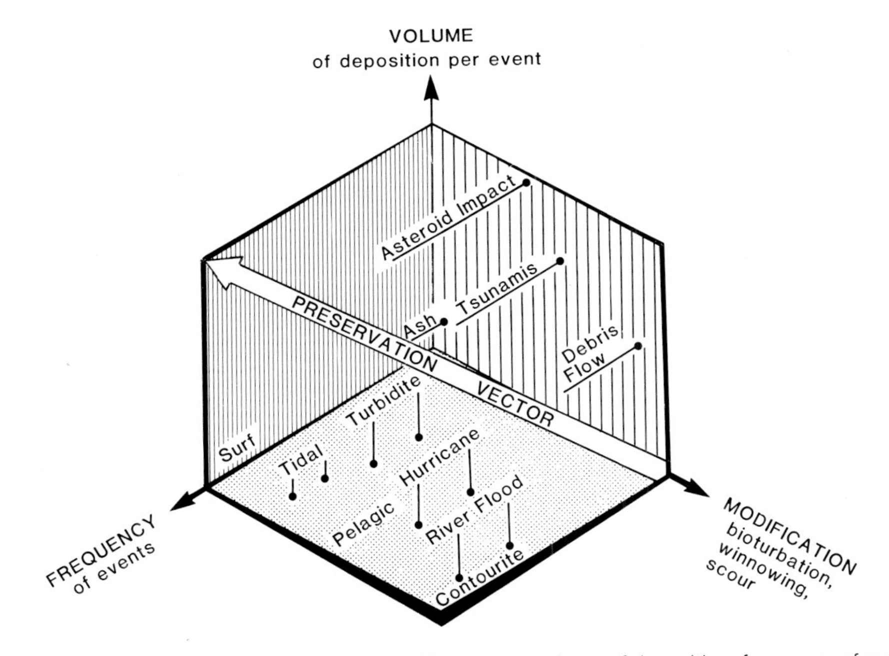

Figure 9.2 Three-dimensional conceptual graph of relationships among volume of deposition, frequency of events, and post-depositional modification for various sedimentary processes. Minimum preservation occurs at lower right corner and maximum at upper left corner. Contourites would have a very low preservation potential, compared with other deep-marine deposits, including pelagic, turbidite and debris flow deposits (modified from Dott 1982).

Table 9.1 Sedimentary characteristics of muddy contourites.

	Notes from marine-based studies	Preservation potential
Occurrence	thick sequences of hemipelagic/contourite sediment ridges in association with turbidites – overlying graded muds or sands on the continental rise.	good, difficult to recognize alone
Structure	dominantly homogeneous, bedding not well defined; bioturbation common; coarse lag concentrations (especially biogenic); primary silt/mud lamination – rare but where present similar to fine-grained turbidites, but lacking Bouma sequences.	good, difficult to distinguish from fine-grained turbidites
Texture	dominantly silty mud: frequently high sand content (0–15%) of biogenic tests; medium to poorly sorted, ungraded, no offshore textural trends; may show marked textural difference from interbedded turbidites.	not good, altered by diagenesis or tectonism
Fabric	mud fabric: smaller, more randomly arranged particle clusters than the large, oriented groupings of mud turbidites; primary silt laminate or coarse lag deposits show grain orientation parallel to the current (along-slope); interbedded, reworked turbidites may show widely biomodal grain orientations.	not good to poor
Composition	generally combination of biogenic and terrigenous material; biogenic material; usually rich in coccoliths, forams, diatoms, radiolaria, etc. (function of productivity, depth in relation to CCD); calcium carbonate usually high (10–40%) except where low productivity or below CCD; general absence of shallow-water material (usually broken tests)	fair

Table 9.2 Sedimentary characteristics of sandy contourites.

	Notes from marine-based studies	Preservation potential
Occurrence	thin lag deposits in muddy contourite successions; reworked tops of sandy turbidites in interbedded sequences; coarse lag in deep-sea channels and straits.	good, but identification is difficult
Structure	frequently disturbed and bioturbated; irregular coarse layers with little primary structure preserved; where undisturbed horizontal and ripple cross-lamination (but not in Bouma sequence), and orientation parallel to the bottom current direction; may show reverse grading near the top and a sharp upper contact.	good, especially current-direction indicators
Texture	silt to sand-sized, rarely gravel; may be well sorted, lacking mud; tendency to low or negative skewness values; few offshore trends.	not good
Fabric	indication of grain orientation parallel to the bottom-current (along-slope); may show secondary mode of turbidite fabric (? at right angles); may show more polymodal or random (disturbed) fabric.	not good
Composition	dependent on region and immediate supply of sandy material; concentration of grains of higher specific gravity (frequently biogenic lag in muddy contourites); original compositional grading in turbidites obscured by reworking; some evidence of along-slope reworking.	fair

Figure 9.3 X-radiographic positive prints of contourites from the Nova Scotian continental rise, Orphan Knoll area, north of Flemish Cap, on the labrador slope. a, structureless, sandy winnowed material within the core of the Western Boundary Undercurrent, water depth 2600 m, 49° 40.0′ N Lat./47° 52.9′ W Long. (Lehigh core 77034–13A); b, laminated sands and muds, alternating with bioturbated mud, near the core of the Western Boundary Undercurrent, water depth 2508 m, 48° 33.3′ N Lat./46° 42.3′ W Long. (Lehigh core 84035–001); c, laminated sands and muds, alternating with bioturbated mud, on the downslope side of the Western Boundary Undercurrent, water depth 3210 m, 49° 30.2′ N Lat./46° 09.6′ W Long. (piston core 79017–005). For location map see Figure 9.5.

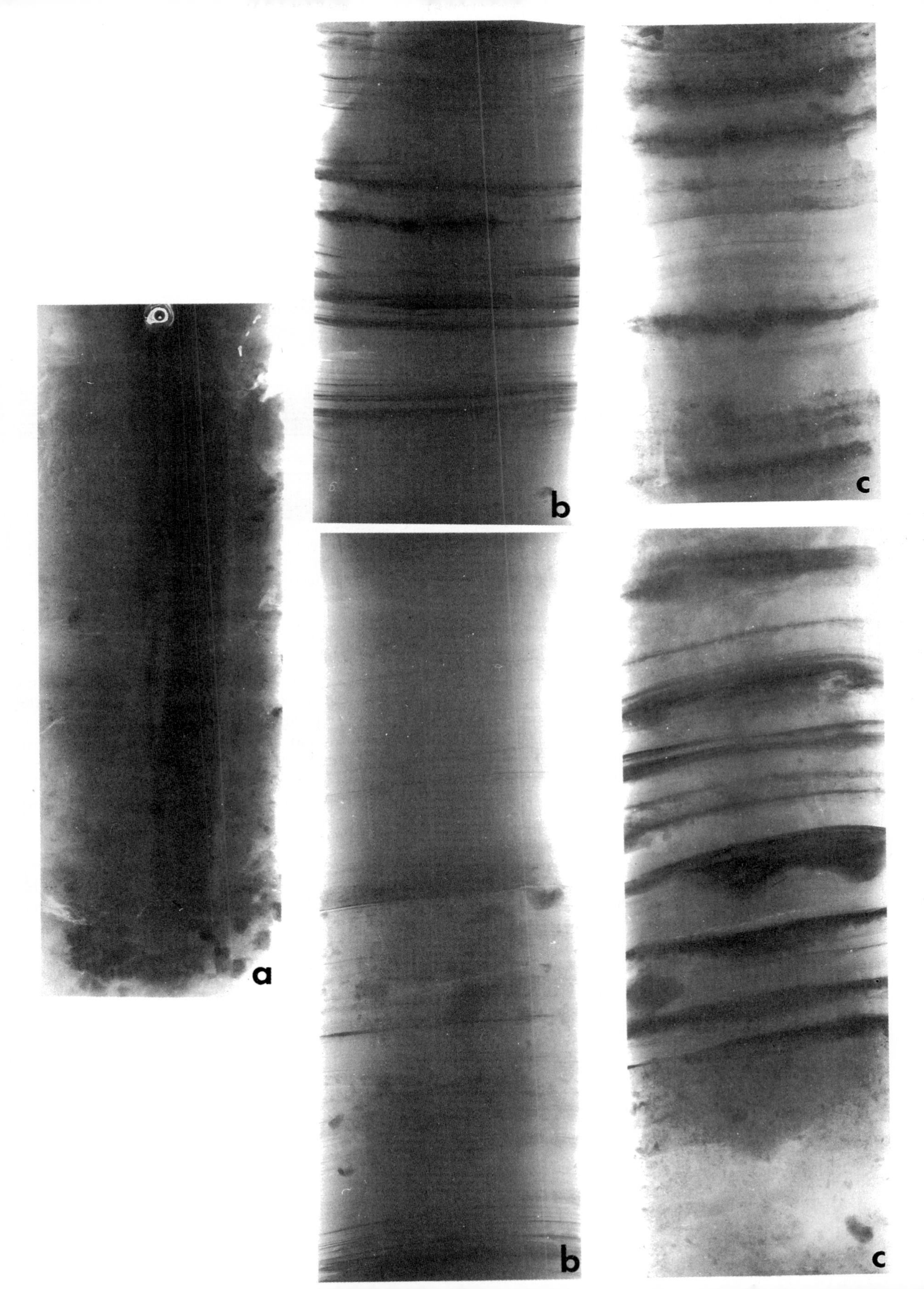
a
b
b
c
c

& Ellwood 1980). Some of these features are difficult to assess because of the fine scale of lamination and grading – commonly sampling must be on the centimetre to millimetre scale to ascertain patterns – and some of these features may not survive diagenesis and tectonic deformation in ancient successions.

Very little information is available on the vertical variations in lithology through sediment-drift successions. This is largely a function of the scale of these features. Piston cores are under-representative of the large scale of the drifts – long core facilities are now being developed to sample 30 m of sediment, while sediment drifts have relief of 200–2000 m (Johnson & Schneider 1969). Thus, seismic data are commonly used because no lithologic data are generally available. Most interpretations of the sedimentology of the drifts are based solely upon interpretations of seismic records and bathymetry of the sea floor.

As discussed in Chapter 5, seismic data have provided considerable information on continental slopes. Various patterns of reflectors, designated as seismic facies, have been recognized (e.g. Sangree *et al.* 1978) and are interpreted mainly as mass-flow deposits on slopes (Fig. 5.4). Patterns observed in seismic records differ in areas of known bottom-current activity and topographic constriction (Mountain & Tucholke 1985, Myers & Piper in press). For example, Myers' study (Myers 1986, Myers & Piper in press) of Late Cenozoic sedimentation in the Labrador Sea identified the following seismic facies as a result of contour-current or bottom-current circulation (Fig. 9.4): (a) mounded stratified facies, dominated by well stratified reflectors that undulate gently or build asymmetric mounds with wavelengths of tens of kilometres; (b) chaotic basin-fill facies, with variable amplitude, irregular to hummocky, discontinuous reflectors – hyperbolic reflectors are present throughout, but most common near the upper bounding surface; (c) mounded-chaotic facies, with locally subparallel to slightly irregular reflectors of a more continuous nature, exhibiting an overall mound-like geometry, with wavelengths of tens of kilometres; (d) wavy-stratified facies, characterized by discontinuous, wavy and overlapping, moderate-amplitude reflectors; and (e) sediment-wave facies, consisting of symmetric to asymmetric, stratified mounds, with relief of 20–250 m and variable wavelengths, commonly 2–10 km. Other facies, including a weakly stratified facies and an irregularly stratified facies, lack sufficient diagnostic seismic characteristics and are interpreted as the result of mixed turbidity current, bottom current and/or hemipelagic influence.

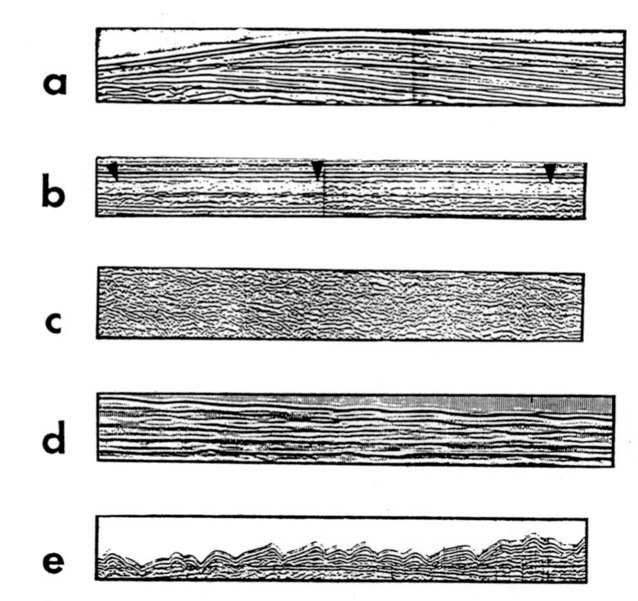

Figure 9.4 Interpretations of seismic facies due to contour-currents or bottom currents. (a) mounded-stratified; (b) chaotic basin-fill (bounded by well-stratified); (c) mounded-chaotic; (d) wavy-stratified; (e) sediment-wave. Modified from Myers (1986).

The distinctions between turbidites and contourites from seismic records is inherently equivocal and interpretive. One classic case involves the contourite mounds or 'depositional anticlines' observed between Flemish Cap and Orphan Knoll, offshore eastern Canada (Mitchum 1985) (Fig. 9.5). In transverse section these deposits could easily be mistaken for submarine fans, in that they have a 'mounded' shape and are associated with other deep-water sediments. They are distinguished from submarine-fan deposits for the following reasons (Mitchum 1985): (a) internal reflectors are very continuous in longitudinal section; (b) in cross section they are asymmetric; (c) they are larger than many fans; and (d) they have possible gentle toplap deposits. On a regional scale these features parallel the strike of the continental margin and they occur where contour currents are focused (Mitchum 1985).

The distinction between turbidites and contourites, and other bottom-current deposits, is very important on continental margins and has a direct bearing on the hydrocarbon potential of a region. As discussed by Damuth & Kumar (1975a) continental-rise successions dominated by turbidity-current and related mass-flow processes contain excellent source and reservoir rocks, whereas continental-rise successions influenced by contour-current and bottom-current processes may contain excellent source rocks, but lack good reservoir rocks.

Major external factors

The major external factors favouring transport and deposition by contour currents are, in general, the opposite of those factors favouring the development of mass-flow facies on continental margins (see Ch. 5). The following factors will be discussed: (a) sea-level variation; (b) stratification of deep-water masses and palaeocirculation; and (c) sediment input. For a more thorough discussion see Chapter 4.

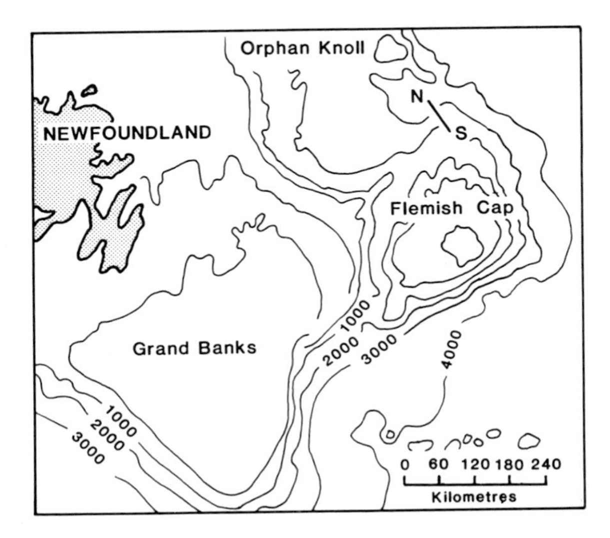

SEA-LEVEL VARIATION

As discussed in Chapters 4 and 5, the importance of eustatic sea-level variations on the development of continental margins has been recognized through seismic stratigraphy. In deep-marine siliciclastic systems, during periods of high sea level, the 'continuous' processes of sedimentation by contourites and hemipelagic draping dominate, whereas during low sea-level stands the 'discontinuous' processes of mass wasting predominate and overwhelm the effects of contour currents (Emery & Uchupi 1972, Gorsline 1980, Sheridan 1986). Shanmugam & Moiola (1982), in their discussion on eustatic control of turbidites and winnowed turbidites, plotted the distribution of contourites and turbidites on the global sea-level curves of Vail *et al.* (1977). In their analysis, they hypothesize that episodes of low sea level also cause vigorous contour currents, which winnow the fines from turbidites. Thus, both turbidites and winnowed turbidites (i.e. some contourites) may correspond to low sea-level stands. Given the low preservation potential of contourites, the lack of good dating control, and the fact that contourites are generally unrecognized in the ancient, such plots of contourite/turbidite distribution versus eustatic sea level must be viewed with caution (see Ch. 4 for controls on deep-marine sedimentation).

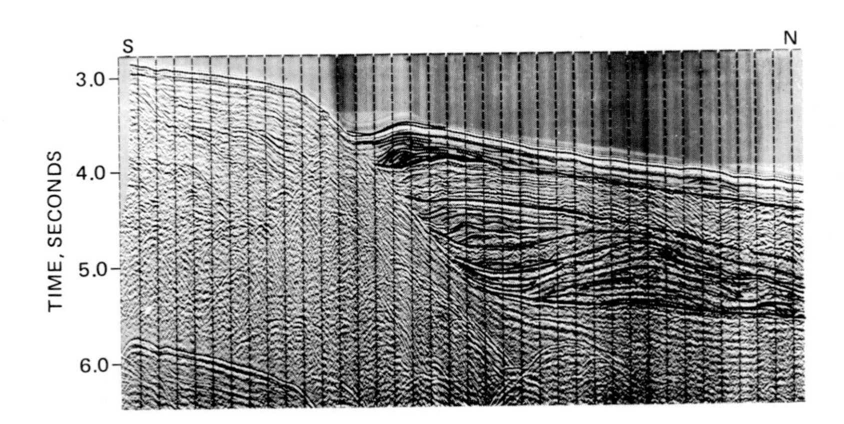

Figure 9.5 Contourite mounds between Flemish Cap and Orphan Knoll, on the Labrador slope, offshore eastern Canada. From Mitchum (1985). Location of profile indicated on map.

STRATIFICATION OF DEEP-MARINE WATER MASSES AND PALAEOCIRCULATION

Below the surface waters of the ocean basins there are distinct water masses with subtly different densities and chemical signatures. These masses can be traced for considerable distances at different levels in the ocean. Stratification of the water column into these distinct water masses is due to density differences caused by temperature and salinity variations. This stratification is a stable feature of the world's oceans, but the magnitude of density contrasts and the effects of surface winds result in a complex circulation pattern (Figs 9.6 & 7).

In the modern oceans, deep thermohaline circulation is mainly driven by the formation of cold, saline waters in the polar regions. The coldest water forms near Antarctica and sinks to form the bottom-water masses of the Indian, South Pacific and Atlantic Oceans. Sources of deep bottom water in the North Atlantic are more complex than in the Indian and Pacific oceans, and include a component of the cold/saline bottom waters formed near Antarctica, cold/saline water formed in the Norwegian and Labrador seas, and warm/saline bottom waters from the Mediterranean Sea (Tolmazin 1985, McCave & Tucholke 1986). The circulation patterns at abyssal depths have been modelled by Stommel (1957) and Stommel & Aarons (1960a & b). In their model, there are two deep-marine sources – a North Atlantic source and a Weddell Sea source in Antarctica (S1 and S2, Fig. 9.6) which diffuse uniformly over the entire ocean area. The model shows western boundary undercurrents and an eastward directed flow in the southern oceans (Fig. 9.7) (McLellan 1965; Tolmazin 1985), in agreement with the observed deep flow patterns in the oceans (Fig. 4.8 & 9) (Tolmazin 1985).

During other geologic time periods, the bottom-water circulations were halokinetic, driven by warm and saline bottom waters descending from shallow shelf seas which bordered the eastern sides of ocean basins in arid zones (Brass *et al.* 1982, Hay 1984, Oberhansli & Hsü 1986). This non-actualistic palaeocirculation was mainly the result of a different arrangement of continental plates, giving different climatic conditions than those found today (Figs 4.11 & 12). Halokinetic circulation was predominant during the Cretaceous, Palaeocene and Early Holocene. Deep thermohaline circulation was established by the Palaeogene, but was still controlled more by salinity rather than by temperature (Oberhansli & Hsü 1986).

Another major feature of earlier ocean circulation was input from the Tethys water mass (Fig. 4.12). The

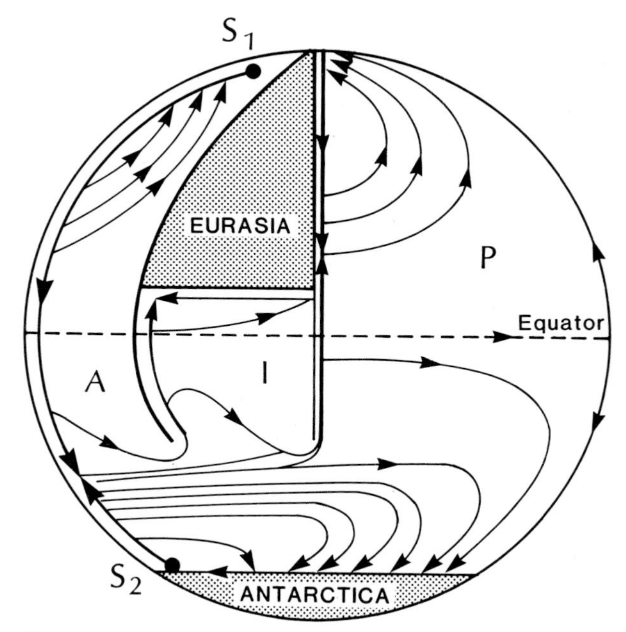

Figure 9.6 The abyssal circulation predicted by the Stommel–Aarons model. S1 and S2 are sinks of deep cold water in the North Atlantic and Weddell Sea. A, Atlantic Ocean; I, Indian Ocean; P, Pacific Ocean. From Tolmazin (1985).

Callovian North Atlantic had an east–west equatorial orientation, was narrow (about 500 km wide) and deepened by about 300 m due to Callovian subsidence. Sluggish bottom currents gave rise to sediment drifts, with most bottom water derived from Tethys. Trade winds favoured extensive upwelling along the southern African margin (Robertson & Ogg 1986). The effect of the Tethys bottom current extended through the Palaeogene.

The response of the deep sea to glaciation is not well understood (Lohmann 1978). Historically, two viewpoints have been presented. According to Worthington (1968), deep-ocean circulation during glacial maxima was similar to that seen today but with bottom waters having a slightly higher density due to the higher salinity of oceans during glacial stages. As deglaciation began, oceans quickly became stratified, with very dense bottom water in the deep-ocean and light, glacial melt-waters near the surface. After development of this strong stratification, there was no longer any high-latitude production of deep bottom waters; i.e. deglaciation led to stagnation of the deep ocean waters, lasting up to 15 000 years. By contrast, in Weyl's (1968) view, the present-day sources of cold, deep, bottom water did not exist during glacial episodes because

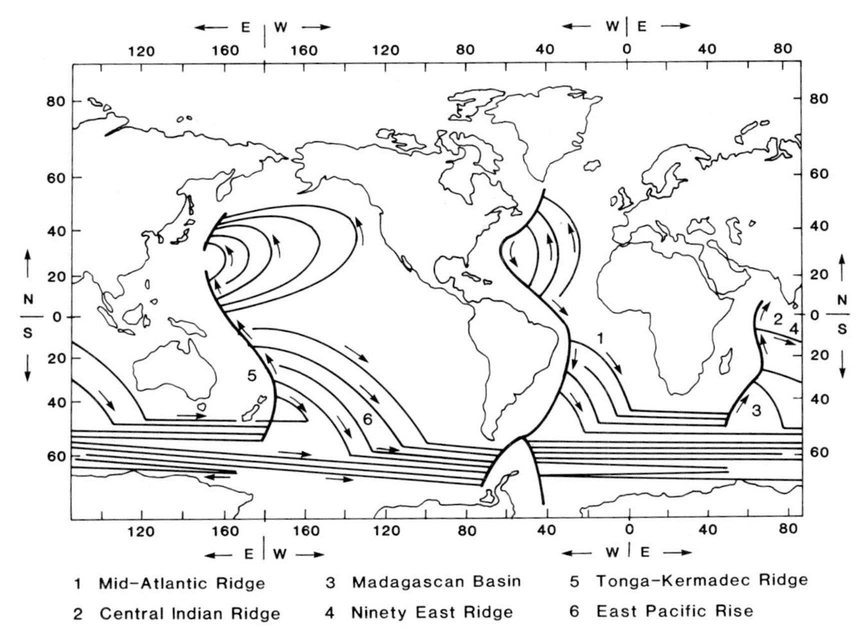

Figure 9.7 The abyssal circulation predicted by the Stommel–Aarons model, superimposed upon a world geographic map. From Tolmazin (1985).

high-latitude seas were covered by glacial ice. Instead, most bottom water originated by sinking of warm, relatively salty waters of the Central Atlantic.

Detailed sedimentologic, palaeontologic and isotopic data from Pleistocene deep-sea sediments do not confirm either the Worthington (1968) or Weyl (1968) hypotheses (Lohmann 1978). The data indicate sluggish movement of North Atlantic bottom waters occurred during glacial times (Schnitker 1974, 1979, Lohmann 1978). Obviously, the detailed palaeoceanographic response of an ocean is a function of the palaeogeographic configuration as well. The present Weddell Sea and Norwegian Sea – Denmark Strait configurations and sill depths are rather critical to the modern circulation pattern; each area may have been affected differently during glacial times (Piper, D.J.W., pers. comm. 1987).

SEDIMENT INPUT

The control of sediment input on the generation of contourite drifts has been well documented only for the western Central and North Atlantic (McCave & Tucholke 1986) (Fig. 4.9). Here, the Bjorn and Gardar Drifts occur downcurrent from the significant volcanic and glacial input of Iceland, whereas in similar oceanographic settings on the west side of Reykjanes Ridge no sediment drifts are formed because of the lack of sediment input. Similarly, the superb development of sediment drifts on the Blake–Bahama Outer Ridge is related to sediment supply from the Blake Plateau. This sediment is carried to the drift region by the Gulf Stream (McCave & Tucholke 1986). If, however, there is continual sediment supply from a lateral (line) source, the formation of a sediment drift is precluded by the continual downslope mass wasting, as evidenced on portions of the Nova Scotian continental rise (McCave & Tucholke 1986). Thus, there appear to be critical minimum and maximum thresholds of sediment input for the development of sediment drifts. The drifts take millions of years to form and, as originally discussed by Heezen *et al.* (1966), there is a delicate balance between along-slope, current-controlled deposition and downslope mass-flow processes in shaping the continental rises.

Styles of contourite deposits

MODERN OCCURRENCES

Few long-term measurements have been made of bottom-current activity associated with deep current-controlled sedimentation, and most of the sites that have been studied were not associated with sediment drifts (McCave & Tucholke 1986). Deep current-controlled sedimentation is best understood in the western Central and North Atlantic where there are numerous sediment drifts (Fig. 4.9). In addition to these large sediment drifts, there are a number of abyssal bedforms superimposed upon many of the drifts. These bedforms show a decreasing hierarchy in size from mud waves → furrows → transverse and longitudinal ripples, and other small current structures that can be observed on bottom photographs (McCave & Tucholke 1986) (Table 9.3) (Fig. 9.8).

Table 9.3 Indicators of relative (increasing from 1 to 8) current strength from sea floor photographs.

Currents		Degree of bedform development	Bottom water
Tranquil	1	Flocs of organic/mineral debris; undisturbed animal tracks.	Clear
↓	2	Subtle smoothing; rare flocs.	↓
	3	Weak lineation; appearance of tool marks (poorly developed); flocs of debris in lee of obstacles.	
	4	Small crags with tails of mineral debris	
	5	Appearance of barchan ripples of unconsolidated silt/sand; crags and tails; weak scour crescents in front of obstacles.	
	6	Mounds and tails; longitudinal ripples; common cornices; crags and tails widespread.	
	7	Well developed crags and tails very common; well developed scour crescents around obstacles; some erosional plucking of seabed and of existing bedforms.	
Strong	8	Strong and widespread development of erosional plucking; tool marks, and scour around obstacles; cohesive sediment exposed and unconsolidated silt/sand absent except in protected areas.	Very cloudy

Note: Scales not linear: the lowest speeds (scale No. 1) are <5 cm/s whereas the fastest (scale No. 8) are probably >40 cm/s.

CASE STUDY: FURROWS, BLAKE–BAHAMA OUTER RIDGE

Intermediate-scale bedforms in the deep sea have long been recognized from bottom photographs (Heezen & Hollister 1971). In the mid-1970s, deep-towed instrument packages showed that previously unexplained hyperbolic echoes were caused by furrows on the seabed (Hollister *et al.* 1974, 1976a). Detailed bathymetric and side-scan surveys of the Blake–Bahama Outer Ridge (Fig. 4.9) showed the occurrence of abyssal mudwaves, with superimposed smaller furrows. Furrows are (a) oriented at about 25–35° to the strike of the mudwave crestlines; (b) spaced 20–125 m apart; and (c) traceable along their length for distances up to 5 km. In plan view, furrows are very straight and show 'tuning fork' junctions opening into the current. In cross-section, furrows are 1–4 m wide, 0.75–2 m deep, and steep-sided with flat floors (Hollister *et al.* 1976b). Bottom current measurements (20 m above the bottom) show maximum current velocities of 8–10 cm/s, oriented parallel to the strike of the furrows. The thickness of the benthic boundary layer at this site is approximately half the spacing of the furrows, suggesting that the furrows are related to secondary helical circulation patterns within the benthic boundary layer. These small-scale furrows were interpreted by Hollister *et al.* (1976b) to be the cause of the fine-scale hyperbolic echos commonly seen on 3.5 kHz and 12 kHz records.

At other sites between the transition from the Blake–Bahama Outer Ridge to the abyssal plain, larger furrows (20 m deep × 50–150 m wide × 50–200 m spacing) are clearly erosional. As with the smaller-scale mudwave furrows, these larger furrows also show 'tuning-fork' patterns in plan view, which open in an upcurrent direction. Bottom current measurements (20 m above the seafloor) reached a maximum of 8 cm/s, compared with a maximum of 4 cm/s above the adjacent abyssal plain. Composite vertical deep-tow temperature profiles in the erosional furrow area show that the boundary layer averages about 60 m thick (Fig. 9.9). The well-mixed boundary layer is believed to result from interactions between deep, steady (? geostrophic) currents and the bottom topography (Hollister *et al.* 1976b). This results in the development of secondary circulations, which maintain the erosional furrow topography.

CASE STUDY: DEEP-MARINE CLIMBING DUNES, ROCKALL TROUGH, (?) LATE MIOCENE ATLANTIC OCEAN

A (?) late Miocene field of dunes or sandwaves has been identified in a 350-km^2 area in 1080–1180 m water depth in the subsurface of the northern part of the

Rockall Trough, offshore Scotland (Fig. 9.10) (Richards *et al.* 1987). The dune deposits are up to 200 m thick (using a sediment velocity of 1.6 km/s), and are relict features which are mantled by a drape of hemipelagic muds. Two seismic reflection profiles show that these features have major internal bounding surfaces, between which the seismic facies vary considerably (Fig. 9.11). The basal Unit 1 (Fig. 9.11) consists of a 105–m thick succession of climbing bedforms, similar to the type A climbing sand ripples of Jopling & Walker (1968). The average angle of climb of the dunes is *c*. 3°. Migration was toward the south, away from the Wyville–Thomson Ridge (Fig. 9.10). Unit 2 (Fig. 9.11) is a transitional phase between climbing dunes and sinusoidal dunes, geometrically resembling type B and transitional ripple drift lamination (Jopling & Walker 1968). The upper Unit 3 (Fig 9.11) only occurs on seismic line A (Figs 9.10 & 11) and is locally eroded below the younger deposits of Unit 4 (Fig. 9.11), interpreted as Plio–Pleistocene slide deposits. Near the top of the dune package, in Unit 3, individual dunes have heights of 18 m and wavelengths of approximately 1 km; in cross-section, these features resemble sinusoidal ripple lamination (Jopling & Walker 1968). There are no long cores from this site. By analogy with other dune bedforms, these deep-sea climbing dunes are interpreted as sand (?) deposits from currents in which the ratio of bedload to suspended load varies with time. Unit 1 represents rapid dune migration (toward the south), with deposition from tractional currents. As the strength of the tractional currents decreased, deposition from suspension increased in importance and transitional (Unit 2) and sinusoidal (Unit 3) bedforms developed.

These time-varying currents are interpreted to reflect the interaction of cold, bottom-water overflows from the Norwegian Sea with the Wyville-Thomson Ridge. Cold, dense bottom currents originate in the Norwegian Sea, and flow over the ridge into the main Atlantic basin. The interface between the cold, bottom overflows and the overlying water mass occurs at *c*. 500 m water depth. During the Miocene, the Wyville–Thomson Ridge was not a barrier to flow, and strong tractional currents deposited the dunes of Unit 1. During the late Miocene, the Faeroe–Iceland Ridge subsided (Roberts 1975), siphoning off most of the overflow water that earlier would have returned to the North Atlantic over the Wyville–Thomson Ridge. This reduction in overflow resulted in diminished current strength, with an increased accumulation of sediment from suspension (Units 2 and 3). At present, the dune system is inactive.

CASE STUDY: MUDWAVES, ATLANTIC OCEAN

Mudwaves mainly comprise hemipelagic mud, with thin bands of current-reworked terrigenous silt or biogenic debris, and locally thin turbidite sands (Allen 1984). Mudwaves are very common in the Atlantic Ocean, particularly along the north-western margin, where there is an influence of the Western Boundary Undercurrent, between 2500 and 3500 m water depth, and on sediment drifts associated with the Blake–Bahama Outer Ridge, the Greater Antilles Outer Ridge and the Feni Drift (Fig. 4.9). These abyssal mudwaves have three dominant morphologies (Allen 1984): (a) steep and irregular (wavelengths: 0.5–2 km, heights: 10–100 m, moderate to steep angles of climb) (Fig. 9.12a & b); (b) steep and regular (wavelengths: 0.5–3 km, heights: 10–50 m; low (4 to 10°) angles of climb) (Fig. 9.12c & d); and (c) flat and regular (wavelengths: 3–10 km; heights: <10 m; very low (2–4°) angles of climb) (Fig. 9.12e & f).

Mudwaves are very long-lived phenomena, and can represent millions of years of accumulation ($>10^5$–10^6) (Roberts & Kidd 1979; Tucholke 1979). They are not, however, 'equilibrium' bedforms, but represent long-term depositional and erosional events associated with bottom current flow. Two mechanisms have been proposed for the origin of abyssal mudwaves. The first is that mudwaves are the response of the seafloor to an instability phenomenon associated with interactions between a deformable bed and a turbid ocean bottom, similar to the origin of antidunes in fluvial systems. In such cases, mudwaves should form under flows with densiometric Froude numbers ≥ 1; a condition supported by Kolla's *et al.* (1980a) estimate of 0.91 for the Froude number on the interface of the benthic thermocline.

The second possible origin is that there is differential deposition on the sea floor due to the influence of large-scale internal lee-waves, caused by perturbations on the sea floor, superimposed on turbid ocean bottom currents (Flood 1978). In this case, a critical Froude number does not have to be reached prior to the formation of the mudwaves. Stationary lee waves have been calculated to be stable for stratified flow over a fixed barrier in a channel with rigid upper and lower boundaries (Allen 1984). The resultant streamline pattern shows the occurrence of a train of waves downstream from the perturbation (Fig. 9.13). Once established, the mudwaves may act as perturbations to subsequent flow, and 'seed' additional leeside waves. According to Allen (1984) this second model is consistent with (a) wavelengths of mudwaves, (b) the develop-

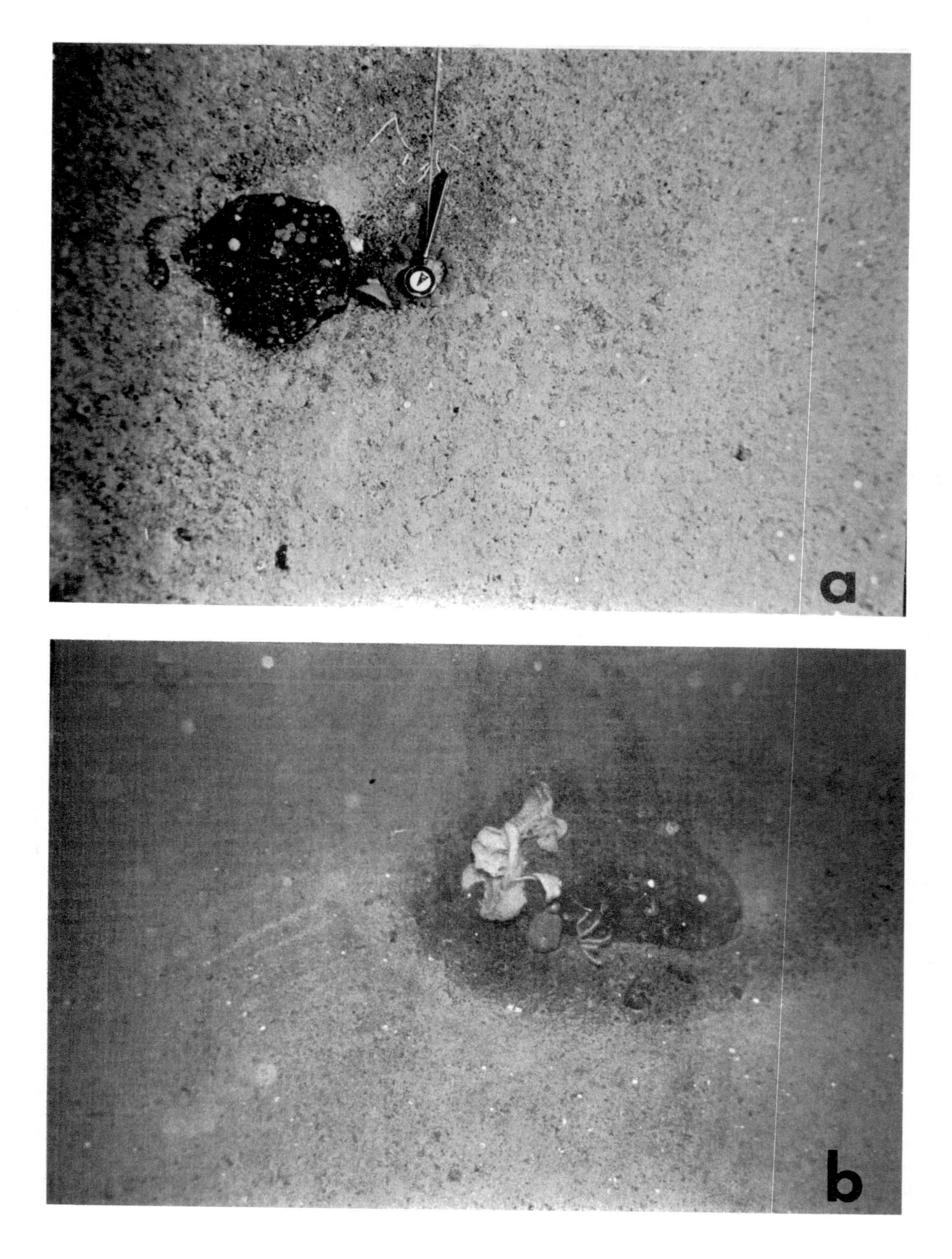
a
b

Figure 9.8 Oblique (a) and vertical (b, c) bottom photographs of the sea floor near the axis of the Western Boundary Undercurrent (WBU), Orphan Knoll region, Labrador slope, eastern Canada. Water depth 2500 m. 49° 30′ N Lat./47° W Long. a, manganese-covered cobble in a bioturbated sandy mud just upslope of the axis of the WBU. The diameter of the compass case is 7.5 cm; b, manganese-covered pebbles and boulder in a foraminiferal sand beneath the axis of the WBU. Benthic organisms are attached on the upstream side of the boulder, note current lineations in the left half of the photograph; c, heavily bioturbated mud, just downslope from the axis of the WBU.

ment of mudwaves on only one side of the levees of submarine channels, and (c) the upslope migration of mudwaves. McCave & Tucholke (1986) also agree that the origin of abyssal mudwaves is best explained by internal lee-waves, triggered by upstream perturbations. It is unlikely that mudwaves would be recognizable in ancient successions, due to insufficient outcrop scale. However, these mudwaves may account for the lateral variations seen in sand/silt : mud ratios in deep-marine sediments (Allen 1984). There are problems with the ultimate preservation potential of such features because they are deposited on oceanic crust.

CASE STUDY: SEDIMENT DRIFTS, CENTRAL AND NORTH ATLANTIC OCEAN

In the western Central and North Atlantic, large sediment drifts occur mainly along the path of the Norwegian Sea Overflow water, and at sites where the Gulf Stream interacts with the deep thermohaline currents. Smaller sediment drifts are associated with the Labrador Current and north-flowing eastern boundary currents at the base of the Iberian continental rise (Fig. 4.9) (McCave & Tucholke 1986). Bottom photographs on sediment drifts show that rippled sands, current crescents and lags occur at sites of active bottom-current flow (Fig. 9.8); muds are deposited at sites marginal to the axis of the boundary flows (McCave 1982, Schafer & Asprey 1982, Carter & Schafer 1983) (Fig. 9.8). Deposits at these sites correspond, respectively, to the 'sandy' and 'muddy' contourites of Stow (1982) and Stow & Lovell (1979) (see Tables 9.1 & 2).

In the Faro Drift, Gulf of Cadiz (Fig. 4.9), muddy contourites are typically homogeneous and bioturbated, with the preservation of only thin, irregular winnowed silt concentrations and rare silt laminae (Fig. 9.14a) (Gonthier *et al.* 1984). In contrast, the sandy contourites show better preservation of primary sedimentary structures, including thin lag concentrations of coarse sand or shell fragments, laminated and cross-

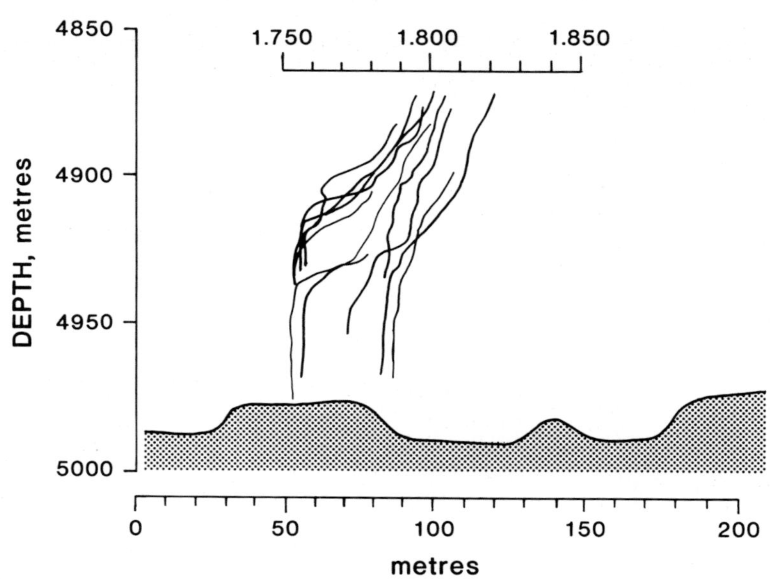

Figure 9.9 Composite of vertical deep-tow temperature profiles showing a well-mixed bottom boundary layer over large furrows near the Blake–Bahama Outer Ridge–Abyssal Plain contact. Note the relationship between the well-mixed layer thickness and furrow spacing (No vertical exaggeration). From Hollister *et al.* (1976).

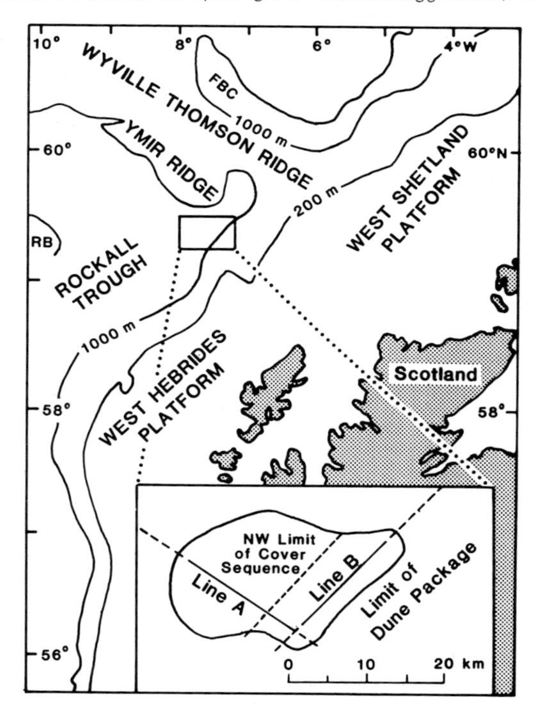

Figure 9.10 Rockall Trough study area showing position of seismic lines A and B (Fig. 9.11). FSC, Faeroe–Shetland Channel; FBC, Faeroe–Bank Channel; RB, Rosemary Bank. The relative positions of seismic lines A and B shown in inset. From Richards *et al.* (1987).

laminated sands and silts (Fig. 9.14b). Bioturbation is still common, but not as pervasive as in the muddy contourites, presumably reflecting higher energy conditions and/or decreased flux of organic matter.

Typical vertical facies trends consist of a 'coupled negative-positive' sequence, in which there is a gradual increase, reaching a maximum, then a gradual decrease in the inferred current strengths at a given site. These vertical sequences may be thick (Fig. 9.15 KC 8221) or thin with sharp erosional boundaries (Fig. 9.15 KC 8220 and KC 8217), reflecting slower and faster temporal changes in current strength, respectively. Other patterns include the rapid and sudden occupation of a site, followed by a gradual abandonment (Fig. 9.15 KC 8226). Such temporal changes are related to variations of current strength as a result of climatic or oceanographic factors, or the lateral migration of the axis of boundary currents through time (Gonthier *et al.* 1984; cf. Ledbetter 1984).

Not all sediment drifts show this variation in current velocity. For example, Kidd & Hill (1986a & b) in their study of the Feni and Gardar sediment drifts (Figs 4.9 & 9.16) found that there were few overall trends throughout the 1600 m-thick accumulation of sediment. These large sediment drifts are inferred to be the result of transport and deposition of material by thermohaline bottom currents, effective in this part of the North Atlantic since the Eocene–Oligocene. High-resolution seismic profiles (3.5 kHz) of the sediment

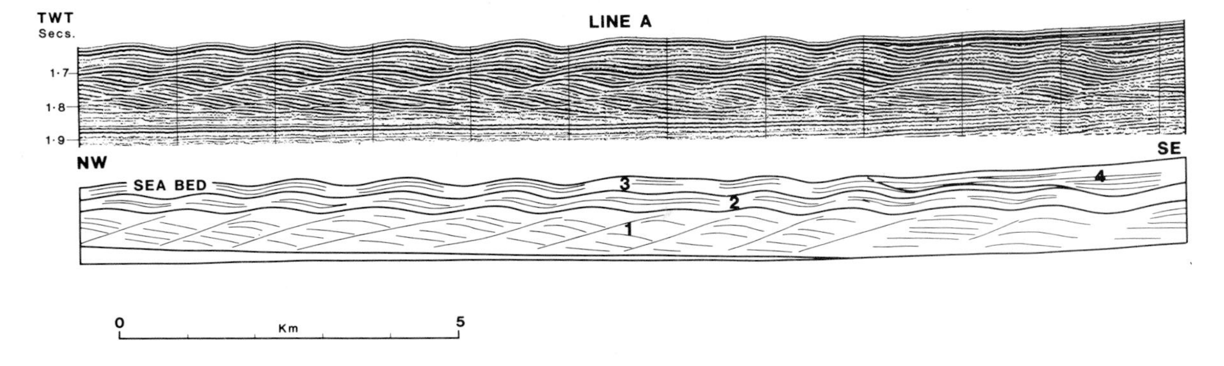

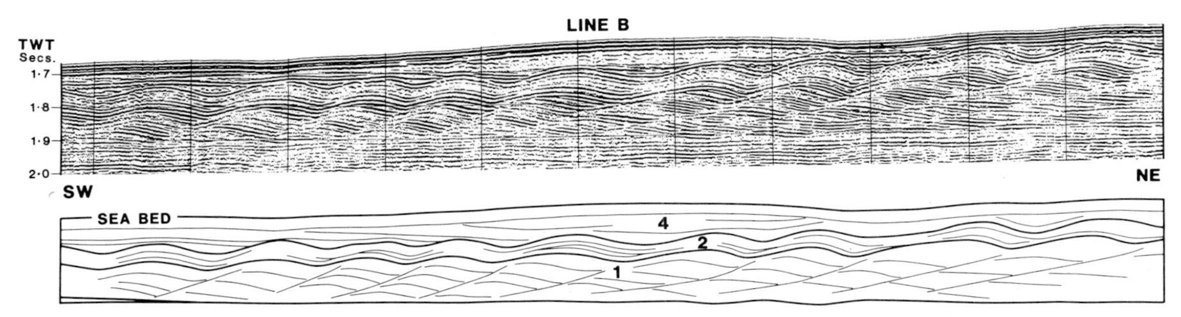

Figure 9.11 Seismic profiles of climbing, transitional and sinusoidal dunes, Rockall Trough. 1, lowest climbing-dune unit; 2, middle or transitional-dune unit; 3, top or sinusoidal dune unit; 4, overlying slump/slide unit. From Richards *et al.* (1987).

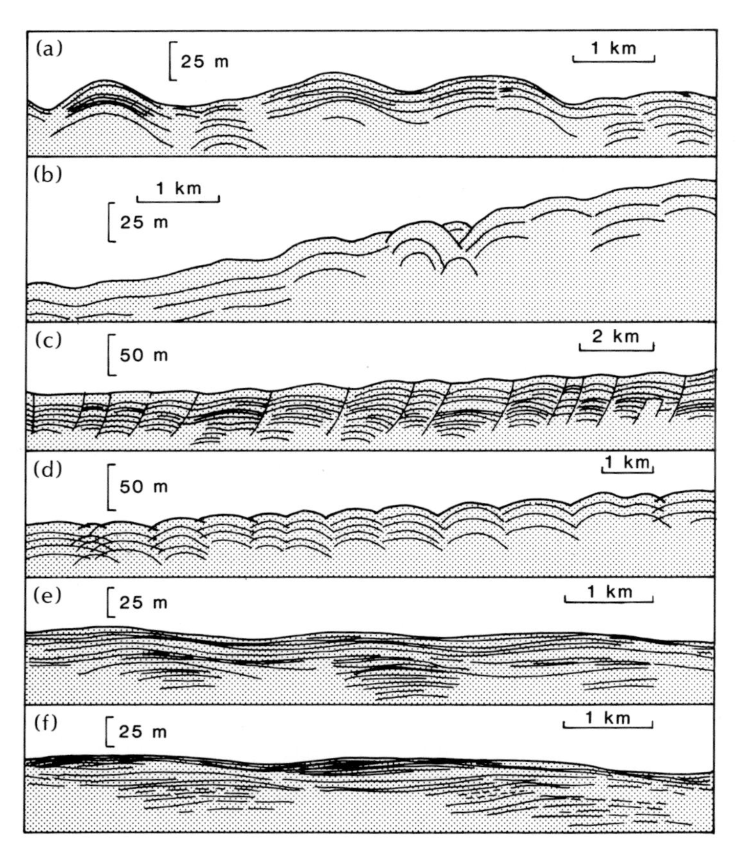

Figure 9.12 Reconstructions of seismic profiler records across different types of mudwaves, showing the seabed bathymetry and principal internal reflectors. a,b, Steep, irregular waves, equatorial Atlantic Ocean; c,d, steep, regular waves, equatorial and southeast Atlantic Oceans; e,f, flat, regular waves, equatorial Atlantic Ocean.

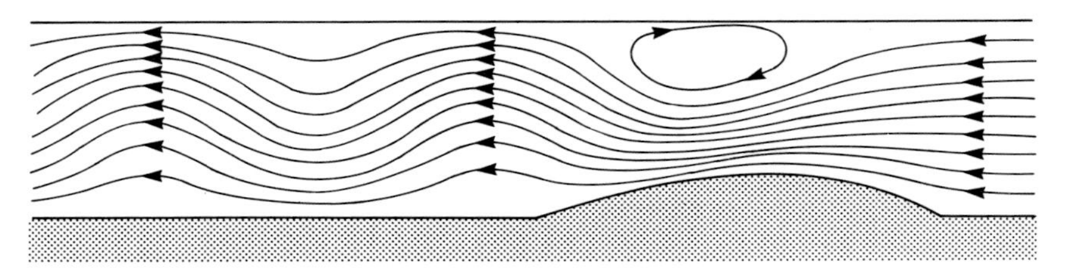

Figure 9.13 Flowlines of lee waves in a stratified flow downstream of a low mound, for densiometric Froude number (Fr′) = 0.2. From Allen (1984).

(a) **MUDDY CONTOURITES: Characteristics**

centimetres 0 1 2 3 4 5 6 7 8 9 10

Structure:
Homogeneous
Bedding poor or absent
Irregular winnowed concentrations
Rare primary silt laminae
Bioturbated and burrowed

Texture:
Dominantly silty-mud
0 – 15% sand-sized
Poorly-sorted

Composition:
Combination of biogenic and terrigenous (i.e. hemipelagic)
Part may be far-travelled
Organic carbon (av. 0.3 – 1.0%)
Carbonate (commonly high)
Absence of shallow-water biogenics

Fabric:
Magnetic anisotropic fabric may parallel bottom current
Small clay-particle (?) clusters with preferred orientation

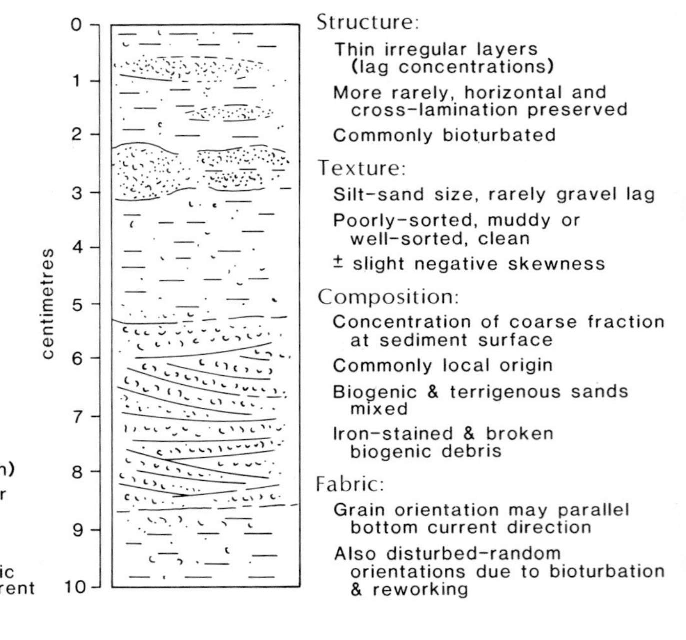

Figure 9.14 Sedimentary characteristics of a, muddy contourites, and b, sandy contourites from Stow & Holbrook (1984).

drifts show sediment waves superimposed on the drift deposits (Fig. 9.17). Detailed DSDP coring of wave crests and troughs failed to show appreciable changes in facies, either spatially across the sediment drifts or temporally through 1600 m of section. In cores from the upper 70 m at least, the lithostratigraphic variations apparently correspond solely to glacial/interglacial stages (Fig. 9.18). Sediment within the waves is mainly pelagic, with little or no current reworking since 2.4 Ma. Broad scale changes in the sedimentary regime of the drifts are due to interactions of major water masses in the Atlantic, including the Norwegian Sea Overflow Water, the Antarctic Bottom Water and the North Atlantic Deep Water. During the preglacial stage, the Norwegian Sea Overflow Water was the main current affecting the area during a period of active sediment wave migration. The influence of this water mass diminished with the onset of glacial–interglacial periods. Holocene sedimentation since that time has mainly deposited a pelagic drape, and bottom currents

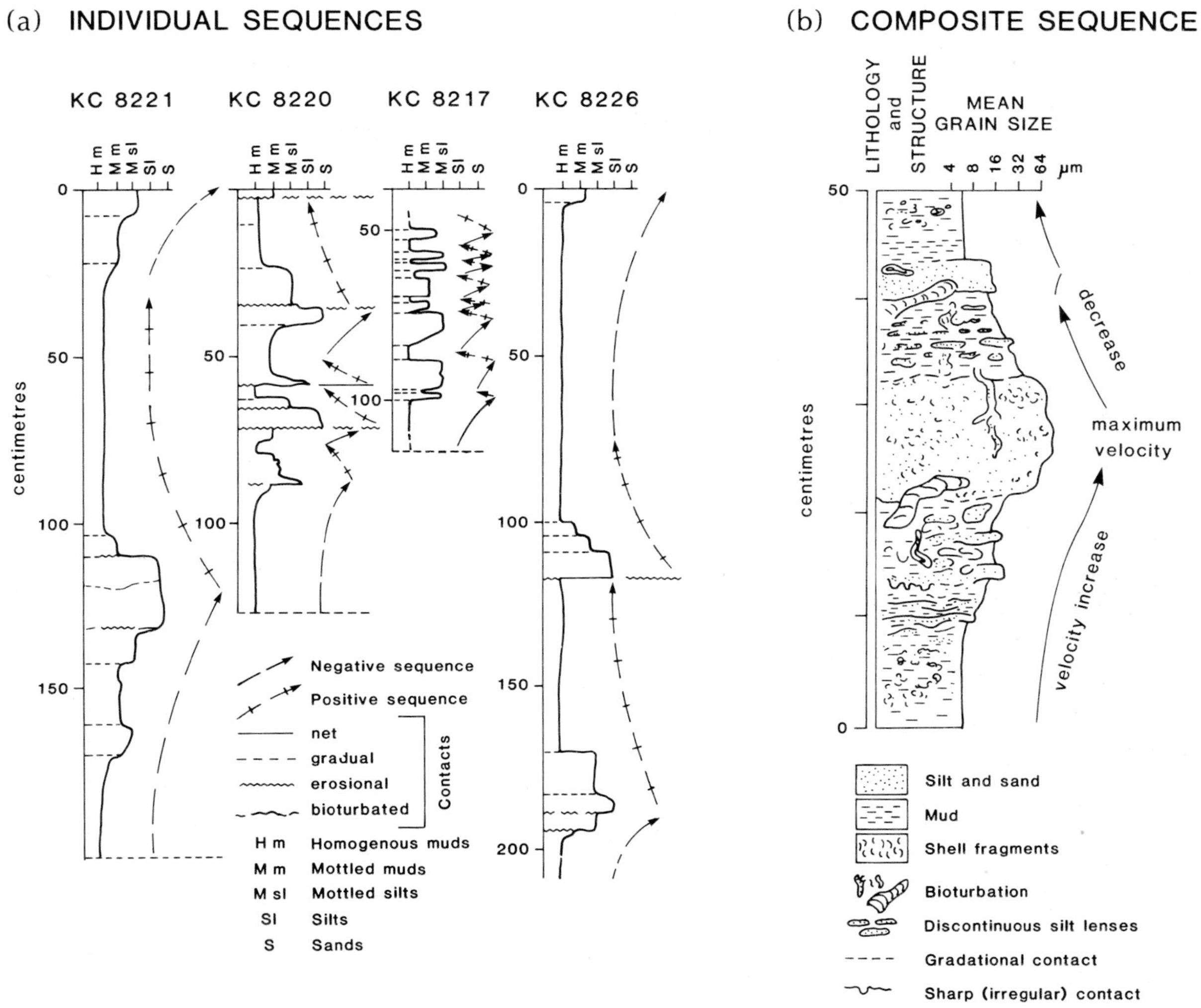

Figure 9.15 a, Sketches of core sections through contourite facies, showing coarsening-upward (negative) and fining-upward (positive) vertical sequences; b, schematic model showing vertical sequence of contourite facies. Faro Drift. From Gonthier *et al.* (1984).

have simply maintained the pre-existing morphology of the now relict sediment drifts.

The large-scale controls on the overall shape of sediment drifts are the form of the pre-existing sea floor and the influence of this topography on the contour-current circulation. The speed of contour currents is increased in zones of steeper bottom gradient and diminished in zones of reduced gradient (McCave & Tucholke 1986) (Fig. 9.19a & b). Sedimentation preferentially occurs in the more tranquil, lower gradient zones. The morphology of the sediment drifts depends on the location of the axis of maximum flow in relation to the pre-existing sea floor topography and on the distribution of suspended sediment matter within the flow (McCave & Tucholke 1986). Depending on the rate of change of the sea floor gradient, double drifts (Gardar & Bjorn, Figs 4.9 & 9.19a) or 'plastered' drifts (Hatton and N. Feni Drifts, Figs 4.9 & 9.19b) occur at the base-of-slope. If the rate of change of the sea floor gradient is very abrupt, the sediment drifts may become separated

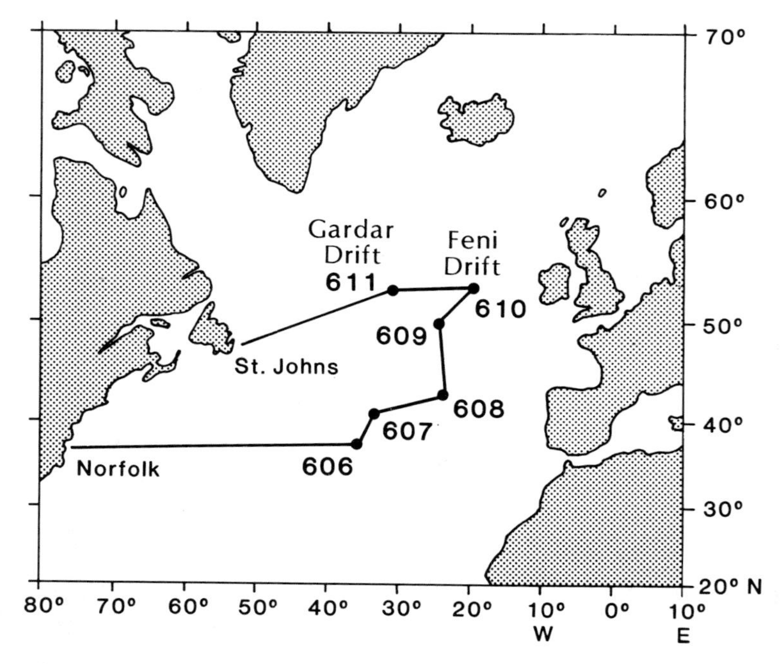

Figure 9.16 Location of sediment drifts Sites 610 and 611 on DSDP Leg 94 palaeoclimate transect. From Kidd & Hill (1986a).

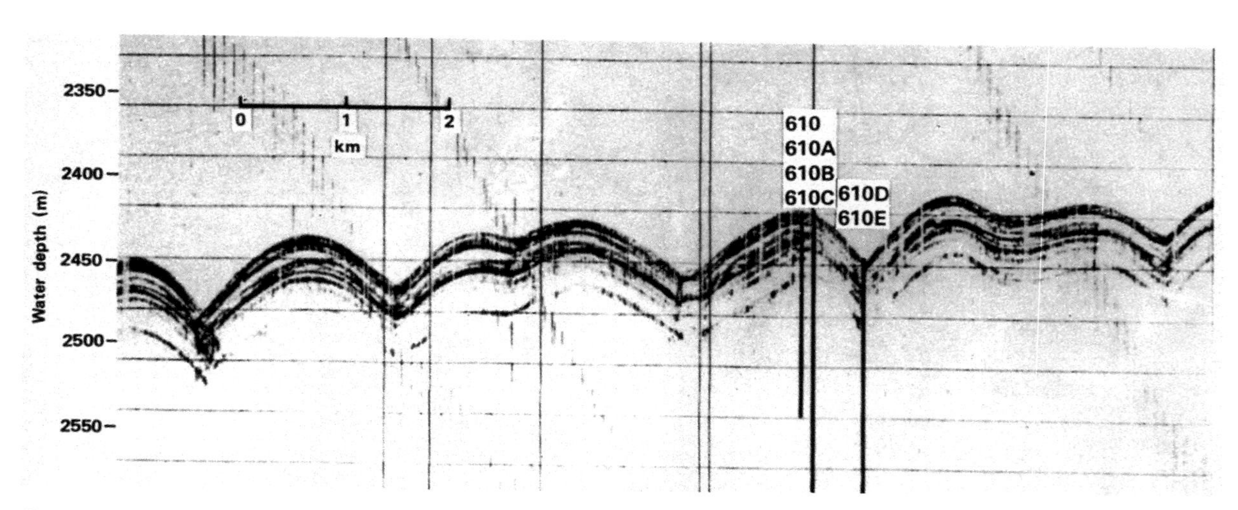

Figure 9.17 High-resolution seismic profile (3.5 kHz) of the sediment waves on Feni Drift, illustrating location of crest and trough drilling Site 610. From Kidd & Hill (1986a).

from the slope (Caicos Outer Ridge, Figs 4.9 & 9.19d). In areas of more complex sea floor morphology (e.g. in areas of combined or reversing flows around a corner or a ridge) 'detached' drifts may form, and grow by progradation in the direction of the initial flows (Greater Antilles Outer Ridge, Blake–Bahama and Eirik Ridge, Figs 4.9 & 9.19c).

Studies of the seismic stratigraphy of contourite drifts and mounds show that the surficial morphologic patterns of sediment drifts can be preserved in the subsurface (Kennard 1984, Mitchum 1985). Seismic profiles of sediment drifts near Orphan Knoll and Flemish Cap, offshore Newfoundland (Figs 4.9, 9.5 & 9.20) show a series of 'mound' and 'moat' topographies (Kennard

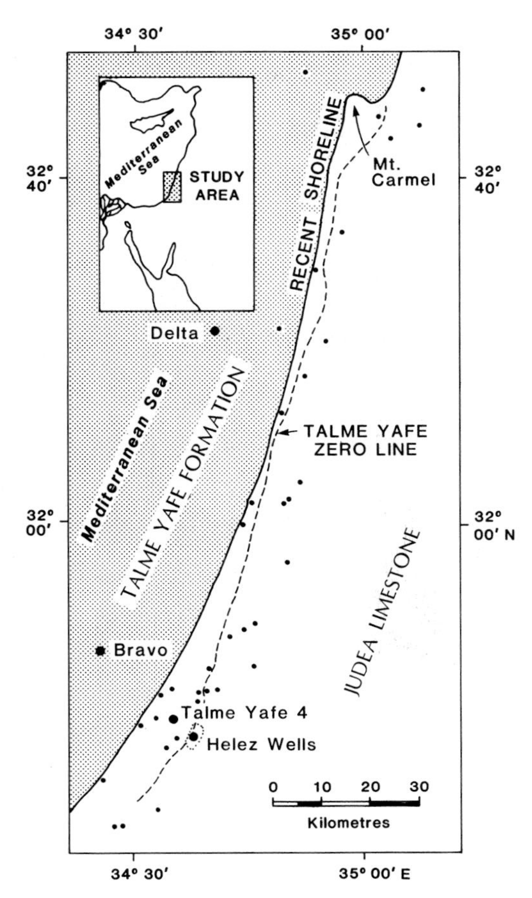

Figure 9.24 Map of the coastal plain of Israel and its continental shelf, showing the location of drill holes (dots). Designated drill holes are mentioned in the text. From Bein & Weiler (1976).

by various mass-wasting processes. This material was redeposited on the slope by contour-following currents. Within the whole succession, planktonic fauna are ubiquitous and especially prevalent in the marls to the west which are interpreted as pelagic deposits (Fig. 9.25). Fine-grained sediment was transported from the shelf and onto the slope as nepheloid flows.

The Talme Yafe deposit consists of calcilaminite (Facies Class D), calcilutite (Facies Class E), calcarenite (Facies Class B), calcirudite (Facies Class F) and marl (Facies Class G) lithofacies. The bulk of the Talme Yafe (perhaps a few hundred metres) consists of calcilaminite (Facies Class D), which is found in most of the drill holes and along 200 m-thick coastal outcrop sections. The calcilaminite shows a frequent alternation of calcilutite and calcarenite (Fig. 9.26). The laminae are parallel, with sharp boundaries and no grading. Micro-scale cross-lamination is less common. The calcilutite laminae are ungraded, consist of fine to coarse calcilutite, with a small proportion of coarser dispersed sand-size grains (Fig. 9.27). The calcarenite laminae are fine to coarse-grained sand, with a poor degree of sorting. Small flute marks, cut-and-fill structures, intraclasts and micro-load structures occur. Some of the units are extensively bioturbated. Erosional channels, up to 20-m wide, occur within the calcilaminite facies (Fig. 9.28).

Calcarenite facies (Facies Class B) occur in the Talme Yafe deposits west of the Helez oil field (Fig. 9.24). Calcarenites mainly comprise unsorted, randomly-oriented skeletal material, up to 0.5 mm in diameter, with the only discrete internal sedimentary structures consisting of flaser structures and intraclasts. Calcirudites (Facies F1.1) occur near the base of the Talme Yafe Formation, attaining a maximum thickness of 145 m in the Talme Yafe 4 drillhole (Fig. 9.24). The calcirudite comprises chaotically oriented blocks, 2–3 m across, and finer gravel clasts set in a calcilutite matrix (Fig. 9.29). No bedding is discernible. Rock fragments are poorly sorted, and angular, and are thought to have been derived from the underlying Aptian Telamim Formation. The marl facies (Facies Class G) occurs only in the western Bravo and Delta drill holes (Fig. 9.24) and is mainly structureless. No sedimentary structures are preserved. The marl comprises fine lutite, with about 25% clay and 5–10% coarse calcilutite and calcarenite. Planktonic foraminifera and nannoplankton are abundant.

The following major features characterize the Talme Yafe deposit: (a) predominant shelf-derived calcareous detritus; (b) periodic alternation of strong and weak currents; (c) a small proportion of turbidites; (d) an elongated, asymmetric prism, in which the isopach contours parallel the trend of the continental rise

Figure 9.22 (opposite top) Seismic section B, North Flemish Cap Double Mound. See Figure 9.21 caption for explanation. Vertical exaggeration is approximately 6:1. From Kennard (1984).

Figure 9.23 (opposite) Seismic section C, Sackville Spur. See Figure 9.21 caption for explanation. Vertical exaggeration is approximately 6:1. From Kennard (1984).

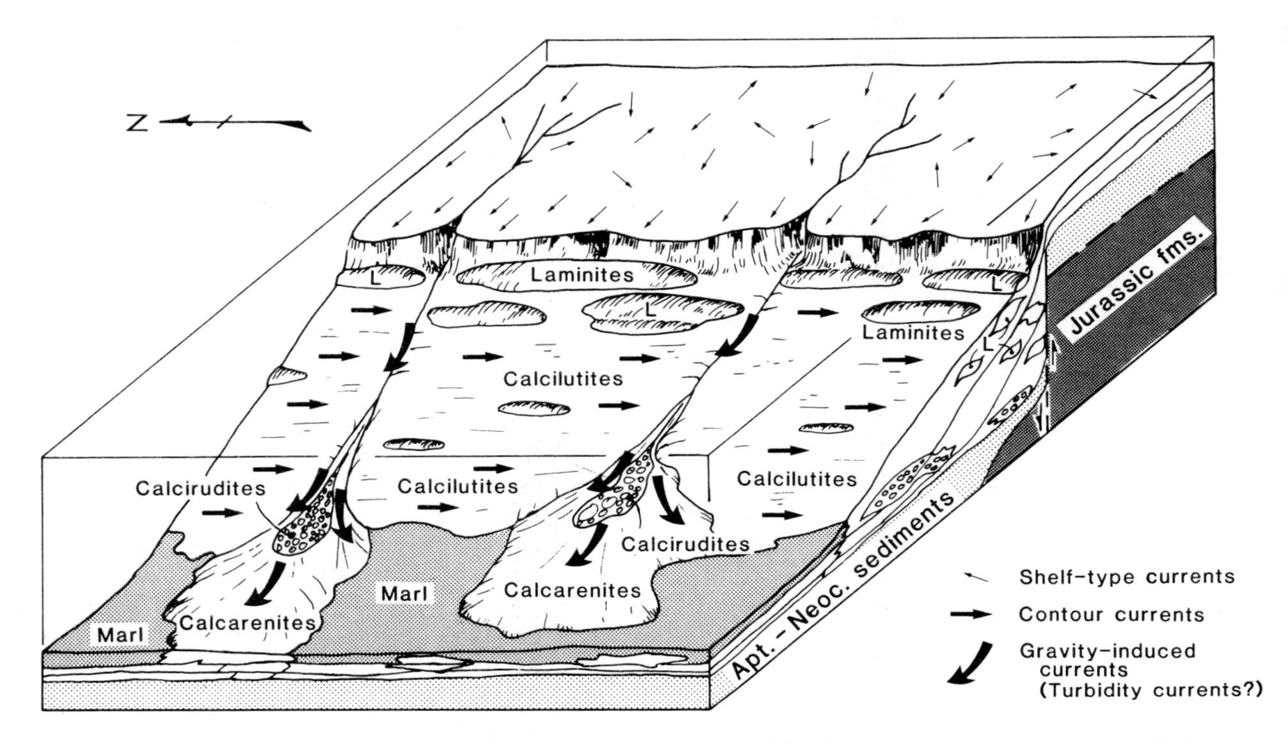

Figure 9.25 Schematic three-dimensional model of palaeogeography for the Talme Yafe Formation, Israel continental shelf. From Bein & Weiler (1976).

(Fig. 9.30); and, (e) a stratigraphic position between shallow neritic carbonate platform deposits and deep-marine pelagic marls.

The overall shape of the formation on isopach maps, palaeogeographic setting, and facies all point clearly to deposition as a sediment drift on the lower continental slope/rise. In this Miocene example, the source of clastic material was from epicontinental platform carbonates located to the east. Downslope movement from the shelf break was by nepheloid flow for the fine-grained detritus, and mass flows (? turbidity currents or debris flows) for the coarser grained material. Coarse-grained sediment gravity flows, which deposited the calcirudites (Facies Class F) were confined to submarine canyons which dissected the continental slope, and to small submarine-fans at the base-of-slope. The main currents which dispersed the sediment and shaped the continental rise were contour currents. The contour-currents emplaced much of the laminated calcilutite (Facies Class D) as muddy contourites, and the flaser-bedded, and intraclast-strewn calcarenites (Facies Class B) as sandy contourites, perhaps beneath the axis of the contour current. Marls (Facies Class G) were due to pelagic sedimentation in areas more removed from the contour-current influence.

Facies model for contourites

It is premature to propose a facies model for contourite-dominated systems. In modern settngs there is a hierarchy, from smallest to largest, of bed-forms: barchan & longitudinal ripples → erosional furrows → sand-waves (dunes) or mudwaves → drifts. Generally speaking, the internal features of the larger bedforms are only well documented from seismic data, for which there is a transition in seismic facies with increasing energy from: hyperbolic reflectors → wavy-stratified → sediment-wave → mounded-stratified facies (Fig. 9.4). In zones of maximum velocity (i.e. along the axis of a contour-current) there may be erosion and non-deposition, with the development of sediment lags, condensed sections and unconformities.

Possible proximal to distal trends away from the current-axis can be hypothesized for sediment drifts, based on results from modern studies. An example of

Figure 9.26 Calcilaminite (Facies Class D), showing alternating calcilutite (light) and calcarenite (dark). Lamina boundaries are sharp and no graded lamination was observed. Length of label 19 mm. Drill hole 9. From Bein & Weiler (1976).

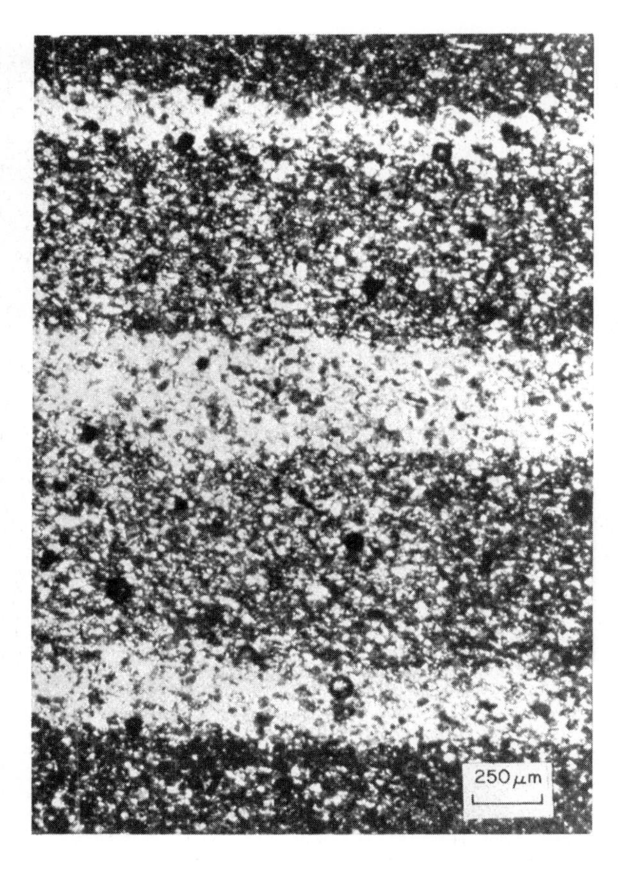

Figure 9.27 Calcilaminite (Facies Class D) detail of Figure 9.26. Note that no clear-cut grain size sorting can be observed: lutite is present in the calcarenite laminae and vice versa. Maximum grain size is 85 μm. Drill hole 9. From Bein & Weiler (1976).

this approach can be summarized from data obtained in Orphan Knoll Basin, where the Western Boundary Undercurrent impinges on the east Labrador lower slope and rise at *c.* 2700 m-water depth (i.e. Carter *et al.* 1979, Schafer & Asprey 1982) (Fig. 9.31). The trends depend upon the type of sediment source, the direction of supply (downslope versus along-slope), the relative strength of the currents beneath the contour current versus along the flanks, organic flux to the bottom, the carbonate compensation depth, range of water depths, the bottom gradient, among other factors. For the proposed summary sequence, water depths are 300–3000 m; current strengths are 0 to 20 cm/s; a zone of upwelling occurs just upslope of the sediment drift system; and sediment supply is ice-rafted debris from floating ice and resedimentation of shelf sediment by mass-wasting processes downslope. Accumulation rates are high on the flanks of the Western Boundary Undercurrent, diminishing towards the axis of the current. Sand percentages vary from 10% on the upslope flank, to 65% beneath the axis of the current, to 35–40% on the downslope flank. Beneath the axis of the Western Boundary Undercurrent, sediment is a thin, sandy gravel lag, with Fe–Mn coatings on pebbles. Organic carbon content is low, bioturbation intensity low, and calcium carbonate content is high, reflecting a high diversity of benthic deep-water calcareous foraminifera.

Proximal/distal trends in sediment character from a contour current axis may vary both spatially and

Figure 9.28 Calcilaminite (Facies Class D): a 20–m wide erosional channel cut through a well-bedded undisturbed laminite succession. Beds filling the channel wedge out towards the margins of the channel. Mount Carmel. From Bein & Weiler (1976).

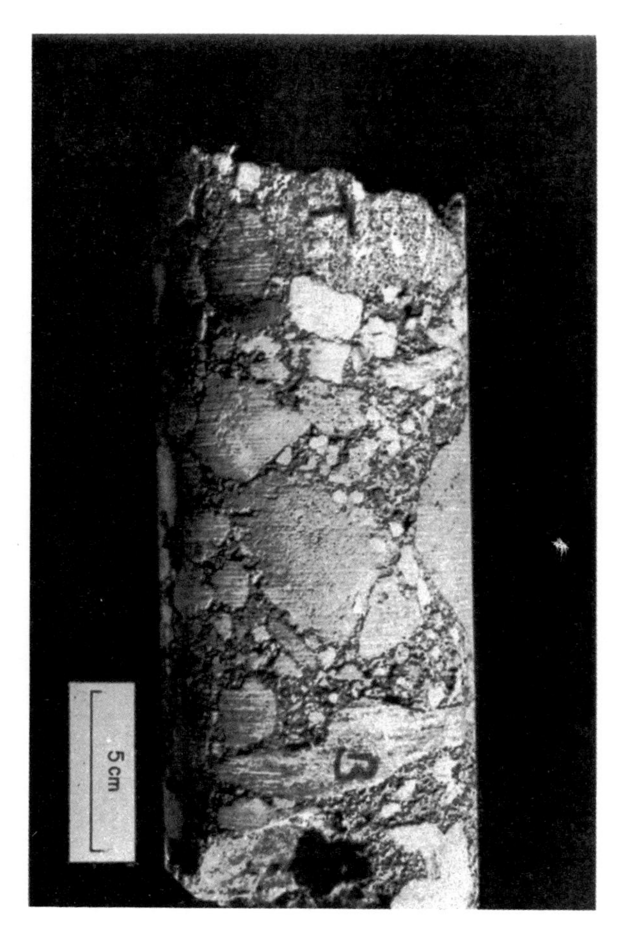

Figure 9.29 Rudite (Facies F1.1) consists of angular, unsorted fragments (maximum of a few decimetres in diameter) with a calcilutite matrix. Talme Yafe 4 drill hole. From Bein & Weiler (1976).

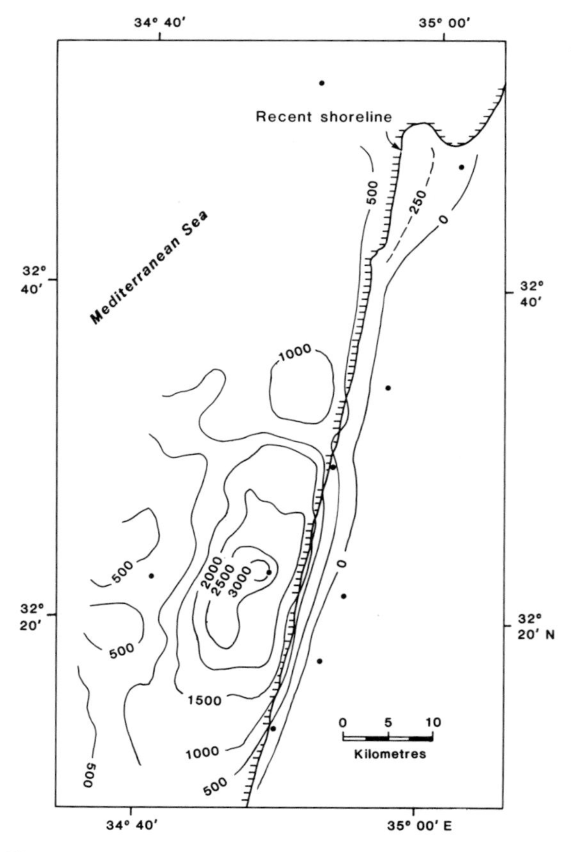

Figure 9.30 Isopach map of the Talme Yafe Formation in the Israel continental shelf and in the northern coastal plain. Contours based on drill hole data (Fig. 9.24) and correlated seismic profiles. From Bein & Weiler (1976).

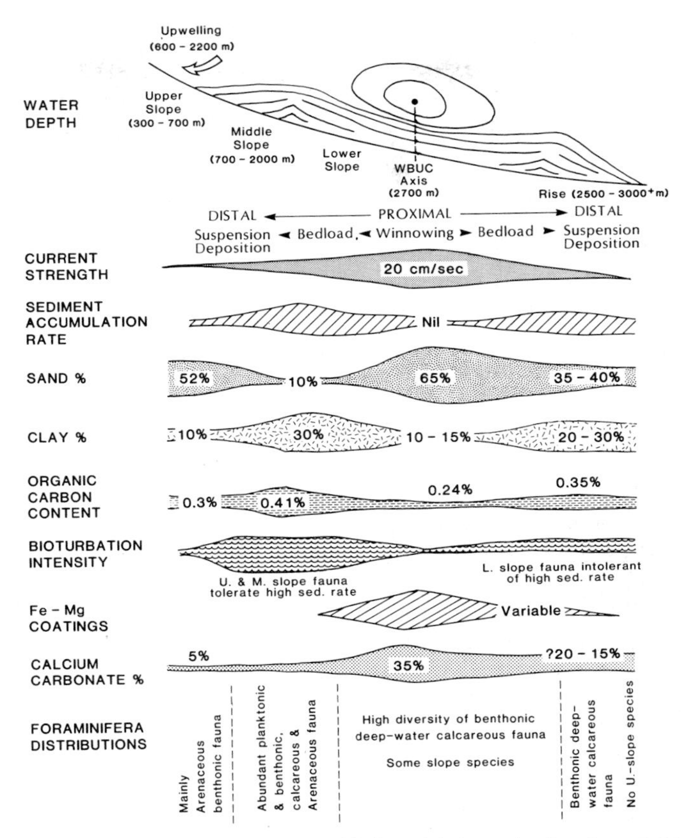

Figure 9.31 Schematic representation of distal-proximal-distal (left to right) trends of sediment accumulation in the vicinity of a strong contour current flowing along a continental slope or rise, with the development of a double mound sequence. Based on data gathered from the sea floor beneath the Western Boundary Undercurrent, near Orphan Knoll, Labrador Slope, eastern Canada (cf. Carter *et al.* 1979, Schafer & Asprey 1982).

temporally, depending upon the rate of shifting of the path of the contour-current axis (Ledbetter 1979). These shifts generate fining–coarsening–fining sequences. The bottom sediment trends across sediment drifts are isochronous, occurring on timelines which can have various 'wave-form' or irregular morphologies (see Fig. 9.19). The identification of such features in the ancient requires stratigraphic preservation of long (? up to km) correlative along-strike units. Finally, disconformities and/or subtle discordances in the seismic sections of modern mudwaves and drifts, and ordered stacking of seismic facies, may be the result of meandering of the axes of deep-water flows through time and space. With more data on the relationship of drifts to the deep circulation, we may begin to understand the internal structures of sediment drifts, and to make predictions of expected trends in ancient successions.

PART 3

Plate tectonics and sedimentation

Small-scale synsedimentary normal faults, Kimmeridge deep-marine 'Boulder Bed' succession, NE Scotland. Note the sandstone clastic dyke intruding along the fault surface at right. These beds and faults are associated with the major phase of late Jurassic extensional tectonics in the northern North Sea.

CHAPTER 10

Evolving and mature passive margins

Introduction

Rifting and subsequent continental breakup to form oceanic crust have been modelled by McKenzie (1978) under a regime of pure shear, with crust and lithosphere both thinning as a result of extension. More recently, Wernicke (1985) has suggested that rifting under simple shear, producing asymmetric crustal thinning and asymmetric half-graben geometry about the rift axis, better explains the geometry of rift systems. In this case, continental failure is along a low-angle detachment surface that reaches the surface as a 'breakaway fault' situated to one side of the rift zone.

The Wernicke model has been applied to the crustal structure of the Viking Graben by Beach (1986), and the Grand Banks – western Iberia rift zone by Boillot *et al.* (1987), and Tankard & Welsink (1987). Whatever model is valid, perhaps a combination of both, crustal thinning to form oceanic crust has important implications for passive margin stratigraphies.

The McKenzie and Wernicke models involve a phase of initial fault-controlled subsidence, followed by exponentially decreasing thermally-controlled subsidence for an evolving margin, so that deep-marine sediments may be deposited locally soon after rift onset, or over a broader area after continental breakup. In some cases, away from the boundary between continental and oceanic crust, rift-stage deep-water sediments may pass gradually upwards into shallow marine deposits resulting from slope/shelf progradation as sediment accumulation rate begins to keep pace with and eventually exceeds the rate of thermal subsidence. Variations in sea level may considerably alter this simplified sequence. Closer to the boundary between continental and oceanic crust, thermal subsidence may take the sea floor into deep water (2–4 km) so that subsequent sedimentation results in no significant shallowing.

Rifting may involve several episodes of lithospheric stretching and thermal subsidence before continental breakup. In general, it is assumed that the lithosphere is subject to an initial phase of attenuation and subsidence, followed by post-breakup thermal subsidence as the lithosphere cools (McKenzie 1978, Steckler & Watts 1978). The longer the time of rifting, the greater will be the proportion of syn-rift sediments in the passive-margin record (Cochran 1983). We recognize five stages in the evolution of passive margins (Table 10.1). The pre-rift sediments are deposited on a relatively stable craton, or are related to an older, unrelated tectonic framework.

Rift (or syn-rift) sediments are deposited during extension and are affected by rotation of upper crustal blocks on listric or planar faults (e.g. Surlyk 1978). The continental breakup phase follows major attenuation of the lower crust and the appearance of the first oceanic crust. Breakup is not an instantaneous event along a margin, but may propagate along the rift zone. Even locally, variable stretching rates (β-factors) between segments of the rift zone, each bounded by transfer faults (Gibbs 1984), can lead to a complex and protracted history of breakup. In the South Atlantic, rifting began in the Oxfordian–Kimmeridgian (136–147 Ma), and ended with diachronous breakup from the Hauterivian near the Agulhas Plateau, to early Albian in the vicinity of the Niger Delta (DSDP Leg 75, Hay, Sibuet *et al.* 1984); the rifting phase varied by about 15 m.y. (Fig. 10.1).

Some researchers refer to a 'post-rift phase' after continental breakup, but this is an unnecessary and redundant term in the case of passive margins. The rift phase ends as a result of breakup, so the adjective 'post-

Table 10.1 Evolution of passive continental margins (based partly on Brice *et al.* 1983 with considerable modifications).

Phase	Tectonic signature	Structural style	Seismic signature	Environments and lithological signature
Pre-rift	Stable craton?	Internally concordant Gentle dips Little faulting	Concordant events Strong, low frequency events	Variable shallow-marine to non-marine sediments
Rift or syn-rift	Lithospheric extension	Block faulting (listric or planar) Angular unconformities Steeply dipping strata	Weak, discontinuous reflections Angular reflection patterns Truncated sequences	Shaly and sandy siliciclastics (marine to non-marine), local high organic-matter concentrations, local deep lakes/narrow deep-marine troughs/hypersaline basins
Breakup	Initiation of sea-floor spreading. Major attenuation of lower crust ends	Internally concordant Gentle dips Little faulting ± halokinesis with associated complex faulting	Concordant events Strong continuous reflections grading into transparent zones ± complex patterns with halokinesis	Sandy siliciclastics, carbonates
Post-breakup or early drift	Rapid thermal subsidence and deepening	Internally concordant, ± deep faulting and shallow folding with halokinesis or shale diapirism	Weak concordant reflections ± complex patterns with strong reflections at base with halokinesis or shale diapirism	Marine siliciclastics and carbonates Transgressive sequence
Mature margin	Slowing thermal subsidence and increasing flexural rigidity of lithosphere	Internally concordant, Diminishing faulting/folding due to halokinesis or shale diapirism ± major sediment slides	Weak discontinuous reflections having progradational offlap patterns ± cut-and-fill patterns with erosion, chaotic patterns with sediment slides	Shaly siliciclastics, turbidites ± slide sheets/debrites, canyon fills. Regressive sequence. Sea-level variation increasingly influential

breakup' is preferred for clarity. We reserve the term 'post-rift' for failed rifts in which extension ceases without continental breakup, allowing most thermal contraction to occur before later events are inititiated. The term 'post-rift' would therefore be appropriate for aulacogens, or for sequels to early failed attempts at continental rupture, as was the Triassic – earliest Jurassic extensional episode along the North Atlantic borderlands.

The post-breakup phase (Table 10.1) involves thermal subsidence, major deepening and the development of onlap sequences. The mature passive margin (Phase 5) is characterized by a prograding slope, and complex erosion/deposition.

The Triassic to Cretaceous history of the North Atlantic involves all five phases of rifting, and includes an additional post-rift phase after aborted rifting in the Triassic; the second and successful rift phase began in the Late Jurassic (Kimmeridgian).

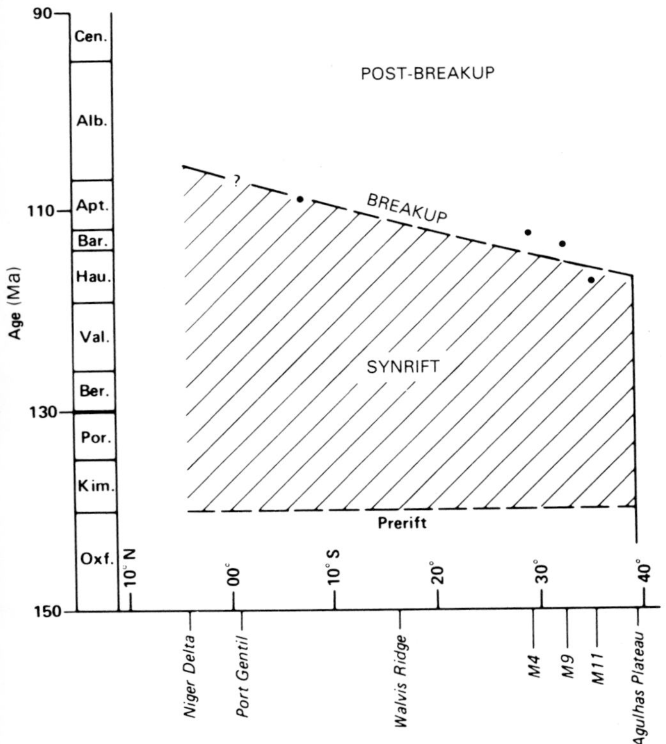

Figure 10.1 Duration of syn-rift sedimentation along the African continental margin from Cape Agulhas to the Niger Delta. Time scale from Odin (1982). From Sibuet *et al.* (1984).

Stratigraphy

The main controls on stratigraphy during the development of either a passive continental margin or an aulacogen, are: (a) thermal contraction of initially warm and thin lithosphere; (b) sedimentary loading (flexure) of a lithosphere that cools and becomes progressively thicker and more rigid after extension has ceased, and (c) compaction, palaeobathymetry, local erosion and global sea-level changes, all of lesser importance (Watts & Thorne 1984), although global sea-level changes exert relatively greater influence as tectonic subsidence rate decreases with time after breakup (Thorne & Watts 1984).

Thermal and mechanical models for the evolution of passive continental margins, taking account of the effects of thermal contraction and sediment loading, have been developed (e.g. Beaumont *et al.* 1982 and see references in Coward *et al.* 1987). These models involve increasing the flexural rigidity of the cooling lithosphere with age and allow a prediction of the stratigraphy for an evolving passive continental margin. A comparison of the predicted and observed stratigraphic profiles permits an assessment of the relative importance of the various factors that controlled the development of the actual stratigraphy (cf. Watts & Thorne 1984).

Although every evolving and mature passive margin will possess a unique stratigraphy related to its own geological development, many tend to show a deep-marine stratigraphy in the post-breakup phases as illustrated by the Mesozoic evolution of the West African margin off Cabinda, Angola (Brice *et al.* 1983). However, during the syn-rift phase of some evolving passive margins, bathymetric depressions between the crests of fault blocks may be sufficiently deep to allow the accumulation of deep-marine sediments. An example of this is found in the Galicia margin, west of Spain and Portugal, where inferred Upper Jurassic – Lower Cretaceous slides and olistostromes of Facies Class F are described by Mauffret & Montadert (1987) banked against normal fault scarps.

As a general rule, the sediments of the post-breakup, or 'drift' phase tend to become finer grained and more biogenic with time, as continental source areas become relatively less important: coarse-grained deep-marine sediments (Facies Classes A, B, C & F), if present, are overlain mainly by finer grained (Facies Classes D & E) turbidites and/or contourites, above which hemipelagic and pelagic deposits (Facies Classes E & G) increase in volume. Submarine canyons and gullies may incise the evolving margin, transporting shelf sediments, together with more locally eroded material, deeper into the basin to form turbidite systems such as submarine fans, either within slope basins or on the developing continental rise.

Kingston *et al.* (1983) recognize four types of passive continental margin (Fig. 10.2) based on the most important control on sedimentation. These four types are: (a) 'normal' siliciclastic margin; (b) carbonate bank (platform); (c) major delta with major gravity-controlled tectonic features, and (d) margin dominated by salt tectonics. Naturally, continental margins may embrace a number of these 'types' within a relatively restricted geographical area, for example the Gulf of Mexico contains carbonate platform-fringed margins (West Florida Slope and Campeche Escarpment), together with a hybrid of the delta- and salt diapir-influenced margin (Texas–Louisiana Slope and Mississippi Fan) (Bouma *et al.* 1978 and references therein).

The reader is referred to other parts of this book for further details on the sedimentary facies that characterize evolving and mature passive continental margins (Ch. 3). and for a general discussion on the controls on slope physiography and sediment types (Ch. 4, 5 & 9).

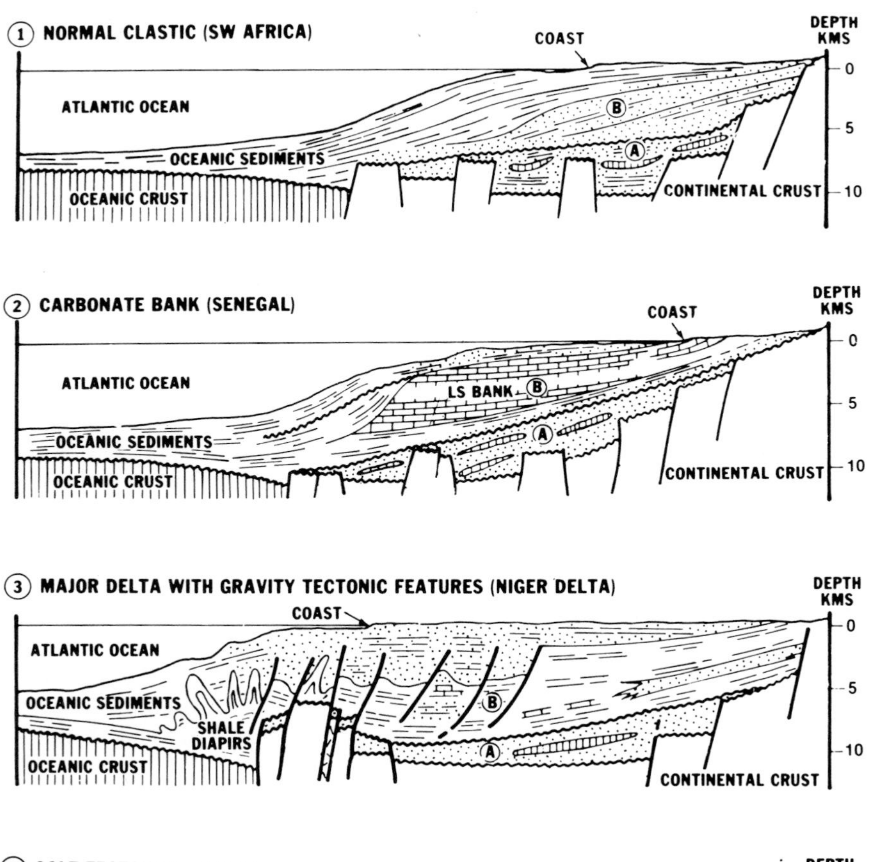

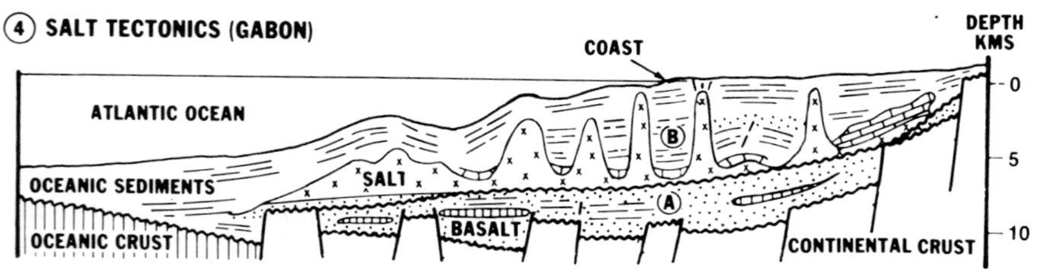

Figure 10.2 End-member classification of passive continental margins from Kingston *et al.* (1983). A = non-marine and B = marine stratigraphy. See text for explanation.

Salt diapirism appears to be common in the slope and rise sediments of many passive continental margins, for example off northern Israel (Almagor & Garfunkel 1979, Almagor & Wiseman 1982, Garfunkel 1984, Garfunkel & Almagor 1985), in the northern Gulf of Mexico (Bouma *et al.* 1978, Humphris 1978, Berryhill 1981, Bouma 1983), and off North Carolina, where salt structures extend for 300 km, parallel to and seaward of the base of the continental slope (Cashman & Popenoe 1985). Diapirism may alter the sediment distribution patterns in deep water by: (a) diverting pre-existing canyons; (b) creating intraslope basins within which sediment ponding can occur; (c) generating slope failure through mass-wasting; (d) providing a local source of sediment; and (e) providing easy-slip horizons for mass-wasting triggered by, for example, earthquakes and other processes (e.g. the Messinian salt horizon in the continental margin off Israel).

Changes in ocean circulation patterns

Continental rifting and the development of ocean basins, leading to important relative changes in plate positions, are linked to major changes in ocean-current circulation patterns and the global mass balance of water, which, in turn, may be associated with important temporal changes in deep-marine facies.

The onset of deep-ocean thermohaline circulation cells may rework the passive margin sediments into contourite drifts and other small- to large-scale bedforms. Neogene–Pleistocene examples of this phenomenon have been documented for the Norwegian–Greenland Sea (DSDP Leg 38, Talwani, Udintsey *et al.* 1976), the Atlantic continental margin off western South Africa (DSDP Leg 75, Hay, Sibuet *et al.* 1984), off the east coast of the USA (DSDP Leg 93, van Hinte, Wise *et al.* 1985a & b; DSDP Leg 95, Poag, Watts *et al.* 1987), and the Gulf of Cadiz as a result of Mediterranean outflow (Faugères *et al.* 1984, Gonthier *et al.* 1984).

The deep-ocean currents may also produce regional erosional surfaces (disconformities) as in the Palaeocene and Oligocene history of the northwest Australian margin (von Rad & Exon 1983). The Exmouth Plateau in this area has subsided to subsea depths of *c.* 2000 m associated with low sediment supply and erosion/non-deposition beneath vigorous bottom currents.

The opening of the Drake Passage in the Antarctic Ocean in the latest Oligocene – early Miocene (25–30 Ma) produced a deep circum-polar current, leading to the thermal isolation of Antarctica and increased global cooling. The cooling is associated with a major turnover in planktonic organisms with significant Neogene extinctions (Keller & Barron 1983). The present circulation in the Antarctic was established 13.5–12.5 Ma.

Failed rift systems

Deep-water sedimentation occurs not only along evolving and mature passive continental margins, but also in 'failed' rift systems (aulacogens), i.e. where rifting ceased before any oceanic crust was generated. Aulacogens such as the Precambrian Athapuscow Aulacogen, Great Slave Lake, northern Canada (Hoffman *et al.* 1974), contain deep-water turbidite deposits that accumulated during a phase of rapid subsidence of the rift associated with crustal extension. At the western end of the Benue Trough, a Cretaceous to Tertiary, 1000 km-long and 100 km-wide aulacogen on the margin of West Africa, there are up to 12 km of submarine-fan deposits beneath fluvio–deltaic sediments of the Niger System (Weber 1971, Burke *et al.* 1972, Maurin *et al.* 1986). The Lower Palaeozoic deep water sediments of the Ouachita Mountains, Oklahoma, are also interpreted as the infill of a failed rift system associated with the development of the Iapetus Ocean; the arms of the rift system involved the South Oklahoma Aulacogen, Reelfoot Rift, early Illinois Basin, Rome Basin and the Ouachita Basin (Lowe 1985). Nearly 12 km cumulative thickness of turbidite facies accumulated in the Ouachita Basin (Lowe 1985).

The North Sea furnishes a superbly documented, large, failed rift system associated with the opening of the North Atlantic (Glennie 1986a & b, Brooks & Glennie 1987). As with other rift systems, early Palaeozoic tectonic lineaments have considerably influenced the location and geometry of the North Sea basins (cf. Glennie 1986b, Cartwright 1987, Dingwall 1987, Dore & Gage 1987, Haszeldine & Russell 1987). Extensive deep-marine turbidite systems developed, particularly during the Late Jurassic and Tertiary history of the northern North Sea (e.g. Lovell 1978, Johnson & Stewart 1985). In contrast to turbidite systems developed on mature passive margins, the North Sea submarine fans typically are small in radius and sand-rich (Stow *et al.* 1982).

Ocean margin basins associated with rifting

While discussions of passive margin stratigraphy often emphasize the continental rise-slope prism, there are important relatively small basins that form in response to the major rifting processes and that may receive considerable thicknesses of deep-marine sediments. Moreover, such basins, developed on attenuated and fractured continental crust, typically record much of the overall tectonic development of the passive margin. For such sediment accumulations, the term 'basin' is used loosely to denote thick depocentres that may not actually be contained within a completely rimmed topographic low.

Examples of these ocean-margin, or marginal-rift, basins occur around the South Atlantic where nine such basins are recognized (Fig. 10.3); one of these, the Santos Basin (Figs 10.4 & 5) of offshore southern Brazil (Williams & Hubbard 1984), was initiated in the Late Jurassic and contains more than 10 km of Aptian to Recent sediments. The basin depocentre is 700 km in length and is bounded to the north and south by basement and volcanic highs; the outer part of the basin, in deeper water, passes laterally into thin sediment cover over the submarine Sao Paulo Plateau. After reflector R5 time (Fig. 10.4), considerable thicknesses of deep-marine sediments accumulated. At times, for example in the Cenomanian and Santonian, this basin had a restricted water circulation, because of tectonic sills, and therefore organic-rich fine-grained sediments preferentially accumulated. Figure 10.5 shows the depositional model for the Upper Cretaceous 'regressive sequence' of the southern Santos Basin, with three discrete progradational wedges bounded by major unconformities. Overall, each sequence shows an upward transition from deep-water turbidite systems into shallow-marine shelf sediments.

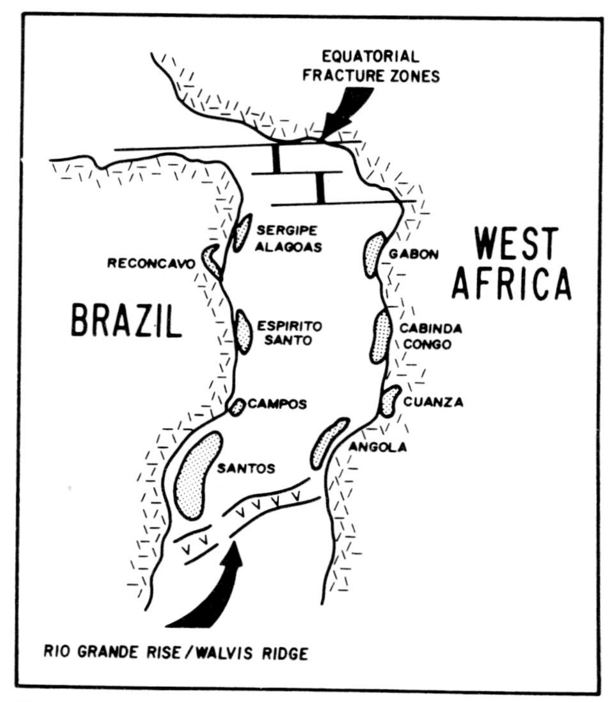

Figure 10.3 Early Cretaceous reconstruction of the equatorial South Atlantic to show main depocentres or basins along the passive continental margin. From Williams & Hubbard (1984).

CASE STUDY: EVOLUTION OF THE MAZAGAN MARGIN OFF NORTHWEST AFRICA

The Mazagan continental margin (Fig. 10.6) is located on the eastern side of the Central Atlantic Ocean. In this book, we use 'North Atlantic' for the region north of the Newfoundland – Azores – Gibraltar Fracture Zone (breakup age = Aptian and younger), 'Central Atlantic' for the region south of this fracture zone to the equatorial fracture zones of Figure 10.3 (breakup age = Bathonian – Callovian, Gradstein & Sheridan 1983), and 'South Atlantic' for the remainder of the Atlantic Ocean (breakup age in Fig. 10.1).

Data from many sources, particularly magnetic reversals in the oceanic crust, have permitted Klitgord & Schouten (1986) to produce a series of maps illustrating the development of the Central Atlantic Ocean. The oldest oceanic crust is Middle Jurassic at about 170 Ma. By the Late Jurassic to Early Cretaceous, a large enclosed ocean basin had formed in the present Central Atlantic (Fig. 10.7).

Rifting along the Mazagan continental margin began in the Triassic by crustal thinning associated with listric faulting, to give a series of basins and highs (DSDP Leg 79, Hinz, Winterer *et al.* 1984). The basins received sediments mainly from granitic basement highs (Fig. 10.8). By the close of the Triassic, halite with minor potash was deposited in shallow marine salt pans across the rift system, perhaps a little below sea level. At this time, the estimated width of the rift basin from the Mazagan margin (at DSDP Site 546) to the Nova Scotian margin to the northwest, is believed to have been approximately 150–200 km.

By the Early Jurassic, marine waters rapidly drowned the basin to a depth of a few hundred metres. During the latter part of the Early Jurassic and the Middle Jurassic, shallow-water carbonate banks occupied parts of the block faults near DSDP Site 544, whereas downslope at DSDP Site 547 redeposited sediments (limestone breccias) of Facies Class A and F accumulated together with hemipelagic radiolarian

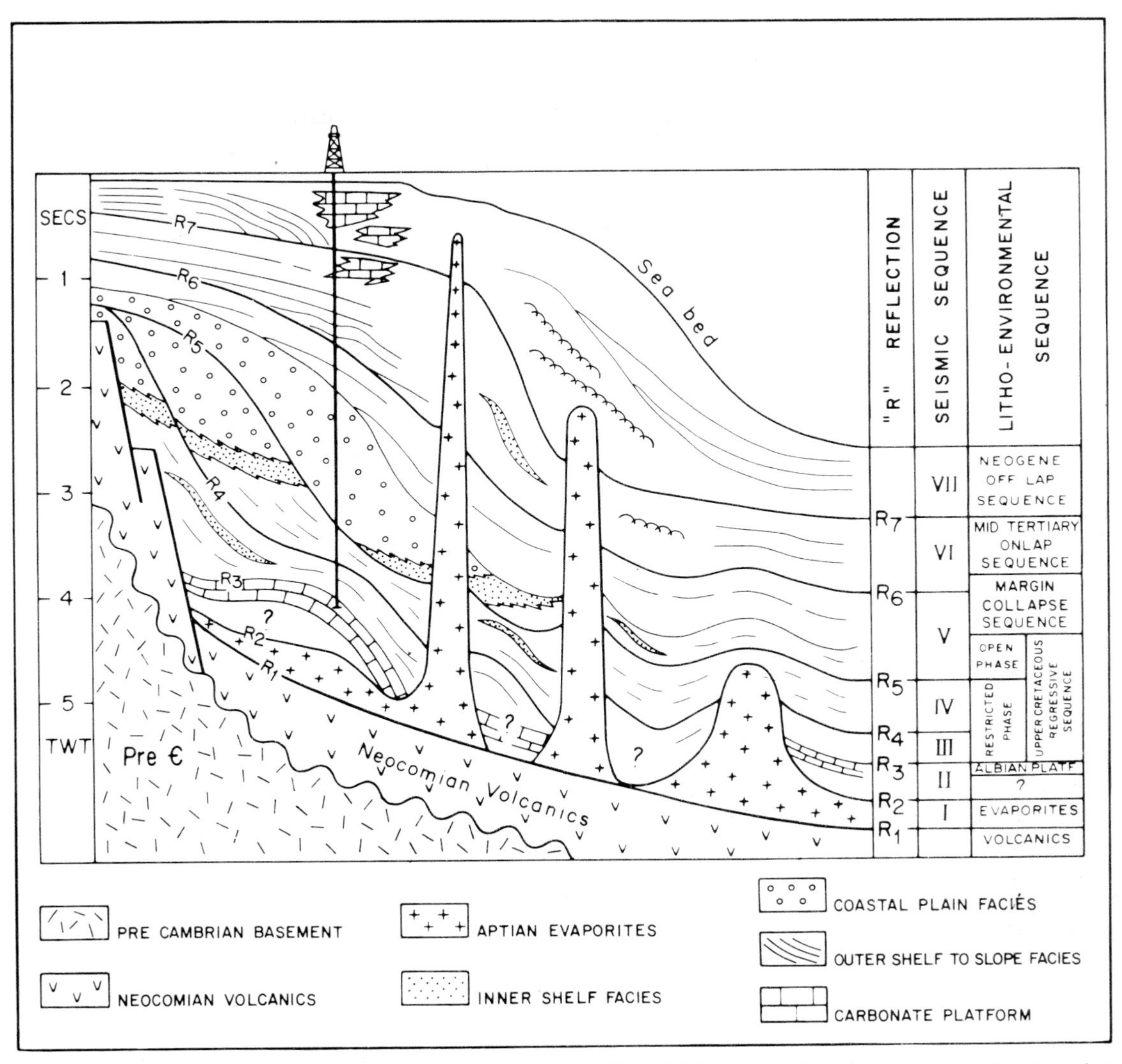

Figure 10.4 Schematic cross-section through the Santos Basin (cf. Fig. 10.3 for location) to show the generalized stratigraphic framework as defined by seismic profiles and drilling. From Williams & Hubbard (1984).

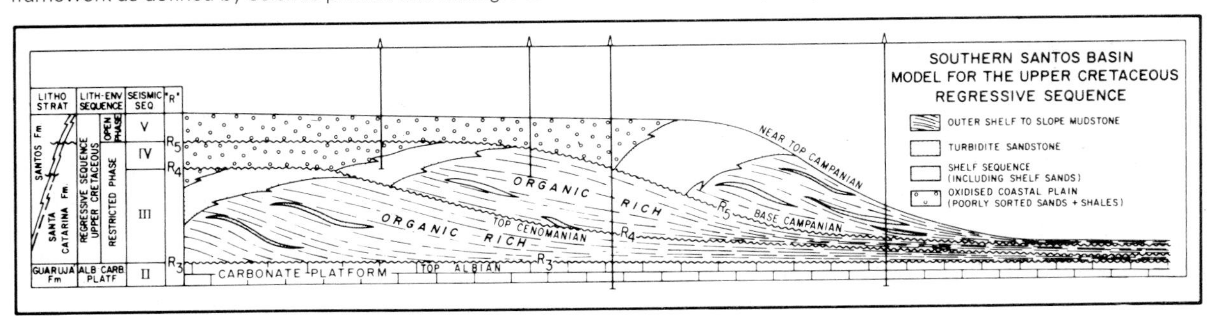

Figure 10.5 Depositional model for the Upper Cretaceous regressive sequence in the Santos Basin, offshore Brazil. Three major progradational sequences are defined between important unconformities. From Williams & Hubbard (1984).

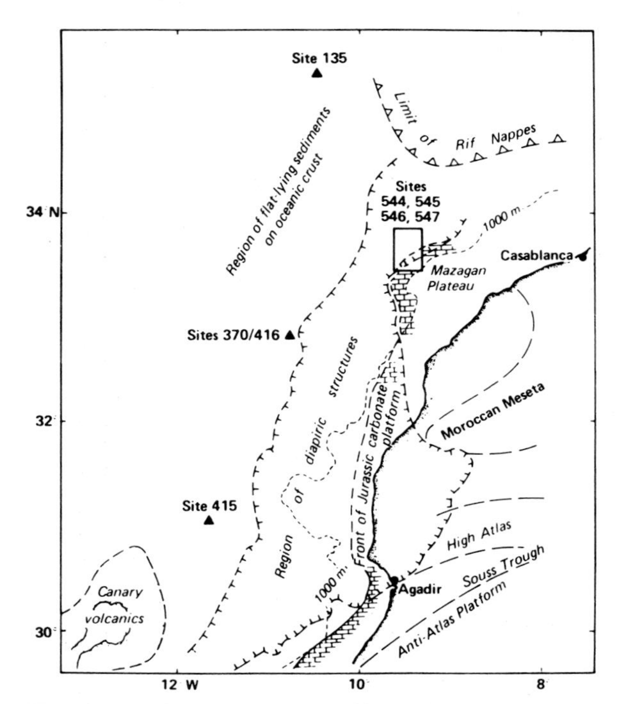

Figure 10.6 Continental margin off NW Africa to show location of DSDP Leg 79 sites. From Deroo *et al.* (1984).

marls (Facies Class G) in slowly subsiding basins a few hundred metres deep and, intermittently, poorly oxygenated.

The Bathonian–Callovian onset of sea-floor spreading was associated with extensive normal faulting and slightly accelerated subsidence rates, possibly with renewed crustal stretching. This tectonism particularly affected the more proximal areas of the Mazagan margin near DSDP Site 545, creating the main present-day structural relief of the Mazagan Escarpment. Gravity-driven faulting in the upper crust continued after breakup, and probably reached maximum intensity in the Late Jurassic, a time of major transgression. By Oxfordian times, carbonate reefs were established along the edge of the Mazagan Plateau (Fig. 10.9), supplying talus, including Facies Classes A and F, to the peripheral deeper water slopes (Steiger & Jansa 1984).

During the Early Cretaceous, large quantities of terrigenous sand (Facies Classes A, B and C) were delivered to the offshore Morocco Basin by turbidity currents flowing down the continental slope south of Mazagan. In the Aptian, hemipelagic clay-rich nannofossil ooze (Facies Class G) was rapidly deposited on the Mazagan Slope; accumulation of this facies continued through the Albian and most of the Cenomanian. Debris flows, slides and other sediment gravity flows or mass flows, depositing Facies Classes A to F, were common on the slope.

Following a phase of erosion and, or, non-deposition in the Turonian and Santonian, debris flows and associated deep-water processes once again led to active accretion of the Mazagan Slope during the Campanian and Maestrichtian, and intermittently in the Palaeocene and Eocene. Also, through the post-breakup and mature phases in the evolution of the margin, erosive events cut progressively deeper into older sediments, forming submarine canyons and other erosional features.

The Oligocene is represented by Facies Class G deposits as a thin incomplete nannofossil chalk record, truncated by a major unconformity, above which hemipelagic clay-rich nannofossil ooze accumulated on the Mazagan Slope in the latest Miocene, with rare coarser-grained sediment gravity flows (Fig. 10.9), continuing up to the present day. A second unconformity that eliminates the upper Miocene Tortonian Stage is believed to be due to vigorous thermohaline currents crossing the Mazagan Slope at that time. At DSDP Site 545, debris flow deposits include Jurassic limestone clasts with a matrix of Miocene nannofossil ooze, the clasts being derived from shallow water oolitic shoals, coral bioherms and deeper shelf deposits (Steiger & Jansa 1984).

The drilling results from DSDP Leg 79, Sites 544, 547 and 545 (Fig. 10.6), have been synthesized into an age-versus-depth diagram (Fig. 10.10) by Winterer & Hinz (1984). These 'subsidence curves' suggest slow rates of subsidence for the post-breakup history of the Mazagan margin. For the continental margin of northwest Africa, Leckie (1984) has summarized the second-order sea-level curves of Vail *et al.* (1977), the relative intensity of oceanic anoxic events (Arthur & Schlanger 1979), the episodes of upwelling along this margin, and

Figure 10.7 Cenozoic to Tertiary reconstructions of the continents around the North Atlantic during: (a) Middle Jurassic to Early Cretaceous, (b overleaf), Early Cretaceous to Early Palaeogene, and (c overleaf), Middle Palaeogene to Present. Large arrows indicate plate motions relative to North America. Note position of Mazagan Plateau and New Jersey margin case studies on the present map. After Klitgord & Schouten (1986).

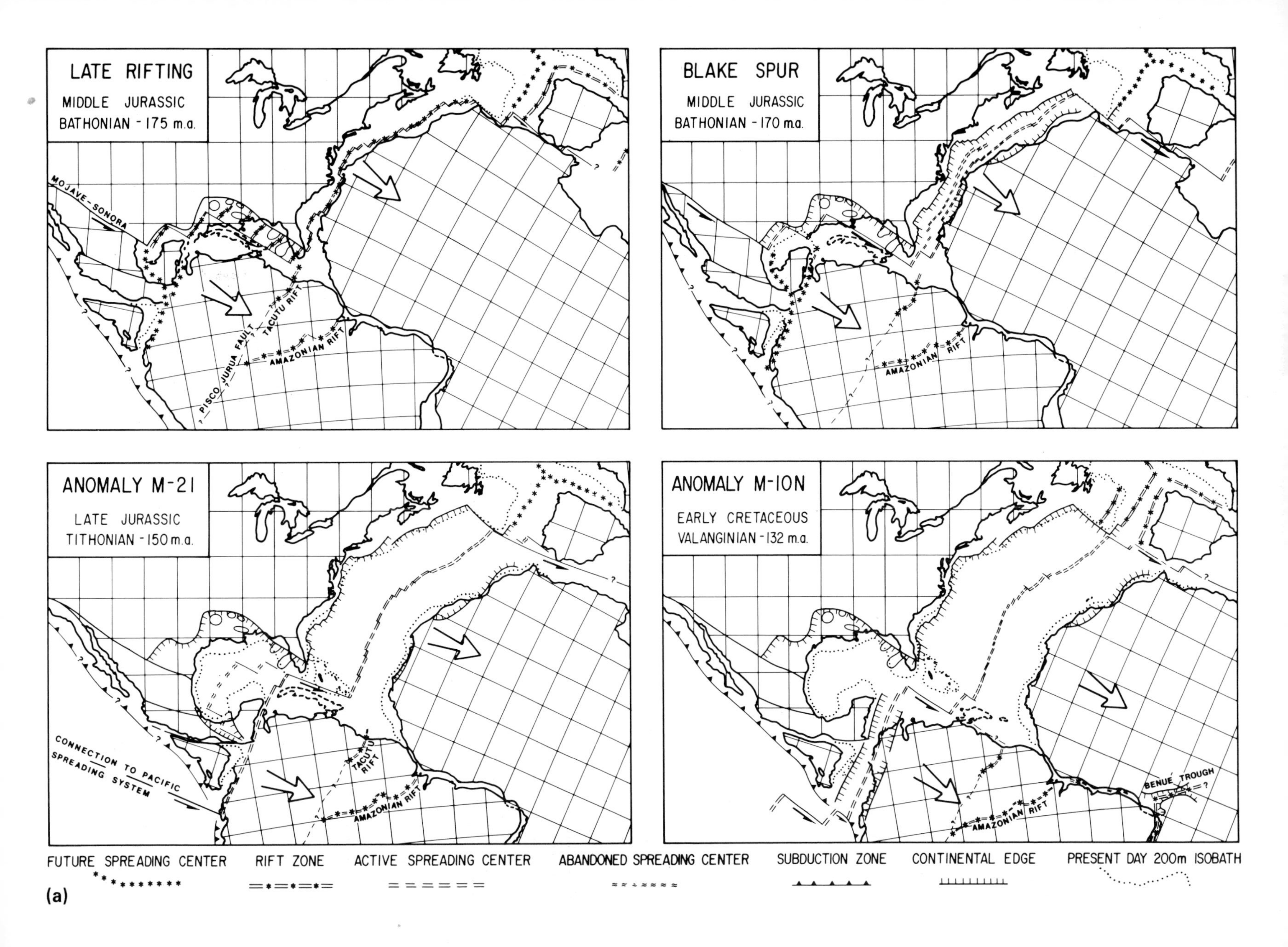
LATE RIFTING
MIDDLE JURASSIC
BATHONIAN - 175 m.a.
MOJAVE-SONORA
PISCO JURUA FAULT
TACUTU RIFT
AMAZONIAN RIFT
BLAKE SPUR
MIDDLE JURASSIC
BATHONIAN - 170 m.a.
AMAZONIAN RIFT
ANOMALY M-21
LATE JURASSIC
TITHONIAN - 150 m.a.
CONNECTION TO PACIFIC SPREADING SYSTEM
TACUTU RIFT
AMAZONIAN RIFT
ANOMALY M-10N
EARLY CRETACEOUS
VALANGINIAN - 132 m.a.
BENUE TROUGH
AMAZONIAN RIFT
FUTURE SPREADING CENTER
RIFT ZONE
ACTIVE SPREADING CENTER
ABANDONED SPREADING CENTER
SUBDUCTION ZONE
CONTINENTAL EDGE
PRESENT DAY 200m ISOBATH
(a)

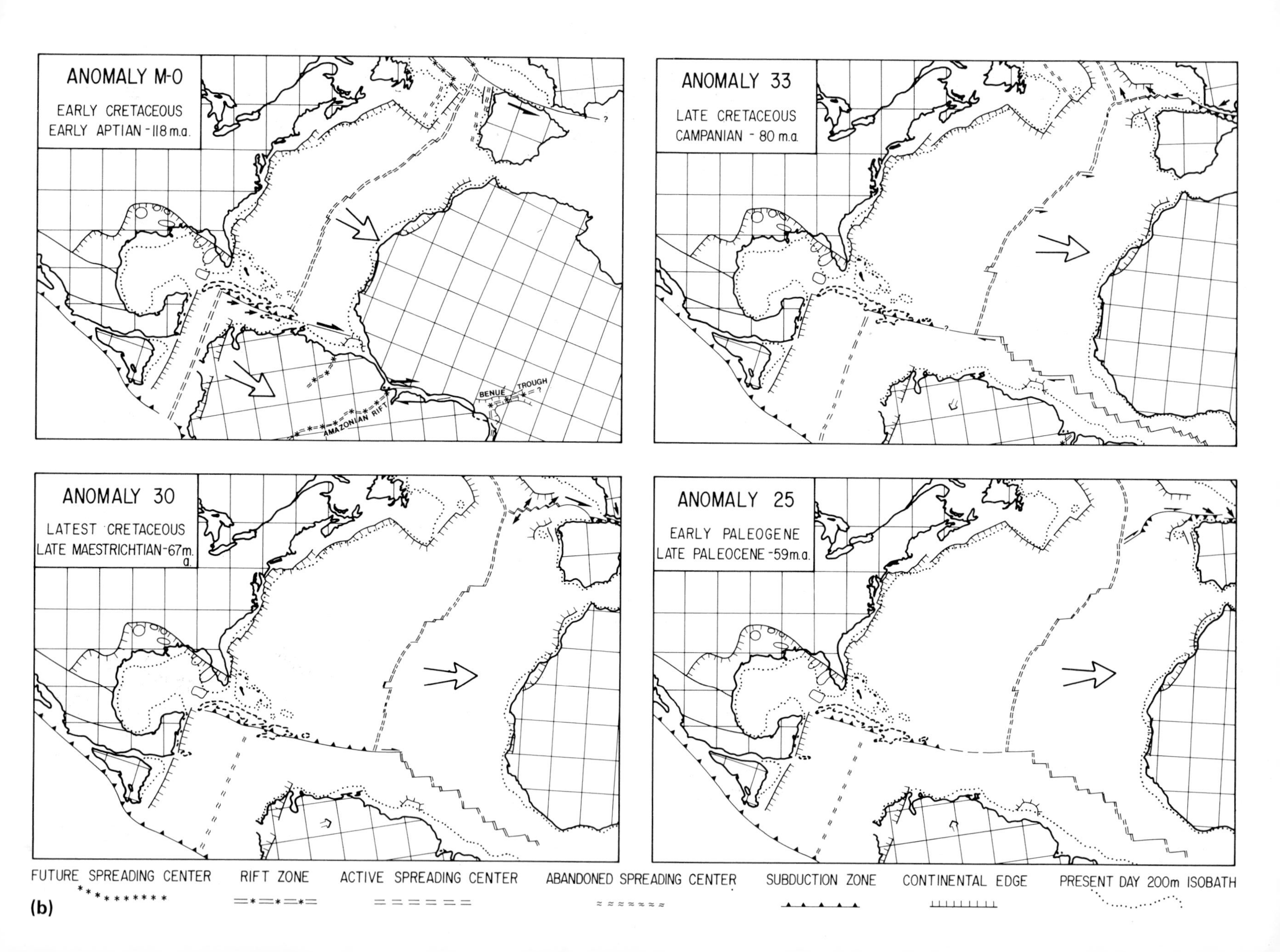

ANOMALY M-0
EARLY CRETACEOUS
EARLY APTIAN - 118 m.a.
BENUE TROUGH
AMAZONIAN RIFT
ANOMALY 33
LATE CRETACEOUS
CAMPANIAN - 80 m.a.
ANOMALY 30
LATEST CRETACEOUS
LATE MAESTRICHTIAN - 67 m.a.
ANOMALY 25
EARLY PALEOGENE
LATE PALEOCENE - 59 m.a.
FUTURE SPREADING CENTER
RIFT ZONE
ACTIVE SPREADING CENTER
ABANDONED SPREADING CENTER
SUBDUCTION ZONE
CONTINENTAL EDGE
PRESENT DAY 200m ISOBATH

(b)

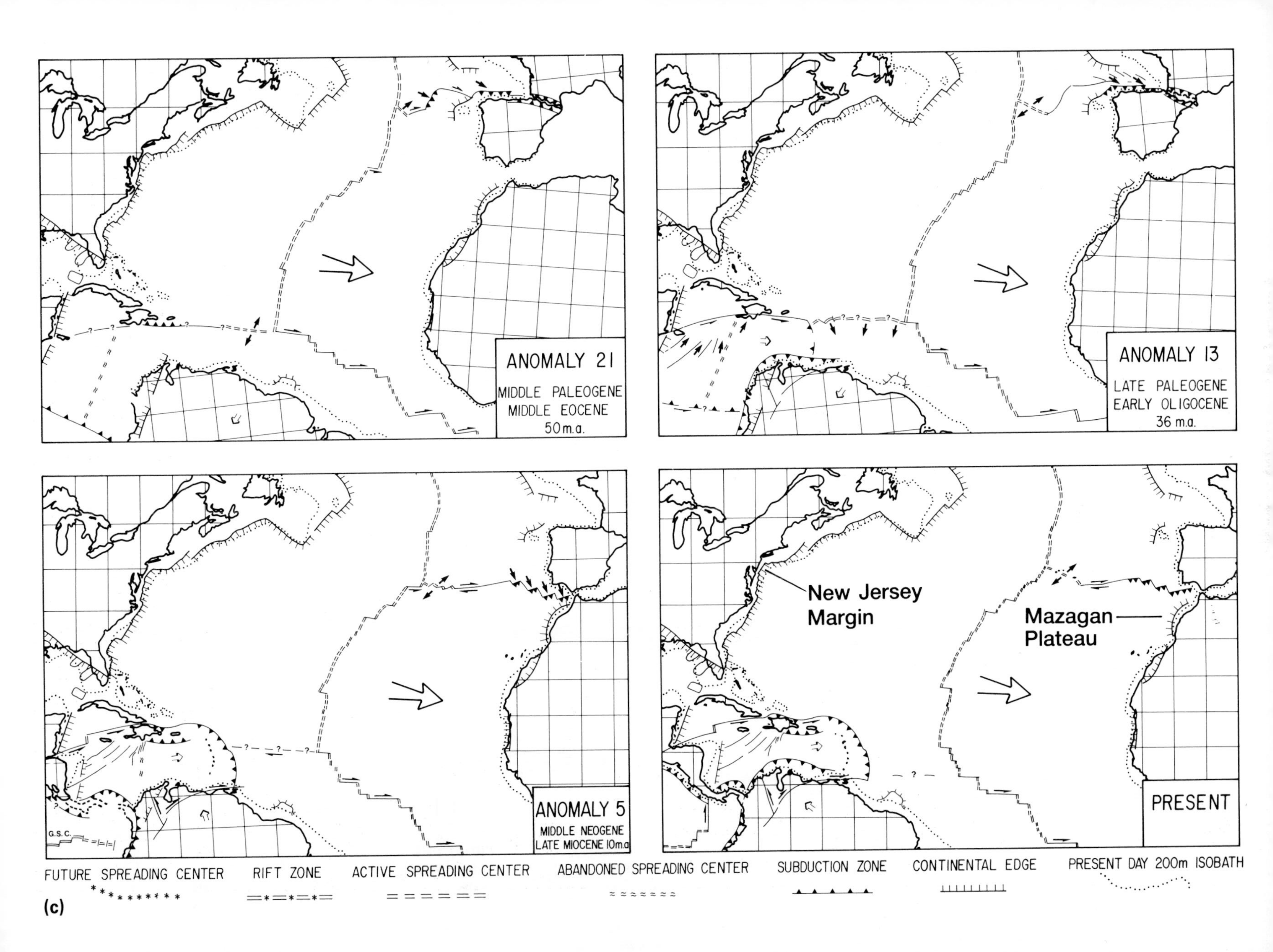
ANOMALY 21
MIDDLE PALEOGENE
MIDDLE EOCENE
50 m.a.
ANOMALY 13
LATE PALEOGENE
EARLY OLIGOCENE
36 m.a.
ANOMALY 5
MIDDLE NEOGENE
LATE MIOCENE 10m.a
G.S.C.
New Jersey
Margin
Mazagan
Plateau
PRESENT
FUTURE SPREADING CENTER
RIFT ZONE
ACTIVE SPREADING CENTER
ABANDONED SPREADING CENTER
SUBDUCTION ZONE
CONTINENTAL EDGE
PRESENT DAY 200m ISOBATH
(c)

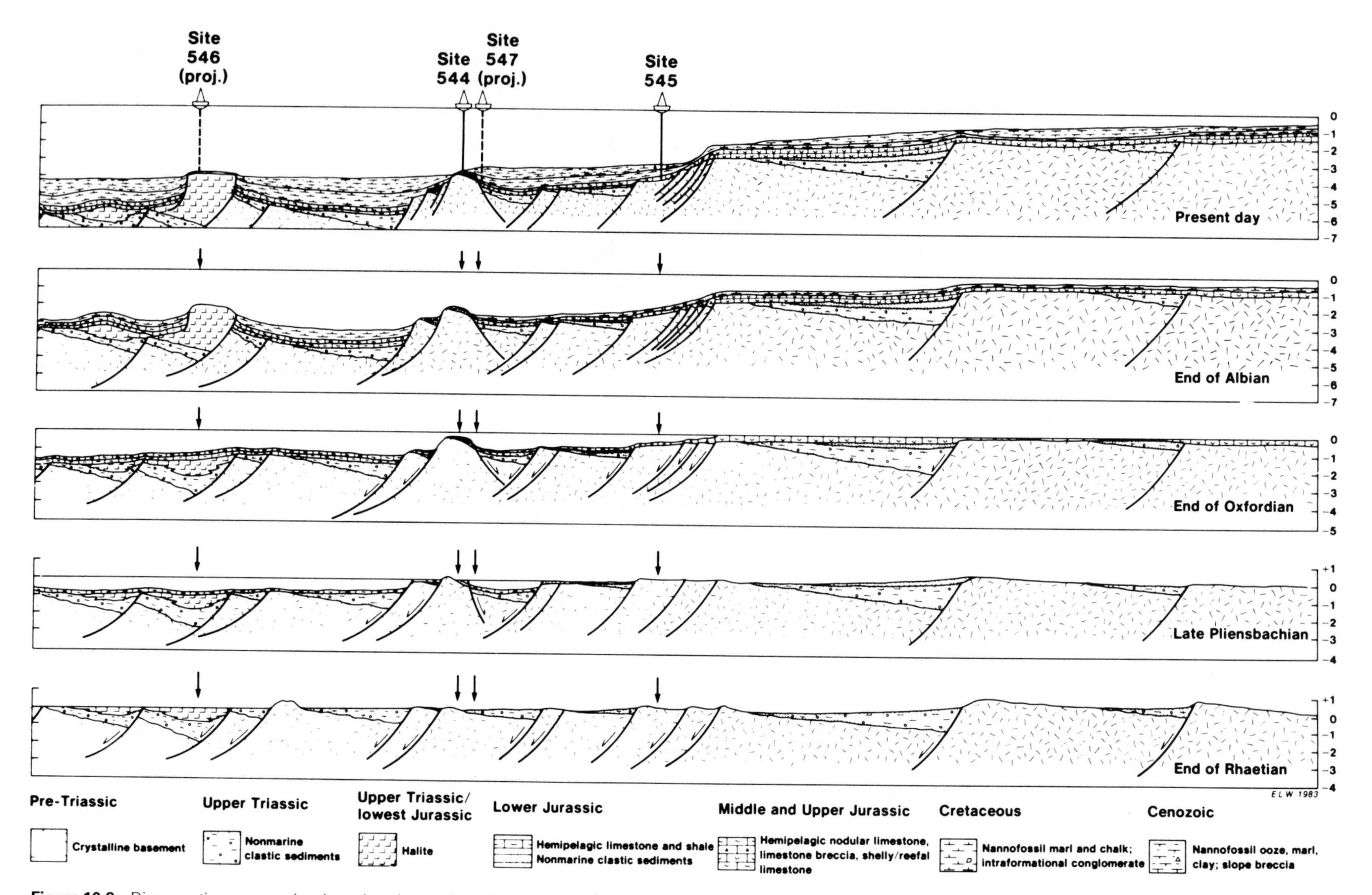

Figure 10.8 Diagramatic cross-section, based on the results of DSDP Leg 79, at five stages in the geological evolution of the Mazagan continental margin. No vertical exaggeration. See Figure 10.6 for location of DSDP sites. The northwest end of the cross-section is very schematic in order to show the salt diapir drilled at Site 546. Major faults inferred to be active during sedimentation are labelled with movement arrows. From Winterer & Hinz (1984).

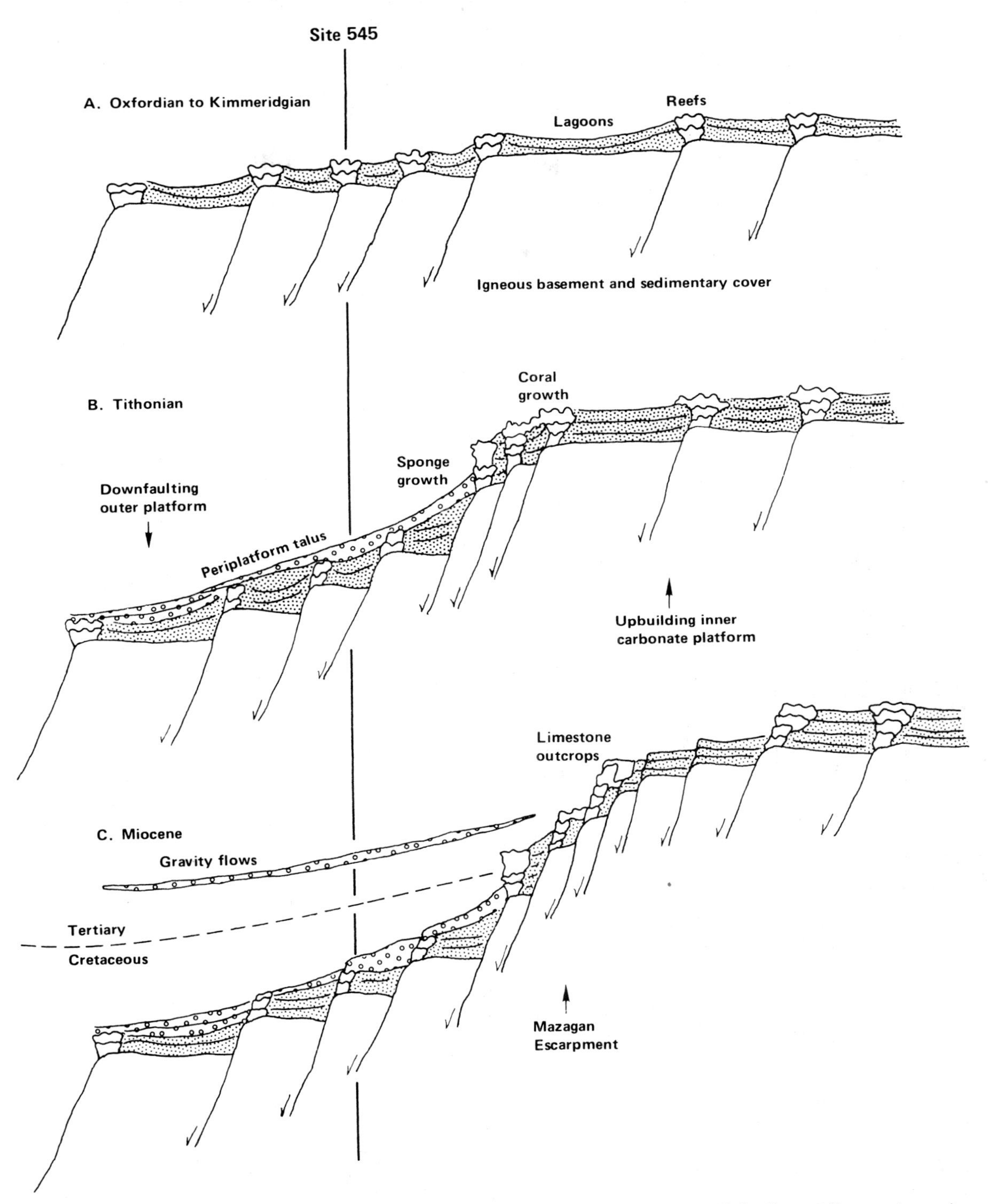

Figure 10.9 Interpreted development of the Mazagan Plateau around DSDP Site 545 from Oxfordian to Miocene times. A, Oxfordian–Kimmeridgian: reef development above back-tilting, slowly subsiding, blocks in a carbonate ramp; B, Tithonian: formation of Mazagan Escarpment; C, Miocene: foundering of established Mazagan Escarpment by mass wasting and burial below various deep-marine sediment gravity flows. After Steiger & Jansa (1984).

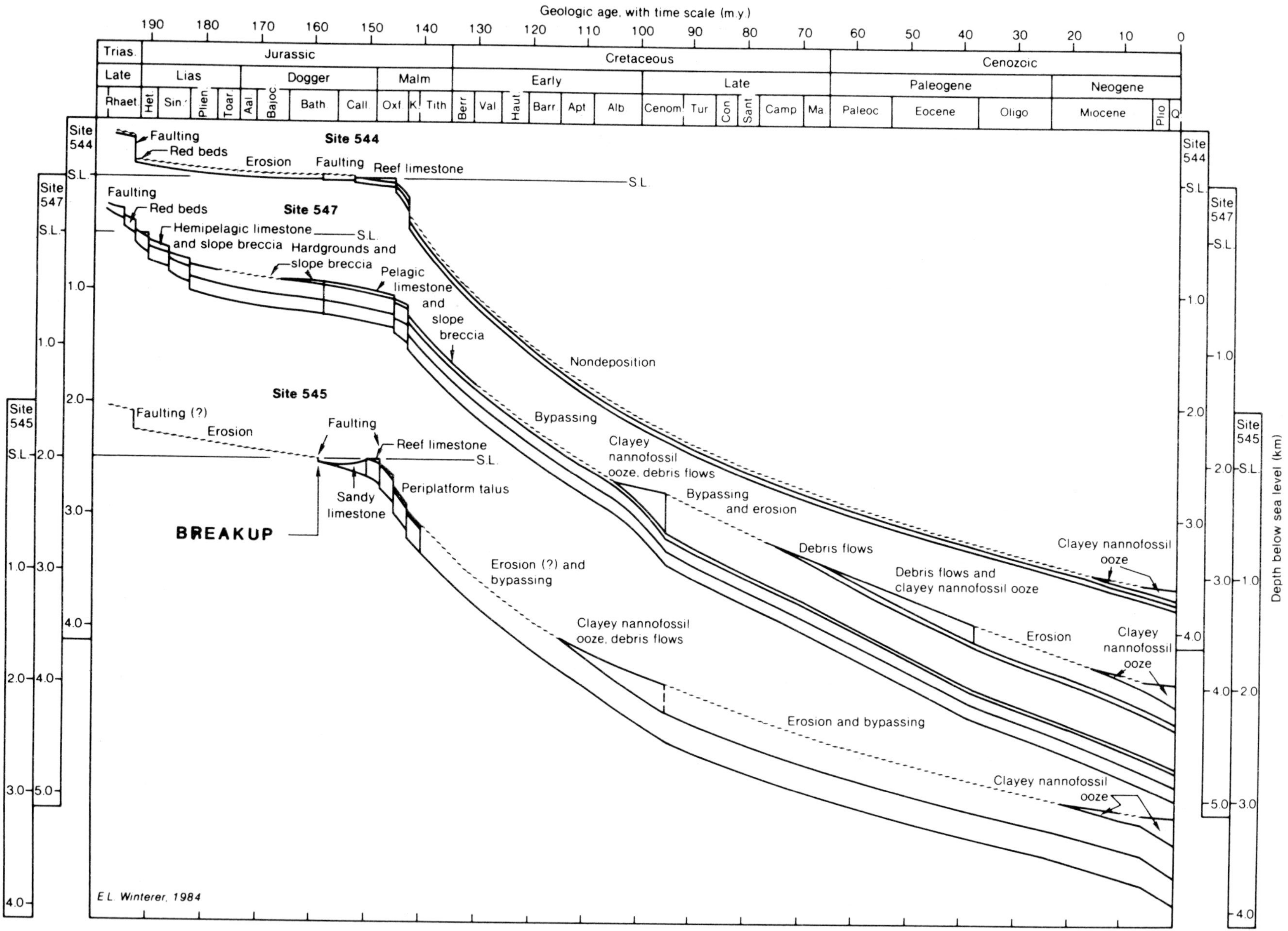

Geologic age, with time scale (m.y.)

190 180 170 160 150 140 130 120 110 100 90 80 70 60 50 40 30 20 10 0

Trias. | Jurassic | Cretaceous | Cenozoic

Late | Lias | Dogger | Malm | Early | Late | Paleogene | Neogene

Rhaet. | Het. | Sin. | Plien. | Toar. | Aal. | Bajoc. | Bath. | Call. | Oxf. | K. | Tith. | Berr. | Val. | Haut. | Barr. | Apt. | Alb. | Cenom. | Tur. | Con. | Sant. | Camp. | Ma. | Paleoc. | Eocene | Oligo. | Miocene | Plio. | Q.

Site 544
S.L.
1.0
2.0
3.0
4.0

Site 547
S.L.
1.0
2.0
3.0
4.0
5.0

Site 545
S.L.-2.0
1.0-3.0
2.0-4.0
3.0-5.0
4.0

Site 544

Faulting
Red beds
Erosion
Faulting
Reef limestone
S.L.
Nondeposition

Site 547

Faulting
Red beds
Hemipelagic limestone and slope breccia
S.L.
Hardgrounds and slope breccia
Pelagic limestone and slope breccia
Bypassing
Clayey nannofossil ooze, debris flows
Bypassing and erosion
Debris flows
Debris flows and clayey nannofossil ooze
Erosion
Clayey nannofossil ooze
Clayey nannofossil ooze

Site 545

Faulting (?)
Erosion
Faulting
Reef limestone
S.L.
Sandy limestone
Periplatform talus
BREAKUP
Erosion (?) and bypassing
Clayey nannofossil ooze, debris flows
Erosion and bypassing
Clayey nannofossil ooze

E.L. Winterer, 1984

Depth below sea level (km)

the periods of enhanced diversity of planktonic foraminfera (Fig. 10.11). Organic-rich black shales accumulated along the Mazagan margin during the Aptian–Albian and Cenomanian–Turonian oceanic anoxic events (OAEs), coincident with times of rising sea level (Fig. 10.11). Furthermore, Leckie (1984) shows that the diversity of planktonic foraminfera decreased during the two OAEs, presumably due to an intensified oxygen minimum zone off central Mazagan in latest Aptian to early Albian, and latest Cenomanian to early Turonian times (Fig. 10.11). Thus global sea-level changes during the evolution of the Mazagan margin exerted a fundamental control on the nature of deep-marine sediment types.

CASE STUDY: US ATLANTIC MARGIN OFF NEW JERSEY

This case study is based mainly on the results of DSDP Leg 93 (van Hinte, Wise *et al.* 1987) and Leg 95 (Poag, Watts *et al.* 1987).

The central Eastern Seaboard of the United States, off New Jersey in the Baltimore Canyon Trough area (Fig. 10.12), is amongst the most intensively studied continental shelves and margins of the world. More than 40 boreholes, thousands of kilometres of seismic reflection profiles and considerable quantities of shallow-core, box-core and grab samples have been recovered. A summary of characteristics of seismic

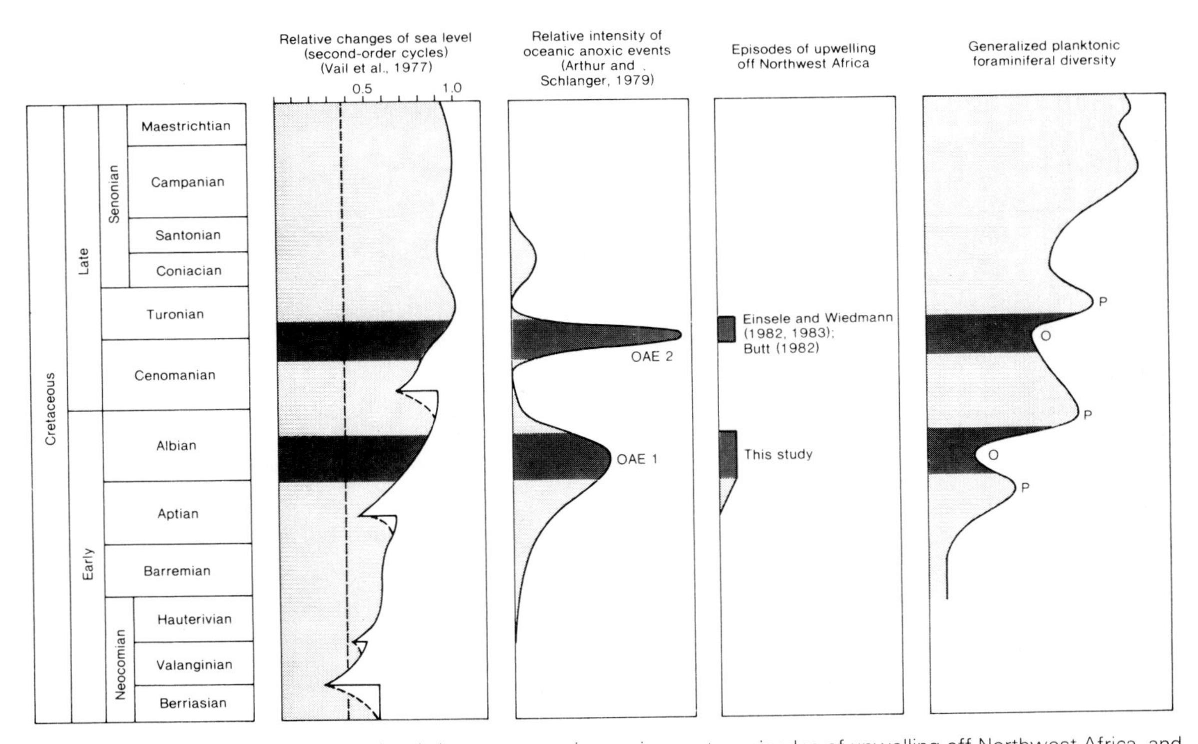

Figure 10.11 Correlation of global sea-level changes, oceanic anoxic events, episodes of upwelling off Northwest Africa, and diversity trends in planktonic foraminifera; O, oligotaxic; P, polytaxic. After Leckie (1984).

Figure 10.10 (opposite) Age-versus-depth geohistory diagram inferred for DSDP Sites 544, 547 and 545. Note that each diagram has its own depth scale. Subsidence curves constrained by sea-level crossings and present-day depths. Exponential subsidence curves are assumed where facies data do not constrain the depths. Isostatic (Airy) corrections made for sediment loading. Wavy lines denote erosion, sediment bypassing or non-deposition. (After Winterer & Hinz 1984; timescale after van Hinte 1978 a & b).

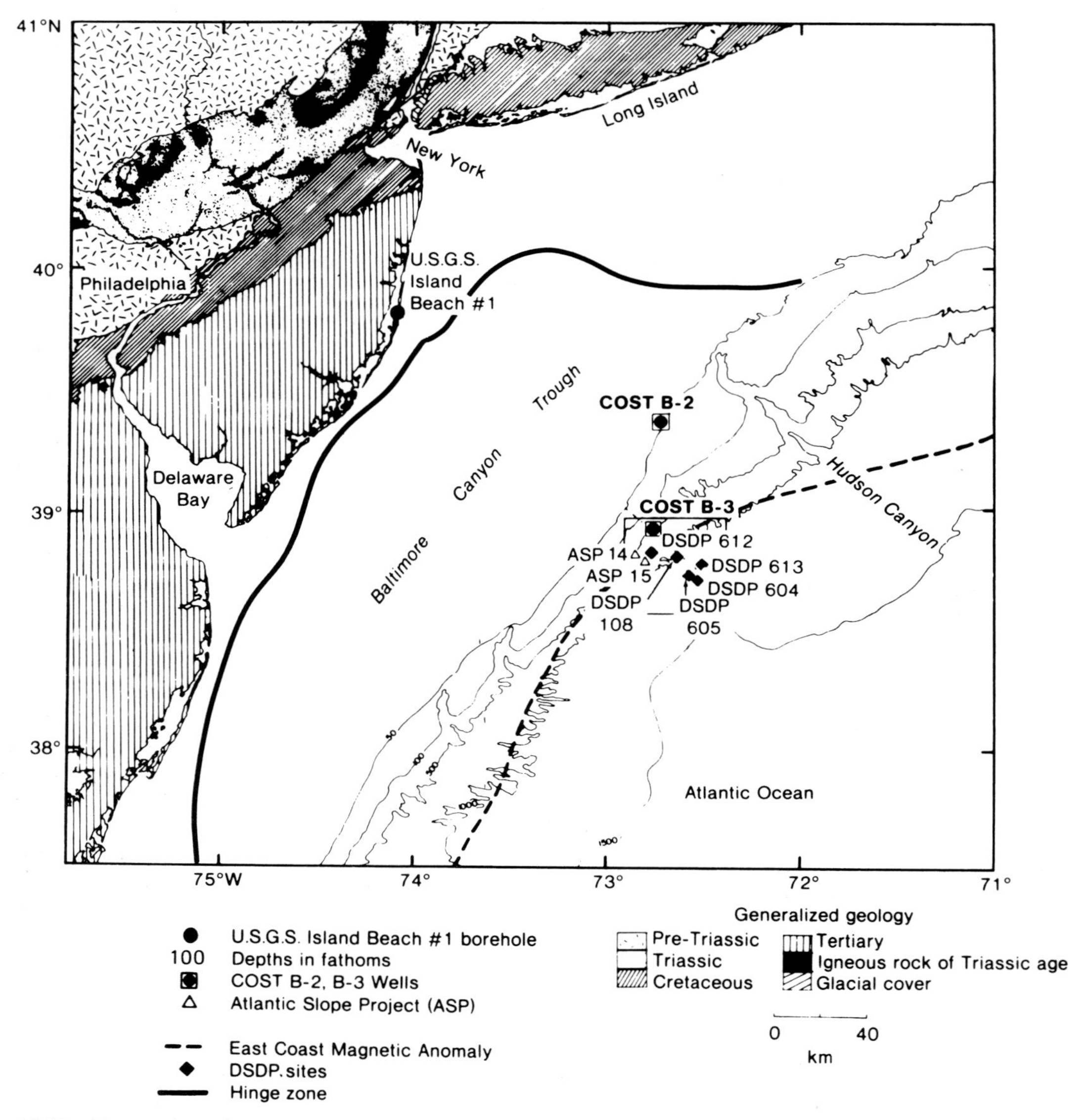

Figure 10.12 Map to show the regional context of the US continental margin off New Jersey together with the location of various DSDP sites and the COST B–2 and B–3 boreholes. After Farre & Ryan (1987).

facies in this area is given in Table 10.2 (Schlee & Hinz 1987). Many geological and geophysical studies have been published from the area following the early work of Drake *et al.* (1959), for example Steckler & Watts (1978), Watts & Steckler (1979), Grow (1980), Poag (1980, 1985), Schlee (1981), Thorne & Watts (1984), Watts & Thorne (1984), van Hinte Wise *et al.* (1987), Poag, Watts *et al.* (1987), Klitgord & Schouten (1986), Klitgord *et al.* (1988) and Poag & Valentine (1988).

The offshore part of the Baltimore Canyon Trough covers about 200 000 km^2 of the continental shelf and slope from Cape Hatteras to Long Island, New York (Fig. 10.12). More than 18 km of sediments occur in the trough (Schlee 1981), with more than 11 km of Jurassic. The standard reference section of the Baltimore Canyon Trough is the US Geological Survey multichannel seismic reflection profile Line 25 (Grow 1980) close to the COST B–2 and COST B–3 boreholes (Figs 10.12 & 13).

Borehole data from the Baltimore Canyon Trough

Table 10.2 Seismic facies present on the continental rise seaward of Baltimore Canyon Trough, eastern USA. After Schlee & Hinz (1987).

Seismic facies unit	Environmental facies interpretation	External form	Internal configuration	Reflection geometry at boundaries
Sigmoid low relief, lens-shaped as a shelf-edge progradation	Fluvial input to broad delta system and active outbuilding of shelf	Low, broad lens	Sigmoid along a dip section with low angle downlap of reflections	Low angle downlap at base and broadly concordant at top
Open marine shelf, parallel, moderate continuity and intensity	Interbedded sand and silt, wave-transported by widespread bottom current activity	Sheet behind palaeoshelf edge	Parallel to divergent	Generally concordant at top and bottom, but can be marked by minor channelling and reflection termination
Slope transition blanket, low continuity, hummocky reflection pattern, variable intensity	Slump fill and mass-wasting deposits	Discontinuous blanket over palaeoslope-upper rise area	Discontinuous reflection of variable amplitude and breadth	Channelling unit of variable thickness, spotty in distribution; grades laterally into basin fill
Slope-front fill, moderate continuity, low–high amplitude, adjacent to platform front	Mass wasting, debris fans, talus that accumulated in front of steep platform edge	Linear wedge, thinning toward palaeoslope and seaward under upper rise	Downlap in seaward direction and onlap toward platform	Channelled at base, generally parallel arrangement of reflections; conformable to low angle disconformity at top
Sheet drape, moderate to high continuity and amplitude of parallel reflections	Hemipelagic clays and oozes; possible sand interlayers	As a unit, tends to follow bottom topography as a broad, thin blanket, usually in upper part of section	Parallel reflections	Usually concordant at base and possibly channelled (minor) or concordant at top
Onlapping basin fill	Interbedded turbidites and hemipelagic deposits (mainly silts, clay, and ooze)	Broad blankets that thin toward ancestral continental slope	Parallel, moderately continuous to discontinuous reflections of variable amplitude	Low angle onlap toward slope at base, minor channelling or concordance at top and for both the base and top over much of rise
Channel basin fill	Lenticular sand and silts, probably part of fan complex	Broad lens-like tabular sheets; difficult to distinguish channels, levees etc, except in surficial deposits	Discontinuous and low continuity reflections, lensing of smaller units within main one	Irregular channelling and cutout of reflections at base; can be concordant at top; spotty blanket
Chaotic basin fill	Mass wasting slump tongues and blocks, turbidites	Irregular blanket marked by hummocky reflections; variable thickness, particularly on slope	Discontinuous hyperbolic returns of variable amplitude and breadth	Tends to be best developed under slope and upper rise
Contour-current basin fill	Deep-water thermo-haline deposits caused by lateral transport and buildup of fine-grained clastic sediment	Lozenge-shaped bodies with low angle downlap of reflectors	Parallel, moderate–low conformity–reflections that downlap and are truncated at top	Broad swalelike erosional depressions toward base; onlapping basin facies intertongue with this facies and overlie it

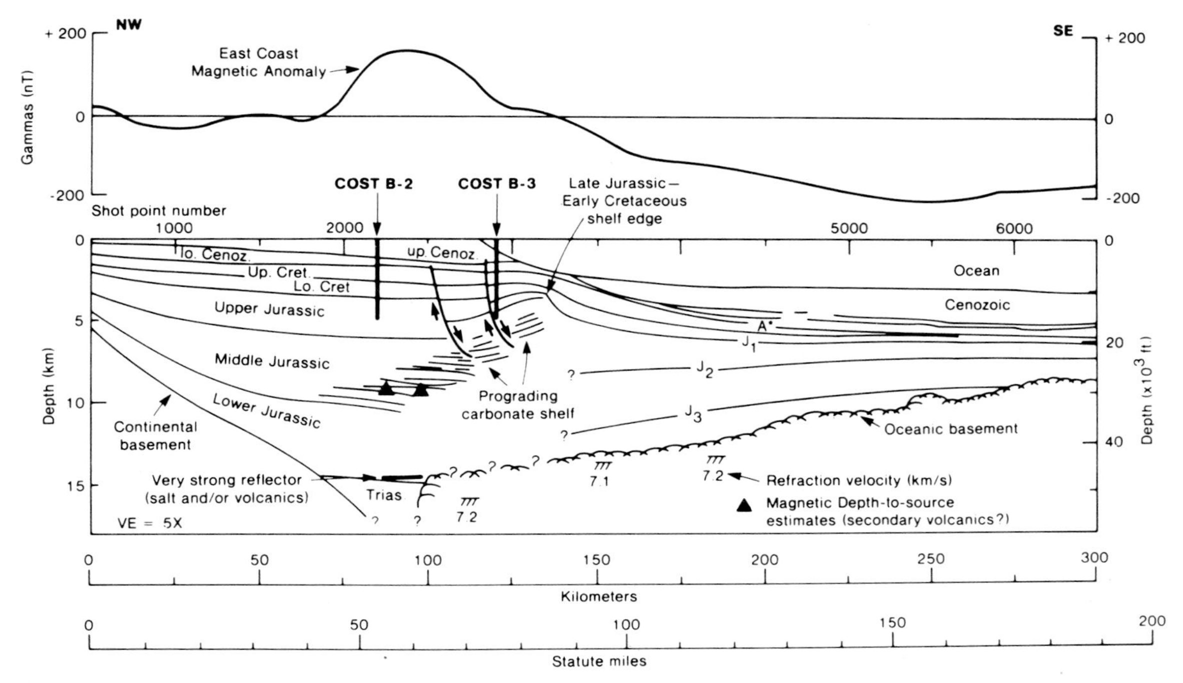

Figure 10.13 Schematic geological section of the Baltimore Canyon Trough along the US Geological Survey multichannel seismic reflection profile 25 showing the location of the COST B–2 and B–3 boreholes. A*, J1, J2 and J3 are prominent seismic reflectors. From Grow (1980) in Poag & Watts (1987).

area suggest that the main causes of subsidence at this margin were thermal contraction and sedimentary loading following the heating and attenuation of the crust and lithosphere prior to and during the Bathonian–Callovian continental breakup phase when North America separated from Africa (Fig. 10.7). Modelling of this margin in the vicinity of the COST B–2 well suggests that the overall geometry of the basement below the trough, the presence of the coastal plain, and the relative stratigraphic highs seaward of the hinge line (Figs 10.12 & 13) can be explained essentially by sediment loading (Steckler & Watts 1978).

The post-breakup subsidence led to deep-marine conditions along the continental margin: DSDP Site 603, on the lower continental rise, drilled an extensive Lower Cretaceous submarine fan system, up to 298 m thick, of turbidite sands, black shales and interbedded limestones of Valanginian to early Aptian age (Fig. 10.14). The history of this deep-marine clastic system in summarized in Figure 10.15 (Wise & van Hinte 1987), and involves the development of a lowstand fan (see Ch. 7). Wise & van Hinte (1987) believe that this and other coeval coarse-grained clastic systems along the Eastern Seaboard of the USA probably formed during more humid climates associated with tectonic rejuvenation of the source area: the uplift being due to rifting and continental breakup in the North Atlantic between North America and Europe.

In the continental slope off New Jersey, there are a number of distinct unconformities, falling into two groups (Poag & Watts 1987): those that are correlated from basin to basin and appear approximately coincident with the 'global' erosional events postulated by Vail *et al.* (1977) or Haq *et al.* (1987), and those of a more local extent, restricted to individual basins and slopes. Periods of non-deposition are extremely variable, for example an 11 m.y. hiatus between upper Eocene and upper Oligocene strata (observed in the COST B–3 and other wells), and an hiatus of 5 m.y. duration between lower and middle Miocene strata. Variations in water depth are recorded across many of these unconformities, for example the 9 m.y. gap from upper Oligocene to middle Miocene strata shows a change from lower to upper bathyal depths in the COST B–3 well, whereas the same hiatus in the COST B–2 well is defined by a change from outer to inner sub-littoral depths, i.e. an overall shallowing upwards.

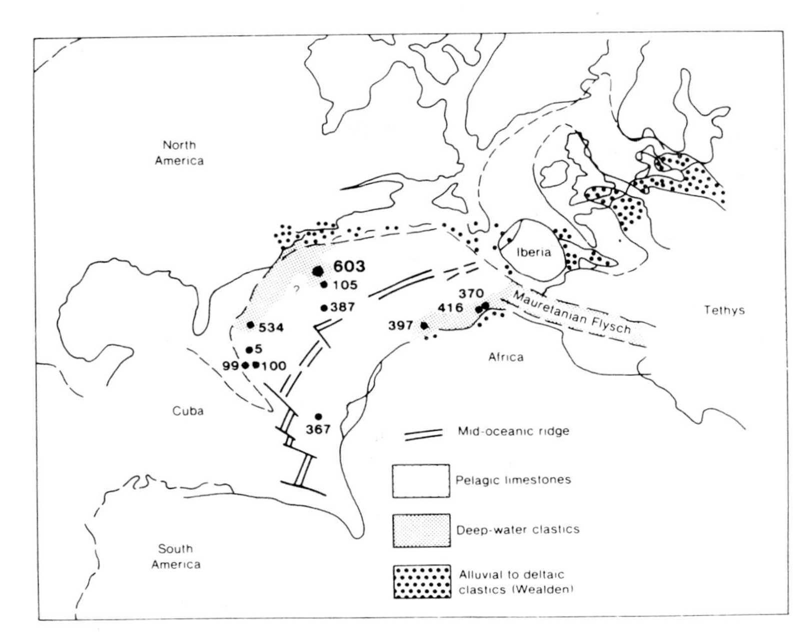

Figure 10.14 Early Cretaceous palaeogeography of the North Atlantic region to show the distribution of pelagic limestones, deep-marine clastics, and fluvio-deltaic ('Wealden') facies associations. After Wise & van Hinte (1987).

Erosive submarine currents are invoked to explain some of these periods of non-deposition (Poag & Watts 1987, p. 21).

For the last 135 m.y. the New Jersey margin has been dominated by large scale gravity sliding (Facies Class F) and other sediment gravity flows (Poag, Watts *et al.* 1987). Most boreholes in the Baltimore Canyon Trough show a substantial deepening of the sedimentary environments from middle shelf to outer shelf and upper slope during the Turonian-Santonian interval, approximately 90–80 Ma. During the Late Jurassic, shelf-edge reefs developed, and were subsequently buried in the Late Cretaceous by turbidites and debris flow deposits. During the lowered sea levels in the Pleistocene, deep incision of the shelf edge produced many submarine canyons (Poag, Watts *et al.* 1987).

The site survey for DSDP Leg 95 included a SeaMARC sidescan sonar survey of the three distinct provinces of the continental margin: (a) upper and middle continental slope; (b) lower continental slope, and (c) upper continental rise (Farre & Ryan 1987). Quaternary terrigenous sediments cut by U- and V-shaped canyons and valleys that die out downslope typify the upper and middle continental slope. Rugged outcrops of middle Eocene bio-siliceous limestones characterize the lower continental slope where gradients are typically 4.5°. In the SW of the survey area (Fig. 10.12), on the lower slope, two canyon systems begin in water depths of 1500–1600 m with steep joint-controlled walls and broad flat floors coincident with bedding surfaces in the Eocene sediments. In the eastern part of the lower slope, there are numerous downslope-oriented gullies up to a few metres in depth and width, interpreted as gouge marks excavated into the Eocene limestone by slides of rock and unconsolidated sediments (Farre & Ryan 1987).

The upper continental rise, with gradients of about 1.5°, is covered by terrigenous sediments that onlap the erosional surface of the lower continental slope. The upper continental rise is cut by 30–70 m-deep, 1–2 km-wide and 4–7 km-long depressions, interpreted by Farre & Ryan (1987) as erosional hollows created by downslope-moving mass flows and rock slides, preferentially excavated in the region of abrupt gradient change. Similar depressions are described from comparable environments, for example off North Carolina by Bunn & McGregor (1980), and seaward of slope canyons at the base of the Malta Escarpment, eastern Mediterranean, where allochthonous slope material from the rim of one of the depressions suggests an erosional origin during sediment mass flow/sliding (Cita *et al.* 1982, Biju-Duval *et al.* 1983).

Observations from the *Alvin* submersible reveal olistoliths of Eocene chalk on parts of the upper continental rise off New Jersey, together with debris flow deposits from 5–15 m thick (Farre & Ryan 1987). One of the olistolith blocks, several metres in diameter, glided downslope for more than 4 km on a gradient of approximately 1.5° to leave a visible trail.

Figure 10.16 is a summary of the geology and surface

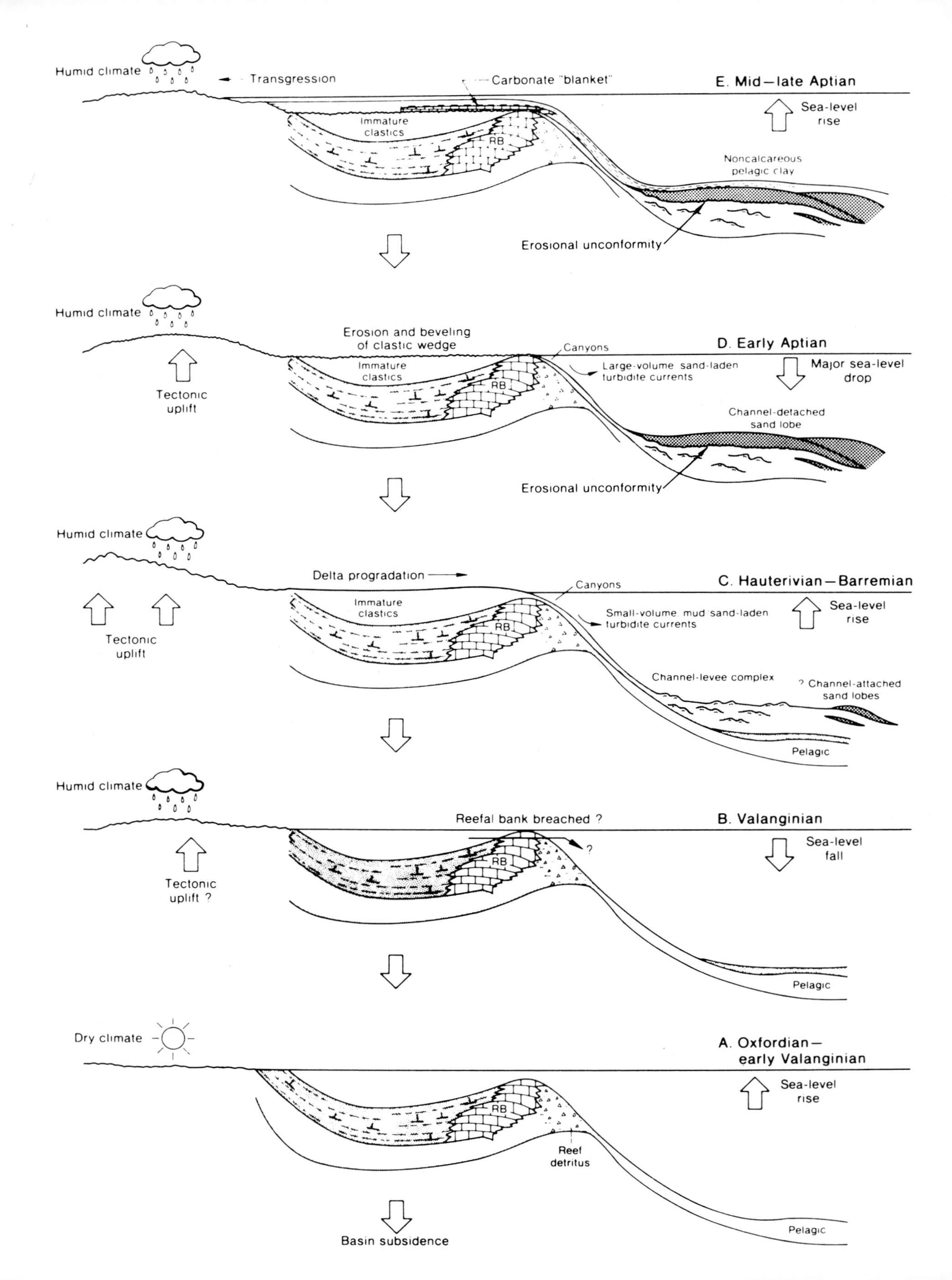

Humid climate
Transgression
Carbonate "blanket"
E. Mid—late Aptian
Sea-level rise
Immature clastics
RB
Noncalcareous pelagic clay
Erosional unconformity
Humid climate
Erosion and beveling of clastic wedge
Canyons
D. Early Aptian
Tectonic uplift
Immature clastics
Large-volume sand-laden turbidite currents
Major sea-level drop
RB
Channel-detached sand lobe
Erosional unconformity
Humid climate
Delta progradation
Canyons
C. Hauterivian—Barremian
Immature clastics
Small-volume, mud/sand-laden turbidite currents
Sea-level rise
RB
Tectonic uplift
Channel-levee complex
? Channel-attached sand lobes
Pelagic
Humid climate
Reefal bank breached ?
B. Valanginian
Sea-level fall
RB
Tectonic uplift ?
Pelagic
Dry climate
A. Oxfordian—early Valanginian
Sea-level rise
RB
Reef detritus
Pelagic
Basin subsidence

morphology for the continental margin downslope from the shelf break off New Jersey (Farre & Ryan 1987), emphasizing the role of mass wasting, the nature of unconformities in passive margins, and the way in which submarine canyons, valleys and gullies may die out downslope. These features may be widely applicable to many lower slope and upper continental rise settings at mature passive margins.

CASE STUDY: NORTHERN GULF OF MEXICO

The Gulf of Mexico provides an example of a semi-enclosed small ocean basin that contains siliciclastic and carbonate deep-marine sediments, with changing importance of the different source areas through time (e.g. Bertagne 1984, fig. 6). While the Mississippi River drainage basin has supplied siliciclastics to the deep Gulf of Mexico, resedimented carbonates have been supplied from the Western Florida Shelf to the east, and the Campeche–Yucatan Bank in the south. The Gulf of Mexico covers an area greater than 1.5 million km^2, with maximum water depths of more than 3700 m (Martin 1978, Martin & Bouma 1978). A useful general review of the geology of the Gulf of Mexico can be found in Bouma *et al.* (1978); Table 10.3 summarizes the geological history (Shaub *et al.* 1984).

The deep Gulf of Mexico contains two abyssal plains, Sigsbee Plain in the west and the Florida Plain in the east (Fig. 10.17), and is underlain by up to 10 km of probable Jurassic to Holocene sediments (Shaub *et al.*

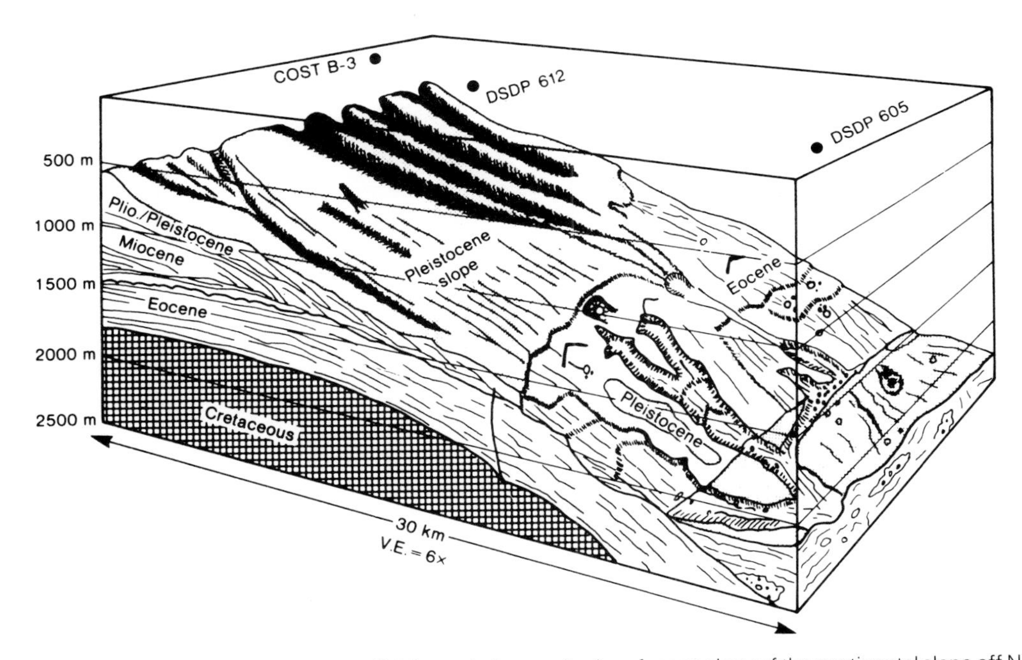

Figure 10.16 Schematic summary of the superficial morphology and subsurface geology of the continental slope off New Jersey in the region of the COST B–3 borehole and DSDP Sites 612 and 605. After Farre & Ryan (1987).

Figure 10.15 Schematic diagrams showing the Early Cretaceus depositional history around DSDP Site 603 and the adjacent shelf of the Baltimore Canyon Trough area, with the oceanic basin of the evolving North Atlantic to the right (east). Interpretations based, in part, on recent commercial drilling. A, Oxfordian–early Valanginian, progradation then aggradation of carbonate reef, culminating in drowning of reef by early Valanginian times. B, Valanginian, regression promotes submarine dissection of reef, perhaps with sand and mud bypassing the reef to be redeposited in deeper water. C, Hauterivian–Barremian, despite a rising sea level, rapid delta progradation across the shelf occurred due to hinterland uplift and enhanced erosion. Siliciclastics accumulate on the shelf and in submarine fan complexes. D, Early Aptian, eustatic fall in sea level causes substantial redeposition of shelf clastics in submarine fan complexes. E, Mid-late Aptian, a rising sea level, and associated transgression, lead to the development of clastic traps on the inner shelf and the re-establishment of carbonate sedimentation towards the outer shelf; only fine-grained hemipelagites and pelagites accumulate in deep water. After Wise & van Hinte (1987).

Table 10.3 Summary of deep Gulf of Mexico seismic units. After Shaub *et al.* (1984).

Unit	Age	Typical reflection characteristics	Suggested depositional environment depocenter source
Challenger	Middle Jurassic(?) to Middle Cretaceous[1,2,3]	Moderate amplitudes, low frequency; generally continuous, parallel and sub-horizontal in central Gulf; discontinuous and gently dipping, deformed and even chaotic along the base of the Florida and Campeche platforms	Unit immediately overlies acoustic basement (oceanic and transitional crust). Predominantly deep marine sediments in central Gulf. Evaporites, shallow then deep marine along Campeche and Florida Escarpments. Eastern depocenter may have source in Tampa Embayment; central Gulf depocenter, apparent source in Campeche region
Campeche	Upper Cretaceous to Early Tertiary(?)[1,2]	Low amplitude, low frequency; continuous, parallel, horizontal or gently dipping	Predominantly deep marine distal clastics in western ⅔ of study area; pelagic in east. Western depocenter source is probably Rio Grande Embayment
Lower Mexican Ridges	Early(?) to Middle(?) Tertiary	Moderate amplitudes and frequencies; generally continuous; commonly parallel and horizontal	Predominantly deep marine distal clastic sediments. Broad western depocenter attributable to ancestral Texas and Mexican rivers. Considered a continuation and progradation of the sedimentation pattern of the underlying Campeche unit. Clastics are distributed throughout the entire deep basin
Upper Mexican Ridges	Middle Tertiary(?) to Late Miocene[3]	High amplitude, high frequency; continuous, parallel, horizontal/ sub-horizontal; minor channels and clinoforms near depocenters.	Predominantly deep marine distal sands, silts, muds. Western margin progradation and sedimentation continues from western margin. A northeastern depocenter is established in the study area for the first time and is attributed to the ancestral Mississippi River
Cinco de Mayo	Late Miocene through Pliocene[3,4]	Generally acoustically transparent; otherwise, variable amplitude and frequency; parallel and sub-horizontal	Abyssal terrigenous and biogenic ooze. No major depocenter in deep Gulf; sediments thicken slightly in northern and southwestern Gulf. Most of clastic supply may be trapped by sedimentary deformation along northern and western margins[5,6,7,8]

Continued

Table 10.3 *continued* Summary of deep Gulf of Mexico seismic units after Shaub *et al.* (1984).

Unit	Age	Typical reflection characteristics	Suggested depositional environment depocenter source
Sigsbee	Pleistocene[3,4]	Mid-fan: variable, but generally high amplitude and frequency; complex, even chaotic reflection configurations interpreted as channels, levees, and channel-fill, inter-channel and overbank strata. Lower fan: high amplitude, high frequency; continuous, parallel and horizontal; in places wavy or distorted with channels. Western and southwestern continental rise: generally acoustically transparent	Abyssal submarine fan and other northern-source mass-transport deposits in eastern ⅔ of Gulf, contributed by Pleistocene Mississippi River[9,10,11], mostly suspension deposits in west; some fine-grained turbidites also derived from Mexican rivers in western basin[10]

[1]Addy & Buffler (1984). [2]Schlager *et al.* (1984). [3]Worzel *et al.* (1973). [4]Ewing (1969). [5]Bryant *et al.* (1968). [6]Buffler *et al.* (1979). [7]Martin (1978). [8]Shaub (1983). [9]Davies (1972). [10]Stuart & Caughey (1976). [11]Moore & Woodbury (1978).

1984). Multichannel seismic reflection profiles reveal sediment onlap onto transitional crust at the margins of the deep basin (Fig. 10.18a), and probable oceanic crust under the central basin (Fig. 10.18b).

Important Late Tertiary halokinesis and shale diapirism in the northern continental slope caused much of the coarse sediment influx to the Gulf to be trapped in upper slope basins (cf. Ch. 5), such that the upper Miocene to Pliocene Cinco de Mayo unit (Table 10.3, Fig. 10.18) represents a phase of relative sediment starvation in the deep Gulf. The youngest Sigsbee unit, which includes the Mississippi Fan, comprises up to about 3 km of deep-marine sediments ranging from coarse sediment gravity flow to pelagic deposits (Shaub *et al.* 1984, Bouma, Coleman *et al.* 1986). In the western and southwestern parts of the deep basin, the Pleistocene succession comprises the thinned distal equivalents of an ancestral Mississippi Fan system, predominantly deposits of low-concentration flows.

In a seismic study of the Veracruz Tongue (western Gulf of Mexico), defined as the region between the Mexican Ridge fold belt to the west and the Campeche Knolls salt province to the east, Bertagne (1984) reconstructed the changing source areas for the deep-marine clastic systems from the mid-Cretaceous to the present (Fig. 10.19). The Triassic to Middle Jurassic rifting in the Gulf was followed by a short period of seafloor spreading in the Late Jurassic, with the associated, post-breakup, subsidence phase being completed by the mid-Cretaceous (Buffler *et al.* 1981): at this time, several kilometres of subsidence had occurred to generate an overall basin morphology and depth somewhat similar to the present-day. Halokinesis in the Campeche Knolls salt province first formed a barrier to turbidity-current flow from the east in the Middle Miocene, and the growth of the Mexican Ridge fold belt gradually impeded turbidity-current flow from the west by the end of the Pliocene (Fig. 10.19) (Bertagne 1981). The result of the shifting of depocentres was that the sand-rich Miocene turbidite systems were covered by hemipelagic clays and clayey oozes. This study from the western Gulf of Mexico is a particularly good example of how structural control, including halokinesis, can modify basin shape, relocate deep-marine depocentres, and cause the abandonment of relatively sand-rich turbidite systems.

GLORIA side-scan sonar imagery of the Gulf of Mexico (USGS Map Series) reveals a complex pattern of sinuous channels on the surface of the Mississippi Fan, and fan sediments abutting the base of the Florida Escarpment. DSDP Leg 96 (Bouma, Coleman *et al.* 1986) drilled sites on the Mississippi Fan with the aim

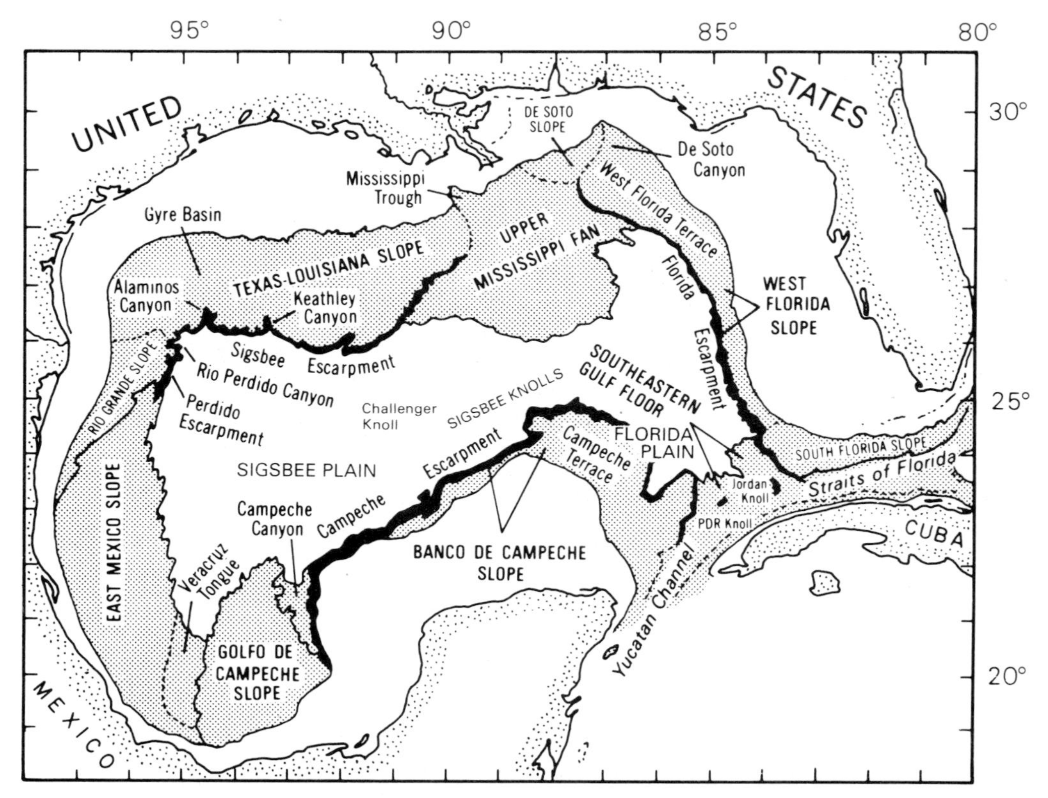

Figure 10.17 Physiographic provinces in the Gulf of Mexico, including principal submarine canyons, sea valleys and escarpments.

of testing and developing the models for fan sedimentation that are derived mainly from ancient systems. (Ch. 7). Also, part of DSDP Leg 96 was devoted to studying two intraslope basins (see Ch. 5 case study on Orca Basin).

In the Florida and Sigsbee Abyssal Plains (Fig. 10.17), there are thick sections of turbidites, hemipelagites and pelagites ranging from Pliocene to Holocene (Burk *et al.* 1969, Martin & Bouma 1978). DSDP Leg 1 (Ewing *et al.* 1969) showed that the Sigsbee Knolls that interrupt the smooth topography of the plain are the surface expression of salt diapirs that pierce and deform thousands of metres of strata. In contrast, the knolls in Florida Plain are associated with positive magnetic anomalies and are interpreted to have an igneous origin (Pyle *et al.* 1969).

Much of the northern continental slope of the Gulf of Mexico contains salt diapirs that, in some cases, have acted as dams to downslope resedimentation processes to produce intraslope basins (see Ch. 5). Bouma (1983) documents substantial sand bodies that were trapped in submarine canyons as a result of halokinesis, and he interprets the main times of sand resedimentation into such structural traps to be during lowstands in sea level.

The Mississipi Fan is a large arcuate wedge of Pleistocene clastic sediments with a radius varying from 330–380 km, and derived mainly from the Mississippi River drainage basin. Gradients typically range from 0.5–0.25° on the middle fan (Moore *et al.* 1978, Bouma *et al.* 1985b) in water depths from 1900–3000 m. DSDP Leg 96 (Bouma, Coleman *et al.* 1986) showed that sediment accumulation rates over the present fan surface are about an order of magnitude slower than in the Pleistocene when rates were up to about 1100–1200 cm/1000 years (Fig. 10.20).

DSDP Sites 621 and 622 are located within a prominent, sinuous, middle-fan channel; Site 617 on an adjacent levee, and Site 620 in overbank deposits approximately 18 km northeast of the channel sites. Four holes were drilled on the lower fan, and one on the lateral fan margin of the youngest 'fanlobe'. The

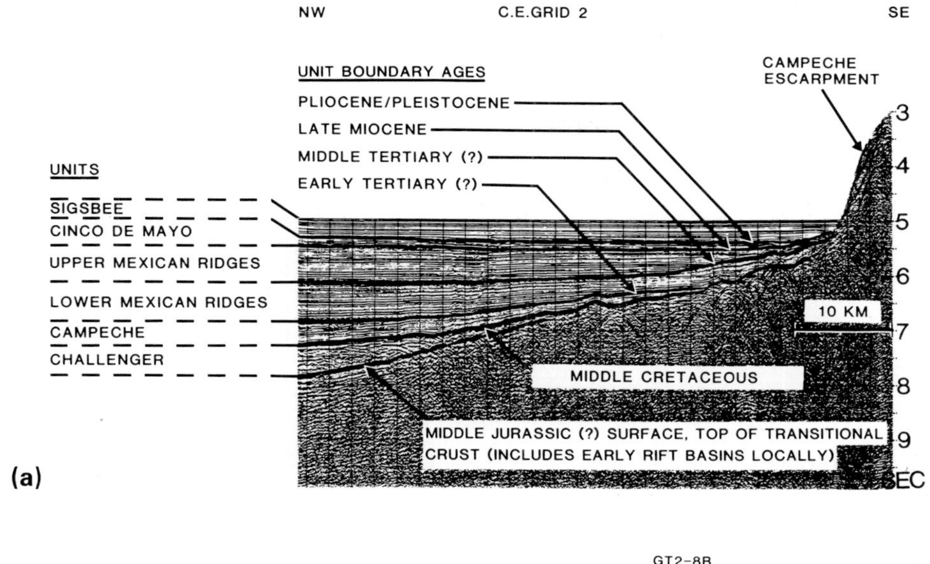

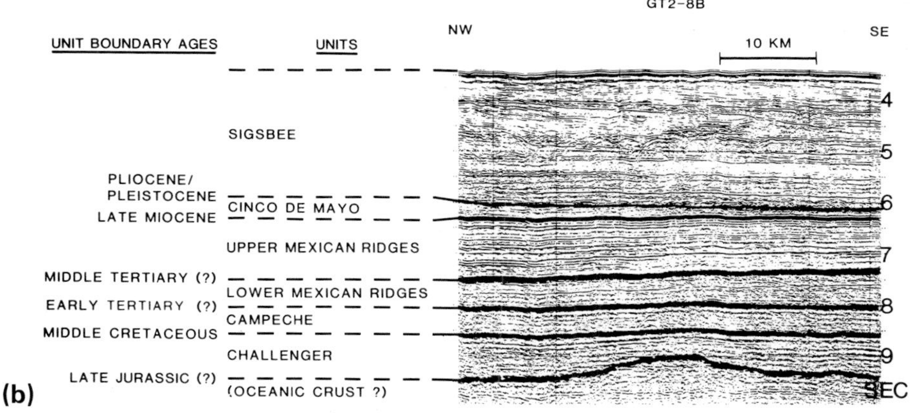

Figure 10.18 Seismic profiles in the southern (a) and central (b) Gulf of Mexico to show seismic units. See Table 10.3 for explanation. After Shaub *et al.* (1984).

maximum cored depths of about 420 m (Sites 620 and 615) only penetrated the upper parts of the Pleistocene stratigraphy. Figure 10.21 shows some representative cored facies from the Pleistocene part of the middle fan; Figure 7.3 (Ch. 7) summarizes cored sequences in the middle-fan channel.

High resolution and multichannel seismic reflection profiles in the middle fan suggest a temporal change in the channel shape and sedimentary character during the late Pleistocene to Holocene (Fig. 10.22) (Stelting *et al.* 1985, Pickering *et al.* 1986b). The reflectors shown in Figure 10.22 were drilled on Leg 96, and are interpreted as follows: type 1 high-amplitude, low continuity reflectors as base-of-channel gravels and sands; type 2 medium-amplitude, short, semi-transparent to transparent reflectors as channel margin sands and silts; type 3 low-amplitude, relatively continuous, curvilinear, semi-transparent to transparent reflectors as upper channel muds and clays, and type 4 medium-amplitude, subparallel, hummocky, discontinuous reflectors as overbank silts, muds and clays. The seismic evidence shown in Figure 10.22, proved by drilling, shows an infilling and fining of the channel from the Pleistocene lowstand to the Holocene raised

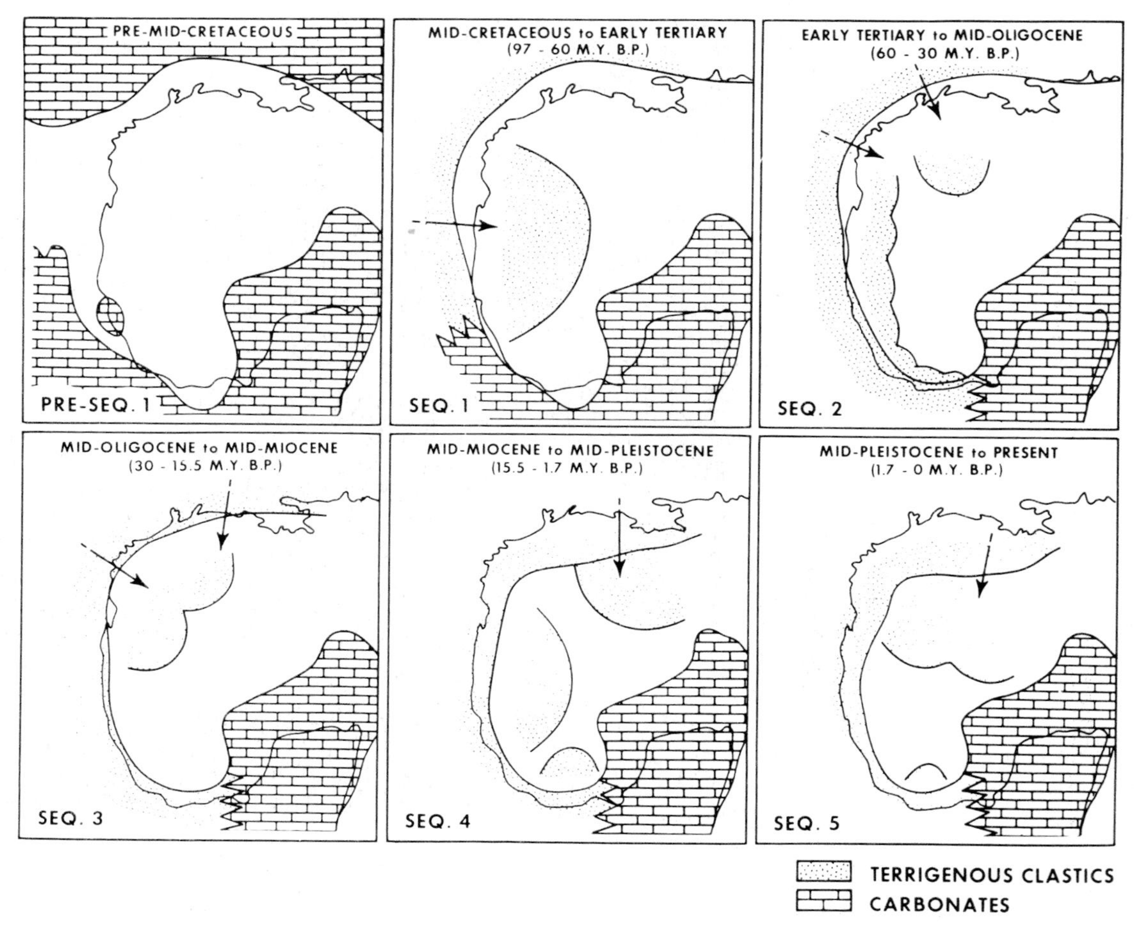

Figure 10.19 Location of shelf edges and major deep-marine depocentres in the western Gulf of Mexico, Veracruz Tongue, from pre–mid Cretaceous to present. Inferred direction of reworked sediments from major deltas shown by arrows. After Bertagne (1984).

sea level, demonstrating the importance of fluctuations in sea level in controlling fan processes (Pickering *et al.* 1986b).

Fragments of ancient passive margins

Many large fragments of ancient passive margins are documented from orogenic belts. Amongst the thicker successions is the 14 km-thick, upper Precambrian (Riphean–Vendian) Barents Sea Group and Lökvikfjell Group, Finnmark, North Norway (Siedlecka & Siedlecki 1967, Johnson *et al.* 1978, Pickering 1985, Siedlecka 1985). The oldest exposed part of this passive margin consists of at least 3200 m of deep-marine turbidite and related deposits (Pickering 1981b, 1985), passing up into 2500–3500 m of upper continental slope and prodelta (Pickering 1981b, 1982a, 1984b), delta-front, delta-top and associated shallow-marine sediments (Siedlecka & Edwards 1980). Up to 1500 m of shallow-marine, intertidal and supratidal carbonates, overlain by 1500 m of essentially fluviatile sediments, form the upper part of the Barents Sea Group (Siedlecka 1978, 1985, Siedlecka *et al.* 1989). 5700 m of upper Precambrian–Eocambrian shelf, marginal marine and continental sediments of the Lökvikfjell Group (Levell & Roberts 1977, Johnson *et al.* 1978, Levell 1980a & b) overlie the Barents Sea Group with disconformity or local angular unconformity. This passive margin stratigraphy, developed facing an ocean basin

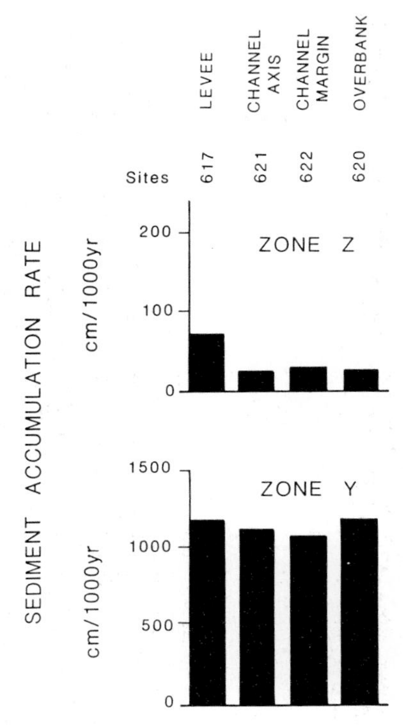

Figure 10.20 Undecompacted sediment accumulation rates for DSDP Leg 96 Sites 617 (levee), 621, 622 (channel) and 620 (overbank). The sediment accumulation rates for the Ericson Zone Y are based on seismic projection to 'horizon 30', the Y/X boundary. Note the dramatic decrease in sediment accumulation rate from the Pleistocene (Zone Y) to the Holocene (Zone Z).

towards the present N–NE, was juxtaposed against the Fennoscandian Shield by major dextral shear between 640–540 Ma (Kjøde *et al.* 1978).

The study of ancient passive margins has generated a better understanding of carbonate-dominated deep-water sedimentation models, such as the carbonate apron models of Mullins & Cook (1986) and Colacicchi & Baldanza (1986). Watts & Garrison (1986), in a study of the evolution of a Mesozoic carbonate slope on the south Tethyan margin (2500 m-thick Sumeini Group, Oman), document carbonate submarine fans, base-of-slope debris aprons and gullied bypass slopes, integrating them into the Permian–Early Cretaceous rifting-subsidence history of this passive margin.

The Cambro–Ordovician of eastern North America provides a well documented example of an ancient passive margin (Williams & Stevens 1974, James & Stevens 1986) with deep-water sediments offshore from a carbonate platform (Walker *et al.* 1983, Stanley & Ratcliffe 1985). Chapter 5 contains a case study from this margin. On the western side of North America, Cook (1979) describes a N-trending Late Cambrian – Early Ordovician margin in central Nevada where 150 m of hemipelagic carbonates (Facies Classes E and G) accumulated on a deep-marine slope, coeval with 600 m of shallow-marine limestones 170 km east of the slope. Seaward of the thin slope succession, in which up to 50% of the section comprises slide, slump, debris-flow and turbidity-current deposits (Facies Classes A to D), there are 1000–2000 m of submarine fan sediments.

CASE STUDY: JURASSIC PASSIVE MARGIN, SOUTHERN ALPS, ITALY

The Southern Alps of Italy preserve, almost intact, a cross-section through a Jurassic mature passive margin (Winterer & Bosellini 1981), although Jenkyns (1980) prefers an interpretation of Tethyan facies deposition primarily in 'intra-continental basins and on continental margins influenced by transcurrent faulting, where abrupt positive and negative motions were the rule rather than the exception.' Regional tectonostratigraphic evidence suggests that strata exposed in the Southern Alps of Italy, together with coeval sediments to the south, were deposited on the southern margins of Tethys, on the Apulian Plate.

After the Hercynian Orogeny, the western part of the present Mediterranean region was part of a single continental block of Africa, Europe and North America, with shallow continental seas over much of this crust. Local crustal fragmentation by block faulting occurred in the Permian and Triassic, contemporaneous with volcanism (e.g. Winterer & Bosellini 1981). Triassic rift basins became the sites for deep-marine sedimentation. The Middle Jurassic opening of the Central Atlantic Ocean followed a second episode of Early Jurassic rifting and basin subsidence, with 'synrift' sedimentation and the foundering of shallow-marine carbonate platforms lasting for about 12–15 m.y. Typical deep-marine sediments are preserved as carbonate turbidites and other mass-flow deposits (Alvarez *et al.* 1985). The earliest dated oceanic crust is not recorded in southern Italy, but in the Sub-Betic Ranges of southern Spain, in post-breakup basins, as pre-Aalenian (earliest Middle Jurassic, approx. 185 Ma) pillow basalts, associated with pelagic sediments (Paquet 1969). Northwest-dipping subduction was occurring in the western Tethys at least by the Late Jurassic accompanied by westward obduction of ophiolitic material onto the Apulian plate in Greece and

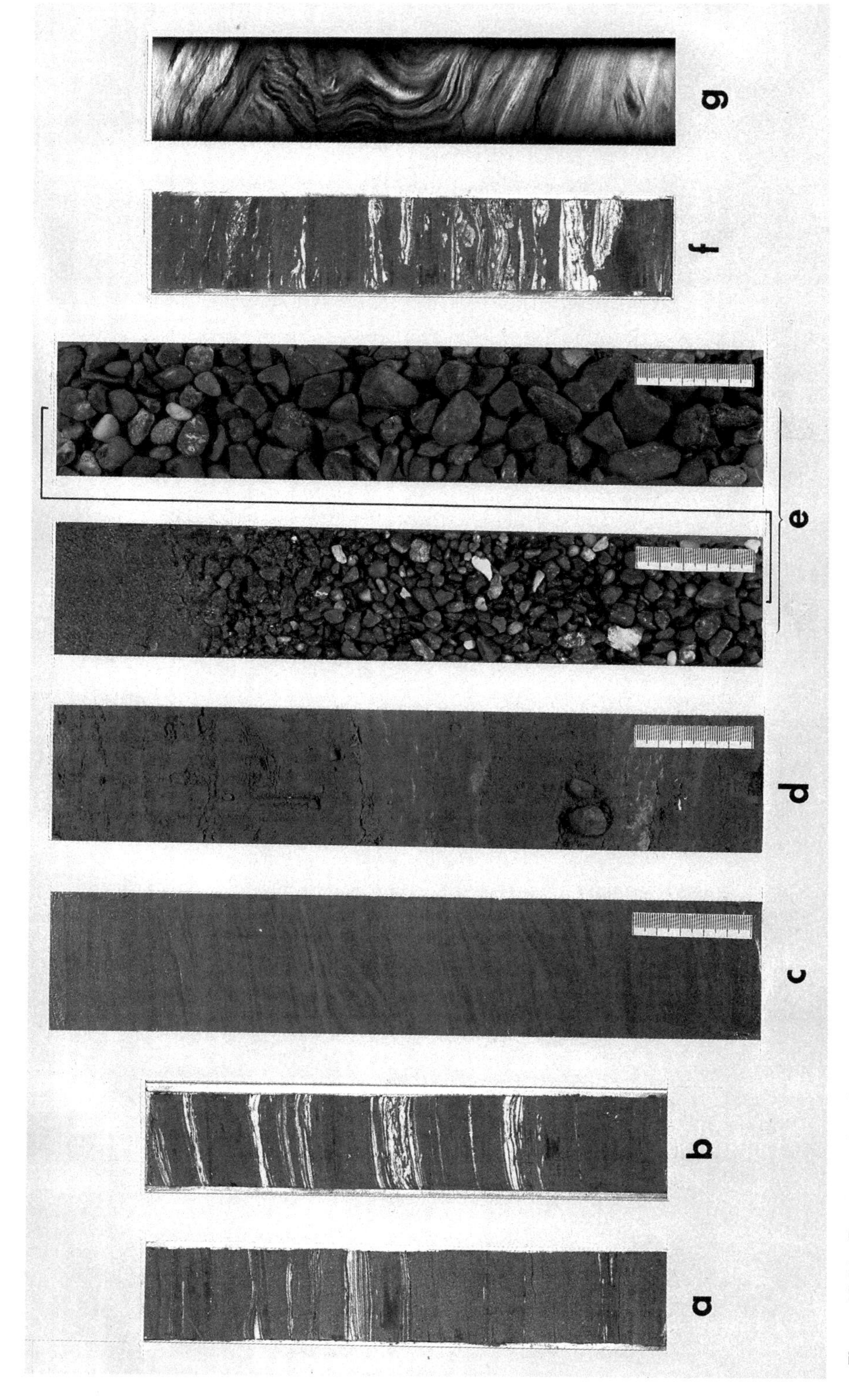

Figure 10.21 Representative photographic plates to show the characteristic sedimentary facies drilled in the channel–levee–overbank system of the Missisipi Fan. Sub-bottom depths correspond to the top of each plate. 5 cm bar scales; plates without scale have a core width of approx. 5.8 cm. a, silt-laminated mud (Site 622, 93.4 m); b, silt-laminated mud (Site 621, 158.7 m); c, sandy silt (Site 622, 156.2 m); d, pebbles in silt laminated mud (Site 622, 197.7 m); e, gravel overlain by sand (Site 621, 213.9 m); f, muddy silts, silty muds and silts (Site 617, approx. 85 m); g, X-ray radiograph of wet-sediment deformation, slide fold, in Site 617 (approx. 47 m). (After Pickering *et al.* 1986b).

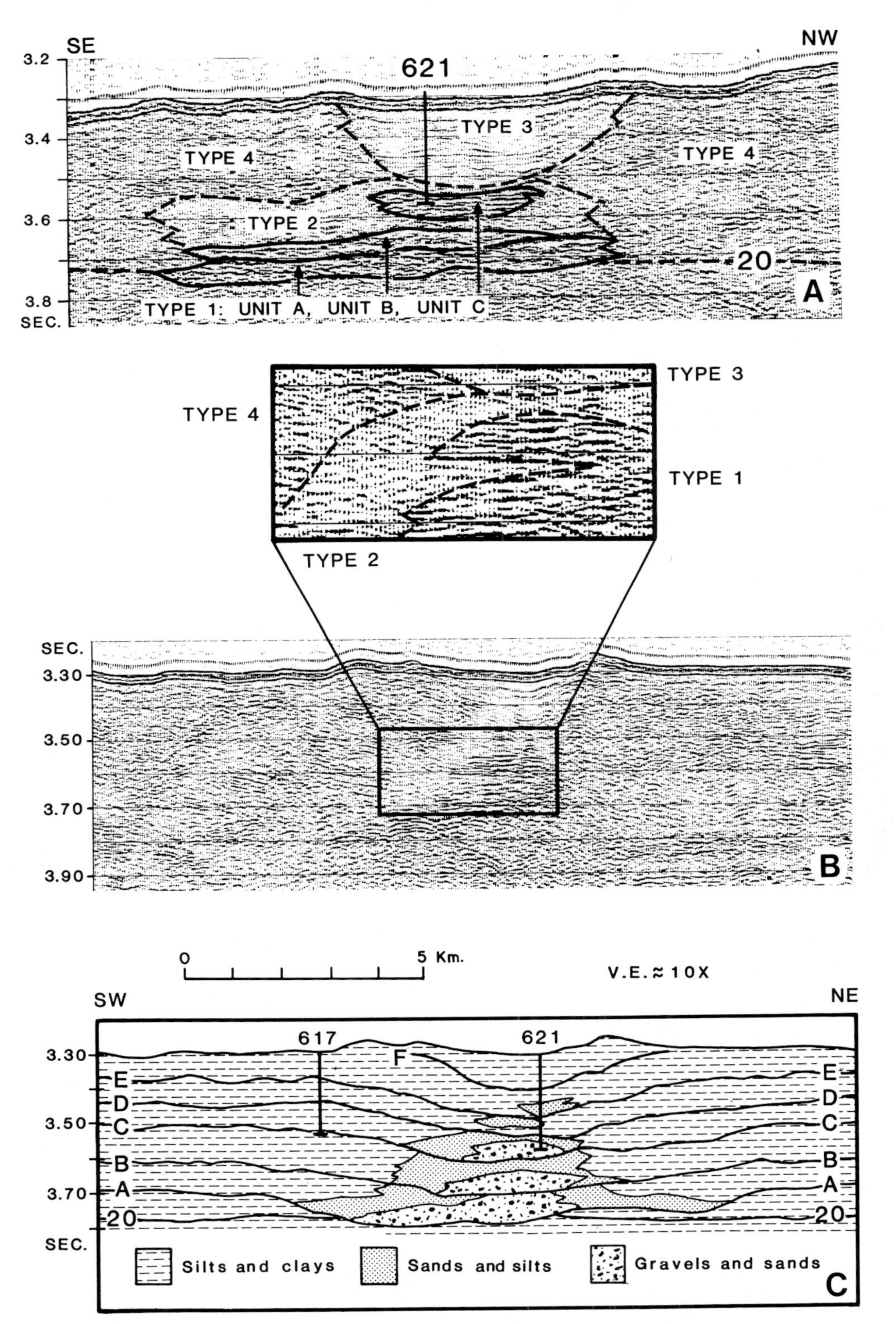

Figure 10.22 Seismic facies and their interpretation, based on tying the data to cored intervals in DSDP Sites 617 and 621. See text for explanation. Note overall fining upwards in the channel due to the eustatic rise in sea level from the Pleistocene to Holocene. (After Pickering *et al.* (1986b).

Yugoslavia (Aubouin *et al.* 1977). Thus, passive margin sedimentation along the western edge of the Apulian Plate probably spanned only the Middle to Late Jurassic.

Figure 10.23 is a palinspastic reconstruction for the Late Jurassic, from Bernoulli (1980), to show the lithologic units in the Southern Alps, together with their relative thicknesses. Early Jurassic rifting (Fig. 10.24) led to the foundering of the carbonate platforms, with rapid subsidence transferring shallow-marine areas into the realm of deep-marine sedimentation. Areas of rapid subsidence probably lay between listric faults, with coarse debris (Facies Classes A, B and F) being shed down the steep fault scarps. The many erosional gaps and the stratigraphic asymmetry between the active faults suggests contemporaneous rotational tilting (Gaetani & Poliana 1978).

During the Middle Jurassic, large oolitic submarine fans prograded westwards into the Belluno Trough (Fig. 10.24) to reach thicknesses up to about 1000 m (Bosellini & Masetti 1972). At the same time, the 'Ammonitico Rosso Inferiore', or red nodular limestone accumulated elsewhere. By the close of Ammonitico Rosso Inferiore times, water depths are estimated at about 1000 m (Winterer & Bosellini 1981).

By the Late Jurassic, slow pelagic sedimentation (Facies Class G) covered most of the region (Fig. 10.24), without evidence of synchronous faulting. Biogenic pelagic sediments, such as siliceous radiolarian oozes, accumulated below the CCD. Typical 'condensed' successions overlying the shallow-marine (Lower Jurassic) platform carbonates include the nodular and marly Ammonitico Rosso with its rare corroded ammonites. In contrast, the Early Cretaceous involved renewed fault activity with subsidence continuing into the Tertiary, and further episodes of deep-marine sedimentation in the northern Italian Apennines, including the 200–400 m thick Coniacian-Eocene Scaglia Rossa Formation of turbidites, hemipelagites and pelagites (Stow *et al.* 1984). Alvarez *et al.* (1985) believe that the Jurassic – Lower Tertiary pelagic carbonates in the Umbria–Marches Apennines, northern peninsular Italy, were deposited on a promontory projecting north from Apulia into Tethys, similar to Florida–Bahamas today.

Winterer & Bosellini (1981) estimate the subsidence and water-depth history, based on the following assumptions: (a) the non-isostatic part of the total subsidence (i.e. the thermal component) is proportional to the square-root of the time elapsed since post-breakup subsidence commenced; (b) isostatic loading by sediments was local; (c) isostatic load corrections can be made with specific gravities of water = 1.0, sediment = 2.3, and mantle = 3.3, and (d) global sea level remained constant. Winterer & Bosellini (1981) calculated rates of platform foundering in the Southern Alps of between 100–300 m/m.y. for the first million years of rifting. Jenkyns (1980) believes that the Mesozoic Tethyan margin did not follow a smooth subsidence curve, and cites the local persistence of shallow-water Jurassic 'oolites' and associated stromatolitic limestones as evidence for local subsidence reversals. Coeval with the oolites, deep-water red nodular limestones, radiolarian cherts or white coccolith-rich calpionellid calcilutites were deposited elsewhere (Bernoulli & Jenkyns 1974). Figure 10.25 is Jenkyns' (1986) proposed subsidence tracks for parts of the Tethyan continental margin based on facies.

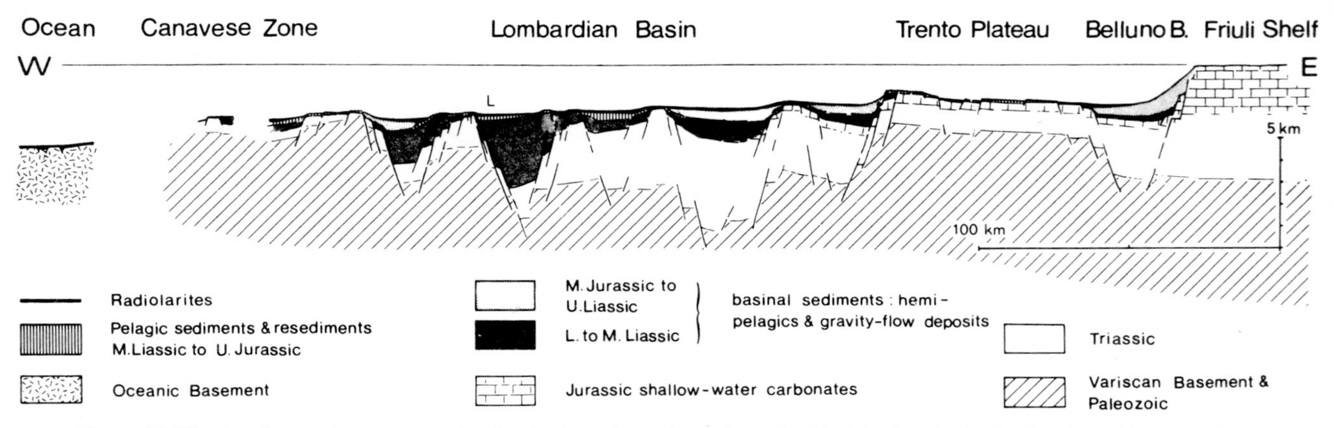

Figure 10.23 A palinspastic reconstruction for the Late Jurassic to show the litholologic units in the Southern Alps, together with their relative thicknesses. From Bernouilli (1980).

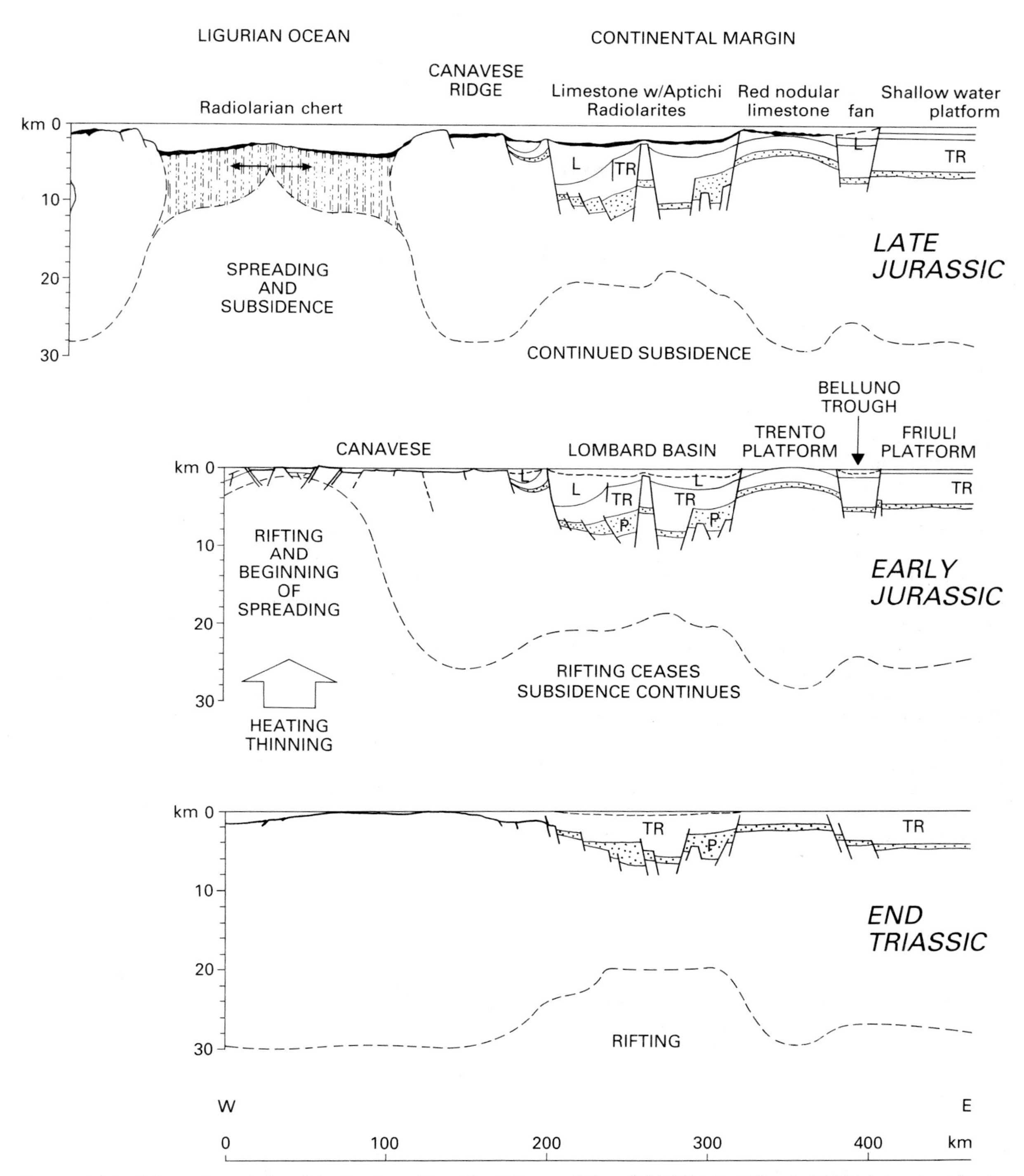

Figure 10.24 Schematic evolution of the Southern Alps. After Winterer & Bosellini (1981). End Triassic (192 Ma): Lombard Basin (centre) development by rifting with thick shallow-marine and non-marine infill. Continental crust is shown as thinned beneath basin. Basement to Trento Platform, as well as parts of the crust to the west of the Lombard Basin, is Permian igneous plutonic and volcanic rocks. In France and Italy, Triassic thins southwestward towards the site of the future Ligurian oceanic basin (see top diagram). Early Jurassic (178 Ma): Rapid subsidence of Lombard Basin and Belluno Trough to depths of about 1000 m, immediately prior to subsidence of Trento Platform. Late Jurassic (143 Ma): Opening of Ligurian oceanic basin to a width probably less than 2000 km. Infill of Ligurian Basin initially by pelagic limestones, then Cretaceous terrigenous flysch. Red nodular limestones accumulate on relatively shallow platforms, with basinal sedimentation as red marly limestones or radiolarites, depending on water depth. Timescale of van Hinte (1978b).

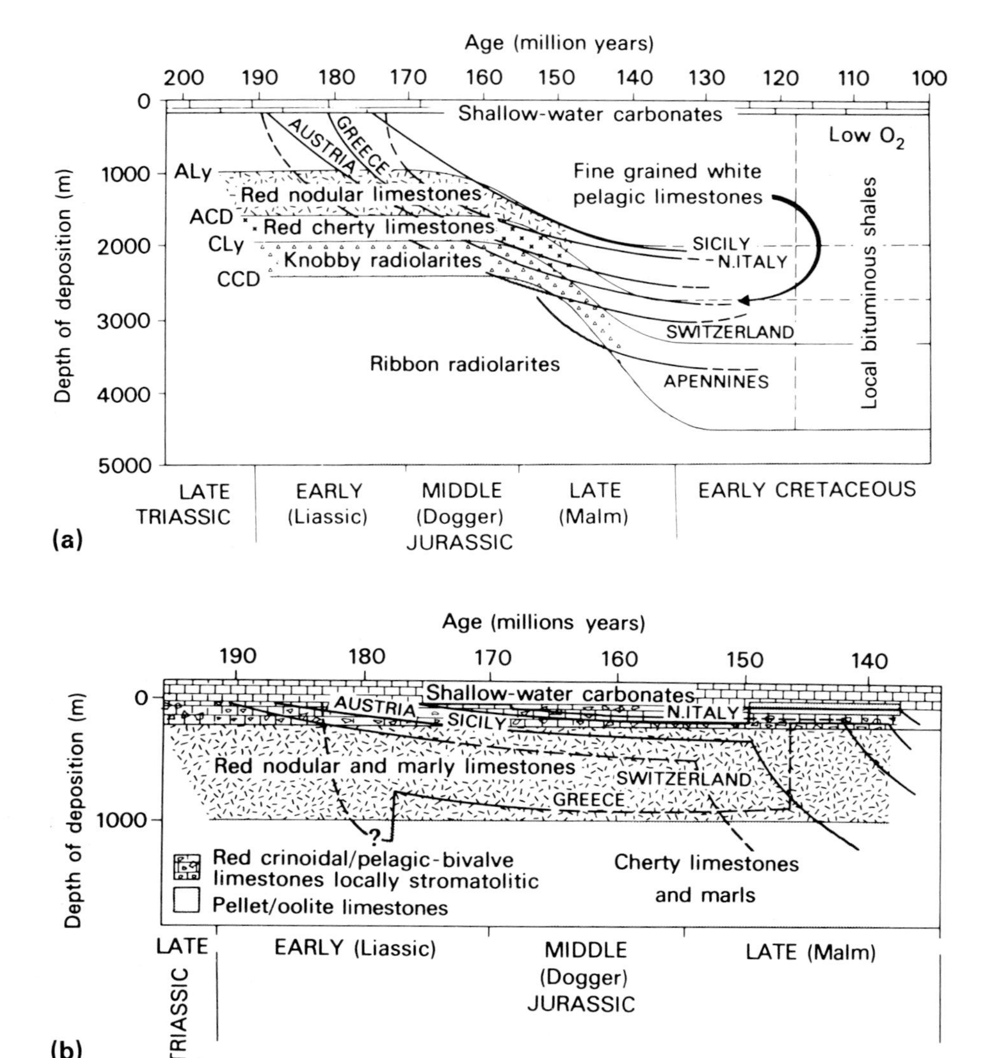

Figure 10.25 Idealized models of subsidence tracks for various parts of the Tethyan continental margin. **a**, ALy, aragonite lysocline; ACD, aragonite compensation depth; CLy, calcite lysocline; CCD, calcite compensation depth. Curve labelled 'Apennines' is proposed subsidence track for former Tethyan ocean floor now represented by Ligurian ophiolites of northern Italy. In this model, all subsidence tracks are thought to be unaffected by syn-sedimentary tectonic activity. The Early Cretaceous was locally characterized by waters of low-oxygen content and favoured the deposition of bituminous shales. **b**, In this model, where blocks with relatively shallow marine histories have been selected, the time-depth profiles do not follow smooth subsidence tracks but were influenced by uplift and/or accelerated differential subsidence of the sea floor. Dotted lines indicate periods of non-deposition. After Jenkyns (1986).

Greece, for example, shows an erratic track with two major periods of uplift, at about 176 and 146 Ma (timescale of van Hinte 1978b).

In summary, the Triassic–Jurassic pelagic successions of the north Italian Tethyan realm are interpreted to have accumulated in relatively small extensional fault-bounded and fault-controlled intra-continental basins of an evolving passive continental margin (Bernoulli & Jenkyns 1974, Jenkyns 1986) (Fig. 10.26), with the local development of oceanic crust in regions external to these basins. Contemporaneous strike-slip tectonics appears to have been important (Dercourt *et*

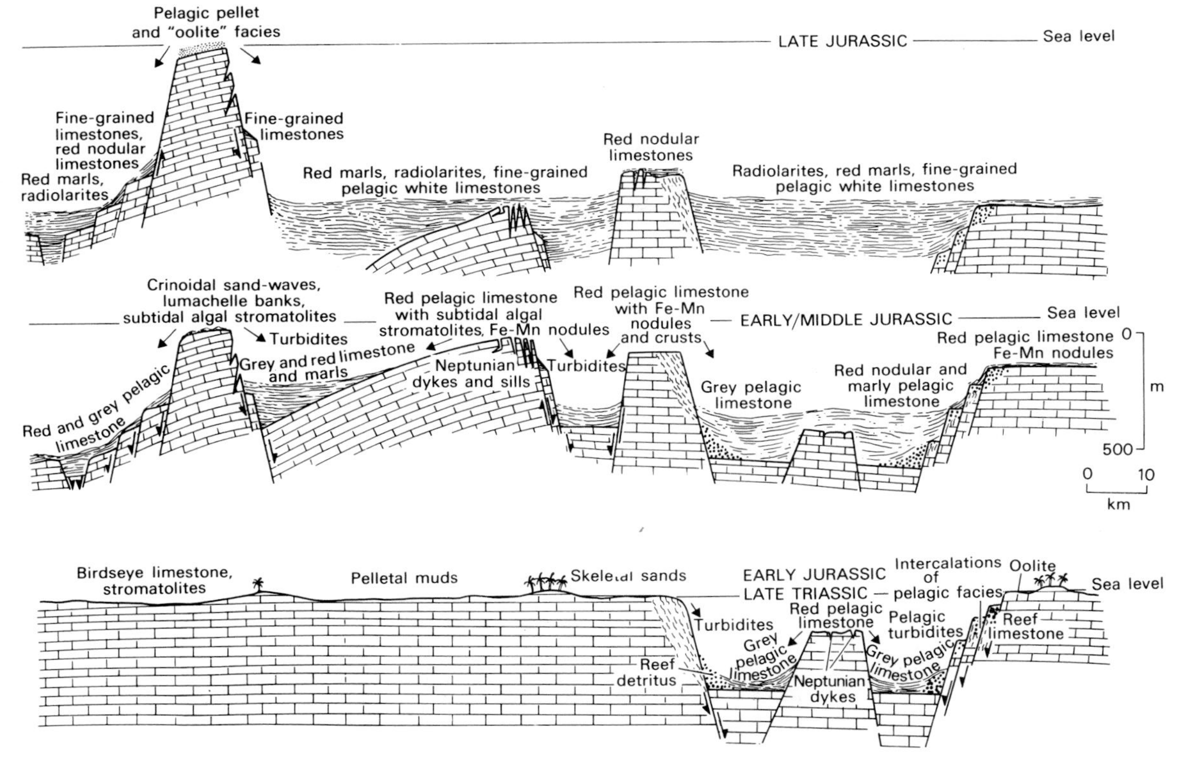

Figure 10.26 Schematic diagram of the development of the Tethyan continental margin during the Triassic and Jurassic. Block faulting and differential subsidence affected most carbonate platforms to produce a 'seamount and basin topography' in which coeval condensed and expanded successions accumulated. An overall levelling and smoothing of the sea floor during deepening characterized the Late Jurassic, except for seamounts on which pelagic 'oolites' accumulated: such seamounts or plateaus, e.g. the Trento Platform, persisted into the Cretaceous. After Jenkyns (1986).

al. 1986, Ricou *et al.* 1986). The Gulf of California may provide a useful plate-tectonic analogue, with transtension giving oceanic crust in a narrow, elongate, and deep ocean basin.

Although this section has focused on the Jurassic Period, the Middle Triassic in northern Italy shows a similar horst and graben topography, and associated shallow/deep marine sedimentation related to extensional tectonics in the western Tethys (Blendinger 1986).

CASE STUDY: FAN-DELTA/SUBMARINE FAN SEDIMENTATION, LATE JURASSIC TO EARLY CRETACEOUS, AND EARLY PALAEOGENE, EAST GREENLAND AND VIKING GRABEN, NORTH SEA

This case study is based mainly on the work of Surlyk (1978, 1984), Stow *et al.* (1982), Stow (1985), Stewart (1987), and Turner *et al.* (1987). Given the similarity in facies and the interpreted tectonic setting for the essentially contemporaneous deep-marine successions in East Greenland and the South Viking Graben, northern North Sea, this section is a combined case study from these areas. In the North Sea, an intracratonic rift system developed with pre-rift, syn-rift (rift) and post-rift phases, but without breakup and generation of a mature passive margin. In East Greenland, the early syn-rift history is followed by continued passive-margin development.

The middle Volgian to Valanginian Wollaston Foreland Group, in eastern Greenland, occurs as a succession of sediment gravity flows and sediment mass flows interpreted as fan-delta and submarine fan deposits. These sediments accumulated in half-grabens formed during the rifting phase close to the Jurassic–Cretaceous boundary, in which the regional basin slope dipped eastwards (Fig. 10.27b). At about the same time, similar sedimentation was taking place in the

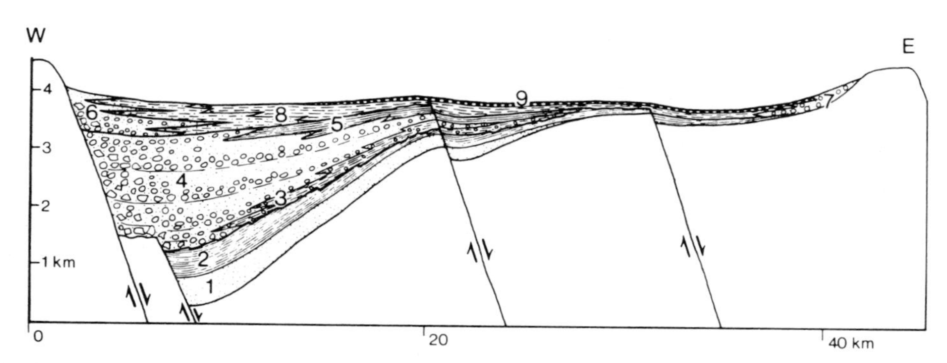

Figure 10.27a East–west palinspastic cross-section through the northern Wollaston Foreland, East Greenland, for the end Valanginian to Early Cretaceous. 1 to 9 are formations or members. 1, Bathonian? – Lower Oxfordian; 2, Upper Oxfordian – Lower Kimmeridgian; 3–5, Middle Volgian – Ryazanian; 6–9, Ryazanian – Valanginian. After Surlyk (1984).

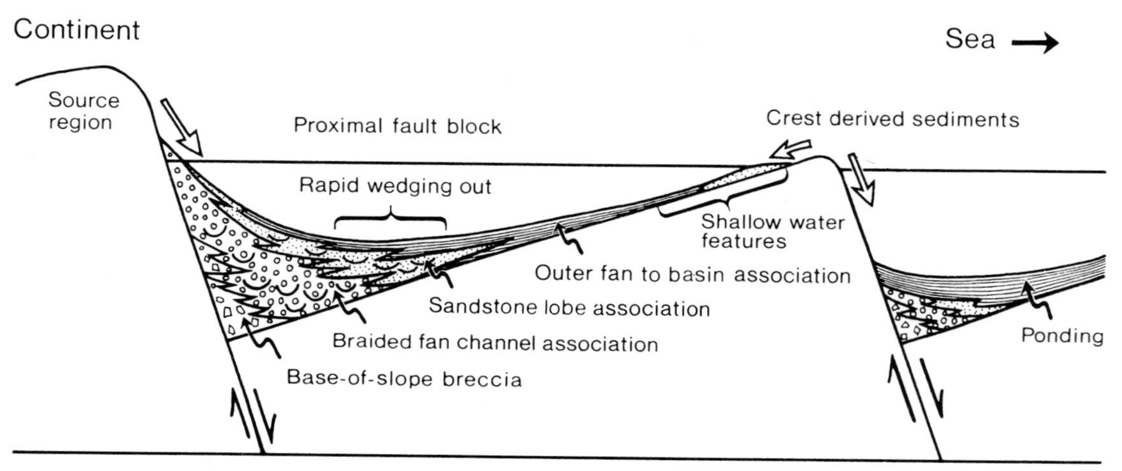

Figure 10.27b Model for fan-delta to submarine fan sedimentation along fault scarps on tilted fault blocks, based on the Late Jurassic of East Greenland. After Surlyk (1978).

North Sea Viking Graben, in the area of the modern Brae, South Brae, Tiffany, Toni and Thelma and other oilfields (Figs 10.28 & 29). The Wollaston Foreland Group, East Greenland, is exposed for tens of kilometres along strike, and thins eastwards, over 30–40 km (Fig. 10.27a), from about 3000 m to a thickness of 0–50 m: the wedge was deposited in approximately 10–12 m.y.

The Brae oilfield succession is a 300–600 m-thick basin-fill. The study by Stow *et al.* (1982) was based on over 2500 m of recovered core from 13 wells, combined with electric log and dipmeter data. The South Brae oilfield succession is up to 760 m thick and consists of graben-margin and basin-fill successions. The most recent studies are based on over 6700 m of recovered core from 41 wells, combined with geophysical logs (mainly gamma ray and sonic) and dipmeter data (Turner *et al.* 1987).

In East Greenland, water depths during sedimentation at the fault scarps are estimated to vary from near sea level or emergent to about 1000 m in the basin axis, 15 km east of and parallel to the scarp (Fig. 10.27). For the Viking Graben fault-controlled fans, water depths are unknown, although sedimentological evidence suggests depths below wave-base.

The sediments in both East Greenland and the South Viking Graben appear similar, with a considerable amount of conglomeratic facies. In the Brae oilfield, Stow *et al.* (1982) estimate the following composition: 45% Facies Classes E and G; 35% Facies Class A; 20% Facies Classes B, C and D. About 3% comprises deformed wet-sediments, such as slides of Facies Class F.

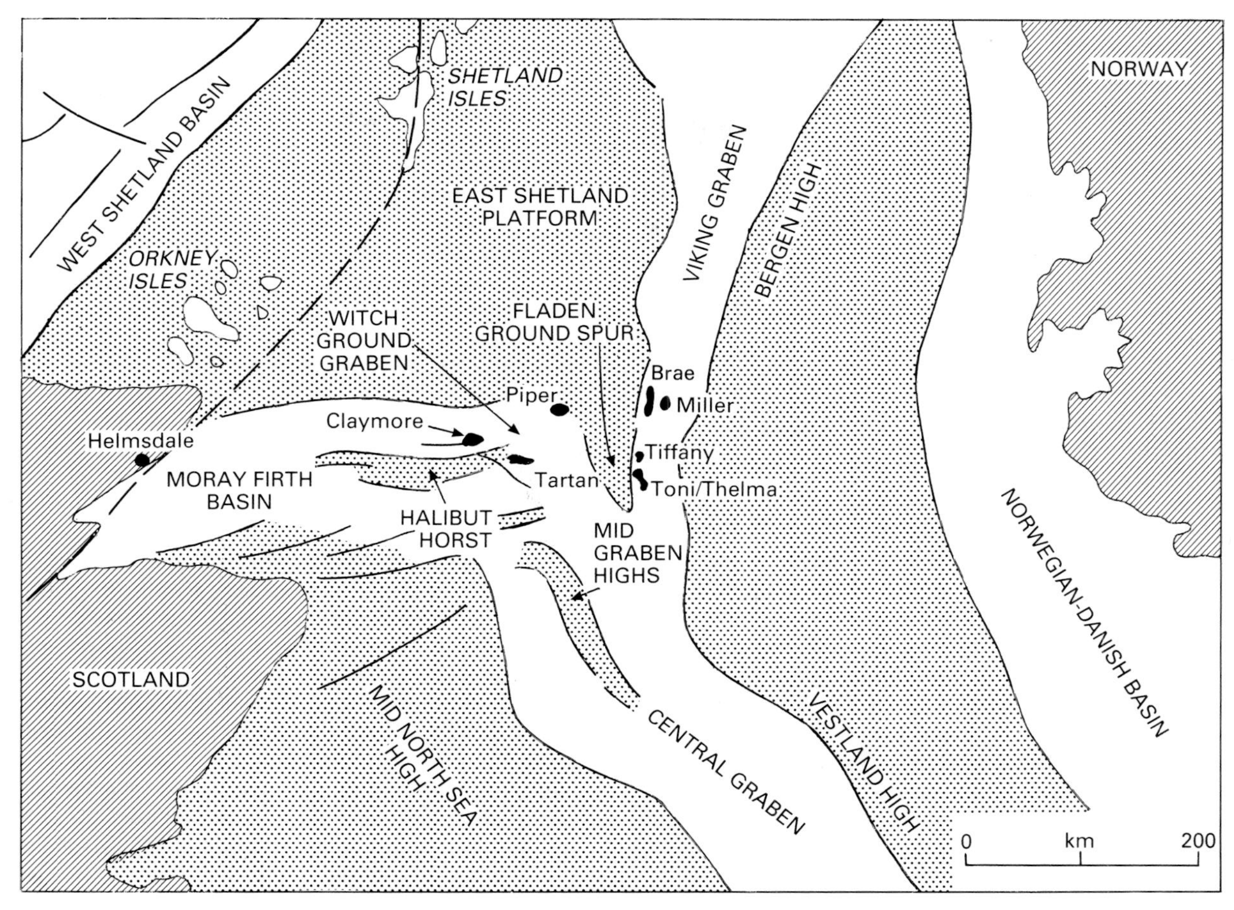

Figure 10.28 Map to show the principal structural features of the northern North Sea with principal oilfields.

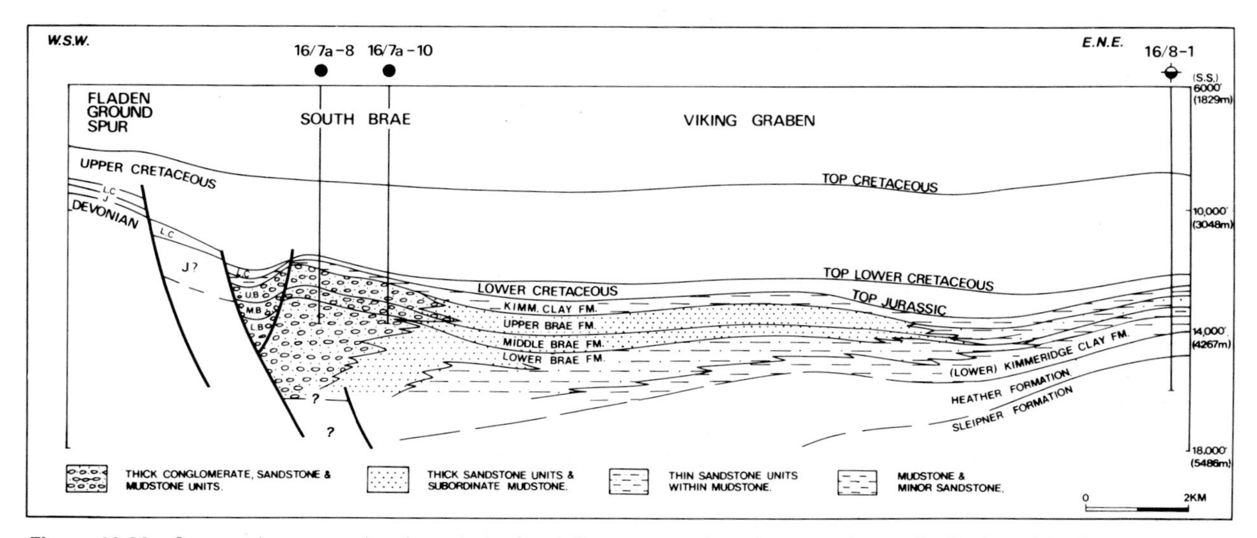

Figure 10.29 Structural cross-section through the South Brae area to show the approximate distribution of the lithologic units. After Turner *et al.* (1987).

In the South Brae oilfield there is a high degree of organization of sedimentary facies, with 30–90 m-thick units of conglomerate and sandstone (Facies Classes A and B) alternating with 15–60 m-thick units of finer grained, thin-bedded sandstone, interlaminated mudstone and laminated mudstone (Facies Classes C, D, E) (Fig. 10.30).

Lateral facies relationships within the South Brae oilfield clearly demonstrate cyclicity in both a lateral and vertical sense (Fig. 10.30). Laterally extensive mudstone units with minor sandstone interbeds are continuous sheets which blanket the coarser-grained facies; these muddy sheets are interpreted as blanketing the submarine fans during periods of reduced sediment supply. Channel-fill conglomerate and sandstone (Facies Classes A, B) are lenticular, scour-based units, with up to 75 m of relief in the northern part of the oilfield (Fig. 10.30). Channel width: depth ratios decrease downfan. Channel-axis facies are mainly conglomeratic and chaotic mudstone breccias (Facies Classes A, F), whereas channel-margin sediments are much finer grained, predominantly sandstone and mudstone (Facies Classes C, D, E).

Coarse-grained sediment gravity flows eroded and infilled submarine channels during low stands of sea-level. Because overlying sandstone units are very continuous and sheet-like, the resultant fining-upwards sequences are interpreted to reflect reduced supply of coarse detritus to the fan system. Sheet-like conglomerates and sandstones (Facies Classes A, B) form laterally adjacent to thick mudstone packets, especially near the margins of the fan. These units are interpreted to be deposited from sediment gravity flows which were unconfined at the lateral margins of the fan fringe. The finest grained mudstones and intercalated sandstones are overbank deposits, emplaced on topographically higher levees and interchannel sites.

The primary sediment source of the Brae and South Brae oilfields was Devonian sediments subaerially exposed on the Fladen Ground Spur to the west. Sediments of the Brae Formation were deposited adjacent to the Fladen Ground Spur, along the western margin of the South Viking Graben. Both the East Greenland and Brae and South Brae oilfield studies suggest small-radius (5–20 km), coarse-grained submarine fans that developed at the base of active fault scarps (Figs 10.27, 28 & 29). About 100–150 km farther onshore from these North Sea examples, to the west in the Moray Firth, coeval deep-marine fault-controlled sedimentation produced base-of-slope talus deposits, with debris-flow deposits (Facies Class A), and sediment slides (Facies Class F) being commonly associated with finer grained 'Kimmeridge Clay' lithologies (Facies Classes E and G, with some D) (Pickering 1983b, 1984a).

Tectonic control of fan development in the Brae oilfield resulted in six fining-upward sequences (50–150 m thick), within the overall fining-upward basin-fill (300–600 m thick). In the South Brae oilfield seven or eight fining-upwards sequences occur (Fig. 10.30). The base-of-slope sediments occur as a breccia-conglomerate facies association, deposited mainly by rock-fall and avalanching down the steep fault scarps. Inner fan sediments are dominated by conglomerates and sandstones, interpreted to have resulted from various sediment gravity flows. Mudstone facies associations are interpreted as mainly turbidites in interchannel, outer-fan, fan-margin and basinal settings.

Deposition of the Brae Formation was terminated at the South Brae and Brae oilfields by the Volgian, corresponding to a eustatic sea-level rise (Vail *et al.* 1984, Haq *et al.* 1987). Elsewhere within the North Sea, fan sedimentation continued. The eustatic sea-level rise had a local effect in cutting off the source area for the Brae and South Brae fans.

The Claymore oilfield on the southwest flank of the Witch Ground Graben, Outer Moray Firth, includes Upper Jurassic fault-controlled turbidite systems (Maher & Harker 1987). Figure 10.31 is a summary history of the development of the Claymore turbidite system, showing a comparable evolution to that of the Viking Graben. Indeed, during the Late Jurassic to Early Cretaceous, the extensive normal faulting in the northern North Sea formed coarse-grained deep-marine fans, commonly associated with nearby emergent source areas and connecting fan-deltas (Fig. 10.32).

In the Early Palaeogene, composite chronostratigraphic diagrams in conjunction with coastal onlap curves (Figs 10.33 & 34) show that on the basis of seismic stratigraphic data, the central North Sea can be divided into ten depositional sequences which are related to relative changes in sea level (Stewart 1987). Major subsidence of the Central Graben of the North Sea occurred in the Danian and was associated with an influx of sand from the Orkney–Shetland platform (Maureen Fan, Fig. 10.33), a consequence of uplift and rejuvenation of that area. The subsidence is associated with a rapid relative sea-level drop and a basinward shift of coastal onlap. Large volume submarine fans were built at this time, constructed in water depths of about 200 m. From the late Danian to early Ypresian, five lowstand fans developed, separated by thin hemipelagic mudstones deposited during highstand events. Thus, sea-level change was an important factor controlling the character of sediment within the North Sea.

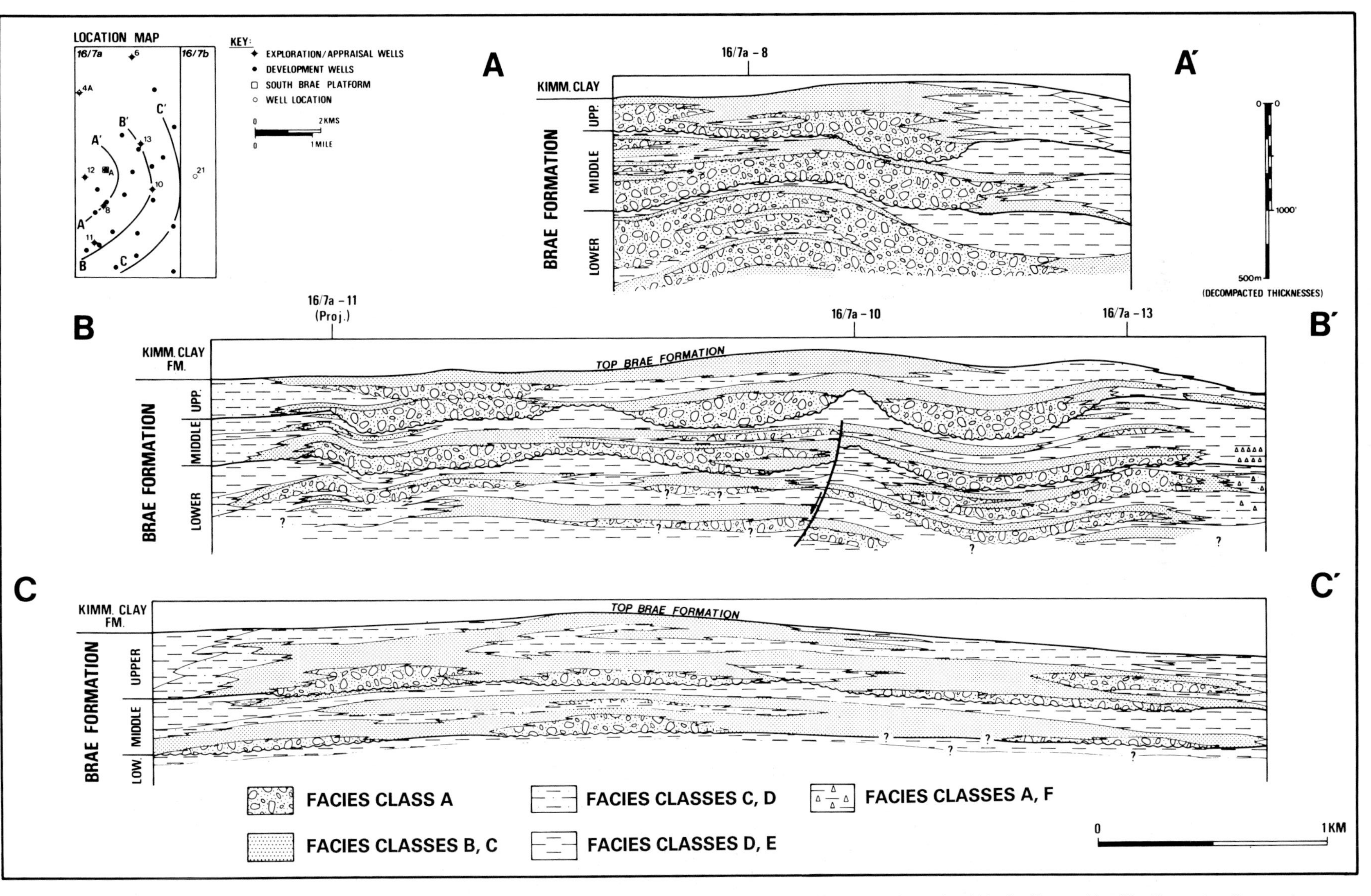

Figure 10.30 Profiles through the South Brae submarine fan to show the distribution of various facies associations. The upper datum is within the Kimmeridge Clay Formation, flattened to horizontal. Well data from decompacted logs have been extrapolated onto the lines of the section from adjacent wells. After Turner *et al.* (1987).

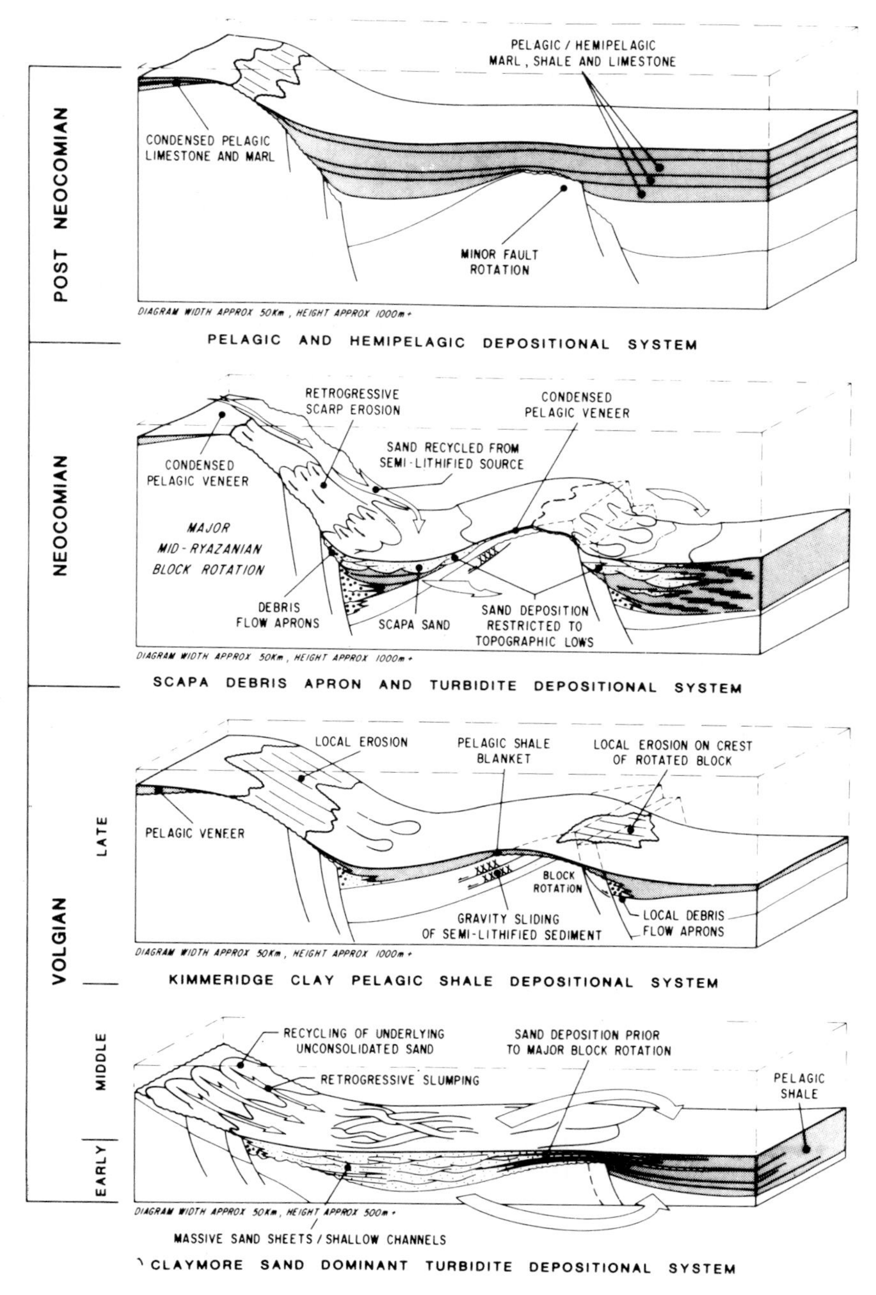

Figure 10.31 Summary diagrams for the geological evolution of the Late Jurassic to Early Cretaceous deep-marine systems of the Witch Ground Graben, northern North Sea. These models are also applicable to the contemporaneous development of the Viking Graben. After Boote & Gustav (1987).

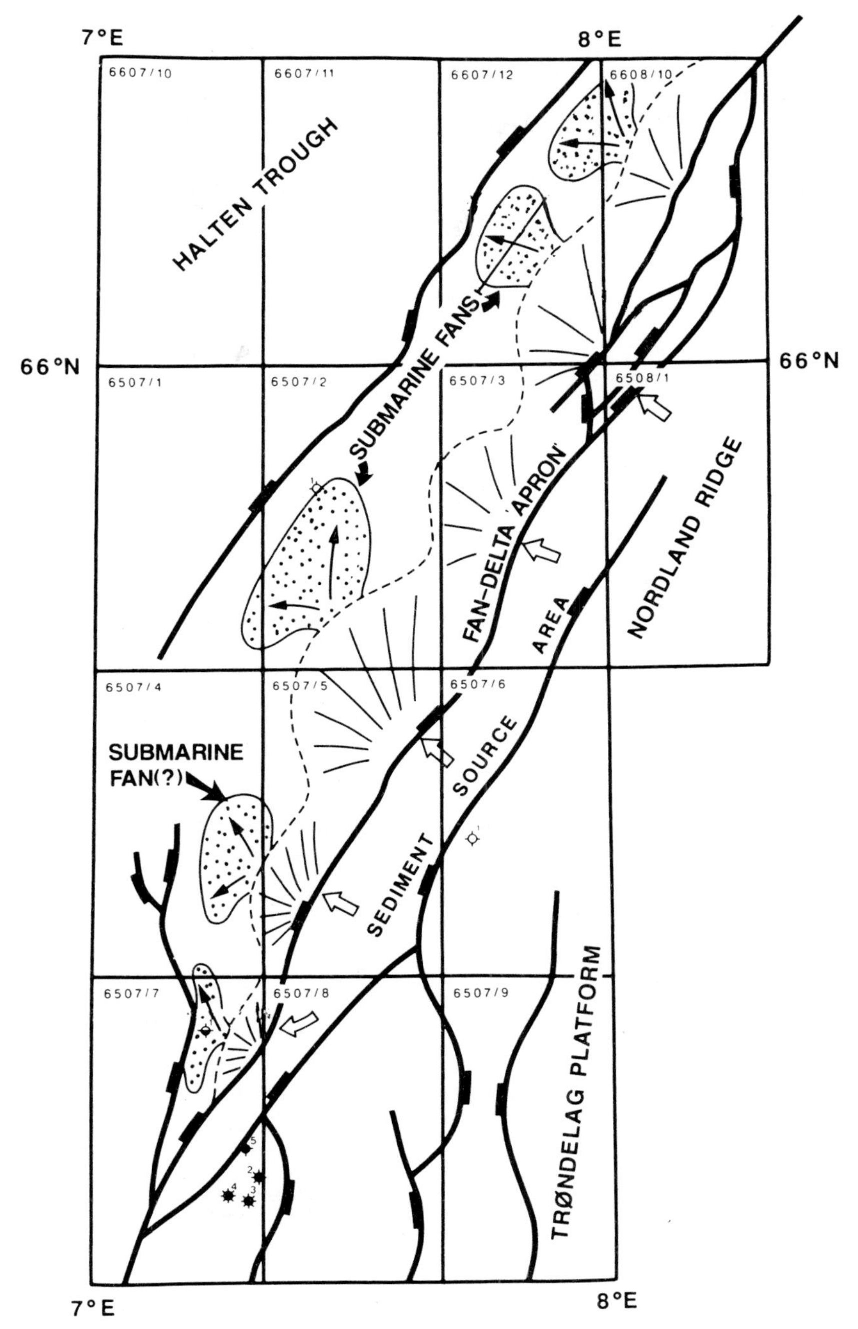

Figure 10.32 Schematic depositional model for the Lower Cretaceous from the northeastern part of the northern North Sea. The sediments of the deep-marine turbidite systems were derived from the erosion of the Nordland Ridge to the east. Compare with models shown in Figures 10.27b and 10.31. After Hastings (1987).

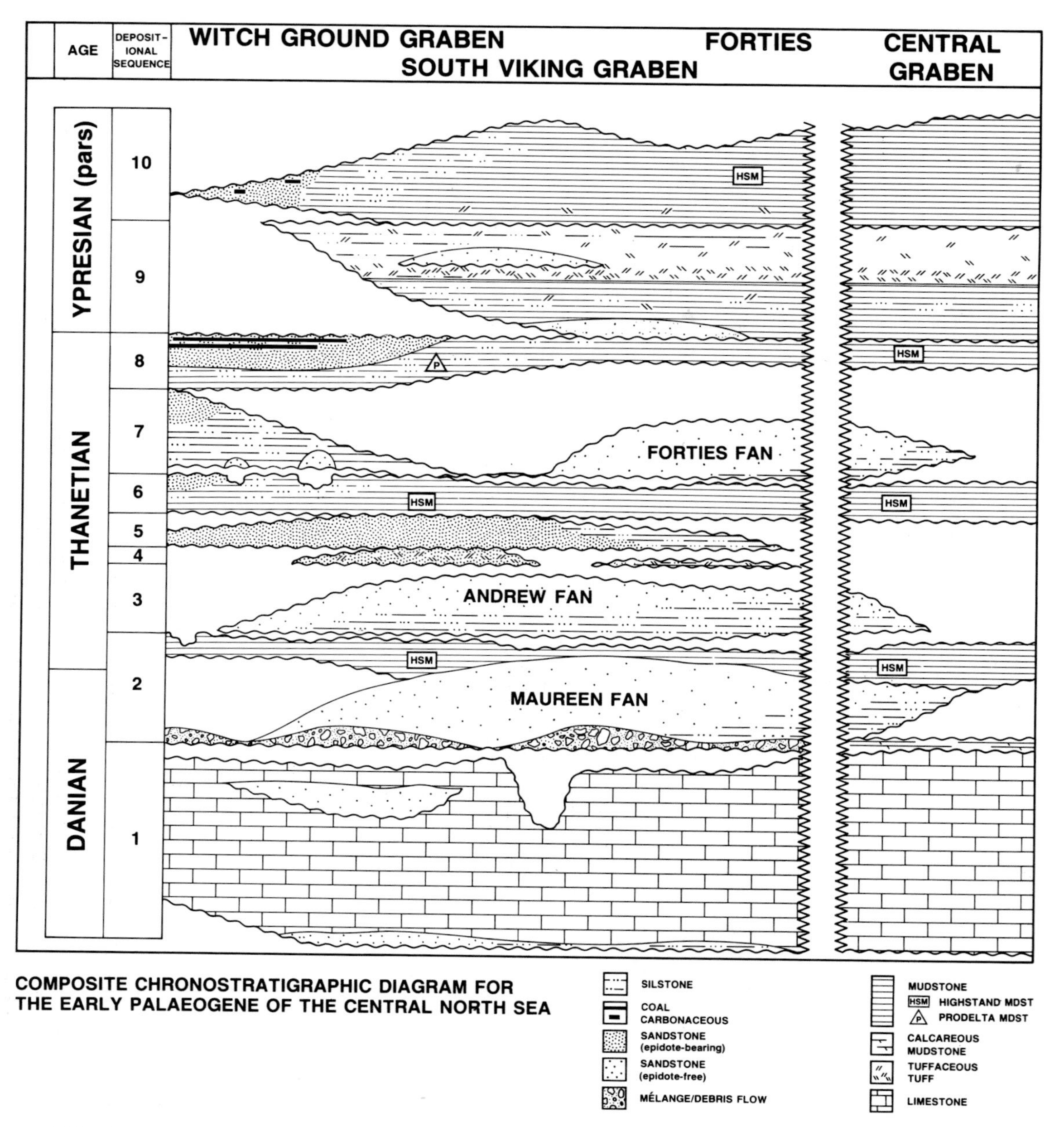

Figure 10.33 Composite chronostratigraphic diagram for the Early Palaeogene of the central North Sea. After Stewart (1987). See Figure 10.34 for coastal onlap curves.

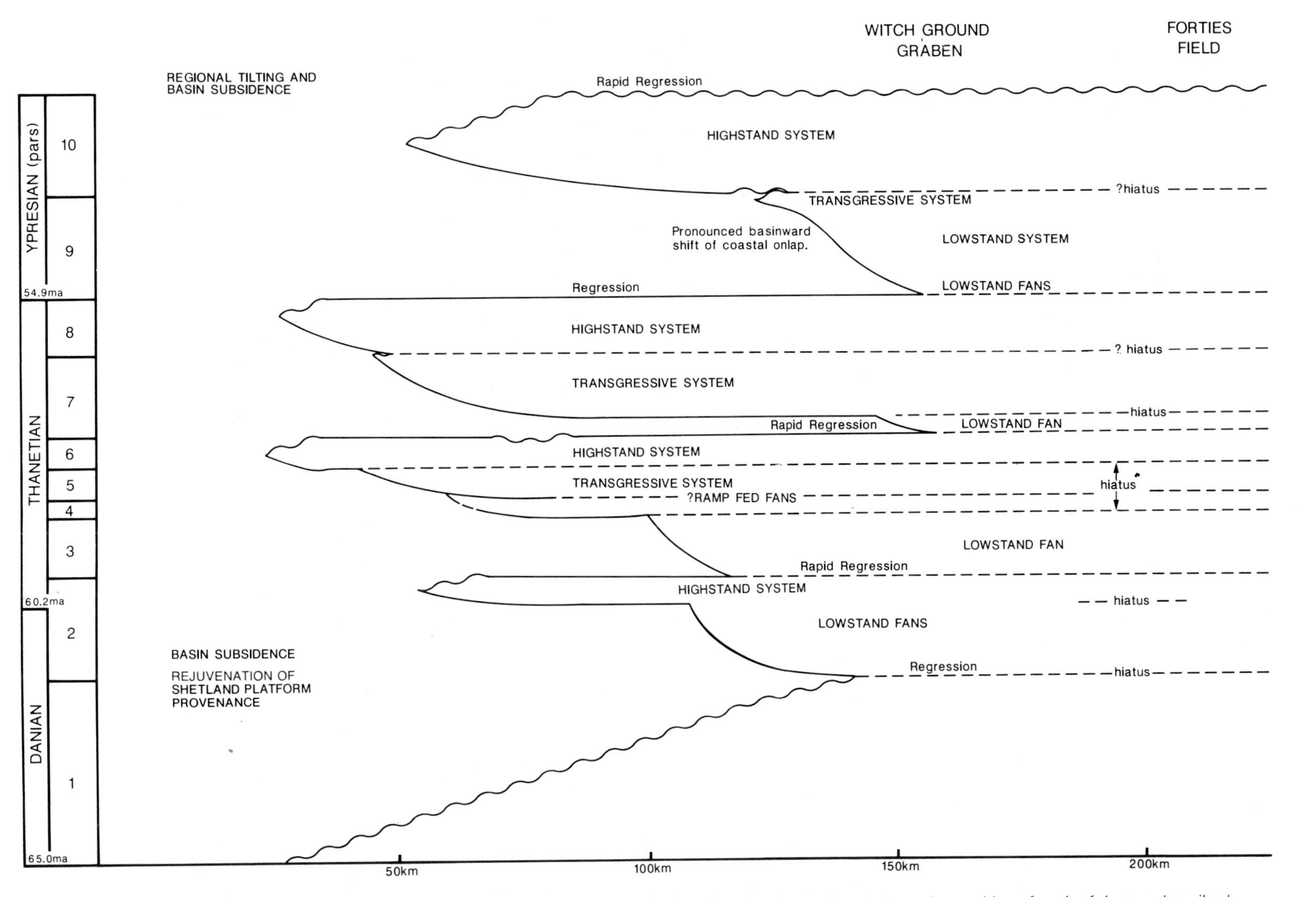

Figure 10.34 Coastal onlap curve for the Early Palaeogene of the Central North Sea. The diagram shows the relative onlap position of each of the ten described depositional sequences and the depositional systems operating within each. The terms highstand and lowstand refer only to the relative position of sea level. After Stewart (1987). See Figure 10.33 for stratigraphy of ten depositional sequences labelled 1 to 10.

Road section in uppermost part of the Caradoc-Ashgill Cloridorme Formation, Quebec Appalachians, Canada, showing thrust imbricate stack (duplex structure) of sandy (middle fan?) turbidites. Note human scale. These beds were deposited in a deep-marine foreland basin (Hiscott *et al.* 1986).

CHAPTER 11

Active Convergent Margins

Introduction

This chapter considers forearcs, backarc/marginal basins, and foreland basins, with reference to deep-marine sedimentation patterns. The modern case studies are chosen to reflect the complex spatial association of such basins associated with active convergent margins. A more wide-ranging treatment of active margins may be found in books edited by Burk & Drake (1974), Talwani & Pitman (1977), Watkins *et al.* (1979), Leggett (1982), Kokelaar & Howells (1984), Nasu *et al.* (1985), Allen & Homewood (1986), Coward & Ries (1986), and Moore (1986).

Zones of plate convergence tend to be dominated by thrust belt tectonics (cf. Boyer & Elliott 1982, Ellis 1988) and sedimentation, together with magmatic arc activity. Where plate convergence is long-lived, orthogonal or oblique, and there is substantial terrigenous sediment supply, then very wide accretionary imbricate systems may form with widths up to several hundred kilometres. Examples include the Lesser Antilles and Makran accretionary prisms. Active convergent margins can involve: (a) subduction–accretion, with the consumption of oceanic crust to generate arc volcanism, or (b) continental (or arc) collision–underplating, typically associated with foreland basin thrust systems where neither subduction nor arc volcanism occurs. Examples include the present-day Timor Trough and Taiwan Foreland Basin.

There are many examples of basins at convergent margins that are not clearly forearc, backarc/marginal, or foreland basins. For example, the deep eastern Mediterranean, characterized by basin-plain sheet sedimentation (see Ch. 8), is still tectonically active even though typical subduction processes probably ceased about 5 Ma with the disappearance of all oceanic crust between Turkey and Africa. Plate convergence continues with only limited under-thrusting of Africa along the Cypriot Arc, but there are regional deformation zones within a possible 300 km- wide band stretching from Herodotus Abyssal Plain to the east along the northern edge of the African and Arabian Plates (Woodside 1977). Also, halokinesis to form salt pillars and anticlines is an active process in Herodotus Basin, and the north–south alignment of many diapirs, oblique to the regional trend of the Mediterranean Ridge and the edge of the Nile Cone, suggests basement control, with possible continuation of the north–south structural grain of the North African crystalline basement (Woodside 1977). Thus, while essentially horizontal basin-plain or abyssal-plain sedimentation is occurring, complex compressional tectonics and halokinesis are controlling the stratigraphy.

Modern subduction-related systems: forearcs

Wherever plates converge towards subduction zones, a range of deep-marine sedimentary basins may form between the trench and the volcanic arc (Fig. 11.1). Deep-marine forearc basins range from relatively small accretionary-prism slope basins, to large forearc basins such as the Barbados and Tobago Basins in the Lesser Antilles forearc, where dimensions are larger than 100 km, and sediment infills thicker than 2000 m and 4000 m respectively. Dickinson & Seely (1979) define the following basins, all of which generally contain substantial thicknesses and volumes of deep-marine sediments: (a) intra-massif basins within and upon basement terranes of the arc massif; (b) residual basins on oceanic or transitional crust between the arc massif and the site of initial subduction; (c) accretionary-prism basins upon the accreted elements of the developing subduction–accretion complex, and (d) basins constructed both upon the arc massif and accreted subduction complex. Basins may also develop that are intermediate between types (a) to (d).

Seely (1979) recognized the importance of irregular plate boundaries, with promontories and re-entrants in continental crust, as a major control on the type of forearc basin that develops together with the associated basement. However, the volume of forearc sediments, and therefore the size of an accretionary prism will be influenced largely by factors such as the magnitude of river drainage into the forearc region and the longevity of subduction–accretion processes.

Theoretical models have been proposed for the development of forearcs by frontal accretion and underplating, by taking account of the internal deformation in accretionary prisms, to explain convergent–extensional wedges (e.g. parts of the Middle America Trench margin) versus convergent–compressional wedges (e.g. parts of the Nankai Trench prism). For example Platt's (1986) 'critical taper model' suggests that if accretionary prisms are disequilibrated by elongation (frontal accretion), then they will tend to an equilibrium geometry and taper by internal folding and thrusting. In contrast, over-shortened prisms will tend to equilibrium by elongation mainly through extensional normal faulting. In the latter case, we may expect thicker successions of undeformed sediments to develop upon the accretionary prism, essentially controlled by extensional, normal, growth faulting at shallow depths. Accretionary prisms may grow both by the addition of sediment at the toe of a prism or from 'underplating'. Underplating is used to describe two different scales of accretion from below a thrust system at active convergent margins: (a) crustal thickening from continental and/or anomalous crust, including arcs, being accreted, and (b) the addition of oceanic plate sediments from a subducting slab, by being plastered onto the underside of an accretionary prism at depth.

In the following sections we consider the tectonics and sedimentation that occur in the main elements of a forearc, i.e. trenches (and the preservation potential of a trench stratigraphy), accretionary prism-top (slope) basins and larger forearc basins. An idealized summary model for forearc sedimentation is presented. There is a section dealing with wet-sediment injection structures in forearcs. There is also a section on backarc/marginal basins before the modern and ancient case studies are considered.

Trench sedimentation

Deep-marine trench sediments may be derived from three separate source areas: (a) oceanic-plate sediments passively conveyed by plate motions into a trench during subduction; (b) lateral sediment input from the forearc, and (c) axial transport of sediments along the trench, possibly from distant areas. Flexure and associated extensional faulting of the subducting oceanic plate will lead to low-angle unconformities, with onlap between oceanic plate sediments and younger strata.

Oceanic-plate sediments may occur as a thin veneer of pelagites, hemipelagites and fine-grained turbidites (Facies Classes D, E & G) above oceanic basaltic basement, associated with Facies Group G3 chemogenic sediments such as cherts and manganiferous sediments, on what has been called the 'pelagic plate' (Schweller & Kulm 1978). Examples of such fine-grained successions, spanning geologically long time intervals, occur on the Pacific and Cocos Plates where sediment accumulation rates are slow, typically from 2–5 mm/1000 years (or 2–5 m/m.y.), and below the carbonate compensation depth (CCD) (Schweller & Kulm 1978). In contrast, in the Indian Ocean, the oceanic plate is buried beneath a thick succession of Bengal Fan sediments that are being

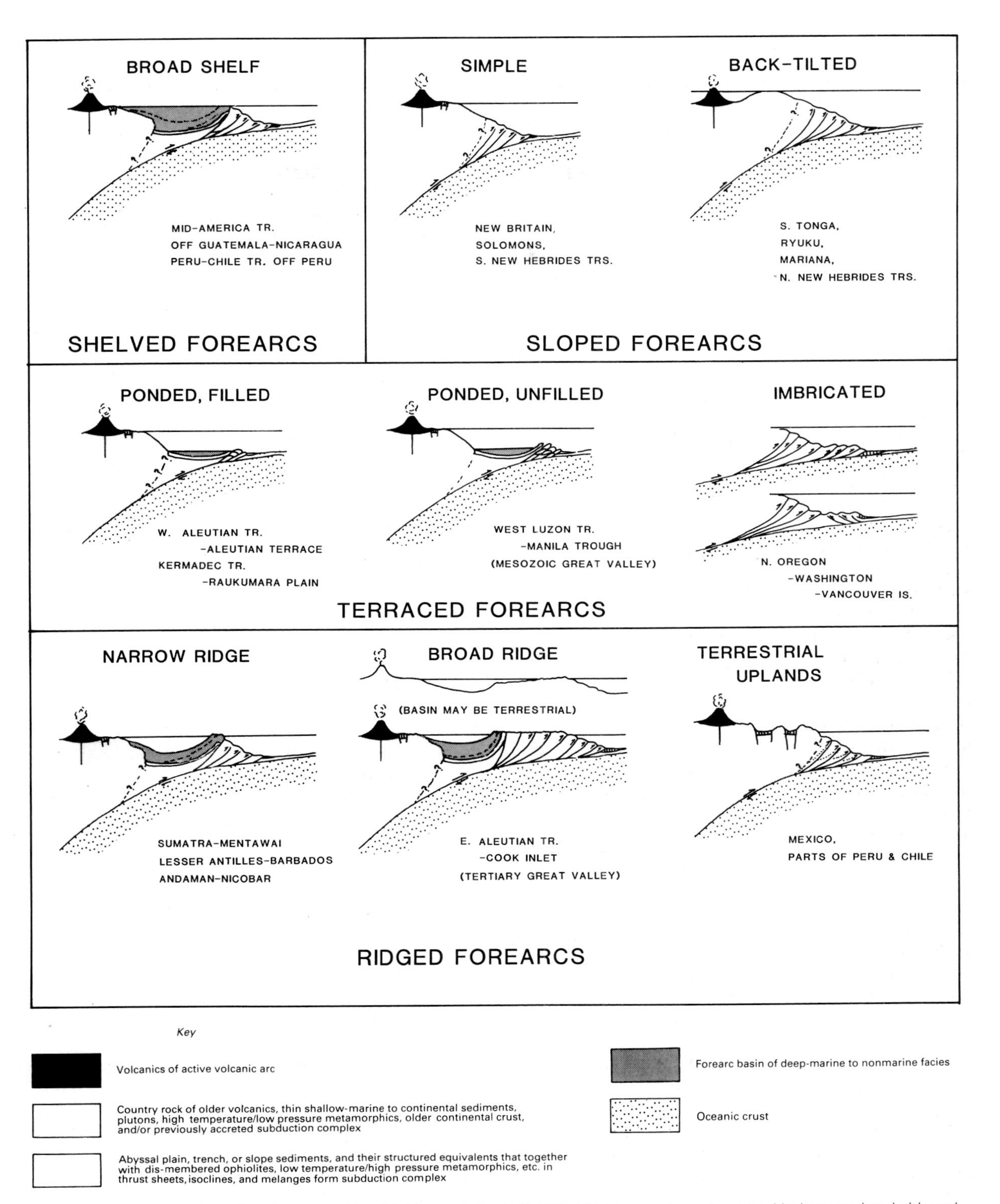

Figure 11.1 Models of modern forearcs. After Dickinson & Seely (1979). NB. arc may be submerged below sea level although shown here as emergent.

conveyed obliquely into the Java Trench. In the north, the Indian Plate is covered by up to several thousand metres of terrigenous turbidites (Curray & Moore 1974, Curray *et al.* 1979, Karig *et al.* 1979, Moore *et al.* 1982). An essentially comparable, thick, clastic succession fed northwards from the South American continental margin onto the westward-subducting Atlantic Plate is at least several hundred metres thick over parts of the Lesser Antilles Trench (Westbrook 1982, Biju-Duval, Moore *et al.* 1984, Brown & Westbrook 1987, Moore, Mascle *et al.* 1987, Mascle, Moore *et al.* 1988). Such thick marine clastic successions on some subducting oceanic plates lead us to prefer the general term 'oceanic plate sediments' rather than 'pelagic plate sediments', because in many cases the sediments are mainly terrigenous turbidites, not pelagites.

Lateral sediment input from the forearc may produce several contrasting trench-floor deposits: (a) a blocky sea floor, due to slides (Facies Group F2) and debris flow deposits (Facies Group A1), either locally-derived from the lower slope and inner trench-slope, or from more distant sources such as the upper slope, a forearc ridge and/or shallow-marine shelf areas; (b) relatively coarse-grained sediments (Facies Classes A, B & C) fed from submarine canyons, channels and gullies to construct trench fans as channel–levee–overbank systems or sheet systems, and (c) smooth-surfaced, relatively fine-grained deposits (Facies Classes D, E & G), for example those supplied by turbidity currents flowing through slope channels, and those deposited from nepheloid layers (see Ch. 2). These latter fine-grained sediments infill and/or mantle uneven topography and tend towards sheet systems with onlap of peripheral slopes.

Large submarine canyons may funnel coarse terrigenous material directly to the trench floor, effectively bypassing the lower slope; examples of such canyons include those documented from the Middle America Trench by Underwood & Bachman (1982). These authors also describe, from the Middle America Trench, tectonic ridges along the trench slope that block and/or deflect smaller canyons so that coarse sediments are trapped within up-slope basins. In the Central Aleutian Trench, not associated with well developed canyons, Holocene volcaniclastic sand-layers cored in the trench suggest that unconfined, non-channelized, turbidity currents had sufficient velocities to climb the trench-slope break and transport sediment across the lower trench-slope into the trench (Underwood 1986). Thus, although canyons provide a means of effective, efficient, slope bypass for coarse sediments, unconfined flows may also lead to forearc bypassing.

Active sediment input from the subducting plate, rather than passive conveyance of open-ocean sediments to the trench, includes seamount-margin sedimentation (slides, debris flows, turbidity currents etc.). Such sediments may include carbonate reef clasts, volcanic material, and mixed shallow- and deep-marine faunas. The nature of any preserved carbonate may be strongly influenced by the position of the CCD.

Continent–continent, arc–continent and arc–arc collision will result in trench sedimentation that may comprise clastic/volcaniclastic material from opposing margins in varying proportions. Such laterally-supplied sediments may then be funnelled axially along the trench. Collision-related deep-marine trenches, or foreland basins, are considered later in this chapter.

Figure 11.2 summarizes the eight principal types of trench fill that we recognize, together with their most characteristic facies classes. The eight types are not mutually exclusive, but serve to emphasize the range of trench fills. Indeed, one trench may contain all eight types at various places along the trench axis. The eight end-member types are: (a) starved trench; (b) sheet system; (c) trench fan; (d) chaotic lateral infill from forearc; (e) axial trench channel; (f) seamount-choked; (g) subducting oceanic-plate thick clastic system, and (h) lateral influx of clastics/volcaniclastics from two opposing continental and/or arc margins in a deep-marine foreland basin (Fig. 11.2).

Axial transport of sediments in trenches may deposit any facies class, although the most common classes are A, B, C, D and E (Table 3.1). Axial channels will tend to develop in topographic lows along the inner trench wall in starved trenches, but on the outer side of a trench if lateral sediment input from the forearc is fast and volumetrically large. Examples of axial channels include the northward axial drainage in the Chile Trench (Thornburg & Kulm 1987), and the westward drainage from the triple junction of the Pacific, Philippine and Eurasian Plates (mainly from the Fuji River), along the Nankai Trench (Shimamura 1986, Taira & Nitsuma 1986, Le Pichon *et al.* 1987a & b). Any substantial lateral influx of sediments, such as a trench fan, will tend to divert an axial channel toward the outer (oceanward) floor of the trench – as observed in parts of the Chile Trench by Thornburg & Kulm (1987), and in the Nankai Trench opposite the Tenryu Fan (Fig. 11.3, Le Pichon *et al.* 1987b).

Facies Classes A to G occur in all eight trench-fill types (above), but in varying proportions that reflect the dominant control on sedimentation. For example, starved trenches, where there is little clastic input, will tend to comprise Classes D, E and G, whereas in

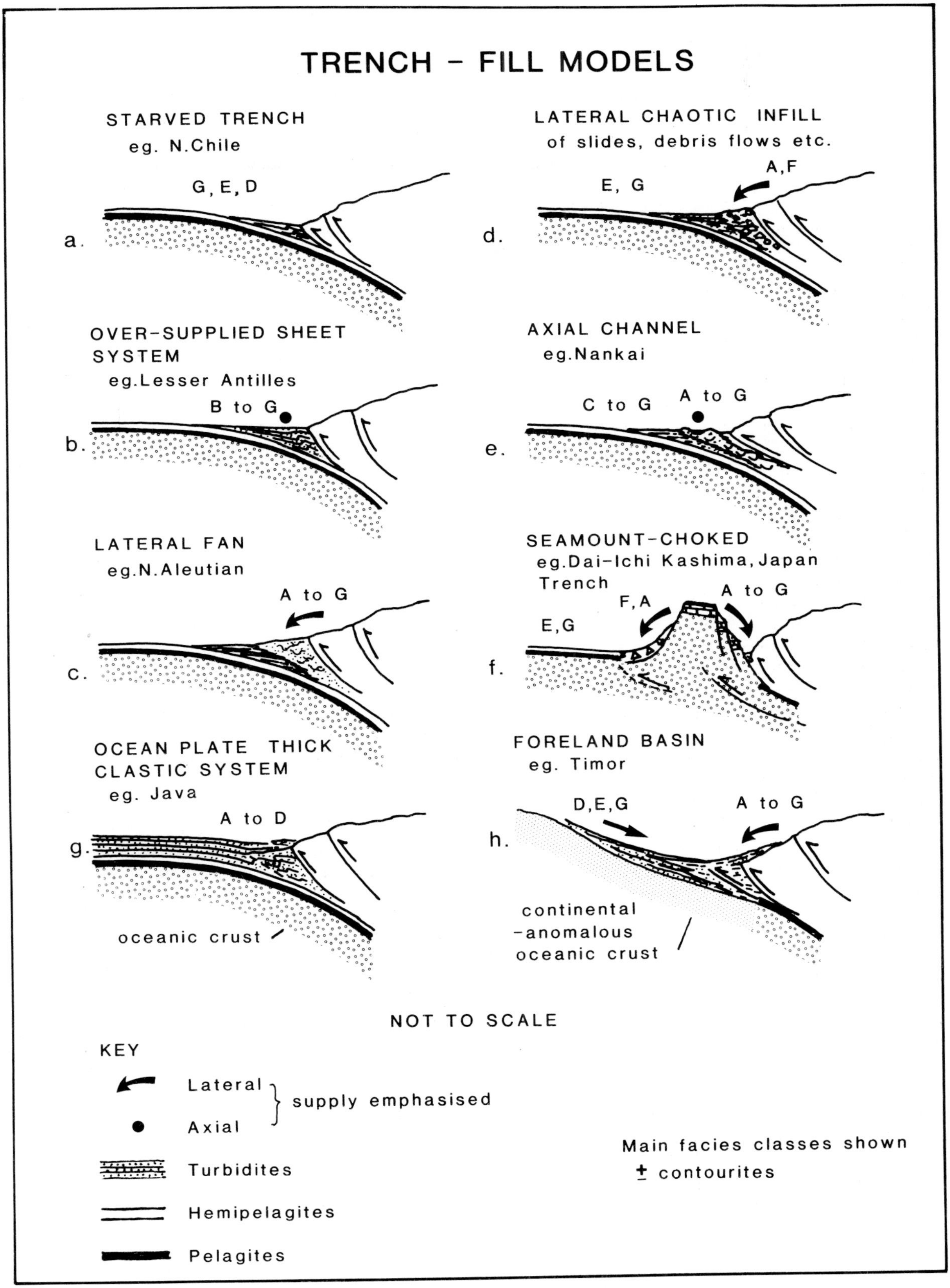

TRENCH – FILL MODELS

a. STARVED TRENCH
eg. N.Chile
G, E, D

b. OVER-SUPPLIED SHEET SYSTEM
eg.Lesser Antilles
B to G

c. LATERAL FAN
eg.N.Aleutian
A to G

d. LATERAL CHAOTIC INFILL
of slides, debris flows etc.
E, G
A, F

e. AXIAL CHANNEL
eg.Nankai
C to G
A to G

f. SEAMOUNT–CHOKED
eg.Dai–Ichi Kashima, Japan Trench
E, G
F, A
A to G

g. OCEAN PLATE THICK CLASTIC SYSTEM
eg. Java
A to D
oceanic crust

h. FORELAND BASIN
eg. Timor
D, E, G
A to G
continental
–anomalous
oceanic crust

NOT TO SCALE

KEY

Lateral } supply emphasised
Axial

Turbidites

Hemipelagites

Pelagites

Main facies classes shown
± contourites

Figure 11.2 Trench-fill models to show the main facies classes and predominant transport paths in various settings. The facies classes are listed in order of volumetric importance. See text for explanation.

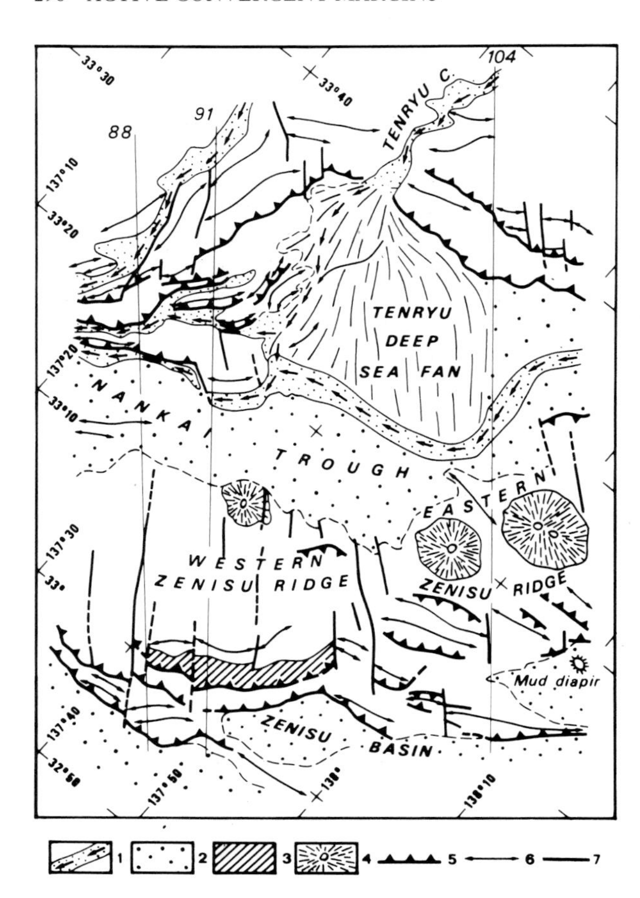

Figure 11.3 Map of the eastern Nankai Trench to show the main structural and sedimentological features, from Box 5 of the Franco–Japanese KAIKO Project and based on Seabeam bathymetry and single-channel seismic reflection profiles. 1, deep-marine channel; 2, trench fill; 3, probable outcrop of basement; 4, igneous volcano; 5, thrust; 6, fold; 7, unspecified fault.

seamount-choked trenches, near the subducting and/or accreting seamount, Classes A and F may predominate as rockfall, debris flow and slide deposits. In such cases, the finer grained matrix for any debris flows/olistostromes and slides may be derived from pelagites, hemipelagites and fine-grained turbidites that mantled the seamount.

Detailed observations by the Franco–Japanese KAIKO Project, near the triple junction of the Japan – Bonin (Ogasawara) and Nankai Trenches off southeast central Honshu, have shown the deposits of rockfall, debris flow, slide and other sediment redeposition processes on the disintegrating margins of Dai-ichi Kashima seamount entering the trench (Fig. 11.4a & b). It is possible that the seamount talus, as an olistostrome, could provide an ideal precursor to mélange during later thrusting and shearing processes as parts of the seamount are incorporated into the accretionary prism. Where siliciclastic influx is large, for example with trench fans (Fig. 11.2c & g), Facies Classes A to E will be volumetrically substantial.

Thermohaline circulation, as ephemeral to semi-permanent ocean currents, may rework trench floor, or forearc slope, sediments into contourites. Thornburg & Kulm (1987) document silt and sand laminae, winnowed from hemipelagic muds and distal turbidites, that are particularly well developed in the sediment-starved, constricted, parts of the Chile Trench where geostrophic currents accelerate between the steep inner and outer trench walls.

Axial gradients within trenches, combined with the linear basin shape, conspire to favour pronounced asymmetry in axial sedimentation patterns away from sediment-entry points such as submarine canyons. Thornburg & Kulm (1987), in a study of the Chile Trench, derive proximal to distal trench stratigraphies from 'depositional' to 'erosional' fans, to sheeted then ponded basins (Fig. 11.5). Although we would avoid defining depositional versus erosional fans, we assume that Thornburg & Kulm (1987) use these terms only to emphasize the predominant interpreted processes operating on such fans. Near sediment entry points, fast rates of sediment accumulation favour fast fan aggradation, together with relatively rapid channel avulsion/migration, and lead to an erosional fan system (Fig. 11.5a & b). More distal trench environments will contain fewer and/or more shallow channel–levee–overbank systems, and will pass into essentially ponded basins, where a sheet system is dominated by fine-grained sediments and, possibly, contourites (Fig. 11.5c & d).

Sufficiently fast sediment supply, with a relatively stable trench topography and gradients, will allow the axial progradation of clastic systems, thereby generating single or stacked coarsening-upward sequences. However, the complexities of changing paths of sediment supply, varying sea level, and tectonic–sedimentary processes that alter the shape and gradient of the trench floor and walls, will ensure a far more complicated stratigraphy.

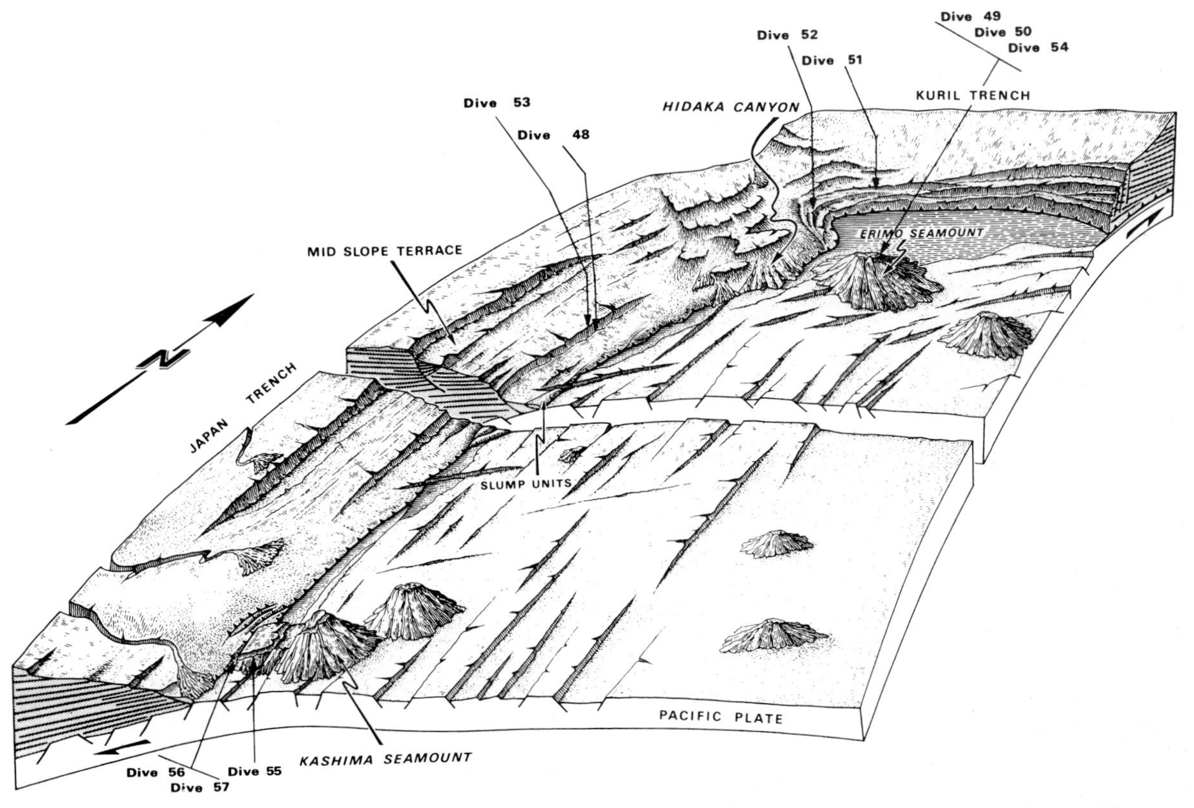

Figure 11.4a Idealized morphology of the Japan and Kuril Trenches based on Seabeam mapping, seismic profiling and submersible dives (locations shown). After Cadet *et al.* (1987).

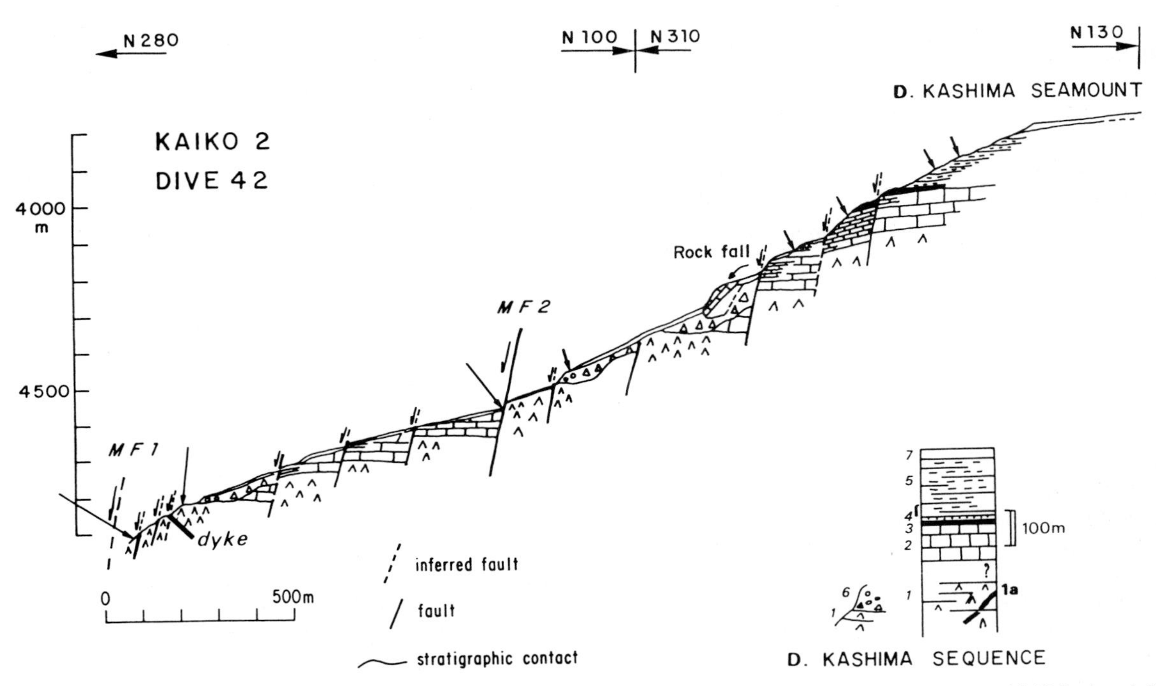

Figure 11.4b Cross-section of Dai-ichi Kashima Seamount scarp based on dive 42 of the Franco–Japanese KAIKO Project (after Pautot *et al.* 1987). MF1 and MF2 are the main normal faults. 1, basaltic lava (1a = dyke); 2, lower Cretaceous shallow-marine limestone; 3, brown marls; 4, chalk; 5, upper Miocene to lower Pliocene yellow marl; 6, sedimentary breccia; 7, Recent hemipelagic mud. Sample locations shown by arrows.

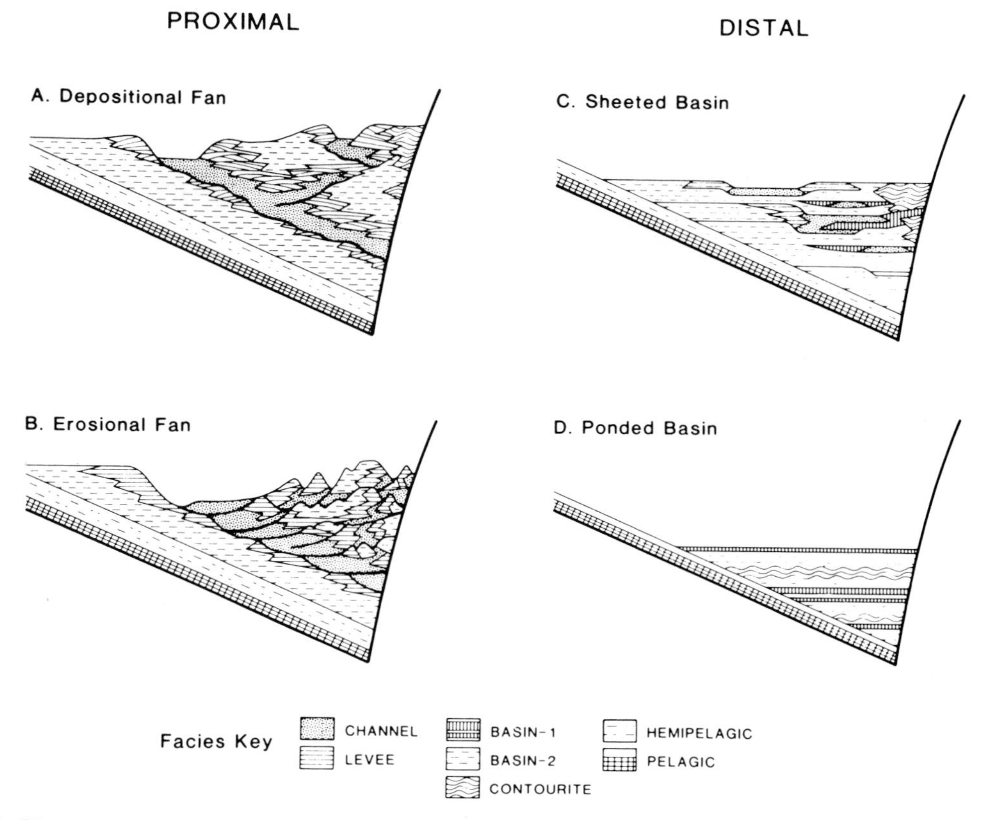

Figure 11.5 Diagrams to show the range in trench-fill stratigraphies and environments based on lithologic, morphologic and seismic data from the Chile Trench (after Thornburg & Kulm 1987). The proximal 'depositional' and 'erosional' fans to distal 'sheeted' systems develop along the axis of the trench. In some cases, submarine canyons may debouch sediments directly into sheet systems without the development of a channelized sediment-dispersal system.

Preservation of trench stratigraphy and recognition

The nature of subduction–accretion processes militates against the common preservation of relatively intact trench stratigraphies in the geological record. Instead, accretionary processes tend to severely deform the prism sediments by flattening and penetrative simple shear during thrusting and folding. However, some mechanisms of subduction–accretion provide a better chance of accreting a relatively complete trench succession, for example during ocean-ridge and seamount subduction.

Basal decollement below the trench fill, for example in oceanic basalts, will greatly increase the probability of preserving a relatively complete trench fill. In many trenches, seismic evidence suggests that the pelagites and hemipelagites immediately overlying oceanic basaltic basement tend to be subducted, whereas it is the trench turbidites and/or oceanic-plate thick clastics that are off-scraped in frontal accretion processes.

Results from the KAIKO Project, in the Nankai Trench, suggest that subduction zones may migrate oceanward in increments, controlled by the position of oceanic basement ridges (Le Pichon *et al.* 1987a & b). In the eastern Nankai Trench, the aseismic Zenisu Ridge is approaching the northward-dipping subduction zone and trench, and multi-channel seismic reflection profiles suggest that the thrust (deformation) front may have jumped to the south and oceanward of the Nankai Trench (Fig. 11.6), due to the mechanical strength of the ridge and upper oceanic crust during flexure. Should such a decollement surface develop into a new subduction zone, thereby de-activating the present trench as the deformation front, then a com-

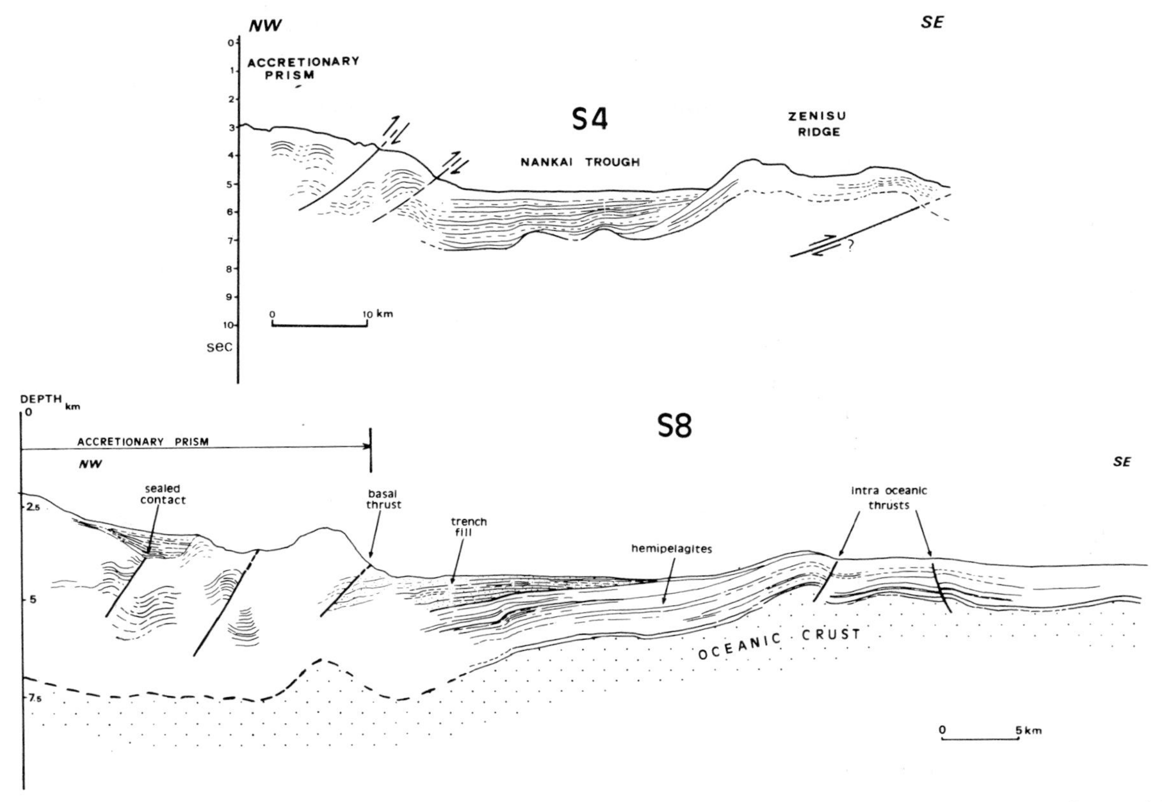

Figure 11.6 Interpreted multichannel seismic profiles S4 and S8 across the eastern part of the Nankai Trench in Box 5 of the Franco–Japanese KAIKO Project. Note the thrusting south of the present Nankai Trench, to the south of the Zenisu Ridge, where a new trench might develop. See text for explanation. After Le Pichon *et al.* (1987b).

plete trench-fill stratigraphy floored by oceanic basalts may be accreted. Furthermore, the strength of any accreted basement could shield the overlying strata from the intense internal deformation characteristic of other ocean-facing accretionary prisms.

Seamount subduction, with the off-scraping of parts of the basaltic edifice together with the sediment veneer and associated trench sediments, may also preserve an identifiable trench stratigraphy. Such successions may comprise a tectonically-sheared association, or juxtaposition, of open-ocean limestones, cherts, pelagites, hemipelagites and thin-bedded turbidites, rockfall and debris-flow deposits, olistostromes, and relatively coarse-grained, siliciclastic, trench-fill turbidites.

Thornburg & Kulm (1987) discuss the importance of the nature of trench-fill sediments in determining the location of successive deformation fronts and, therefore, the nature of stratigraphic duplication. They suggest that sediment-starved trenches will tend to decouple within the upper part of the oceanic basaltic layer (although it is hard to see why such decollement should not continuously occur only immediately above the oceanic crust in trench sediments); also, they suggest that where sediment accumulation rates are fast enough to generate thick trench successions then no, or very little, oceanic crust is accreted (cf. Thornburg & Kulm 1987, fig. 18). Where a prominent trench axial channel is developed, they suggest that successive decollement will tend to occur on the oceanward (outer trench) side of such clastics. While the modelling of Thornburg & Kulm (1987) contains many useful elements, we believe that the controls on preserving certain stratigraphic successions are far more complex and should await more detailed comparative research.

In the absence of any oceanic basalts associated with deep-marine sediments, it is probably impossible to differentiate a forearc basin, accretionary-prism slope basin, and trench fill. Apart from the depositional site, there appear to be no unique sedimentary characteristics of trench-fill siliciclastics.

Forearc basins/slope basins

Within a subduction–accretion complex, forearc basins of various sizes develop, controlled by compressional, extensional, and in some cases dominated by oblique-slip tectonics. Many forearc basins do not show typical submarine fan facies-associations and sequences, but contain essentially sheet-like deep-water deposits, for example the 20 × 30 km upper Miocene Makara Basin (van der Lingen 1982) that is divided by a narrow zone of deformed Cretaceous–Palaeogene mélange. Thus, it is difficult to make any generalizations about the deep-water stratigraphy that develops in such basins. An example of the complex interaction between tectonics and sedimentation can be appreciated by considering the small slope basins that commonly develop on accretionary prisms such as the Nias Ridge (Fig. 11.7a), with dimensions from 3 × 1 km and a few hundred metres of sediment fill, to 21 × 4 km with greater than 600 m of sediment fill (Stevens & Moore 1985) (Figs. 11.7b, c & d). Other good examples revealed on seismic profiles are documented from the Columbia forearc basin, SW Caribbean Plate (Lu & McMillen 1983), and from the Makran forearc in the Gulf of Oman, NW Indian Ocean (White & Louden 1983).

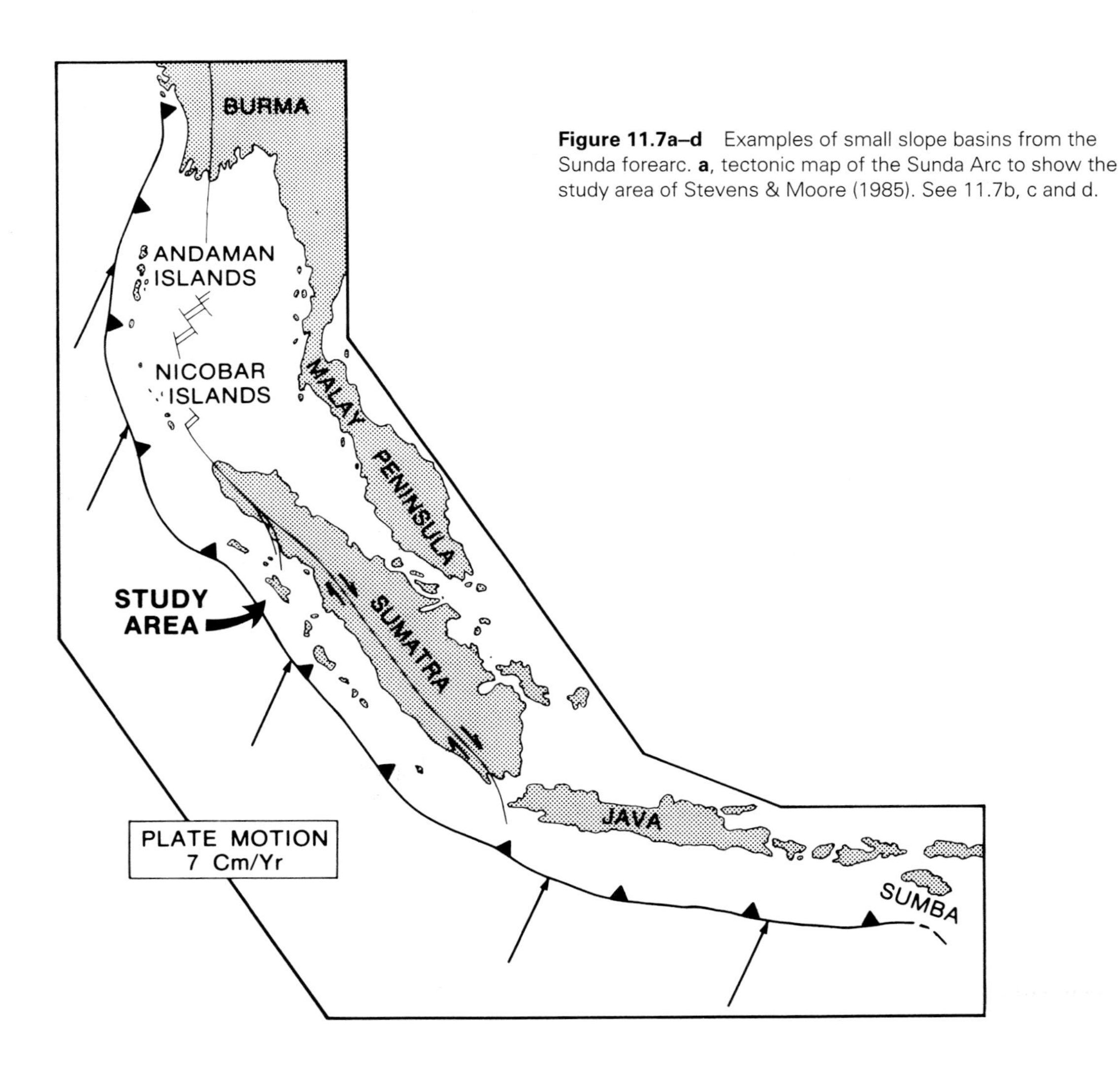

Figure 11.7a–d Examples of small slope basins from the Sunda forearc. **a**, tectonic map of the Sunda Arc to show the study area of Stevens & Moore (1985). See 11.7b, c and d.

Many of the slope basins are controlled by thrust faults, and growing folds in the deforming accretionary prism sediments, revealed as ridges at the surface. In the Nias Ridge examples, the thrusts define the arcward side of a basin, while the trenchward margins show progressive onlap by the basin fill (Fig. 11.7b & c). Within individual basins, the development of subsurface folds is contemporaneous with thrusting, and major fold hinges are parallel to oblique to the strike of the forearc slope (Fig. 11.7d). The orientation of major structural elements parallel to the strike of the forearc slope tends to favour sediment feeding along the axes of many basins, with geographically restricted lateral input (Fig. 11.7d). These slope basins, controlled by compressional tectonics, are also developed in backarc and foreland basins, wherever thrust-imbricate systems form during sedimentation.

In the Columbia Basin, the western part of the accretionary prism is smoothed and masked by the Magdalena Fan, with fast rates of sediment accumulation, whereas farther east there is near-surface complex deformation, including a mid-slope structural high and numerous thrust-top basins (Lu & McMillen 1983). Towards the coast of Makran, White & Louden (1983) document slope basins becoming progressively filled, and ridge tops being buried by sediments, thereby smoothing the forearc slope adjacent to areas with fast rates of sediment accumulation. Also, deeply-incised canyons and gullies in the Makran upper slope run southwards down the regional slope, but upon reaching exposed fold ridges they are deflected through 90° to run along the axes of the slope terraces/basins.

The clarity of tectonic processes in forearc slope-basins generally is best seen where rates of sediment accumulation are relatively slow. For example, Gnibidenko *et al.* (1985) show that the location of many submarine canyons on the forearc upper-slope of the Tonga-Kermadec Trench is strongly influenced by two fault systems, one parallel and the other transverse to the main trend of the trench. Also, horst-and-graben

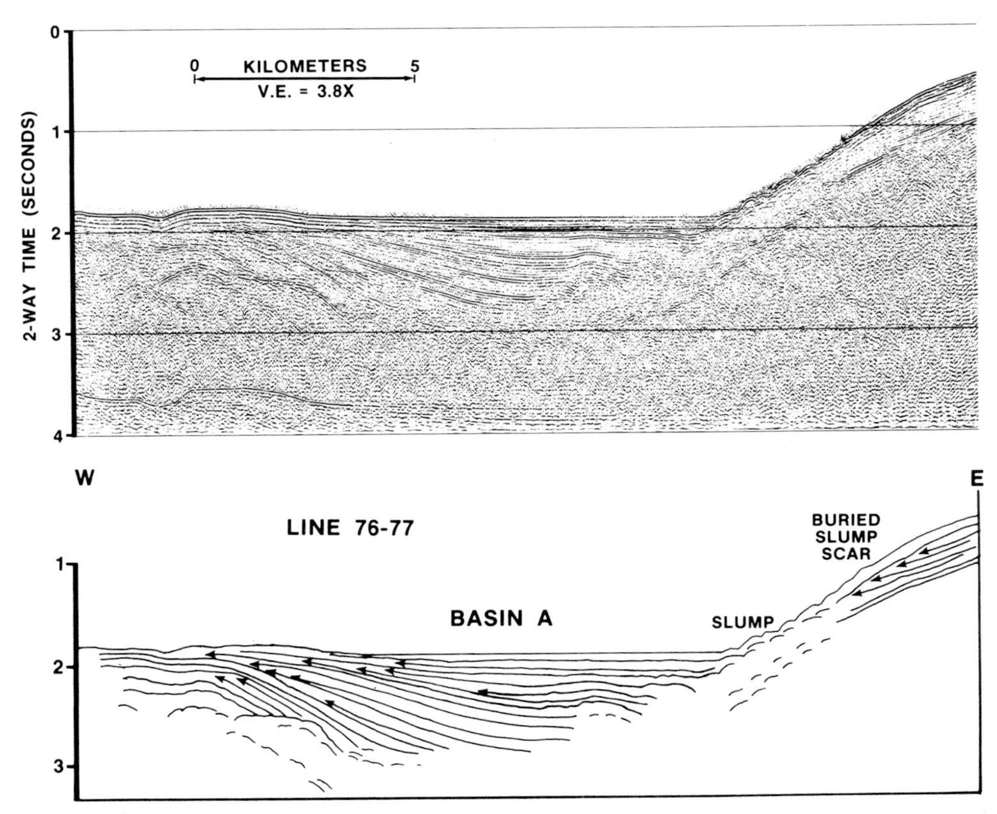

Figure 11.7b Migrated seismic reflection profile across basin, and interpretation (below) to show asymmetry of basin infill, complex onlap/offlap relationships and unconformities at the western (trenchward) side and slide-dominated eastern margin.

structures, with numerous normal faults, attest to the importance of extensional tectonics at least locally in this forearc where vertical uplift is believed to have been 5000–7000 m (Gnibidenko *et al.* 1985).

During forearc deformation, either compressional or extensional, the rate of back-tilting of the basin floor may be faster than the rate of sediment accumulation and/or slope-incision by canyons, gullies and other channels. In such circumstances, the sediment transport paths may be deflected along the back (arcward) side of these basins until conditions favour re-establishment of regional downslope flow. This process will result in linear facies development along the strike of the forearc slope, and is observed in many slope basins such as the Muroto Basin, a forearc basin off southeast Shikoku on the Nankai accretionary prism. Here, the main and tributary submarine canyons have been deflected eastwards along the back of the basin for a distance of about 50 km before turning down into the Nankai Trench (Fig. 11.8).

The tectonic telescoping of many deep-marine forearcs leads to an overall shallowing-up succession in the basins, ultimately to terrestrial sedimentation unconformably on the vestigial forearc. Such a tectonically-controlled shallowing-up will be expressed as a major coarsening-upward sequence in the forearc basin, and/or slope-basin fill. These sequences are generated because additional off-scraping and frontal accretion of trench sediments causes relatively large lower-slope basins (receiving mainly fine-grained sediments) to be translated gradually toward the arc and into shallower water depths characterized by coarser grained siliciclastics and/or carbonates (Fig. 11.9). An example of such shallowing-up sequences has been described from

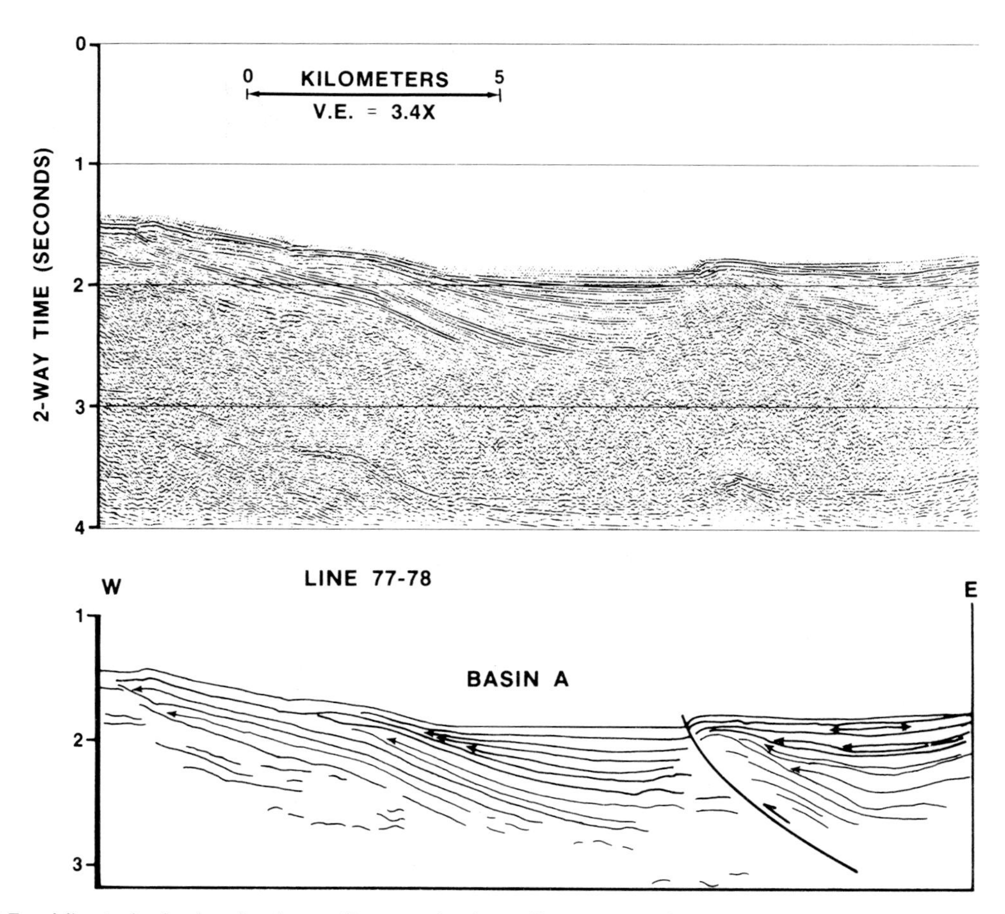

Figure 11.7c Migrated seismic reflection profile across basin, and interpretation (below), to show asymmetry of basin infill, with arcward-dipping thrust defining and controlling the ridge between basins.

the Makran accretionary prism by Platt *et al.* (1985), who interpret a mid-Miocene to Early Pliocene slope and shelf succession as having been 'deposited directly on abyssal-plain turbidites' (Panjgur turbidites) 'without any detectable stratigraphic or structural discordance' (Fig. 11.9a & b). Up to 4 km of shelf deposits (Talar shelf sandstones) pass laterally, southwards, into thin-bedded, fine-grained, deep-marine turbidites over a distance of only a few kilometres (Fig. 11.9a). It is hard to see, however, what evidence unequivocally indicates that this supra-accretionary prism basin developed upon completely undeformed abyssal-plain sediments and not simply upon a deep-marine slope-basin, turbiditic, sheet system. Thrust faulting with stratigraphic duplication is common on all scales in the Makran accretionary prism (e.g. duplex structure in thin-bedded turbidites, Fig. 11.10), but apparently does not similarly affect the Panjgur turbidites, contrary to what might be expected if an early compressional deformation had predated the main shallowing-upward sequence in the prism-top basin. For a discussion of this problem, see Platt *et al.* (1985, p. 509).

Figure 11.11 is a conceptual model to show how supra-accretionary prism basins may develop a shallowing- and coarsening-upward sequence as such a basin is translated gradually away from the trench. However, unlike the explanation favoured by Platt *et al.* (1985) for the Makran slope-basin fill (above), our model suggests that: (a) where slope basins show large-scale shallowing-upward sequences with relatively little internal deformation, it is most likely that they were initiated as slope basins and not trench-fills or open-ocean, abyssal-plain, systems; (b) although difficult to appreciate at outcrop, there may be a series

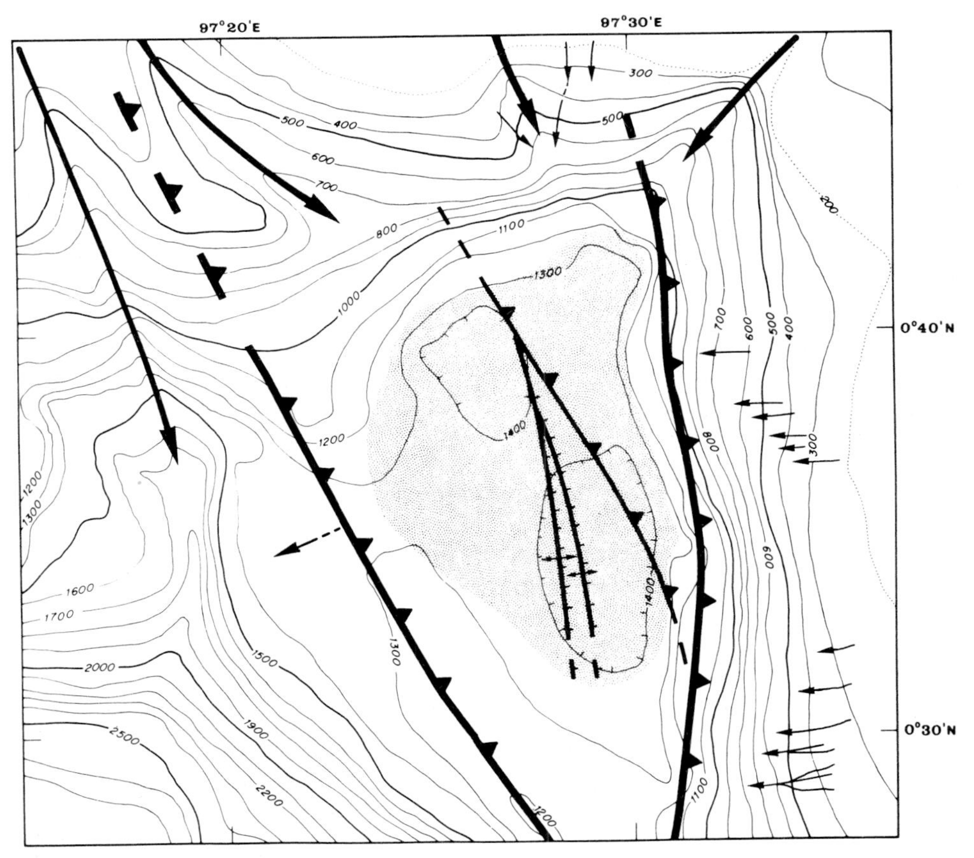

Figure 11.7d Map of basin shown in **b** and **c** to show the main structural features of the slope basin (stipple) and the sediment-transport routes. Thrusts and folds shown using standard legend. Thick arrows show principal conduits, and smaller arrows subordinate, sediment-transport paths. From Stevens & Moore (1985).

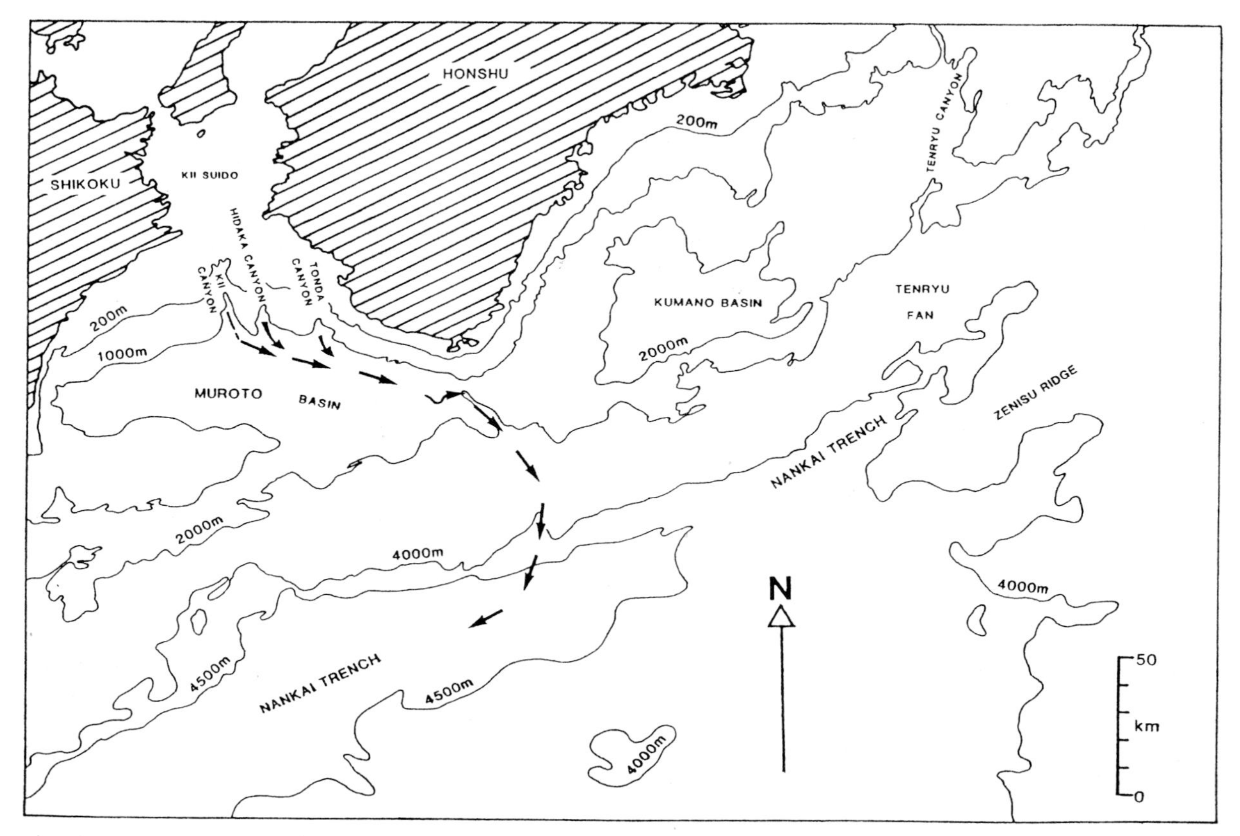

Figure 11.8 Submarine topography of the eastern Nankai Trench and surrounding area to show principal sediment-transport paths in the large Muroto forearc basin and trench. Note that the canyons supplying sediments to the Muroto Basin are deflected eastward along the arcward side of the basin, for more than 50 km, to feed into a large canyon that supplies sediment to the eastern part of the Nankai Trench. See text for explanation.

of internal disconformities and/or unconformities, possibly with a consistent overall dip (Fig. 11.11), that reflects accelerated pulses of tectonic activity, such as uplift events, and (c) the internal deformation of such a basin-fill, although gentle, should increase in intensity (and variability?) both with depth and towards the margins of the basin. Naturally, some internal deformation of the slope-basin sediments is expected, for example the margins of these basins, if fault-controlled, are susceptible to tectonic dislocation, but the overall stratigraphy has a high preservation potential, especially in thick successions. High-resolution, multi-channel, seismic reflection profiles through active accretionary prisms may provide the only means of seeing the scale of such subtle features.

In the slope landward of the Middle America Trench prism, off Guatemala, Lundberg (1982) identified a number of aproximately 1200 m-thick coarsening-upward sequences, but ascribed them to submarine fan progradation. However, such sequences are of a scale and tectonic location similar to the shallowing-up sequences described here. Clearly, if a slope-basin is translated from the lower to upper slope, but still in deep water, then any shallowing-up may be difficult to identify except on the basis of micropalaeontological evidence and, perhaps, relatively subtle overall grain-size trends. The Great Valley Sequence of California, at present subaerial, evolved from a deep-marine forearc basin that developed during the Mesozoic–Early Cenozoic (Ingersoll 1978a & b), and represents a good example of the shallowing-up (and coarsening-up) sequences that can form in forearc basins.

Large-scale coarsening-upward sequences generated simply by progradation should not produce unconformities in a forearc basin-fill. However, if the coarsening-upward sequences were formed during

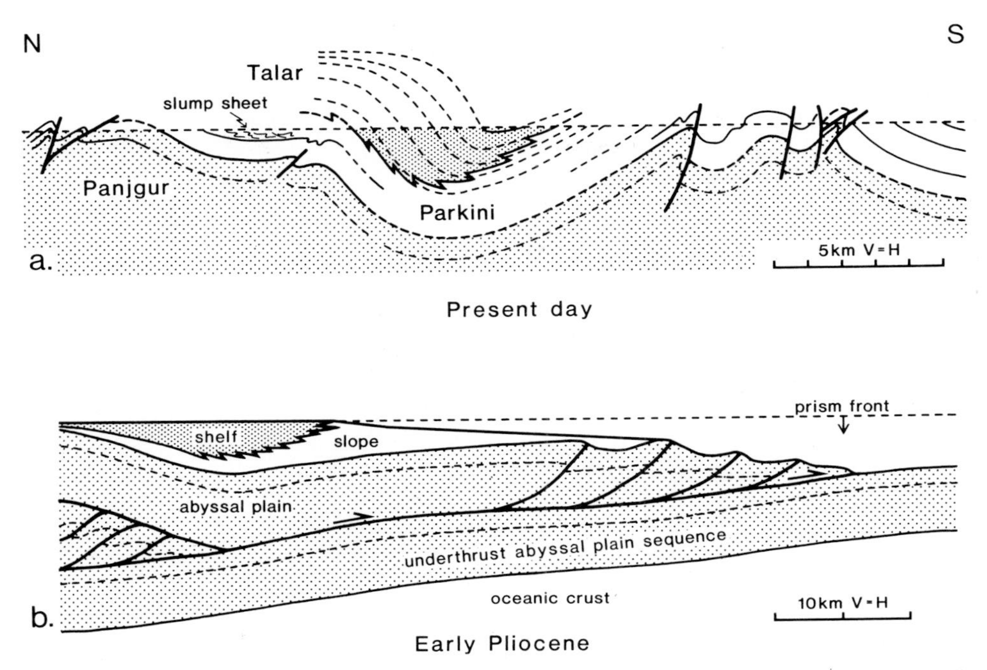

Figure 11.9 (a) Section across eastern termination of the Kulanch syncline and nearby structures. Schematic time lines are drawn to emphasize the structural and stratigraphic relationships. (b) Reconstruction of frontal part of early Pliocene accretionary prism to illustrate the idea that the shelf and slope successions were deposited upon undisturbed abyssal-plain deposits that were uplifted by duplication along major, bedding-parallel, thrusts. The structure at depth is entirely speculative. See text for discussion. After Platt *et al.* (1985).

uplift and back-tilting of forearc slope-basins, then low-angle unconformities should be common, possibly rotated in a consistent direction, and representing phases or pulses of uplift and infill (Fig. 11.11). Unfortunately, the angle of such unconformities may be so small as to escape detection in ancient successions.

Forearc summary model

Figure 11.12 is a summary of the main large-scale sedimentary features found in forearcs. Inevitably, the complex interplay between tectonics and sedimentation may make such environments very difficult to reconstruct accurately from the geological record, especially as the final geological relationships will commonly bear little, if any, similarity to the original spatial and temporal association of environments due to subduction related compression/extension, out-of-sequence thrusting and, possibly, major strike-slip. Furthermore, Figure 11.12 does not distinguish between mud- and sand-rich forearcs, where different processes and deformation styles may prevail. We would expect mud-rich forearcs to contain a greater proportion of mud diapirs, mud-ridges and mud-volcanoes. At this time, the data-base from active margins precludes a useful comparison between mud- and sand-rich forearcs.

Wet-sediment injections

The tectonics and sedimentation patterns of active convergent margins, especially forearcs and accretionary prisms, provide ideal environments in which overpressured sediments may become liquefied and fluidized to produce wet-sediment structures and injections such as mud diapirs or seepages. While all scales of sandy injections may occur, the largest diapirs appear to comprise a mud or shale matrix with incorporated blocks of lithified or semi-lithified rock. Examples of mud and shale diapirs (as pipes or ridges) and volcanoes are well documented from the Lesser Antilles accretionary prism (Higgins & Saunders 1967, Biju-Duval *et al.* 1982, Westbrook & Smith 1983, Brown & Westbrook 1987), and from the Timor Trough area

Figure 11.10 Footwall duplex structure in turbidites in the Garuk Kaur (thrust) fault zone, Makran accretionary prism. The light-coloured 30 cm-thick turbidite bed has been shortened from at least 88 m to its present 29.6 m (from a line balance), representing a 66% shortening. After Platt & Leggett (1986).

(Barber *et al.* 1986, Karig *et al.* 1987). Interpretation of a seismic profile across a mud volcano in the Barbados Basin (Fig. 11.13a), by Brown & Westbrook (1988), suggests that there are lateral, layer-parallel, injections stretching for more than 2 km from the main feeder pipe (Fig. 11.13b). In ancient successions with limited exposure, such chaotic layers, up to hundreds of metres thick, could easily be misinterpreted as debris flows, or if severely deformed as entirely 'tectonic' mélanges.

Shale diapirism may develop where sediments become over-pressured at depth (above), leading to intrusion into overlying and surrounding lithologies in a fluid state. The vertical escape velocities of the mud–water slurries may reach sufficient magnitudes to hydraulically fracture consolidated and semi-lithified lithologies, incorporate them into the slurry as blocks, and even extrude them from clastic volcanoes on the sea bed, or subaerially on islands. Islands formed from mud volcanoes are documented, for example Chatham Island, south of Trinidad in the West Indies (Adams 1908, Bower 1951, Arnold & Macready 1956, Birchwood 1965, Higgins & Saunders 1967), and from Timor (Barber *et al.* 1986). An onland eruption of a mud volcano this century occurred near Gisbourne, New Zealand (Strong 1931, Stoneley 1962).

Liquefaction of over-pressured sediments at depth and subsequent wet-sediment injection may result from: (a) seismically-induced stresses due to faulting; (b) loading or overburden stresses induced by tectonic thrust-thickening and/or fast rates of sediment accumulation, including slides; (c) hydrocarbon build-up, for example methane or gas hydrates, during the decay of organic matter within shales (Hedberg 1974); (d) escape of deep crustal or upper mantle thermogenic gases and fluids; (e) dehydration reactions during compaction and/or mineralogical transformations in

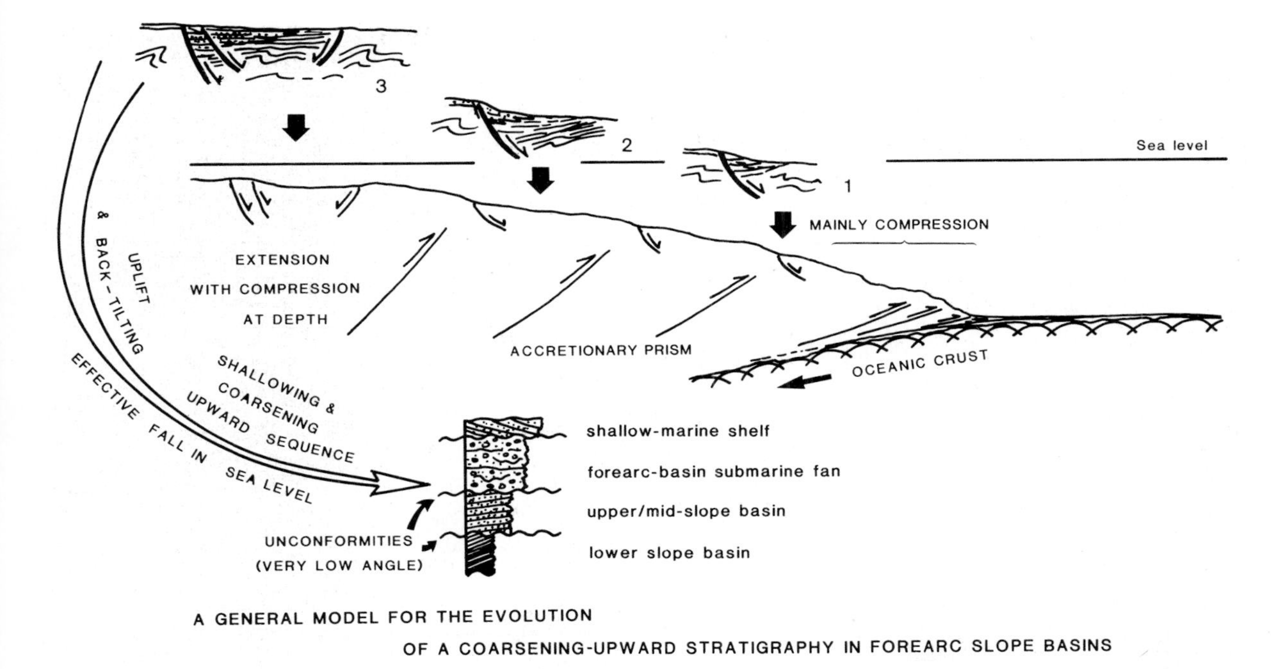

Figure 11.11 A general model for the evolution of the stratigraphy in some accretionary-prism top basins that are progressively uplifted and incorporated further back from the trench with time. Compare with Figure 11.9a & b, and see text for explanation.

clays, and (f) unstable density inversions, possibly caused by thrusting relatively dense lithologies, such as off-scraped oceanic basalts, over less dense, water-saturated muddy sediments.

Following shale diapirism, the reduced pore pressure at depth causes a volume decrease in the 'donor horizon', with subsidence peripheral to the intrusion, as seen on seismic profiles (Fig. 11.13a from Biju-Duval *et al.* 1982). Although such intrusions will tend to inject vertically or subvertically, the easiest local escape paths may be horizontal to produce a complex wet-sediment sill and dyke system. In all cases, any obliquely-aligned bedding, layering or foliation immediately adjacent to an intrusion will tend to be deformed into alignment with the margins. Vertical injection through horizontal strata may upturn bedding against a dyke, a feature commonly observed in field exposures.

Mud diapirs may occur as near-circular pipes or as sheets with more irregular geometries. They may: (a) extrude as mud volcanoes and ridges; (b) appear related to fault zones and joint patterns, or (c) occur in an apparently random distribution. Wet-sediment extrusion on the sea floor may locally change sediment-transport paths and provide a local, exotic, source of sediment whose petrography, faunal content and age, are anomalous with respect to the surrounding sediment on the sea floor.

In the Sunda Arc accretionary prism, south of Sumba, Indonesia, Breen *et al.* (1986) used SeaMARC II side-scan sonar mapping to define three distinct structural styles within 15–25 km of the thrust front: (a) lower-slope folds and thrusts; (b) mud volcanoes, and mud ridges paralleling the thrust front, and (c) conjugate strike-slip faults 10–15 km from the thrust front. The mud injections occur in water depths greater than 4000 m where sea-floor gradients range from 2.5°–5.3°. The mud ridges are 200–300 m high, 1–2 km wide, 5–18 km long, and trend east to west slightly oblique to the thrust front. The mud volcanoes have similar relief above the sea floor but are more symmetrical. Reflections from the mud injections suggest a blocky, rubble-

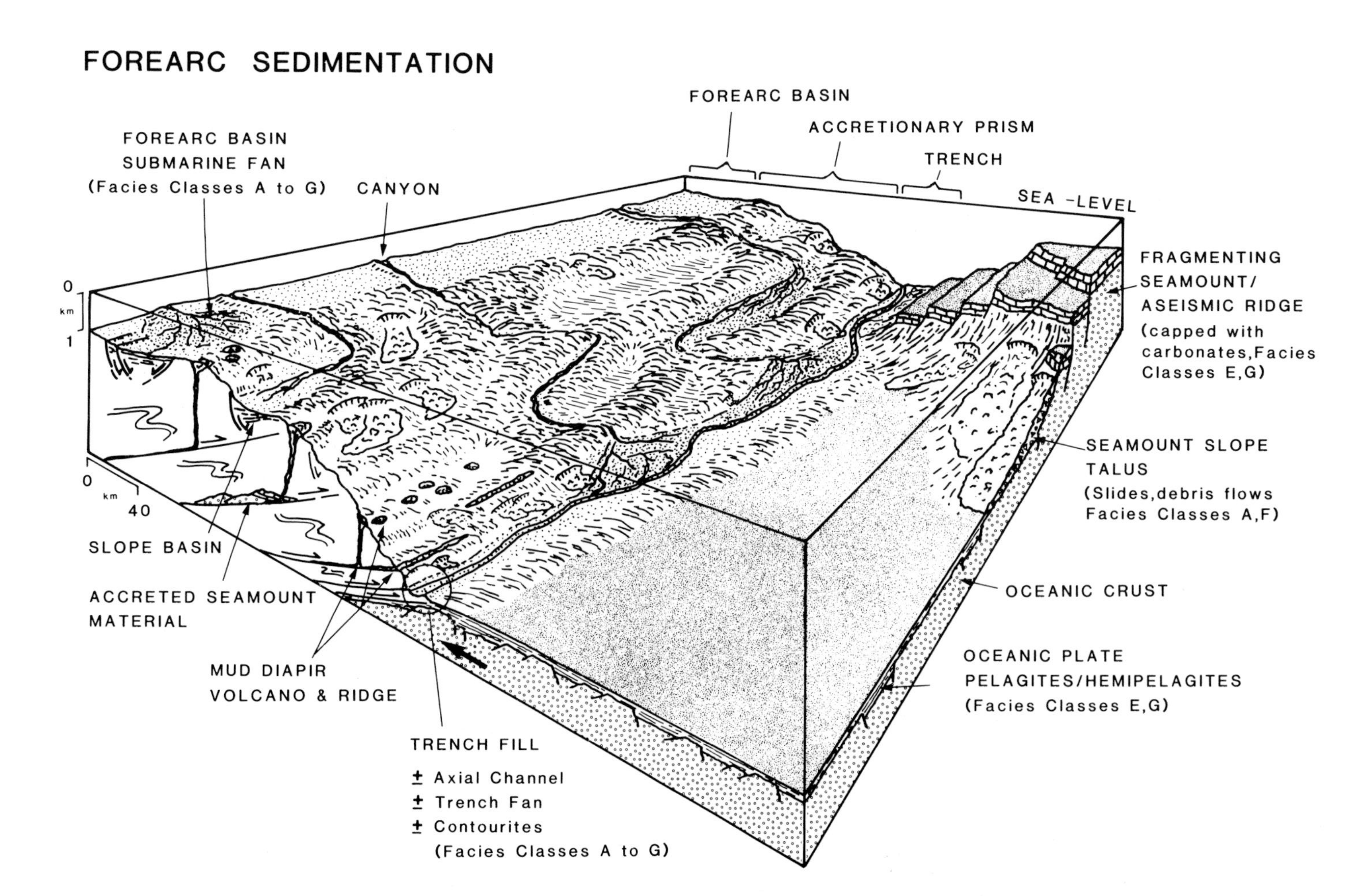

Figure 11.12 A general model for the sedimentary features of many forearcs, particularly accretionary prisms, and subducting oceanic plates. See text for explanation.

strewn relief. Deformation of the wet sediments in the lower slope occurs by folding and thrusting, wrench-faulting and mud intrusion/extrusion. This deformation has altered sediment-transport paths such that submarine feeder channels are sinuous, branching, and locally parallel or oblique to the strike of the slope.

One of the principal objectives of ODP Leg 110 (Mascle, Moore *et al.* 1988), in the northern Barbados Ridge, was to investigate the role of pore fluids in an accretionary prism. Pore-fluid data from the various holes in an east–west transect (along approximately 15°32′ N Latitude), with Site 674 about 17 km west of

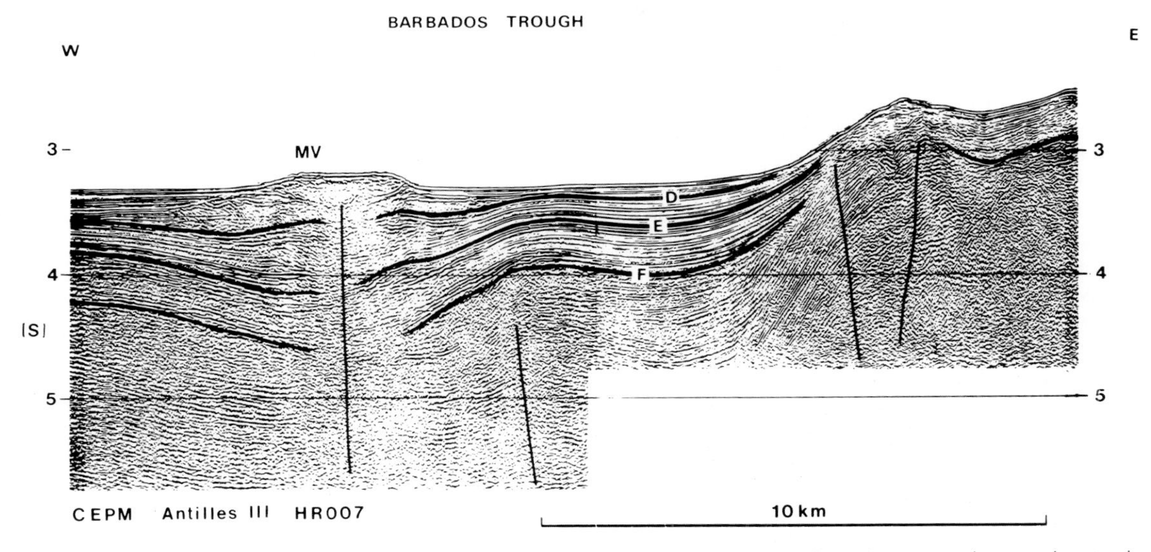

Figure 11.13a Migrated seismic reflection profile from the eastern side of the Barbados Trough, an accretionary-prism top basin in the Lesser Antilles to show a mud diapir. The diapir has penetrated the sea floor to form an 80 m high and 2 km wide mud volcano (MV) that has also intruded laterally into the sediments (chaotic reflectors). Uplift of the margin of the basin is shown by the thinning and upwarping of sedimentary horizons D, E and F. From DSDP Leg 78, Biju-Duval *et al.* (1984). Profile is located on Figure 11.19.

Figure 11.13b Interpretation of the seismic line in (a) by Brown & Westbrook (1988) to show the shape of the mud diapir and volcano; note the presence of layer-parallel injections more than 2 km from the main feeder pipe.

the deformation front, the maximum distance from this front, revealed two distinct fluid realms (Fig. 11.14). The methane-bearing region is restricted to the decollement zone and subjacent underthrust sediment, whereas the methane-free region, with chloride anomalies occurs in the accretionary prism (Fig. 11.14). Methane content decreases below the decollement zone but increases again towards the permeable sand layer at the base of the hole in Site 671. Similar anomalies are observed at Sites 676 and 674. Preliminary C isotope data on methane samples from Leg 110 (Mascle, Moore *et al.* 1988) suggests that it has a thermogenic origin in physio-chemical conditions not found in the drilled holes. Negative chloride anomalies occur along faults, for example the chloride minimum at the decollement zone in Site 671 (Fig. 11.14). Chloride anomalies are associated with three out of five faults in Site 674 (Fig. 11.14). The chloride anomalies may occur because of: (a) ultrafiltration processes in the buried clays that act as semipermeable membranes to facilitate the flow of water but inhibit the migration of various ions; (b) dehydration of smectites at depth, or (c) melting of gas hydrates as documented from the Middle America Trench by Gieskes *et al.* (1984). Methane is much more sensitive than the chlorides in indicating fluid flow. These important results from ODP Leg 110 demonstrate the chemical partitioning that occurs in progressively dewatering sediments due to fluid flow within subduction systems. The stress system within accretionary prisms tends to favour the development of hydrofractures that approximately parallel the major decollement rather than having steep dips. However, the occurrence of mud volcanoes shows that during times of relatively violent escape of pore fluids, subvertical hydrofractures form. Thus, pore-fluid activity, and their preferential fluid migration along fault surfaces, provide a fundamental key to understanding wet-sediment deformation within accretionary prisms. Mud diapirism as pipes or sills, seepage and mélange formation, including microstructures such as vein arrays (e.g. Knipe 1986), are all governed by pore-fluid behaviour.

It is important to appreciate that mud or shale diapirism, together with debris flow, slide and rockfall processes, may produce deposits that appear similar to one another in limited outcrops or in cores from the subsurface. Furthermore, the deposits from all these processes may be tectonically sheared to form mélanges

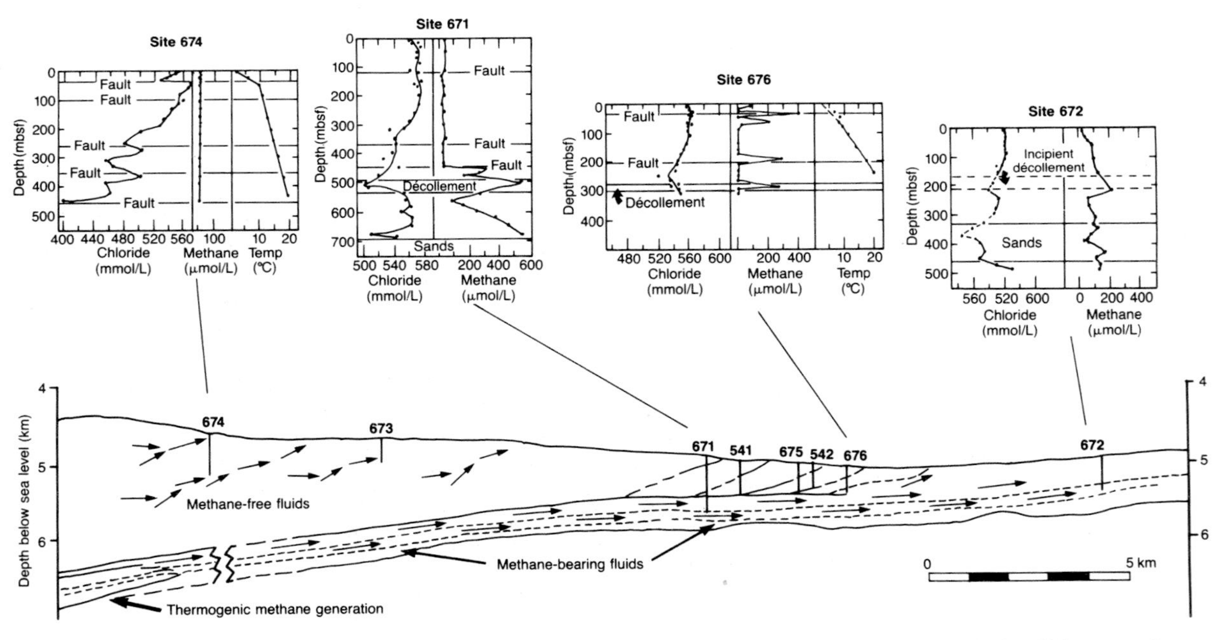

Figure 11.14 Plot of methane and chloride concentrations in pore fluids from selected sites on the Barbados Ridge accretionary prism complex, and definition of two pore-fluid realms: a methane and chloride, and a methane-free, realm. Over-pressured sediments may fail at depth and, lubricated by such fluids, can form mud injections as diapirs, seepages and volcanoes. From ODP Leg 110, Mascle, Moore *et al.* (1988).

(Cowan 1985, Pickering *et al.* 1988a). Thus, the correct interpretation for the origin of chaotic deposits with a mud matrix depends upon very accurate field mapping, detailed small-scale observations and regional considerations. In ancient successions, the deeper part of a mud or shale diapir has a high chance of being preserved whereas the surface and near-surface part probably will be eroded. A corollary of this difference in preservation potential is that while the deeper part of a diapir will tend to be preserved, it is most susceptible to severe deformation. Figure 11.15 shows a likely deformation path for the deeper part of a mud or shale diapir, where folding, flattening and shearing may produce a mélange in which highly-sheared country rock encloses lozenges of mud-shale diapir. The original diapiric relationships will only be preserved within the tectonically-formed lozenges (Fig. 11.15). Mélanges with these features, and interpreted as highly deformed diapirs, are documented from the Shimanto Belt of southern Japan (Pickering *et al.* 1988a).

Marginal/backarc basins

Marginal basins are amongst the more complex of deep-water sedimentary basins. Stratigraphies in these basins are extremely variable, and there are few general models. Marginal basins have been defined as 'relatively small semi-isolated oceanic basins, spatially associated with active or inactive volcanic-arc and trench systems' (Kokelaar & Howells 1984), but that exclude forearc settings. Included in the term 'marginal basin' are those basins that form by extension and thinning of the crust behind magmatic arcs (i.e. backarc basins). A common feature of marginal basins is that their development is controlled by a phase of major crustal extension, involving oceanic, continental or transitional crust. As a general term, 'marginal basin' is preferred for such basins because there is no connotation of an association with an emergent volcanic island arc.

The nature of the basement crust beneath marginal basins may be oceanic, continental or transitional.

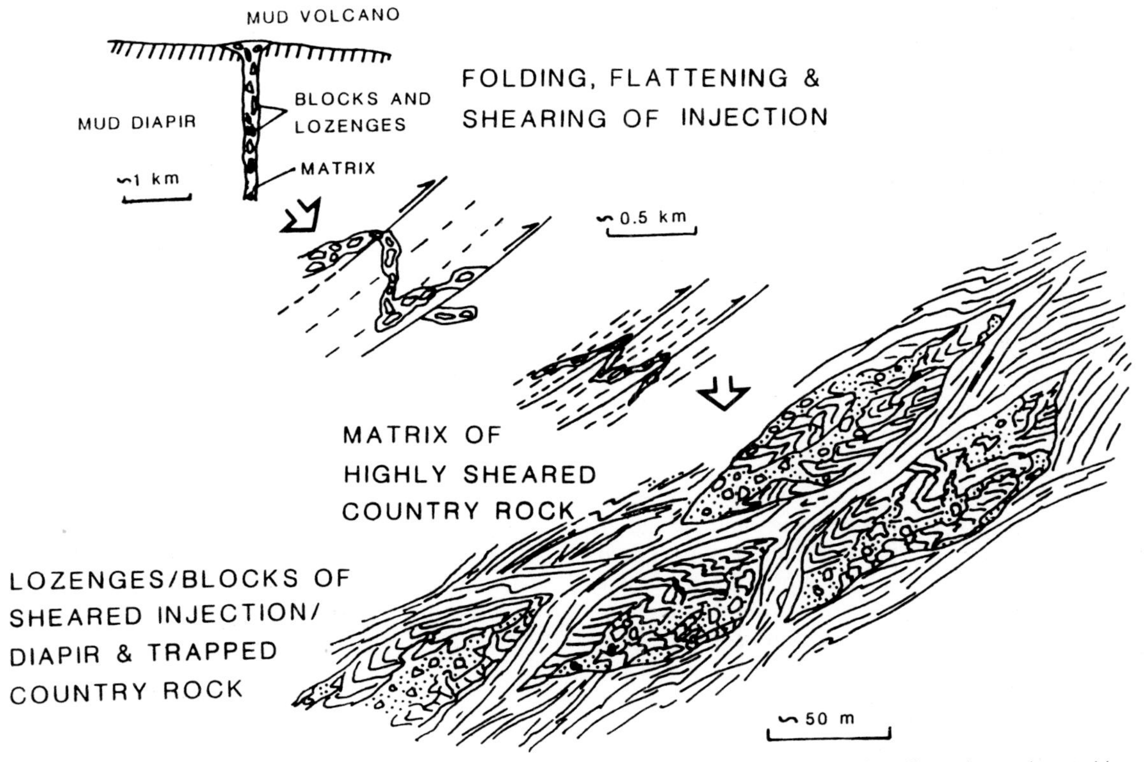

Figure 11.15 Conceptual model for the deformation of a mud diapir in a forearc setting by folding, flattening and stretching to produce a mélange. Note that the mélange may only show original wet-sediment deformation features within the 'lozenges' of material, enclosed within entirely tectonically-deformed thin-bedded, fine-grained lithologies. Blocks within the mélange may range from igneous and metamorphic to sedimentary, including biogenic, rocks. After Pickering *et al.* (1988a).

Furthermore, large thicknesses of deep-marine sediments can accumulate in such basins. Examples are: (a) the Sarawak Basin (South China Sea), floored by pre-Oligocene oceanic crust, now covered by 8000 m of relatively undisturbed sediments, and (b) the Sunda shelf edge that has advanced about 300 km north due to post-Eocene progradation in a marginal basin (Houtz & Hayes 1984).

The initial rifting phase in the development of marginal basins is commonly associated with a large influx of relatively coarse-grained clastics that tends to infill the irregular basin topography. As rifting continues, thick successions of relatively coarse-grained deep-water clastics (Facies Classes A, B & C) become restricted to the margins of the basin. Fine-grained sedimentation (Facies Classes D, E and G) dominates the central parts of the basin. Large abyssal plains may eventually develop within 'mature' marginal basins. The final infill of marginal basins is extremely variable and will depend upon the subsidence or uplift history, together with the manner in which the basin is destroyed.

Marginal/backarc-basin sedimentation may include considerable volumes of volcaniclastic deep-water sediments (see Cas & Wright 1987). Carey & Sigurdsson (1984) have related the distribution of volcaniclastic sediments to stages in the evolution of a marginal (backarc) basin (Fig. 11.16). They recognize three important contributary sources to the sedimentary apron bordering marginal basins: (a) a primary volcaniclastic influx from subaerial and/or subaqueous arc eruptions; (b) an epiclastic influx from the erosion of the subaerial and/or subaqueous arc complex, and (c) a continuous accumulation of pelagic biogenic and aeolian debris.

Carey & Sigurdsson (1984) define four phases in the evolution of a backarc (marginal) basin (Fig. 11.16). The first phase represents the initial rifting and development of an intra-arc basin, with steep unstable basin margins typically mantled by various sediment mass-flow deposits (Facies Classes A & F). The second phase involves backarc spreading and island-arc volcanism: initially very unstable faulted basin margins tend to be smoothed by a veneer of sediment gravity flows (Facies Classes B, C & D). Minor amounts of Facies Classes E and G also accumulate. The third phase, basin maturity, involves a decrease of backarc spreading rates, with increased preservation of fine-grained sediment gravity flows and hemipelagites/pelagites (Facies Classes D, E & G). Since the rate of aggradation of this apron is directly linked to the magnitude of the arc volcanic activity, an assessment of the rate of sediment accumulation should provide a qualitative estimate of the pace of arc volcanism. Pelagic sediment accumulation rates will depend, amongst other factors, upon the depth of the CCD and the biological productivity in the water column.

Amongst the best-documented marginal basins are those of the SW Pacific (cf. Leitch 1984) shown in Figure 11.17a. Leitch (1984) summarizes some of the main features of the marginal basins of the SW Pacific as: (a) a spreading history of short duration; (b) initial rifting of relatively thick crust (the South Rennell Trough being an exception), and (c) rifting either associated with arc volcanism, as in the SW Pacific marginal basins (Bismarck, Lau, Havre and North Fiji basins), or rifting with little or no arc volcanism, as in the NW Pacific marginal basins. Figure 11.17b shows typical DSDP deep-marine stratigraphies in the SW Pacific marginal basins. Hundreds of metres of deep-marine, essentially fine-grained, sediments were drilled and show considerable variation, particularly in the Palaeogene–Neogene sediments. Site 206, for example, in the New Caledonia Basin (Fig. 11.17b), comprises mainly siliceous oozes, whereas Site 286 (now a forearc setting), in the New Hebrides Basin, contains abundant volcanigenic conglomerates, sands and silts, with relatively little siliceous ooze (Fig. 11.17b). Such differences reflect proximity to an active volcanic centre furnishing large volumes of sediments – Site 286 being near the New Hebrides Arc, with Site 206 on oceanic crust hundreds of kilometres from any sediment source.

The Mesozoic to Early Tertiary development of the Black Sea and Caspian Sea provides a useful ancient example of backarc/marginal basins. Zonenshain & Le Pichon (1986) believe that the marginal sea reached its maximum size in the Eocene at about 3000 km in length by 900 km in width, the central part of the basin being subducted during collision between the Arabian Promontory and the Eurasian margin. The large marginal sea consisted of four deep basins floored by oceanic crust, namely the Great Caucasian Sea, South Caspian Sea, and the western and eastern Black Sea basins. An estimated 14–15 km of sediments accumulated in the eastern and western Black Sea basins, upon 5–6 km thick oceanic crust (Zonenshain & Le Pichon 1986).

CASE STUDY: NORTH FIJI BASIN

The interaction of extensional, compressional and oblique-slip tectonics along the perimeters of marginal basins can be extremely complex. In the North Fiji Basin, a major submarine mountain chain formed by

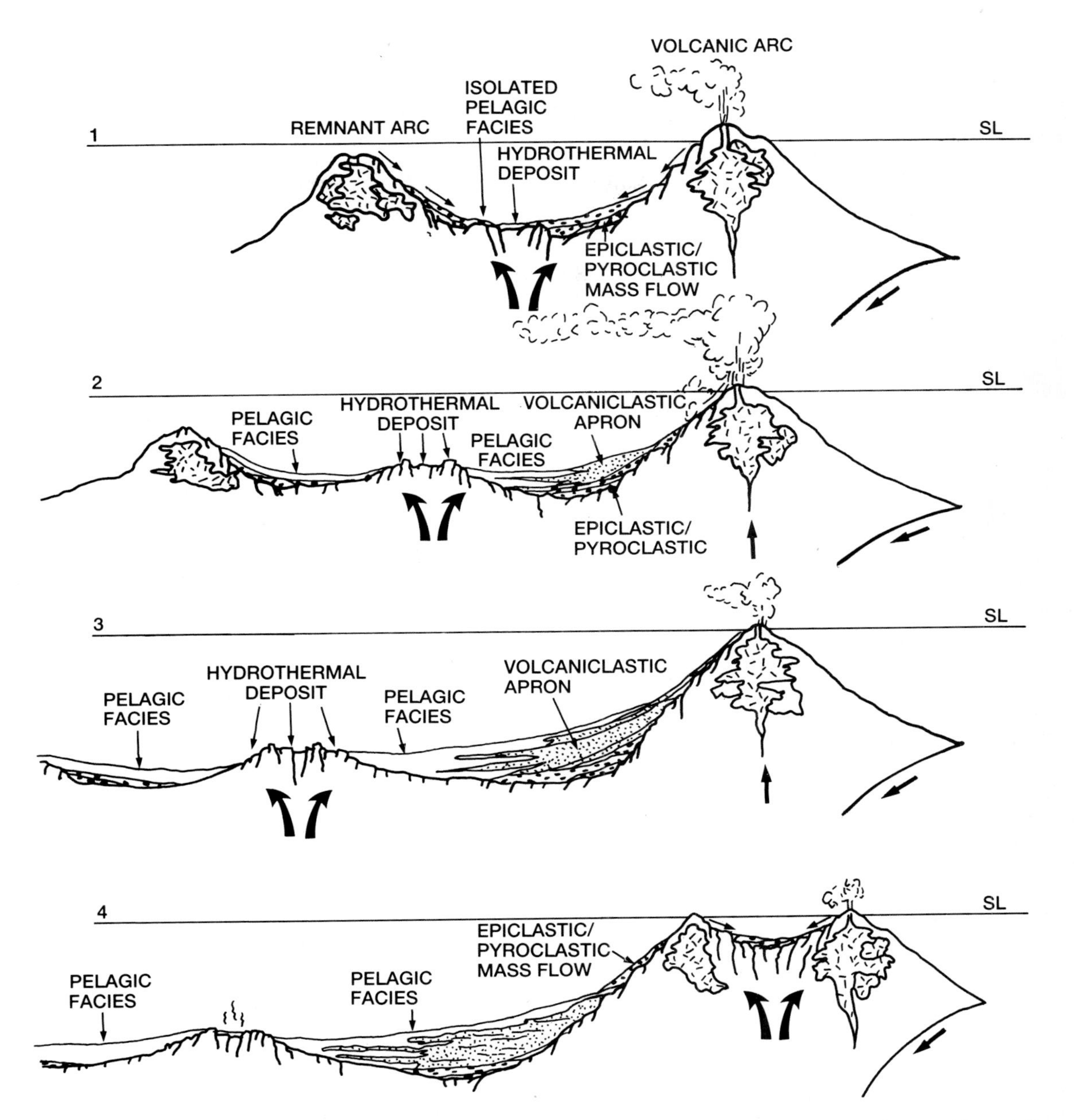

Figure 11.16 Distribution of volcanigenic sediments in an evolving marginal basin. Stage 1, early rifting and large influx of volcaniclastics; Stage 2, basin widening by backarc spreading with active island arc volcanism. Thick volcaniclastic apron developed marginal to active arc; Stage 3, basin maturity, decreasing arc volcanism favours transgression with finer grained, hemipelagic and pelagic, sedimentation over coarse volcaniclastic apron; Stage 4, basin inactivity, backarc spreading ceases and splitting of volcanic arc initiates a new cycle of basin formation. After Carey & Sigurdsson (1984).

two parallel ridges, the d'Entrecasteaux Zone (DEZ) and the West Torres Massif (WTM) of the New Hebrides island arc, is causing both a complex plate interaction and indentation of the arc (Collot *et al.* 1985, Fisher *et al.* 1986).

The islands of the New Hebrides form a NNW–SSE trending intra-oceanic island arc associated with the eastward subduction of the Indo–Australian Plate beneath the North Fiji Basin; the subduction zone dips toward the Pacific Ocean (Figs. 11.18a & b). The DEZ is an area of high relief, up to 100 km wide and with a mean depth of about 3500 m, composed of MORB-

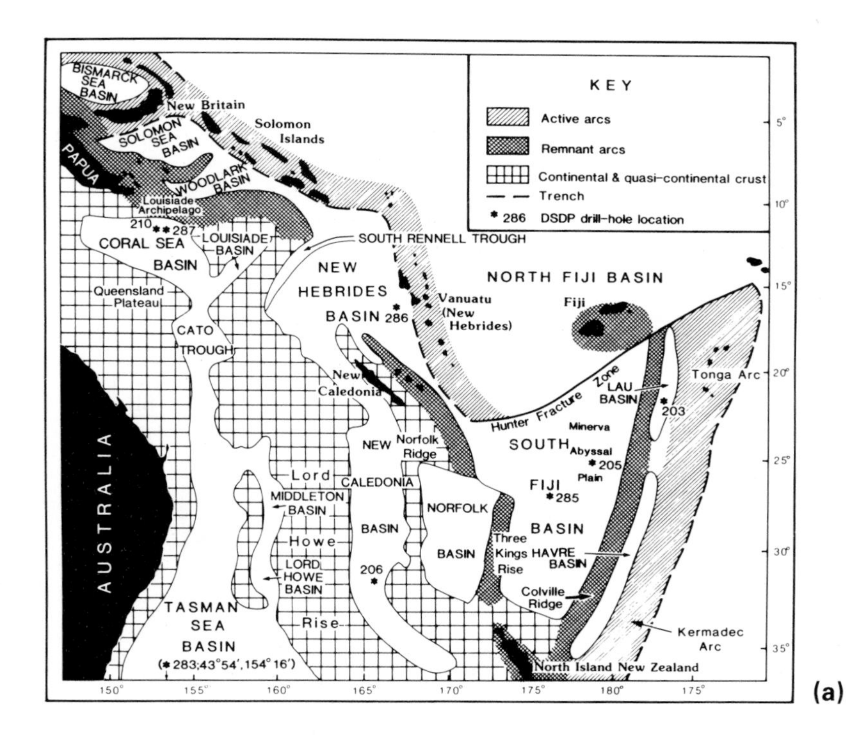

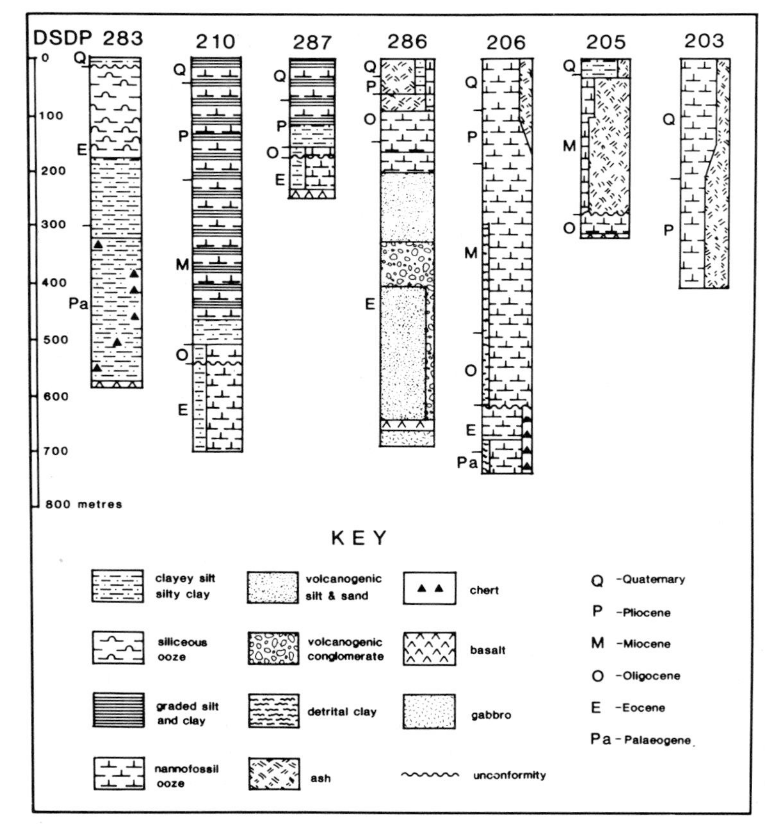

Figure 11.17 Marginal basins and their stratigraphies. **a**, Location of marginal basins in the SW Pacific. 5° of longitude is about 550 km. **b**, DSDP stratigraphies from SW Pacific marginal basins. Tasman Basin, Site 283; Coral Sea, Sites 210 & 287; New Hebrides Basin, Site 286; New Caledonia Basin, Site 206; South Fiji Basin, Site 205; Lau Basin, Site 203. After Leitch (1984).

affinity basalt. The indentation and oblique subduction (10°–15°) of the DEZ against the New Hebrides, at least since 2 Ma, has led to a mean Quaternary uplift rate of 1 mm/year in the arc. This uplift can be explained by radial horizontal stresses resulting from the collision of the DEZ with the New Hebrides. The resultant compressive stresses have modified the width of the arc in this region, with the resulting deformation closely resembling the theoretical stress field produced by the indentation of a long narrow plastic body (arc) by a rigid flat plate (DEZ) (Fig. 11.18c).

The New Hebrides block is both elastically bent under vertical stresses and plastically deformed and pushed eastwards by the resultant horizontal stresses (Collot *et al.* 1985). The recent uplift of the eastern chain is well explained by this model, with the north and south sections of the arc being influenced mainly by a tensional stress field (Fig. 11.18b). Thus, in this case, deep-marine sedimentation is taking place in a range of environments and tectonic regimes from convergence with subduction (New Hebrides forearc), extensional (North Fiji Basin backarc), and oblique-slip (north and south of the central block of Fig. 11.18b). Furthermore, the rapid Quaternary uplift, due to collision of the DEZ with the New Hebrides Arc, has led to the accumulation of a thick succession of deep-marine sediments near the southern margin of the DEZ, probably eroded from Espiritu Santa Island (Fisher *et al.* 1986).

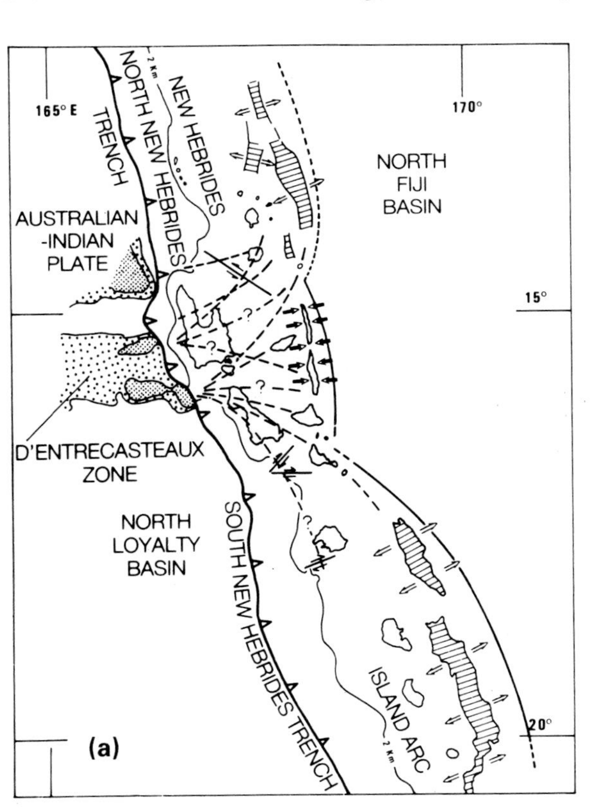

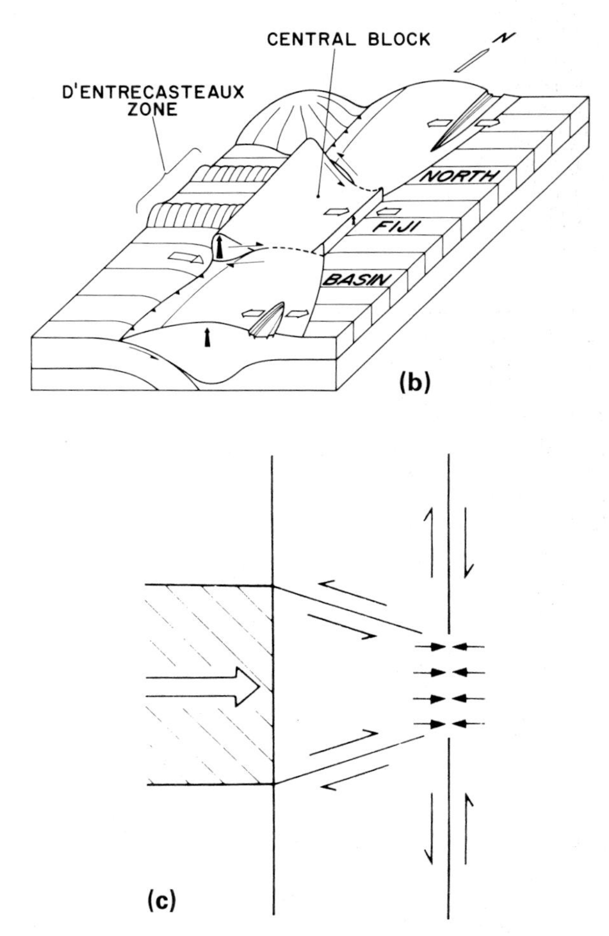

Figure 11.18 Complex plate deformation in the New Hebrides island arc, adapted from Collot *et al.* (1985) and Fisher *et al.* (1986). **a**, Map of the New Hebrides arc region to show the stress pattern along the arc. d'Entrecasteaux Ridge; ruled legends are backarc basins; solid lines show western and eastern limits of arc; dashed lines indicate inferred radiating stress pattern in plate; strike-slip faults interpreted from earthquake mechanism solutions; 2000 m isobath emphasizes the protruding shape of the Central Block (see 11.18b). **b**, Three-dimensional model to illustrate the main tectonic features of the New Hebrides arc in the region of the d'Entrecasteaux Ridge (not to scale). **c**, Diagramatic strain pattern for the general case of indentation of a finite-width body by a flat rigid die. After Molnar & Tapponnier (1975). Compare this strain pattern with that in (b). See text for explanation.

CASE STUDY: LESSER ANTILLES SUBDUCTION COMPLEX, FOREARC–BACKARC

This study is based mainly on the results of DSDP Leg 78A (1981), in Biju-Duval, Moore *et al.* (1984), and Brown & Westbrook (1987).

The Lesser Antilles intra-oceanic arc is situated between the backarc Grenada Basin to the west, and the westward-subducting Atlantic Ocean Plate to the east (Fig. 11.19); ocean plate subduction is occurring at about 2 cm/year. The Lesser Antilles subduction complex provides a good example of the variation and complexity in tectonics and sedimentation along active convergent plate margins. The forearc of the Lesser Antilles contains the accretionary prism of the Barbados Ridge complex, together with two prominent forearc basins, the Tobago and Barbados Basins (Fig. 11.20), the former basin containing more than 4000 m of sediment fill that becomes progressively more deformed eastwards. The Barbados Ridge Complex is the widest example of an accretionary prism associated with an island arc; even the outer trench rise, defined by a positive isostatic gravity anomaly 150 km seaward of the plate boundary, lies buried beneath the subduction complex. At the deformation front, sediment thicknesses beneath the Atlantic Abyssal Plain vary from more than 4000 m in the south to about 700 m in the DSDP Leg 78A area.

The Lesser Antilles forearc ranges in width from more than 450 km in the south, to less than 150 km in the north. The considerable width of the system, together with the very rapid rates of subsidence and fast sediment accumulation rates, at least since the Pliocene, have conspired to mask the trench as a bathymetric feature (Fig. 11.20). Thus, the seaward boundary of the forearc is defined by a deformation front of folding and thrust faulting. In the region of DSDP Leg 78A drilling, the accretionary prism (Barbados Ridge Complex) is about 260 km wide, and to the west of the emergent Barbados Ridge seismic evidence suggests back-thrusting in the deep marine Tobago Basin (Fig. 11.20). Most along-strike structural variation can be related to the influence of basement relief.

The Lesser Antilles volcanic arc has been active since at least the middle Eocene and possibly since Cretaceous times. North of Martinique opposite the Puerto Rico Basin, the arc divides into an outer arc (Aves Ridge), extinct since the middle Miocene, and an inner arc that has been active from at least about 5 or 6 Ma. In the south, west of Tobago Basin, there is only one arc, probably active since 55 Ma. The crust beneath the arc is about 30 km thick. Between the arc and the subduction zone, there is a 100 km-wide segment of oceanic crust overlain by a thick succession of deep-marine forearc deposits, the Tobago Basin and Lesser Antilles Basin.

The backarc Grenada Basin, underlain by an anomalously thick oceanic crust (Boynton *et al.* 1979) similar to that of the Venezuela Basin, in part contains more than 6000 m of essentially arc-derived sheet-like sediments. The Grenada Basin almost certainly had an extensional origin due to separation of the Lesser Antilles from the Aves Ridge, probably in the Late Cretaceous to Early Tertiary. In a study of 29 piston cores (Pleistocene–Holocene) from the Grenada Basin, Carey & Sigurdsson (1978), and Sigurdsson *et al.* (1980), describe coarse-grained, volcaniclastic, sediment gravity flow deposits in beds up to 4.5 m thick (Facies Classes A, B & C), interbedded with predominantly green–grey hemipelagic muds and clays (Facies Classes E & G). The upper part of the cores comprise Holocene, brown, pelagic clay of Facies Class G. Their study of the turbidites showed that the Bouma T_a division is present in 76% of the beds, T_b in 41%, T_c in 9%, T_d in 6% and T_e in 33%. The debris flow deposits (Facies Class A or F) contain pumice clasts up to 6.5 cm in diameter in a fine-grained matrix of glass shards, crystals, lithic and biogenic fragments, and clay. Petrographic studies suggest that the volcaniclastic material was derived from multiple sources, including subaerial eruptions on the islands of St Vincent, St Lucia, Martinique, Dominica and Guadeloupe.

Associated with the accretionary prism, there are a number of 'supra-complex' basins, the largest of which is the Barbados Basin (Fig. 11.19) with up to 2000 m of sediment only deformed near widely-spaced fault zones. The eastern margin of the Barbados Basin is currently being deformed by east-dipping reverse faults and thrusts with horizontal separations of 10–15 km; the related folds produce ridges 500–600 m above the basin floors (Brown & Westbrook 1987). Locally, such basins have been subject to considerable vertical tectonics with undeformed sediments over 1000 m thick uplifted on ridges between supra-complex basins.

The Tobago Basin is a major forearc basin that was probably formed by the upward growth of the accretionary complex. At the southern end of the basin, sediments show overstep of the deformed prism sediments. This stratigraphic relationship could be the result of pulses of deformation at the basin-margin, each followed by relatively stable phases when sediment overstep occurs. However, at the southeastern margin of the Tobago Basin, where overstep occurs,

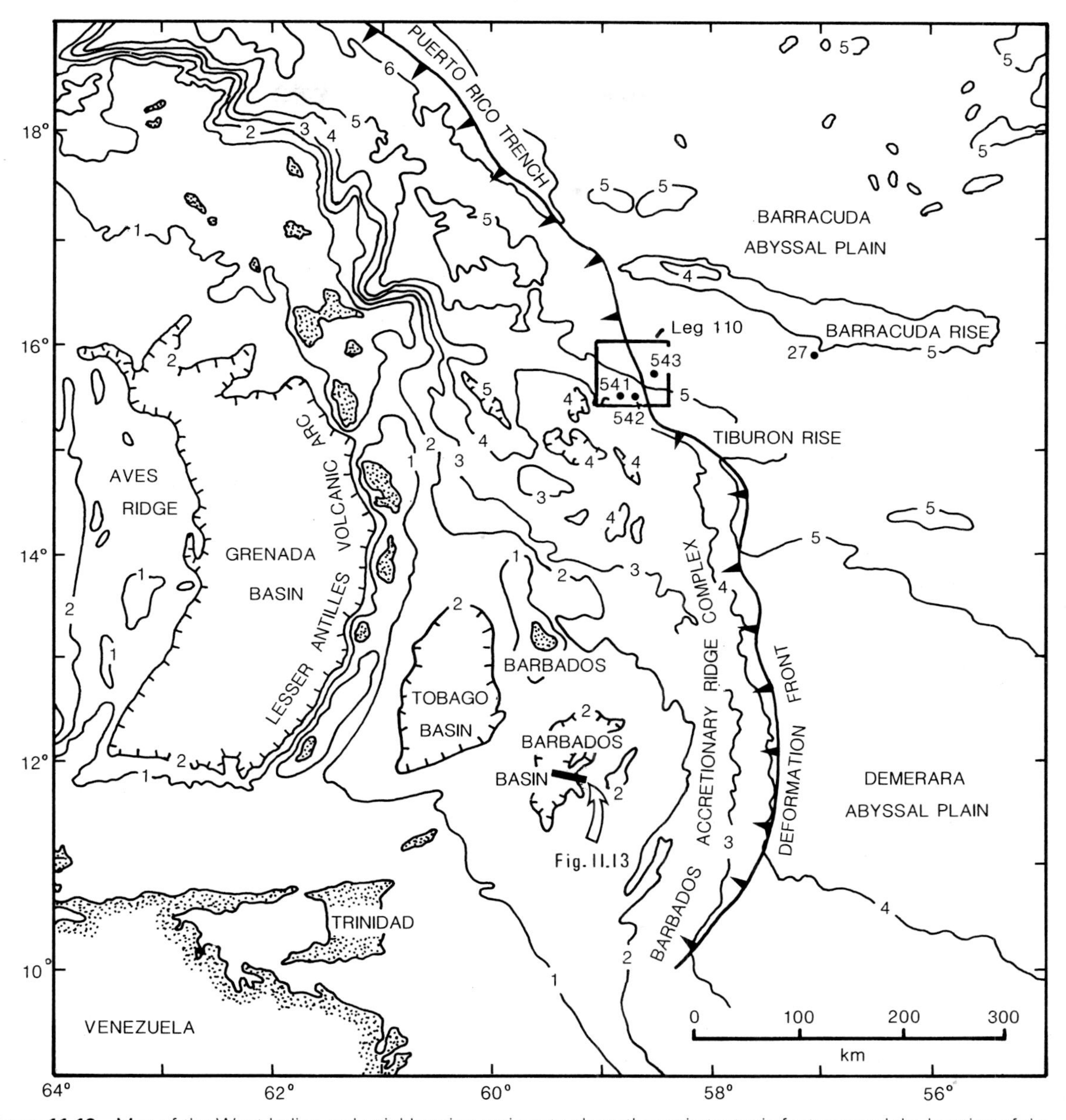

Figure 11.19 Map of the West Indies and neighbouring regions to show the main tectonic features and the location of the principal sedimentary basins, together with boxed ODP Leg 110 area. Location of seismic line is shown for Figure 11.13a. Sites 541–3 were drilled on DSDP Leg 78A.

the basin sediments have been shielded from compression since the Pliocene by a spur of South American metamorphic basement, also seen on the island of Tobago.

Important, large-scale, forearc mud diapirism has been recognized both from onshore outcrops in Trinidad (Higgins & Saunders 1974), and from seismic profiles and sidescan sonar surveys in the accretionary complex offshore (Fig. 11.13a & b) (Michelson 1976, Biju-Duval *et al.* 1982, Stride *et al.* 1982, Brown & Westbrook 1987). It appears that such diapirs preferentially develop along fault zones, especially those along the

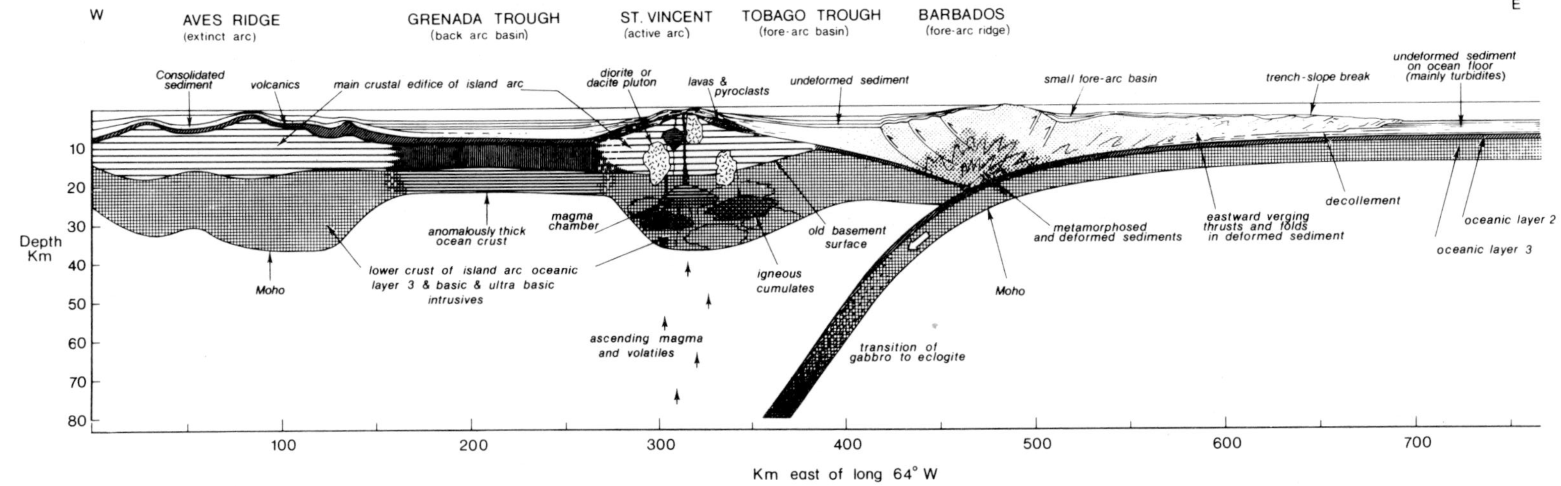

Figure 11.20 Cross-sectional model of the Lesser Antilles arc system through Barbados and St Vincent, derived from all available geophysical data, but primarily gravity and seismic refraction data. After Westbrook *et al.* (1984).

margins of sedimentary basins. The mud diapirism appears to be most abundant in the southern part of the accretionary complex where turbidites from the Orinoco submarine fan are accreted rapidly (Brown & Westbrook 1987). The origin of these diapirs, up to 17 km long by 1 km wide, has been ascribed to enhanced pore-fluid pressures created by load-induced stress as the accretionary complex thickens (Westbrook & Smith 1983).

Sediments toward the toe of the Barbados Ridge Complex, drilled on Leg 78A (Sites 541–543), comprise lower Miocene to Quaternary hemipelagic/pelagic marly calcareous oozes (Facies Group G1), structureless muds (Facies Class E), radiolarian muds (Facies Group G2) and ash bands. These sediments occur at Sites 541 and 543, the latter site also penetrating 44 m of basaltic pillow lavas at 411 m sub-bottom. The paucity of terrigenous turbidites at these sites probably reflects deposition on a topographic high. GLORIA long-range side-scan sonar has shown the importance of gravity sliding at the deformation front, opposite the Tiburon Rise (Fig. 11.19) where a debris flow and its chute cover an area of 100 km^2.

Sedimentation patterns are principally governed by: (a) long-lived relatively slow subduction–accretion; (b) a major southern source of terrigenous sediments, axially-funnelled from the South American margin (Damuth 1977), although the finer grained silt and clay are carried by the Guiana Current system, and the coarser grained silt and sand transported by turbidity currents down the South American continental slope onto the Atlantic Abyssal Plain; (c) mass wasting and sediment gravity flows derived from the forearc accretionary prism; (d) volcanigenic sediments derived from the volcanic island chain; and (e) production of pelagic, calcareous, biogenic material, at least in areas above the CCD.

CASE STUDY: MIDDLE AMERICA AND PERUVIAN CONVERGENT MARGIN

This study is based mainly on results from DSDP Leg 66 (Watkins, Moore *et al.* 1982), Leg 67 (Aubouin, von Huene *et al.* 1982), Leg 84 (von Huene, Aubouin *et al.* 1985), and Bourgois *et al.* (1988).

The active convergent continental margin off Guatemala, containing the Middle America Trench, is one of the classic examples of subduction–accretion, involving the offscraping of sediments and rocks from the downgoing oceanic Cocos Plate (Fig. 11.21). This margin differs from the Lesser Antilles area (above) because sedimentation is associated with an arc developed on continental crust, with the only substantial deep-marine deposition occurring in the forearc.

Subduction began in the Early Tertiary (von Huene, Aubouin *et al.* 1985). By the Neogene, a well defined forearc basin had developed in middle to upper bathyal depths with slope and trench sedimentation occurring at abyssal depths, greater than 3500 m. Along the axis of the Middle America Trench, hemipelagic silt and clay, together with arc-derived turbidites, are ponded on the descending oceanic plate with its veneer of pelagic sediments. Minor amounts of sand also occur, with displaced shelf and slope faunas. Beneath this trench-fill, the Cocos Plate is covered by Miocene, pelagic white foraminiferal/nannoplanktonic chalk and red–brown clays, overlain by Pliocene/Quaternary hemipelagic silt and clay, with rare turbidites derived from the inner slope (Fig. 11.22). In contrast, the trench slope, locally incised by submarine canyons, comprises more varied facies, including pebbly mudstones of Facies Group A1 (Fig. 11.22). There is evidence to suggest that 50–200 m-thick lobes of slope sediments are subject to non-episodic, geologically continuous, downslope plastic creep processes (Baltuck *et al.* 1985). Sediment accumulation rates are high, for example DSDP Site 565: 165 m/m.y. from 0–80 m sub-bottom; 13 m/m.y. from 80–90 m sub-bottom, and 123 m/m.y. from 90–328 m sub-bottom.

Despite plate convergence, little compressional deformation is observed off Guatemala in the forearc. Instead, considerable seismic evidence exists for extensional tectonics (Fig. 11.23). A model based on underplating as the main accretionary mechanism can satisfactorily account for the extension. This active margin off Guatemala has been termed a 'convergent–extensional active margin' by Aubouin *et al.* (1982a & b, 1984), and Bourgois *et al.* (1988), to contrast it with convergent–compressional margins such as the Lesser Antilles forearc (Biju-Duval, Moore *et al.* 1984). The convergent–extensional active margins appear to contain a thin sediment cover compared to convergent–compressional margins.

In a Seabeam, multi- and single-channel seismic reflection survey of the continental slope off Peru, between 4°–10° S latitude, Bourgois *et al.* (1988) recognize three distinct morpho-structural domains: (a) an upper slope, to a depth of 2500 m, dipping generally oceanward at about 5° with a slight upward convexity, and cut mainly by straight V-shaped canyons; (b) a middle slope, characterized by many curvilinear scarps with large offsets of the sea floor of up to 1200 m toward the trench, interpreted as the result of major mass failure and 'tectonic collapse' of the slope, particularly in the upper parts, and (c) a lower slope to the relatively

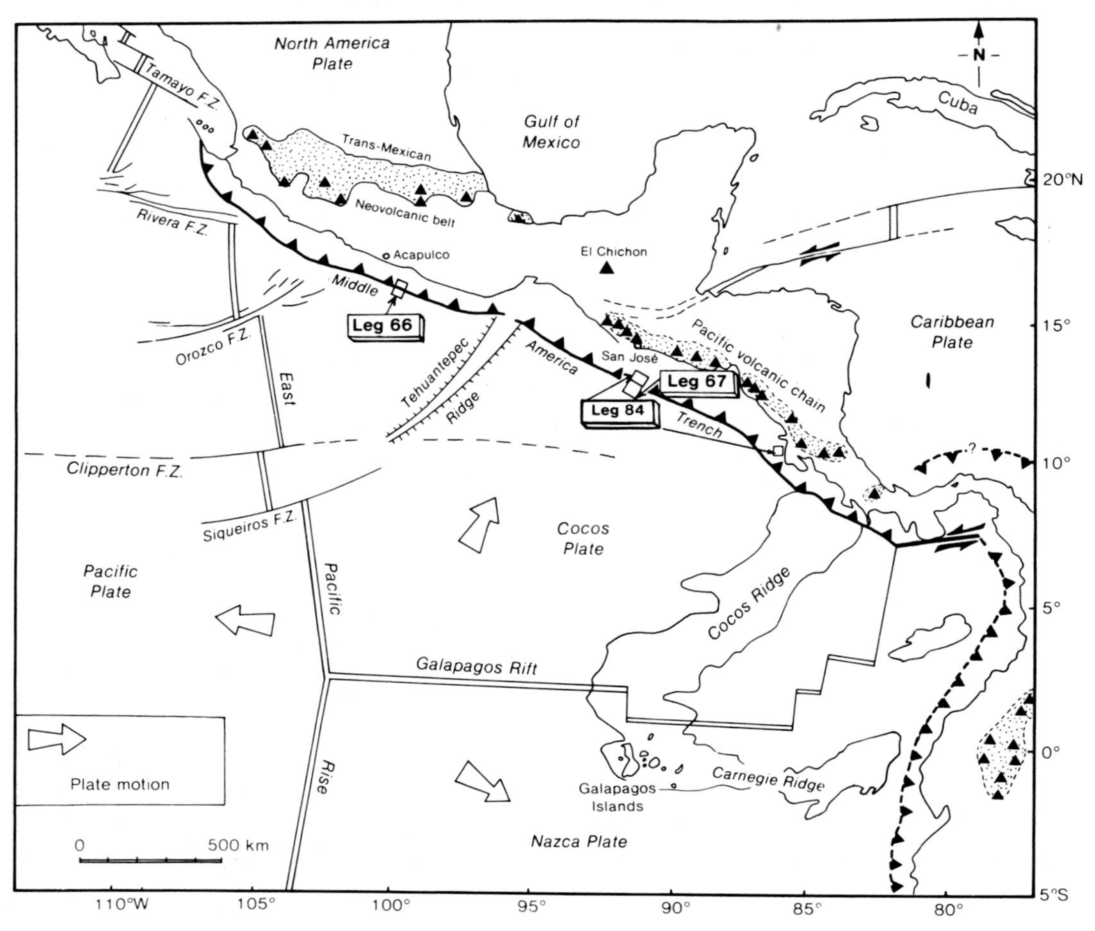

Figure 11.21 Map of Central America and the Middle America Trench (after Baltuck *et al.* 1985). Sites of DSDP Legs 66, 67 and 84 are shown; plate motions shown by open arrows. Solid triangles are Quaternary volcanoes.

flat trench floor at about 5000 m depth, showing a hummocky topography and interpreted as the deposits of major slope failure from higher up the continental slope. The lower slope shows ridges that are interpreted as the surface expression of major thrusts in the frontal and toe regions of the accretionary prism. The compressional part of the accretionary prism is a minimum 15 km and maximum 85 km width normal to the trench axis; the change from the predominantly compressional to extensional part of the margin is defined at the middle slope–lower slope boundary, at least where it can be recognized in the region of 5° S latitude (Bourgois *et al.* 1988). The Peruvian margin, like the Middle America forearc, studied on DSDP Leg 84 (von Huene, Aubouin *et al.* 1985), comprises a young accretionary complex stacked against continental crust and also represents a 'convergent–extensional margin' (Aubouin *et al.* 1984).

CASE STUDY: WESTERN PACIFIC 10°–45° NORTH

The Western Pacific Ocean, between about 10° and 45° North latitude, contains a number of major arc systems and associated basins (Fig. 11.24). Some aspects of the Japan Trench and Nankai Trench, and forearcs, are considered together with the Ryukyu Arc and associated Okinawa Basin, a marginal basin developed by backarc rifting. The tectonic development of this region since 25 Ma is summarized by Letouzey & Kimura (1985, 1986), and is shown in Figure 11.25.

The Japanese or Honshu Arc, associated with the triple junction between the Philippine, Eurasian and

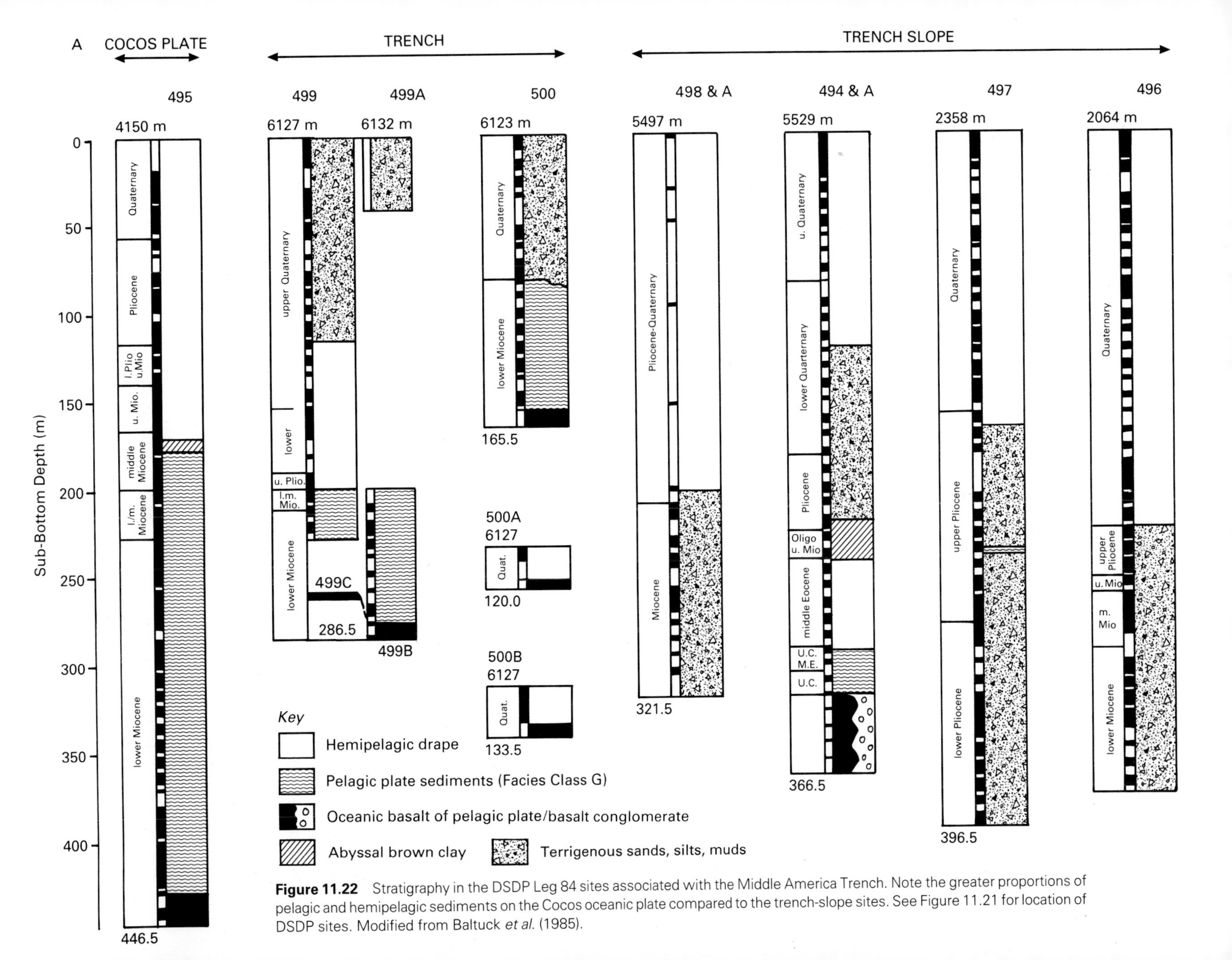

Figure 11.22 Stratigraphy in the DSDP Leg 84 sites associated with the Middle America Trench. Note the greater proportions of pelagic and hemipelagic sediments on the Cocos oceanic plate compared to the trench-slope sites. See Figure 11.21 for location of DSDP sites. Modified from Baltuck *et al.* (1985).

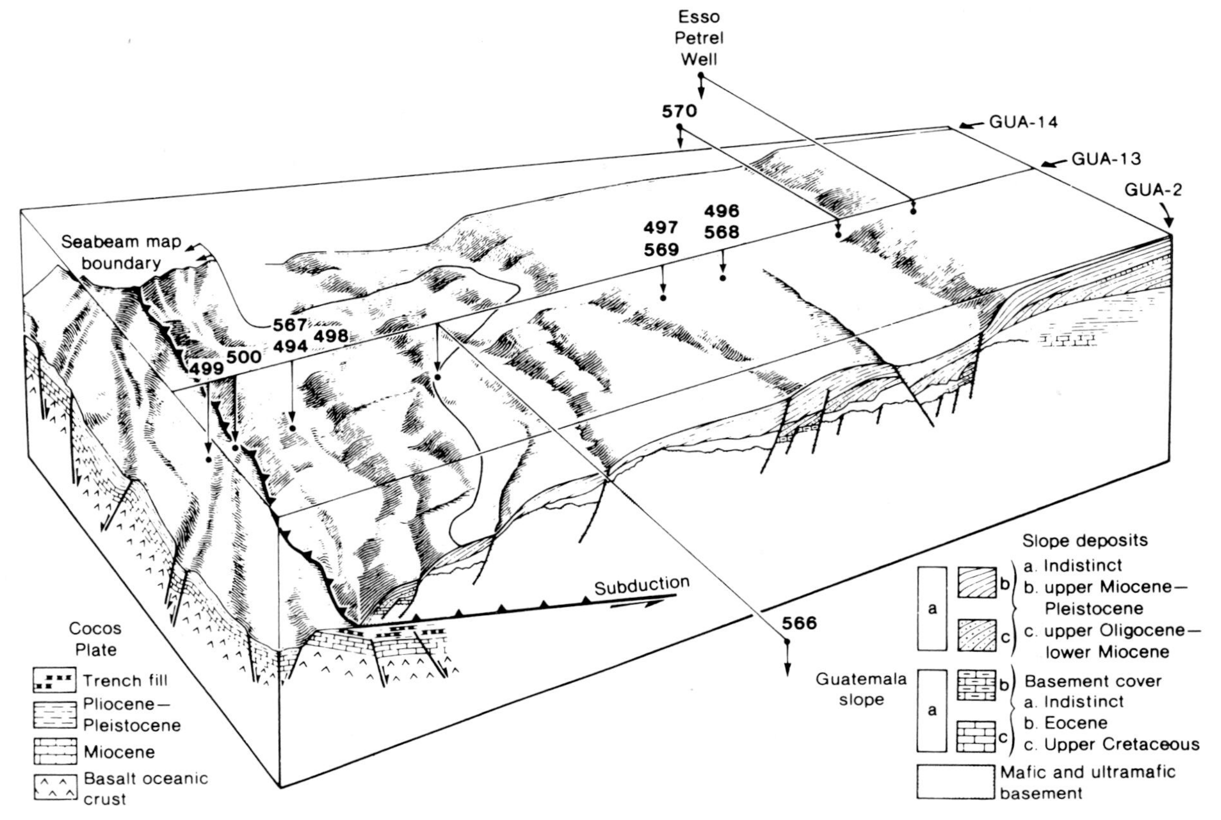

Figure 11.23 Block diagram showing the convergent zone between the Cocos Plate and the Caribbean (Central America) Plate in the Middle America Trench off Guatemala, together with the location of the DSDP Leg 84 sites. Note the importance of extensional tectonics in the forearc region. After Aubouin *et al.* (1985).

Pacific Plates, contains two discrete trenches–the northeastern Japan Trench and the southwestern Nankai Trench. Along the margin of the Pacific Plate, the Japan Trench passes northward into the Kuril Trench, and southward into the Izu–Bonin (Ogasawara) Trench. The Japan Trench–Honshu Arc, is perhaps the most extensively studied Pacific convergent margin (for a review, see Shiki & Misawa 1982).

All trenches in this region have different sedimentary fills. In the Nankai Trench, large amounts of terrigenous mud and sand have accumulated, fed mainly from the Fuji River via submarine canyons into the trench axis to cover the trench floor. Mud diapirism has been reported in this area (Shipboard Scientific Staff of the KAIKO Project 1985). In the Japan Trench, there is a relatively thin veneer of terrigenous sediments that appears to have been redeposited in the trench by slope instability (mainly sliding) from the forearc lower trench slope (Facies Class F). In contrast, there is a thick sediment cover in the western part of the more northerly Kuril Trench. A very thick, subsiding accumulation of terrigenous sediments is derived from Boso Canyon and covers the triple junction of the Izu – Bonin (Ogasawara) Trench and Sagami and Japan Trenches. The surrounding ocean floors incline towards the centre of the triple-junction triangle. At the triple junction, the rate of subsidence is much faster than the rate of sediment accumulation.

There is considerable variation in the style of accretion in the trenches, with extensional tectonics and uplift also controlling sedimentation. For example, DSDP Leg 87 in the forearc of the Japan Trench has shown that Late Cretaceous and Early Palaeogene uplift created the Oyashio landmass that developed in the outer forearc (Karig, Kagami *et al.* 1983). Subsequent late Oligocene to late Pliocene major subsidence of the forearc led to submergence, with post-early Pleistocene convergence causing renewed uplift and accretion.

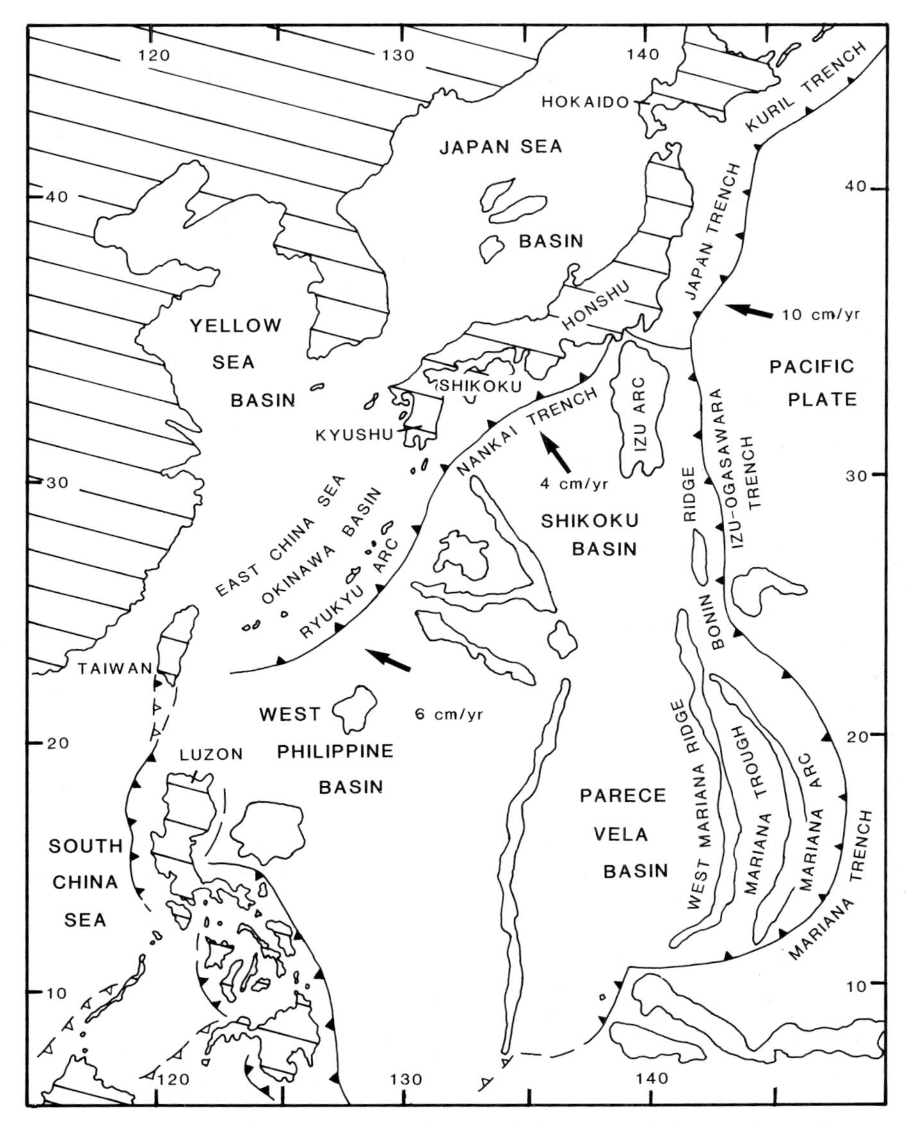

Figure 11.24 Map of the western Pacific to show the main tectonic features and plate vectors with approximate rates of convergence (black arrows). Map redrawn and modified after Letouzey & Kimura (1985).

Thus, a feature of the stratigraphy is the alternation of deep- and shallow-marine sedimentation, associated with local emergence of parts of the forearc.

Subduction style is variable in the 4000–4800 m deep, 10–20 km wide, Nankai Trench (Fig. 11.26a & b) (Aoki *et al.* 1983). Here, the angle of subduction is steeper along the eastern part of the trench compared to the west. Leggett *et al.* (1985) show that accretion along the Nankai Trench changes from: (a) subduction–accretion domains with off-scraping of trench deep-marine sediments near the toe of the lower trench slope, to (b) domains of underplating. Where off-scraping and frontal accretion occur, imbricate thrusts are resolvable at intervals of 1–6 km on high resolution, multi-channel, seismic reflection profiles (Fig. 11.26). These thrusts have possible lateral, along-strike, continuity up to 70 km. An area of underplating is proposed beneath the growing Minami–Muroto Knoll, an area of uplift associated with contemporaneous extensional growth faults. Upslope from the knoll, an estimated

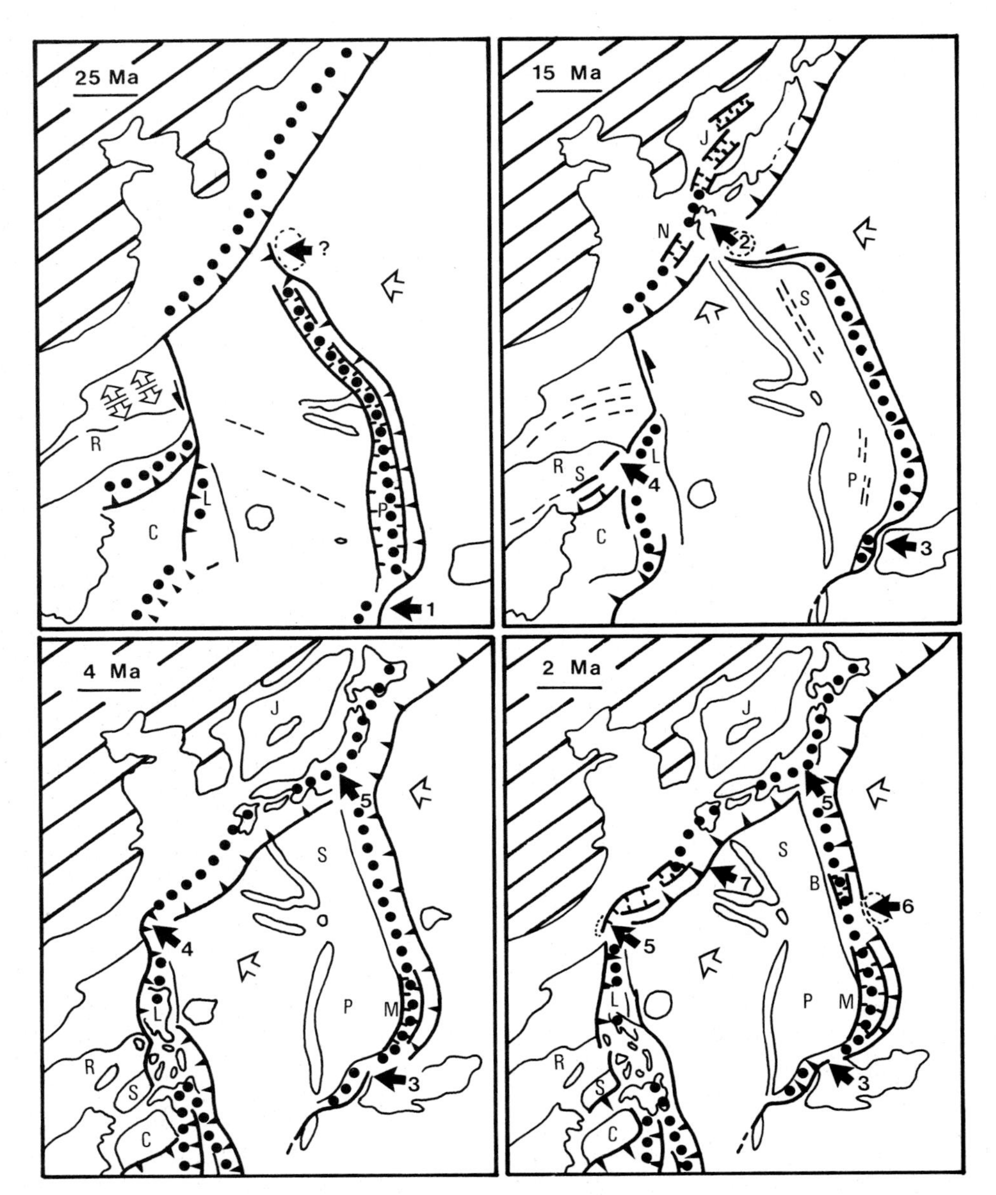

Figure 11.25 Palaeogeographic reconstructions for the Philippine Sea at 25 Ma, 15 Ma, 4 Ma and 2 Ma. Symbols are: R, Reed Bank–North Palawan continental block; L, Luzon Island; S, Sulu Basin; C, Celebes Sea; N, North Okinawa Basin; J, Japan Sea; P, Parece Vela Basin; SH, Shikoku Basin; M, Manana Basin; B, Bonin Basin. Solid arrows indicate collisions involving: 1, Mapia Ridge; 2, Palau Kyushu Ridge; 3, Caroline Ridge; 4, North Luzon island arc; 5, Izu–Bonin Ridge or Izu–Ogasawara Arc; 6, Ogasawara Plateau; 7, Amami Plateau Palau Kyushu Ridge; 8, 'Philippine Mobile Belt'. Graben symbols indicate backarc basin opening. After Letouzey & Kimura (1985). Large dots indicate volcanic arcs.

Figure 11.26 Segments from time-migrated multi-channel seismic reflection profile Lines 55-1 & 55-8 across the eastern part of the Nankai Trench and lower slope to show internal and superficial deformation in the accretionary prism. Note the slope basins (SB), structural inversion, or pop-up, feature (IS), trench-fill (TF) with onlap against the subducting oceanic-plate sediments, the top of the oceanic-plate sediments (P) and the top of the ocean crust (O), are all marked. Two-way travel time shown in seconds. Plates courtesy of JAPEC and A. Taira.

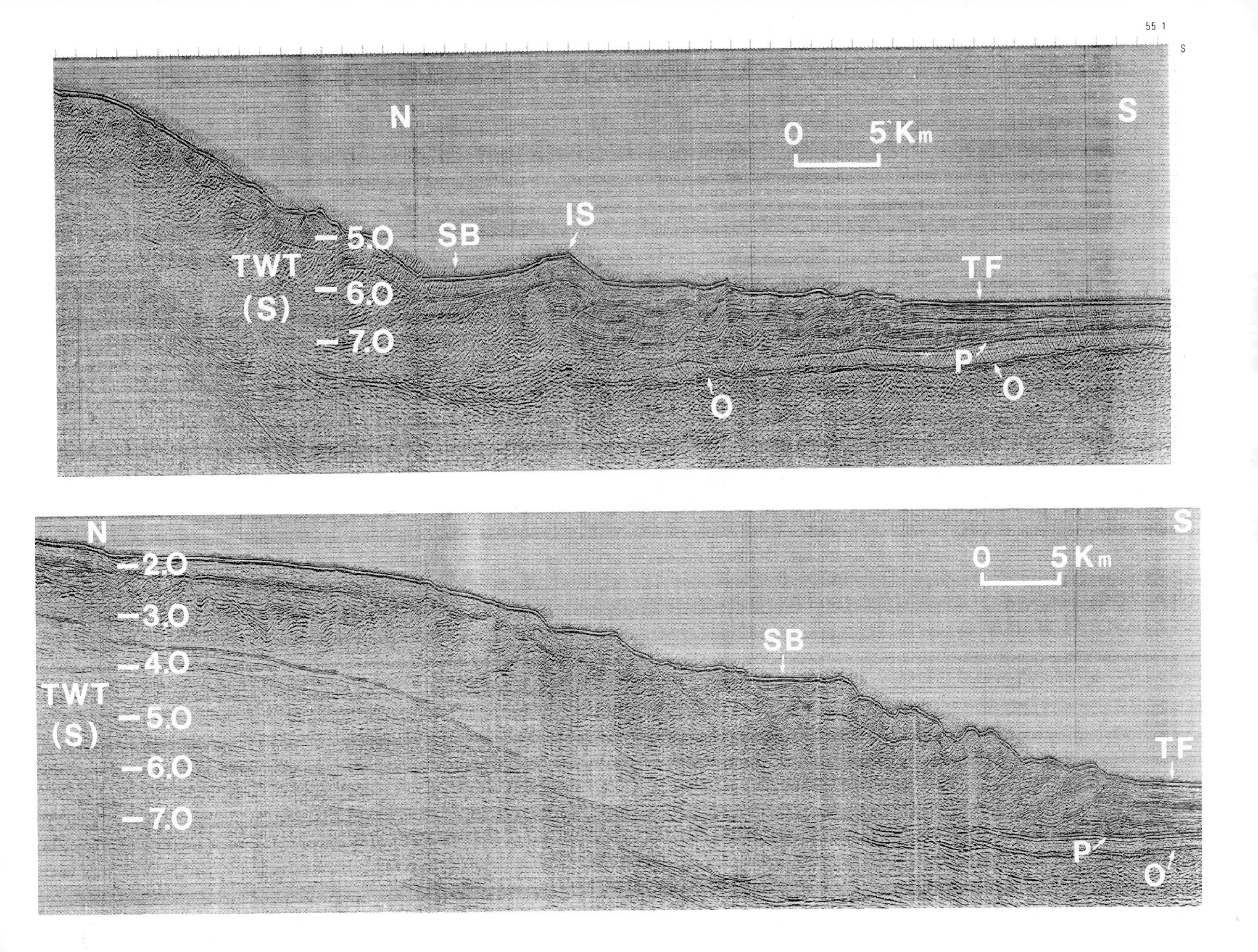
N
S
0
5 Km
IS
SB
TF
TWT
(S)
— 5.0
— 6.0
— 7.0
P
O
O
N
S
0
5 Km
SB
TF
TWT
(S)
— 2.0
— 3.0
— 4.0
— 5.0
— 6.0
— 7.0
P
O

1200 m of deep-marine sediments have accumulated as a footwall-block succession.

The Japan and Nankai Trenches and associated forearc areas demonstrate the complexity and variability in tectonic style associated with subduction–accretion. Not only is there considerable mass wasting of the inner trench slopes by sliding and other processes, but small sedimentary basins have developed in response to both compressional and extensional tectonics: some of these basins having considerable thicknesses of sediment (Fig. 11.27).

Forearc sedimentation and tectonics may show many similar features to large backarc/marginal basins, where extensional tectonics may be of primary importance in generating stratigraphy. As an example, the Okinawa Basin, a marginal basin behind the Ryukyu Trench (Fig. 11.24) is considered using data from Letouzey & Kimura (1985) and Kimura (1985). The Okinawa Basin has a width of about 230 km, a length of 1300 km, and an 'inner' graben 50–100 km wide. About 3000–4000 m of mainly deep-marine sediments have accumulated in the south, compared to 7000–8000 m in the north. The history of the Okinawa Basin may be summarized as follows: (a) rifting within the volcanic arc beginning in the Neogene and still active today; (b) synchronous opening and subsidence of the Okinawa Basin with tilting and subsidence of the forearc terrain – the late Miocene erosional surface now being 4000 m

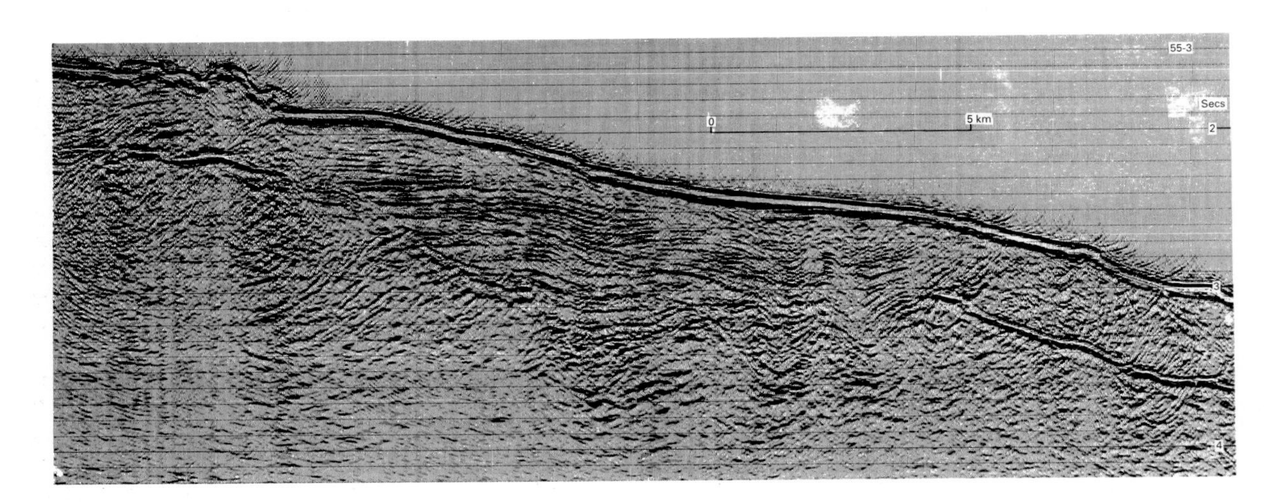

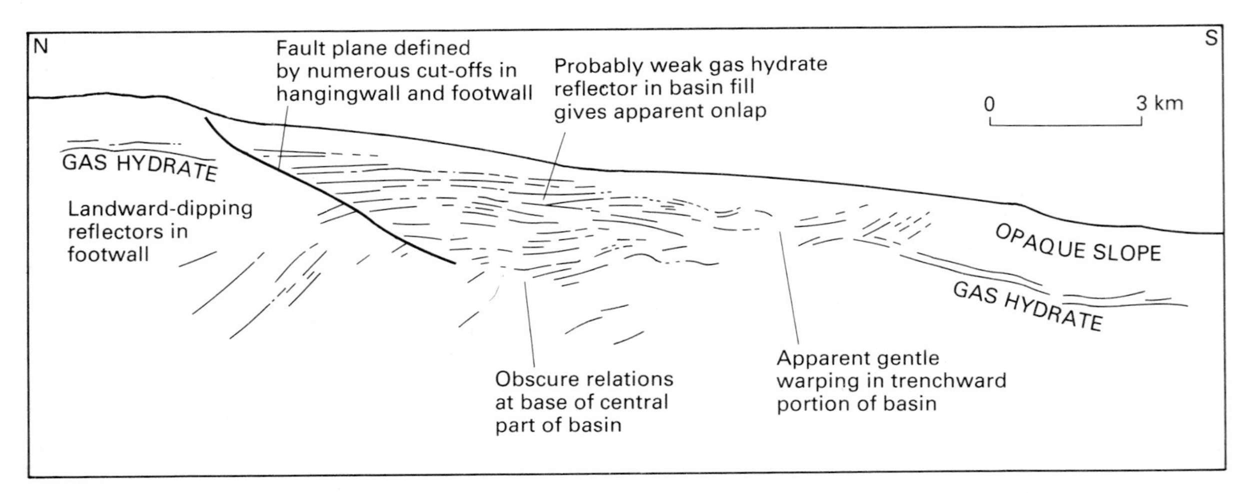

Figure 11.27 Segment from time-migrated seismic reflection profile Line 55-3 across the accretionary prism facing the eastern part of the Nankai Trench to show details in slope basin. From Leggett *et al.* (1985).

below sea level in the forearc terrace above the trench slope, and (c) a Pliocene, 1.9 Ma, major phase of crustal extension in the southern and central Okinawa Basin.

Palaeomagnetic studies suggest a 45°–50° clockwise rotation of the southern Ryukyu Arc since the late Miocene. This rotation was probably related to the collision of Taiwan and the north Luzon Arc with the Chinese mainland margin, provoking the lateral 'extrusion', clockwise rotation and buckling of the southern Ryukyu non-volcanic arc. Crustal extension has produced, on the arc side of the Okinawa Basin, Pleistocene–Recent half-grabens that are 10–80 km wide with floors that dip up to 20°–30° to the southeast. Near Taiwan, the slopes are cut by many canyons and channels that obscure the half-graben structure.

Ancient subduction–accretion systems

There are many examples of ancient subduction–accretion systems, interpreted with varying degrees of confidence. Some are very extensive, for example the Cretaceous Chugach Terrane, southern Alaska (Fig. 11.28), with a linear continuity of about 2000 km (Nilsen & Zuffa 1982), whereas other systems have considerably less present-day along-strike continuity such as the Ordovician–Silurian Southern Uplands prism, Scotland, extending for about 120 km (Leggett *et al.* 1982). However, if this system continues into the Northern Appalachians, for example into Central Newfoundland, as seems likely, then its length is considerably greater. The interpretation of ancient subduction–accretion systems ideally relies upon the recognition of consanguineous arc volcanism with identifiable forearc and backarc tectono-sedimentary environments. If there has been later major oblique-slip tectonics, then reconstructing the ancient arc-related system relies upon correlating far-displaced terranes.

Ideally, in reconstructing ancient subduction–accretion systems, it is important to map out a complete arc-trench system. In the case of the Cretaceous Chugach Terrane, Nilsen and Zuffa (1982) identify: (a) forearc slope deposits, including slope basins, with south-trending submarine canyons; (b) a major zone of mélange against the flysch zone of the Chugach Terrane, the contact being a major landward-dipping fault zone; (c) southward-transported forearc-basin deposits forming a belt parallel to other Chugach Terrane tectono-stratigraphic belts, and (d) a major magmatic arc of granitic intrusions bounded by dextral strike-slip fault zones (Fig. 11.28b). Also, most fault-bounded blocks dip towards the arc, with the style of faulting, folding and metamorphism typical of modern subduction–accretion systems. Sediment dispersal was mainly along the trench from east to west, with lateral infilling from the forearc area to the north (Fig. 11.28b).

The deep-water sedimentary fill of arc-related basins can be extremely variable both in petrography and grain size. The lower Eocene Tyee and Flournoy Formations of western Oregon have been interpreted as deep forearc-basin sediments (Chan & Dott 1983), and comprise a 2000 m-thick succession that includes voluminous sands that were transported into deep water along a broad front from the shelf, rather than being fed from point sources such as canyons. Parts of the sand-rich fan system, however, were fed from complex nested channels up to 350 m wide and 40 m deep incised into the shelf/slope at shallow depths (see Ch. 7 case study for more details). Chan & Dott (1983) recognize an overall shallowing-up trend in this forearc basin, with deltaic systems prograding across the narrow shelf and out over the basin-fill in early Eocene times. Minimum sediment accumulation rates were 67 cm/1000 years with an estimated separation of 750–1700 years between sediment gravity flows (Chan & Dott 1983). As with many forearc basins, fans are unlike those on mature passive margins because of the relatively small, elongate basins, i.e. they tend to be more confined at their lateral margins in forearcs.

The template upon which forearc basin sedimentation develops may comprise an accretionary prism and/or volcanic arc basement. The Cenozoic West Sumatran forearc basin, for example, shows a deep, partially-filled basin between the shelf and an outer arc ridge, with flat-lying seismic facies showing onlap over: (a) the arc-massif on the landward side, and (b) either the subsiding accretionary prism, or attenuated continental crust, on the trench side (Beaudry & Moore 1985). Deposition presently occurs in water depths of 600–1000 m.

CASE STUDY: CRETACEOUS–NEOGENE, SOUTHERN JAPAN

The Cretaceous–Neogene Shimanto Belt, extending for more than 1800 km from the Nansei Islands in the southwest to Boso Peninsula southeast of Tokyo (Fig. 11.29a), is one of the best studied of any ancient subduction–accretion systems (Kanmera 1976a & b, Suzuki & Hada 1979, Teraoka 1979, Taira *et al.* 1980, 1982, Tazaki *et al.* 1980, Sakai & Kanmera 1981, Ogawa 1982, 1985, Taira 1985). In southern Japan, Taira (1985) defines three major tectono-stratigraphic

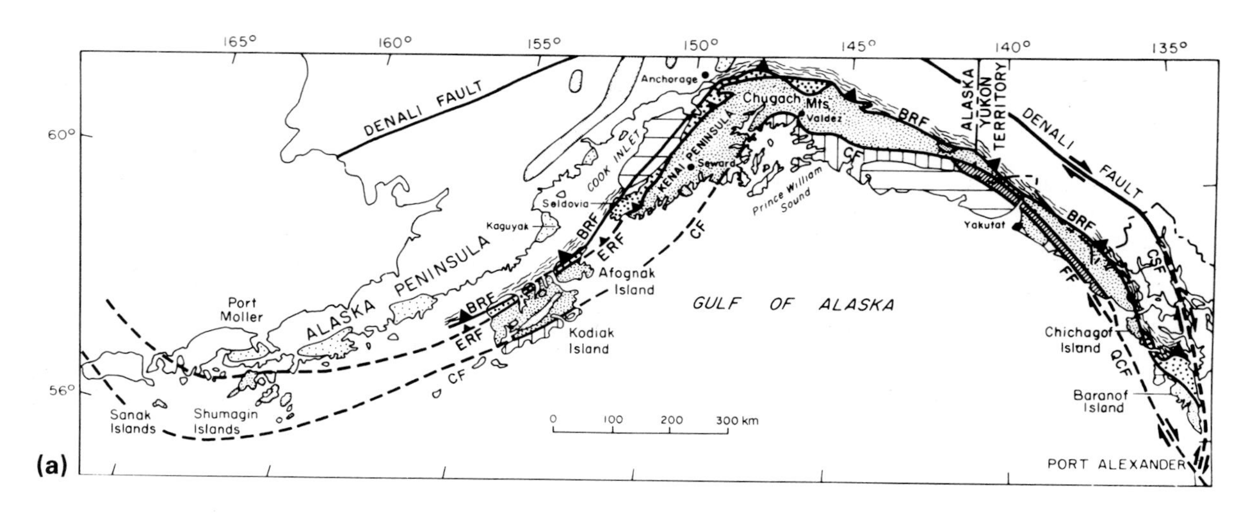

165°
160°
155°
150°
145°
140°
135°
60°
56°
DENALI FAULT
Anchorage
Chugach Mts.
Valdez
ALASKA
YUKON TERRITORY
COOK INLET
KENAI PENINSULA
Seward
Seldovia
Prince William Sound
Kaguyak
BRF
ERF
CF
CSF
FF
QCF
Yakutat
Afognak Island
GULF OF ALASKA
ALASKA PENINSULA
Port Moller
Kodiak Island
Chichagof Island
Baranof Island
Sanak Islands
Shumagin Islands
0 100 200 300 km
PORT ALEXANDER
(a)

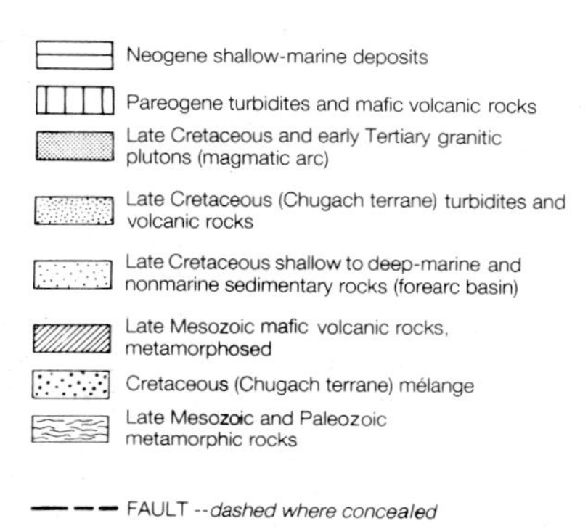

Neogene shallow-marine deposits
Pareogene turbidites and mafic volcanic rocks
Late Cretaceous and early Tertiary granitic plutons (magmatic arc)
Late Cretaceous (Chugach terrane) turbidites and volcanic rocks
Late Cretaceous shallow to deep-marine and nonmarine sedimentary rocks (forearc basin)
Late Mesozoic mafic volcanic rocks, metamorphosed
Cretaceous (Chugach terrane) mélange
Late Mesozoic and Paleozoic metamorphic rocks
FAULT --dashed where concealed

BRF BORDER RANGES FAULT
CSF CHATHAM STRAIT FAULT
CF CONTACT FAULT
ERF EAGLE RIVER FAULT
FF FAIRWEATHER FAULT
QCF QUEEN CHARLOTTE FAULT

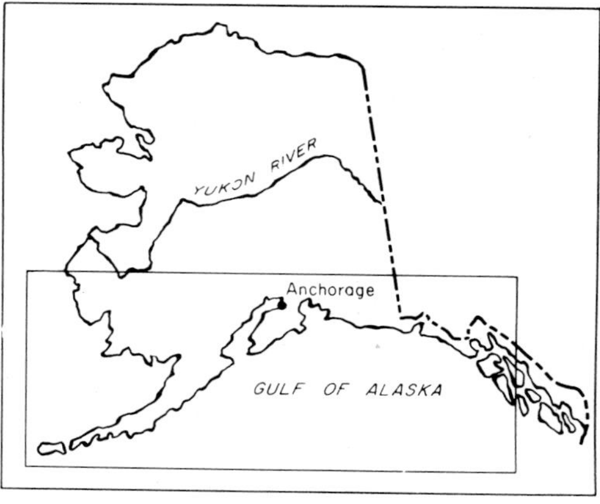

YUKON RIVER
Anchorage
GULF OF ALASKA

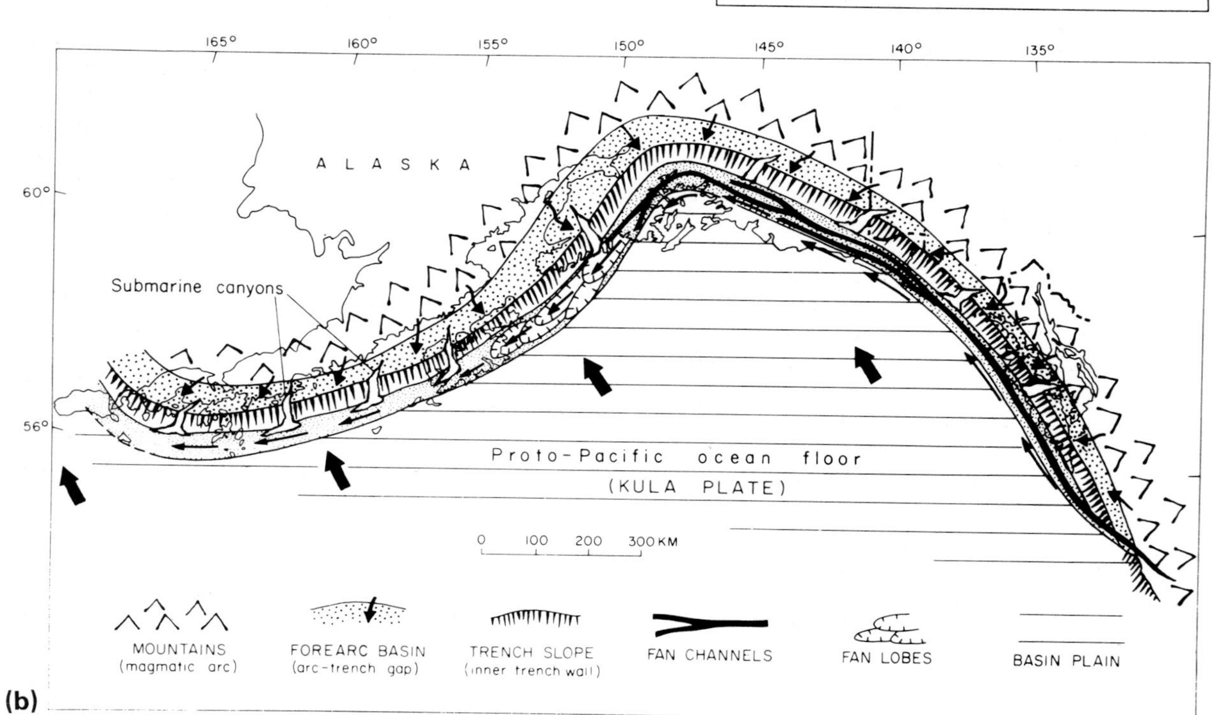

165°
160°
155°
150°
145°
140°
135°
60°
56°
ALASKA
Submarine canyons
Proto-Pacific ocean floor
(KULA PLATE)
0 100 200 300 KM
MOUNTAINS (magmatic arc)
FOREARC BASIN (arc-trench gap)
TRENCH SLOPE (inner trench wall)
FAN CHANNELS
FAN LOBES
BASIN PLAIN
(b)

'terranes' together with several zones of substantial strike-slip displacement (Fig. 11.29): (a) a northern composite terrane of pre-Jurassic rocks accreted to the Asian margin prior to the opening of the Sea of Japan, including the Hida, Sangun and Yamaguchi belts; (b) a Jurassic subduction complex of the Chyugoku, Chichibu, Tamba, Mino and Ashio belts, and the associated metamorphic Sambagawa and Ryoke belts, and (c) the Cretaceous–Neogene Shimanto Belt, interpreted as forearc and intra-arc environments. In general, the Shimanto Belt, dislocated by major faults that outcrop as northerly-dipping high-angle reverse faults, youngs towards the south. Oblique subduction and strike-slip tectonics have been important in the evolution of Japan since at least Late Palaeozoic times (Taira *et al.* 1983).

The Shimanto Belt comprises five main belts (Taira 1985) from north to south: (a) extrusive andesites and rhyolites, granitic intrusives and terrestrial sediments of the magmatic arc, with 130–30 Ma ages but clustering between 95–70 Ma; (b) Upper Cretaceous, mainly deep-marine coarse-grained clastics fed from east to west into a series of small basins in the forearc; (c) south of the Median Tectonic Line (MTL), a Carboniferous–Jurassic non-magmatic outer arc of basalts, limestones, cherts, sandstones and mudrocks; (d) an Early–Late Cretaceous outer-arc shelf-basin of mainly shallow-marine and fluvial to brackish-marine sediments, and (e) to the south of the Butsuzo Tectonic Line (BTL), the Cretaceous–Neogene Shimanto subduction–accretion complex. In structural sections, from north to south, across central Honshu and Shikoku to the Nankai Trench, it is possible to appreciate the longevity of Japan as an active convergent margin (Fig. 11.29 & 30). The modern analogue is the present subduction system.

In Shikoku, the Shimanto Belt consists of a northern Cretaceous belt, and a southern Eocene to lower Miocene, belt (Taira *et al.* 1982, Taira 1985). The Cretaceous Shimanto is divisible into four major tectono-stratigraphic units bounded by steeply-dipping reverse faults: (a) Neocomian–Coniacian brackish to shallow-marine sediments; (b) Aptian–Cenomanian turbidites; (c) Coniacian–Campanian turbidites and mélanges, and (d) upper Campanian–Maastrichtian shallow-marine deposits, including sediment slides. Figure 11.31 (Taira *et al.* 1982) summarizes the palaeogeography for the Cretaceous (Lower) Shimanto Group, and emphasizes the range of deep-marine environments from forearc-basin submarine fans, to accretionary-prism ridged basins in deeper water, to trench sedimentation. Wet-sediment deformation is recognized, both as surface gravity-slides and as tectonic, thrust-generated, mélange. Work by Pickering *et al.* (1988a) suggests that many of these mélanges may have originated as mud or shale diapirs that were subsequently folded, sheared and flattened.

The Coniacian–Campanian turbidite-mélange unit comprises tectonic slivers of pillow basalts, radiolarian cherts (Facies Class G), hemipelagic mudrocks (Facies Classes E & G), acidic tuffs and sandstone turbidites (Facies Classes B, C & D) in a shale matrix. Studies by Taira *et al.* (1980), Taira (1981) and Kodama *et al.* (1983), suggest that they were formed in equatorial latitudes, while the Campanian turbidites accumulated essentially at present-day latitudes. There is a *c.* 50 m.y. age difference between the oceanic plate material (including basalt and limestone) and the matrix of the mélange. The difference in latitude represents *c.* 3000 km, which can be used with age data to calculate a plate speed of about 6 cm/year. The age and lithology of the facies within the mélanges show a systematic southward change (Fig. 11.32), thereby allowing successive reconstructions of the trench stratigraphy. They show that progressively younger oceanic crust was being subducted and partially obducted.

The Southern Shimanto comprises an Eocene and lower Miocene belt. The lithologies are predominantly deep-marine clastics and igneous rocks, interpreted as deposits of forearc accretionary prism environments (Taira 1985). The offshore geology (Fig. 11.30) shows modern subduction–accretion processes, and in this case study highlights the problem of separating the 'ancient' Upper (Southern) Shimanto Belt from the 'modern' subduction system.

The evolution of the Shimanto Belt has been episodic, as seen by the discrete development of the various 'sub-belts', each during intervals of *c.* 10–20 m.y. This episodic evolution is probably the result of cyclic subduction of the oceanic plate in 10–20 m.y. cycles (Taira *et al.* 1980). Structural and stratigraphic data point to the importance of movement on transform faults

Figure 11.28 Location of Chugach Terrane, Gulf of Alaska, showing the major tectono-stratigraphic features Fig. 11.28(a), and a palaeogeographic reconstruction for the Late Cretaceous. In Figure 11.28(b), Large arrows indicate approximate direction of sea floor motion, small arrows show principal directions of sediment transport, and palinspastic restoration shows a more southerly position for the Chugach Terrane based on palaeomagnetic data. After Nilsen & Zuffa (1982).

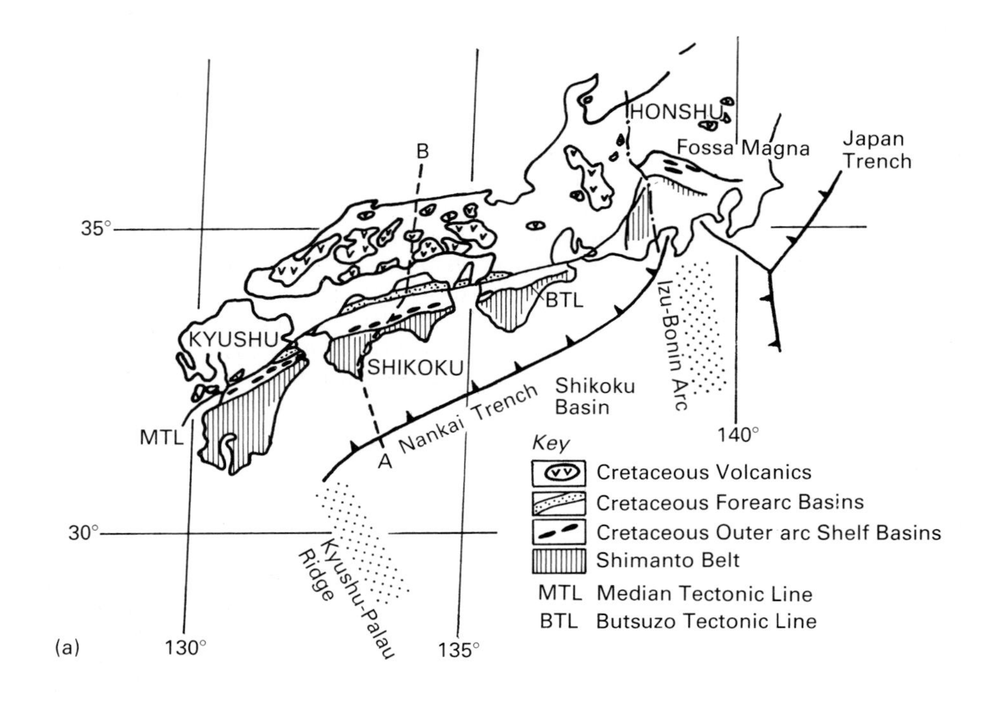

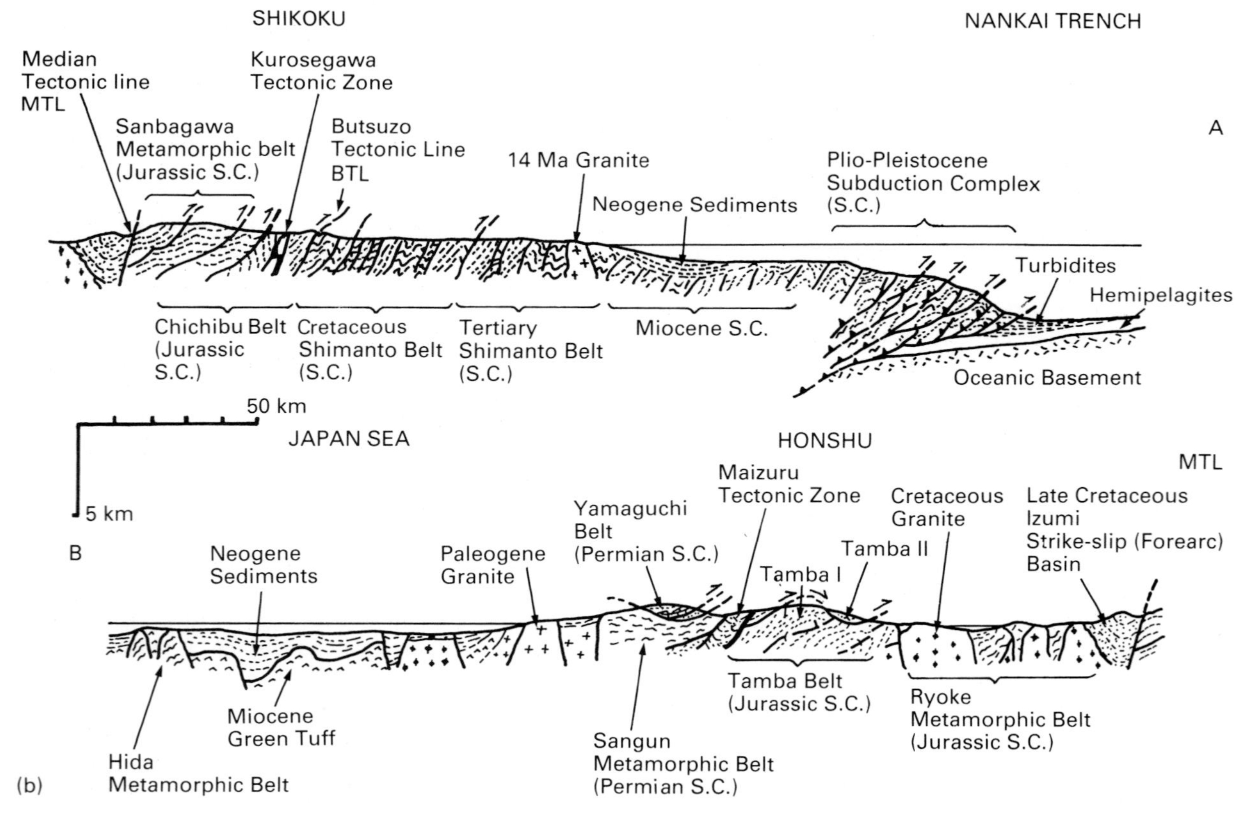

Figure 11.29 Schematic geological cross-section of SW Japan, located on Figure 11.29(a), to show the long-lived history of Japan as part of an active convergent margin. Main tectono-stratigraphic zones or belts, together with major bounding lineaments, shown in Figures 11.29 (a & b). After Taira (1985). See text for explanation.

during the late Oligocene to early Miocene (Ogawa 1985). For example, the Miura–Boso Terrane of SE central Honshu was transferred to the Honshu Arc from the Izu–Bonin Arc in this manner. The accreted material comprises: (a) Miocene–Pliocene, volcaniclastic, deep- to shallow-marine sediments; (b) dismembered ophiolitic rocks, and (c) some rocks of the Shimanto Group. All were accreted under predominant dextral transpressive tectonics. This stress is believed to have been the result of oblique subduction of the NE Philippine Plate, with a present average annual relative displacement northwestwards of about 3.5 cm/year (Ogawa 1985).

The dextral transpressive shear is responsible for the Neogene accretion of the Mineoka Ophiolite Belt, a chaotic volcanic and sedimentary succession, including olistostromes, that accumulated in the proto Izu–Bonin Ridge (Izu–Ogasawara Arc) and was later accreted onto the Honshu Arc (Ogawa 1983). The Mineoka Group, probably Lower Tertiary to lower Miocene, comprises an ophiolitic and pelagic/terrigenous clastic succession. The lower siliceous and calcareous claystones/mudstones are overlain by arkosic turbidites, above which there is a shallowing-upward sequence into possible shallow-marine, tidally-dominated clastics (Ogawa 1983)–a stratigraphy that may reflect the uplift of the basin during sustained arc–arc collision.

To explain the origin of some mélanges in the Shimanto Belt, Ogawa (1985) proposed a model involving subduction–accretion processes operating in submarine trenches, whereby subducting seamounts are at least partially off-scraped at the leading edge of the accretionary prism (Fig. 11.33). Such off-scraping leads to the incorporation of seamount material, along with oceanic-plate sediments, into the accretionary prism. This model has a superb modern analogue in the Dai-ichi Kashima Seamount that is currently being dismembered, partially subducted and accreted in the southern part of the Japan Trench (Mogi & Nishizawa 1980).

The Pleistocene Ashigara Group in central Honshu is a further example of a sedimentary basin formed between the colliding Izu–Ogasawara and Honshu Arcs (Huchon & Kitazato 1984). Figure 11.34 is a summary of the stratigraphy of the group. The arc–arc collision led to the uplift of a source for the voluminous conglomerates of the Miocene Tanzawa Group. Palaeobathymetric studies, based on benthic foraminifera, suggest that the lower to middle Ashigara Group was deposited in water depths of 1000–2000 m (Fig. 11.34). Above this deep-marine clastic system, there is a coarsening-upward sequence, interpreted to reflect a shallowing-up, from the 1500 m-thick Neishi Formation (deep-sea plain) and 1300 m-thick Seto Formation (submarine fan) into the Hata Formation (shelf edge) and Shiozawa Formation (alluvial fan). Soon after the deposition of the upper Ashigara Group, the area was overthrust by the Tanzawa mountains along the Kannawa Fault and folded under NW–SE compression. At about 0.3 Ma, the direction of compression changed considerably to a N–S or NE–SW compression when the colling Izu–Ogasawara Arc finally locked against central Japan (Huchon & Kitazato 1984).

Deep-marine foreland basins

In recent years, active convergent plate margins have been divided into those in which oceanic crust is consumed by subduction processes, generally associated with arc magmatism and volcanism, and margins where continents and arcs collide, resulting in underplating without subduction. Clearly, many remnant sutures mark the site of complex collisional events that may have initially reflected subduction–accretion, finally being replaced by underplating as continent–continent collision locked the system.

Of the few modern deep-marine foreland basins that have been recognized, the only two that have been described in any detail are: (a) the western Taiwan foreland basin that resulted from the collision of the Luzon volcanic arc with mainland China *c.* 4 Ma, and is 400 km long, 100 km wide and, in the southern part, has water depths up to about 1500 m (Covey 1986), and (b) the Timor–Tanimbar foreland basin in the Banda Arc region, formed because of collision between the Banda volcanic arc and the northern Australian continental margin (Audley-Charles 1986a & b). The Timor–Tanimbar foreland basin is described as a case study below. In the western Taiwan foreland basin, plate convergence in the Taiwan area is about 7 cm/year (70 km/m.y.), shows a diachronous (oblique) north to south closure such that the Manila Trench dips overall towards the south, involves 160 km of shortening to date, involves uplift rates of about 5 mm/year and denudation rates for the entire Taiwan of 5.5 mm/year (Covey 1986).

Foreland basins develop as a response to load-induced stresses, for example thrust sheets flexing the lithosphere. Quinlan & Beaumont (1984) consider a continuous visco-elastic plate to be a reliable rheological model of the lithosphere, and they conclude that a load applied to a thick lithosphere results in a relatively wide and shallow basin, whereas the same load applied to a

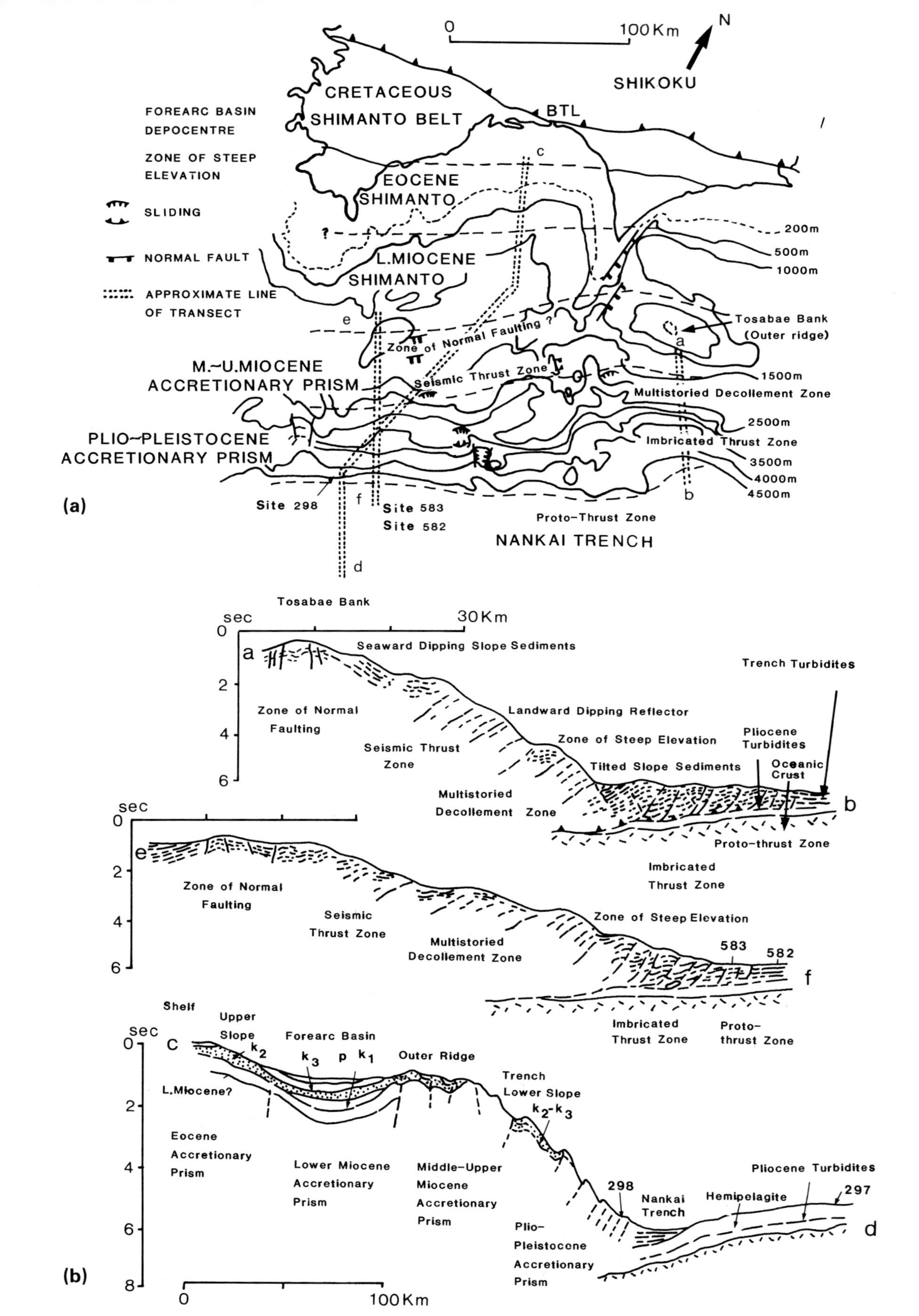
0
100 Km
N
SHIKOKU
CRETACEOUS
SHIMANTO BELT
BTL
FOREARC BASIN
DEPOCENTRE
ZONE OF STEEP
ELEVATION
SLIDING
NORMAL FAULT
APPROXIMATE LINE
OF TRANSECT
EOCENE
SHIMANTO
L.MIOCENE
SHIMANTO
200m
500m
1000m
Tosabae Bank
(Outer ridge)
Zone of Normal Faulting ?
M.~U.MIOCENE
ACCRETIONARY PRISM
Seismic Thrust Zone
1500m
Multistoried Decollement Zone
2500m
PLIO~PLEISTOCENE
ACCRETIONARY PRISM
Imbricated Thrust Zone
3500m
4000m
4500m
(a)
Site 298
Site 583
Site 582
Proto-Thrust Zone
NANKAI TRENCH
Tosabae Bank
sec
30Km
Seaward Dipping Slope Sediments
Trench Turbidites
Zone of Normal
Faulting
Landward Dipping Reflector
Seismic Thrust
Zone
Zone of Steep Elevation
Pliocene
Turbidites
Tilted Slope Sediments
Oceanic
Crust
Multistoried
Decollement Zone
Proto-thrust Zone
Imbricated
Thrust Zone
Zone of Normal
Faulting
Seismic
Thrust Zone
Multistoried
Decollement Zone
Zone of Steep Elevation
583
582
Imbricated
Thrust Zone
Proto-
thrust Zone
Shelf
Upper
Slope
Forearc Basin
Outer Ridge
L.Miocene?
Trench
Lower Slope
Eocene
Accretionary
Prism
Lower Miocene
Accretionary
Prism
Middle-Upper
Miocene
Accretionary
Prism
Pliocene Turbidites
298
Nankai
Trench
Hemipelagite
297
Plio-
Pleistocene
Accretionary
Prism
(b)
100Km

thin lithosphere generates a relatively narrow and deep basin. Thus, thin lithosphere provides the most favourable template upon which to develop deep-marine foreland basins. The recognition of ancient foreland basin successions relies upon the ability to recognize tectonic loading as the primary control on subsidence.

An important type of foreland basin, and one for which there are few well documented case studies, are those that develop in front of obducting ophiolitic basement. A good example is the Upper Cretaceous foreland basin in the Oman Mountains, associated with the obduction of the Semail ophiolite (Robertson 1987). The impingement of the advancing thrust-load on the edge of the Tethys Ocean down-flexed the crust and passive margin in the late Coniacian to Campanian (88.5 to 73 Ma) to create a foredeep, or foreland basin, that migrated towards the foreland. This foreland basin was partly infilled by terrigenous turbidites and olistostromes, in a deep-marine environment below the CCD (Robertson 1987).

Ori *et al.* (1986), in a study of the Central Adriatic post-Oligocene foreland basin, make a number of very interesting observations that summarize many of the tectono-stratigraphic attributes of foreland basin development: (a) the foreland basins (foredeeps) migrated in front of the advancing thrust pile, in this case towards the northeast; (b) thrust highs, with vertical displacements up to 1000 m, can be eroded to

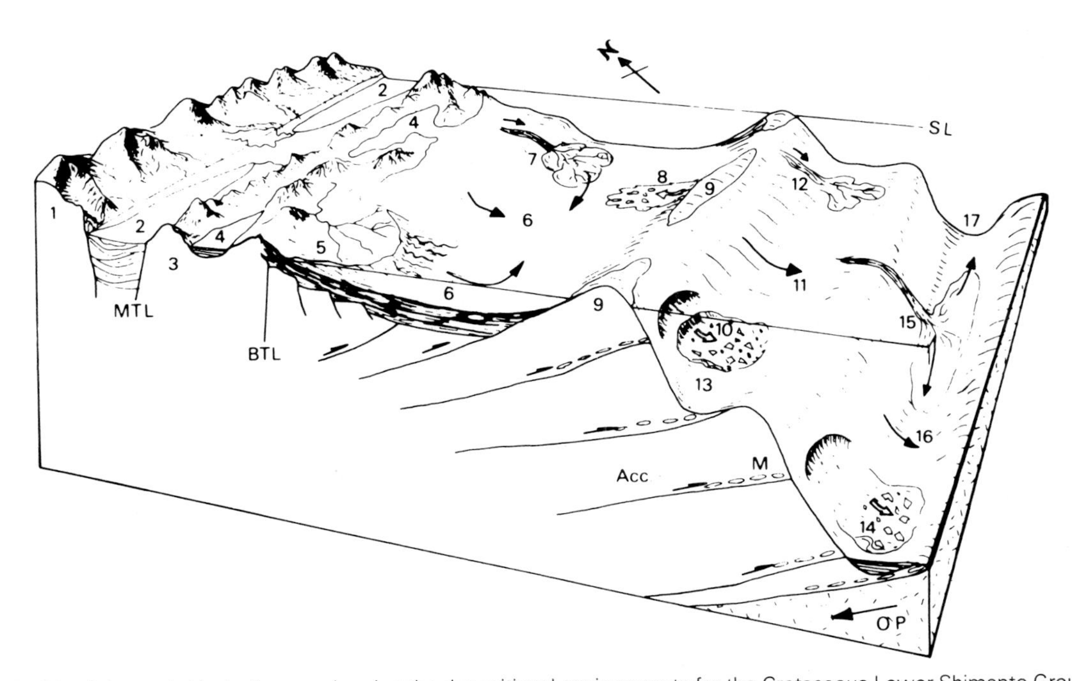

Figure 11.31 Schematic block diagram showing the depositional environments for the Cretaceous Lower Shimanto Group (not to scale), after Taira *et al.* (1982). 1, seaward limit of Ryoke high P–T belt; 2, intra-arc basins; 3, Sambagawa low P–T and Chichibu Belts, with; 4, mainly Lower Cretaceous small, forearc, shelf basins with thin fluvial and shallow-marine sands and mudrocks; 5, deltaic deposits; 6, deep-marine, non-fan, turbidite systems; 7 submarine canyon, fan-channel and fan system; 8, slide deposits, including olistostromes, on upper inner-trench wall; 9, trench-slope break, partly emergent; 10, slide deposits including olistostromes on upper inner-trench wall; 11, non-fan turbidite systems; 12, canyon-fan system contributing sediments to: 13, 'perched' accretionary-prism top basins; 14, slides including olistostromes on lower inner-trench wall; 15, canyon-fan system; 16, lateral supply of turbidites to: 17, trench floor. MTL = Median Tectonic Line; BTL = Butsuzo Tectonic Line; Acc = Accretionary prism wedge separated by thrusts with mélange (M); OP = intermittently subducting oceanic plate. The Palaeogene Shimanto Belt environments are regarded as similar to this Lower Cretaceous reconstruction, but with the deformed Lower Shimanto Group being exposed and eroded, and a new accretionary prism and associated environments having developed further southeast.

Figure 11.30 (opposite) Offshore geological features of the Shikoku subduction zone, southern Japan (11.30a), together with interpretation from seismic profiles of transects A–B, E–F and C–D (11.30b) located in Figure 11.30(a). After Taira (1985).

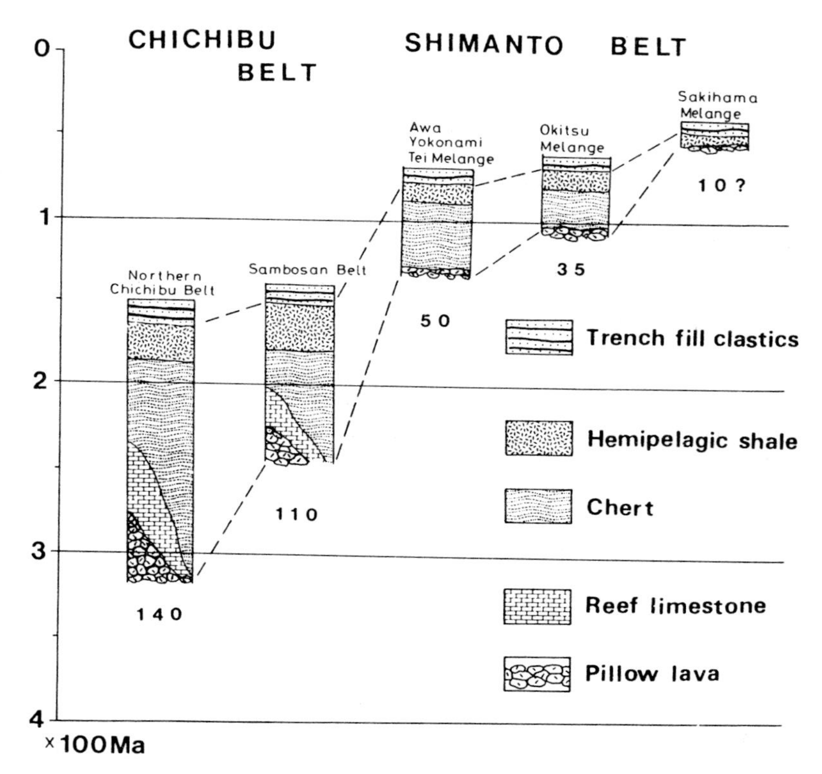

Figure 11.32 Reconstruction of oceanic plate stratigraphy in the Shikoku subduction system during Jurassic to Eocene times. Note that younger oceanic plate had been subducted with time. Numbers shown below columns represent age difference between the oldest oceanic-plate material and trench-fill clastics in millions of years. After Taira (1985).

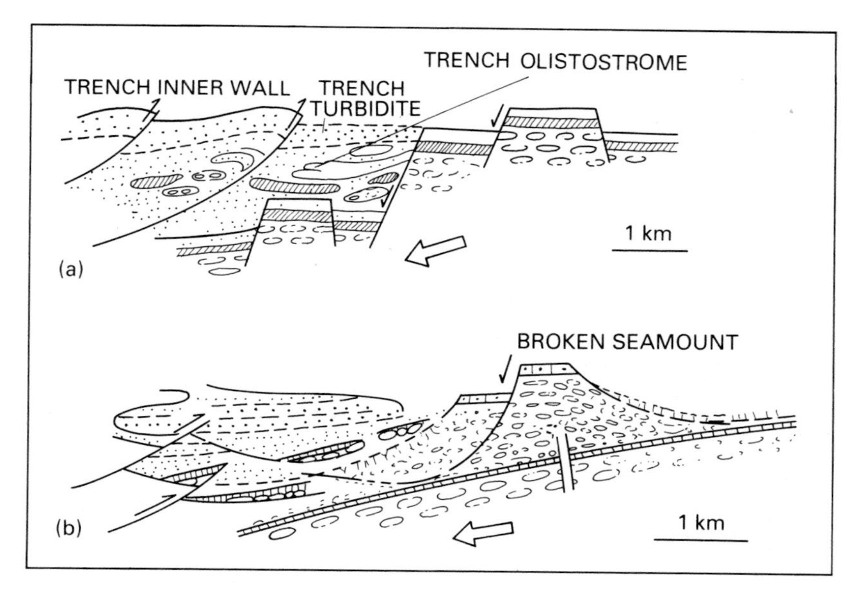

Figure 11.33 Conceptual model for the accretion and off-scraping of subducting oceanic plate material in a trench, after Ogawa (1985). (a) Formation of trench olistostrome from the normally-faulted oceanic floor, including basaltic material. (b) Off-scraping of parts of a subducting seamount.

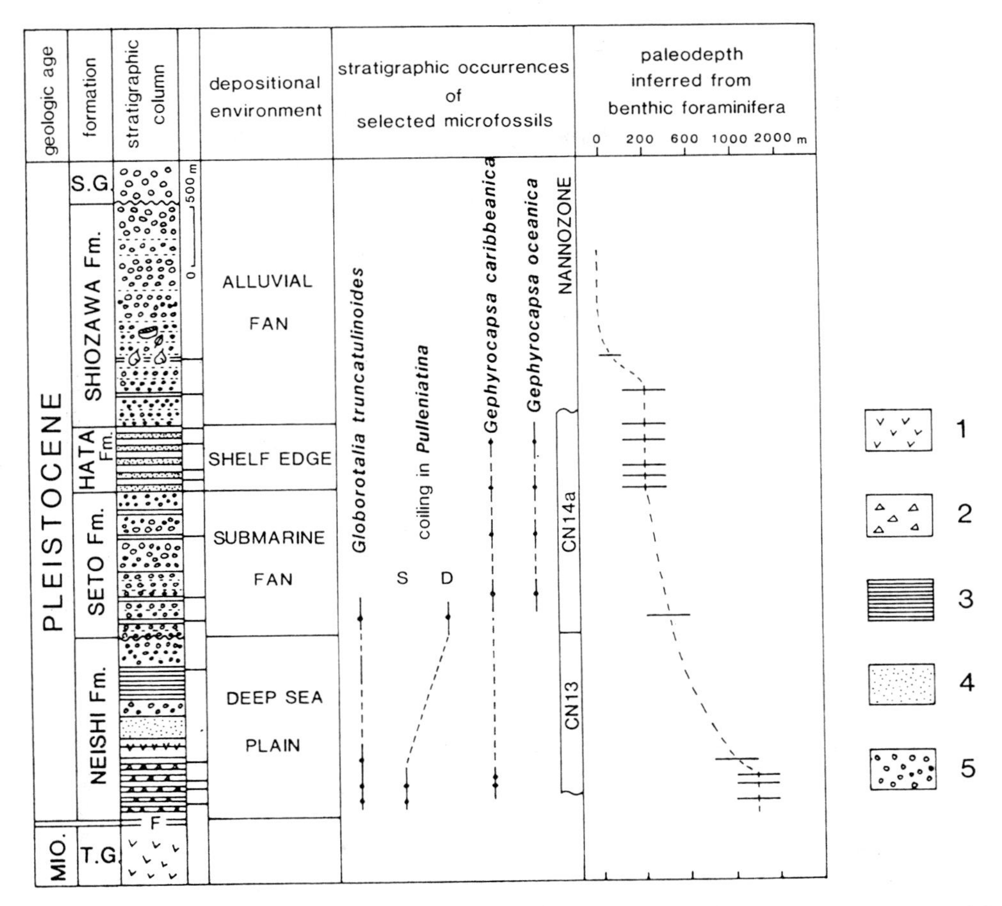

Figure 11.34 Summary of the stratigraphy and palaeoenvironments of the Pleistocene Ashigara Group, SE Japan (after Huchon & Kitazato 1984). 1, volcaniclastics; 2, lapilli tuffs; 3, siltstones; 4, sandstones; 5, conglomerates, S.G. = Suruga gravels; T.G. = Tanzawa Group. See text.

produce a local sediment source; (c) sediment dispersal patterns are mainly parallel to the long axis of the basin, although lateral supply is important, and (d) the foredeep was segmented into discrete depocentres as a result of contemporaneous tectonism and diapirism (probably due to basement control). In the Central Adriatic case, the Apennines are a complex thrust belt of tectonic units that are still moving, and have been emplacing thrust sheets since the Oligocene.

The deep-water Miocene onshore parts of the Periadriatic foreland basin are very well documented by Ricci Lucchi (1975a & b, 1981, 1986) and Ricci Lucchi & Ori (1984, 1985). Ricci Lucchi & Ori (1984, 1985) recognize both: (a) major clastic bodies with volumes of 3000–30 000 km^3 that represent the progressive infilling of northeasterly migrating foredeeps, together with (b) minor basin-fills of 50–500 km^3. Amongst the smaller basins, thrust-based (thrust-top) basins marginal to the foredeep, or 'piggyback basins' (Ori & Friend 1984), are assigned to a general class of 'satellite basins' (Ricci Lucchi & Ori 1984, 1985). Rates of depocentre (foredeep) shifting from the Oligocene to the present day have been as great as about 7.5 cm/year (Miocene), with fast rates of sediment accumulation tending to occur during phases of most rapid depocentre shifting, equivalent to periods of greatest tectonic activity: for example, during the fast depocentre migration of the Miocene, sediment accumulation rates reached about 87 cm/1000 years (Ricci Lucchi & Ori 1984).

The main style of deep-water sedimentary fill of the Miocene foreland basins described by Ricci Lucchi and others involves essentially sheet systems with many ponded flows, including Facies C2.4 megaturbidites (Ricci Lucchi & Valmori 1980). The basins were effectively over-supplied with respect to their size, a feature that appears common to many foreland basin successions. A detailed sedimentological case study of the Miocene Marnoso-arenacea Formation from the Periadriatic area is given in Chapter 8 on sheet systems.

CASE STUDY: NEOGENE–QUATERNARY SOUTHERN BANDA ARC

The Java Trench marks the boundary between the northward moving Indian Ocean Plate and the southeast part of the Asian Plate (Fig. 11.35a). At the eastern end of the Java Trench there is northward convergence and underplating of the Australian continental margin beneath the Banda Arc, without any intervening oceanic crust (Jacobson *et al.* 1979, Bowin *et al.* 1980, Audley-Charles 1986a & b, Karig *et al.* 1987).

The non-volcanic Outer Banda Arc can be interpreted as an emergent part of the imbricate thrust system formed in response to the underplating of the Australian margin beneath the Timor–Babar–Tanimbar–Kai island chain. The Southern Banda Arc region, therefore, may contain one of the few identified modern, deep-marine foreland basins, with the Timor Trough as a foreland basin whose subsidence has been controlled, at least in part, by loading of the North Australian margin by thrust sheets of the Outer Banda Arc. The Australian shelf represents the distal part of the foreland.

Prior to foreland-basin development, subduction–accretion processes similar to those now occurring farther west along the Java Trench are believed to have operated. The forearc was then converted into a foreland basin as the nature of the crust that was subducted/underplated changed. Before collision and emplacement of the allochthonous nappes in Timor, micropalaeontological studies suggest that sedimentation on both the Asian and Australian margins of the Java Trench, from at least the late Miocene to early Pliocene, occurred in water depths greater than 2000 m.

Micropalaeontological studies in Timor suggest post-collision uplift rates varying from 1.5–3 mm/year, associated with crustal shortening by nappe emplacement at rates possibly greater than 62.5–125 mm/year (Audley-Charles 1986a). Tectonic thickening of the Australian continental margin and shelf has occurred by imbrication (mainly by southerly-transported thrust sheets) together with other crustal shortening processes. Figure 11.35b shows some of the major thrust slices under the northern margin of the Tanimbar Trough, a deep marine basin representing the along-strike continuation of the Timor Trough and the axis of the present foreland basin. The period of overthrusting is estimated at 0.4–0.8 Ma (Audley-Charles 1986a). Figure 11.36 summarizes the stratigraphy on the north Australian passive continental margin (foreland), the colliding island arc complex of Timor, and DSDP Site 262 within the deep-marine foreland basin, or foredeep.

Seismic reflection and SeaMARC II (side-scan and swath bathymetry) reveal a tectonic pattern analogous to that of typical oceanic subduction zones, with a deformation front in the Timor Trough discontinuously advancing southward, as successive thrust sheets of sediment are assembled from the underplating Australian margin (Karig *et al.* 1987). Furthermore, Karig *et al.* (1987) postulate oblique underplating, leading to dextral shear along a NE-trending fault system that offsets the outer arc between the islands of Savu and Roti. Within Timor, sediments range up to more than 1000 m thick, with rapid lateral thickness changes reflecting the local growth of folds and/or thrusts, at least some of which are controlled by the reactivation of normal faults in the down-bending Australian margin. Within Timor Trough, at DSDP Site 262 (Veevers, Hietzler *et al.* 1974), fine-grained siliciclastics and biogenic carbonates were recovered (Fig. 11.36), suggesting that coarser-grained sediments are trapped on the inner trough slope. Sediment failures on the lower slope provide the main source of material for the trough floor. A large amount of Facies Class F occurs, particularly as sediment slides.

Figure 11.35 Location and major tectonic features of the Southern Banda Arc and Timor Trough (a), and an interpreted NW–SE cross-section (b), after Audley-Charles (1986b), to show the similarity between this foreland-basin and an ocean-facing accretionary prism. See text.

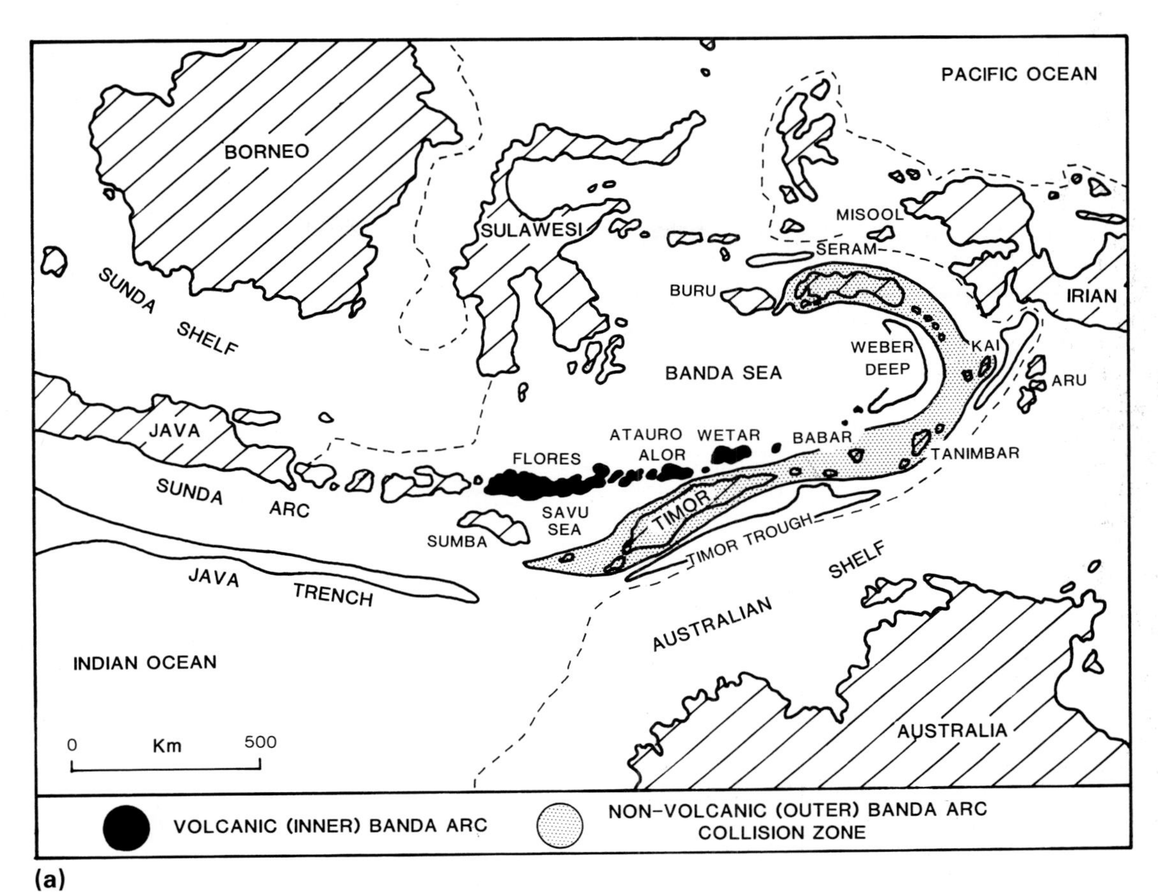
PACIFIC OCEAN
BORNEO
SULAWESI
MISOOL
SERAM
BURU
IRIAN
SUNDA
SHELF
WEBER
DEEP
KAI
BANDA SEA
ARU
JAVA
ATAURO
WETAR
BABAR
ALOR
FLORES
TANIMBAR
SUNDA
ARC
SAVU
SEA
TIMOR
SUMBA
TIMOR TROUGH
SHELF
JAVA
TRENCH
AUSTRALIAN
INDIAN OCEAN
AUSTRALIA
0
Km
500
VOLCANIC (INNER) BANDA ARC
NON-VOLCANIC (OUTER) BANDA ARC
COLLISION ZONE

(a)

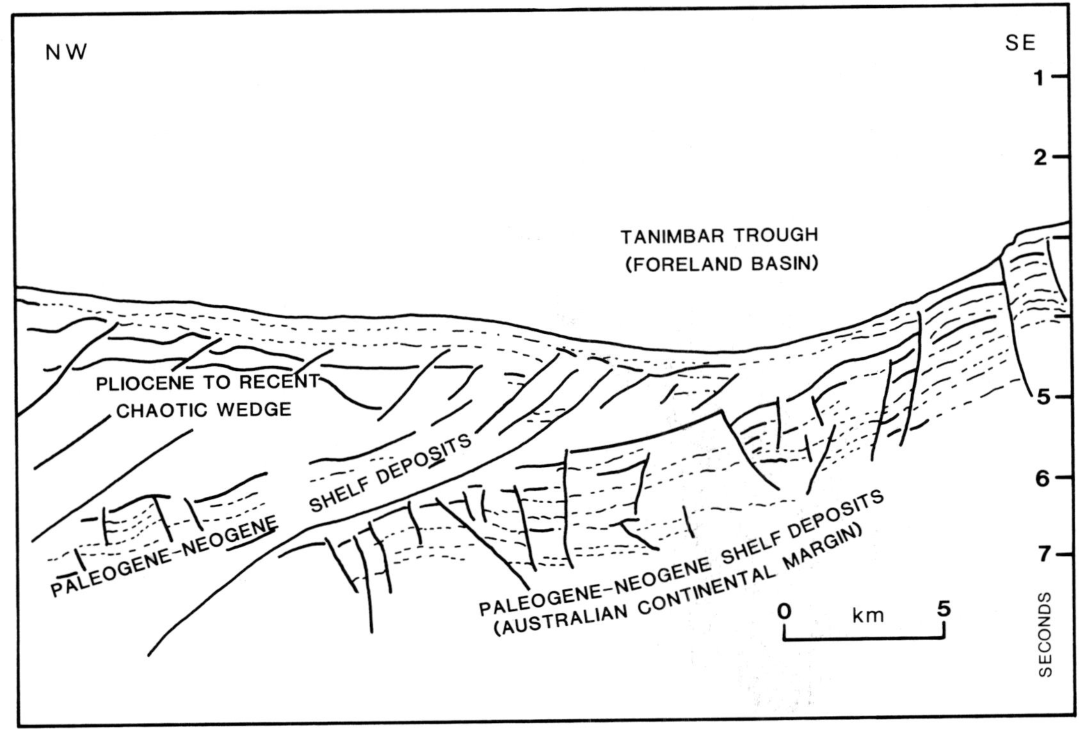
NW
SE
1
2
TANIMBAR TROUGH
(FORELAND BASIN)
PLIOCENE TO RECENT
CHAOTIC WEDGE
5
SHELF DEPOSITS
6
PALEOGENE-NEOGENE
7
PALEOGENE-NEOGENE SHELF DEPOSITS
(AUSTRALIAN CONTINENTAL MARGIN)
0
km
5
SECONDS

(b)

CASE STUDY: TERTIARY SOUTH PYRENEAN FORELAND BASIN

The South Pyrenean east–west trending foreland basin developed from early Eocene to Miocene times during north–south compression (Labaume *et al.* 1985). Within this foreland basin, a well-preserved turbidite system accumulated as the lower/middle Eocene Hecho Group (Mutti 1977, 1984, 1985, Mutti & Johns 1979, Mutti *et al.* 1984) (Figs 11.37a & b).

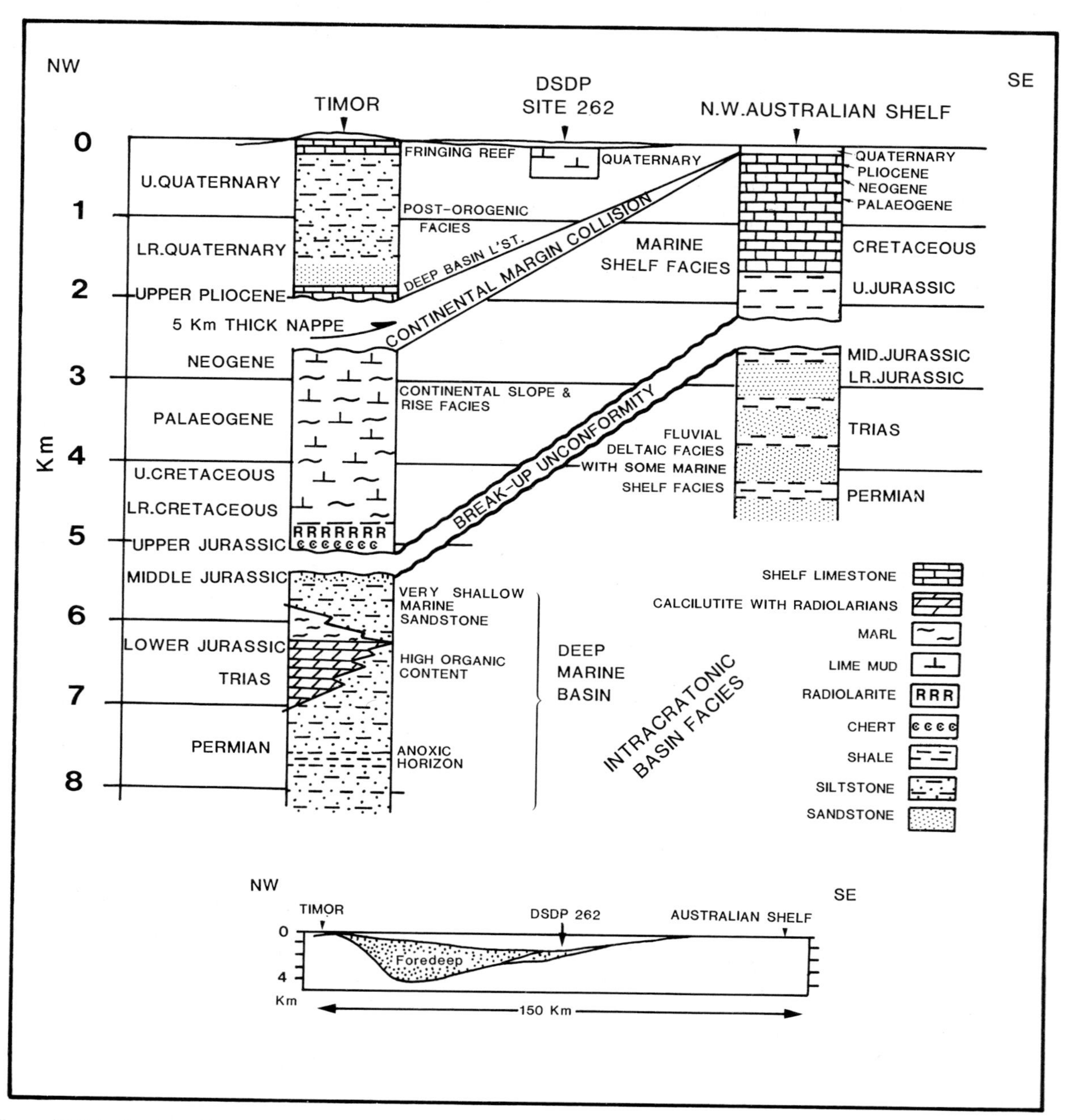

Figure 11.36 Stratigraphic columns at the margins of the Timor Trough (Timor and the North Australian continental shelf) and within the foreland basin. After Audley-Charles (1986b). DSDP Site 262 data from Veevers *et al.* (1978).

During the Late Cretaceous, 100–75 Ma, the Iberian Plate rotated approximately 400 km southeast with respect to a stable Europe, resulting in sinistral shear along the European–Iberian plate boundary. In the Late Cretaceous to Eocene, the relative plate motions changed to a NNW–SSE convergence of about 130 km to produce a north–south compressive regime that was active throughout the Tertiary (Fig. 12.17 & 18). Two foreland basins formed during the compressional phase – a North and South Pyrenean Tertiary Basin. Estimated minimum average southward migration of the border of the southern foreland basin was 5 mm/year, with the advance of the deformation front from early Eocene to early Miocene (Fig. 11.38) at about 3.5 mm/year, based on palinspastic reconstructions (Labaume *et al.* 1985).

Within the southern foreland basin, the main transport path for the clastics was axial and from east to west (Fig. 11.37a). Today the preserved basin dimensions are 250 km long and 15–45 km width, from the Tremp region in the east to the Pamplona region in the west. Fluvio-deltaic sedimentation predominated in the east,

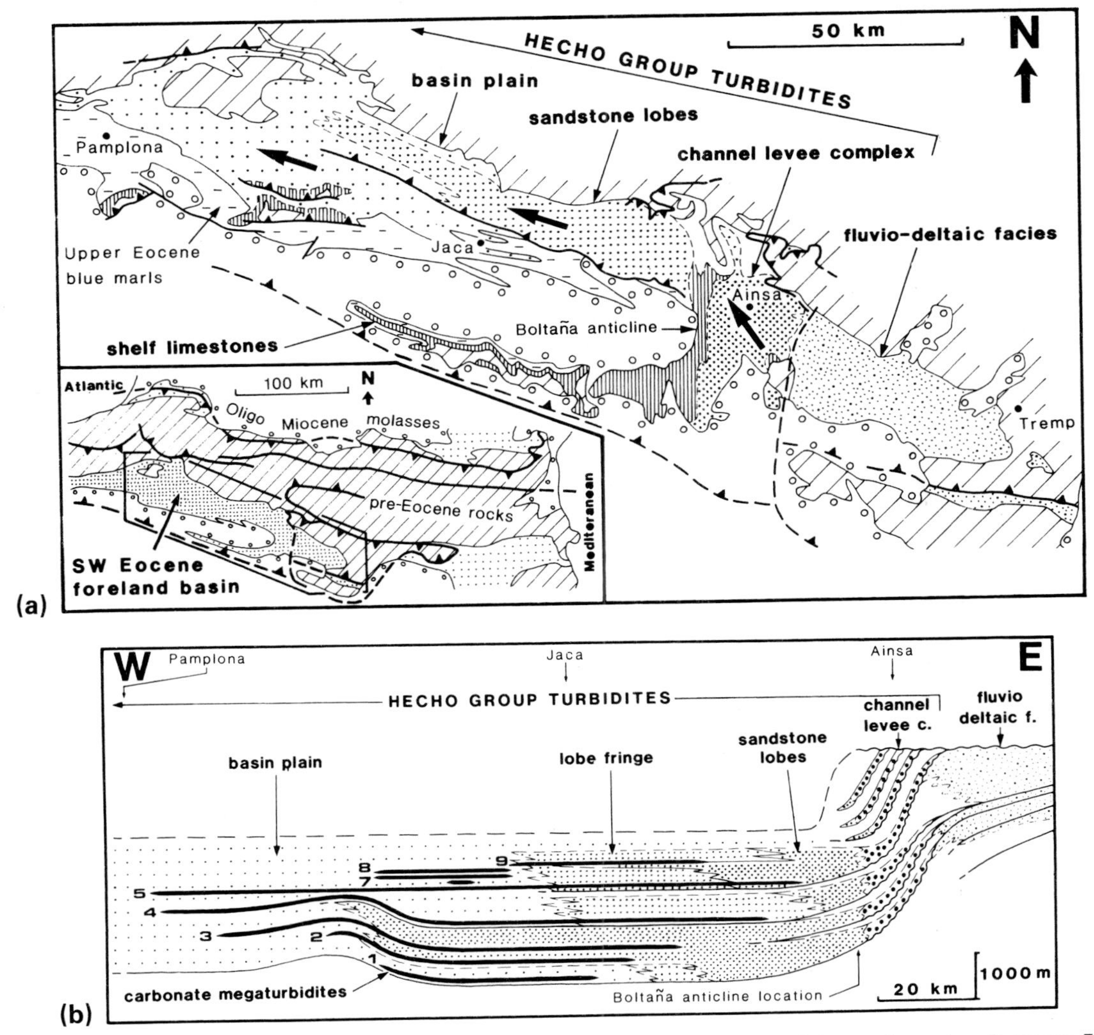

Figure 11.37a Gross facies distribution of the Lower and Middle Eocene sediments of the south-western Pyrenees Foreland Basin. After Labaume *et al.* (1985). Arrows are palaeocurrents in the Hecho Group, siliciclastic, turbidite system. Inset map shows the location of the basin in the Pyrenees. (b) Schematic location of the nine carbonate megaturbidites in a restored longitudinal section of the Hecho Group. After Mutti *et al.* (1985) and Labaume *et al.* (1985).

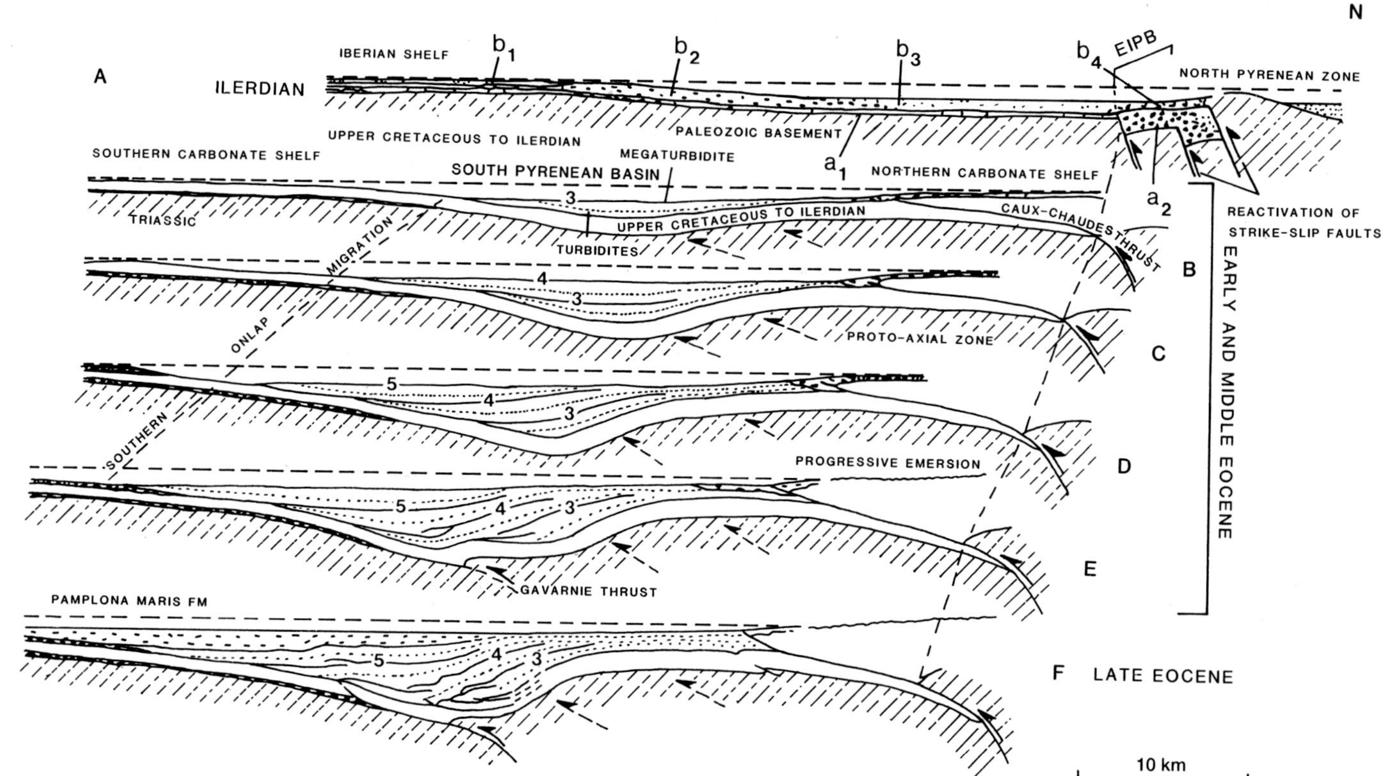

Figure 11.38 Evolutionary model for the Early and Middle Eocene of the southern Pyrenees Navarra–Aragon turbidite systems. In 11.38A: a = Cenomanian to Early Campanian, a1 shelf limestones, a2 base-of-fault-scarp breccias; b = Late Campanian to Ilerdian, b1 shelf sediments, b2 slope sediments, b3 turbidites, b4 base-of-slope breccias. Time lines are shown by megaturbidites (thick black lines 3 to 5). Panels E to F show Eocene evolution. After Labaume *et al.* (1985).

while to the west a deep-marine turbidite basin formed that was of the order of 1000 m deep and from 15–45 km wide (Mutti *et al.* 1984). The palaeogeographic boundary between the deltaics and the turbidite basin was controlled by the Cinca Fault. Contemporaneous with foreland basin sedimentation, extensive carbonate shelves formed on both the northern and southern basin margins. The northern carbonate shelf is not preserved but has been inferred from palaeocurrent and sediment composition studies. Along the length of the foreland basin, structural highs, like the Boltana and Mediano anticlines, were active during sedimentation. These syn-sedimentary highs acted as dams for deep-marine sediments to become ponded against. The most significant of these structures is the Boltana anticline that began to grow in the early Eocene, and divided the Hecho Group into an eastern and western basin (Mutti 1984).

The lower and middle Eocene Hecho Group (Figs 11.37a & b) is a turbidite system that accumulated in the foreland basin, and has a length of 175 km, a width of 40–50 km, a maximum thickness of 3500 m, and an estimated volume of 21 000–26 000 km^3 (Mutti 1984). The deep-marine turbidite facies are well documented in many papers, amongst which the best known are by Mutti (1977, 1985) and Labaume *et al.* (1985). The eastern portion of the Hecho Group comprises six, deep marine, channel–levee complexes, and the western part, with an aggregate thickness of 3500 m, consists of fan-fringe and basin-plain deposits. In the western part of the Hecho Group, Mutti & Johns (1979) found that the mean thickness of turbidites in the basin plain is approximately 11.8 cm, whereas in the fan-fringe deposits the mean value is only 4.9 cm. They interpreted the thicker, mud-rich, turbidites as a result of flows which bypassed the sand-rich fans.

A particularly interesting feature of the basin-plain deposits is the occurrence of 'megaturbidites' in beds up to 200 m thick. The megaturbidites ideally show a five-fold division (Fig. 11.39) and, based on criteria for slope failure and assuming a seismic origin for these beds with earthquakes of at least magnitude 7, Seguret *et al.* (1984) estimated that the sediment flows involved volumes up to 200 km^3.

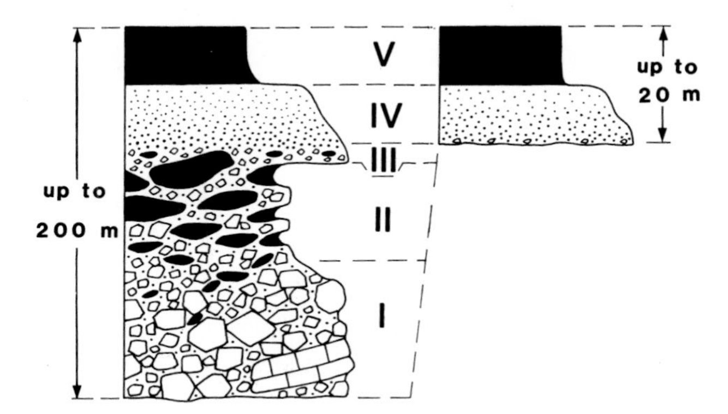

Figure 11.39 Example of a 200 m thick 'megabed' from the Tertiary foreland basin of the southern Pyrenees. Division I is mainly clast-supported and poorly organized megabreccia with limestone blocks up to 100 m thick; Division II & III is carbonate megabreccia with rip-up clasts of slope mudstone (II) and graded calcirudite with a calcarenite matrix (III); Division IV comprises coarse- to fine-grained calcarenite of bioclastic material with rare detrital quartz grains; Division V is an homogeneous marlstone. After Seguret *et al.* (1984) and Labaume *et al.* (1985).

Cliff-top view of the Awa mélange, Cretaceous Northern Shimanto Belt, south of Kubotsu and Tosa Shimizu, SW Shikoku, Japan. The tight-to-isoclinally folded sandstones and shales define a ghost stratigraphy involving dismembered folds of wavelengths from 20–30 m. Mélanges such as this one, showing some evidence of wet-sediment deformation, may have originated as mud diapirs in a forearc slope/accretionary prism, subsequently suffering (enjoying) intense deformation by progressive flattening, folding and shearing (cf. Pickering *et al.* 1988a).

CHAPTER 12

Oblique-slip continental margin basins

Introduction

Plate boundaries along which oblique-slip tectonics predominate may be the site of deep-marine sedimentary basins. While such margins may show tectonostratigraphic features in common with evolving and mature passive or convergent margins, it is generally the complexity of oblique-slip basins that warrants special consideration. Furthermore, the associated igneous and volcanic stratigraphy appears to be unique to such plate tectonic settings and this latter factor may prove to be one of the best means of interpreting possible ancient oblique-slip margins.

One of the most important aspects of oblique-slip continental margins has been the recognition of 'suspect terranes' or unrelated crustal and geological elements that have been juxtaposed through terrane accretion events (see edited volume by Howell, 1985). A suspect terrane is defined as an 'area characterized by an internal continuity of geology, including stratigraphy, faunal provinces, structure, metamorphism, igneous petrology, metalogeny, and palaeomagnetic record, that is distinct from that of neighbouring terranes and cannot be explained by facies changes' (Keppie 1986). In the Circum-Pacific, palaeobiogeographic and palaeomagnetic studies indicate terrane displacement from a few hundred to more than 6000 km! (Howell *et al.* 1985). Terrane accretion may occur over long time periods, for example China has a complex history of such events from the Late Precambrian, in Mongolia, continuing up until the collision of India with Asia about 45 Ma. Oblique subduction associated with strike-slip, or oblique-slip, has been a feature of terrane accretion in the evolution of Japan, at least throughout the Mesozoic and Cenozoic (cf. Taira *et al.* 1983).

A simple shear couple (Fig. 12.1) appears to provide an appropriate tectonic model for many cases of basin evolution controlled by oblique-slip (Harding 1974). However, a more complex modified simple shear model like that described by Aydin & Page (1984) for the San Francisco Bay region, California, may be more useful (Fig. 12.2). Ideally, the first set of strike-slip faults, the R (synthetic) and R′ (antithetic) shears, make angles of $\phi/2$ and $90° - \phi/2$ with the applied shear direction, respectively, where ϕ, the angle of internal friction, is typically around 30°. With progressive deformation, first the R′ and then the R faults are deactivated to become passive strain markers. Eventually, P shears form that are parallel to subparallel with respect to the principal applied shear direction and it is these faults that tend to accommodate the largest displacements.

Oblique-slip tectonics commonly result in complex facies changes over relatively short distances. Synclines and anticlines tend to develop synchronous with sedimentation, to generate elongate sedimentary depocentres that preferentially concentrate coarser grained clastics as elongate submarine fans and channel systems. The development of syn-sedimentary anticlines can generate intrabasinal highs that may act as local sediment sources, including gravity-controlled sediment mass flows. Normal growth faults and thrusts may develop along different margins of the same basin, deforming wet sediment at depth as well as creating areas of local uplift and subsidence. Thus, along the same oblique-slip zone, deep-marine clastic systems can develop under both transpressional and transtensional regimes, the latter producing sag and pull-apart basins. Such pull-apart basins commonly are deep relative to their width, resulting in abrupt lateral facies changes. Pull-apart basins tend to be floored by oceanic crust whereas sag basins have thinned with subsided continental crust.

The increasing availability of seismic sections has led to the recognition of flower structures as a common feature of oblique-slip zones. Flower structures represent the shallow expression of deeper, near-vertical, strike-slip fault zones where either compressive (transpressional) stresses produce splayed reverse faults as a positive flower structure, or tensional forces (transtension) generate curved normal faults as a negative flower structure. Well-illustrated positive flower structures occur beneath lower-upper Miocene clastics on the shelf off northwest Palawan, Philippines (Fig. 12.3) (Roberts 1983), as the result of compression with sinistral shear during the development of a thrust imbricate system.

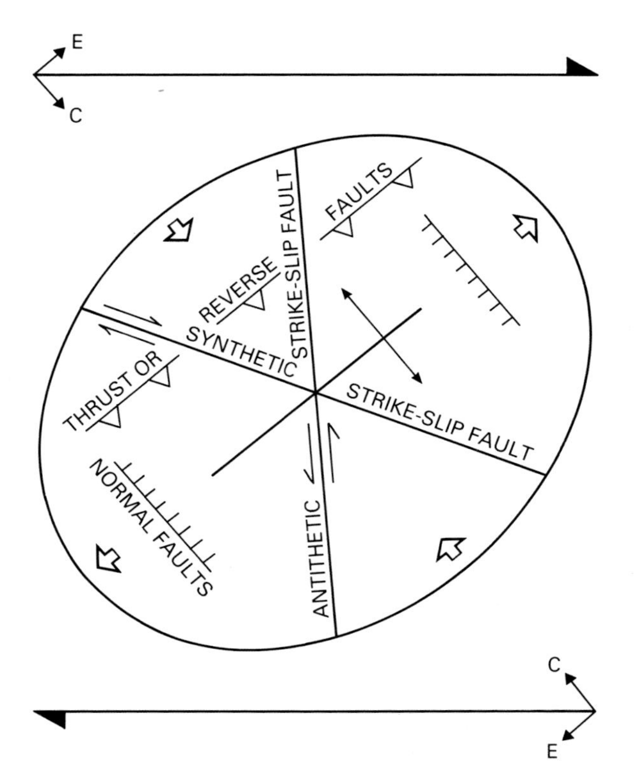

Figure 12.1 Summary diagram to show the structures produced from simple shear under a dextral shear couple.

Modern oblique-slip mobile zones

Harding (1983) and Harding *et al.* (1985) document the characteristics of dextral transtension in the Andaman Sea where divergent strike-slip has produced a single major fault bordered by discontinuous oblique faults (Figs 12.4 & 5). Water depths in this region are greater than 1000 m. The Andaman Sea is a marginal basin where spreading to produce oceanic crust began about 13.5 Ma, based on the magnetic anomalies. The central spreading axis strikes northeast, and is linked by NNW–N-striking dextral faults (Fig. 12.4). In this region: (a) *en echelon* folds occur with dextral offsets from 2.4–3.2 km; (b) abundant normal faults occur, formed contemporaneously with wrench faults that are disclosed in a simple *en echelon* pattern, and (c) there are localized reverse faults near an S-shaped closing or restraining bend at the southern end of the principal fault (Fig. 12.5) (Harding 1983, Harding *et al.* 1985). These observations are consistent with major dextral shear along the West Andaman Fault zone. The fault density and abundance outside of the principal West Andaman Fault is asymmetric, with more numerous faults to the east (Fig. 12.5). In the southern part of the study area, transtension has produced negative flower structures in Miocene–Pleistocene sediments (Fig. 12.6). Thus, although the deformation patterns are complex and occur with sedimentation, overall there is a predictable geometry consistent with major transtensional dextral shear.

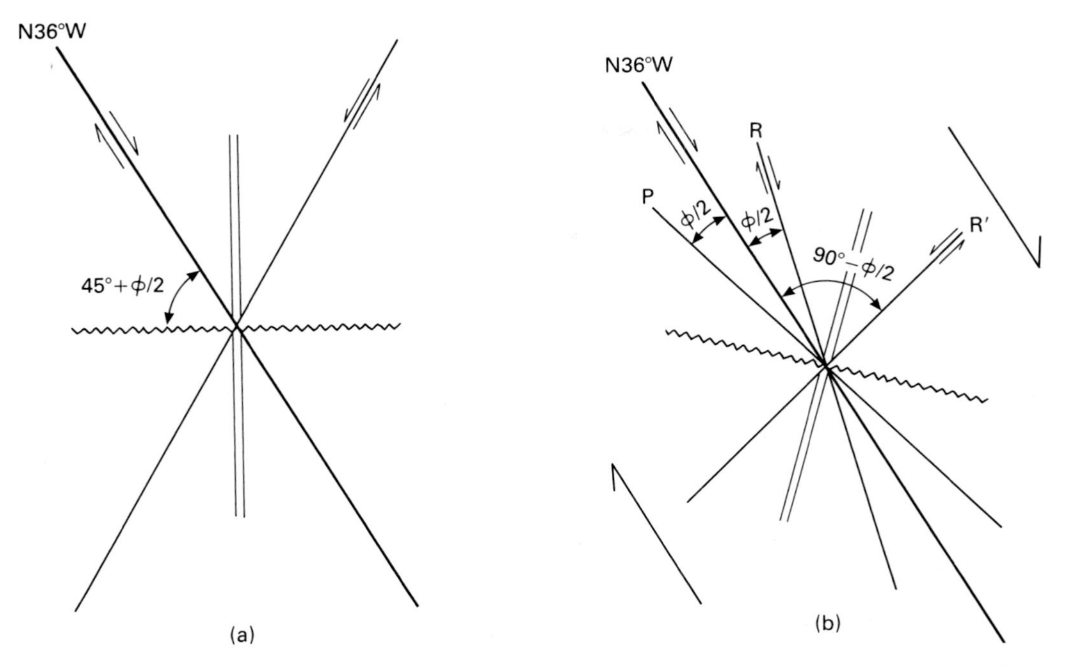

Figure 12.2 Models for structures produced by a dextral shear couple applied to the San Andreas Fault zone and associated structures, using (a) Coulomb–Anderson, and (b) simple shear. Both cases are drawn with the major dextral strike-slip as N36° W, parallel to the Pacific–North American plate boundary and some of the main strike-slip faults. After Aydin & Page (1984). See text for explanation of the P, R and R′ shears.

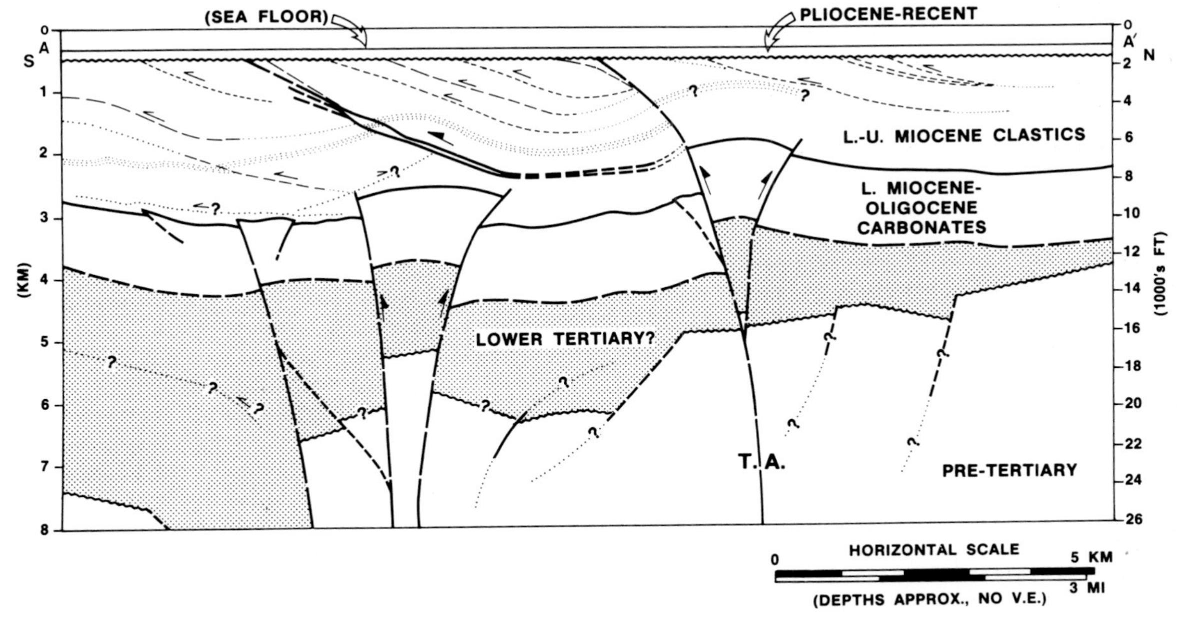

Figure 12.3 Positive (transpression-produced) flower structure interpreted from seismic reflection profile from the northwest shelf of Palawan, Philippines. Relative motion towards observer shown by 'T' and away 'A'. Such structures typically develop along zones of transpression. After Roberts (1983).

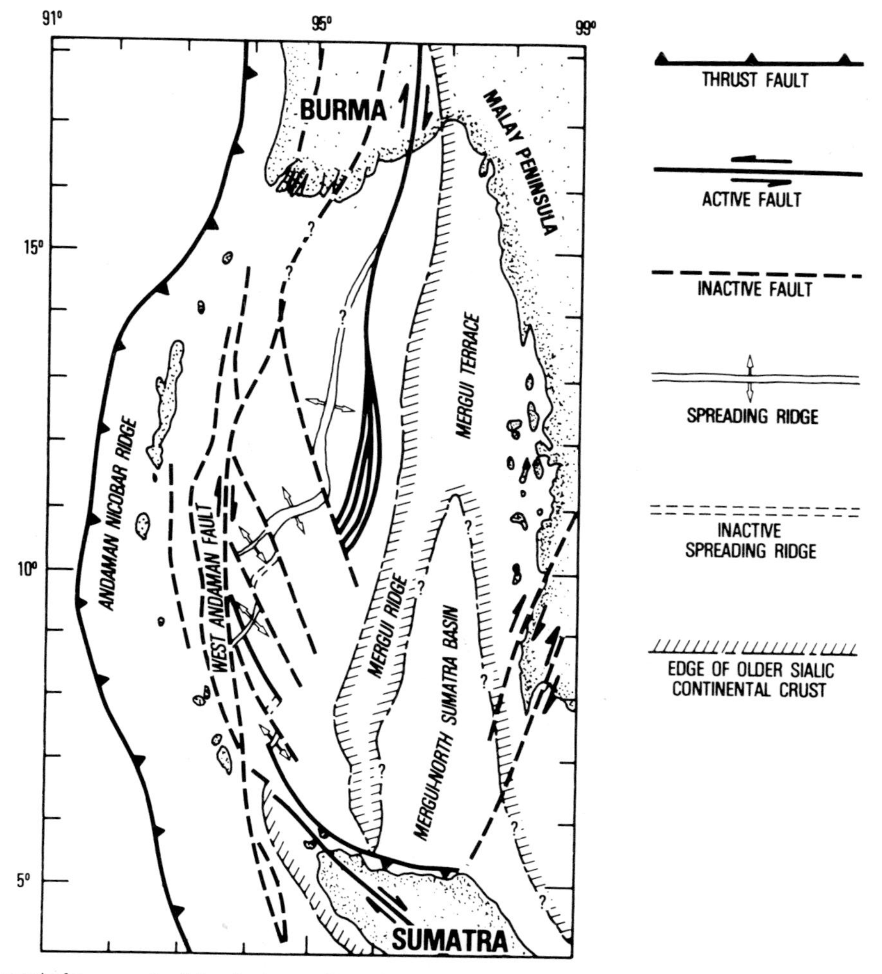

Figure 12.4 Tectonic framework of the Andaman Sea after Harding *et al.* (1985). Boxed area shown in detail in Fig. 12.5.

Another well-known deep marine environment where oblique-slip tectonics are important occurs in the Hikurangi Trough at the southern end of the Kermadec Trench, off North Island, New Zealand, where the westward-subducting Pacific Plate changes from roughly orthogonal collision with the Indian Plate to approximately transcurrent dextral relative displacement (Fig. 12.7a). DSDP Leg 90 was partly concerned with this area (Kennett, von der Borch *et al.* 1986), particularly the Hikurangi Trough and Chatham Rise areas. Structures in the Hikurangi Trough are consistent with a major dextral shear couple between the Pacific and Indian plates (Fig. 12.7b). The Alpine Fault of South Island, New Zealand, separates blocks of continental crust. Its extension in the Hikurangi Trough separates continental crust to the west from subducting oceanic crust to the east. The Alpine Fault, along the west of South Island, is an intracontinental trench–trench transform that was initiated in the latest Eocene–Oligocene (Norris *et al.* 1978, Lewis *et al.* 1986). In the late Oligocene–early Miocene, a phase of extension was replaced by the present pattern of transpression (Carter & Norris 1976) due to the southward migration of the Indo–Pacific pole of rotation (Walcott 1978).

The net result of the plate motions off northeast New Zealand is that the northwestern edge of Chatham Rise (Fig. 12.7a) is being dragged under the Kaikoura Ranges (Fig. 12.7b) in northeast South Island. Also, the down-warped edge of Chatham Rise can be traced under the southern end of Hikurangi Trough where a thick succession of turbidites has accumulated. The

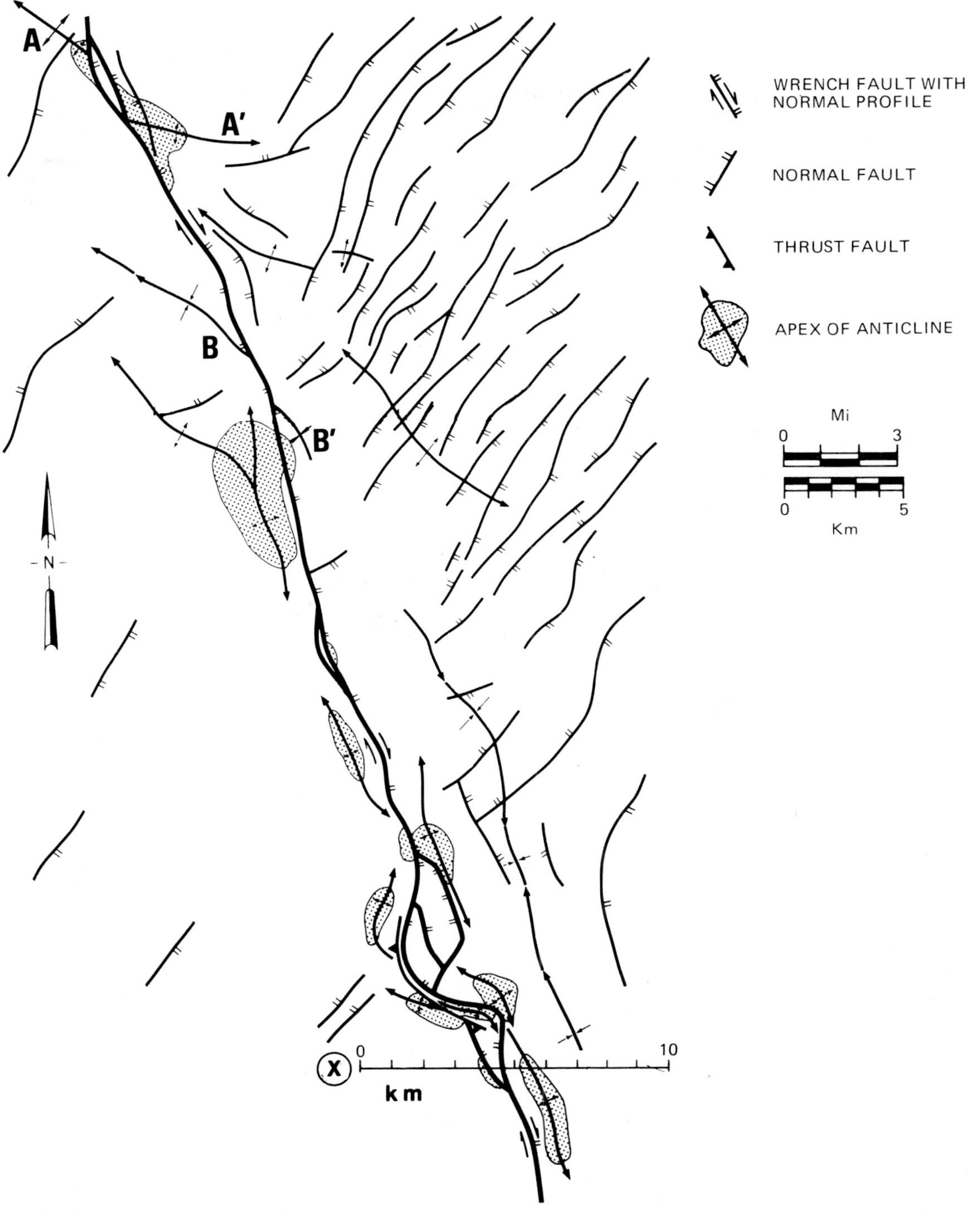

Figure 12.5 Detailed structural map of part of the Andaman Sea (see Fig. 12.4 for location). Note that many of these (Tertiary) structures are compatible with having developed under a dextral shear couple. AA′ and BB′ represent offset of structural features. After Harding (1983).

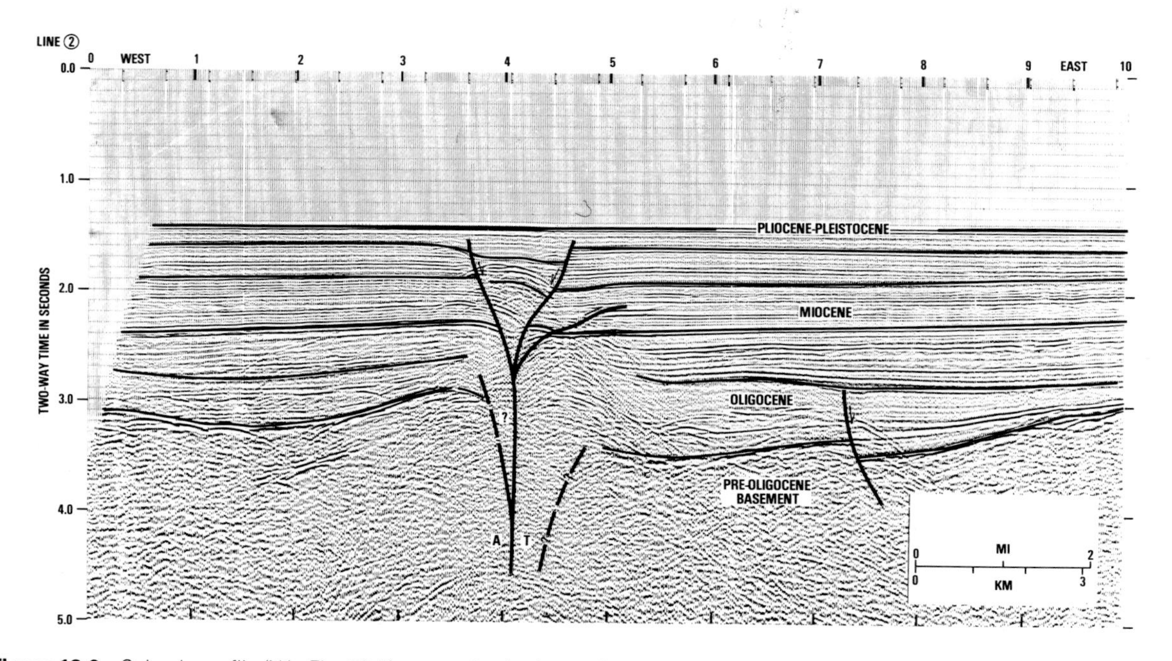

Figure 12.6 Seismic profile (X in Fig. 12.5) across the Andaman Sea wrench fault to show interpreted negative flower structure (after Harding *et al.* 1985). 'A' = relative motion away, and 'T' towards, observer. These structures typically develop along zones of transtension.

slope between Mernoo Saddle and the southwestern Hikurangi Trough (Fig. 12.7b) shows evidence of Neogene tectonic activity (Fig. 12.8a) and superficial slope instability to produce large-scale sedimentary slides of Facies Group F2 (Fig. 12.8b). The 'tectonic' faults are late Neogene in origin, based on the acoustic units they cut, and generally dip southward; their genesis is ascribed to seismic activity along the boundary, North Mernoo Fault zone (Lewis *et al.* 1986).

Dextral strike-slip tectonics has been important in controlling the development of deep-marine clastics since at least 38 Ma in South Island, New Zealand (Norris *et al.* 1978); for example, there are hundreds of metres of Oligocene–Miocene calcareous, sand-rich and mud-rich sediments in the Moonlight Zone of southwest South Island.

Norris *et al.* (1978) have shown that the Moonlight Zone, with more than 2000 m of Cainozoic flysch, is an Oligocene–Miocene infill at the northern extension of the NNE–SSW Solander Trough. In the Oligocene, fault-controlled deep-marine sandy systems (probably small submarine fans) accumulated from sediment sources to the west and east in the approximately N–S basin (Carter & Lindqvist 1975, 1977). By the early Miocene, the dominant transport direction of sandy flysch had changed from lateral supply to southward, axial, progradation except off parts of the eastern fault-controlled margin, where calcareous sediment was being deposited as small submarine fans (cf. Norris *et al.* 1978, Figs 7A & B). By late middle Miocene time, fluvial conglomerates had replaced the deep-marine sedimentation. The critical position of the Moonlight Zone along the boundary of the Indo–Australian and Pacific plates has led to considerable post-Miocene vertical tectonics, with large amounts of uplift, especially in the last few million years. Features that Norris *et al.* (1978) regard as typical of strike-slip tectonics and sedimentation in the Moonlight Zone include: (a) the narrowness of the zone of subsidence, and (b) the rapidity of lateral and vertical facies changes. Thus, throughout the Cainozoic, southwest South Island and the northern part of the Solander Trough (eastern edge of the Tasman Basin) have been tectonically and sedimentologically a continuous belt of flysch basins controlled by oblique-slip processes (Summerhayes 1979).

Figure 12.7a Map of the southwest Pacific, east of North Island New Zealand, to show main topographic features, boundary between Indian and Pacific Plates, direction and rates of plate convergence. After Davey *et al.* (1986).

Figure 12.7b Major faults in the area of the southern Hikurangi Trough (basin) offshore from North and South Islands, New Zealand. Tags on faults show downthrown block; triangles indicate thrust/reverse faults. Inset simple shear strain ellipse in appropriate orientation for major dextral-shear couple between the Indian and Pacific Plates. See text. After Lewis *et al.* (1986).

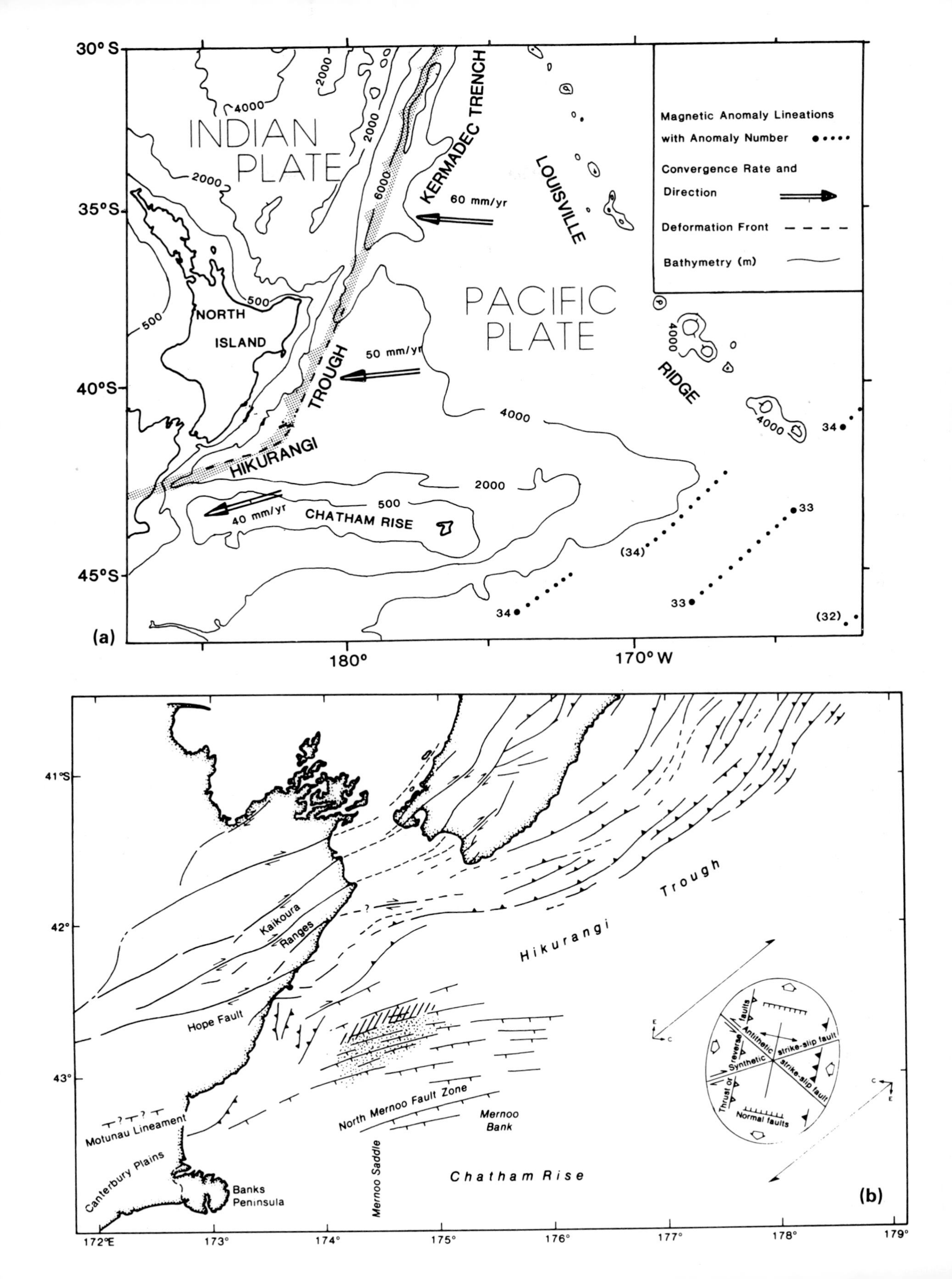

30° S
35°S
40°S
45°S
180°
170° W
(a)
INDIAN PLATE
PACIFIC PLATE
KERMADEC TRENCH
LOUISVILLE
RIDGE
HIKURANGI
TROUGH
NORTH ISLAND
CHATHAM RISE
60 mm/yr
50 mm/yr
40 mm/yr
2000
4000
6000
500
34
(34)
33
(32)
Magnetic Anomaly Lineations with Anomaly Number
Convergence Rate and Direction
Deformation Front
Bathymetry (m)
41°S
42°
43°
172°E
173°
174°
175°
176°
177°
178°
179°
Kaikoura Ranges
Hope Fault
Hikurangi Trough
Motunau Lineament
North Mernoo Fault Zone
Mernoo Bank
Mernoo Saddle
Chatham Rise
Canterbury Plains
Banks Peninsula
Thrust or reverse faults
Antithetic strike-slip fault
Synthetic strike-slip fault
Normal faults
(b)

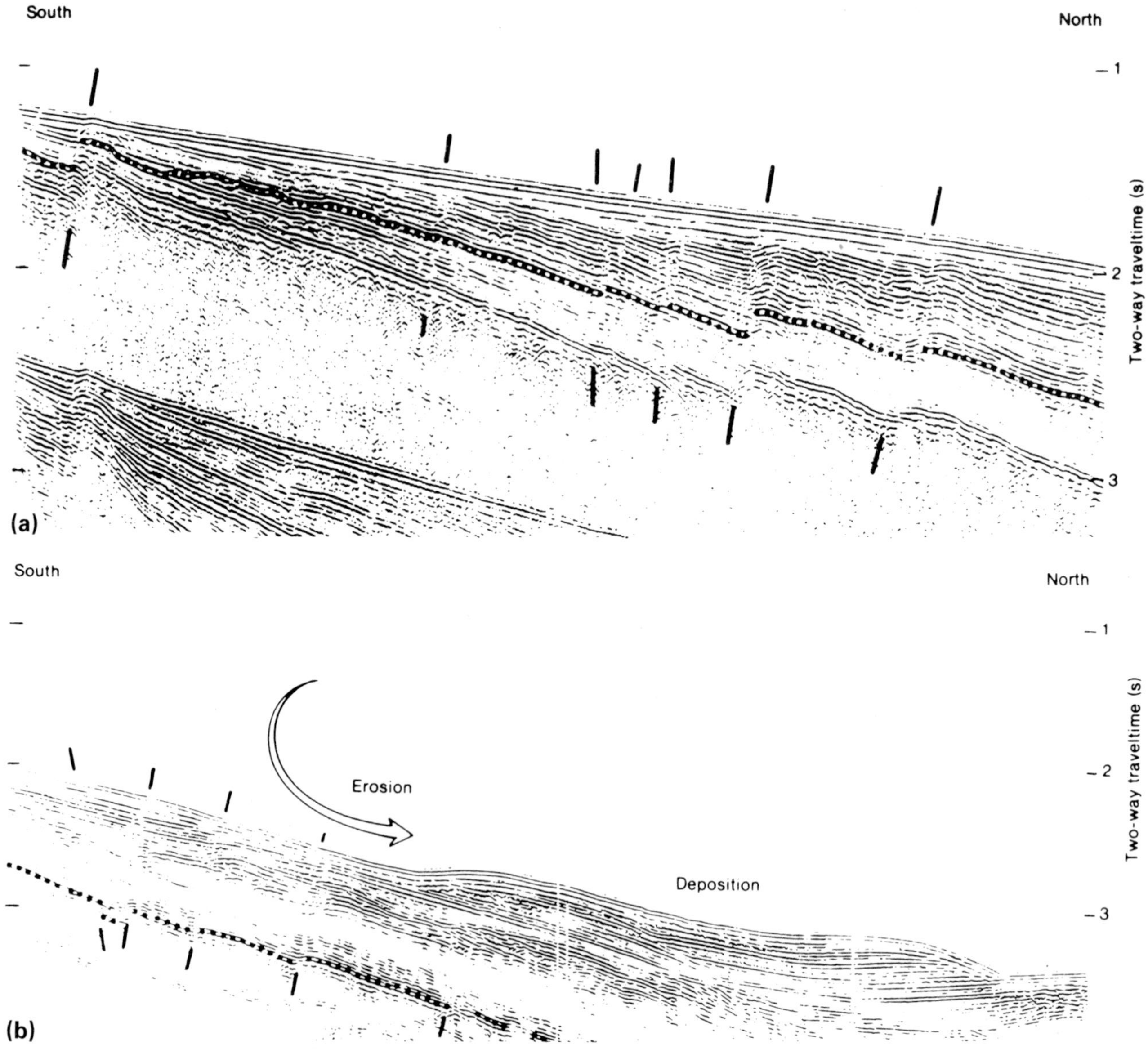

South

North

(a)

Two-way traveltime (s)

South

North

Erosion

Deposition

(b)

Two-way traveltime (s)

Figure 12.8 Seismic profiles, Magellan G–9, across (a) North Mernoo Fault Zone (water depth 800–1450 m), and (b) across lower North Chatham Slope and adjacent Hikurangi Trough (water depth 1500–2400 m). TWT time in 1 second intervals approximately equivalent to 750 m in seawater. Stippled line is unconformity of uppermost Miocene to Recent (pure and impure nannofossil oozes) upon Middle-Upper Miocene pure nannofossil oozes. Note in Fig. 12.8a that most normal faults dip towards the crest of Chatham Rise, and in 12.8b the zone of erosion and downslope redeposition. After Lewis *et al.* (1986).

CASE STUDY: CALIFORNIA CONTINENTAL MARGIN AND GULF OF CALIFORNIA

The continental margin of both offshore and onshore California is an active transform plate margin dominated by right-lateral or dextral strike-slip motions (Emery 1960, Crouch 1981). Figure 12.9 is a summary of the geological evolution of this margin since 30 Ma (Saunders *et al.* 1987). Subduction of oceanic lithosphere below western North America occurred from the Cretaceous to Miocene (Atwater & Molnar 1973). The Pacific–Guadalupe ridge intersected the continental margin about 29 Ma to create two triple junctions, the Mendocino transform–transform–trench and the Rivera ridge–trench–transform that migrated towards the north and south respectively. The San Andreas transform fault system developed between the two triple junctions. From about 29–12.5 Ma, the

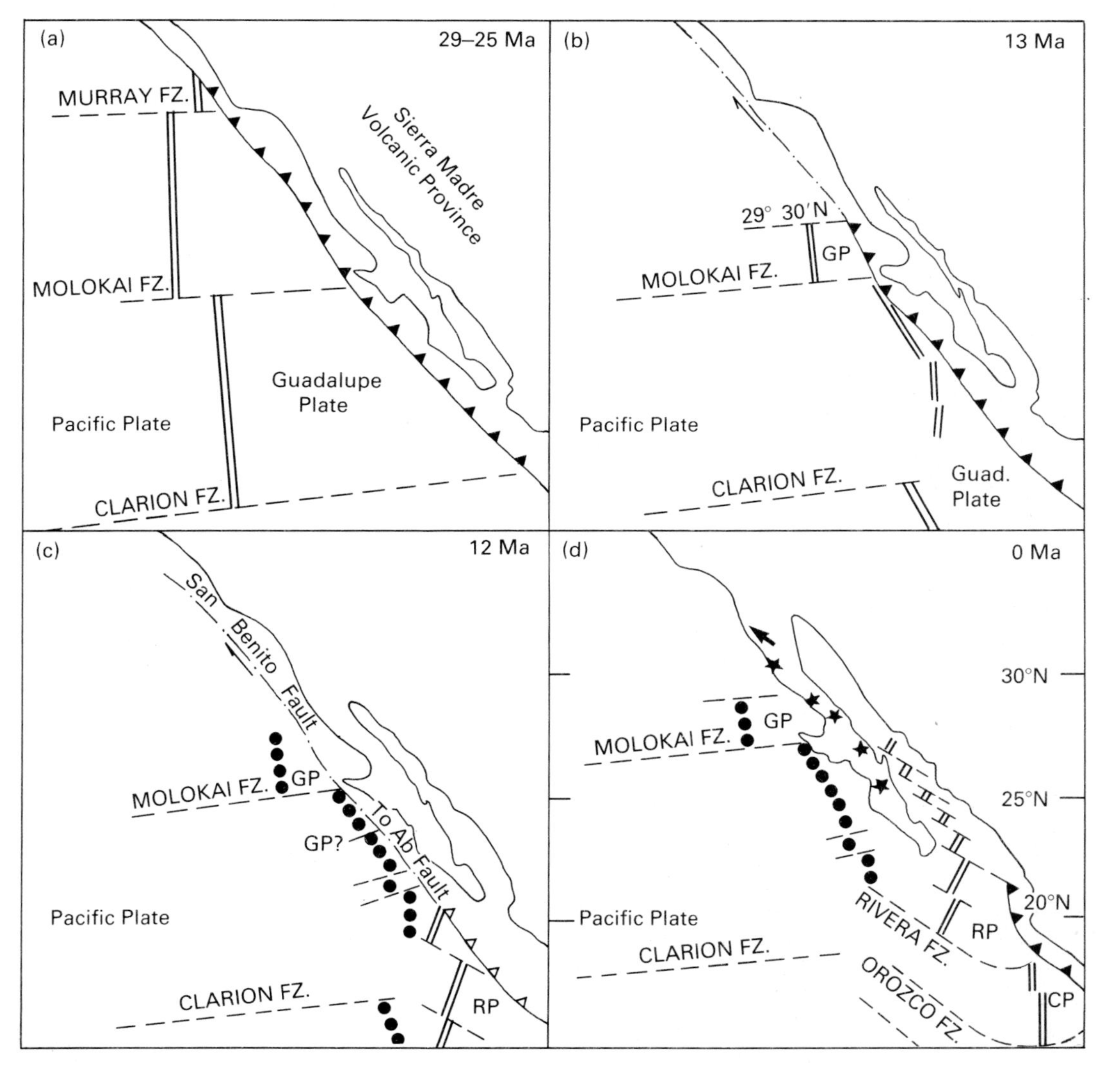

Figure 12.9 Generalized plate-tectonic development of the eastern Pacific from 29–0 Ma. Present-day outlines of Baja California and the Gulf of California shown for reference. GP, Guadalupe Plate; RP, Rivera Plate; CP, Cocos Plate; To–Ab Fault, Tosco–Abreojos Fault; circles, abandoned spreading centre; stars, Late Cenozoic volcanic fields; 29° 30′ N = unnamed fracture zone at this latitude. After Saunders *et al.* (1987).

Pacific–Guadalupe spreading centre was progressively consumed southwards at the trench. By about 12 Ma, the entire continental margin as far south as Baja California was converted into a dextral-slip transform system. In gross terms, the northern sector of the San Andreas fault system is dominated by transpression in contrast to the Gulf of California where transtension has occurred to generate a narrow ocean basin. The two case studies on the California margin can be considered as contrasting oblique-slip plate boundaries where deep-marine sedimentation forms a major component of the stratigraphic record.

NORTH-CENTRAL CALIFORNIA CONTINENTAL MARGIN

The California continental margin has evolved through the Neogene as part of a broad zone of oblique-slip plate tectonics (Fig. 12.9). There are many small deep-marine basins (Fig. 12.10), controlled by a complex array of compressive and/or tensional tectonics. To demonstrate the complexity of the deep-marine sedimentation in these basins, a brief review of Santa Monica and San Pedro basins is presented, based mainly on Malouta (1981) and Nardin (1983).

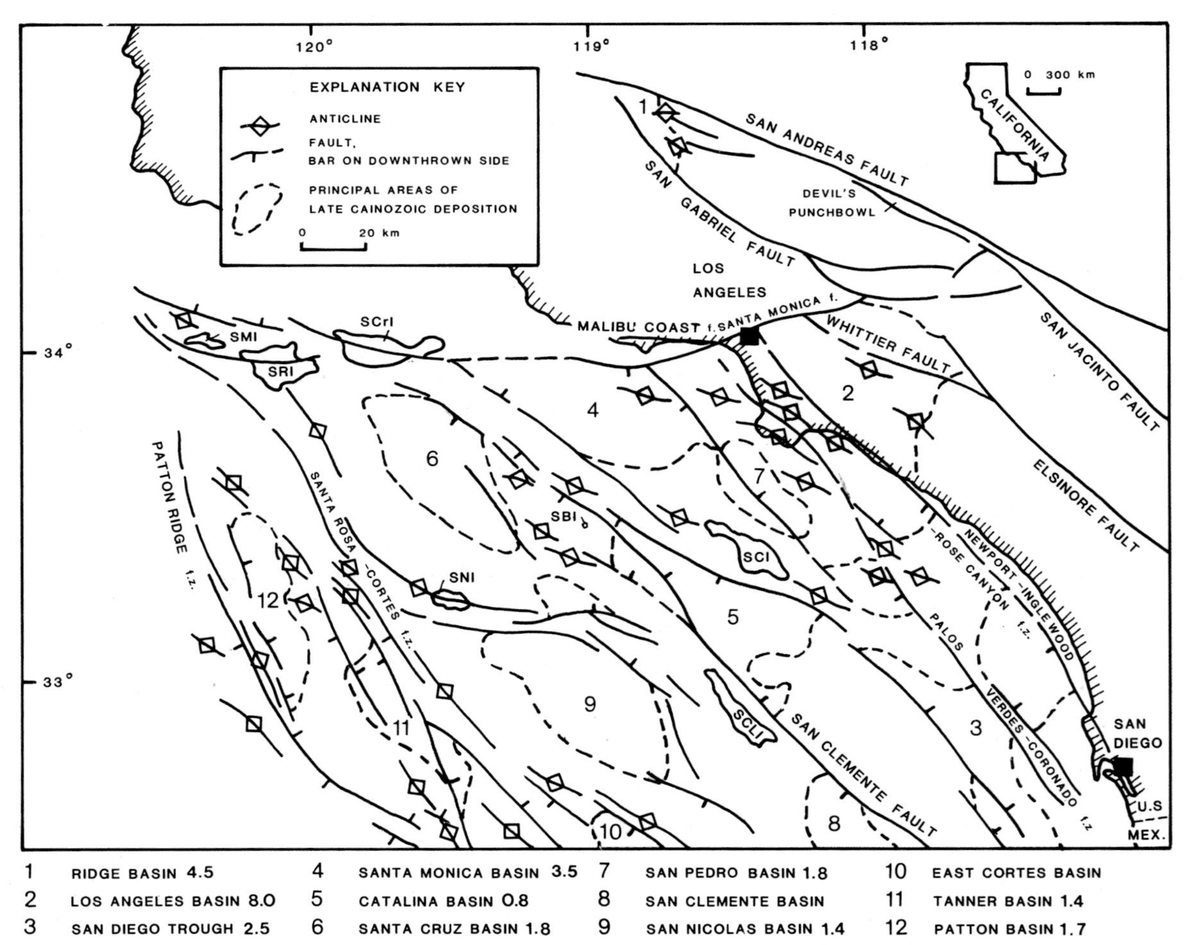

Figure 12.10 Sedimentary basins and main structural features of the southern California Borderland and immediately adjacent onland areas. SMI, San Miguel Island; SRI, Santa Rosa Island; SCrl, Santa Cruz Island; SCI, Santa Catalina Island; SBI, Santa Barbara Island; SNI, San Nicolas Island; SCLI, San Clemente Island. Approximate thickness of basin fills in kilometres as follows: Ridge Basin, 4.5; Los Angeles Basin, 8.0; Santa Monica Basin, 3.5; San Diego Basin, 2.5; San Pedro Basin, 1.8; Santa Cruz Basin, 1.8; Catalina Basin, 0.8; San Nicolas Basin, 1.4; Patton Basin, 1.7; Tanner Basin, 1.4. NW-trending faults are dextral and synthetic to the San Andreas Fault. Santa Monica Fault is sinistral. Anticline trend is mainly WNW. Adapted from Junger (1976) Moore (1969) and Howell *et al.* (1980).

Santa Monica and San Pedro Basins are NW-trending, structurally-controlled, depressions with maximum depths of 938 m and 912 m respectively. These depressions are now partially filled with sediments. The adjacent shelves and shelf-slopes are the limbs of large anticlinoria that acted as barriers to sediment transport from the Los Angeles Basin in pre-late Quaternary times (Emery 1960, Nardin & Henyey 1978). Basin slopes range up to about 18° on the San Pedro escarpment. To the southwest, the basins are bounded by the structurally-controlled Santa Cruz–Catalina Ridge. Submarine canyons that head near the shoreline feed sediments, mainly by trapping long-shore drift cells, into the basins (Hueneme, Mugu, Dume, Santa Monica, Redondo and San Pedro Canyons). Other processes that provide sediment to the basins include shelf bypassing and resuspension (nepheloid layer) of fine-grained sediments, and mass movements due to slope failure. Much of the slope instability and failure is due to both fast rates of sediment accumulation and the regional seismic activity. The seismic characteristics of the sediments that are filling the California Borderland basins are shown in Table 12.1.

Small radius submarine fans cover large areas of the basins, most of which are well described by the Normark (1970, 1978) 'suprafan model'. Figure 12.11 demonstrates the lateral variation in facies associations over relatively small distances. The geometry of the clastic infills of these basins is primarily governed by local tectonic elements, with developing folds and active faults creating sediment sinks and structural barriers to sediment transport pathways. For example, Santa Monica Canyon now is filled to the crest of a growing (Dume) anticline that has been a sediment dam at the mouth of the canyon (Junger & Wagner 1977, Nardin 1983). A channel now has breached this anticline, but as yet has not achieved an equilibrium gradient with the basin floor.

Stable slopes are characterized by onlapping slope reflectors that thin towards the upper slope; locally at the base of the slope such deposits interdigitate with basin-floor sediments, and are locally onlapped by them. Slope sedimentation, in part, has been strongly influenced by the late Quaternary glacio-eustatic changes in sea level, linked to fluctuating rates of sediment input. Slope instability has created a 140 km^2 composite slide mass (Facies Class F) between Hueneme and Mugu Fans. The slide mass has controlled fan growth patterns by restricting and deflecting turbidity currents (Nardin 1983).

Tectonic deformation also has affected basin-floor/plain deposits (mainly Facies Class E & G). An example where such tectonic effects are well-developed is near Avalon Sill, at the southern margin of San Pedro Basin, where late Pleistocene deformation has upturned and faulted onlapping basin floor/margin sediments (Nardin 1983).

The detailed analysis of Santa Monica and San Pedro Basins led Nardin (1983) to develop a basin-fill model to characterize the sedimentation in the California Borderland basins that includes the following: (a) small fault-controlled basins; (b) both structural and sedimentological (mass movement) barriers to sediment transport paths; (c) small radius, generally coarse-grained, submarine fans fed by (d) canyons that tend to intercept littoral drift currents, to feed sediment into deeper water, and (e) fluctuating sea levels exerting considerable influence on the overall growth pattern of the deep-marine clastic systems.

Strike-slip tectonics can cause the truncation of submarine canyon heads. This has been well documented for Ascension Canyon in San Clemente Basin by Nagel *et al.* (1986), on which the following description is largely based (also see Ch. 6).

The Outer Santa Cruz Basin, between Monterey Bay and San Francisco, began to form in the Miocene by extension associated with dextral shear along faults of the San Andreas system (Hoskins & Griffiths 1971, Howell *et al.* 1980). This northwest-dipping basin contains over 3000 m of Neogene deep-marine sediments (Hoskins & Griffiths 1971), and is bounded to the northeast by the granitic Farallon Ridge–Pigeon Point High, and to the southwest by Franciscan rocks of the Santa Cruz High (Nagel & Mullins 1983) (Fig. 12.12).

The NNW-trending San Gregorio fault zone, 1–2 km wide and comprising a complex zone of highly fractured rocks, has controlled the location and offset of canyons. *En echelon* folds and associated faults are compatible with dextral simple shear. Three canyon-cutting episodes are recognized at about 6.6–2.8 Ma, 750 000 years BP and 18 000 years BP (Nagel *et al.* 1986). Initial canyon excavation appears to have been associated with relative lowstands of sea level, for example at about 3.8 Ma (Pliocene) the NW headward region of Ascension Canyon was juxtaposed against the seaward part of Monterey Canyon (Fig. 12.13), coincident with a lowstand of sea level on the Haq *et al.* (1987) curve (Fig. 4.3). Thus, palaeogeographic reconstructions for this time suggest that Ascension Canyon formed the distal extension of nearby Monterey Canyon, and was subsequently offset along the San Gregorio fault zone along which 110 km of dextral displacement have been recorded (Graham & Dickinson

Table 12.1 Principal seismic facies for Santa Monica and San Pedro Basins.

Seismic Facies	Facies parameters Amplitude	Frequency	Continuity	Configuration	Reflection geometry at base	Lateral relationships	Depositional environment
Broad low-relief mound	Variable	Variable	Discontinuous: can be continuous locally	Hummocky, subparallel; on 1-s records relief and gradient decrease from upper to lower mid-fan and reflections become more parallel and even	Onlap toward basin margin and topographic irregularities; downlap locally	Gradational to complex mound, onlap basin fill and onlap slope facies	Mid-fan variable velocity turbidity currents; low-relief channels resolvable on 3-s records; upper and lower mid-fan can be distinguished on 1-s and 0.25-s records and by relief and gradient
High-relief complex mound	Variable	Variable	Discontinuous	Hummocky, subparallel, contorted to chaotic locally; pronounced discordance among some reflections	Onlap toward basin margin and topographic irregularities; downlap	Occurs near submarine canyon and grades to broad low-relief mound facies	Upper fan and canyon; high relief channels and levees
Chaotic mound	Variable	Variable	Discontinuous	Chaotic, distorted or hummocky; high relief surface	Variable	Commonly associated with onlap slope facies and truncated and contorted slope reflections; intercalated or gradational with onlap basin-fill facies	Mass movement—slides, slumps, mass flows; degree of reflection disruption depends on types of mass movement process
Onlap basin fill	Variable on 3-s records continuous reflections tend to be high; discontinuous reflections tend to be low	Tends to be uniform	Mainly continuous on 3-s records, variable on 1-s records	Even, parallel	Onlap toward basin margin and topographic irregularities; concordant at centre of basin	Gradational to broad low-relief mound facies; continuous with onlap slope facies along some horizons	Lower fan and basin plain (can be distinguished on 0.25-s records); relatively low-velocity turbidity current flows

Channel fill	Usually high on 1-s records	Variable	Discontinuous on 1-s records	Parallel to hummocky	Concordant or discordant	May be gradational with levees on fan or basin plain if depositional type; onlaps if erosional	Turbidity current channels
Onlap slope	Variable but tends to be low on 1-s records	Variable	Continuous	Convergent upslope; may be wavy, disrupted or contorted due to tectonic deformation or mass movement	Onlap	Onlapped by mound or fill facies; gradational with some mound facies; continuous with basin fill along some horizons	Hemipelagic slope deposition; particle settling from low energy turbid layer or nepheloid flows resuspended from shelf and upper slope
Sheet drape	Relatively low on 1-s records	Relatively uniform on 1-s records	Continuous	Parallel and conforms to underlying topography	Concordant on irregular topography; locally onlaps along basin margin	Gradational with onlap slope front fill facies; alternates with onlap basin-fill facies	Basin-plain hemipelagic particle settling from suspended sediment concentrations (nepheloid layers)
Migrating wave	Variable	Relatively high on 0.25-s records	Continuous	Wavy, subparallel; asymmetric; resembles climbing ripples on 0.25-s records; mounded, wavy to hummocky on 3-s records	Concordant	Subfacies with broad, low-relief mound facies (fan); relief decreases and wave length decrease toward margin of fan	Turbidity current phenomenon on large levee complex in mid-fan environment

Nardin, T.R. (1983), Table 1 p. 1107).

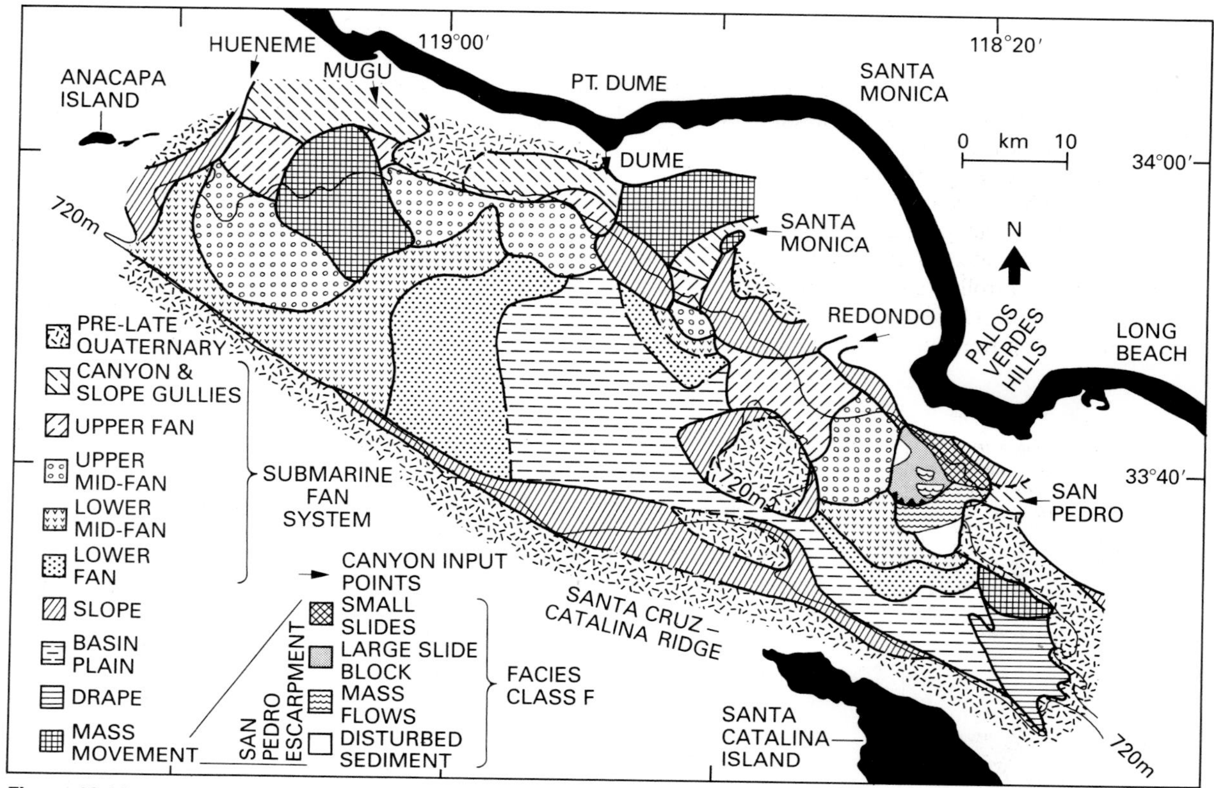

Figure 12.11

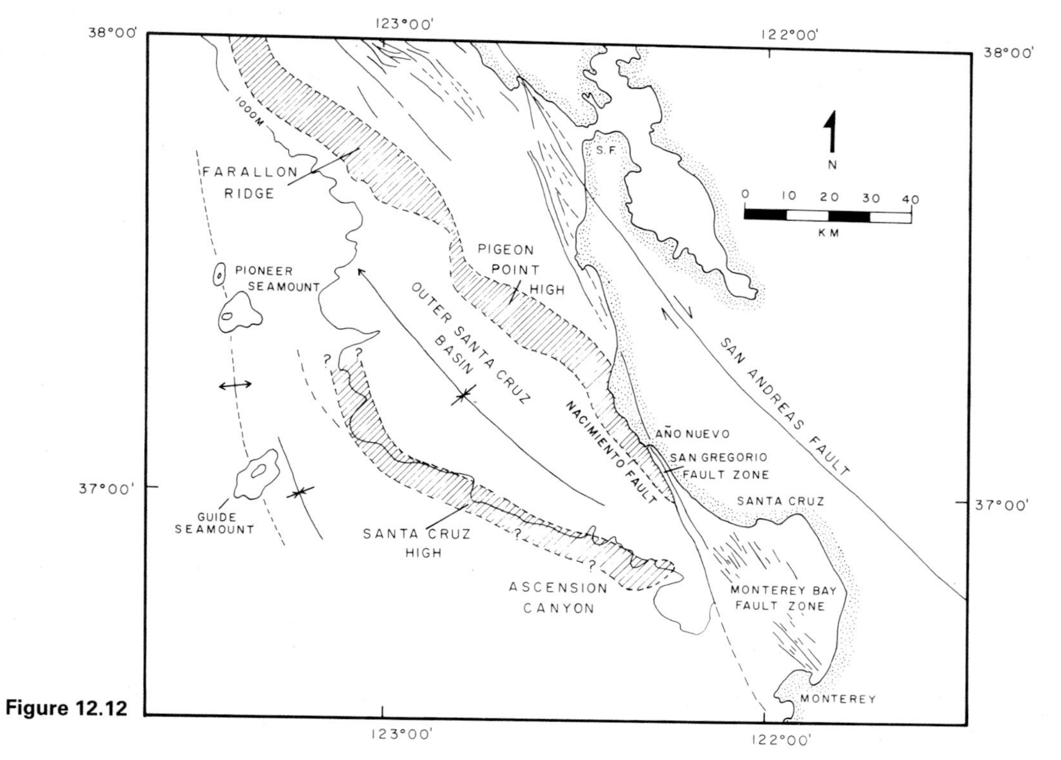

Figure 12.12

1978). Some of the SE heads of Ascension Canyon may have been formed during the late Pliocene and early Pleistocene sea-level lowstands at about 2.8 Ma and 1.75 Ma, respectively, and then offset to the NNW. There appear to have been at least two canyon-cutting events in the last 750 000 years after the entire Ascension Canyon system migrated to the NNW past Monterey Canyon – again controlled by lowstands of sea level.

Nagel & Mullins (1983) estimate about 105 km dextral offset along the San Gregorio fault zone since the initial displacement 12 Ma, based on cross-fault offset pairs in upper Miocene Santa Cruz mudstone. Seventy kilometres of this offset occurred since 6.6 Ma (late Miocene), giving an average displacement rate of 1.06 cm/year. The last 35 km offset occurred at a slower average rate of 0.65 cm/year. Today, Ascension Canyon appears to be an inactive shelf-edge canyon, with the associated fan valley receiving only rare Holocene sand/silt turbidites and mainly hemipelagic muds (Hess & Normark 1976).

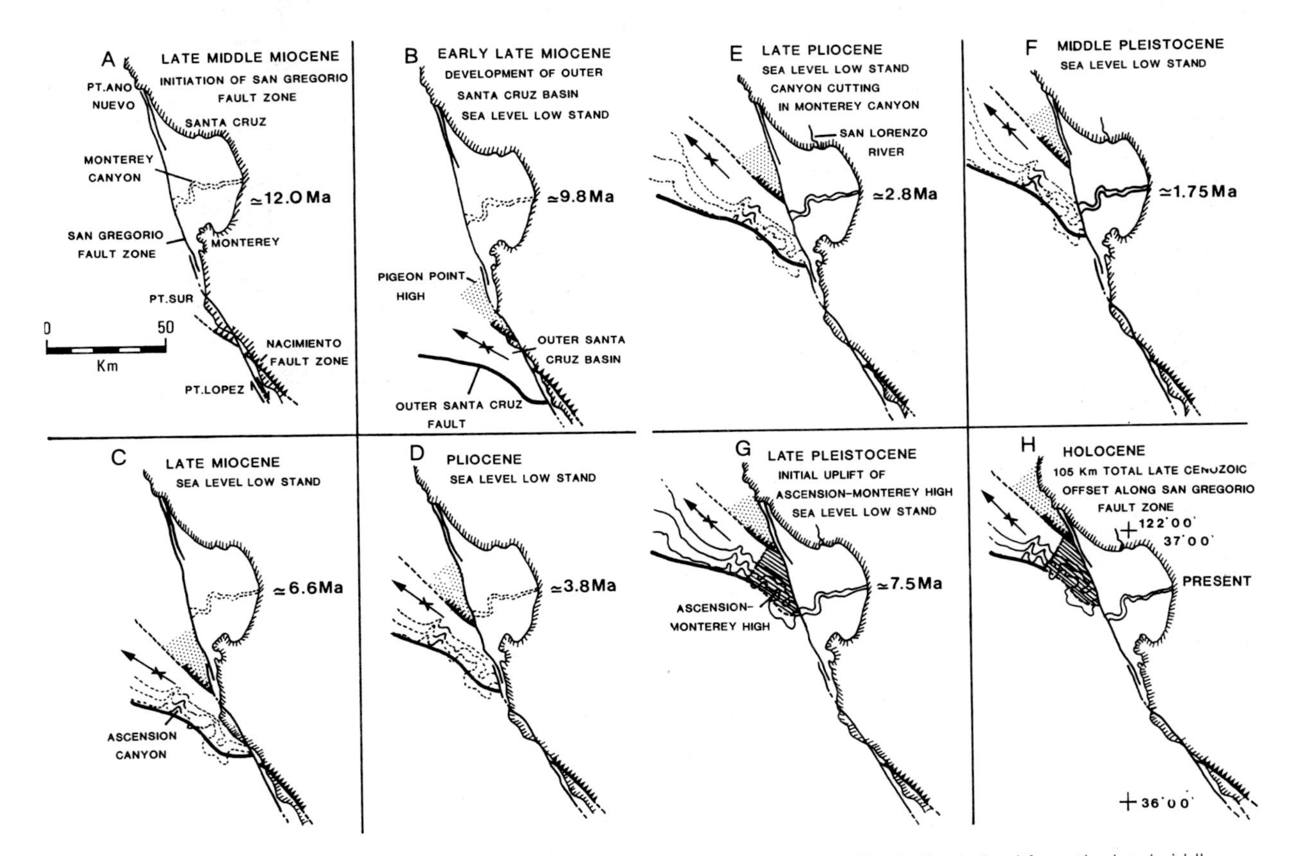

Figure 12.13 Palinspastic palaeogeographic reconstructions of the south-central California Borderland from the late/middle Miocene to Present (12–0 Ma). Reference coastline shown in Fig. 12.12. Maps show the possible development of the Ascension Submarine Canyon system and adjacent continental shelf in relation to Monterey Canyon. Most tectonic movement is assumed to have occurred along the San Gregorio Fault zone. See text. After Nagel *et al.* (1986).

Figure 12.11 (opposite) Late Quaternary depositional environments in Santa Monica and San Pedro Basins. Note the extreme lateral heterogeneity of deep-marine environments and, therefore, changes in facies. After Nardin (1983).

Figure 12.12 (opposite) Generalized structural map of the Outer Santa Cruz Basin, also showing the San Gregorio and Monterey fault zones. The area shown in this figure is the reference frame for the palaeogeographies in Fig. 12.13. After Nagel *et al.* (1986).

GULF OF CALIFORNIA, A TRANSTENSIONAL OCEAN BASIN

This study is based mainly on the results of DSDP Leg 64 (Curray, Moore *et al.* 1982).

The Gulf of California may be considered both as a young rift system with opposing juvenile passive margins, and as a transtensional oblique-slip zone. Approximately 300 km of dextral slip along the San Andreas and related faults in central and southern California (Ehlig 1981) favours the discussion of this case study here rather than in the passive margin chapter.

The Gulf of California can be considered as a deep-marine transtensional zone located between two triple junctions (Mendocino and Rivera), that passes northwards into a region of predominantly dextral transpressive plate interactions. South of the Rivera triple junction, the plate boundary is characterized by subduction–accretion processes (Fig. 12.14E).

The results of DSDP Leg 64 suggest that at about 5.5 Ma, dextral transform motion between the North American and Pacific Plates jumped from offshore to the eastern side of the Peninsular Ranges batholith, thereby initiating: (a) movement along the present San Andreas Fault, and (b) opening of the Gulf of California. Relative displacement over the last 5.5 m.y., at an estimated rate of 5.6 cm/year, is consistent with the matching of geological markers across the plate boundaries and indicates about 300 km offset. By about 3.2 Ma, the first lineated magnetic anomalies and oceanic crust had formed at the mouth to the Gulf. Developing spreading centres erupted basaltic magmas into wet sediments (Einsele *et al.* 1980). Figure 12.14 shows the proposed plate tectonic scenario for the evolution of the deep marine Gulf of California. The oldest marine sediments are inferred to be younger than 5.5 Ma.

An intriguing aspect of the palaeogeographic reconstruction, and something that was supported and refined by DSDP Leg 63 results (Yeats, Haq *et al.* 1981) is that about 13 Ma the rate of sediment accumulation on the Magdalena submarine fan was substantially reduced – a change that appears to correspond to the cessation of delivery of quartzo-feldspathic sandy turbidites to the fan. DSDP Leg 64 results indicate that the source area to Magdalena Fan was cut off because of movement along a transform, the Tosco–Abriejos Fault, combined with the beginning of substantial crustal extension and associated subsidence causing a regional apparent sea-level rise. The ensuing 7 m.y., from 12.5–5.5 Ma, was associated with offshore transform faulting at the plate edge, during which time the triple junction southeast of Cabo San Lucas probably became increasingly unstable because of changes in the plate shear vectors between the North American and Pacific plates (Blake *et al.* 1978, Spencer & Normark 1979).

At about 5.5 Ma, when the transform motion between the North American and Pacific plates jumped towards the northeast, offshore major transform faulting ceased. The magnetic anomalies in the Gulf of California indicate a diachronous transition from extended continental crust to oceanic crust from 4.9–3.2 Ma. The last 3.5 m.y. have witnessed crustal thinning by block and listric faulting, together with igneous sill injection about well-defined spreading centres.

DSDP Sites 474, 475 and 476 straddle the boundary from continental to oceanic crust (Fig. 12.15). Sites 475 and 476 penetrated hemipelagic muddy sediments (mainly Facies Group G), a thin interval of sediments containing phosphates and glauconite, and a cobble conglomerate of metamorphic clasts, Site 476 ending in weathered granite (Fig. 12.15). Site 474 penetrated mainly muddy turbidites (Facies Group D and G), ending in middle Pliocene 3 Ma oceanic crust. Depth indicators at these sites suggest water depths of about 1000 m by the time that oceanic crust was first generated.

The range of deep-marine sediments encountered on DSDP Leg 64 is well illustrated by Site 474 (see Table 12.2). Overall, the succession shows a deepening trend from middle-fan to outer-fan and fan-fringe deposits. The upper part of Site 474 is notable for the development of Facies Class F and A deposits, interpreted mainly as deposits of sediment slides and debris flows, respectively. The most rapid sediment accumulation rates are recorded above the Pliocene–Quaternary contact, reaching about 40 cm/1000 years. An important feature of Site 474 is that it represents an entirely deep-marine succession developed on oceanic crust in contrast to Sites 475 and 476 that show an overall deepening stratigraphy from subaerial alluvial plain sediments (e.g. cobble conglomerates) to isolated offshore bank deposition in an oxygen minimum zone (phosphatic sediments with glauconite grains), to deep-marine, diatomaceous, muddy and silty turbidites and hemipelagites.

The second area of investigation on DSDP Leg 64 was in the Guaymas Basin in the central Gulf of California. Sites 477, 478 and 481 in the Guaymas Basin indicate faster rates of sediment accumulation than at the mouth of the Gulf, up to 270 cm/1000 years (van Andel 1964). An interesting aspect of the sedi-

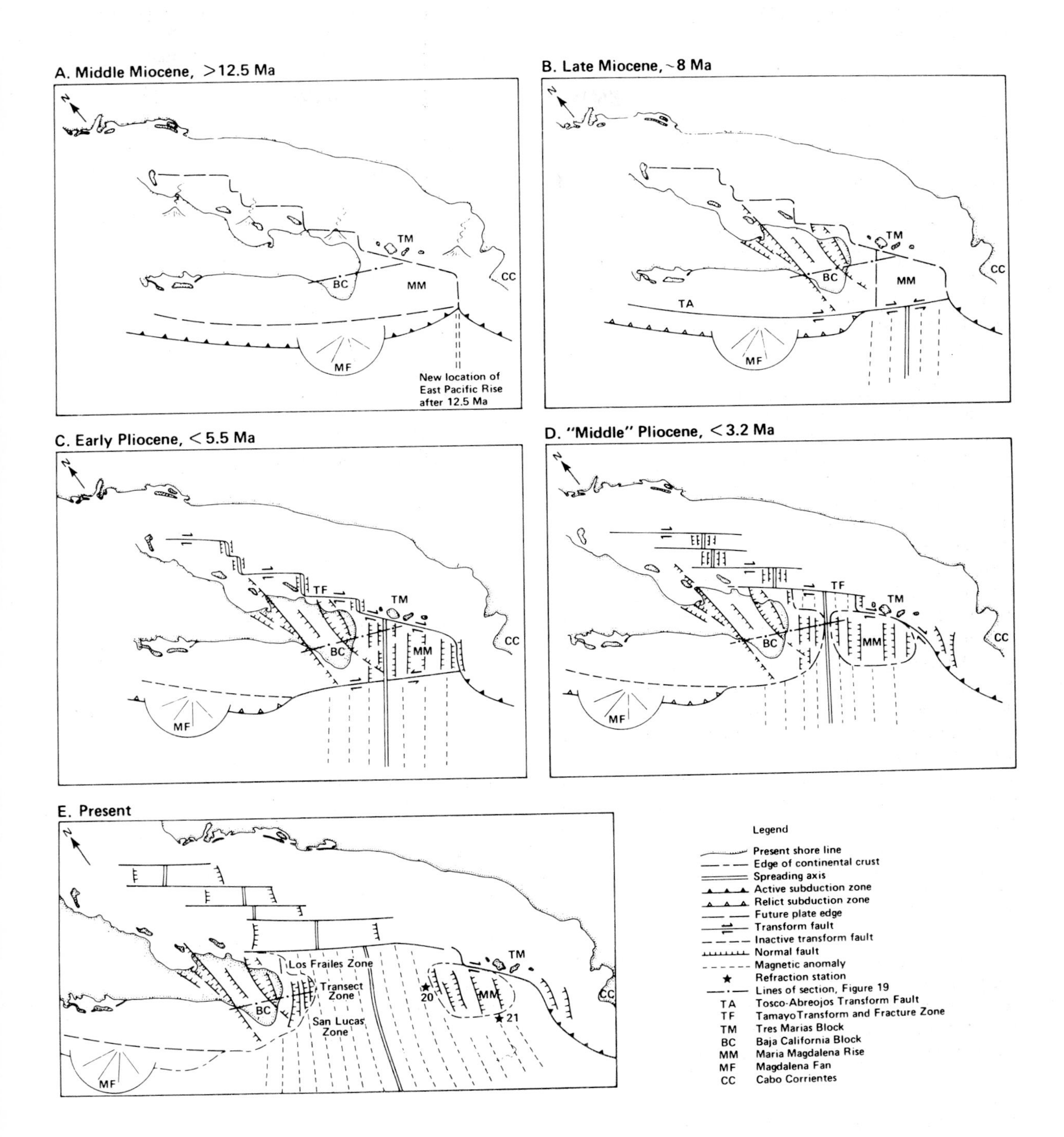

Figure 12.14 Geological history of the Baja California margin based on the results of DSDP Leg 64. After Curray, Moore *et al.* (1982) and various other sources. See text.

mentology was the identification of hydrothermally altered, basin-plain, mud turbidites in Site 477: the silt/sand turbidites contain an epidote–chlorite–quartz–albite–pyrite–pyrrhotite–dolomite–anhydrite mineral assemblage. Particularly well displayed in Site 480 are Upper Quaternary laminated diatomites (Fig. 12.16) deposited in water depths of *c*. 2000 m. The sediments comprise mainly planktonic diatoms, with minor silicoflagellates, radiolarians and variable amounts of benthic and planktonic foraminifera; the terrigenous component is olive brown silty clay. Metre-thick intervals of varve-like sediments occur between non-laminated units. The varves are attributed to annual diatom blooms associated with oceanic upwelling

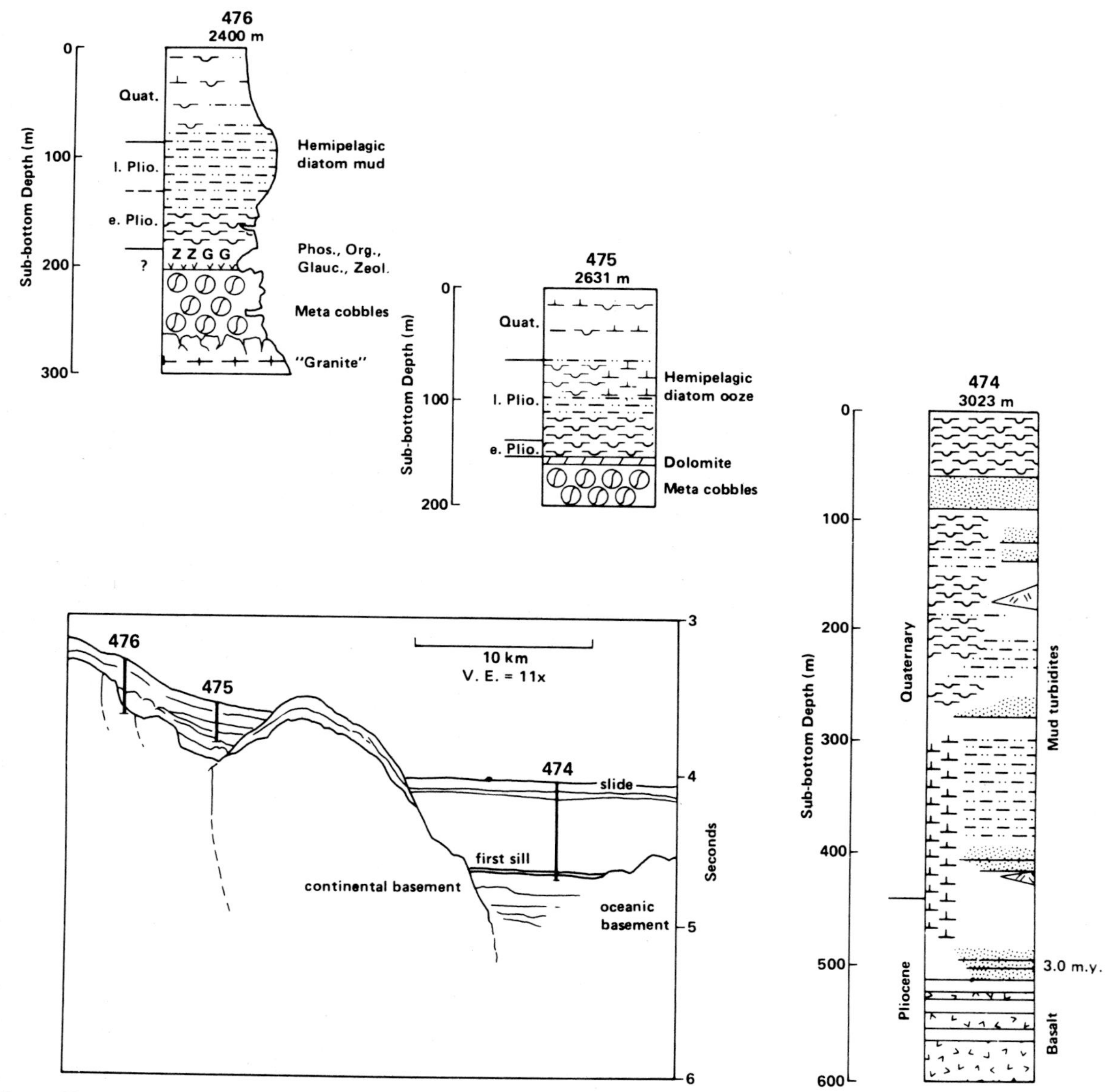

Figure 12.15 Summary sections through DSDP Leg 64 Sites 474, 475 & 476 in the Gulf of California, located on section. For explanation of lithologic patterns refer to Table 12.2. Note the early Pliocene rapid deepening associated with the opening of this part of the Gulf. After Curray, Moore *et al.* (1982).

Table 12.2 Sedimentary lithological units, DSDP Site 474.

Unit	Interval	Sub-bottom Depth (m)	Lithology	Sedimentary environment	Age	Estimated Sedimentation Rate (m/m.y.)	Thickness (m)	Contact
I	Cores 474–1–474–3	0.0–21.0	dusky yellow green to greyish olive diatomaceous muds to oozes and diatomaceous nannofossil marls with downward increasing clayey silts	hemipelagic/distal fan	to NN21/20	47	30.5	transitional
II	Core 474–4–Section 474–10,CC	21.0–87.5	olive brown to greyish olive; firm nannofossil diatomaceous muds and a coarse arkose sand to conglomerate	redeposited slump-debris flow	(NN19)	very high(?)	76.0	top = transitional; bottom = sharp, abrupt
III	Sections 474–11–1–474A–8, CC	87.5–239.0	clayey silts to silty clays, nannofossil diatomaceous muds, nannofossil marls, and scattered arkose muds	hemipelagic mud and mud turbidites on outer fan	NN20	395	142.0	top = sharp; bottom = transitional to Core 474–12
IV	Sections 474A–8,CC–474A–28,CC	239.0–420.0	greyish olive silty claystone to clayey siltstone		early Pleistocene	395–86	181.0	top and bottom transitional
IVA	Sections 474–9–1–474A–10,CC	(239.0–258.5)	mostly uniform silty claystones with siliceous fossils and nannofossils	hemipelagic with some mud turbidites	NN19	395	19.5	
IVB	Sections 474A–10,CC–474A–28,CC	(258.5–420.0)	mostly cycles of thick, clayey quartzose siltstone beds and some arkose sands; mud flows;siliceous fossils diminished	middle fan and mud turbidites		395–86	(161.5)	gradational
V	Sections 474A–28,CC–474A–41.5 (25 cm) and 474A–45–1	420.0–533.0	olive grey clayey siltstone with thick mass flows and cemented arkose silty claystones between dolerite sills	middle fan, mud turbidites, and hemipelagic mud	late Pliocene	86	(~115)	interflow claystones

Figure 12.16 Organic-rich, laminated, hemipelagic, sediments from DSDP Leg 64, Site 480–29–3, 145–150 cm'. Photograph by K. Kelts. These laminites are interpreted as a result of annual phytoplankton and algal blooms. Scale to left graduated in mm and cm.

events in warm climatic periods, whereas the more homogeneous units are correlated with cooler and drier climatic periods (Kelts & Niemitz 1982).

The Gulf of California case study emphasizes the complexity of deep-marine sedimentation patterns in a large transtensional ocean basin that forms the southward extension of the San Andreas Fault system. Sustained and protracted dextral-slip plate motions have governed the history of western North America since at least 12.5 Ma.

Ancient deep-marine oblique-slip mobile zones

The identification of ancient oblique-slip plate margins is one of the most difficult aspects of reconstructing palaeogeographies. By inspection of modern plate margins, some degree of oblique extension or compression is the norm rather than the exception. However, ancient plate margins where suspect terrain accretion has occurred are most probably associated with considerable strike-slip. Examples include the Lower Palaeozoic northern Appalachians–southern Caledonides in Newfoundland and Britain, and the Mesozoic–Tertiary of the Pyrenees. In both cases, deep-marine sediments constitute a large part of the stratigraphy, suggesting sedimentation on thin lithosphere subject either to transpression or transtension.

CASE STUDY: MESOZOIC PYRENEES

The Mesozoic history of the Pyrenees was dominated by sea-floor spreading in the Ligurian Tethys, Central and North Atlantic and Bay of Biscay together with the anticlockwise rotation of the Iberian plate. These plate motions resulted in an overall sinistral shear couple between the Iberian and southern European plates (Fig. 12.17). Puigdefabregas & Souquet (1986)

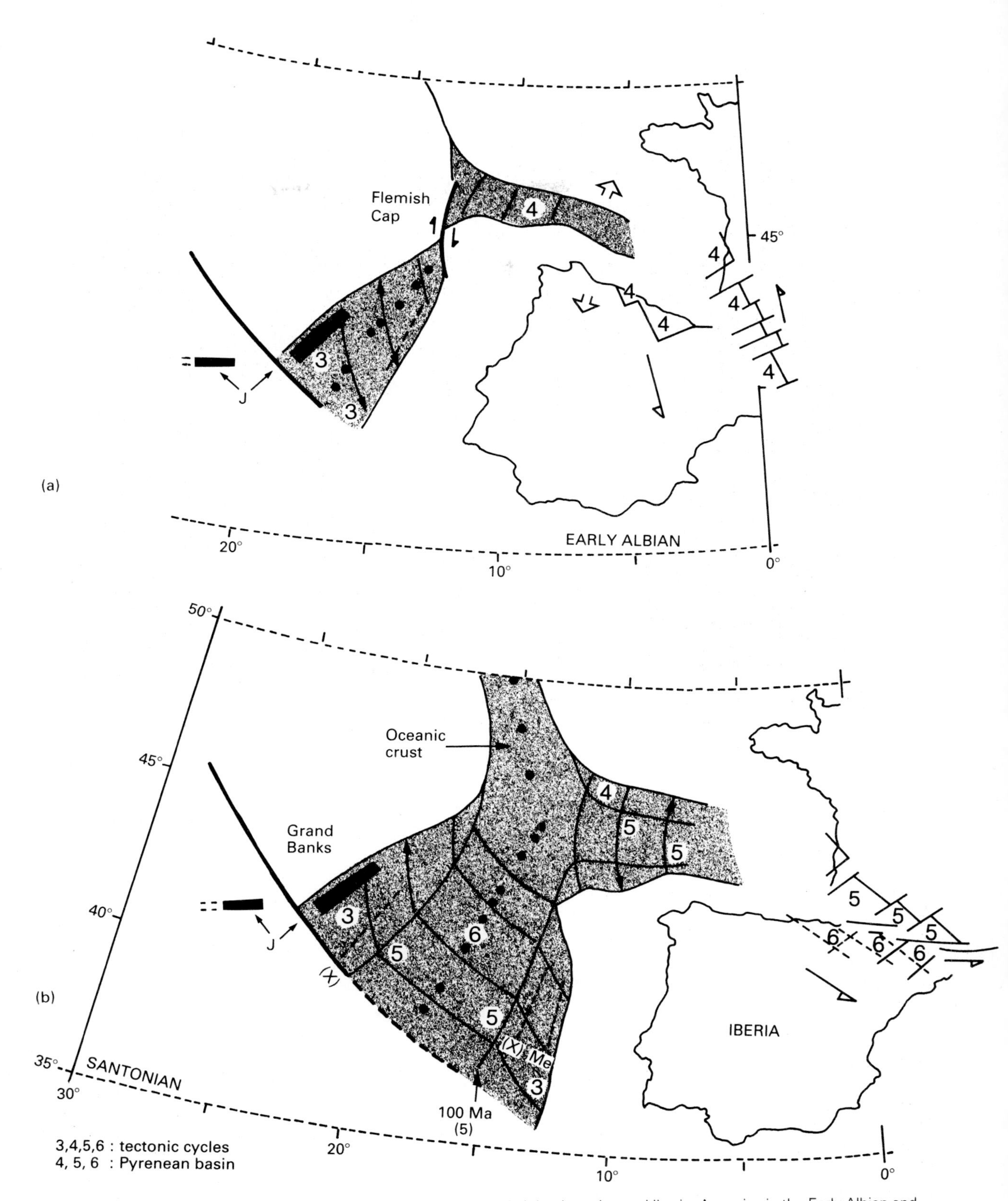

Figure 12.17 Hypothetical reconstruction of the southernmost North Atlantic region and Iberia–Armorica in the Early Albian and Santonian to show the development of the Mesozoic–Tertiary deep-marine basins in the Pyrenees. The basins are shown as due to major sinistral transtension along the Iberian and European Plate margins. After Puigdefabregas & Souquet (1986).

summarize the transtensional, wrench-faulting phase in the development of the Pyrenees in the following way. In the middle Albian–Cenomanian, transtension was associated with alkaline magmatism, thermal metamorphism and diapirism of the Triassic evaporites. The deeper basins were filled by deep-marine slope-apron and basin turbidites of the 'Flysch Noir' or Pyrenean flysch (Souquet *et al.* 1985). Figure 12.18 shows the present-day outcrop of the Flysch Noir, a reconstruction of the lozenge-shaped, deep-marine

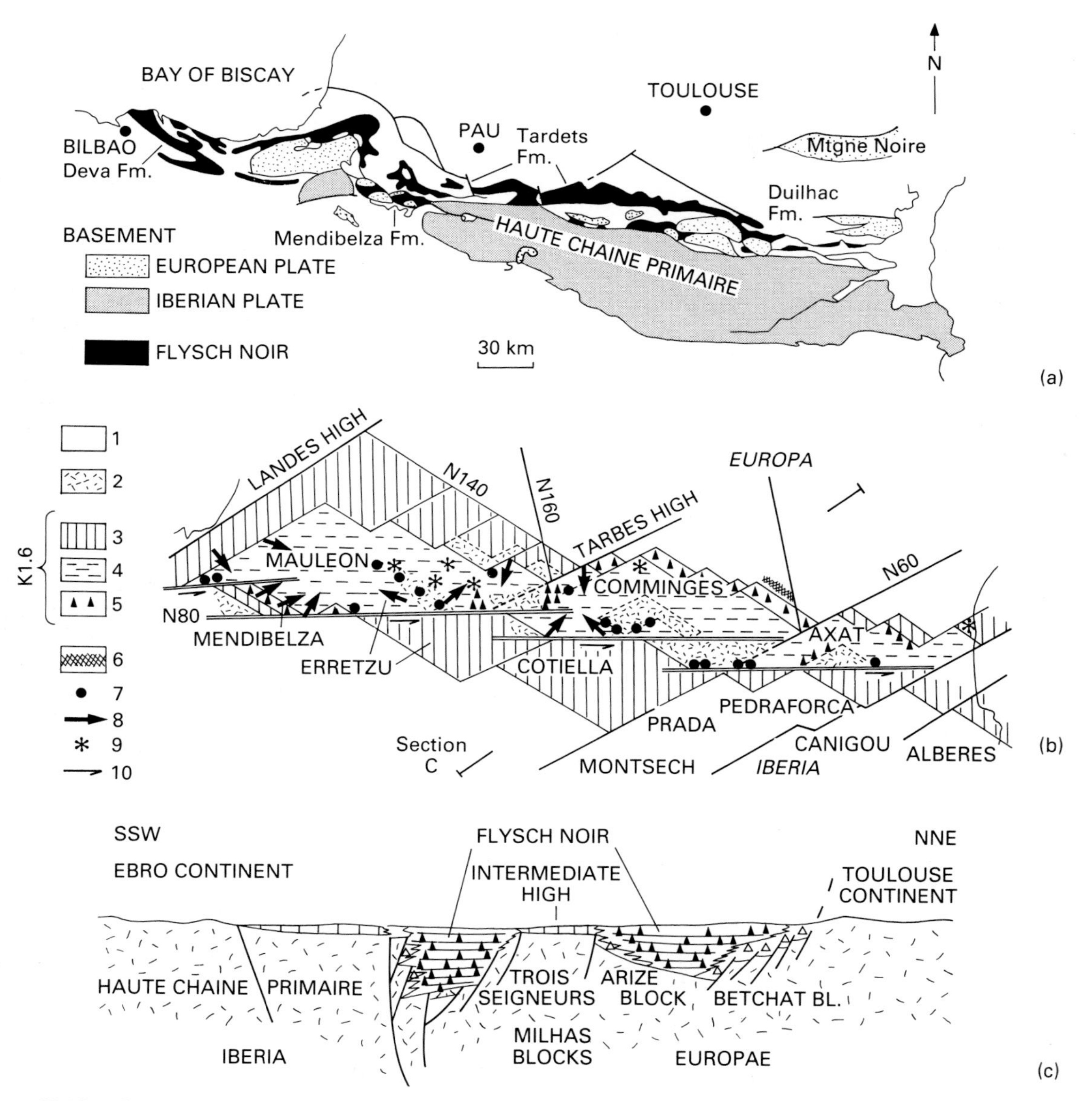

Figure 12.18a Generalized geological map of the North Pyrenean Fault zone in the northeastern Pyrenees to show the outcrop of the oldest Pyrenean flysch (Middle Albian–Cenomanian) that infilled the deeper parts of the developing basins, the 'Flysch Noir' Formation. **12.18b**: Palaeogeography for the 'Flysch Noir' Formation times, with **12.18c** showing hypothetical cross-section along Section C shown in 12.18b. Key as follows: 1, emergent areas; 2, eroded highs and/or over-riding blocks, 3, shelf deposits, 4, Albian turbidites; 5, fault-controlled debris apron; 6, bauxites; 7, lherzolites; 8, turbidite transport directions; 9, mid-Cretaceous magmatic rocks; 10, inferred strike-slip faults. After Puigdefabregas & Souquet (1986).

pull-apart basins, and a schematic cross-section through the basins. An example of slide folds in the Flysch Noir is shown in Figure 3.10a.

During the Cenomanian–middle Santonian, a global rise in sea level effectively widened the basin, and the site of shelf-carbonate sedimentation retreated landward. Carbonate turbidites from the drowned former shelf were shed into deeper water. In the middle Santonian, there was a relative fall in sea level, together with a tectonic tilting of the margin fault blocks; this resulted in widespread unconformities, and in deeper water the accumulation of slope breccias (Facies Class F) and megaturbidites (Facies Class A and Facies C2.4 & Class F). The late Santonian–Maastrichtian was associated with further normal faulting and subsidence, under sinistral transtension, and the development of a carbonate debris sheet extending downslope into the basin to interfinger with siliciclastic submarine fans showing axial sediment transport (Van Hoorn 1970). There was a major rise in sea level in the Santonian to give coastal onlap successions, and the development of progressive unconformities (Simo *et al.* 1985) shows that sedimentation was synchronous with folding. Sinistral transpression became increasingly more important into the Tertiary. This plate convergence caused uplift in the NE Pyrenees so that the area became a source for the clastics that were shed into the evolving foreland basins (see Ch. 11, Fig. 11.36).

Thus, the Mesozoic–Tertiary history of the Pyrenees provides an insight into plate margins that were initially dominated by transtension and then transpression as plate convergence finally closed the remaining seaways between the European and Iberian plates.

CASE STUDY: LOWER PALAEOZOIC NORTH-CENTRAL NEWFOUNDLAND AND BRITAIN

The large Cambrian Atlantic-type Iapetus Ocean (Wilson 1966) was transformed in the Early Ordovician into a Pacific-type ocean with subduction zones, sites of ophiolite obduction, and complex terrane accretion particularly along the Laurentian margin of North America and its continuation into the British Caledonides (Williams & Hatcher 1982, 1983, Williams 1984, 1985, Bluck 1985). Terrane accretion occurred progressively outwards from the Laurentian margin (Williams 1985) until the Iapetus Ocean was finally closed in Devonian (Emsian) times (Soper *et al.* 1987). Although the demise of any vestigial ocean did not occur until the Emsian, palaeontologic, tectono-stratigraphic and palaeomagnetic evidence suggests initial closure of the Iapetus Ocean with complete subduction of intervening oceanic crust between Laurentia (North America) in the region of Western Newfoundland and Eastern Avalonia (Britain and Eire south of the Iapetus Suture) probably by the late Ashgill (Pickering 1987a & b, Pickering *et al.* 1988b). Western Avalonia (eastern Newfoundland, Acadia and Appalachian Piedmont terrane) was accreted in the Early–Middle Devonian (Keppie 1986). Figure 12.19, from Pickering *et al.* 1988b summarizes the history of the Iapetus Ocean from Arenig to Llandovery times.

Suspect terranes that accreted to the Laurentian margin are widely reported from the Lower Palaeozoic Appalachians and Caledonides (Fig. 12.19) (Williams & Hatcher 1982, 1983, Williams 1984, 1985, Curray *et al.* 1982, Soper & Hutton 1984, Bluck 1985, Keppie 1986, Ziegler 1986a, b & c, Anderson & Oliver 1986, Bluck & Leake 1986, Hutton & Dewey 1986, Elders 1987, Pickering 1987b, Soper *et al.* 1987, Pickering *et al.* 1988b). The Laurentian margin appears to have been a site of complex terrane accretion throughout the Ordovician until the Middle Devonian, with most evidence supporting terrane docking by sinistral shear (e.g. Soper & Hutton 1984, Bluck 1985, Keppie 1986, Hutton & Dewey 1986, Elders 1987, Pickering 1987a & b, Soper *et al.* 1987, Blewett & Pickering 1988, Pickering *et al.* 1988b).

Although it is difficult to assess the cumulative amount of strike slip along the Laurentian margin during terrane accretion, the radiometric age range and petrography of granodiorite clasts in the Southern Uplands of Scotland and the Long Range area of Newfoundland, together with palaeocurrent data, suggest that at some time during the Caradoc these terranes may have been juxtaposed (Elders 1987). If so, then at least 1500 km of cumulative sinistral displacement must have occurred from Caradoc to Middle Devonian (Emsian) along various fault zones when the British Isles basement was finally assembled. Both in the British Isles and Newfoundland, deep-marine turbidite systems (mainly slope and slope basin fills) occur throughout the Ordovician and Lower Silurian successions (e.g. Watson 1981, Arnott 1983a & b, Arnott *et al.* 1985, Pickering 1987a), and are overlain by shallow shelf to non-marine deposits.

Regional considerations, together with available palaeomagnetic data, suggest a palaeogeography for the latest Ordovician–earliest Silurian (Fig. 12.19) in which: (a) the once contiguous Iapetus Ocean was finally destroyed by continent–continent collision between Laurentia and Eastern Avalonia to eliminate any intervening oceanic crust; (b) destruction of oceanic crust continued both to the north and south of this collisional site with westward to northwestward subduction as Baltica and Western Avalonia approached Laurentia; (c) oblique collision of Eastern Avalonia with Laurentia, probably first occurring opposite western Newfoundland at a promontory in the continental margin, leading to sustained sinistral oblique-slip, creating transtensional pull-apart, and transpressional thrust-controlled, deep-marine basins.

In north-central Newfoundland, structural and sedimentological studies suggest Ashgill-Wenlock active fault-controlled sedimentation, mainly in relatively small, deep-marine basins controlled by NW-

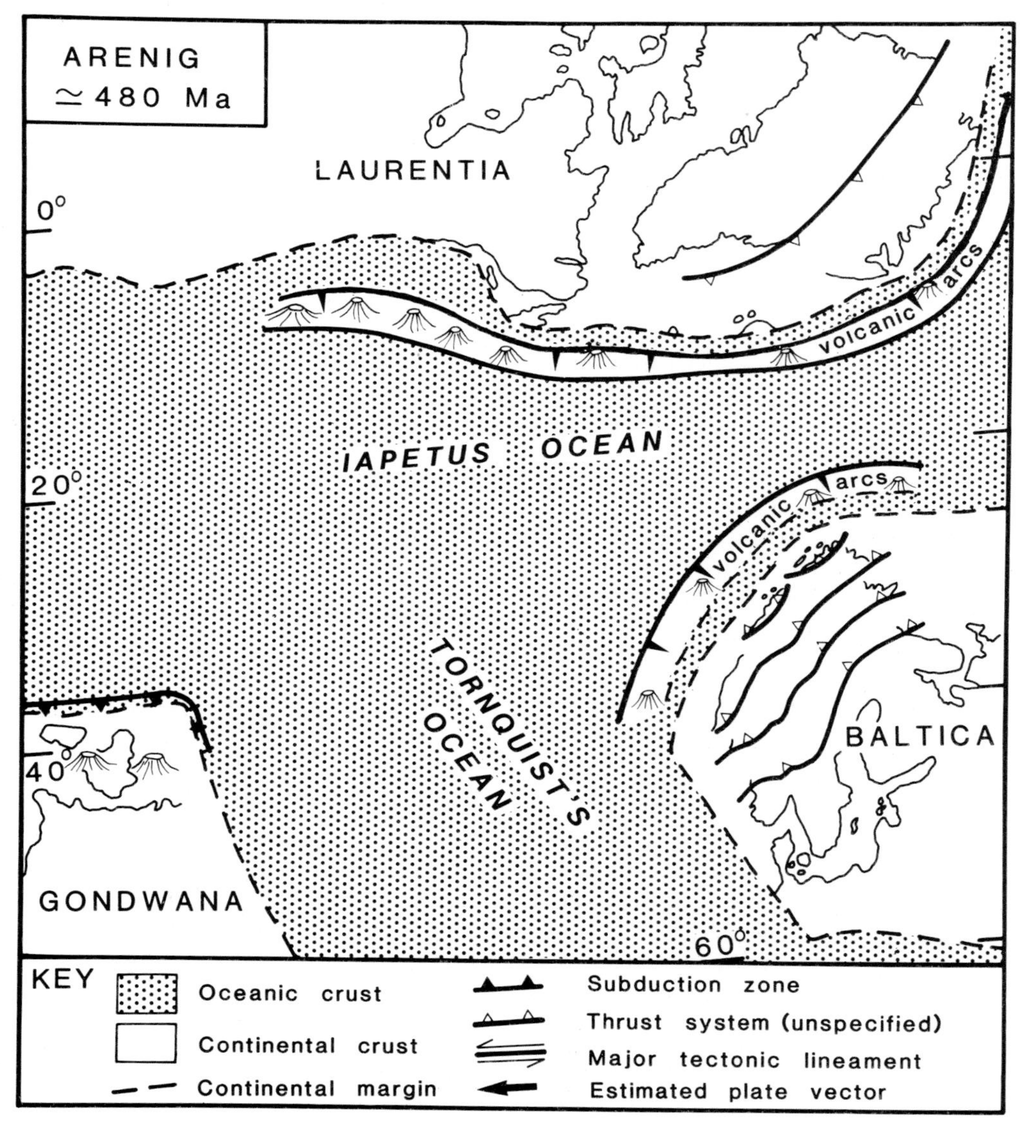

Figure 12.19 Three palaeogeographic maps for the Lower Palaeozoic Iapetus Ocean, Tornquist's Ocean and the Rheic Ocean. Note that the closure of the Iapetus Ocean is interpreted as involving protracted, sinistral-shear dominated, transpression and transtension between the Eastern Avalonian continent and Laurentia. The Ashgill–Wenlock deep-marine sediments in Central Newfoundland are believed to have accumulated in a foreland basin analogous to the present-day Timor Trough or the Taiwan Foreland Basin. After Pickering *et al.* (1988b).

dipping thrust faults, commonly associated with tectonized olistostromes (mélanges) and growth faults (Nelson 1981, Arnott 1983a & b, Arnott *et al.* 1985, Pickering 1987a & b). In New World Island, Newfoundland, Arnott (1983a & b) has demonstrated that essentially contemporaneous successions, separated by major syn-sedimentary faults, probably developed in discrete neighbouring basins. A syn-sedimentary Late Ordovician–Silurian D1 deformation is recognized both in the Gander and Dunnage terranes. Major tectonic elements associated with the Dunnage–Gander terranes that may have been active, as ancestral structures, during the Late Ordovician–Silurian include the Hermitage Flexure (Brown & Colman-Sadd 1976), the Lobster Cove–Chanceport and Lukes Arm–Sops Head faults across north-central Newfoundland (Nelson 1981, Watson 1981, Arnott 1983a & b, Arnott *et al.* 1985), with possible NW-dipping thrust faults including the ancestral Lukes Arm, Toogood, Cobbs Arm, Byrne Cove and Boyds Island faults on New

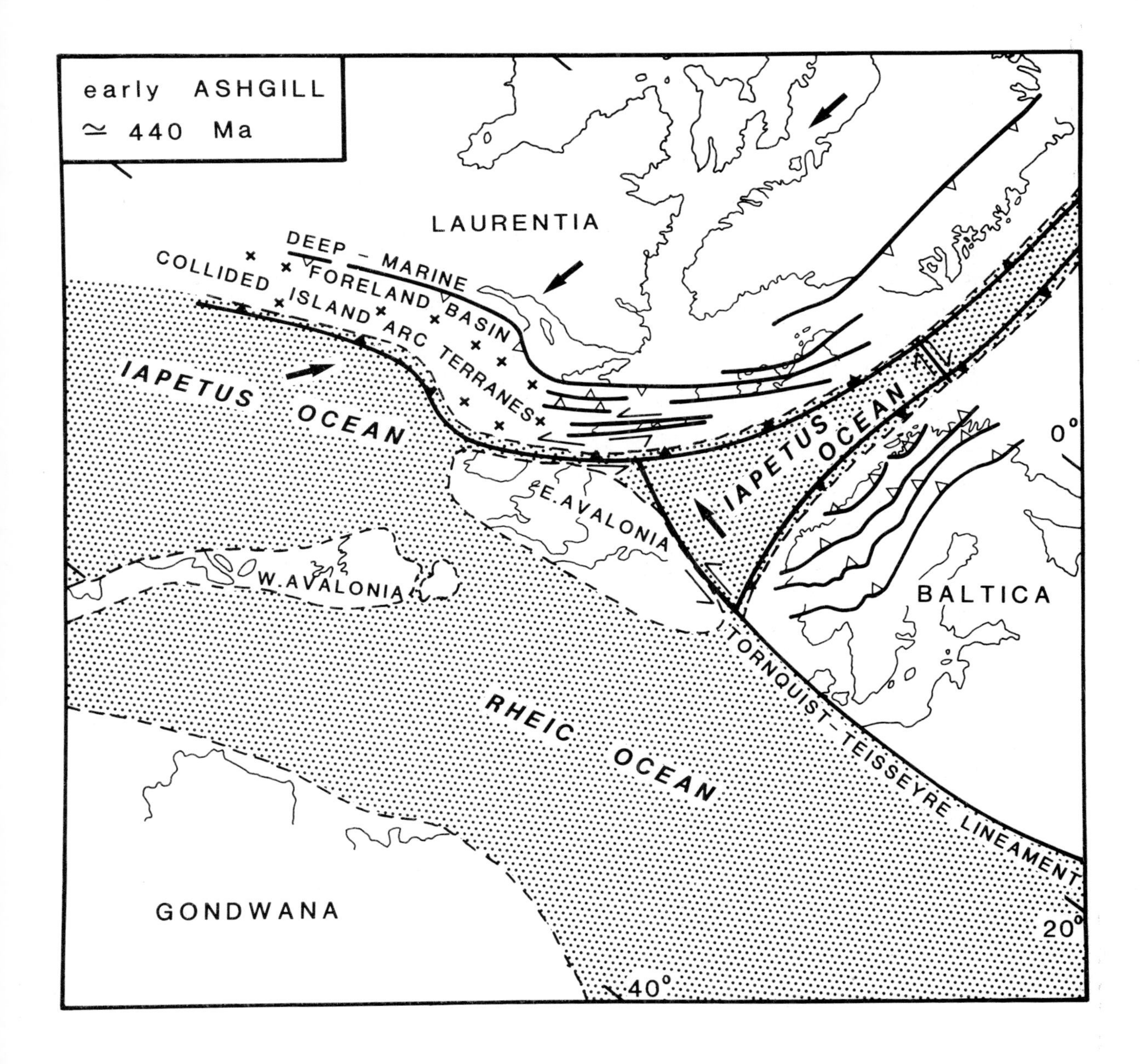

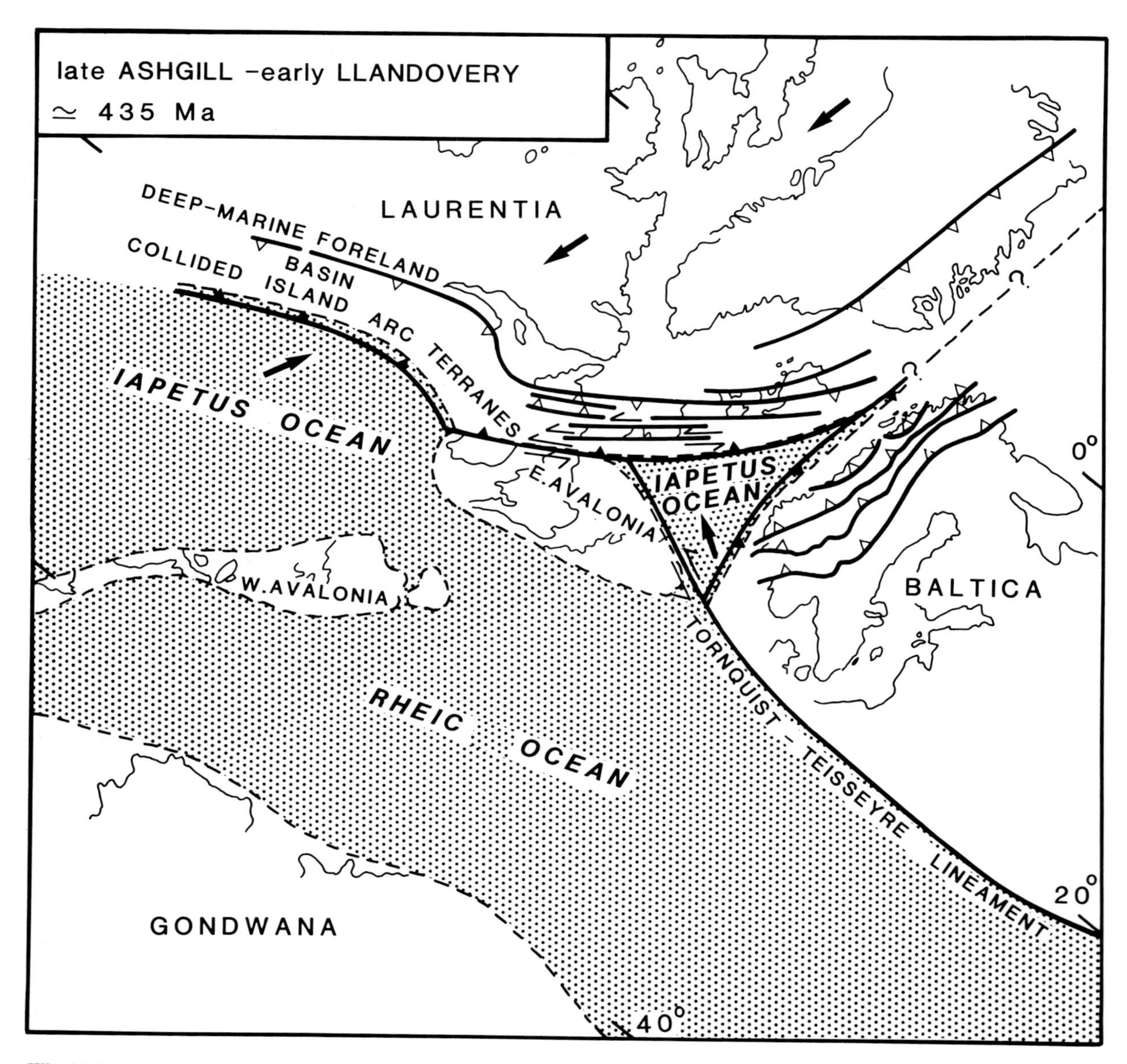

World Island (Nelson 1981, Arnott 1983a & b, Arnott *et al.* 1985), and the ancestral New Bay Fault (Pickering 1987a).

Figure 12.20 shows a palaeogeographic reconstruction of north-central Newfoundland in the late Caradoc – Ashgill (Blewett pers. comm.). This map shows a number of small, fault-controlled, deep-marine basins separated by structural highs that are commonly at an obtuse angle to the main basin-bounding faults. The sediment source for the basins was from the deeply dissected arc terrane to the north, with conglomeratic and sandy facies classes (A,B,C,F) typically occurring in submarine channels and valleys up to several hundred metres deep and a few kilometres wide; the surrounding lithologies mainly being silty or muddy as Facies Classes D and E, with minor G (cf. Nelson 1981, Watson 1981, Arnott 1983a & b, Pickering 1987a). The sedimentology in these deep-marine basins in Newfoundland varies considerably from thick coarse-grained submarine fan deposits, as in the Milliners Arm Formation, New World Island (Watson 1981 and Ch. 7 case study), to thick, fine-grained,

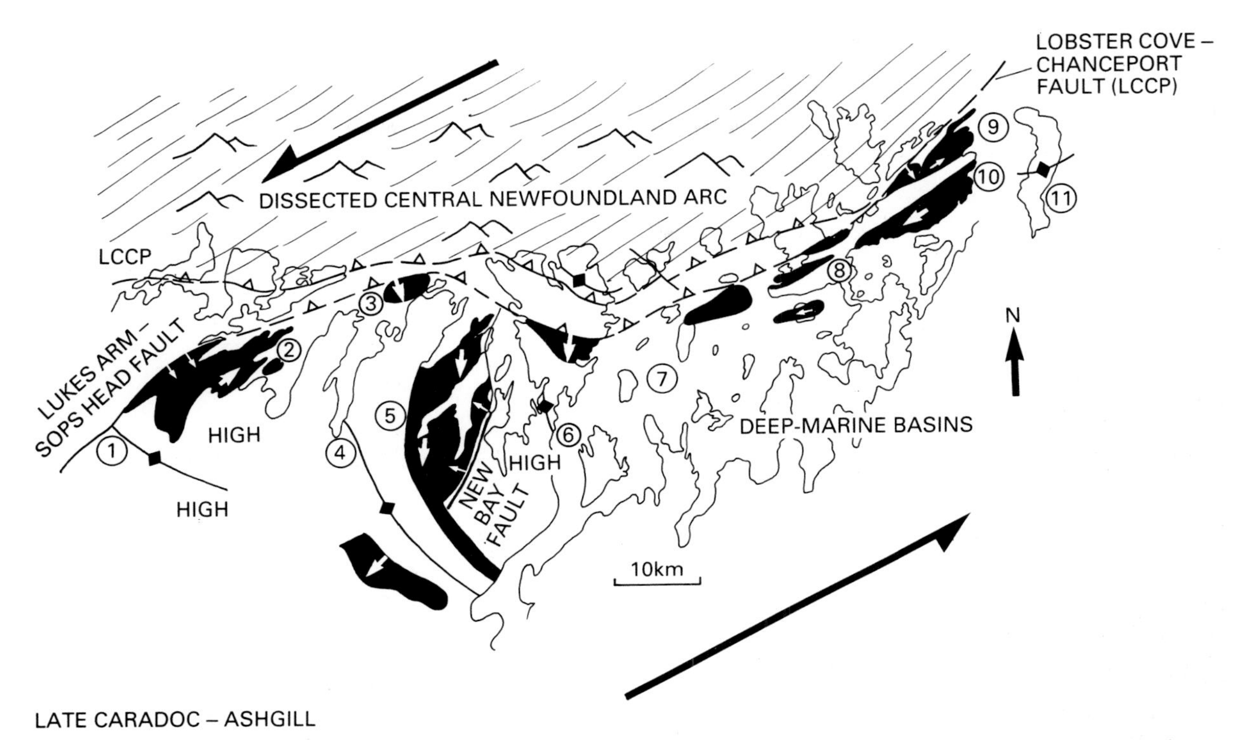

Figure 12.20 Late Caradoc–Ashgill palaeogeographic reconstruction for the environments of deposition of the deep-marine successions in north Central Newfoundland, based on unpublished work by R. Blewett and various other sources (see text). Black legend represents basin depocentres between structural highs; white arrows are schematized palaeocurrent directions; main faults interpreted as active during sedimentation are: LCCP, Lobster Cove–Chanceport Fault; LASH, Lukes Arm–Sops Head Fault, both as thrusts/oblique-slip faults. Note the deeply dissected Central Newfoundland Island Arc to the north, that collided with Laurentia in the Llandeilo, and provided the main source of sediments for the basins to the south on the plate margin. Also, note the inferred importance of major sinistral shear at that time. By the late Ashgill, Eastern Avalonia probably was sited hundreds of kilometres to the southeast and the intervening environments are interpreted by Pickering *et al.* (1988b) as within a major deep-marine foreland basin. Numbers refer to areas from which stratigraphic and structural data have been integrated to produce this palaeogeographic interpretation (from R. Blewett).

slope-basin fills cut by coarse-grained channel successions, as in the Point Leamington Formation, New Bay area (Pickering 1987a). Furthermore, there is abundant evidence of wet-sediment deformation as sediment slides (Pickering 1987a).

The California Borderland provides a useful modern analogue, with a complex series of fault-controlled basins oceanward of a major zone of strike-slip. A major difference, however, is that during the Silurian, in Newfoundland, there was no major ocean basin to the south/southeast of these marine basins unlike off California today (see Fig. 12.19). The seaway to the southeast was probably more like the present day Timor Trough. In the example from north-central Newfoundland, the major strike-slip/oblique-slip appears to have occurred under a sinistral shear couple along faults such as the Lobster Cove–Chanceport, and Lukes Arm–Sops Head faults (Fig. 12.20). Sinistral shear associated with deep-marine sedimentation appears to have occurred from Late Ordovician to Middle Silurian times, followed by the infilling of remaining marine seaways by shallowing-upward successions.

References

Aalto, K. R. 1976. Sedimentology of a mélange: Franciscan of Trinidad, California. *J. Sed. Petrol.* **46**, 913–29.

Abbate, E., V. Bortolotti & P. Passerini 1970. Olistosomes and olistoliths. *Sed. Geology* **4**, 521–57.

Adams, J. H. 1908. The eruption of the Waimata mud spring. *New Zealand Mines Record* **12**, 908–12.

Addy, S. K. & R. T. Buffler 1984. Seismic stratigraphy of the shelf and slope; northeastern Gulf of Mexico. *Bull. Am. Assoc. Petrolm Geol.* **68**, 1782–9.

Addy, S. K. & H. Kagami 1979. Sedimentation in a closed trough north of the Iberia abyssal plain in Northeast Atlantic. *Sedimentology* **26**, 561–75.

Aksu, A. E. 1984. Subaqueous debris flow deposits in Baffin Bay. *Geo-Marine Letts* **4**, 83–90.

Allen, J. R. L. 1960. The Mam Tor Sandstones: a 'turbidite' facies of the Namurian deltas of Derbyshire, England. *J. Sed. Petrol.* **30**, 193–208.

Allen, J. R. L. 1969. Some recent advances in the physics of sedimentation. *Proc. Geol. Assoc.* **80**, 1–42.

Allen, J. R. L. 1970. The sequence of sedimentary structures in turbidites, with special reference to dunes. *Scottish J. Geology* **6**, 146–61.

Allen, J. R. L. 1982. Sedimentary structures: their character and physical basis. *Developments in sedimentology* **30** (parts I and II). Amsterdam: Elsevier.

Allen, J. R. L. & N. L. Banks 1972. An interpretation and analysis of recumbent-folded deformed cross-bedding. *Sedimentology* **19**, 257–83.

Allen, P. A. & P. Homewood (eds) 1986. *Foreland basins*. Int. Assoc. Sedimentologists Spec. Publn 8. Oxford: Blackwell Scientific.

Almagor, G. & Z. Garfunkel 1979. Submarine slumping in continental margin of Israel and northern Sinai. *Bull. Am. Assoc. Petrolm Geol.* **63**, 324–40.

Almagor, G. & G. Wiseman 1982. Submarine slumping and mass movements on the continental slope of Israel. In *Marine slides and other mass movements*, S. Saxov & J. K. Nieuwenhuis (eds), 95–128. New York: Plenum.

Almgren, A. A. 1978. Timing of Tertiary submarine canyons and marine cycles of deposition in the southern Sacramento Valley, California. In *Sedimentation in submarine canyons, fans, and trenches*, D. J. Stanley & G. Kelling (eds), 276–91. Stroudsburg, PA: Dowden, Hutchinson & Ross.

Alvarez, W., R. Colacicchi & A. Montanari 1985. Synsedimentary slides and bedding formation in Apennine pelagic limestones. *J. Sed. Petrology* **55**, 720–34.

Anderson, J. B., D. D. Kurtz & F. M. Weaver 1979. Sedimentation on the Antarctic continental slope. In *Geology of continental slopes*, L. J. Doyle & O. H. Pilkey (eds), 265–83. Soc. Econ. Paleont. Mineral. Spec. Publn 27.

Anderson, R. N. 1986. *Marine geology*. New York: Wiley.

Anderson, T. B. & G. T. H. Oliver 1986. The Orlock Bridge Fault: a major Late Caledonian sinistral fault in the Southern Uplands terrane, British Isles. *Trans. R. Soc. Edinburgh: Earth Sci.* **77**, 203–22.

Andreson, A. & L. Bjerrum 1967. Slides in subaqueous slopes in loose sand and silt. In *Marine geotechnique*, A. F. Richards (ed.), 221–39. Urbana: University of Illinois Press.

Angelier, J., B. Colletta, J. Chorowicz, L. Ortlieb & C. Rangin 1981. Fault tectonics of the Baja California Peninsula and the opening of the Sea of Cortez, Mexico. *J. Structural Geology* **3**, 347–57.

Aoki, Y., T. Tamano & S. Kato 1983. Detailed structure of the Nankai Trough from migrated seismic sections. In *Studies in continental margin geology*, J. S. Watkins & C. L. Drake (eds), 309–22. Am. Assoc. Petrolm Geol. Mem. 34.

Arnold, R. & G. A. Macready 1956. Island-forming mud volcano in Trinidad, British West Indies. *Bull. Am. Assoc. Petrolm Geol.* **40**, 2748–58.

Arnott, R. J. 1983a. Sedimentology, structure and stratigraphy of north-east New World Island, Newfoundland. Unpubl. Ph.D. Thesis, Oxford University.

Arnott, R. J. 1983b. Sedimentology of Upper Ordovician–Silurian sequences on New World Island, Newfoundland: separate fault-controlled basins? *Can. J. Earth Sci.* **20**, 345–54.

Arnott, R. J., W. S. McKerrow & L. R. M. Cocks 1985. The tectonics and depositional history of the Ordovician and Silurian rocks of Notre Dame Bay, Newfoundland. *Can. J. Earth Sci.* **22**, 607–18.

Arnott, R. W. & F. J. Hein 1986. Submarine canyon fills of the Hector Formation, Lake Louise, Alberta: Late Precambrian syn-rift deposits of the proto-Pacific miogeocline. *Bull. Can. Petrolm Geol.* **34**, 395–407.

Arrhenius, G. 1963. Pelagic sediments. In *The sea*, Vol. 3, M. N. Hill (ed.), 655–727. New York: Wiley.

Arthur, M. A. & J. H. Natland 1979. Carbonaceous sediments in

North and South Atlantic: the role of salinity in stable stratification of Early Cretaceous basins. In *Deep drilling results in the Atlantic Ocean: continental margins and paleoenvironment*, M. Talwani, M. W. Hay & W. B. F. Ryan (eds), 375–401. Maurice Ewing Series 3. Washington: Am. Geophys. Union.

Arthur, M. A. & S. O. Schlanger 1979. Cretaceous 'oceanic anoxic events' as causal factors in development of reef-reservoired giant oil fields. *Bull. Am. Assoc. Petrolm Geol.* **63**, 870–85.

Arthur, M. A., W. E. Dean & D. A. V. Stow 1984. Models for the deposition of Mesozoic–Cenozoic fine-grained organic-carbon-rich sediment in the deep sea. In *Fine-grained sediments: deep-water processes and facies*, D. A. V. Stow & D. J. W. Piper (eds), 527–60. Geol Soc. (London) Spec. Publn 15. Oxford: Blackwell Scientific.

Armi, L. 1978. Mixing in the deep oceans: the importance of boundaries. *Oceanus* **21**, 14–19.

Armi, L. & E. d'Asaro 1980. Flow structures of the benthic ocean. *J. Geophys. Res.* **85**, 469.

Athanasiou-Grivas, D. 1978. Reliability analysis of earth slopes. *Proc. Soc. Engineering Sci. 15th Annual Meeting*, Gainesville, University of Florida, 453–8.

Atwater, T. & P. Molnar 1973. Relative motion of the Pacific and North American Plates deduced from seafloor spreading in the Atlantic, Indian and South Pacific Oceans. In *Proceedings of the conference on tectonic problems in the San Andreas Fault system*, R. L. Kovach & A. Nur (eds), 136–48. Stanford University: Geol Sci. Publn 13.

Aubouin, J. & R. von Huene 1985. Summary: Leg 84, Middle America Trench transect off Guatemala and Costa Rica. In *Initial Reports Deep Sea Drilling Project 84*, R. von Huene, J. Aubouin *et al.*, 936–57. Washington, DC: US Government Printing Office.

Aubouin, J. *et al.* 1977. Les Hellenides dans l'optique de la technique des plaques. *Proc. 6th Colloquium on the Geology of the Aegean Region, Athens*, 1335–54.

Aubouin, J., J. Bourgois & J. Azema 1984. A new type of active margin: the convergent–extensional margin, as exemplified by the Middle America Trench off Guatemala. *Earth Planet. Sci. Letts* **67**, 211–18.

Aubouin, J., J. Bourgois, J. Azema & R. von Huene 1985. Guatemala margin: a model of convergent extensional margin. In *Initial Reports Deep Sea Drilling Project 84*, R. von Huene, J. Aubouin *et al.*, 911–17. Washington, DC: US Government Printing Office.

Aubouin, J., J. Bourgois, R. von Huene & J. Azema 1982a. La marge pacifique du Guatemala: un modéle de marge extensive en domaine convergent. *C.R. Acad. Sci. Paris* **295**, 607–14.

Aubouin, J., J. F. Stephan, J. Roump & V. Renard 1982b. The Middle America Trench as an example of a subduction zone. *Tectonophysics* **86**, 113–32.

Aubouin, J., R. von Huene *et al.* 1982c. *Initial Reports Deep Sea Drilling Project* 67. Washington, DC: US Government Printing Office.

Audley-Charles, M. G. 1986a. Rates of Neogene and Quaternary tectonic movements in the Southern Banda Arc based on micropalaeontology. *J. Geol Soc. (London)* **143**, 161–75.

Audley-Charles, M. G. 1986b. Timor–Tanimbar Trough: the foreland basin of the evolving Banda orogen. In *Foreland basins*, P. A. Allen & P. Homewood (eds), 91–102. Int. Assoc. Sedimentologists Spec. Publn **8**. Oxford: Blackwell Scientific.

Aydin, A. & B. M. Page 1984. Diverse Pliocene–Quaternary tectonics in a transform environment, San Francisco Bay region, California. *Bull. Geol Soc. Am.* **95**, 1303–17.

Baba, J. & P. D. Komar 1981. Measurements and analysis of settling velocities of natural quartz sand grains. *J. Sed. Petrol.* **51**, 631–40.

Bagnold, R. A. 1956. The flow of cohesionless grains in fluids. *Phil. Trans. R. Soc. London (A)* **249**, 235–97.

Bagnold, R. A. 1962. Auto-suspension of transported sediment: turbidity currents. *Proc. R. Soc. London (A)* **265**, 315–19.

Bagnold, R. A. 1966. An approach to the sediment transport problem from general physics. *Prof. Pap. US Geol Surv.* 422–I.

Bagnold, R. A. 1973. The nature of saltation and 'bed-load' transport in water. *Proc. R. Soc. London (A)* **332**, 473–504.

Bailey, E. B. & J. Weir 1932. Submarine faulting in Kimmeridgian times: east Sutherland. *Trans. R. Soc. Edinburgh* **47**, 431–67.

Baker, E. T. & B. M. Hickey 1986. Contemporary sedimentation processes in and around an active west coast submarine canyon. *Marine Geology* **71**, 15–34.

Ballance, P. F. & H. G. Reading (eds) 1980. Sedimentation in oblique-slip mobile zones. *Int. Assoc. Sedimentologists Spec. Publn* **4**. Oxford: Blackwell Scientific.

Bally, A. W. 1980. Basins and subsidence – a summary. In *Dynamics of plate interiors. Am. Geophys. Union, Geodynamics Series* **1**, 5–20.

Bally, A. W. 1982. Musings over sedimentary basin evolution. *Phil. Trans. R. Soc. London (A)* **305**, 325–38.

Bally, A. W. (ed.) 1983. *Seismic expression of structural styles*. Vol. 1–3. Am. Assoc. Petrolm Geol. Studies in Geology 15.

Baltuck, M., E. Taylor & K. McDougall 1985. Mass movement along the inner wall of the Middle America Trench, Costa Rica. In *Initial Reports Deep Sea Drilling Project 84*, R. von Huene, J. Aubouin *et al.*, 551–70. Washington, DC: US Government Printing Office.

Barber, A. J., S. Tjokrosapoetro & T. R. Charlton 1986. Mud volcanoes, shale diapirs, wrench faults, and mélanges in accretionary complexes, Eastern Indonesia. *Bull. Am. Assoc. Petrolm Geol.* **70**, 1729–41.

Barnes, N. E. & W. R. Normark 1984. Diagnostic parameters for comparing modern submarine fans and ancient turbidite systems. *Geo-Marine Letts* **3**, map following p. 224.

Barron, E. J. 1983. A warm, equable Cretaceous: the nature of the problem. *Earth-Sci. Rev.* **19**, 305–38.

Barron, E. J. & J. M. Whitman 1981. Ocean sediments in space and time. In *The sea*, Vol. 7, C. Emiliani (ed.), 689–733. New York: Wiley Interscience.

Bartolini, C., C. Gehin & D. J. Stanley 1972. Morphology and recent sediments of the Western Alboran Basin in the Mediterranean Sea. *Marine Geology* **13**, 159–224.

Bartow, J. A. 1966. Deep submarine channel in upper Miocene, Orange County, California. *J. Sed. Petrol.* **36**, 700–5.

Baturin, G. N. 1982. Phosphorites on the seafloor: origin, composition and distribution. *Developments in sedimentology* **33**. Amsterdam: Elsevier.

Beach, A. 1986. A deep seismic reflection profile across the northern North Sea. *Nature* **323**, 53–5.

Beach, A. 1987. A regional model for linked tectonics in NW Europe. In *Petroleum geology of north west Europe*. J. Brooks & K. W. Glennie (eds), 43–8. London: Graham & Trotman.

Beaumont, C., C. E. Keen & R. Bantillier 1982. A comparison of foreland and rift margin sedimentary basins. In *The evolution of sedimentary basins*, P. Kent, M. H. P. Bott, D. P. McKenzie & C. A. Williams (eds), 295–318. London: The Royal Society.

Beaudry, D. & G. F. Moore 1985. Seismic stratigraphy and Cenozoic evolution of West Sumatra forearc basin. *Bull. Am.*

Assoc. Petrolm Geol. **69**, 742–59.
Bein, A. & Y. Weiler 1976. The Cretaceous Talme Yafe Formation, a contour current shaped sedimentary prism of carbonate debris at the continental margin of the Arabian craton. *Sedimentology* **23**, 511–32.
Belderson, R. H. & A. H. Stride 1969. The shape of submarine canyon heads revealed by Asdic. *Deep-Sea Res.* **16**, 103–4.
Belderson, R. H., N. H. Kenyon, A. H. Stride & C. D. Pelton 1984. A 'braided' distributary system on the Orinoco deep-sea fan. *Marine Geology* **56**, 195–206.
Belt, E. S. & L. Brussiers 1981. Upper Middle Ordovician submarine fans and associated facies, northeast of Quebec City. *Can. J. Earth Sci.* **18**, 981–94.
Bennett, R. H. & T. A. Nelson 1983. Seafloor characteristics and dynamics affecting geotechnical properties at shelfbreaks. In *The shelfbreak: critical interface on continental margins.*, D. J. Stanley & G. T. Moore (eds), 333–55. Soc. Econ. Paleont. Mineral. Spec. Publn 33.
Bennetts, K. R. W. & O. H. Pilkey 1976. Characteristics of three turbidites, Hispaniola–Caicos Basin. *Bull. Geol Soc. Am.* **87**, 1291–300.
Bentor, Y. K. (ed.) 1980. *Marine phosphorites–geochemistry, occurrence, genesis.* Soc. Econ. Paleont. Mineral. Spec. Publn 29.
Berger, W. H. 1974. Deep-sea sedimentation. In *The geology of continental margins*, C. A. Burk & C. L. Drake (eds), 213–41. New York: Springer.
Berger, W. H. & D. J. W. Piper 1972. Planktonic foraminifera: differential settling, dissolution and redeposition. *Limn. Oceanogr.* **17**, 275–87.
Berger, W. H., E. Vincent & H. R. Thierstein 1981. The deep-sea record: major steps in Cenozoic ocean evolution. In *The deep sea drilling project: a decade of progress*, J. E. Warme, R. G. Douglas & E. L. Winterer (eds), 489–504. Soc. Econ. Paleont. Mineral. Spec. Publn 32.
Berggren, W. A. 1982. Role of ocean gateways in climatic change. In *Climate in Earth history, studies in geophysics, Panel on pre-Pleistocene climates*, 118–38. Washington, DC: National Academy Press.
Berkland, J. O. 1972. Paleogene 'frozen' subduction zone in the Coast Ranges of northern California. *24th Int. Geol Congress, Montreal*, Section 3, 99–105.
Bernouli, D. (contributor) 1980. Southern Alps of Ticino. In *Geology of Switzerland*, a guide-book (Pt. A), R. Trumpy, 80–2. Basel–New York: Wepf.
Bernoulli, D. & H. C. Jenkyns 1974. Alpine, Mediterranean, and Central Atlantic Mesozoic facies in relation to the early evolution of the Tethys. In *Modern and ancient geosynclinal sedimentation*, R. H. Dott & R. H. Shaver (eds), 129–60. Soc. Econ. Paleont. Mineral. Spec. Publn 19.
Berryhill, H. L. Jr 1981. Ancient buried submarine trough, northwest Gulf of Mexico. *Geo-Marine Letts* **1**, 105–10.
Bertagne, A. J. 1984. Seismic stratigraphy of Veracruz Tongue, deep southwestern Gulf of Mexico. *Bull. Am. Assoc. Petrolm Geol.* **68**, 1894–907.
Berthois, L. & Y. le Calves 1960. Etude de la vitesse de chute des coquilles de foraminifères planctoniques dans un fluid comparativement a celle de grains de quartz. *Inst. Peches Mar.* **24**, 293–301.
Bertrand, M. 1897. Structure des Alps française et recurrence de certaines facies sédimentaires. *VIe. Int. Geol Congress Zurich*, 161–77.
Bhatia, M. R. 1983. Plate tectonics and geochemical composition of sandstones. *J. Geology* **91**, 611–27.
Bhatia, M. R. & K. A. W. Crook 1986. Trace element characteristics of graywackes and tectonic setting discrimination of sedimentary basins. *Contrib. Mineral. Petrol.* **92**, 181–93.
Biju-Duval, B., S. Charier, M. Taviani, Y. Morel, A. Baudrimont, P. F. Burollet, P. Clairfond, G. Clavzon, P. Colantoni, G. Mascle, L. Montadert, R. Perrier, P. Orsoloni, C. Ravenne & E. Winnock 1983. Dépressions circulaires au pied de l'escarpement de Malte et morphologie des escarpements sous-marine: problems d'interprétation. *Rev. Inst. France Pet.* **38**, 605–19.
Biju-Duval, B., P. Le Quellec, A. Mascle, V. Renard & P. Valery 1982. Multibeam bathymetric survey and high resolution seismic investigation on the Barbados Ridge Complex (eastern Caribbean): a key to the knowledge and interpretation of an accretionary wedge. *Tectonophysics* **86**, 275–304.
Biju-Duval, B., J. Letouzey & L. Montadert 1979. Variety of margins and deep basins in the Mediterranean. In *Geological and geophysical investigations of continental margins*, J. S. Watkins, L. Montadert & P. W. Dickerson (eds), 293–317. Tulsa, Oklahoma: Am. Assoc. Petrolm Geol.
Biju-Duval, B., J. C. Moore *et al.* 1984. *Initial Reports Deep Sea Drilling Project 78A.* Washington, DC: US Government Printing Office.
Birchwood, K. M. 1965. Mud volcanoes in Trinidad. *Inst. Petrolm Geol Rev.* **19**, 164–7.
Biscaye, P. E. & S. L. Eittreim 1977. Suspended particulate loads and transports in the nepheloid layer of the abyssal Atlantic Ocean. *Marine Geology* **23**, 155–72.
Biscaye, P. E., W. D. Gardner, J. R. V. Zaneveld, H. Pak & B. Tucholke 1980. Nephels! Have we got nephels! *EOS, Trans. Am. Geophys. Union* **61**, 1014.
Blake, M. C. Jr, R. H. Campbell, T. W. Dibblee, D. G. Howell, T. H. Nilsen, W. R. Normark, J. C. Vedder & E. A. Silver 1978. Neogene basin formation in relation to plate-tectonic evolution of San Andreas Fault system, California. *Bull. Am. Assoc. Petrolm Geol* **62**, 344–72.
Blatt, H. G., G. V. Middleton & R. C. Murray 1980. *Origin of sedimentary rocks*, 2nd edn. Englewood Cliffs, New Jersey: Prentice-Hall.
Blendinger, W. 1986. Isolated stationary carbonate platforms: the Middle Triassic (Ladinian) of the Marmolada area, Dolomites, Italy. *Sedimentology* **33**, 159–83.
Blewett, R. S. & K. T. Pickering 1988. Sinistral shear during Acadian deformation in north-central Newfoundland, based on transecting cleavage. *J. Structural Geology* **10**, 125–7.
Blow, W. H. & W. A. Berggren 1972. A Cenozoic time scale – some implications for regional geology and paleobiogeography. *Lethaia* **5**, 195–215.
Bluck, B. J. 1985. The Scottish paratectonic Caledonides. *Scottish J. Geology* **21**, 437–64.
Bluck, B. J. & B. E. Leake 1986. Late Ordovician to Early Silurian amalgamation of the Dalradian and adjacent Ordovician rocks in the British Isles. *Geology* **14**, 917–19.
Boillot, G., P. A. Dupeuble, I. Hennequin-Marchand, I. Lamboy, J. P. Lepretre & P. Musellec 1974. Le role des décrochements "tardi-hiercyniens" dans l'évolution structurale de la marge continentale et dans la localisation des grands canyons sous-marins a l'ouest et au nord de la Péninsule Iberique. *Rev. Geogr. Phys. Geol. Dyn.* **16**, 75–86.
Boillot, G. & Leg 103 Scientific Party 1987. Tectonic denudation of the upper mantle along passive margins: a model based on drilling results (ODP Leg 103, western Galicia margin, Spain). *Tectonophysics* **132**, 335–42.

Boltunov, V. A. 1970. Certain earmarks distinguishing glacial and moraine-like glaciomarine sediments, as in Spitsbergen. *Int. Geol Rev.* **12**, 204–11.

Boote, D. R. D. & S. H. Gustav 1987. Evolving depositional systems within an active rift, Witch Ground Graben, North Sea. In *Petroleum geology of north west Europe*, J. Brooks & K. W. Glennie (eds), 819–33. London: Graham & Trotman.

Booth, J. S., D. A. Sangrey & J. K. Fugate 1985. A nomogram for interpreting slope stability of fine-grained deposits in modern and ancient marine environments. *J. Sed. Petrol.* **55**, 29–36.

Bosellini, A. & D. Masetti 1972. Ambiente e dinamica depozionale del Calcare del Vajont (Giurassico medio-Prealpi Bullenesi e Friulane). *Ann. University of Ferrara* **5**, 87–100.

Bouma, A. H. 1962. *Sedimentology of some flysch deposits: a graphic approach to facies interpretation*. Amsterdam: Elsevier.

Bouma, A. H. 1964. Ancient and recent turbidites. *Geol. Mijnb.* **43**, 375–9.

Bouma, A. H. 1972. Recent and ancient turbidites and contourites. *Trans. Gulf Coast Ass. Geol Soc.* **22**, 205–21.

Bouma, A. H. 1973. Contourites in Niessenflysch, Switzerland. *Ecol. Geol. Helv.* **66/2**, 315–23.

Bouma, A. H. 1979. Continental slopes. In *Geology of continental slopes*, L. J. Doyle & O. H. Pilkey (eds), 1–15. Soc. Econ. Paleont. Mineral. Spec. Publn 27.

Bouma, A. H. 1983. Intraslope basins in northwest Gulf of Mexico: a key to ancient submarine canyons and fans. In *Studies in continental margin geology*, J. S. Watkins & C. L. Drake (eds), 567–81. Am. Assoc. Petrolm Geol. Mem. 34.

Bouma, A. H. & A. Brouwer (eds) 1964. *Turbidites. Developments in Sedimentology* **3**. Amsterdam: Elsevier.

Bouma, A. H. & C. D. Hollister 1973. Deep ocean basin sedimentation. In *Turbidites and deep water sedimentation*, G. V. Middleton & A. H. Bouma (eds), 79–118. Pacific Section, Soc. Econ. Paleont. Mineral. Short Course Notes, Anaheim.

Bouma, A. H. & T. H. Nilsen 1978. Turbidite facies and deep-sea fans–with examples from Kodiak Island, Alaska. *Proc. 10th Offshore Technology Conference*, Houston, 559–70.

Bouma, A. H., N. E. Barnes & W. R. Normark (eds) 1985a. *Submarine fans and related turbidite systems*. New York: Springer.

Bouma, A. H., J. M. Coleman & DSDP Leg 96 Shipboard Scientists 1985b. Mississippi Fan: Leg 96 program and principal results. In *Submarine fans and related turbidite systems*, A. H. Bouma, N. E. Barnes & W. R. Normark (eds), 247–52. New York: Springer.

Bouma, A. H., J. M. Coleman *et al.* 1986. *Initial Reports Deep Sea Drilling Project* 96. Washington, DC: US Government Printing Office.

Bouma, A. H., G. T. Moore & J. M. Coleman (eds) 1978. Framework, facies, and oil-trapping characteristics of the upper continental margin. *Am. Assoc. Petrolm Geol. Studies in Geology* **7**.

Bouma, A. H., C. E. Stelting & J. M. Coleman 1984. Mississippi Fan: internal structure and depositional processes. *Geo-Marine Letts* **3**, 147–54.

Bouma, A. H., C. E. Stelting & DSDP Leg 96 Shipboard Scientists 1983. Seismic stratigraphy and sedimentary processes in Orca and Pigmy basins. In *Initial Reports Deep Sea Drilling Project 96*, A. H. Bouma, J. M. Coleman, A. W. Meyer *et al.*, 563–76. Washington, DC: US Government Printing Office.

Bourcart, J. 1938. La marge continentale: essai sur les regressions et les transgressions marines. *Bull. Soc. Geol. France* **8**, 393–474.

Bourgois, J., G. Pautot, W. Bandy, T. Boinet, P. Chotin, P. Huchon, B. Mercier de Lepinay, F. Monge, J. Monlau, B. Pelletier, M. Sosson & R. von Huene 1988. Seabeam and seismic reflection imaging of the neotectonic regime of the Andean continental margin off Peru (4°S to 10°S). *Earth Planet. Sci. Letts* **87**, 111–26.

Bowen, A. J., W. R. Normark & D. J. W. Piper 1984. Modelling of turbidity currents on Navy Submarine Fan, California Continental Borderland. *Sedimentology* **31**, 169–85.

Bower, T. H. 1951. Mudflow occurrence in Trinidad, British West Indies. *Bull. Am. Assoc. Petrolm Geol.* **35**, 908–12.

Bowin, C., G. M. Purdy, C. Johnson, G. G. Shor, L. Lawver, H. M. S. Hartono & P. Jezek 1980. Arc continent collision in the Banda Sea. *Bull. Am. Assoc. Petrolm Geol.* **64**, 868–915.

Bowles, F. A., W. F. Ruddiman & W. H. Jahn 1978. Acoustic stratigraphy, structure, and depositional history of the Nicobar Fan, western Indian Ocean. *Marine Geology* **26**, 269–88.

Boyer, S. E. & D. Elliott 1982. Thrust systems. *Bull. Am. Assoc. Petrolm Geol.* **66**, 1196–1230.

Boynton, C. H., G. K. Westbrook, M. H. P. Bott & R. E. Long 1979. A seismic refraction investigation of crustal structure beneath the Lesser Antilles island arc. *Geophys. J. R. Astronomical Soc.* **58**, 371–93.

Braithwaite, C. J. R. 1973. Settling behaviour related to sieve analysis of skeletal sands. *Sedimentology* **20**, 251–62.

Brass, G. W., E. Saltzman, J. L. Sloan II., J. R. Southam, W. W. Hay, W. T. Holser & W. H. Peterson 1982. Ocean circulation, plate tectonics and climate. In *Climate in Earth history, studies in geophysics, Panel on pre-Pleistocene climates*, 83–9. Washington, DC: National Academy Press.

Breen, N. A., E. A. Silver & D. M. Hussong 1986. Structural styles of an accretionary wedge south of the island of Sumba, Indonesia, revealed by SeaMARC II side scanner. *Bull. Geol Soc. Am.* **97**, 1250–61.

Brice, S. E., M. D. Cochran, G. Pardo & A. D. Edwards 1983. Tectonics and sedimentation of the South Atlantic rift sequence: Cabinda, Angola. In *Studies in continental margin geology*, J. S. Watkins & C. L. Drake (eds), 5–18. Am. Assoc. Petrolm Geol. Mem. 34.

Broecker, W. S. 1974. *Chemical oceanography*. New York: Harcourt Brace Javanovich.

Brooks, J. & K. W. Glennie (eds) 1987. *Petroleum geology of north west Europe*. London: Graham & Trotman.

Broster, B. E. & S. R. Hicock 1985. Multiple flow and support mechanisms and the development of inverse grading in a subaquatic glacigenic debris flow. *Sedimentology* **32**, 645–57.

Brown, K. M. & G. K. Westbrook 1987. The tectonic fabric of the Barbados Ridge accretionary complex. *Marine Petrolm Geology* **4**, 71–81.

Brown, K. M. & G. K. Westbrook 1988. Mud diapirism and subcretion in the Barbados Ridge accretionary prism: the role of fluids in accretionary processes. *Tectonics* **7**, 613–40.

Brown, L. F. & W. L. Fisher 1977. Seismic-stratigraphic interpretation of depositional systems: examples from Brazilian rift and pull-apart basins. In *Seismic stratigraphy–applications to hydrocarbon exploration*, C. E. Payton (ed.), 213–48. Am. Assoc. Petrolm Geol. Mem. 26.

Brown, P. A. & S. P. Colman-Sadd 1976. Hermitage flexure: figment or fact? *Geology* **4**, 561–4.

Bryant, W. R., J. Antoine, M. Ewing & B. Jones 1968. Structure of Mexican continental shelf and slope, Gulf of Mexico. *Bull. Am. Assoc. Petrolm Geol.* **52**, 1204–28.

Bucher, W. H. 1940. Submarine valleys and related geologic

problems of the North Atlantic. *Bull. Geol Soc. Am.* **51**, 489–512.
Buffler, R. T., F. J. Shaub, R. Huerta, A. B. K. Ibrahim & J. S. Watkins 1981. A model for the early evolution of the Gulf of Mexico basin. *Oceanologica Acta* **4**, 129–36.
Buffler, R. T., F. J. Shaub, J. S. Watkins & J. L. Worzel 1979. Anatomy of the Mexican Ridge, southwestern Gulf of Mexico. In *Geological and geophysical investigations of continental margins*, J. S. Watkins, L. Montadert & P. W. Dickerson (eds), 319–27. Am. Assoc. Petrolm Geol. Mem. 29.
Bulfinch, D. L. & M. T. Ledbetter 1984. Deep Western Boundary Undercurrent delineated by sediment texture at base of North American continental rise. *Geo-Marine Letts* **3**, 31–6.
Bunn, A. R. & B. A. McGregor 1980. Morphology of the North Carolina continental slope, Western North Atlantic shaped by deltaic sedimentation and slumping. *Marine Geology* **37**, 253–66.
Burk, C. A. & C. L. Drake (eds) 1974. *The geology of continental margins*. New York: Springer.
Burk, C. N. *et al.* 1969. Deep-sea drilling into the Challenger Knoll, central Gulf of Mexico. *Bull. Am. Assoc. Petrolm Geol.* **53**, 1338–47.
Burke, K. 1967. The Yallahs Basin: a sedimentary basin southeast of Kingston, Jamaica. *Marine Geology* **5**, 45–60.
Burke, K. C., T. F. J. Dessauvagie & A. J. Whiteman 1972. Geological history of the Benue valley and adjacent areas. In *African geology*, T. F. J. Dessauvagie & A. J. Whiteman (eds), 187–205. Ibadan: Ibadan University Press.
Burkry, D. 1975. Coccolith and siliciflagellate stratigraphy, northwestern Pacific Ocean. In *Initial Reports Deep Sea Drilling Project 32*, R. L. Larson, R. Moberly *et al.*, 677–701. Washington, DC: US Government Printing Office.
Burne, R. V. 1970. The origin and significance of sand volcanoes in the Bude Formation (Cornwall). *Sedimentology* **15**, 211–28.
Butt, A. 1982. Micropalaeontological bathymetry of the Cretaceous pelagic sedimentation of Western Morocco. *Palaeogeog. Palaeoclimatol. Palaeoecol.* **37**, 235–75.
Byers, C. W. 1977. Biofacies patterns in euxinic basins: a general model. In *Deep-water carbonate environments*. H. E. Cook & P. Enos (eds), 5–17. Soc. Econ. Paleont. Mineral. Spec. Publn 25.

Cacchione, D. A. & J. B. Southard 1974. Incipient sediment movement by shoaling internal gravity waves. *J. Geophys. Res.* **70**, 2237–42.
Cacchione, D. A., G. T. Rowe & A. Malahoff 1978. Submersible investigation of outer Hudson submarine canyon. In *Sedimentation in submarine canyons, fans, and trenches*, D. J. Stanley & G. Kelling (eds), 42–50. Stroudsburg, PA: Dowden, Hutchinson & Ross.
Cadet, J. P., K. Kobayashi, L. Jolivet, J. Aubouin, J. Boulegue, J. Dubois, H. Hotta, T. Ishii, K. Konishi, N. Nitsuma & H. Shimamura 1987. Deep scientific dives in the Japan and Kuril Trenches. *Earth Planet. Sci. Letts* **83**, 313–28.
Calvert, S. E. 1966. Accumulation of diatomaceous silica in the sediments of the Gulf of California. *Bull. Geol Soc. Am.* **77**, 569–96.
Campbell, J. S. & D. L. Clark 1977. Pleistocene turbidites of the Canada Abyssal Plain of the Arctic Ocean. *J. Sed. Petrol.* **47**, 657–70.
Cant, D. J. & F. J. Hein 1986. Depositional sequences in ancient shelf sediments: some contrasts in style. In *Shelf sands and sandstones*, R. J. Knight & J. R. McLean (eds), 303–12. Can. Soc. Petrolm Geol. Mem. 11.
Carey, D. L. & D. C. Roy 1985. Deposition of laminated shale: a field and experimental study. *Geo-Marine Letts* **5**, 3–9.
Carey, S. & H. Sigurdsson 1978. Deep-sea evidence for distribution of tephra from the mixed magma eruption of the Soufriere of St Vincent, 1902: ash turbidites and air fall. *Geology* **6**, 271–4.
Carey, S. & H. Sigurdsson 1984. A model of volcanogenic sedimentation in marginal basins. In *Marginal basin geology*, B. P. Kokelaar & M. F. Howells (eds), 37–58. Geol Soc. (London) Spec. Publn 16. Oxford: Blackwell Scientific.
Carpenter, G. 1981. Coincident sediment slump/clathrate complexes on the US Atlantic continental slope. *Geo-Marine Letts* **1**, 29–32.
Carpenter, R., M. L. Peterson & J. T. Bennett 1982. Pb-210 derived sediment accumulation and mixing rates for the Washington continental slope. *Marine Geology* **48**, 135–64.
Carson, B. & N. P. Arcaro 1983. Control of clay-mineral stratigraphy by selective transport in Late Pleistocene-Holocene sediments of Northern Cascadia Basin–Juan de Fuca Abyssal Plain: implications for studies of clay-mineral provenance. *J. Sed. Petrol.* **53**, 395–406.
Carson, B., E. T. Baker, B. M. Hickey, C. A. Nittrouer, D. J. DeMaster, K. W. Thorbjarnarson & G. W. Snyder 1986. Modern sediment dispersal and accumulation in Quinault submarine canyon–a summary. *Marine Geology* **71**, 1–13.
Carter, L. & C. T. Schafer 1983. Interaction of the Western Boundary Undercurrent with the continental margin off Newfoundland. *Sedimentology* **30**, 751–68.
Carter, L., C. T. Schafer & M. A. Rashid 1979. Observations on depositional environments and benthos of the continental slope and rise, east of Newfoundland. *Can. J. Earth Sci.* **16**, 831–46.
Carter, R. M. 1975. A discussion and classification of subaqueous mass-transport with particular application to grain-flow, slurry-flow and fluxoturbidites. *Earth Sci. Rev.* **11**, 145–77.
Carter, R. M. 1988. The nature and evolution of deep-sea channel systems. *Basin Res.* **1**, 41–54.
Carter, R. M. & J. K. Lindqvist 1975. Sealers Bay submarine fan complex, Oligocene, southern New Zealand. *Sedimentology* **22**, 465–83.
Carter, R. M. & J. K. Lindqvist 1977. Balleny Group, Chalky Island, southern New Zealand; an inferred Oligocene submarine canyon and fan complex. *Pacific Geology* **12**, 1–46.
Carter, R. M. & R. J. Norris 1977. Redeposited conglomerates in Miocene flysch sequence at Blackmount, western Southland, New Zealand. *Sed. Geology* **18**, 289–319.
Cartwright, J. 1987. Transverse structural zones in continental rifts–an example from the Danish sector of the North Sea. In *Petroleum geology of north west Europe*, J. Brooks & K. W. Glennie (eds), 441–52. London: Graham & Trotman.
Cas, R. 1979. Mass-flow arenites from a Paleozoic interarc basin, New South Wales, Australia: mode and environment of emplacement. *J. Sed. Petrol.* **49**, 29–44.
Cas, R. A. F. & J. V. Wright 1987. *Volcanic successions: modern and ancient: a geological approach to processes, products and successions*. London: Unwin-Hyman.
Cashman, K. V. & P. Popenoe 1985. Slumping and shallow faulting related to the presence of salt on the continental slope and rise off North Carolina. *Marine Petrolm Geology* **2**, 260–71.
Chan, M. A. & R. H. Dott 1983. Shelf and deep-sea sedimentation in Eocene forearc basin, western Oregon – fan or non-fan? *Bull. Am. Assoc. Petrolm Geol.* **67**, 2100–16.

Chipping, D. H. 1972. Sedimentary structures and environment of some thick sandstone beds of turbidite type. *J. Sed. Petrol.* **42**, 587–95.

Chough, S. & R. Hesse 1976. Submarine meandering thalweg and turbidity currents flowing for 4000 km in the Northwest Atlantic Mid-Ocean Channel, Labrador Sea. *Geology* **4**, 529–34.

Chough, S. & R. Hesse 1980. Northwest Atlantic Mid-Ocean Channel of the Labrador Sea: III. Head spill vs body spill deposits from turbidity currents on natural levees. *J. Sed. Petrol.* **50**, 227–34.

Cita, M. B., C. Beghi, A. Camerlenghi, K. A. Karstens, F. W. McCoy, A. Nosetto, E. Parisi, F. Scolari & L. Tomadin 1984. Turbidites and megaturbidites from the Herodotus Abyssal Plain (eastern Mediterranean) unrelated to seismic events. *Marine Geology* **55**, 79–101.

Cita, M. B., F. Benelli, B. Bigioggero, A. Bossio, C. Broglia, H. Chezar, G. Clauzon, A. Colombo, M. Giambastiani, A. Malinverno, E. L. Miller, E. Parisi, G. Salvatorino & P. Vercesi 1982. Unusual debris flow deposits from the base of the Malta Escarpment (Eastern Mediterranean). In *Marine slides and other mass movements*, S. Saxov & J. Nieuwenhuis (eds), 305–22. New York: Plenum Press.

Clari, P. & G. Ghibaudo 1979. Multiple slump scars in the Tortonian type area (Piedmont Basin, northwestern Italy). *Sedimentology* **26**, 719–30.

Cleary, W. J., O. H. Pilkey & M. W. Ayers 1977. Morphology and sediments of three ocean basin entry points, Hatteras Abyssal Plain. *J. Sed. Petrol.* **47**, 1157–70.

Cloetingh, S. 1986. Tectonics of passive margins: implications for the stratigraphic record. *Geol. Mijnb.* **65**, 103–17.

Cochran, J. R. 1983. Effects of finite rifting times on the development of sedimentary basins. *Earth Planet. Sci. Letts* **66**, 289–302.

Colacicchi, R. & A. Baldanza 1986. Carbonate turbidites in a Mesozoic pelagic basin: Scaglia Formation, Apennines – comparison with siliciclastic depositional models. *Sed. Geology* **48**, 81–105.

Collette, B. J., J. J. Ewing, R. A. Lagaay & M. Truchan 1969. Sediment distribution in the oceans: the Atlantic between 10° and 19°N. *Marine Geology* **7**, 279–345.

Collinson, J. D. 1969. The sedimentology of the Grindslow Shales and the Kinderscout Grit: a deltaic complex in the Namurian of northern England. *J. Sed. Petrol.* **39**, 194–221.

Collinson, J. D. 1970. Deep channels, massive beds and turbidity current genesis in the Central Pennine Basin. *Proc. Yorkshire Geol Soc.* **37**, 495–520.

Collot, J. Y., J. Daniel & R. V. Burne 1985. Recent tectonics associated with the subduction/collision of the d'Entrecasteaux Zone in the central New Hebrides. *Tectonophysics* **112**, 325–56.

Coniglio, M. 1986. Synsedimentary submarine slope failure and tectonic deformation in deep-water carbonates. Cow Head Group, western Newfoundland. *Can. J. Earth Sci.* **23**, 476–90.

Conolly, J. R. & M. Ewing 1967. Sedimentation in the Puerto Rico Trench. *J. Sed. Petrol.* **37**, 44–59.

Cook, H. E. 1979. Ancient continental slope sequences and their value in understanding modern slope development. In *Geology of continental slopes*, L. J. Doyle & O. H. Pilkey (eds), 287–305. Soc. Econ. Paleont. Mineral. Spec. Publn 27.

Cook, H. E. & P. Enos (eds) 1977. *Deep-water carbonate environments*. Soc. Econ. Paleont. Mineral. Spec. Publn 25.

Cook, H. E. & H. T. Mullins 1983. Basin margin environment. In *Carbonate depositional environments*, P. A. Scholle, D. G. Bebout & C. H. Moore (eds), 540–617. Am. Assoc. Petrolm Geol. Mem. 33.

Cook, H. E., M. E. Field & J. V. Gardner 1982. Characteristics of sediments on modern and ancient continental slopes. In *Sandstone depositional environments*, P. A. Scholle & D. Spearing (eds), 329–64. Tulsa: Am. Assoc. Petrolm Geol. Mem. 31.

Covey, M. 1986. The evolution of foreland basins to steady state: evidence from the western Taiwan foreland basin. In *Foreland basins*, P. A. Allen & P. Homewood (eds), 77–90. Int. Assoc. Sedimentologists Spec. Publn 8. Oxford: Blackwell Scientific.

Cowan, D. S. 1985. The origin of some common types of melange in the Western Cordillera of North America. In *Formation of active ocean margins*, N. Nasu *et al.* (eds), 257–72. Tokyo: Terrapub.

Cowan, D. S. & B.M. Page 1975. Recycled Franciscan material in Franciscan melange west of Pass Robles, California. *Bull. Geol. Soc. Am.* **86**, 1089–95.

Coward, M. P. & A. C. Ries (eds) 1986. *Collision tectonics*. Geol Soc. (London) Spec. Publn 19. Oxford: Blackwell Scientific.

Coward, M. P., J. F. Dewey & P. L. Hancock (eds) 1987. *Continental extensional tectonics*. Geol Soc. (London) Spec. Publn 28. Oxford: Blackwell Scientific.

Crouch, J. K. 1979. Neogene tectonic evolution of the California Continental Borderland and western Transverse Ranges. *Bull. Geol Soc. Am.* **90**, 338–45.

Crouch, J. K. 1981. Northwest margin of California Continental Borderland: marine geology and tectonic evolution. *Bull. Am. Assoc. Petrolm Geol.* **65**, 191–218.

Crowell, J. C. 1957. Origin of pebbly mudstones. *Bull. Geol Soc. Am.* **68**, 993–1010.

Curray, J. R. & D. G. Moore 1974. Sedimentary and tectonic processes in the Bengal deep-sea fan and geosyncline. In *The Geology of continental margins*, C. A. Burk & C. L. Drake (eds), 617–27. New York: Springer.

Curray, J. R., D. G. Moore, L. A. Lawver, F. J. Emmel, R. W. Raitt, M. Henry & R. Kieckhefer 1979. Tectonics of the Andaman Sea and Burma. In *Geological and geophysical investigations of continental margins*, J. S. Watkins, L. Montadert & P. W. Dickerson (eds), 189–98. Am. Assoc. Petrolm Geol. Mem. 29.

Curray, J. R., D. G. Moore *et al.* 1982. *Initial Reports Deep Sea Drilling Project 64*. Washington, DC: US Government Printing Office.

Curray, J. R., D. G. Moore, K. Kelts & G. Einsele 1982. Tectonics and geological history of the passive continental margin at the tip of Baja California. In *Initial Reports Deep Sea Drilling Project 64*, J. R. Curray, D. G. Moore *et al.*, 1089–116. Washington, DC: US Government Printing Office.

Cutshall, N. H., I. L. Larsen, C. R. Olsen, C. A. Nittrouer & D. J. DeMaster 1986. Columbia River sediment in Quinault Canyon, Washington–evidence from artificial radionuclides. *Marine Geology* **71**, 125–36.

Daly, R. A. 1936. Origin of submarine canyons. *Am. J. Sci.* **31**, 410–20.

Damuth, J. E. 1975. Echo-character of the Western Equatorial Atlantic floor and its relationship to the dispersal and distribution of terrigenous sediments. *Marine Geology* **18**, 17–45.

Damuth, J. E. 1977. Late Quaternary sedimentation in the western equatorial Atlantic. *Bull. Geol Soc. Am.* **88**, 695–710.

Damuth, J. E. & R. W. Embley 1981. Mass-transport processes on Amazon Cone: western equatorial Atlantic. *Bull. Am. Assoc. Petrolm Geol.* **65**, 629–43.

Damuth, J. E. & R. D. Flood 1984. Morphology, sedimentation processes and growth pattern of the Amazon deep-sea fan. *Geo-Marine Letts* **3**, 109–17.

Damuth, J. E. & R. D. Flood 1985. Amazon Fan, Atlantic Ocean In *Submarine fans and related turbidite systems*, A. H. Bouma, W. R. Normark & N. E. Barnes (eds), 97–106. New York: Springer.

Damuth, J. E. & M. A. Gorini 1976. The equatorial Mid-Atlantic Canyon: a relict deep-sea channel on the Brazilian continental margin. *Bull. Geol. Soc. Am.* **87**, 340–6.

Damuth, J. E. & N. Kumar 1975a. Amazon Cone: morphology, sediments, age, and growth pattern. *Bull. Geol. Soc. Am.* **86**, 863–78.

Damuth, J. E. & N. Kumar 1975b. Late Quaternary depositional processes on the continental rise of the western equatorial Atlantic: comparison with the western North Atlantic and implications for reservoir rock distribution. *Bull. Am. Assoc. Petrolm Geol.* **59**, 2171–81.

Damuth, J. E., V. Kolla, R. D. Flood, R. O. Kowsmann, M. C. Monteiro, M. A. Gorino, J. J. C. Palma & R. H. Belderson 1983a. Distributary channel meandering and bifurcation patterns on the Amazon deep-sea fan as revealed by long-range side-scan sonar (GLORIA). *Geology* **11**, 94–8.

Damuth, J. E., R. O. Kowsmann, R. D. Flood, R. H. Belderson & M. A. Gorino 1983b. Age relationships of distributary channels on Amazon Deep-Sea Fan: implications for fan growth pattern. *Geology* **11**, 470–3.

Davey, F. J., M. Hampton, J. Childs, M. A. Fisher, K. Lewis & J. R. Pettinga 1986. Structure of a growing accretionary prism, Hikurangi margin, New Zealand. *Geology* **14**, 663–6.

Davies, D. K. 1972. Deep sea sediments and their sedimentation, Gulf of Mexico. *Bull. Am. Assoc. Petrolm Geol.* **56**, 2212–39.

Davies, G. R. 1977. Turbidites, debris sheets and truncation structures in upper Paleozoic deep-water carbonates of the Svedrup Basin, Arctic Archipelago. In *Deep-water carbonate environments*, H. E. Cook & P. Enos (eds), 221–49. Soc. Econ. Paleont. Mineral. Spec. Publn 25.

Davies, I. C. & R. G. Walker 1974. Transport and deposition of resedimented conglomerates: the Cap Enrage Formation, Cambro-Ordovician, Gaspe, Quebec. *J. Sed. Petrol.* **44**, 1200–16.

Degens, E. T. & D. A. Ross (eds) 1974. *The Black Sea–geology, chemistry and biology*. Am. Assoc. Petrolm Geol. Mem. 20.

Deichmann, N., J. Ansorge & St Mueller 1986. Crustal structure of the Southern Alps beneath the intersection with the European Geotraverse. *Tectonophysics* **126**, 57–83.

Dercourt, J. *et al.* 1986. Geological evolution of the Tethys Belt from the Atlantic to the Pamirs since the Lias. *Tectonophysics* **123**, 241–315.

Deroo, G., J. P. Herbin & J. Roucache 1984. Organic geochemistry of Cenozoic and Mesozoic sediments from deep sea drilling Sites 544 to 547, Leg 79, eastern North Atlantic. In *Initial Reports Deep Sea Drilling Project* 79, 721–42. Washington, DC: US Government Printing Office.

Dickinson, W. R. & D. R. Seely 1979. Structure and stratigraphy of forearc regions. *Am. Assoc. Petrolm Geol.* **63**, 2–31.

Dickinson, W. R. & C. A. Suczek 1979. Plate tectonics and sandstone composition. *Bull. Am. Assoc. Petrolm Geol.* **63**, 2164–82.

Dickinson, W. R. & R. Valloni 1980. Plate settings and provenance of sands in modern ocean basins. *Geology* **8**, 82–6.

Dill, R. F. 1964. Sedimentation and erosion in Scripps submarine canyon head. In *Papers in marine geology*, R. L. Miller (ed.), 23–41. New York: McMillan.

Dimberline, A. J. & N. H. Woodcock 1987. The southeast margin of the Wenlock turbidite system, Mid-Wales. *Geol J.* **22**, 61–71.

Dingwall, R. G. 1987. Ocean-floor spreading and the Mesozoic deltaic systems of the North Atlantic rim. In *Petroleum geology of north west Europe*, J. Brooks & K. W. Glennie (eds), 1177–88. London: Graham & Trotman.

Donovan, D. T. & E. J. W. Jones 1979. Causes of world-wide changes in sea level. *J. Geol Soc. (London)* **136**, 187–92.

Dore, A. G. & M. S. Gage 1987. Crustal alignments and sedimentary domains in the evolution of the North Sea, north-east Atlantic margin and the Barents Shelf. In *Petroleum geology of north west Europe*, J. Brooks & K. W. Glennie (eds), 1131–48. London: Graham & Trotman.

Doreen, J. M. Jr 1951. Rubble bedding and graded bedding in Talara Formation of northwestern Peru. *Bull. Am. Assoc. Petrolm Geol.* **35**, 1829–49.

Dott, R. H. Jr 1961. Squantum "Tillite", Massachusetts – evidence of glaciation or subaqueous mass movements? *Bull. Geol Soc. Am.* **72**, 1289–306.

Dott, R. H. Jr 1963. Dynamics of subaqueous gravity depositional processes. *Bull. Am. Assoc. Petrolm Geol.* **47**, 104–29.

Dott, R. H. Jr 1978. Tectonics and sedimentation a century later. *Earth Sci. Rev.* **14**, 1–34.

Dott, R. H. Jr 1982. S.E.P.M. Presidential Address: Episodic sedimentation–How normal is average? How rare is rare? Does it matter? *J. Sed. Petrol.* **53**, 5–23.

Douglas, R. G. & H. L. Heitman 1979. Slope and basin benthic foraminifera of the California Borderland. In *Geology of continental slopes*, L. J. Doyle & O. H. Pilkey (eds), 231–46. Soc. Econ. Paleont. Mineral. Spec. Publn 27.

Doyle, L. J. & O. H. Pilkey (eds) 1979. *Geology of continental slopes*. Soc. Econ. Paleont. Mineral. Spec. Publn 27.

Drake, C. L., M. Ewing & G. L. Sutton 1959. Continental margins and geosynclines: the East Coast of North America north of Cape Hatteras. *Phys. Chem. Earth* **3**, 110–98.

Drake, D. E., P. G. Hatcher & G. H. Keller 1978. Suspended particulate matter and mud deposition in Upper Hudson submarine canyon. In *Sedimentation in submarine canyons, fans, and trenches*, D. J. Stanley & G. Kelling (eds), 33–41. Stroudsburg, PA: Dowden, Hutchinson & Ross.

Driscoll, M. L., B. E. Tucholke & I. N. McCave 1985. Seafloor zonation in sediment texture on the Nova Scotian lower continental rise. *Marine Geology* **66**, 25–41.

Droz, L. & G. Bellaiche 1985. Rhône deep-sea fan: morphostructure and growth pattern. *Bull. Am. Assoc. Petrolm Geol.* **69**, 460–79.

Droz, L. & D. Mougenot 1987. Mozambique Upper Fan: origin of depositional units. *Bull. Am. Assoc. Petrolm Geol.* **71**, 1355–65.

Dunbar, R. B. & W. H. Berger 1981. Fecal pellet flux to modern bottom sediment of Santa Barbara Basin (California) based on sediment trapping. *Bull. Geol Soc. Am.* **92**, 212–18.

Dzułynski, S. & J. E. Sanders 1962. Current marks on firm mud bottoms. *Trans. Connecticut Academy Sci.* **42**, 57–96.

Dzułynski, S. & E. K. Walton 1965. Sedimentary features of flysch and greywackes. *Developments in sedimentology* **7**. Amsterdam: Elsevier.

Dzułynski, S., M. Ksiaskiewicz & Ph. H. Kuenen 1959. Turbidites in flysch of the Polish Carpathian Mountains. *Bull. Geol Soc. Am.* **70**, 1089–118.

Edwards, M. B. 1986. Glacial environments. In *Sedimentary environments and facies*, 2nd edn, H. G. Reading (ed.), 445–70. Oxford: Blackwell Scientific.

Egan, J. A. & D. A. Sangrey 1978. Critical state model of cyclic load pore pressures. *Am. Soc. Civil Engrs Spec. Conference Earthquake Engineering Soil Dynamics* **1**, 410–24.

Egloff, J. & G. L. Johnson 1975. Morphology and structure of the southern Labrador Sea. *Can. J. Earth Sci.* **12**, 2111–33.

Ehlig, P. L. 1981. Origin and tectonic history of the basement terrane of the San Gabriel Mountains, central Transverse Region. In *The geotectonic development of California*, W. G. Ernst (ed.), 254–83. New Jersey: Prentice-Hall.

Einsele, G. & J. Wiedmann 1982. Turonian black shales in the Moroccan coastal basins: first upwelling in the Atlantic Ocean? In *Geology of the Northern African continental margin*, U. von Rad *et al.* (eds), 396–414. Berlin: Springer.

Einsele, G. & J. Wiedmann 1983. Cretaceous upwelling off Northwest Africa: a summary. In *Coastal upwelling*, J. Thiede & W. Suess (eds), 485–99. New York: Plenum Press.

Einsele, G., J. M. Gieskes, J. M. Curray, D. Moore, E. Aguayo, M.-P. Aubry, D. Fornari, J. Guerrero, M. Kastner, K. Kelts, M. Lyle, M. Matola, A. Molina-Cruz, J. Niemitz, J. Rueda, A. Saunders, H. Schrader, B. Simoniet & V. Vacquier 1980. Intrusion of basaltic sills into highly porous sediments, and resulting hydrothermal activity. *Nature* **283**, 441–5.

Eittreim, S., A. Grantz & J. Greenberg 1982. Active geologic processes in Barrow Canyon, northeast Chukchi Sea. *Marine Geology* **50**, 61–76.

Elders, C. 1987. The provenance of granite boulders in conglomerates of the Northern and Central Belts of the Southern Uplands of Scotland. *J. Geol Soc. (London)* **144**, 853–63.

Ellis, M. 1988. Lithospheric strength in compression: initiation of subduction, flake tectonics, foreland migration of thrusting, and an origin of displaced terranes. *J. Geology* **96**, 91–100.

Elmore, R. D., O. H. Pilkey, W. J. Cleary & H. A. Curran 1979. Black Shell turbidite, Hatteras Abyssal Plain, western Atlantic Ocean. *Bull. Geol Soc. Am.* **90**, 1165–76.

Embley, R. W. 1976. New evidence for occurrence of debris flow deposits in the deep sea. *Geology* **4**, 371–4.

Embley, R. W. 1980. The role of mass transport in the distribution and character of deep-ocean sediments with special reference to the North Atlantic. *Marine Geology* **38**, 23–50.

Embley, R. W. & R. Jacobi 1986. Mass wasting in the western North Atlantic. In *The geology of North America Volume M, The western North Atlantic region*, Ch. 29, 479–90. Geol Soc. Am.

Embley, R. W., J. I. Ewing & M. Ewing 1970. The Vidal deep-sea channel and its relationship to the Demerara and Barracuda abyssal plain. *Deep-Sea Res.* **17**, 539–52.

Emery, K. O. 1960a. *The sea off southern California.* New York: Wiley.

Emery, K. O. 1960b. Basin plains and aprons off southern California. *J. Geology* **68**, 464–79.

Emery, K. O. 1981. Geological limits of the 'continental shelf'. *Ocean Dev. Int. Law J.* **10**, 1–11.

Emery, K. O. & J. D. Milliman 1978. Suspended matter in surface waters: influence of river discharge and of upwelling. *Sedimentology* **25**, 125–40.

Emery, K. O. & E. Uchupi 1972. Western North Atlantic Ocean: topography, rocks, structure, water, life and sediments. Am. Assoc. Petrolm Geol. Mem. 17.

Enos, P. 1969. Anatomy of a flysch. *J. Sed. Petrol.* **39**, 680–723.

Enos, P. 1977. Flow regimes in debris flow. *Sedimentology* **24**, 133–42.

Enos, P. & C. H. Moore 1983. Fore-reef slope environment. In *Carbonate depositional environments*, P. A. Scholle, D. G. Bebout & C. H. Moore (eds), 508–37. Am. Assoc. Petrolm Geol. Mem. 33.

Ericson, D. B., M. Ewing & B. C. Heezen 1951. Deep-sea sands and submarine canyons. *Bull. Geol Soc. Am.* **62**, 961–5.

Ericson, D. B., M. Ewing & B. C. Heezen 1952. Turbidity currents and sediments in North Atlantic. *Bull. Am. Assoc. Petrolm Geol.* **36**, 489–511.

Ericson, D. B., M. Ewing, G. Wollin & B. C. Heezen 1961. Atlantic deep-sea sediment cores. *Bull. Geol Soc. Am.* **72**, 193–286.

Ewing, J. & C. H. Hollister 1972. Regional aspects of deep-sea drilling in the western North Atlantic. in *Initial Reports Deep Sea Drilling Project 11*, 951–73. Washington, DC: US Government Printing Office.

Ewing, M. & E. M. Thorndike 1965. Suspended matter in deep ocean water. *Science* **147**, 1291–4.

Ewing, M. *et al.* 1969. *Initial Reports Deep Sea Drilling Project 1.* Washington, DC: US Government Printing Office.

Ewing, M., B. C. Heezen, D. B. Ericson, J. Northrop & J. Dorman 1953. Exploration of the northwest Atlantic Mid-Ocean Canyon. *Bull. Geol Soc. Am.* **64**, 865–8.

Ewing, M., W. J. Ludwig & J. I. Ewing 1965. Oceanic structural history of the Bering Sea. *J. Geophys. Res.* **70**, 4593–600.

Eyles, C. H. 1987. Glacially influenced submarine-channel sedimentation in the Yakataga Formation, Middleton Island. *J. Sed. Petrol.* **57**, 1004–17.

Farre, J. A. & W. B. F. Ryan 1987. Surficial geology of the continental margin offshore New Jersey in the vicinity of Deep Sea Drilling Project Sites 612 and 613. In *Initial Reports DSDP 95*, C. W. Poag, A. B. Watts *et al.*, 725–58. Washington DC: US Government Printing Office.

Farre, J. A., B. A. McGregor, W. B. F. Ryan & J. M. Robb 1983. Breaching the shelfbreak: passage from youthful to mature phase in submarine canyon evolution. In *The shelfbreak: critical interface on continental margin*, D. J. Stanley & G. T. Moore (eds), 25–39. Soc. Econ. Paleont. Mineral. Spec. Publn 33.

Farrell, S. G. 1984. A dislocation model applied to slump structures, Ainsa Basin, south central Pyrenees. *J. Structural Geology* **6**, 727–36.

Faugeres, J.-C., D. A. V. Stow & E. Gonthier 1984. Contourite drift moulded by deep Mediterranean outflow. *Geology* **12**, 296–300.

Feeley, M. H. 1982. Structural and depositional relationships of intraslope basins, northern Gulf of Mexico. Unpubl. M.Sc. Thesis, Texas A&M University, College Station.

Felix, D. W. & D. S. Gorsline 1971. Newport submarine canyon, California; an example of the effects of shifting loci of sand supply upon canyon position. *Marine Geology* **10**, 177–98.

Fisher, M. A., J.-Y. Collot & G. L. Smith 1986. Possible causes for structural variation where the New Hebrides island arc and the d'Entrecasteaux zone collide. *Geology* **14**, 951–4.

Fisher, R. V. & H.-U. Schmincke 1985. *Pyroclastic rocks.* Berlin: Springer.

Flood, R. D. 1978. *Studies of deep sea sedimentary microtopography in the North Atlantic Ocean.* Ph.D. Thesis. Massachusetts Institute of Technology Woods Hole Oceanographic Institution, Woods Hole Oceanographic Institution Report WHO1–78–64.

Flores, G. 1955. Discussion of 'Les resultats des études pour la recherche petrolifère en Sicilie (Italie)' by E. Beneo. *4th World*

Petrolm Congress Rome, *Proc. Sect.* **1**, 121–2.

Fralick, P. W. & A. D. Miall 1981. Grant 84: sedimentology of the Matinenda Formation. In *Geoscience Research Grant Program, Summary of Research 1980–1981*, E. G. Pye (ed.), 80–9. Misc. Papers Ontario Geol Surv. 98.

Fruth, L. S. Jr 1965. The 1929 Grand Banks turbidite and the sediments of the Sohm Abyssal Plain. Columbia University, New York, unpubl. Ph.D. Thesis.

Fukushima, Y., G. Parker & H. M. Pantin 1985. Prediction of ignitive turbidity currents in Scripps Submarine Canyon. *Marine Geology* **67**, 55–81.

Gaetani, M. & G. Poliani 1978. Il Toarciano e il Giurassico medio in Albenza (Bergamo). *Riv. Italiana Paleontologia e Stratigrafia* **84**, 349–82.

Galloway, W. E. & D. K. Hobday 1983. *Terrigenous clastic depositional systems: applications to petroleum, coal and uranium exploration*. New York: Springer.

Garcia-Mondejar, J., F. M. Hines, V. Pujalte & H. G. Reading 1985. Sedimentation and tectonics in the western Basque-Cantabrian area (northern Spain) during Cretaceous and Tertiary times. In *Excursion guidebook*, M. D. Mila & J. Rosell (eds), 6th European Regional Meeting, Lleida, Spain. Int. Assoc. Sedimentologists.

Gardiner, S. & R. N. Hiscott 1988. Deep-water facies and depositional setting of the lower Conception Group (Hadrynian), southern Avalon Peninsula, Newfoundland. *Can. J. Earth Sci.* **25**, 1579–94.

Gardner, W. D., P. E. Biscaye, J. R. V. Zaneveld & M. J. Richardson 1985. Calibration and comparison of the LDGO nephelometer and the OSU transmissometer on the Nova Scotian rise. *Marine Geology* **66**, 323–44.

Garfunkel, Z. 1984. Large-scale submarine rotational slumps and growth faults in the Eastern Mediterranean. *Marine Geology* **55**, 305–24.

Garfunkel, Z. & G. Almagor 1985. Geology and structure of the continental margin off northern Israel and the adjacent part of the Levantine Basin. *Marine Geology* **62**, 105–31.

Garrison, L. E., N. H. Kenyon & A. H. Bouma 1982. Channel systems and lobe construction in the Mississippi Fan. *Geo-Marine Letts* **2**, 31–9.

Gates, O. & W. Gibson 1956. Interpretation of the configuration of the Aleutian Ridge. *Bull. Geol Soc. Am.* **67**, 127–46.

Gawthorpe, R. L. & H. Clemmey 1985. Geometry of submarine slides in the Bowland Basin (Dinantian) and their relation to debris flows. *J. Geol Soc. (London)* **142**, 555–65.

Gennesseaux, M., P. Guibout & H. Lacombe 1971. Enregistrement de courants de turbidite dans la vallée sous-marine du Var (Alpes-Maritimes). *C.R. Acad. Sci. Paris* **273**, 2456–9.

Gennesseaux, M., A. Mauffret & G. Pautot 1980. Les glissements sous-marins de la pente continentale niçoise et la rupture de cables en mer Ligure (Mediterranée occidentale). *Comptes Rendus Academy Sci., Paris (D)* **290**, 959–62.

Ghibaudo, G. 1980. Deep-sea fan deposits in the Macigno Formation (Middle-Upper Oligocene) of the Gordana Valley, northern Apennines. *J. Sed. Petrol.* **50**, 723–42.

Gibbs, A. D. 1984. Structural evolution of extensional basin margins. *J. Geol Soc. (London)* **141**, 609–20.

Gibson, R. E. 1958. The progress of consolidation in a clay layer increasing in thickness with time. *Geotechnique* **8**, 171–82.

Gieskes, J. M., K. Johnston & M. Boehm 1984. Appendix. Interstitial water studies, Leg 66. In *Initial Reports Deep Sea Drilling Project 66*, R. von Huene, J. Aubouin *et al.*, 961–7. Washington, DC: US Government Printing Office.

Glasby, G. P. (ed.) 1977. *Marine manganese deposits. Oceanography Ser.* **15**. Amsterdam: Elsevier.

Glennie, K. W. (ed.) 1986a. *Introduction to the petroleum geology of the North Sea*, 2nd edn. Oxford: Blackwell Scientific.

Glennie, K. W. 1986b. Structural framework and pre-Permian history of the North Sea. In *Introduction to the petroleum geology of the North Sea*, K. W. Glennie (ed.), 853–64. Oxford: Blackwell Scientific.

Gloppen, T. G. & R. J. Steel 1981. The deposits, internal structure and geometry of six alluvial fan–fan delta bodies (Devonian, Norway)–a study in the significance of bedding sequences in conglomerates. In *Recent and ancient non-marine depositional environments: models for exploration*, F. G. Ethridge & R. M. Flores (eds), 49–69. Soc. Econ. Paleont. Mineral. Spec. Publn 31.

Gnibidenko, H. S., G. A. Anosov, V. V. Argentov & I. K. Pushchin 1985. Tectonics of the Tonga–Kermadec Trench and Ozbourn Seamount junction area. *Tectonophysics* **112**, 357–83.

Gonthier, E. G., J.-C. Faugeres & D. A. V. Stow 1984. Contourite facies of the Faro Drift, Gulf of Cadiz. In *Fine-grained sediments: deep-water processes and facies*, D. A. V. Stow & D. J. W. Piper (eds), 275–92. Geol Soc. (London) Spec. Publn 15. Oxford: Blackwell Scientific.

Gorsline, D. S. 1978. Anatomy of margin basins. *J. Sed. Petrol.* **48**, 1055–68.

Gorsline, D. S. 1980. Deep-water sedimentologic conditions and models. *Marine Geology* **38**, 1–21.

Gorsline, D. S. 1984. A review of fine-grained sediment origins, characteristics, transport and deposition. In *Fine-grained sediments: deep-water processes and facies*, D. A. V. Stow & D. J. W. Piper (eds), 17–34. Geol Soc. (London) Spec. Publn 15. Oxford: Blackwell Scientific.

Gorsline, D. S. & K. O. Emery 1959. Turbidity current deposits in San Pedro and Santa Monica basins off southern California. *Bull. Geol Soc. Am.* **70**, 279–90.

Gorsline, D. S., R. L. Kolpack, H. A. Karl, D. E. Drake, P. Fleischer, S. E. Thornton, J. R. Schwalbach & C. E. Svarda 1984. Studies of fine-grained sediment transport processes and products in the California Continental Borderland. In *Fine-grained sediments: deep-water processes and facies*, D. A. V. Stow & D. J. W. Piper (eds), 395–415. Geol Soc. (London) Spec. Publn 15. Oxford: Blackwell Scientific.

Govean, F. M. & R. E. Garrison 1981. Significance of laminated and massive diatomites in the upper part of the Monterey Formation, California. In *The Monterey Formation and related siliceous rocks of California*, R. E. Garrison & R. G. Douglas (eds), 181–98. Pacific Section Soc. Econ. Paleont. Mineral. Spec. Publn.

Gradstein, F. M. & R. E. Sheridan 1983. On the Jurassic Atlantic Ocean and a synthesis of results of Deep Sea Drilling Project Leg 76. In *Initial Reports Deep Sea Drilling Project 76*, R. E. Sheridan, F. M. Gradstein *et al.*, 913–43. Washington, DC: US Government Printing Office.

Graham, S. A. & W. R. Dickinson 1978. Evidence for 115 kilometres of right-slip on the San Gregorio–Hosgri fault trend. *Science* **199**, 179–81.

Gravenor, C. P., V. von Brunn & A. Dreimanis 1984. Nature and classification of waterlain glaciogenic sediments, exemplified by Pleistocene, Late Paleozoic and late Precambrian deposits. *Earth Sci. Rev.* **20**, 105–84.

Griggs, G. B., A. G. Carey Jr & L. D. Kulm 1969. Deep-sea

sedimentation and sediment-fauna interaction in Cascadia Channel and on Cascadia Abyssal Plain. *Deep-Sea Res.* **16**, 157–70.

Grim, P. J. & F. P. Naugler 1969. Fossil deep-sea channel on the Aleutian Abyssal Plain. *Science* **163**, 383–6.

Grow, J. A. 1980. Deep structure and evolution of the Baltimore Canyon Trough in the vicinity of the COST No. B-3 well. In *Geological studies of the COST No. B-3 Well, United States Mid-Atlantic continental slope area*, P. A. Scholle (ed.), 117–25. US Geol Surv. Circular 833.

Guiraud, M. & M. Seguret 1987. Soft-sediment microfaulting related to compaction within the fluvio-deltaic infill of the Soria strike-slip basin (northern Spain). In *Deformation of sediments and sedimentary rocks*, M. E. Jones & R. M. F. Preston (eds), 123–36. Geol Soc. (London) Spec. Publn 29. Oxford: Blackwell Scientific.

Hacquebard, P. A., D. E. Buckley & G. Vilks 1981. The importance of detrital particles of coal in tracing the provenance of sedimentary rocks. *Bull. Centres Rech. Explor.-Prod. Elf-Aquitaine* **5**, 555–72.

Hall, J. K. 1979. Sediment waves and other evidence of palaeo-bottom currents at two locations in the deep Arctic Ocean. *Sed. Geol.* **23**, 269–99.

Hamilton, E. L. 1967. Marine geology of abyssal plains in the Gulf of Alaska. *J. Geophys. Res.* **72**, 4189–213.

Hamilton, E. L. 1973. Marine geology of the Aleutian Abyssal Plain. *Marine Geology* **14**, 295–325.

Hamilton, E. L. 1976. Variations of density and porosity with depth in deep-sea sediments. *J. Sed. Petrol.* **46**, 280–300.

Hampton, M. A. 1972. The role of subaqueous debris flow in generating turbidity currents. *J. Sed. Petrol.* **42**, 775–93.

Hampton, M. A. 1975. Competence of fine-grained debris flows. *J. Sed. Petrol.* **45**, 834–44.

Hampton, M. A. 1979. Buoyancy in debris flows. *J. Sed. Petrol.* **49**, 753–8.

Hampton, M. A., A. H. Bouma, P. R. Carlson, B. F. Molnia, E. C. Clukey & D. A. Sangrey 1978. Quantitative study of slope instability in the Gulf of Alaska. *Proc. 10th Offshore Tech. Conference OTC 3314*, 2307–18.

Hancock, J. M. & E. G. Kauffman 1979. The great transgressions of the Late Cretaceous. *J. Geol Soc. London* **136**, 175–86.

Hand, B. M. & J. B. Ellison 1985. Inverse grading in density-current deposits. *Abstracts, 1985 Mid-year meeting*, Soc. Econ. Paleont. Mineral., Golden, Colorado.

Hand, B. M. & K. O. Emery 1964. Turbidites and topography of north end of San Diego Trough, California. *J. Geol.* **72**, 526–42.

Hand, B. M., G. V. Middleton & K. Skipper 1972. Antidune cross-stratification in a turbidite sequence, Cloridorme Formation, Gaspe, Quebec. *Sedimentology* **18**, 135–8.

Haner, B. E. 1971. Morphology and sediments of Redondo submarine fan, southern California. *Bull. Geol Soc. Am.* **82**, 2413–32.

Haq, B. U. 1981. Paleogene palaeoceanography: Early Cenozoic ocean revisited. *Oceanologica Acta Proc. Int. Geol Congress, Geology of Oceans Symposium Paris*, 71–82.

Haq, B. U., J. Hardenbol & P. R. Vail 1987. Chronology of Fluctuating Sea Levels Since the Triassic. *Science* **235**, 1156–67.

Harding, T. P. 1974. Petroleum traps associated with wrench faults. *Bull. Am. Assoc. Petrolm Geol.* **58**, 1290–304.

Harding, T. P. 1983. Divergent wrench fault and negative flower structure, Andaman Sea. In *Seismic expression of structural styles*, Vol. 3, A. W. Bally (ed.), 4.2-1 to 4.2-8. Am. Assoc. Petrolm Geol. Studies in Geology Series 15.

Harding, T. P., R. C. Vierbuchen & N. Christie-Blick 1985. Structural styles, plate-tectonic settings, and hydrocarbon traps of divergent (transtensional) wrench faults. In *Strike-slip deformation, basin formation, and sedimentation*, K. T. Biddle & N. Christie-Blick (eds), 51–77. Soc. Econ. Paleont. Mineral. Spec. Publn 37.

Harms, J. C. 1974. Brushy Canyon Formation, Texas: a deep-water density current deposit. *Bull. Geol Soc. Am.* **85**, 1763–84.

Harms, J. C. & R. K. Fahnestock 1965. Stratification, bed forms and flow phenomena (with an example from the Rio Grande). In *Primary sedimentary structures and their hydrodynamic interpretation*, G. V. Middleton (ed.), 84–115. Soc. Econ. Paleont. Mineral. Spec. Publn 12.

Harms, J. C., J. B. Southard & R. G. Walker 1982. *Structures and sequences in clastic rocks.* Soc. Econ. Paleont. Mineral. Short Course 9.

Harms, J. C., P. Tackenberg, E. Pickles & R. E. Pollock 1981. The Brae oilfield area. In *Petroleum geology of the continental shelf of North-West Europe*, L. V. Illing & G. D. Hobson (eds), 352–7. London: Heyden.

Harper, C. W. Jr 1984. Facies models revisited: an examination of quantitative methods. *Geoscience Canada* **11**, 203–7.

Harris, I. McK. & P. E. Schenk 1968. A study of sedimentary structures in the Goldenville Formation, eastern Nova Scotia. *Maritime Sediments* **4**, 1–3.

Hastings, D. S. 1987. Sand-prone facies in the Cretaceous of Mid-Norway. In *Petroleum geology of north west Europe*, J. Brooks & K. W. Glennie (eds), 1065–78. London: Graham & Trotman.

Haszeldine, R. S. & M. J. Russell 1987. The Late Carboniferous northern North Atlantic Ocean: implications for hydrocarbon exploration from Britain to the Arctic. In *Petroleum geology of north west Europe*, J. Brooks & K. W. Glennie (eds), 1163–75. London: Graham & Trotman.

Hay, A. E. 1983. On the frontal speeds of internal gravity surges on sloping boundaries. *J. Geophys. Res.* **88**, 751–4.

Hay, A. E. 1987a. Turbidity currents and submarine channel formation in Rupert Inlet, British Columbia, Part I: Surge observations. *J. Geophys. Res.* **92**, 2875–82.

Hay, A. E. 1987b. Turbidity currents and submarine channel formation in Rupert Inlet, British Columbia, Part II: the roles of continuous and surge-type flows. *J. Geophys. Res.* **92**, 2883–900.

Hay, A. E., R. W. Burling & J. W. Murray 1982. Remote acoustic detection of a turbidity current surge. *Science* **217**, 833–5.

Hay, W. M. 1984. The breakup of Pangea: climatic, erosional and sedimentological effects. In *Geology of ocean basins, Proceedings of 27th International Geologic Congress* **6**, 15–58.

Hay, W. W., J.-C. Sibuet *et al.* (1984). *Initial Reports Deep Sea Drilling Project* 75. Washington, DC: US Government Printing Office.

Hedberg, H. D. 1974. Relation of methane generation to under-compacted shale, shale diapirs, and mud volcanoes. *Bull. Am. Assoc. Petrolm Geol.* **58**, 661–73.

Hedstrom, B. O. A. 1952. Flow of plastic materials in pipes. *Indust. Engr. Chem.* **44**, 651–6.

Heezen, B. C. 1959. Dynamic processes of abyssal sedimentation: erosion, transportation, and redeposition on the deep-sea floor. *Geophys. J.* **2**, 142–63.

Heezen, B. C. (ed.) 1977. Influence of abyssal circulation on sedimentary accumulations in space and time. *Marine Geology* **23** (special issue).
Heezen, B. C. & C. L. Drake 1964. Grand banks slump. *Bull. Am. Assoc. Petrolm Geol.* **48**, 221–33.
Heezen, B. C., D. B. Ericson & M. Ewing 1954. Further evidence for a turbidity current following the 1929 Grand Banks earthquake. *Deep-Sea Res.* **1**, 193–202.
Heezen, B. C. & M. Ewing 1952. Turbidity currents and submarine slumps and the 1929 Grand Banks earthquake. *Am. J. Sci.* **250**, 849–73.
Heezen, B. C. & C. D. Hollister 1971. *The face of the deep.* New York: Oxford University Press.
Heezen, B. C. & A. S. Laughton 1963. Abyssal plains. In *The sea*, M. N. Hill (ed.) **3**, 312–64. New York: Wiley.
Heezen, B. C. & H. W. Menard 1963. Topography of the deep-sea floor. In *The sea*, Vol. 3, M. N. Hill (ed.), 233–80. New York: Wiley Interscience.
Heezen, B. C., R. Coughlin & W. C. Beckman 1960. Equatorial Atlantic Mid-Ocean Canyon (abstract). *Bull. Geol Soc. Am.* **71**, 1886.
Heezen, B. C., M. Ewing & D. B. Ericson 1951. Submarine togography in the North Atlantic. *Bull. Geol Soc. Am.* **62**, 1407–9.
Heezen, B. C., M. Ewing & D. B. Ericson 1954. Reconnaissance survey of the abyssal plain south of Newfoundland. *Deep-Sea Res.* **2**, 122–33.
Heezen, B. C., C. D. Hollister & W. F. Ruddiman 1966. Shaping of the continental rise by deep geostrophic contour currents. *Science* **152**, 502–8.
Heezen, B. C., G. L. Johnson & D. C. Hollister 1969. The Northwest Atlantic Mid-Ocean Canyon. *Can. J. Earth Sci.* **6**, 1441–53.
Heezen, B. C., R. J. Menzies, E. D. Schneider, W. M. Ewing & N. C. L. Granelli 1964. Congo submarine canyon. *Bull. Am. Assoc. Petrolm Geol.* **48**, 1126–49.
Heezen, B. C., M. Tharp & M. Ewing 1959. The floors of the oceans: I. The North Atlantic. *Geol Soc. Am. Spec. Paper* **65**.
Hein, F. J. 1979. Deep-sea valley-fill sediments, Cap Enrage Formation, Quebec. Ph.D. Thesis. Hamilton, Ontario: McMaster University.
Hein, F. J. 1982. Depositional mechanisms of deep-sea coarse clastic sediments, Cap Enrage Formation, Quebec. *Can. J. Earth Sci.* **19**, 267–87.
Hein, F. J. 1985. Fine-grained slope and basin deposits, California Continental Borderland: facies, depositional mechanisms and geotechnical properties. *Marine Geology* **67**, 237–62.
Hein, F. J. & R. W. Arnott 1983. Precambrian Miette conglomerates, Lower Cambrian Gog quartzites and modern braided outwash deposits, Kicking Horse Pass area. *Can. Soc. Petrolm Geol. Field Trip Guidebook.*
Hein, F. J. & D. S. Gorsline 1981. Geotechnical aspects of fine-grained mass flow deposits: California Continental Borderland. *Geo-Marine Letts* **1**, 1–5.
Hein, F. J. & J. P. M. Syvitski 1987. Sedimentology of Itirbilung Fiord, Baffin Island, Canada. *Abstracts 16th Arctic Workshop, Boreal Inst. for Northern Studies*, 53–5. Edmonton: University Alberta.
Hein, F. J. & R. G. Walker 1982. The Cambro-Ordovician Cap Enrage Formation, Quebec, Canada: conglomeratic deposits of a braided submarine channel with terraces. *Sedimentology* **29**, 309–29.
Heller, P. L. & W. R. Dickinson 1985. Submarine ramp facies model for delta-fed, sand-rich turbidite systems. *Bull. Am. Assoc. Petrolm Geol.* **69**, 960–76.
Hendry, H. E. 1972. Breccias deposited by mass flow in the Breccia Nappe of the French pre-Alps. *Sedimentology* **8**, 277–92.
Hendry, H. E. 1973. Sedimentation of deep water conglomerates in Lower Ordovician rocks of Quebec–composite bedding produced by progressive liquefaction of sediment? *J. Sed. Petrol.* **43**, 125–36.
Hendry, H. E. 1978. Cap des Rosiers Formation at Grosses Roches, Quebec–deposits in the mid-fan region on an Ordovician submarine fan. *Can. J. Earth Sci.* **15**, 1472–88.
Herman, B. M., R. N. Anderson & M. Truchan 1979. Extensional Tectonics in the Okinawa Trough. In *Geological and geophysical investigations of continental margins*, J. S. Watkins, L. Montadert & P. W. Dickerson (eds), 199–208. Am. Assoc. Petrolm Geol. Mem. 29.
Hersey, J. B. 1965a. Sediment ponding in the deep sea. *Bull. Geol Soc. Am.* **76**, 1251–60.
Hersey, J. B. 1965b. Sedimentary basins of the Mediterranean Sea. In *Submarine geology and geophysics*, W. F. Whittard & R. Bradshaw (eds), 75–89. London: Butterworths.
Herzer, R. H. & D. W. Lewis 1979. Growth and burial of a submarine canyon off Motunau, north Canterbury, New Zealand. *Sed. Geol.* **24**, 69–83.
Hess, G. R. & W. R. Normark 1976. Holocene sedimentation history of the major fan valleys of Monterey Fan. *Marine Geology* **22**, 233–51.
Hesse, R. 1965. Herkunfe und Transport der Sedimente im Bayerischen Flyschtrog. *Z. Deutsch. Gesell.* **116**, 147–70.
Hesse, R. 1974. Long-distance continuity of turbidites: possible evidence for an early Cretaceous trench-abyssal plain in the East Alps. *Bull. Geol Soc. Am.* **85**, 859–70.
Hesse, R. 1975. Turbiditic and non-turbiditic mudstone of Cretaceous flysch sections of the East Alps and other basins. *Sedimentology* **22**, 387–416.
Hesse, R. 1982. Cretaceous-Palaeogene Flysch Zone of the East Alps and Carpathians: identification and plate-tectonic significance of 'dormant' and 'active' deep-sea trenches in the Alpine-Carpathian Arc. In *Trench–forearc geology*, J. K. Leggett (ed.), 471–94. Geol Soc. (London) Spec. Publn 10. Oxford: Blackwell Scientific.
Hesse, R. & A. Butt 1976. Palaeobathymetry of Cretaceous turbidite basins of the East Alps relative to the calcite compensation level. *J. Geol.* **84**, 505–33.
Hesse, R. & S. K. Chough 1980. The Northwest Atlantic Mid-Ocean Channel of the Labrador Sea: II. Deposition of parallel laminated levee-muds from the viscous sublayer of low density turbidity currents. *Sedimentology* **27**, 697–711.
Hesse, R., S. K. Chough & A. Rakofsky 1987. The Northwest Atlantic Mid-Ocean Channel of the Labrador Sea. V. Sedimentology of a giant deep-sea channel. *Can. J. Earth Sci.* **24**, 1595–624.
Hicks, D. M. 1981. Deep-sea fan sediments in the Torlesse zone, Lake Ohau, South Canterbury, New Zealand. *New Zealand J. Geol. Geophys.* **24**, 209–30.
Hickey, B., E. Baker & N. Kachel 1986. Suspended particle movement in and around Quinault submarine canyon. *Marine Geology* **71**, 35–83.
Hieke, W. 1984. A thick Holocene homegenite from the Ionian Abyssal Plain (eastern Mediterranean). *Marine Geology* **55**, 63–78.
Higgins, G. E. & J. B. Saunders 1967. Report on 1964 Chatham

Mud Island, Erin Bay, Trinidad, West Indies. *Bull. Am. Assoc. Petrolm Geol.* **51**, 55–64.

Higgins, G. E. & J. B. Saunders 1974. Mud volcanoes, their nature and origin. In *Contributions to the geology and palaeobiology of the Caribbean and adjacent areas. Verhandlungen Naturforschenden Gesellschaft in Basel* **84**, 101–52.

Hilde, T. W. C., J. M. Wageman & W. T. Hammond 1969. The structure of the Tosa Terrace and Nankai Trough off southwestern Japan. *Deep-Sea Res.* **16**, 67–75.

Hill, P. R. 1984a. Facies and sequence analysis of Nova Scotian Slope muds: turbidite vs 'hemipelagic' deposition. In *Fine-grained sediments: deep-water processes and facies*, D. A. V. Stow & D. J. W. Piper (eds), 311–18. Geol Soc. (London) Spec. Publn 15. Oxford: Blackwell Scientific.

Hill, P. R. 1984b. Sedimentary facies of the Nova Scotian upper and middle continental slope, offshore eastern Canada. *Sedimentology* **31**, 293–309.

Hill, P. R. & A. J. Bowen 1983. Modern sediment dynamics at the shelf-slope boundary off Nova Scotia. In *The shelfbreak: critical interface on continental margins*, D. J. Stanley & G. T. Moore (eds), 265–76. Soc. Econ. Paleont. Mineral. Spec. Publn 33.

Hill, P. R., K. M. Moran & S. M. Blasco 1982. Creep deformation of slope sediments in the Canadian Beaufort Sea. *Geo-Marine Letts* **2**, 163–70.

Hinz, K., E. L. Winterer *et al.* 1984. *Initial Reports Deep Sea Drilling Project 79.* Washington, DC: US Government Printing Office.

Hiscott, R. N. 1979. Clastic sills and dikes associated with deep-water sandstones, Tourelle Formation, Ordovician, Quebec. *J. Sed. Petrol.* **49**, 1–10.

Hiscott, R. N. 1980. Depositional framework of sandy mid-fan complexes of Tourelle Formation, Ordovician, Quebec. *Bull. Am. Assoc. Petrolm Geol.* **64**, 1052–77.

Hiscott, R. N. 1981. Deep-sea fan deposits in the Macigno Formation (middle–upper Oligocene) of the Gordana Valley, northern Apennines, Italy–Discussion. *J. Sed. Petrol.* **51**, 1015–21.

Hiscott, R. N. & N. P. James 1985. Carbonate debris flows, Cow Head Group, western Newfoundland. *J. Sed. Petrol.* **55**, 735–45.

Hiscott, R. N. & G. V. Middleton 1979. Depositional mechanics of thick-bedded sandstones at the base of a submarine slope, Tourelle Formation (Lower Ordovician), Quebec, Canada. In *Geology of continental slopes*, L. J. Doyle & O. H. Pilkey (eds), 307–26. Soc. Econ. Paleont. Mineral. Spec. Publn 27.

Hiscott, R. N. & G. V. Middleton 1980. Fabric of coarse deep-water sandstones, Tourelle Formation, Quebec, Canada. *J. Sed. Petrol.* **50**, 703–22.

Hiscott, R. N. & K. T. Pickering 1984. Reflected turbidity currents on an Ordovician basin floor, Canadian Appalachians. *Nature* **311**, 143–5.

Hiscott, R. N., K. T. Pickering & D. R. Beedon 1986. Progressive filling of a confined Middle Ordovician foreland basin associated with the Taconic Orogeny, Quebec, Canada. In *Foreland basins*, P. A. Allen & P. Homewood (eds), 309–25. Int. Assoc. Sedimentologists Spec. Publn 8. Oxford: Blackwell Scientific.

Hoffert, M. 1980. Les 'argiles rouges des grands fonds' dans le Pacifique centre-est: authigenese, transport, diagenese. Thesis, University Louis Pasteur Strasbourg, Mem. 61.

Hoffman, P., J. F. Dewey & K. Burke 1974. Aulacogens and their genetic relation to geosynclines, with a Proterozoic example from Great Slave Lake, Canada. In *Modern and ancient geosynclinal sedimentation*, R. H. Dott Jr & R. H. Shaver (eds), 38–55. Soc. Econ. Paleont. Mineral. Spec. Publn 19.

Hogg, N. G. 1983. A note on the deep circulation of the western North Atlantic: its nature and causes. *Deep-Sea Res.* **30**, 945–61.

Hollister, C. D. 1967. *Sediment distribution and deep circulation in the western North Atlantic.* Ph.D. Thesis. New York: Columbia University Press.

Hollister, C. D. & B. C. Heezen 1972. Geologic effects of ocean bottom currents: western North Atlantic. In *Studies in physical oceanography 2*, A. L. Gordon (ed.), 37–66. New York: Gordon & Breach.

Hollister, C. D. & I. N. McCave 1984. Sedimentation under deep-sea storms. *Nature* **309**, 220–5.

Hollister, C. D., C. Craddock *et al.* 1976a. *Initial Reports Deep Sea Drilling Project 35.* Washington, DC: US Government Printing Office.

Hollister, C. D., R. D. Flood, D. A. Johnson, P. F. Lonsdale & J. B. Southard 1974. Abyssal furrows and hyperbolic echo traces on the Bahama Outer Ridge. *Geology* **2**, 395–400.

Hollister, C. D., R. D. Flood & I. N. McCave 1978. Plastering and decorating in the North Atlantic. *Oceanus* **21**, 5–13.

Hollister, C. D., J. B. Southard, R. D. Flood & P. F. Lonsdale 1976b. Flow phenomena in the benthic boundary layer and bed forms beneath deep-current systems. In *The benthic boundary layer*, I. N. McCave (ed.), 183–204. New York: Plenum.

Homewood, P. & C. Caron 1983. Flysch of the Western Alps. In *Mountain building processes*, K. J. Hsü (ed.), 157–68. New York: Academic Press.

Hopfinger, E. J. 1983. Snow avalanche motion and related phenomena. *Ann. Review Fluid Mechanics* **15**, 47–76.

Horn, D. R. (ed.) 1972. *Ferromanganese deposits on the ocean floor.* New York: Arden House, Harriman and Lamont-Doherty Geol Obs.

Horn, D. R., J. I. Ewing & M. Ewing 1972. Graded-bed sequences emplaced by turbidity currents north of 20°N in the Pacific, Atlantic and Mediterranean. *Sedimentology* **18**, 247–75.

Horn, D. R., M. Ewing, B. M. Horn & M. N. Delach 1971. Turbidites of the Hatteras and Sohm Abyssal Plains, western North Atlantic. *Marine Geology* **11**, 287–323.

Hoskins, E. G. & J. R. Griffiths 1971. Hydrocarbon potential of northern and central California offshore. *Am. Assoc. Petrolm Geol. Mem.* **15**, 212–28.

Houtz, R. E. & D. E. Hayes 1984. Seismic refraction data from Sunda shelf. *Bull. Am. Assoc. Petrolm Geol.* **68**, 1870–8.

Howell, D. G. (ed.) 1985. *Tectonostratigraphic terranes of the circum-Pacific region.* Houston, Texas: Circum-Pacific Council for Energy & Mineral Resources, Earth Science Series 1.

Howell, D. G. & M. H. Link 1979. Eocene conglomerate sedimentology and basin analysis, San Diego and the southern California Borderland. *J. Sed. Petrol.* **49**, 517–40.

Howell, D. G. & W. R. Normark 1982. Sedimentology of submarine fans. *Am. Assoc. Petrolm Geol. Mem.* **31**, 365–404.

Howell, D. G. & R. von Huene 1980. *Tectonics and sediment along active convergent margins.* Soc. Econ. Paleont. Mineral. Short Course San Francisco 1980.

Howell, D. G., J. K. Crouch, H. G. Greene, D. S. McCulloch & J. G. Vedder 1980. Basin development along the Late Mesozoic and Cenozoic California margin: a plate tectonic margin of subduction, oblique subduction, and transform tectonics. In *Sedimentation in oblique-slip mobile zones*, P. F. Ballance & H. G. Reading (eds), 43–62. Int. Assoc. Sedimentologists

Spec. Publn 4. Oxford: Blackwell Scientific.

Howell, D. G., D. L. Jones & E. R. Schermer 1985. Tectonostratigraphic terranes of the circum-Pacific region. In *Tectonostratigraphic terranes of the circum-Pacific*, D. G. Howell (ed.), 3–30. Houston, Texas: Circum-Pacific Council for Energy & Mineral Resources.

Hoyt, W. H. & P. J. Fox 1977. Long-distance turbidite correlations in Horseshoe Abyssal Plain (abstract). *Bull. Am. Assoc. Petrolm Geol.* **61**, 797.

Hsü, K. J. 1974. Mélanges and their distinction from olistostromes. In *Modern and ancient geosynclinal sedimentation*, R. H. Dott Jr & R. H. Shaver (eds), 321–33. Soc. Econ. Paleont. Mineral. Spec. Publn 19.

Hsü, K. J. 1977. Studies of Ventura field, California, 1: facies geometry and genesis of lower Pliocene turbidites. *Bull. Am. Assoc. Petrolm Geol.* **61**, 137–68.

Hsü, K. J. & H. C. Jenkyns (eds) 1974. *Pelagic sediments: on land and under the sea.* Int. Assoc. Sedimentologists Spec. Publn 1. Oxford: Blackwell Scientific.

Hubert, C., J. Lajoie & M. A. Leonard 1970. Deep sea sediments in the Lower Paleozoic Quebec Supergroup. In *Flysch sedimentology in North America*, J. Lajoie (ed.), 103–25. Geol. Assoc. Can. Spec. Paper 7. Toronto: Business & Economic Service.

Hubert, J. F. 1966a. Modification of the model for internal structures in graded beds to include a dune division. *Nature* **211**, 614–15.

Hubert, J. F. 1966b. Sedimentation history of Upper Ordovician geosynclinal rocks, Girvan, Scotland. *J. Sed. Petrol.* **36**, 677–99.

Hubert, J. F. 1967. Sedimentology of pre-Alpine flysch sequences, Switzerland. *J. Sed. Petrol.* **37**, 885–907.

Hubert, J. F., R. K. Suchecki & R. K. M. Callahan 1977. The Cow head breccia: sedimentology of the Cambro-Ordovician continental margin, Newfoundland. In *Deep-water carbonate environments*, H. E. Cook & P. Enos (eds), 125–54. Soc. Econ. Paleont. Mineral. Spec. Publn 25.

Huchon, P. & H. Kitazato 1984. Collision of the Izu block with central Japan during the Quaternary and geological evolution of the Ashigara area. *Tectonophysics* **110**, 201–10.

Humphris, C. C. Jr 1978. Salt movement on the continental slope, northern Gulf of Mexico. In *Framework, facies, and oil-trapping characteristics of the upper continental margin*, A. H. Bouma, G. T. Moore & J. M. Coleman (eds), 69–85. Am. Assoc. Petrolm Geol. Studies in Geology 7.

Hutton, D. H. W. & J. F. Dewey 1986. Palaeozoic terrane accretion in the western Irish Caledonides. *Tectonics* **5**, 1115–24.

Imbrie, J. & K. P. Imbrie 1979. *Ice ages: solving the mystery*, 113–22. London: Macmillan Press.

Ingersoll, R. V. 1978a. Petrofacies and petrologic evolution of the Late Cretaceous fore-arc basin, northern and central California. *J. Geology* **86**, 335–52.

Ingersoll, R. V. 1978b. Submarine fan facies of the Upper Cretaceous Great Valley Sequence, northern and central California. *Sed. Geology* **21**, 205–30.

Ingle J. C. Jr 1981. Origin of Neogene diatomites around the North Pacific rim. In *The Monterey Formation and related siliceous rocks of California*, R. E. Garrison & R. G. Douglas (eds), 159–79. Pacific Section Soc. Econ. Paleont. Mineral. Spec. Publn.

Ingram, R. L. 1954. Terminology for the thickness of stratification and parting units in sedimentary rocks. *Bull. Geol Soc. Am.* **65**, 937–8.

Inman, D. L., C. E. Nordstrom & R. E. Flick 1976. Currents in submarine canyons: an air–sea–land interaction. *Ann. Rev. Fluid Mechanics* **8**, 275–310.

Isaacs, C. M. 1981. Lithostratigraphy of the Monterey Formation, Coleta to Point Conception, Santa Barbara Coast, California. In *Guide 4, Am. Assoc. Petrolm Geol. Annual Meeting*, 9–24.

Isaacs, C. M. 1984. Hemipelagic deposits in a Miocene basin, California: toward a model of lithologic variation and sequence. In *Fine-grained sediments: deep-water processes and facies*, D. A. V. Stow & D. J. W. Piper (eds), 481–96. Geol Soc. (London) Spec. Publn 15. Oxford: Blackwell Scientific.

Jacka, A. D., R. H. Beck, L. C. St Germain & S. G. Harrison 1968. Permian deep-sea fans of the Delaware Mountain Group (Guadalupian), Delaware Basin. In *Guadalupian facies, Apache mountain area, west Texas*, B. A. Silver (ed.), 49–90. Permian Basin Section Soc. Econ. Paleont. Mineral. Publn. 68–11.

Jackson, T. A. 1965. Power-spectrum analysis of two 'varved' argillites in the Huronian Cobalt Series (Precambrian) of Canada. *J. Sed. Petrol.* **35**, 877–86.

Jacobi, R. D. 1984. Modern submarine sediment slides and their implications for mélange and the Dunnage Formation in north-central Newfoundland. *Geol Soc. Am. Spec. Paper* **198**, 81–102.

Jacobson, R. S., G. G. Shor, R. M. Kieckhefer & G. M. Purdy 1979. Seismic refraction and reflection studies in the Timor–Aru Trough system and Australian continental shelf. In *Geological and geophysical investigations of continental margins*, J. S. Watkins, L. Montadert & P. W. Dickerson (eds), 209–22. Am. Assoc. Petrolm Geol. Mem. 29.

James, N. P. & R. K. Stevens 1986. Stratigraphy and correlation of the Cambro-Ordovician Cow Head Group, western Newfoundland. *Bull. Geol Surv. Canada* **366**.

Jeffery, G. B. 1922. The motion of ellipsoidal particles immersed in a viscous fluid. *Proc. R. Soc. London (A)* **102**, 161–79.

Jenkyns, H. C. 1980. Tethys: past and present. *Proc. Geol. Assoc.* **91**, 107–18.

Jenkyns, H. C. 1986. Pelagic environments. In *Sedimentary environments and facies*, 2nd edn, H. G. Reading (ed.), 343–97. Oxford: Blackwell Scientific.

Jipa, D. & R. S. Kidd 1974. Sedimentation of coarser grained interbeds in Arabian Sea and sedimentation processes of the Indus Cone. In *Initial Reports Deep Sea Drilling Project 23*, R. B. Whitmarsh, O. E. Weser, D. A. Ross *et al.*, 471–95. Washington, DC: US Government Printing Office.

Johnson, A. M. 1970. *Physical processes in geology*. San Francisco: Freeman, Cooper.

Johnson, A. M. 1984 (with contributions by J. R. Rodine). Debris flow. In *Slope instability*, D. Brunsden & D. B. Prior (eds), 257–362. New York: Wiley.

Johnson, B. A. & R. G. Walker 1979. Paleocurrents and depositional environments of deep water conglomerates in the Cambro-Ordovician Cap Enrage Formation, Quebec Appalachians. *Can. J. Earth Sci.* **16**, 1375–87.

Johnson, D. 1939. The origin of submarine canyons. *J. Geomorphology* **2**, 42–60, 133–58, 213–36.

Johnson, D. A. & K. A. Rasmussen 1984. Late Cenozoic turbidite and contourite deposition in the southern Brazil Basin. *Marine Geology* **58**, 225–62.

Johnson, D. W. 1967 (originally published 1925). *The New England–Acadian shoreline*. New York: Hafner.

Johnson, G. L. & E. D. Schneider 1969. Depositional ridges in the North Atlantic. *Earth Planet. Sci. Letts* **6**, 416–22.

Johnson, H. D. & D. J. Stewart 1985. Role of clastic sedimentology in the exploration and production of oil and gas in the North Sea. In *Sedimentology: recent developments and applied aspects*, P. J. Brenchley & B. P. J. Williams (eds), 249–310. Geol Soc. (London) Spec. Publn 18. Oxford: Blackwell Scientific.

Johnson, H. D., B. K. Levell & S. Siedlecki 1978. Late Precambrian sedimentary rocks in East Finnmark, north Norway and their relationship to the Trollfjord–Komagelv fault. *J. Geol Soc. (London)* **135**, 517–33.

Jones, M. E. & R. M. F. Preston (eds) 1987. *Deformation of sediments and sedimentary rocks.* Geol Soc. (London) Spec. Publn 29. Oxford: Blackwell Scientific.

Jopling, A. V. & R. G. Walker 1968. Morphology and origin of ripple-drift cross-lamination, with examples from the Pleistocene of Massachusetts. *J. Sed. Petrol.* **38**, 971–84.

Jordan, T. E. 1981. Enigmatic deep-water depositional mechanisms, upper part of the Oquirrh Group, Utah. *J. Sed. Petrol.* **51**, 879–94.

Junger, A. 1976. Tectonics of the southern California Borderland. In *Aspects of the geologic history of the California Continental Borderland*, D. G. Howell (ed.), 486–98. Am. Assoc. Petrolm Geol. Pacific Section Misc. Publn 24.

Junger, A. & H. C. Wagner 1977. *Geology of the Santa Monica and San Pedro basins, California Continental Borderland.* US Geol Surv. Map MF–820.

KAIKO Project, Shipboard Scientific Party 1985. Japanese deep-sea trench survey. *Nature* **313**, 432–3.

Kanmera, K. 1976a. Comparison between past and present geosynclinal sedimentary bodies I. *Kagaku (Science)* **46**, 284–91.

Kanmera, K. 1976b. Comparison between past and present geosynclinal sedimentary bodies II. *Kagaku (Science)* **46**, 371–8.

Karig, D. E., A. J. Barber, T. R. Charlton, S. Klemperer & D. M. Hussong 1987. Nature and distribution of deformation across the Banda arc–Australian collision zone at Timor. *Bull. Geol Soc. Am.* **98**, 18–32.

Karig, D. E., H. Kagami & DSDP Leg 87 Scientific Party 1983. Varied response to subduction in Nankai Trough and Japan Trench forearcs. *Nature* **304**, 148–51.

Karig, D. E., S. Suparka, G. F. Moore & P. E. Hehanussa, 1979. Structure and Cenozoic evolution of the Sunda Arc in the Central Sumatra region. In *Geological and geophysical investigations of continental margins*, J. S. Watkins, L. Montadert & P. W. Dickerson (eds), 223–37. Am. Assoc. Petrolm Geol. Mem. 29.

Karl, H. A. 1976. Processes influencing transportation and deposition of sediments on the continental shelf, southern California. University of S. California, Los Angeles: Unpubl. Ph.D. Thesis.

Kastens, K. A. & A. N. Shor 1985. Depositional processes of a meandering channel on Mississippi Fan. *Bull. Am. Assoc. Petrolm Geol.* **69**, 190–202.

Katz, H.-R. 1974. Margins of the Southwest Pacific. In *The geology of continental margins*, C. A. Burk & C. L. Drake (eds), 549–65. New York: Springer.

Keith, B. D. & G. M. Friedman 1977. A slope–fan–basin–plain model, Taconic sequence, New York and Vermont. *J. Sed. Petrol* **47**, 1220–41.

Keller, G. H. 1982. Organic matter and the geotechnical properties of submarine sediments. *Geo-Marine Letts* **2**, 191–8.

Keller, G. & J. A. Barron 1983. Paleoceanographic implications of Miocene deep-sea hiatuses. *Bull. Geol Soc. Am.* **94**, 590–613.

Kelling, G. 1961. The stratigraphy and structure of the Ordovician rocks of the Rhinns of Galloway. *Quarterly J. Geol Soc. London* **117**, 37–75.

Kelling, G. & J. Holroyd 1978. Clast size, shape, and composition in some ancient and modern fan gravels. In *Sedimentation in submarine canyons, fans and trenches*, D. J. Stanley & G. Kelling (eds), 138–59. Stroudsburg, PA: Dowden, Hutchinson & Ross.

Kelling, G. & D. J. Stanley (eds) 1978. *Sedimentation in submarine canyons, fans, and trenches.* Stroudsburg, PA: Dowden, Hutchinson & Ross.

Kelts, K. & M. A. Arthur 1981. Turbidites after ten years of deep-sea drilling – wringing out the mop? In *The deep sea drilling project: a decade of progress*, R. G. Douglas & E. L. Winterer (eds), 91–127. Soc. Econ. Paleont. Mineral. Spec. Publn 32.

Kelts, K. & J. Niemitz 1982. Preliminary sedimentology of Late Quaternary diatomaceous muds from Deep-Sea Drilling Project Site 480, Guyamas Basin slope, Gulf of California. In *Initial Reports Deep Sea Drilling Project 64*, J. R. Curray, D. G. Moore *et al.*, 1191–210. Washington, DC: US Government Printing Office.

Kendall, M. G. 1969. *Rank correlation methods.* London: Charles Griffin.

Kendall, M. G. 1976. *Time series.* New York: Hafner Press (Macmillan).

Kennett, J. 1982. *Marine geology.* Englewood Cliffs, New Jersey: Prentice-Hall.

Kennett, J. P., R. E. Houtz, P. V. Andrews, A. R. Edwards, V. A. Gostin, N. Hajos, M. Hampton, D. G. Jenkins, S. V. Margolis, A. T. Ovenshine & K. Perch-Nielsen 1975. Cenozoic paleo-oceanography in the southwest Pacific Ocean and the development of the Circumpolar current. In *Initial Reports Deep Sea Drilling Project 29*, J. P. Kennett, R. E. Houtz *et al.*, 1155–69. Washington, DC: US Government Printing Office.

Kenyon, N. H., R. H. Belderson & A. H. Stride 1978. Channels, canyons and slump folds on the continental slope between south west Ireland and Spain. *Oceanol. Acta* **1**, 369–80.

Keppie, J. D. 1986. The Appalachian collage. In *The Caledonide orogen–Scandinavia and related areas*, D. G. Gee & B. A. Sturt (eds), 1217–26. New York: Wiley.

Kessler II, L. G. & K. Moorhouse 1984. Depositional processes and fluid mechanics of Upper Jurassic conglomerate accumulations, British North Sea. In *Sedimentology of gravels and conglomerates*, E. H. Koster & R. J. Steel (eds), 383–97. Calgary, Alberta: Can. Soc. Petrolm Geol. Mem. 10.

Kidd, R. B. & P. R. Hill 1986. Sedimentation on mid-ocean sediment drifts. In *North Atlantic palaeoceanography*, C. P. Summerhayes & N. J. Shackleton (eds), 87–102. Geol Soc. (London) Spec. Publn 21. Oxford: Blackwell Scientific.

Kidd, R. B. & P. R. Hill 1987. Sedimentation on Feni and Gardar sediment drifts. In *Initial Reports Deep Sea Drilling Project 94*, W. F. Ruddiman, R. B. Kidd, E. Thomas *et al.*, 1217–44. Washington DC: US Government Printing Office.

Kidd, R. B. & D. G. Roberts 1982. Long-range sidescan sonar studies of large-scale sedimentary features in the North Atlantic. *Bull. Int. Bassin d'Aquitaine, Bordeaux*, **31**, 11–29.

Kimura, M. 1985. Back-arc rifting in the Okinawa Trough. *Marine Petrolm Geology* **2**, 222–40.

Kimura, T. 1966. Thickness distribution of sandstone beds and cyclic sedimentation turbidite sequence at two localities in Japan. *Earthquake Res. Inst.* **44**, 561–607.

Kingston, D. R., C. P. Dishroon & P. A. Williams 1983. Global basin classification system. *Bull. Am. Assoc. Petrolm Geol.* **67**, 2175–93.

Kjøde, J., K. H. Storetvedt, D. Roberts & A. Gidskehaug 1978. Palaeomagnetic evidence for large-scale dextral displacement along the Trollfjord–Komagelv fault, Finnmark, north Norway. *Phys. Earth Planet. Interiors* **16**, 132–44.

Klitgord, K. D. & H. Schouten 1986. Plate kinematics of the central Atlantic. In *The geology of North America, Vol. M, The western North Atlantic region*, P. R. Vogt & B. E. Tucholke (eds), 351–78. Boulder, Colorado: Geol Soc. Am.

Klitgord, K. D., D. R. Hutchinson & H. Schouten 1988. US Atlantic continental margin; structural and tectonic framework. In *The geology of the North Atlantic, Vol. 1–2, The Atlantic continental margin*, R. E. Sheridan & J. A. Grow (eds), 19–55. Boulder, Colorado: Geol Soc. Am.

Knipe, R. J. 1986. Microstructural evolution of vein arrays preserved in Deep Sea Drilling Project cores from the Japan Trench, Leg 57. In *Structural fabric in Deep Sea Drilling Project cores from forarcs*, J. C. Moore (ed.), 75–87. Geol Soc. Am. Mem. 166.

Kodama, K., A. Taira, M. Okamura & Y. Saito 1983. Paleomagnetisation of the Shimanto Belt in Shikoku, southwest Japan. In *Accretion tectonics in the circum-Pacific regions*, M. Hashimoto & S. Uyeda (eds), 231–41. Tokyo: Terrapub.

Kokelaar, B. P. & M. F. Howells (eds) 1984. *Marginal basin geology*. Geol Soc. (London) Spec. Publn 16. Oxford: Blackwell Scientific.

Kolla, V. & F. Coumes 1987. Morphology, internal structure, seismic stratigraphy, and sedimentation of Indus Fan. *Bull. Am. Assoc. Petrolm Geol.* **71**, 650–77.

Kolla, V., S. Eittreim, L. Sullivan, J. A. Kostecki & L. H. Burckle 1980a. *Marine Geology* **34**, 171–206.

Kolla, V., J. A. Kostecki, L. Henderson & L. Hess 1980b. Morphology and Quaternary sedimentation of the Mozambique Fan and environs, southwestern Indian Ocean. *Sedimentology* **27**, 357–78.

Komar, P. D. 1969. The channelized flow of turbidity currents with application to Monterey deep-sea fan channel. *J. Geophys. Res.* **74**, 4544–58.

Komar, P. D. 1971. Hydraulic jumps in turbidity currents. *Bull. Geol Soc. Am.* **82**, 1477–88.

Komar, P. D. 1977. Computer simulation of turbidity current flow and the study of deep-sea channels and fan sedimentation. In *The sea, Vol. 6, marine modelling*, E. D. Goldberg, I. N. McCave, J. J. O'Brien & J. H. Steele (eds), 603–21. New York: Wiley.

Kranck, K. 1984. Grain-size characteristics of turbidites. In *Fine-grained sediments: deep-water processes and facies*, D. A. V. Stow & D. J. W. Piper (eds), 83–92. Geol Soc. (London) Spec. Publn 15. Oxford: Blackwell Scientific.

Krissek, L. A. 1984. Continental source area contributions to fine-grained sediments on the Oregon and Washington continental slope. In *Fine-grained sediments: deep-water processes and facies*, D. A. V. Stow & D. J. W. Piper (eds), 363–75. Geol Soc. (London) Spec. Publn 15. Oxford: Blackwell Scientific.

Ksiazkiewicz, M. 1954. Graded and laminated bedding in the Carpathian flysch. *Ann. Soc. Geol. Pologne* **22** (1952), 399–449.

Ksiazkiewicz, M. 1960. Pre-orogenic sedimentation in the Carpathian geosyncline. *Geol. Rdsch.* **50**, 8–31.

Kuenen, Ph. H. 1951. Properties of turbidity currents of high density. In *Turbidity currents*, J. L. Hough (ed.), 14–33. Soc. Econ. Paleont. Mineral. Spec. Publn 2.

Kuenen, Ph. H. 1953. Significant feature of graded bedding. *Bull. Am. Assoc. Petrolm Geol.* **37**, 1044–66.

Kuenen, Ph. H. 1964. Deep-sea sands and ancient turbidites. In *Turbidites*, A. H. Bouma & A. Brouwer (eds), 3–33. Developments in Sedimentology 3. Amsterdam: Elsevier.

Kuenen, Ph. H. 1966. Matrix of turbidites: experimental approach. *Sedimentology* **7**, 267–97.

Kuenen, Ph. H. & C. I. Migliorini 1950. Turbidity currents as a cause of graded bedding. *J. Geology* **58**, 91–127.

Kulm, L. D. & G. A. Fowler 1974. Oregon continental margin structure and stratigraphy: a test of the imbricate thrust model. In *The geology of continental margins*, C. A. Burk & C. L. Drake (eds), 261–83. New York: Springer.

Kulm, L. D., R. E. von Huene *et al.* 1973. *Initial Reports Deep Sea Drilling Project 18*. Washington, DC: US Government Printing Office.

Kurtz, D. D. & J. B. Anderson 1979. Recognition and sedimentologic description of recent debris flow deposits from the Ross and Weddell Seas, Antarctica. *J. Sed. Petrol.* **49**, 1159–69.

Labaume, P., E. Mutti, M. Seguret & J. Rosell 1983. Megaturbidites carbonatées du bassin turbiditique de l'Eocene inferieur et moyen sud-pyrénéen. *Bull. Soc. Geol. France* **25**, 927–41.

Labaume, P., E. Mutti & M. Seguret 1985. Megaturbidites: a depositional model from the Eocene of the SW-Pyrenean foreland basin, Spain. *Geo-Marine Letts* **7**, 91–101.

Laird, M. G. 1968. Rotational slumps and slump scars in Silurian rocks, western Ireland. *Sedimentology* **10**, 111–20.

Laird, M. G. 1970. Vertical sheet structures–a new indication of sedimentary fabric. *J. Sed. Petrol.* 40, 428–34.

Laval, A., M. Cremer, P. Beghin & C. Ravenne 1988. Density surges: two-dimensional experiments. *Sedimentology* **35**, 73–84.

Le Pichon, X., T. Iiyama, H. Chamley, J. Charvet, M. Faure, H. Fujimoto, T. Furuta, Y. Ida, H. Kagami, S. Lallemant, J. Leggett, A. Murata, H. Okada, C. Rangin, V. Renard, A. Taira & H. Tokuyama 1987a. Nankai Trough and the fossil Shikoku Ridge: results of Box 6 Kaiko survey. *Earth Planet. Sci. Letts* **83**, 186–98.

Le Pichon, X., T. Iiyama, H. Chamley, J. Charvet, M. Faure, H. Fujimoto, T. Furuta, Y. Ida, H. Kagami, S. Lallemant, J. Leggett, A. Murata, H. Okada, C. Rangin, V. Renard, A. Taira & H. Tokuyama 1987b. The eastern and western ends of Nankai Trough: results of Box 5 and Box 7 Kaiko survey. *Earth Planet. Sci. Letts* **83**, 199–213.

Leckie, R. M. 1984. Mid-Cretaceous planktonic foraminiferal biostratigraphy off central Morocco, Deep Sea Drilling Project Leg 79, Sites 545 and 547. In *Initial Reports Deep Sea Drilling Project* 79, K. Hinz, E. L. Winterer *et al.*, 579–620. Washington, DC: US Government Printing Office.

Ledbetter, M. T. 1979. Fluctuations of Antarctic bottom water velocity in the Vema Channel during the last 160,000 years. *Marine Geology* **33**, 71–89.

Ledbetter, M. T. 1984. Bottom-current speed in the Vema Channel recorded by particle size of sediment fine-fraction. *Marine Geology* **58**, 137–49.

Ledbetter, M. T. & B. B. Ellwood 1980. Spatial and temporal changes in bottom water velocity and direction from analyses of particle size and alignment in deep-sea sediment. *Marine Geology* **38**, 245–61.

Leeder, M. R. 1983. On the dynamics of sediment suspension by residual Reynolds stresses–confirmation of Bagnold's theory. *Sedimentology* **30**, 485–92.

Lehner, P. 1969. Salt tectonics and Pleistocene stratigraphy on continental slope of northern Gulf of Mexico. *Bull. Am. Assoc. Petrolm Geol.* **53**, 2431–79.

Leitch, E. C. 1984. Marginal basins of the SW Pacific and the preservation and recognition of their ancient analogues: a review. in *Marginal basin geology*, B. P. Kokelaar & M. F. Howells (eds), 97–108. Geol Soc. (London) Spec. Publn 16. Oxford: Blackwell Scientific.

Leggett, J. K. 1980. The sedimentological evolution of a Lower Palaeozoic accretionary fore-arc in the Southern Uplands of Scotland. *Sedimentology* **27**, 401–17.

Leggett, J. K. (ed.) 1982. *Trench–forearc geology*. Geol Soc. (London) Spec. Publn 10. Oxford: Blackwell Scientific.

Leggett, J. K. 1985. Deep-sea pelagic sediments and palaeo-oceanography: a review of recent progress. In *Sedimentology recent developments and applied aspects*, P. J. Brenchley & B. P. J. Williams (eds), 95–121. Geol Soc. (London) Spec. Publn 18. Oxford: Blackwell Scientific.

Leggett, J. K., Y. Aoki & T. Toba 1985. Transition from frontal accretion to underplating in a part of the Nankai Trough accretionary complex off Shikoku (SW Japan) and extensional features on the lower trench slope. *Marine Petrolm Geology* **2**, 131–41.

Leggett, J. K., N. Lundberg, C. J. Bray, J. P. Cadet, D. E. Karig, R. J. Knipe & R. von Huene 1987. Extensional tectonics in the Honshu fore-arc, Japan: integrated results of DSDP Legs 57, 87 and reprocessed multichannel seismic reflection profiles. In *Continental extensional tectonics*, M. P. Coward, J. F. Dewey & P. L. Hancock (eds), 593–609. Geol Soc. (London) Spec. Publn 28. Oxford: Blackwell Scientific.

Leggett, J. K., W. S. McKerrow & D. M. Casey 1982. The anatomy of a Lower Palaeozoic accretionary forearc: the Southern Uplands of Scotland. In *Trench–forearc geology*, J. K. Leggett (ed.), 495–520. Geol Soc. (London) Spec. Publn 10. Oxford: Blackwell Scientific.

Letouzey, J. & M. Kimura 1985. Okinawa Trough genesis: structure and evolution of a backarc basin developed in a continent. *Marine Petrolm Geology* **2**, 111–30.

Letouzey, J. & M. Kimura 1986. The Okinawa Trough: genesis of a back-arc basin developing along a continental margin. *Tectonophysics* **125**, 209–30.

Levell, B. K. 1980a. A late Precambrian tidal shelf deposit, the Lower Sandfjord Formation, Finnmark, North Norway. *Sedimentology* **27**, 539–57.

Levell, B. K. 1980b. Evidence for currents associated with waves in Late Precambrian shelf deposits from Finnmark, North Norway. *Sedimentology* **27**, 153–66.

Levell, B. K. & D. Roberts 1977. A re-interpretation of the geology of north-west Varanger Peninsula, East Finnmark, North Norway. *Norges Geologiske Undersokelse* **334**, 83–90.

Leventer, A., D. F. Williams & J. P. Kennett 1983. Relationships between anoxia, glacial meltwater and microfossil preservation in the Orca Basin, Gulf of Mexico. *Marine Geology* **53**, 23–40.

Lewis, K. B. 1971. Slumping on a continental slope inclined at 1°–4°. *Sedimentology* **16**, 97–110.

Lewis, K. B., D. J. Bennett, R. H. Herzer & C. C. von der Borch 1986. Seismic stratigraphy and structure adjacent to an evolving plate boundary, western Chatham Rise, New Zealand. In *Initial Reports Deep Sea Drilling Project 90*, J. P. Kennett, C. C. von der Borch *et al.*, 1325–37. Washington, DC: US Government Printing Office.

Lindsay, J. F. 1968. The development of clast fabric in mudflows. *J. Sed. Petrol.* **38**, 1242–53.

Lindsey, D. A. 1969. Glacial sedimentology of the Precambrian Gowganda Formation, Ontario, Canada. *Bull. Geol Soc. Am.* **80**, 1685–702.

Lindsey, D. A. 1971. Glacial marine sediments in the Precambrian Gowganda Formation at Whitefish Falls, Ontario (Canada). *Palaeogeog. Palaeoclimatol. Palaeoecol.* **9**, 7–25.

Lisitzin, A. P. (ed.) 1972. *Sedimentation in the world ocean*. Soc. Econ. Paleont. Mineral. Spec. Publn 17.

Lohmann, G. P. 1978a. Response of the deep sea to ice ages. *Oceanus* **21**, 58–64.

Lohmann, G. P. 1978b. Abyssal benthonic foraminifera as hydrographic indicators in the western South Atlantic Ocean. *J. Foraminiferal Res.* **8**, 6–34.

Long, D. G. F. 1977. Resedimented conglomerate of Huronian (Lower Aphebian) age, from the north shore of Lake Huron, Ontario, Canada. *Can. J. Earth Sci.* **14**, 2495–509.

Lonsdale, P. 1978. Ecuadorian Subduction System. *Bull. Am. Assoc. Petrolm Geol.* **62**, 2454–77.

Lovell, J. P. B. 1969. Tyee Formation: a study of proximality in turbidites. *J. Sed. Petrol.* **39**, 935–53.

Lovell, J. P. B. 1978. Cenozoic. In *Introduction to the petroleum geology of the North Sea*, 2nd edn, K. W. Glennie (ed.), 179–96. Oxford: Blackwell Scientific.

Lovell, J. P. B. & D. A. V. Stow 1981. Identification of ancient sandy contourites. *Geology* **9**, 347–9.

Lowe, D. R. 1975. Water escape structures in coarse grained sediments. *Sedimentology* **22**, 157–204.

Lowe, D. R. 1976a. Grain flow and grain flow deposits. *J. Sed. Petrol.* **46**, 188–99.

Lowe, D. R. 1976b. Subaqueous liquefied and fluidized sediment flows and their deposits. *Sedimentology* **23**, 285–308.

Lowe, D. R. 1982. Sediment gravity flows: II. Depositional models with special reference to the deposits of high-density turbidity currents. *J. Sed. Petrol.* **52**, 279–97.

Lowe, D. R. 1985. Ouachita trough: part of a Cambrian failed rift system. *Geology* **13**, 790–3.

Lowe, D. R. & R. D. LoPiccolo 1974. The characteristics and origins of dish and pillar structures. *J. Sed. Petrol.* **44**, 484–501.

Lu, R. S. & K. J. McMillen 1983. Multichannel seismic survey of the Columbia Basin and adjacent margins. In *Studies in continental margin geology*, J. S. Watkins & C. L. Drake (eds), 395–410. Am. Assoc. Petrolm Geol. Mem. 34.

Lundberg, N. 1982. Evolution of the slope landward of the Middle America Trench, Nicoya Peninsula, Costa Rica. In *Trench–forearc geology*, J. K. Leggett (ed.), 131–47. Geol Soc. (London) Spec. Publn 10. Oxford: Blackwell Scientific.

Lundegard, P. D., N. D. Samuels & W. A. Pryor 1980. *Sedimentology, petrology and gas potential of the Brallier Formation–Upper Devonian turbidite facies of the central and southern Appalachians*. Rept. DOE/METC/5201-5, US Dept. Energy.

Luskin, B., B. C. Heezen, M. Ewing & M. Landisman 1954. Precision measurement of ocean depth. *Deep-Sea Res.* **1**, 131–40.

Lüthi, S. 1981. Experiments on non-channelized turbidity currents and their deposits. *Marine Geology* **40**, M59–M68.

Macpherson, B. A. 1978. Sedimentation and trapping mechanism in upper Miocene Stevens and older turbidite fans of southeastern San Joaquin Valley, California. *Bull. Am. Assoc. Petrolm Geol.* **62**, 2243–74.

Maher, C. E. & S. D. Harker 1987. Claymore oil field. In *Petroleum geology of north west Europe*, J. Brooks & K. W. Glennie (eds), 835–45. London: Graham & Trotman.

Maiklem, W. C. 1968. Some hydraulic properties of bioclastic carbonate grains. *Sedimentology* **10**, 101–9.

Malouta, D. N., D. S. Gorsline & S. E. Thornton 1981. Processes and rates of Recent (Holocene) basin filling in an active transform margin: Santa Monica Basin, California Continental Borderland. *J. Sed. Petrol.* **51**, 1077–95.

Mammerickx, J. 1970. Morphology of the Aleutian Abyssal Plain. *Bull. Geol Soc. Am.* **81**, 3457–64.

Mantyla, A. W. & J. L. Reid 1983. Abyssal characteristics of the World Ocean waters. *Deep-Sea Res.* **30**, 805–33.

Marjanac, T. 1985. Composition and origin of the megabed containing huge clasts, flysch formation, middle Dalmatia, Yugoslavia. In *Abstracts and poster abstracts*, 270–3. 6th European Regional Meeting, Lleida, Spain. Int. Assoc. Sedimentologists.

Marschalko, R. 1964. Sedimentary structures and paleocurrents in the marginal lithofacies of the central-Carpathian flysch. In *Turbidites*, A. H. Bouma & A. Brouwer (eds), 106–26. Developments in Sedimentology 3. Amsterdam: Elsevier.

Marschalko, R. 1975. Depositional environment of conglomerate as interpreted from sedimentological studies (Paleogene of Klippen Belt and adjacent tectonic units in East Slovakia). *Nauka o Zemi, Geologica* **10** (English summary).

Martin, R. G. 1978. Northern and eastern Gulf of Mexico continental margin: stratigraphic and structural framework. In *Framework, facies, and oil-trapping characteristics of the upper continental margin*, A. H. Bouma, G. T. Moore & J. M. Coleman (eds), 21–42. Am. Assoc. Petrolm Geol. Studies in Geology 7.

Martin, R. G. & A. H. Bouma 1978. Physiography of Gulf of Mexico. In *Framework, facies, and oil-trapping characteristics of the upper continental margin*, A. H. Bouma, G. T. Moore & J. M. Coleman (eds), 3–19. Am. Assoc. Petrolm Geol. Studies in Geology 7.

Martini, I. P. & M. Sagri 1977. Sedimentary fillings of ancient deep-sea channels: two examples from Northern Apennines (Italy). *J. Sed. Petrol.* **47**, 1542–53.

Mascle, A. & P. A. Biscarrat 1979. The Sulu Sea: A marginal basin in southeast Asia. In *Geological and geophysical investigations of continental margins*, J. S. Watkins, L. Montadert & P. W. Dickerson (eds), 373–81. Am. Assoc. Petrolm Geol. Mem. 29.

Mascle, G. & J. Mascle 1983/1984. Accretion and tectono-sedimentary progradation: the Taranto Gulf (southern Italy). *Geo-Marine Letts* **3**, 9–12.

Mascle, A., J. C. Moore *et al.* 1988. *Proc. ODP, Initial Reports (Pt. A), Leg 110.* College Station, Texas, Ocean Drilling Program.

Mauffret, A. & L. Montadert 1987. Rift tectonics on the passive continental margin off Galicia (Spain). *Marine Petrolm Geology* **4**, 49–70.

Maurin, J. C., J. Benkhelil & R. Robineau 1986. Fault rocks of the Kaltunga lineament, NE Nigeria, and their relationship with Benue Trough tectonics. *J. Geol. Soc. (London)* **143**, 587–99.

May, J. A., J. E. Warme & R. A. Slater 1983. Role of submarine canyons on shelfbreak erosion and sedimentation: modern and ancient examples. In *The shelfbreak: critical interface on continental margins*, D. J. Stanley & G. T. Moore (eds), 315–32. Soc. Econ. Paleont. Mineral. Spec. Publn 33.

Maynard, J. B., R. Valloni & H. Yu 1982. Composition of modern deep sea sands from arc-related basins. In *Trench–forearc geology: sedimentation and tectonics on modern and ancient active plate margins*, J. K. Leggett (ed.), 551–61. Geol Soc. (London) Spec. Publn 10. Oxford: Blackwell Scientific.

McCabe, P. J. 1978. The Kinderscoutian Delta (Carboniferous) of northern England; a slope influenced by density currents. In *Sedimentation in submarine canyons, fans, and trenches*, D. J. Stanley & G. Kelling (eds), 116–26. Stroudsburg, PA: Dowden, Hutchinson & Ross.

McCave, I. N. 1972. Transport and escape of fine-grained sediment from shelf areas. In *Shelf sediment transport: process and pattern*, D. J. P. Swift, D. B. Duane & O. H. Pilkey (eds), 225–48. Stroudsburg, PA: Hutchinson & Ross.

McCave, I. N. 1982. Erosion and deposition by currents on submarine slopes. *Bull. Institute Geol. Bassin d'Aquitaine* **31**, 47–55.

McCave, I. N. 1983. Particulate size spectra, behaviour and origin of nepheloid layers over the Nova Scotia continental rise. *J. Geophys. Res.* **88**, 7647–66.

McCave, I. N. 1984. Erosion, transport and deposition of fine-grained marine sediments. In *Fine-grained sediments: deep-water processes and facies*, D. A. V. Stow & D. J. W. Piper (eds), 35–69. Geol Soc. (London) Spec. Publn 15. Oxford: Blackwell Scientific.

McCave, I. N. 1985. Recent shelf clastic sediments. In *Sedimentology: recent developments and applied aspects*, P. J. Brenchley & B. P. J. Williams (eds) 49–65. Geol Soc. (London) Spec. Publn 18. Oxford: Blackwell Scientific.

McCave, I. N. & P. N. Jones 1988. Deposition of ungraded muds from high-density non-turbulent turbidity currents. *Nature* **333**, 250–2.

McCave, I. N., P. F. Lonsdale, C. D. Hollister & W. D. Gardner 1980. Sediment transport over the Hatton and Gardar contourite drifts. *J. Sed. Petrol.* **50**, 1049–62.

McCave, I. N. & B. E. Tucholke 1986. Deep current-controlled sedimentation in the western North Atlantic. In *The Geology of North America. Volume M, the western North Atlantic region*, P. R. Vogt & B. E. Tucholke (eds), 451–68. Boulder, Colorado: Geol Soc. Am.

McGrail, D. W. & M. Carnes 1983. Shelfedge dynamics and the nepheloid layer in the northwestern Gulf of Mexico. In *The shelfbreak: critical interface on continental margins*, D. J. Stanley & G. T. Moore (eds), 251–64. Soc. Econ. Paleont. Mineral. Spec. Publn 33.

McGregor, B. A., W. L. Stubblefield, W. B. F. Ryan & D. C. Twichell 1982. Wilmington submarine canyon: a marine fluvial-like system. *Geology*, **10**, 27–30.

McIlreath, I. A. & N. P. James 1984. Carbonate slopes. In *Facies models*, R. G. Walker (ed.), 2nd edn, 245–57. Geoscience Canada Reprint Series 1.

McKenzie, D. P. 1978. Some remarks on the development of sedimentary basins. *Earth Planet. Sci. Letts* **40**, 25–32.

McKenzie, J. A., H. C. Jenkyns & G. G. Bennet 1979. Stable isotope study of the cyclic diatomite-claystones from the Tripoli Formation, Sicily: a prelude to the Messinian salinity crisis. *Palaeogeog. Palaeoclimatol. Palaeoecol.* **29**, 125–41.

McLean, H. & D. G. Howell 1984. Miocene Blanca Fan, northern Channel Islands, California: small fans reflecting tectonism and

volcanism. *Geo-Marine Letts* **3**, 161–6.
McMillen, K. J., R. H. Enkeboll, J. C. Moore, T. H. Shipley & J. W. Ladd 1982. Sedimentation in different environments of the Middle America Trench, southern Mexico and Guatemala. In *Trench–forearc geology*, J. K. Leggett (ed.), 107–19. Geol Soc. (London) Spec. Publn 10. Oxford: Blackwell Scientific.
Menard, H. W. 1955. Deep-sea channels, topography, and sedimentation. *Bull. Am. Assoc. Petrolm Geol.* **39**, 236–55.
Menard, H. W., S. M. Smith & R. M. Pratt 1965. The Rhône deep-sea fan. In *Submarine geology and geophysics*, W. F. Whittard & R. Bradshaw (eds), 271–85. London: Butterworths.
Miall, A. D. 1983. Glaciomarine sedimentation in the Gowganda Formation (Huronian), northern Ontario. *J. Sed. Petrol.* **53**, 477–91.
Miall, A. D. 1985. Sedimentation on an early Proterozoic continental margin under glaical influence: the Gowganda Formation (Huronian), Elliott Lake area, Ontario, Canada. *Sedimentology*, **32**, 763–88.
Miall, A. D. 1986. Eustatic sea level changes interpreted from seismic stratigraphy: a critique of the methodology with particular reference to the North Sea Jurassic record. *Bull. Am. Assoc. Petrolm Geol.* **70**, 131–7.
Michelson, J. E. 1976. Miocene deltaic oil habitat, Trinidad. *Bull. Am. Assoc. Petrolm Geol.* **60**, 1502–19.
Middleton, G. V. 1965. Antidune cross-bedding in a large flume. *J. Sed. Petrol.* **35**, 922–7.
Middleton, G. V. 1966a. Experiments on density and turbidity currents: I. Motion of the head. *Can. J. Earth Sci.* **3**, 523–46.
Middleton, G. V. 1966b. Experiments on density and turbidity currents: II. Uniform flow of density currents. *Can. J. Earth Sci.* **3**, 627–37.
Middleton, G. V. 1966c. Small scale models of turbidity currents and the criterion for auto-suspension. *J. Sed. Petrol.* **36**, 202–8.
Middleton, G. V. 1967. Experiments on density and turbidity currents: III. Deposition of sediment. *Can. J. Earth Sci.* **4**, 475–505.
Middleton, G. V. 1970. Experimental studies related to problems of flysch sedimentation. In *Flysch sedimentology in North America*, J. Lajoie (ed.), 253–72. Geol. Assoc. Can. Spec. Paper 7. Business & Economic Service.
Middleton, G. V. 1973. Johannes Walther's Law of correlation of facies. *Bull. Geol. Soc. Am.* **84**, 979–88.
Middleton, G. V. 1976. Hydraulic interpretation of sand size distributions. *J. Geology* **84**, 405–26.
Middleton, G. V. & A. H. Bouma (eds) 1973. *Turbidites and deep-water sedimentation.* Los Angeles, California: Pacific Section Soc. Econ. Paleont. Mineral.
Middleton, G. V. & M. A. Hampton 1973. Sediment gravity flows: mechanics of flow and deposition. In *Turbidites and deep water sedimentation*, G. V. Middleton & A. H. Bouma (eds), 1–38. Short course notes, Pacific Sect. Soc. Econ. Paleont. Mineral.
Middleton, G. V. & J. B. Southard 1984. *Mechanics of sediment transport.* 2nd edn Soc. Econ. Paleont. Mineral. Eastern Section Short Course No. 3, Providence.
Milankovitch, M. 1941. Kannon der Erdbestrahlung und seine Anwendung auf das Eiszeitenproblem. *Royal Serb. Acad. Spec. Publn* **133**, 1–633. Belgrade. English translation published 1969, Israel Program for Scientific Translation, US Department Comm.
Mitchell, A. & H. G. Reading 1986. Sedimentation and tectonics. In *Sedimentary environments and facies*, H. G. Reading (ed.), 471–519. Oxford: Blackwell Scientific.
Mitchum, R. M. Jr 1985. Seismic stratigraphic expression of submarine fans. In *Seismic stratigraphy II: an integrated approach to hydrocarbon exploration*, O. R. Berg & D. G. Woolverton (eds), 117–36. Am. Assoc. Petrolm Geol. Mem. 39.
Mogi, A. & K. Nishizawa 1980. Breakdown of a seamount on the slope of the Japan trench. *Proc. Japanese Acad.* **56**, 257–9.
Molnar, P. & P. Tapponnier 1975. Cenozoic tectonics of Asia: effects of continental collision. *Science* **189**, 419–26.
Moore, D. G. 1961. Submarine slumps. *J. Sed. Petrol.* **31**, 343–57.
Moore, D. G. 1965. Erosional channel wall in La Jolla sea-fan valley seen from bathyscope Trieste II. *Bull. Geol Soc. Am.* **76**, 385–92.
Moore, D. G. 1969. Reflection profiling studies of the California Continental Borderland: structure and Quaternary turbidite basins. *Geol. Soc. Am. Spec. Paper* **107**.
Moore, G. F. & K. E. Karig 1976. Development of sedimentary basins on the lower trench slope. *Geology* **4**, 693–7.
Moore, G. F., J. R. Curray & F. J. Emmel 1982. Sedimentation in the Sunda Trench and forearc region. In *Trench–forearc geology*, J. K. Leggett (ed.), 245–58. Geol Soc. (London) Spec. Publn 10. Oxford: Blackwell Scientific.
Moore, G. H. & W. A. Wallis 1943. Time series tests based on signs of differences. *J. Am. Statist. Assoc.* **38**, 153–64.
Moore, G. T. & H. O. Woodbury 1978. Mississippi Fan – morphology, sedimentational history, petroleum potential. *10th Offshore Technology Conference Paper* **3093**, 391–8.
Moore, G. T., G. W. Starke, L. C. Bonham & H. O. Woodbury 1978. Mississippi fan, Gulf of Mexico – physiography, stratigraphy and sedimentational patterns. In *Framework, facies and oil-trapping characteristics of the upper continental margin*, A. H. Bouma, G. T. Moore & J. M. Coleman (eds), 155–91. Am. Assoc. Petrolm Geol. Studies in Geology 7.
Moore, J. C. 1972. Uplifted trench sediments: southwestern Alaska-Bering shelf edge. *Science* **175**, 1103–5.
Moore, J. C. 1974. Turbidites and terrigenous muds, DSDP Leg 25. In *Initial Reports Deep Sea Drilling Project 25*, E. S. W. Simpson, R. Schlich *et al.*, 441–79. Washington, DC: US Government Printing Office.
Moore, J. C. (ed.) 1986. *Structural fabric in Deep Sea Drilling Project cores from forearcs.* Geol Soc. Am. Mem. 166.
Moore, J. C., B. Biju-Duval, J. H. Natland & the Leg 78A Shipboard Scientific Party 1984. Offscraping and underthrusting of sediment at the deformation front of the Barbados Ridge: an introduction to the drilling results of Leg 78A and explanatory notes. In *Initial Reports Deep Sea Drilling Project 78A*, B. Biju-Duval, J. C. Moore *et al.*, 5–22. Washington, DC: US Government Printing Office.
Moore, J. C., A. Mascle, E. Taylor, P. Andreieff, F. Alvarez, R. Barnes, C. Beck, J. Behrmann, G. Blanc, K. Brown, M. Clark, J. Dolan, A. Fisher, J. Gieskes, M. Hounslow, P. McClellan, K. Moran, Y. Ogawa, T. Sakai, J. Schoomaker, P. Vrolijk, R. Wilkens & C. Williams 1987. Expulsion of fluids from depth along a subduction-zone decollement horizon. *Nature* **326**, 785–8.
Moore, J. C., S. Roeske, N. Lundberg, J. Schoomaker, D. S. Cowan, E. Gonzales & S. E. Lucas 1986. Scaly fabrics from Deep Sea Drilling Project cores from forearcs. In *Structural fabric in Deep Sea Drilling Project cores from forearcs*, J. C. Moore (ed.), 55–73. Geol Soc. Am. Mem. 166.
Moore, J. C., J. S. Watkins, K. J. McMillen, S. B. Bachman, J. K. Leggett, N. Lundberg, T. H. Shipley, J.-F. Stephan, F. W. Beghtel, A. Butt, B. M. Didyk, M. Nitsuma, L. E.

Shephard & H. Stradner 1982. Facies belts of the Middle America Trench and forearc region, southern Mexico: results from Leg 66 DSDP. In *Trench–forearc geology*, J. K. Leggett (ed.), 77–94. Geol Soc. (London) Spec. Publn 10. Oxford: Blackwell Scientific.

Morgan, S. R. & K. M. Campion 1987. Eustatic controls on stratification and facies associations in deep-water deposits, Great Valley Sequence, Sacramento Valley, California, abstract. *Bull. Am. Assoc. Petrolm Geol.* **71**, 595.

Morgenstern, N. R. 1967. Submarine slumping and the initiation of turbidity currents. In *Marine geotechnique*, A. F. Richards (ed.), 189–220. Urbana: Illinois University Press.

Morris, R. C. 1971. Classification and interpretation of disturbed bedding types in the Jackfork flysch rocks (Upper Mississippian), Ouachita Mountains, Arkansas. *J. Sed. Petrol.* **41**, 410–24.

Mosher, D. C. & R. N. Hiscott (in press.) Wisconsian proglacial sedimentation west of Verrill Canyon, Scotian slope. *Marine Geology*.

Mountain, G. S. & B. E. Tucholke 1983. Abyssal sediment waves. In *Seismic expression of structural styles*, Vol. 1, A. W. Bally (ed.), 1.2.5–22 to 1.2.5–24. Am. Assoc. Petrolm Geol. Studies in Geology 15.

Mountain, G. S. & B. E. Tucholke 1985. Mesozoic and Cenozoic geology of the U.S. Atlantic continental slope and rise. In *Geological evolution of the United States Atlantic margin*, C. W. Poag (ed.), 293–341. New York: Van Nostrand Reinhold.

Mudie, J. D., J. A. Grow & J. S. Bessey 1972. A near-bottom survey of lineated abyssal hills in the equatorial Pacific. *Marine Geophys. Res.* **1**, 397–411.

Mullins, H. T. & H. E. Cook 1986. Carbonate apron models: alternatives to the submarine fan model for paleoenvironmental analysis and hydrocarbon exploration. *Sed. Geology* **48**, 37–79.

Mullins, H. T. & A. C. Neumann 1979. Deep carbonate bank margin structure and sedimentation in the northern Bahamas. In *Geology of continental margins*, L. J. Doyle & O. H. Pilkey (eds), 165–92. Soc. Econ. Paleont. Mineral. Spec. Publn 27.

Mullins, H. T., G. H. Keller, J. W. Kofoed, D. N. Lambert, W. L. Stubblefield & J. E. Warme 1982. Geology of Great Abaco submarine canyon (Blake Plateau): observations from the research submersible 'Alvin'. *Marine Geology* **48**, 239–57.

Murray, J. & A. F. Rennard 1891. Report on deep-sea deposits based on the specimens collected during the voyage of HMS *Challenger* in the years 1872–1876. In *Challenger Reports*, London: Government Printer.

Mutti, E. 1974. Examples of ancient deep-sea fan deposits from circum-Mediterranean geosynclines. In *Modern and ancient geosynclinal sedimentation*, R. H. Dott Jr & R. H. Shaver (eds), 92–105. Soc. Econ. Paleont. Mineral. Spec. Publn 19.

Mutti, E. 1977. Distinctive thin-bedded turbidite facies and related depositional environments in the Eocene Hecho Group (south-central Pyrenees, Spain). *Sedimentology* **24**, 107–31.

Mutti, E. 1979. Turbidites et cones sous-marins profonds. In *Sedimentation detritique* (fluviatile, littorale et marine), P. Homewood (ed.), 353–419. Switzerland: Institut Geologique University Fribourg.

Mutti, E. 1984. The Hecho Eocene submarine-fan system, south-central Pyrenees, Spain. *Geo-Marine Letts* **3**, 199–202.

Mutti, E. 1985. Turbidite systems and their relations to depositional sequences. In *Provenance of arenites*, G. G. Zuffa (ed.), 65–93. NATO Advanced Scientific Institute. Dordrecht, Holland: D. Reidel.

Mutti, E. & G. Ghibaudo 1972. Un esempio di torbiditi di conoide sottomarina esterna: le Arenarie di San Salvatore (Formazione di Bobbio, Miocene) nell 'Appennino di Piacenza. *Mem. Acc. Sci. Torino Classe Sci. Fis. Nat. Ser.* **4**, no. 16.

Mutti, E. & D. R. Johns 1979. The role of sedimentary by-passing in the genesis of basin plain and fan fringe turbidites in the Hecho Group System (South-Central Pyrenees). *Mem. Soc. Geol. Italy* **18**, 15–22.

Mutti, E. & W. R. Normark 1987. Comparing examples of modern and ancient turbidite systems: problems and concepts. In *Marine clastic sedimentology*, J. K. Leggett & G. G. Zuffa (eds), 1–38. London: Graham & Trotman.

Mutti, E. & F. Ricci Lucchi 1972. Le torbiditi dell'Apennino settentrionale: introduzione all'analisi di facies. *Mem. Soc. Geol. Italy* **11**, 161–99. (1978 English translation by T. H. Nilsen, *Int. Geol. Rev.* **20**, 125–66).

Mutti, E. & F. Ricci Lucchi 1974. La signification de certaines unites sequentielles dans les séries à turbidites. *Bull. Soc. Geol. France* **16**, 577–82.

Mutti, E. & F. Ricci Lucchi 1975. Turbidite facies and facies associations. In *Examples of turbidite facies and facies associations from selected formations of the Northern Apennines, field trip guidebook A–11, 21–36.* IX Int. Congress Sedimentologists, Nice, France. Int. Assoc. Sedimentologists.

Mutti, E. & F. Ricci Lucchi 1978. Turbidites of the Northern Apennines: introduction to facies analysis. *Int. Geol. Rev.* **20**, 125–66.

Mutti, E. & M. Sonnino 1981. Compensation cycles: a diagnostic feature of turbidite sandstone lobes. In *Abstracts Volume*, 120–123. 2nd European Regional Meeting, Bologna, Italy. Int. Assoc. Sedimentologists.

Mutti, E. *et al.*, 1972. Schema stratigrafico e lineamenti di facies del Paleogene marino dell zona centrale sudpirenaica tra Tremp (Catalogna) e Pamplona (Navarra). *Memorie Società Geologica Italiana* **11**, 391–416.

Mutti, E., T. H. Nilsen & F. Ricci Lucchi 1978. Outer fan depositional lobes of the Laga Formation (upper Miocene and lower Pliocene), east-central Italy. In *Sedimentation in submarine canyons, fans, and trenches*, D. J. Stanley & G. Kelling (eds), 210–23. Stroudsburg, PA: Dowden, Hutchinson & Ross.

Mutti, E., E. Remacha, M. Sgavetti, J. Rosell, R. Valloni & M. Zamorano 1985. Stratigraphy and facies characteristics of the Eocene Hecho Group turbidite systems, south-central Pyrenees. In *Field-trip guide book of the 6th European Meeting Int. Assoc. Sedimentologists, Lleida, Spain*, excursion no. 12, 521–600.

Mutti, E., M. Sgavetti & E. Remacha 1984. Le relazioni tra piattaforme deltizie e sistemi torbiditici nel Bacino Eocenico Sud-pirenaico di Tremp-Pamplona. *Giorn. Geol.* **46**, 3–32.

Myers, R. A. 1986. Late Cenozoic sedimentation in the Northern Labrador Sea: a seismic-stratigraphic analysis. M.Sc. thesis. Halifax: Dalhousie University.

Myers, R. A. & D. J. W. Piper (in press). Late Cenozoic sedimentation in the Northern Labrador Sea: a history of bottom circulation and glaciation. *Can. J. Earth Sci.*

Nagel, D. K. & H. T. Mullins 1983. Late Cenozoic offset and uplift along the San Gregorio Fault zone: central California continental margin. In *Tectonics and sedimentation along faults of the San Andreas system*, D. W. Anderson & M. J. Rymer (eds), 91–103. Soc. Econ. Paleont. Mineral. Symp. Pacific Section.

Nagel, D. K., H. T. Mullins & H. G. Greene 1986. Ascension submarine canyon, California – evolution of a multi-head canyon system along a strike-slip continental margin. *Marine Geology* **73**, 285–310.

Nardin, T. R. 1983. Late Quaternary depositional systems and sea level changes – Santa Monica and San Pedro Basins, California Continental Borderland. *Bull. Am. Assoc. Petrolm Geol.* **67**, 1104–24.

Nardin, T. R. & T. L. Henyey 1978. Pliocene-Pleistocene diastrophism of Santa Monica and San Pedro shelves, California Continental Borderland. *Bull. Am. Assoc. Petrolm Geol.* **62**, 247–72.

Nardin, T. R., B. D. Edwards & D. S. Gorsline 1979. Santa Cruz Basin, California Borderland: dominance of slope processes in basin sedimentation. In *Geology of continental slopes*, L. J. Doyle & O. H. Pilkey (eds), 209–21. Soc. Econ. Paleont. Mineral. Spec. Publn 27.

Nardin, T. R., F. J. Hein, D. S. Gorsline & B. D. Edwards 1979. A review of mass movement processes, sediment and acoustic characteristics and contrasts in slope and base-of-slope systems versus canyon-fan-basin floor systems. In *Geology of continental slopes*, L. J. Doyle & O. H. Pilkey (eds), 61–73. Soc. Econ. Paleont. Mineral. Spec. Publn 27.

Nasu, N. *et al.* (eds) 1985. *Formation of active ocean margins.* Tokyo: Terrapub.

Natland, M. L. & Ph. H. Kuenen 1951. Sedimentary history of the Ventura Basin, California, and the action of turbidity currents. In *Turbidity currents and the transportation of coarse sediment into deep water*, J. L. Hough (ed.), 76–107. Soc. Econ. Paleont. Mineral. Spec. Publn 2.

Naylor, M. A. 1980. The origin of inverse grading in muddy debris flow deposits – a review. *J. Sed. Petrol.* **50**, 1111–16.

Naylor, M. A. 1982. The Casanova Complex of the northern Apennines: a mélange formed on a distal passive continental margin. *J. Structural Geology* **4**, 1–18.

Nederlof, F. H. 1959. Structure and sedimentology of the Upper Carboniferous of the upper Pisuega valleys, Cantabrian Mountains, Spain. *Leidse Geol. Meded.* **24**, 603–703.

Nelson, C. H. 1976. Late Pleistocene and Holocene depositional trends, processes, and history of Astoria deep-sea fan, northeast Pacific. *Marine Geology* **20**, 129–73.

Nelson, C. H. & L. D. Kulm 1973. Submarine fans and deep-sea channels. In *Turbidites and deep-water sedimentation*, G. V. Middleton & A. H. Bouma (eds), 39–78. Soc. Econ. Paleont. Mineral. Pacific Section Short Course Notes, Anaheim.

Nelson, C. H., A. Maldonado, F. Coumes, H. Got & A. Monaco 1984. The Ebro deep-sea fan system. *Geo-Marine Letts* **3**, 125–32.

Nelson, C. H., E. Mutti & F. Ricci Lucchi 1975. Comparison of proximal and distal thin-bedded turbidites with current-winnowed deep-sea sands. *9th Int. Congress Sedimentology, Nice, France* **2**, 317–24.

Nelson, C. H., W. R. Normark, A. H. Bouma & P. R. Carlson 1978. Thin-bedded turbidites in modern submarine canyons and fans. In *Sedimentation in submarine canyons, fans, and trenches*, D. J. Stanley & G. Kelling (eds), 177–89. Stroudsburg, PA: Dowden, Hutchinson & Ross.

Nelson, K. D. 1981. Mélange development in the Boones Point Complex, north-central Newfoundland. *Can. J. Earth Sci.* **18**, 433–42.

Nelson, T. A. & D. J. Stanley 1984. Variable depositional rates on the slope and rise off the Mid-Atlantic states. *Geo-Marine Letts* **3**, 37–42.

Nemec, W., S. J. Porebski & R. J. Steel 1980. Texture and structure of resedimented conglomerates – examples from Ksaiz Formation (Famennian–Tournaisian), southwestern Poland. *Sedimentology* **27**, 519–38.

Nemec, W. & R. J. Steel 1984. Alluvial and coastal conglomerates: their significant features and some comments on gravelly mass-flow deposits. In *Sedimentology of gravels and conglomerates*, E. H. Koster & R. J. Steel (eds), 1–31. Calgary: Can. Soc. Petrolm Geol. Mem. 10.

Niem, A. R. 1976. Patterns of flysch deposition and deep-sea fans in the lower Stanley Group (Mississippian), Ouachita Mountains, Oklahoma and Arkansas. *J. Sed. Petrol.* **46**, 633–46.

Nilsen, T. H. 1978. Turbidites of the northern Apennines: Introduction to facies analysis; translation of Mutti, E. & F. Ricci Lucchi 1972. Le torbiditi dell'Appennino settentrionale: introduzione all'analisi di facies. Mem. dell'Soc. Geol. Italy 11, 161–99. *Am. Geol. Institute Reprint Ser.* **3**, 127–66.

Nilsen, T. H. & M. H. Link 1975. Stratigraphy, sedimentology and offset along the San Andreas fault of Eocene to Lower Miocene strata of the northern Santa Lucia Range and the San Emigdio Mountains, Coast Ranges, central California. In *Paleogene symposium, conference in future horizons of the Pacific Coast*, D. S. Weaver, G. R. Hornaday & A. Tipton (eds), 367–400.

Nilsen, T. H. & G. G. Zuffa 1982. The Chugach Terrane, a Cretaceous trench-fill deposit, southern Alaska. In *Trench–forearc geology*, J. K. Leggett (ed.), 213–27. Geol Soc. (London) Spec. Publn 10. Oxford: Blackwell Scientific.

Normark, W. R. 1970. Growth patterns of deep-sea fans. *Bull. Am. Assoc. Petrolm Geol.* **54**, 2170–95.

Normark, W. R. 1978. Fan valleys, channels, and depositional lobes on modern submarine fans: characters for recognition of sandy turbidite environments. *Bull. Am. Assoc. Petrolm Geol.* **62**, 912–31.

Normark, W. R. & F. H. Dickson 1976. Man-made turbidity currents in Lake Superior. *Sedimentology* **23**, 815–32.

Normark, W. R. & D. J. W. Piper 1972. Sediments and growth pattern of Navy deep-sea fan, San Clemente Basin, California Borderland. *J. Geology* **80**, 192–223.

Normark, W. R. & D. J. W. Piper 1984. Navy Fan, California Borderland: growth pattern and depositional processes. *Geo-Marine Letts* **3**, 101–8.

Normark, W. R., N. E. Barnes & F. Coumes 1984. Rhône deep sea fan: a review. *Geo-Marine Letts* **3**, 155–60.

Normark, W. R., G. R. Hess, D. A. V. Stow & A. J. Bowen 1980. Sediment waves on the Monterey Fan levee: a preliminary physical interpretation. *Marine Geology* **37**, 1–18.

Normark, W. R., D. J. W. Piper & G. R. Hess 1979. Distributary channels, sand lobes, and mesotopography of Navy Submarine Fan, California Borderland, with applications to ancient fan sediments. *Sedimentology* **26**, 749–74.

Norris, R. J., R. M. Carter & I. M. Turnbull 1978. Cainozoic sedimentation in basins adjacent to a major continental transform boundary in southern New Zealand. *J. Geol Soc. (London)* **135**, 191–205.

Nowell, A. R. M. & C. D. Hollister (eds) 1985. Deep ocean sediment transport – preliminary results of the high energy benthic boundary layer experiment. *Spec. Issue, Marine Geology* **66**.

Oberhansli, H. & K. J. Hsü 1986. Paleocene-Eocene paleoceanography. In *Mesozoic and Cenozoic oceans*, K. J. Hsü (ed.), 85–100. Geodynamics Ser. 15. Boulder, Colorado: Geol Soc. Am.

Odin, G. S. 1982. *Numerical dating in stratigraphy.* Chichester, England: Wiley.

Ogawa, Y. 1982. Tectonics of some forearc fold belts in and around the arc–arc crossing area in central Japan. In *Trench–*

forearc geology, J. K. Leggett (ed.), 49–61. Geol Soc. (London) Spec. Publn 10. Oxford: Blackwell Scientific.

Ogawa, Y. 1983. Mineoka ophiolite belt in the Izu forearc area–Neogene accretion of oceanic and island arc assemblages on the northeastern corner of the Philippine Sea Plate. In *Accretion tectonics in the circum-Pacific regions*, M. Hashimoto & S. Uyeda (eds), 245–60. Tokyo: Terrapub.

Ogawa, Y. 1985. Variety of subduction and accretion processes in Cretaceous to Recent plate boundaries around southwest and central Japan. *Tectonophysics* **112**, 493–518.

Opdyke, N. D. & J. H. Foster 1970. Palaeomagnetism of cores from the North Pacific. In *Geological investigations of the North Pacific*, J. D. Hayes (ed.), 83–120. Geol Soc. Am. Mem. 126.

Ori, G. G. & P. F. Friend 1984. Sedimentary basins, formed and carried piggyback on active thrust sheets. *Geology* **12**, 475–8.

Ori, G. G., M. Roveri & F. Vannoni 1986. Plio-Pleistocene sedimentation in the Apenninic–Adriatic foredeep (central Adriatic Sea, Italy). In *Foreland basins*, P. A. Allen & P. Homewood (eds), 183–98. Int. Assoc. Sedimentologists Spec. Publn 8. Oxford: Blackwell Scientific.

Orwig, T. L. 1981. Channeled turbidites in the Eastern Central Pacific. *Marine Geology* **39**, 33–57.

Ovenshine, A. T. 1970. Observations of iceberg rafting in Glacier Bay, Alaska, and the identification of ancient ice-rafted deposits. *Bull. Geol Soc. Am.* **81**, 891–4.

Page, B. M. & J. Suppe 1981. The Pliocene Lichi mélange of Taiwan: its plate-tectonic and olistostromal origin. *Am. J. Sci.* **281**, 193–227.

Pantin, H. M. 1979. Interaction between velocity and effective density in turbidity flow: phase plane analysis, with criteria for autosuspension. *Marine Geology* **31**, 59–99.

Pantin, H. M. & M. R. Leeder 1987. Reverse flow in turbidity currents: the role of internal solitons. *Sedimentology* **34**, 1143–55.

Parea, G. C. 1975. The calcareous turbidite formations of the Northern Apennines. In *Examples of turbidite facies and facies associations from selected formations of the Northern Apennines, field trip guidebook A–11*, 52–62. IXth Int. Congress Sedimentologists, Nice, France.

Paola, C. & J. B. Southard 1983. Autosuspension and the energetics of two-phase flows: reply to comments on 'experimental test of autosuspension' by J. B. Southard & M. E. Mackintosh. *Earth Surface Processes Landforms* **8**, 273–9.

Paquet, J. 1969. Etude géologique de l'ouest de la province de Murcie (Espagne). *Soc. Geol. France Mem.* **48**, No. 111.

Parker, G. 1982. Conditions for the ignition of catastrophically erosive turbidity currents. *Marine Geology* **46**, 307–27.

Pautot, G., K. Nakamura, P. Huchon, J. Angelier, J. Bourgois, K. Fujioka, T. Kanazawa, Y. Nakamura, Y. Ogawa, M. Seguret & A. Takeuchi 1987. Deep-sea submersible survey in the Suruga, Sagami and Japan Trenches: preliminary results of the 1985 Kaiko cruise, Leg 2. *Earth Planet. Sci. Letts* **83**, 300–12.

Pettijohn, F. J. & P. E. Potter 1964. *Atlas and glossary of primary sedimentary structures*. Berlin: Springer.

Picha, F. 1979. Ancient submarine canyons of Tethyan continental margins, Czechoslovakia. *Bull. Am. Assoc. Petrolm Geol.* **63**, 67–86.

Pickering, K. T. 1979. Possible retrogressive flow slide deposits from the Kongsfjord Formation; a Precambrian submarine fan, Finnmark, N. Norway. *Sedimentology* **26**, 295–306.

Pickering, K. T. 1981a. Two types of outer fan lobe sequence, from the late Precambrian Kongsfjord Formation submarine fan, Finnmark, north Norway. *J. Sed. Petrol.* **51**, 1277–86.

Pickering, K. T. 1981b. The Kongsfjord Formation – a late Precambrian submarine fan in north-east Finnmark, north Norway. *Norges Geologiske Unders.* **367**, 77–104.

Pickering, K. T. 1982a. Middle-fan deposits from the late Precambrian Kongsfjord Formation submarine fan, northeast Finnmark, northern Norway. *Sed. Geology* **33**, 79–110.

Pickering, K. T. 1982b. A Precambrian upper basin-slope and prodelta in northeast Finnmark, north Norway – a possible ancient upper continental slope. *J. Sed. Petrol.* **52**, 171–86.

Pickering, K. T. 1982c. The shape of deep-water siliciclastic systems: a discussion. *Geo-Marine Letts* **2**, 41–6.

Pickering, K. T. 1983a. Transitional submarine fan deposits from the late Precambrian Kongsfjord Formation submarine fan, NE Finnmark, N. Norway. *Sedimentology* **30**, 181–99.

Pickering, K. T. 1983b. Small scale syn-sedimentary faults in the Upper Jurassic 'Boulder Beds'. *Scottish J. Geology* **19**, 169–81.

Pickering, K. T. 1984a. The Upper Jurassic 'Boulder Beds' and related deposits: a fault-controlled submarine slope, NE Scotland. *J. Geol Soc. (London)* **141**, 357–74.

Pickering, K. T. 1984b. Facies, facies-associations and sediment transport/deposition processes in a late Precambrian upper basin-slope/pro-delta, Finnmark, N. Norway. In *Fine-grained sediments: deep-water processes and facies*, D. A. V. Stow & D. J. W. Piper (eds), 343–62. Geol Soc. (London) Spec. Publn 15. Oxford: Blackwell Scientific.

Pickering, K. T. 1985. Kongsfjord turbidite system, Norway. In *Submarine fans and related turbidite systems*, A. H. Bouma, N. E. Barnes & W. R. Normark (eds), 237–44. New York: Springer.

Pickering, K. T. 1987a. Wet-sediment deformation in the Upper Ordovician Point Leamington Formation: an active thrust-imbricate system during sedimentation, Notre Dame Bay, north-central Newfoundland. In *Deformation of sediments and sedimentary rocks*, M. E. Jones & R. M. F. Preston (eds), 213–39. Geol Soc. (London) Spec. Publn 29. Oxford: Blackwell Scientific.

Pickering, K. T. 1987b. Deep-marine foreland basin and forearc sedimentation: a comparative study from the Lower Palaeozoic Northern Appalachians, Quebec and Newfoundland. In *Marine clastic sedimentology*, J. K. Leggett & G. G. Zuffa (eds), 190–211. London: Graham & Trotman.

Pickering, K. T. & R. N. Hiscott 1985. Contained (reflected) turbidity currents from the Middle Ordovician Cloridorme Formation, Quebec, Canada: an alternative to the antidune hypothesis. *Sedimentology* **32**, 373–94.

Pickering, K. T., S. Agar & Y. Ogawa 1988a. Genesis and deformation of mud injections containing chaotic basalt–limestone–chert associations: examples from the southwest Japan forearc. *Geology*, **16**, 881–5.

Pickering, K. T., M. G. Bassett & D. J. Siveter 1988b. Late Ordovician–early Silurian destruction of the Iapetus Ocean: Newfoundland, British Isles and Scandinavia: a discussion. *Trans. R. Soc. Edinburgh: Earth Sci*, **79**, 361–82.

Pickering, K. T., J. Coleman, M. Cremer, L. Droz, B. Kohl, W. Normark, S. O'Connell, D. Stow & A. Meyer-Wright 1986b. A high sinuosity, laterally migrating submarine fan channel-levee-overbank: results from DSDP Leg 96 on the Mississippi Fan, Gulf of Mexico. *Marine Petrolm Geology* **3**, 3–18.

Pickering, K. T., D. A. V. Stow, M. Watson & R. N. Hiscott 1986a. Deep-water facies, processes and models: a review and classification scheme for modern and ancient sediments. *Earth Sci. Rev.* **22**, 75–174.

Pierson, T. C. 1981. Dominant particle support mechanisms in debris flows at Mt. Thomas, New Zealand, and implications for flow mobility. *Sedimentology* **28**, 49–60.

Piper, D. J. W. 1970. A Silurian deep-sea fan deposit in western Ireland and its bearing on the nature of turbidity currents. *J. Geology* **78**, 509–22.

Piper, D. J. W. 1972a. Turbidite origin of some laminated mudstones. *Geol. Mag.* **109**, 115–26.

Piper, D. J. W. 1972b. Sediments of the Middle Cambrian Burgess Shale, Canada. *Lethaia* **5**, 169–75.

Piper, D. J. W. 1973. The sedimentology of silt turbidites from the Gulf of Alaska. In *Initial Reports Deep Sea Drilling Project 18*, L. D. Kulm, R. von Huene *et al.*, 847–67. Washington, DC: US Government Printing Office.

Piper, D. J. W. 1975. A reconnaissance of the sedimentology of the lower Silurian mudstones, English Lake District. *Sedimentology* **22**, 623–30.

Piper, D. J. W. 1978. Turbidite muds and silts on deep-sea fans and abyssal plains. In *Sedimentation in submarine canyons, fans, and trenches*, D. J. Stanley & G. Kelling (eds), 163–76. Stroudsburg, PA: Dowden, Hutchinson & Ross.

Piper, D. J. W. & D. C. Brisco 1975. Deep-water continental-margin sedimentation, DSDP Leg 28, Antarctica. In *Initial Reports Deep Sea Drilling Project 28*, D. E. Hayes, L. A. Frakes *et al.*, 727–55. Washington, DC: US Government Printing Office.

Piper, D. J. W. & W. R. Normark 1982. Effects of the 1929 Grand Banks earthquake on the continental slope off eastern Canada. *Geol Surv. Canada Current Res. (Pt. B) Paper 82–1B*, 147–51.

Piper, D. J. W. & W. R. Normark 1983. Turbidite depositional patterns and flow characteristics, Navy Submarine Fan, California Borderland. *Sedimentology* **30**, 681–94.

Piper, D. J. W., W. R. Normark & J. C. Ingle 1976. The Rio Dell Formation: a Plio–Pleistocene basin slope deposit in northern California. *Sedimentology* **23**, 309–28.

Piper, D. J. W., A. G. Panagos & G. G. Pe 1978. Conglomeratic Miocene flysch, western Greece. *J. Sed. Petrol.* **48**, 117–25.

Piper, D. J. W., A. N. Shor, J. A. Farre, S. O'Connell & R. Jacobi 1985. Sediment slides and turbidity currents on the Laurentian Fan: sidescan sonar investigations near the epicenter of the 1929 Grand Banks earthquake. *Geology* **13**, 538–41.

Pitman III, W. C. 1978. Relationship between eustacy and stratigraphic sequences of passive margins. *Bull. Geol Soc. Am.* **89**, 1389–403.

Pitman III, W. C. 1979. The effect of eustatic sea level changes on stratigraphic sequences of Atlantic margins. *Am. Assoc. Petrolm Geol. Mem.* **29**, 453–60.

Pitman III, W. C. & X. Golovochenko 1983. The effect of sealevel change on the shelfedge and slope of passive margins. In *The shelfbreak: critical interface on continental margins*, D. J. Stanley & G. T. Moore (eds), 41–58. Soc. Econ. Paleont. Mineral. Spec. Publn 33.

Platt, J. P. 1986. Dynamics of orogenic wedges and the uplift of high-pressure metamorphic rocks. *Bull. Geol Soc. Am.* **97**, 1037–53.

Platt, J. P. & J. K. Leggett 1986. Stratal extension in thrust footwalls, Makran accretionary prism: implications for thrust tectonics. *Bull. Am. Assoc. Petrolm Geol.* **70**, 191–203.

Platt, J. P., J. K. Leggett, J. Young, H. Raza & S. Alam 1985. Large-scale sediment underplating in the Makran accretionary prism, southwest Pakistan. *Geology* **13**, 507–11.

Poag, C. W. 1980. Foraminiferal stratigraphy, palaeoenvironments, and depositional cycles in the outer Baltimore Canyon Trough. In *Geological studies of the COST No. B-3 well, United States mid-Atlantic continental slope area*, P. A. Scholle (ed.), 44–65. US Geol. Surv. Circular 833.

Poag, C. W. 1985. Depositional history and stratigraphic reference section for central Baltimore Canyon Trough. In *Geologic evolution of the United States Atlantic margin*, C. W. Poag (ed.), 343–65. New York: Van Nostrand-Reinhold.

Poag, C. W. 1987. The New Jersey transect: stratigraphic framework and depositional history of a sediment-rich passive margin. In *Initial Reports Deep Sea Drilling Project 95*, C. W. Poag, A. B. Watts *et al.*, 763–816. Washington, DC: US Government Printing Office.

Poag, C. W. & P. C. Valentine 1988. Mesozoic and Cenozoic stratigraphy of the United States Atlantic continental shelf and slope. In *The geology of North America*, Vol. 1–2, *The Atlantic continental margin*, R. E. Sheridan & J. A. Grow (eds), 67–85. Geol Soc. Am.

Poag, C. W. & A. B. Watts 1987. Background and objectives of the New Jersey transect: continental slope and upper rise. In *Initial Reports Deep Sea Drilling Project 95*, C. W. Poag, A. B. Watts *et al.*, 15–27. Washington, DC: US Government Printing Office.

Poag, C. W., A. B. Watts *et al.* 1987. *Initial Reports Deep Sea Drilling Project 95*. Washington, DC: US Government Printing Office.

Pond, S. & G. L. Picard 1978. *Introduction to dynamic oceanography*. Oxford: Pergamon.

Postma, G. 1984. Slumps and their deposits in fan delta front and slope. *Geology* **12**, 27–30.

Powers, D. W. & R. G. Easterling 1982. Improved methodology for using embedded Markov chains to describe cyclical sediments. *J. Sed. Petrol.* **52**, 913–23.

Prince, R. A. & L. D. Kulm 1975. Crustal rupture and the initiation of imbricate thrusting in the Peru-Chile Trench. *Bull. Geol Soc. Am.* **86**, 1639–53.

Prior, D. B. & J. B. Coleman 1982. Active slides and flow in underconsolidated marine sediments on the slopes of the Mississippi Delta. In *Submarine slides and other mass movements*. S. Saxov & J. K. Nieuwenhuis (eds), 21–50. New York: Plenum.

Prior, D. B., B. D. Bornhold & M. W. Johns 1984. Depositional characteristics of a submarine debris flow. *J. Geol.* **92**, 707–27.

Prior, D. B. & E. H. Doyle 1985. Intra-slope canyon morphology and its modification by rockfall processes, US Atlantic continental margin. *Marine Geology* **67**, 177–96.

Prior, D. B., W. J. Wiseman & R. Gilbert 1981. Submarine slope processes on a fan delta, Howe Sound, British Columbia. *Geo-Marine Letts* **1**, 85–90.

Puigdefabregas, C. & P. Souquet 1986. Tectono–sedimentary cycles and depositional sequences of the Mesozoic and Tertiary from the Pyrenees. *Tectonophysics* **129**, 173–203.

Pyle, T. E., J. W. Antoine, D. A. Fahlquist & W. R. Bryant 1969. Magnetic anomalies in Straits of Florida. *Bull. Am. Assoc. Petrolm Geol.* **73**, 196–9.

Quinlan, G. M. & C. Beaumont 1984. Appalachian thrusting, lithospheric flexure, and the Paleozoic stratigraphy of the Eastern Interior of North America. *Can. J. Earth Sci.* **21**, 973–96.

Ramsay, A. T. S. 1977. Sedimentological clues to palaeo-oceanography. In *Oceanic micropalaeontology*, A. T. S. Ramsay (ed.), 1371–453. London: Academic Press.

Rees, A. I. 1968. The production of preferred orientation in a concentrated dispersion of elongated and flattened grains. *J. Geol.* **76**, 457–65.

Reimnitz, E. 1971. Surf-beat origin for pulsating bottom currents in the Rio Balsas submarine canyon, Mexico. *Bull. Geol Soc. Am.* **82**, 81–90.

Reimnitz, E. & K. F. Bruder 1972. River discharge into an ice covered ocean and related sediment dispersal, Beaufort Sea, coast of Alaska. *Bull. Geol Soc. Am.* **83**, 861–66.

Renz, O., R. Lakeman & E. van der Meulen 1955. Submarine sliding in western Venezuela. *Bull. Am. Assoc. Petrolm Geol.* **39**, 2053–67.

Ricci Lucchi, F. 1969. Channelized deposits in the middle Miocene flysch of Romagna (Italy). *Giorn. Geol.* **36**, 203–82.

Ricci Lucchi, F. 1975a. Miocene palaeogeography and basin analysis in the Periadriatic Apennines. In *Geology of Italy*, C. Squyres (ed.), 5–111. Tripoli: Petrolm Exploration Society Libya.

Ricci Lucchi, F. 1975b. Depositional cycles in two turbidite formations of northern Apennines. *J. Sed. Petrol.* **45**, 1–43.

Ricci Lucchi, F. 1978. Turbidite dispersal in a Miocene deep-sea plain: the Marnoso-arenacea of the Northern Apennines. *Geol. Mijnb.* **57**, 550–76.

Ricci Lucchi, F. 1981. The Marnoso-arenacea: a migrating turbidite basin 'over-supplied' by a highly efficient dispersal system. In *Excursion guidebook with contributions on sedimentology of some Italian basins*, F. Ricci Lucchi (ed.), 231–75. 2nd European Regional Meeting, Bologna, Italy. Int. Assoc. Sedimentologists.

Ricci Lucchi, F. 1984. The deep-sea fan deposits of the Miocene Marnoso-arenacea Formation, northern Apennines. *Geo-Marine Letts* **3**, 203–10.

Ricci Lucchi, F. 1986. The Oligocene to Recent foreland basins of the northern Apennines. In *Foreland basins*, P. A. Allen & P. Homewood (eds), 105–39. Int. Assoc. Sedimentologists Spec. Publn 8. Oxford: Blackwell Scientific.

Ricci Lucchi, F. & G. G. Ori 1984. Orogenic clastic wedges of the Alps and Apennines (abstract). *Bull. Geol Soc. Am.* **64**, 798.

Ricci Lucchi, F. & G. G. Ori 1985. Field excursion D: syn-orogenic deposits of a migrating basin system in the NW Adriatic foreland: examples from Emilia–Romagna region, northern Apennines. In *International symposium on foreland basins, excursion guidebook*, P. A. Allen, P. Homewood & G. Williams (eds), 137–76. Cardiff, Wales: CSP.

Ricci Lucchi, F. & G. Pialli 1973. Apporti secondari nella Marnoso-arenacea; 1. Torbiditi di conoide e di pianura sottomarine a Est-Nordest di Perugia. *Bull. Soc. Geol. Italy* **92**, 669–712.

Ricci Lucchi, F. & E. Valmori 1980. Basin-wide turbidites in a Miocene, over-supplied deep-sea plain: a geometrical analysis. *Sedimentology* **27**, 241–70.

Richards, P. C., J. D. Ritchie & A. R. Thomson 1987. Evolution of deep-water climbing dunes in the Rockall Trough – implications for overflow currents across the Wyville–Thomson Ridge in the (?)Late Miocene. *Marine Geology* **76**, 177–83.

Richardson, M. J., M. Wimbush & L. Mayer 1981. Exceptionally strong near-bottom flows on the continental rise of Nova Scotia. *Science* **213**, 887–8.

Ricou, L. E., J. Dercourt, J. Geyssant, C. Grandjacquet, C. Lepvrier & B. Biju-Duval 1986. Geological constraints on the Alpine evolution of the Mediterranean Tethys. *Tectonophysics* **123**, 83–122.

Riddihough, R. P. 1984. Recent movements of the Juan de Fuca plate system. *J. Geophys. Res.* **89**, 6980–94.

Riedel, W. R. 1963. The preserved records: paleontology of pelagic sediments. In *The sea*, M. N. Hill (ed.), **3**, 866–87. New York: Wiley-Interscience.

Roberts, D. G. 1975. Marine geology of the Rockall Plateau and Trough. *Phil. Trans. R. Soc. London (A)* **278**, 447–509.

Roberts, D. G. & R. B. Kidd 1979. Abyssal sediment-wave fields on Feni Ridge, Rockall Trough: long range sonar studies. *Marine Geology* **33**, 175–91.

Roberts, H. H., D. W. Cratsley & T. Whelan 1976. *Stability of Mississippi delta sediments as evaluated by analysis of structural features in sediment borings*. Offshore Technical Conference Pap. OTC 2425.

Roberts, M. T. 1983. Seismic example of complex faulting from northwest shelf of Palawan, Philippines. In *Seismic expression of structural styles*, Vol. 3, A. W. Bally (ed.), 4.2–18 to 4.2–24. Am. Assoc. Petrolm Geol. Studies in Geology Series 15.

Robertson, A. F. 1987. The transition from a passive margin to an Upper Cretaceous foreland basin related to ophiolite emplacement in the Oman Mountains. *Bull. Am. Assoc. Petrolm Geol.* **99**, 633–53.

Robertson, A. H. F. 1976. Pelagic chalks and calciturbidites from the Lower Tertiary of the Troodos Massif, Cyprus. *J. Sed. Petrol.* **46**, 1007–16.

Robertson, A. H. F. 1984. Origin of varve-type lamination, graded claystones and limestone–shale 'couplets' in the Lower Cretaceous of the western North Atlantic. In *Fine-grained sediments: deep-water processes and facies*, D. A. V. Stow & D. J. W. Piper (eds), 437–52. Geol Soc. (London) Spec. Publn 15. Oxford: Blackwell Scientific.

Robertson, A. H. F. & J. G. Ogg 1986. Palaeoceanographic setting of the Callovian North Atlantic. In *North Atlantic palaeoceanography*, C. P. Summerhayes & N. J. Shackleton (eds), 283–98. Geol Soc. (London) Spec. Publn 21. Oxford: Blackwell Scientific.

Robertson, A. H. F. & N. H. Woodcock 1981. Bilelyeri Group, Antalya Complex: deposition on a Mesozoic passive continental margin, south-west Turkey. *Sedimentology* **28**, 381–99.

Rocheleau, M. & J. Lajoie 1974. Sedimentary structures in resedimented conglomerate of the Cambrian flysch, L'Islet, Quebec Appalachians. *J. Sed. Petrol.* **44**, 826–36.

Rodine, J. D. & A. M. Johnson 1976. The ability of debris, heavily freighted with coarse clastic materials, to flow, on gentle slopes. *Sedimentology* **23**, 213–34.

Rona, P. A. 1973. Relationships between rates of sediment accumulations on continental shelves, sea floor spreading and eustacy inferred from the central North Atlantic. *Bull. Geol Soc. Am.* **84**, 2851–72.

Ross, D. A. 1978. Summary of results of Black Sea drilling. In *Initial Reports Deep Sea Drilling Project 42 (Pt. 2)*, D. A. Ross, Y. P. Neprochnov *et al.*, 1149–78. Washington, DC: US Government Printing Office.

Ross, D. A., Y. P. Neprochnov *et al.* 1978. *Initial Reports Deep Sea Drilling Project 42 (Pt. 2)*. Washington, DC: US Government Printing Office.

Ross, D. A., E. Uchupi & C. O. Bowin 1974. Shallow structure of Black Sea. In *The Black Sea – geology, chemistry and biology*, E. T. Degens & D. A. Ross (eds), 11–34. Am. Assoc. Petrolm Geol. Mem. 20.

Ross, D. A., E. Uchupi & E. S. Trimonis 1974. Modern sedimentation in Black Sea. In *The Black Sea – geology, chemistry and biology*, E. T. Degens & D. A. Ross (eds), 249–78. Am. Assoc. Petrolm Geol. Mem. 20.

Rouse, H. 1937. Modern conceptions of the mechanics of turbulence. *Trans. Am. Soc. Civil Engineers* **102**, 436–505.

Rupke, N. A. 1975. Deposition of fine-grained sediments in the abyssal environment of the Algero-Balearic Basin, western Mediterranean Sea. *Sedimentology* **22**, 95–109.

Sadler, P. M. 1982. Bed-thickness and grain size of turbidites. *Sedimentology* **29**, 37–51.

Sagri, M. 1972. Rhythmic sedimentation in the turbidite sequences of the northern Apennines (Italy). *Proc. 24th Int. Geol. Congress* **6**, 82–7.

Sagri, M. 1974. Rhythmic sedimentation in the deep-sea carbonate turbidites (Monte Antola Formation, northern Apennines). *Bull. Soc. Geol. Italy* **93**, 1013–27.

Sakai, T. & K. Kanmera 1981. Stratigraphy of the Shimanto Terrain and tectono-stratigraphic setting of greenstones in the northern part of Miyazaki Prefecture, Kyushu. *Sci. Repts Dept. Geology, Kyushu University* **14**, 31–48.

Sanders, J. E. 1965. Primary sedimentary structures formed by turbidity currents and related resedimentation mechanisms. In *Primary sedimentary structures and their hydrodynamic interpretation*, G. V. Middleton (ed.), 192–219. Soc. Econ. Paleont. Mineral. Spec. Publn 12.

Sangree, J. B., D. C. Waylett, D. E. Frazier, G. B. Amery & W. J. Fennessy 1978. Recognition of continental-slope seismic facies, offshore Texas-Louisiana. In *Framework, facies and oil-trapping characteristics of the upper continental margin*, A. H. Bouma, G. T. Moore & J. M. Coleman (eds), 87–116. Am. Assoc. Petrolm Geol. Studies in Geology 7.

Saunders, A. D., G. Rogers, G. F. Marriner, D. J. Terrell & S. P. Verma 1987. Geochemistry of Cenozoic volcanic rocks, Baja California, Mexico: implications for the petrogenesis of post-subduction magmas. *J. Volcanol. Geotherm. Res.* **32**, 223–45.

Saxov, S. & J. K. Nieuwenhuis (eds) 1982. *Marine slides and other mass movements*. New York: Plenum.

Schafer, C. T. & K. W. Asprey 1982. Significance of some geotechnical properties of continental slope and rise sediments off northeast Newfoundland. *Can. J. Earth Sci.* **19**, 153–61.

Schlager, W., R. T. Buffler & Scientific Party of DSDP Leg 77 1984. DSDP Leg 77 – early history of the Gulf of Mexico. *Bull. Geol Soc. Am.* **95**, 226–36.

Schlanger, S. O., H. C. Jenkyns & I. Premoli-Silva 1981. Volcanism and vertical tectonics in the Pacific basin related to global Cretaceous transgressions. *Earth Planet. Sci. Letts* **52**, 435–49.

Schlee, J. S. 1981. Seismic stratigraphy of Baltimore Canyon Trough. *Bull. Am. Assoc. Petrolm Geol.* **65**, 26–53.

Schlee, J. S. & K. Hinz 1987. Seismic stratigraphy and facies of continental slope and rise seaward of Baltimore Canyon Trough. *Bull. Am. Assoc. Petrolm Geol.* **71**, 1046–67.

Schnitker, D. 1974. West Atlantic abyssal circulation during the past 12 000 years. *Nature* **248**, 385–7.

Schnitker, D. 1979. The deep waters of the western North Atlantic during the past 24 000 years, and the re-initiation of the Western Boundary Undercurrent. *Marine Micropaleontology* **4**, 265–80.

Scholl, D. W. 1974. Sedimentary sequences in the North Pacific trenches. In *The geology of continental margins*, C. A. Burk & C. L. Drake (eds), 493–504. New York: Springer.

Scholl, D. W., E. C. Buffington & D. M. Hopkins 1968. Geologic history of the continental margin of North America in the Bering Sea. *Marine Geology* **6**, 297–330.

Scholl, D. W. & J. S. Creager 1971. Deep Sea Drilling Project, Leg 19. *Geotimes*, November, 12–15.

Scholle, P. A. 1971. Sedimentology of fine-grained deep-water carbonate turbidites, Monte Antola Flysch (Upper Cretaceous), Northern Apennines, Italy. *Bull. Geol Soc. Am.* **82**, 629–58.

Scholle, P. A. & A. A. Ekdale 1983. Pelagic environments. In *Carbonate depositional environments*, P. A. Scholle, D. G. Bebout & C. H. Moore (eds), 620–91. Am. Assoc. Petrolm Geol Mem. 33.

Scholle, P. A., D. G. Bebout & C. H. Moore (eds) 1983. *Carbonate depositional environments*. Am. Assoc. Petrolm Geol Mem. 33.

Schrader, H. J. 1971. Fecal pellets: role in sedimentation of pelagic diatoms. *Science* **174**, 55–7.

Schwan, W. 1980. Geodynamic peaks in Alpinotype orogenies and changes in ocean-floor spreading during Late Jurassic–Late Tertiary time. *Bull. Am. Assoc. Petrolm Geol.* **64**, 350–72.

Schweller, W. J. & L. D. Kulm 1978. Depositional patterns and channelized sedimentation in active eastern Pacific trenches. In *Sedimentation in submarine canyons, fans, and trenches*, D. J. Stanley & G. Kelling (eds), 311–24. Stroudsburg, PA: Dowden, Hutchinson & Ross.

Scott, K. M. 1966. Sedimentology and dispersal pattern of a Cretaceous flysch sequence, Patagonian Andes, southern Chile. *Bull. Am. Assoc. Petrolm Geol.* **50**, 72–107.

Seed, H. B. & K. L. Lee 1966. Liquefaction of saturated sands during cyclic loading. *J. Soil Mechanics Found. Div., Am. Soc. Civil Engineers* **92**, 105–34.

Seely, D. R. 1979. The evolution of structural highs bordering major forearc basins. In *Geological and geophysical investigations of continental margin*, J. S. Watkins, L. Montadert & P. W. Dickerson (eds), 245–60. Am. Assoc. Petrolm Geol. Mem. 29.

Seely, D. R., P. R. Vail & G. G. Walton 1974. Trench slope model. In *The geology of continental margins*, C. A. Burk & C. L. Drake (eds), 249–60. New York: Springer.

Seilacher, A. 1964. Biogenic sedimentary structures. In *Approaches to palaeoecology*, J. Imbrie & N. D. Newell (eds), 296–313. New York: Wiley.

Seilacher, A. 1967. Bathymetry of trace fossils. *Marine Geology* **5**, 413–28.

Sequret, M., P. Labaume & R. Madariago 1984. Eocene seismicity in the Pyrenees from megaturbidites in the South Pyrenean basin (North Spain). *Marine Geology* **55**, 117–31.

Sestini, G. 1970. Flysch facies and turbidite sedimentology. *Sed. Geology* **4**, 559–97.

Shanmugam, G. 1980. Rhythms in deep sea, fine-grained turbidite and debris flow sequences, Middle Ordovician, eastern Tennessee. *Sedimentology* **27**, 419–32.

Shanmugam, G. & R. J. Moiola, 1982. Eustatic control of turbidites and winnowed turbidites. *Geology* **10**, 231–5.

Shanmugam, G. & R. J. Moiola 1985. Submarine fan models: problems and solutions. In *Submarine fans and related turbidite systems*, A. H. Bouma, W. R. Normark & N. E. Barnes (eds), 29–34. New York: Springer.

Shanmugam, G., J. E. Damuth & R. J. Moiola 1985. Is the turbidite facies association scheme valid for interpreting ancient submarine fan environments? *Geology* **13**, 234–7.

Shaub, F. J. 1983. Growth faults on the southwestern margin of the Gulf of Mexico. In *Seismic expression of structural styles 2*, A. W. Bally (ed.), 2.3.3–3. Am. Assoc. Petrolm Geol. Studies in Geology 15.

Shaub, F. J., R. T. Buffler & J. G. Parsons 1984. Seismic stratigraphic framework of deep central Gulf of Mexico Basin. *Bull. Am. Assoc. Petrolm Geol.* **68**, 1790–802.

Sheldon, P. G. 1928. Some sedimentation conditions in the Middle Portage rocks. *Am. J. Sci.* **15**, 243–52.

Shepard, F. P. 1933. Canyons beneath the seas. *Scientific Monthly* **37**, 31–9.

Shepard, F. P. 1951. Mass movements in submarine canyon heads. *Trans. Am. Geophys. Union* **32**, 405–18.

Shepard, F. P. 1955. Delta-front valleys bordering the Mississippi distributaries. *Bull. Geol Soc. Am.* **66**, 1489–98.

Shepard, F. P. 1963. *Submarine geology* (2nd edn). New York: Harper & Row.

Shepard, F. P. 1966. Meander in a valley crossing a deep-sea fan. *Science* **154**, 385–6.

Shepard, F. P. 1975. Progress of internal waves along submarine canyons. *Marine Geology* **19**, 131–8.

Shepard, F. P. 1976. Tidal components of currents in submarine canyons. *J. Geology* **84**, 343–50.

Shepard, F. P. 1977. *Geological oceanography: evolution of coasts, continental margins, and the deep-sea floor.* New York: Crane, Russak & Co.

Shepard, F. P. 1979. Currents in submarine canyons and other sea valleys. In *Geology of continental slopes*, L. J. Doyle & O. H. Pilkey (eds), 85–94. Soc. Econ. Paleont. Mineral. Spec. Publn 27.

Shepard, F. P. 1981. Submarine canyons: multiple causes and long-time persistence. *Bull. Am. Assoc. Petrolm Geol.* **65**, 1062–77.

Shepard, F. P. & R. F. Dill 1966. *Submarine canyons and other sea valleys.* Chicago: Rand McNally.

Shepard, F. P. & M. Einsele 1962. Sedimentation in San Diego Trough and contributing submarine canyons. *Sedimentology* **1**, 81–133.

Shepard, F. P. & K. O. Emery 1973. Congo submarine canyon and fan valley. *Bull. Am. Assoc. Petrolm Geol.* **57**, 1679–91.

Shepard, F. P. & N. F. Marshall 1969. Currents in La Jolla and Scripps submarine canyons. *Science* **165**, 177–8.

Shepard, F. P. & N. F. Marshall 1973a. Storm-generated current in a La Jolla submarine canyon, California. *Marine Geology* **15**, M19–M24.

Shepard, F. P. & N. F. Marshall 1973b. Currents along floors of submarine canyons. *Bull. Geol Soc. Am.* **57**, 244–64.

Shepard, F. P. & N. F. Marshall 1978. Currents in submarine canyons and other sea valleys. In *Sedimentation in submarine canyons, fans, and trenches*, D. J. Stanley & G. Kelling (eds), 3–14. Stroudsburg, PA: Dowden, Hutchinson & Ross.

Shepard, F. P., N. F. Marshall & P. A. McLoughlin 1974. Currents in submarine canyons. *Deep-Sea Res.* **21**, 691–706.

Shepard, F. P., N. F. Marshall & P. A. McLoughlin 1975. Pulsating turbidity currents with relationship to high swell and high tides. *Nature* **258**, 704–6.

Shepard, F. P., N. F. Marshall, P. A. McLoughlin & G. G. Sullivan 1979. *Currents in submarine canyons and other sea valleys.* Am. Assoc. Petrolm Geol. Studies in Geology 8.

Sheridan, R. E. 1986. Pulsation tectonics as the control on North Atlantic palaeoceanography. In *North Atlantic palaeoceanography*, C. P. Summerhayes & N. J. Shackleton (eds), 255–75. Geol Soc. (London) Spec. Publn 21. Oxford: Blackwell Scientific.

Sheridan, R. E., F. M. Gradstein *et al.* 1983. *Initial Reports Deep Sea Drilling Project 76.* Washington, DC: US Government Printing Office.

Shiki, T. & Y. Misawa 1982. Forearc geological structure of the Japanese Islands. In *Trench–forearc geology*, J. K. Leggett (ed.), 63–73. Geol Soc. (London) Spec. Publn 10. Oxford: Blackwell Scientific.

Shimamura, K. 1986. Topography and geological structure in the bottom of the Suruga Trough: a geological consideration of the subduction zone near the collisional plate boundary. *J. Geography Tokyo Geog. Soc.* **95**, 317–38.

Shimkus, K. M. & E. S. Trimonis 1974. Modern sedimentation in Black Sea. In *The Black Sea – geology, chemistry and biology*, E., T. Degens & D. A. Ross (eds), 249–78. Am. Assoc. Petrolm Geol. Mem. 20.

Shultz, A. W. 1984. Subaerial debris-flow deposition in the Upper Paleozoic Cutter Formation. *J. Sed. Petrol.* **54**, 759–72.

Sibuet, J.-C., W. W. Hay, A. Prunier, L. Montadert, K. Hinz & J. Fritsch 1984. Early evolution of the South Atlantic Ocean: role of the rifting episode. In *Initial Reports Deep Sea Drilling Project 75*, W. W. Hay, J.-C. Sibuet *et al.*, 469–81. Washington, DC: US Government Printing Office.

Siedlecka, A. 1978. Late Precambrian tidal-flat deposits and algal stromatolites in the Båtsfjord Formation, East Finnmark, North Norway. *Sed. Geology* **21**, 177–310.

Siedlecka, A. 1985. Development of the Upper Proterozoic sedimentary basins of the Varanger Peninsula, East Finnmark, North Norway. *Bull. Geol Survey Finland* **331**, 175–85.

Siedlecka, A. & M. B. Edwards 1980. Lithostratigraphy and sedimentation of the Riphean Båsnaering Formation, Varanger Peninsula, North Norway. *Norges Geologiske Undersokelse* **355**, 27–47.

Siedlecka, A. & S. Siedlecki 1967. Some new aspects of the geology of Varanger peninsula (Northern Norway). *Norges Geologiske Undersokelse* **247**, 288–306.

Siemers, C. T., R. W. Tillman & C. R. Williamson (eds) 1981. *Deep-water clastic sediments: a core workshop.* Core workshop 2. Tulsa, Oklahoma. Soc. Econ. Paleont. Mineral.

Sigurdsson, H., R. S. J. Sparks, S. Carey & T. C. Huang 1980. Volcanogenic sedimentation in the Lesser Antilles Arc. *J. Geology* **88**, 523–40.

Simons, D. B., E. V. Richardson & C. F. Nordin Jr 1965. Sedimentary structures generated by flow in alluvial channels. In *Primary sedimentary structures and their hydrodynamic interpretation*, G. V. Middleton (ed.), 34–52. Soc. Econ. Paleont. Mineral. Spec. Publn 27.

Simm, R. W. & R. B. Kidd 1984. Submarine debris flow deposits detected by long range side-scan sonar 1000 km from source. *Geo-Marine Letts* **3**, 13–16.

Simo, A., C. Puigdefabregas & E. Gili 1985. Transition from shelf to basin on an active slope. Upper Cretaceous Tremp area, Southern Pyrenees. *Int. Assoc. Sedimentologists 6th European Meeting Lleida Spain, Excursion guidebook 2*, 61–108.

Simpson, J. E. 1982. Gravity currents in the laboratory, atmosphere, and ocean. *Ann. Rev. Fluid Mechanics* **14**, 213–34.

Skipper, K. 1971. Antidune cross-stratification in a turbidite sequence, Cloridorme Formation, Gaspe, Quebec. *Sedimentology* **17**, 51–68.

Skipper, K. & S. B. Bhattacharjee 1978. Backset bedding in turbidites: a further example from the Cloridorme Formation (Middle Ordovician), Gaspe, Quebec. *J. Sed. Petrol.* **48**, 193–202.

Skipper, K. & G. V. Middleton 1975. The sedimentary structures and depositional mechanics of certain Ordovician turbidites, Cloridorme Formation, Gaspe Peninsula, Quebec. *Can. J. Earth Sci.* **12**, 1934–52.

Snyder, G. W. & B. Carson 1986. Bottom and suspended particle sizes: implications for modern sediment transport in Quinault submarine canyon. *Marine Geology* **71**, 85–105.

Socci, A. D. & G. W. Smith 1987a. Recent sedimentological interpretations in the Avalon Terrane of the Boston Basin, Massachusetts. *Maritime Sediments and Atlantic Geology* **23**.

Socci, A. D. & G. W. Smith 1987b. Evolution of the Boston Basin: a sedimentological perspective. In *Sedimentary basins and basin-forming mechanisms*, C. Beaumont & A. Tankard (ed.), 87–99. Can. Soc. Petrolm Geol. Mem. 12.

Soper, N. J. & D. H. W. Hutton 1984. Late Caledonian sinistral displacements in Britain: implications for a three-plate collision model. *Tectonics* **3**, 781–94.

Soper, N. J., B. C. Webb & N. H. Woodcock 1987. Late Caledonian transpression in north west England: timing, geometry and geotectonic significance. *Proc. Yorkshire Geol Soc.* **46**, 175–92.

Souquet, P., E. Debroas, J. M. Boire, Ph. Pons, C. Fixari, J. C. Roux, J. Dol, J. P. Thieuloy, M. Bonnemaison, H. Marivit & B. Peybernes 1985. Le Groupe du Flysch Noir (Albo-Cenomanien) dans les Pyrénées. *Bull. Cent. Rech. Explor. Prod. Elf-Aquitaine* **9**, 183–252.

Soutar, A., S. R. Johnson & T. R. Baumgarter 1981. In search of modern depositional analogs to the Monterey Formation. In *The Monterey Formation and related siliceous rocks of California*, R. E. Garrison & R. G. Douglas (eds), 123–47. Pacific Section Soc. Econ. Paleont. Mineral. Spec. Publn.

Southard, J. B. & M. E. Mackintosh 1981. Experimental test of autosuspension. *Earth Surface Processes Landforms* **6**, 103–11.

Spencer, J. W. 1903. Submarine valleys off the American coast and in the North Atlantic. *Bull. Geol Soc. Am.* **14**, 207–26.

Spencer, J. E. & W. R. Normark 1979. Tosco–Abreojos fault zone: a Neogene transform plate boundary within the Pacific margin of southern Baja California, Mexico. *Geology* **7**, 554–7.

Srivastava, S. P., M. A. Arthur, B. Clement *et al.*, 1987. Site 645. In *Proc., Initial Reports ODP (Pt A)* **105**, 61–418. College Station, Texas: Ocean Drilling Program.

Stanley, D. J. 1973. Basin plains in the Eastern Mediterranean: significance in interpreting ancient marine deposits, 1. Basin depth and configuration. *Marine Geology* **15**, 295–307.

Stanley, D. J. 1974. Pebbly mud transport in the head of Wilmington Canyon. *Marine Geology* **16**, M1–M8.

Stanley, D. J. 1981. Unifites: structureless muds of gravity-flow origin in Mediterranean basins. *Geo-Marine Letts* **1**, 77–83.

Stanley, D. J. 1982. Welded slump-graded sand couplets: evidence for slide generated turbidity currents. *Geo-Marine Letts* **2**, 149–55.

Stanley, D. J. 1988. *Turbidites reworked by bottom currents: Upper Cretaceous examples from St Croix, US Virgin Islands.* Smithsonian Contrib. *Marine Sci.* **33**, 1–79.

Stanley, D. J. & G. Kelling (eds) 1978. *Sedimentation in submarine canyons, fans, and trenches.* Stroudsburg, PA: Dowden, Hutchinson & Ross.

Stanley, D. J. & A. Maldonado 1981. Depositional models for fine-grained sediments in the western Hellenic Trench, eastern Mediterranean. *Sedimentology* **28**, 273–90.

Stanley, D. J. & G. T. Moore (eds) 1983. *The shelfbreak: critical interface on continental margins.* Soc. Econ. Paleont. Mineral. Spec. Publn 33.

Stanley, D. J. & R. Unrug 1972. Submarine channel deposits, fluxoturbidites and other indicators of slope and base-of-slope environments in modern and ancient marine basins. In *Recognition of ancient sedimentary environments*, J. K. Rigby & W. K. Hamblin (eds), 287–340. Soc. Econ. Paleont. Mineral. Spec. Publn 16.

Stanley, D. J., P. Fenner & G. Kelling 1972a. Currents and sediment transport at Wilmington Canyon shelf-break, as observed by underwater television. In *Shelf sediment transport: process and pattern*, D. J. P. Swift, D. B. Duane & O. H. Pilkey (eds), 630–41. Stroudsburg, PA: Dowden, Hutchinson & Ross.

Stanley, D. J., F. W. McCoy & L. Diester-Haass 1974. Balearic Abyssal Plain: an example of modern basin plain deformation by salt tectonism. *Marine Geology* **17**, 183–200.

Stanley, D. J., H. D. Palmer & R. F. Dill 1978. Coarse sediment transport by mass flow and turbidity current processes and downslope transformation in Annot Sandstone canyon-fan valley systems. In *Sedimentation in submarine canyons, fans and trenches*, D. J. Stanley & G. Kelling (eds), 85–115. Stroudsburg, PA: Dowden, Hutchinson & Ross.

Stanley, D. J., D. J. P. Swift, N. Silverberg, N. P. James & R. G. Sutton 1972b. *Late Quaternary progradation and sand 'spillover' on the outer continental margin off Nova Scotia, Southeast Canada:* Smithsonian Contrib. Earth Sci. 8.

Stanley, D. J., P. T. Taylor, H. Sheng & R. Stuckenrath 1981. *Sohm Abyssal Plain: evaluating proximal sediment provenance.* Smithsonian Contrib. Marine Sci. 11, 1–48.

Stanley, R. S. & N. M. Ratcliffe 1985. Tectonic synthesis of the Taconian Orogeny in western New England. *Bull. Geol Soc. Am.* **96**, 1227–50.

Stauffer, P. H. 1967. Grain flow deposits and their implications, Santa Ynez Mountains, California. *J. Sed. Petrol.* **37**, 487–508.

Steckler, M. S. & A. B. Watts 1978. Subsidence of the Atlantic-type continental margin off New York. *Earth Planet. Sci. Letts* **41**, 1–13.

Steiger, T. & L. F. Jansa 1984. Jurassic limestones of the seaward edge of the Mazagan carbonate platform, northwest African continental margin, Morocco. In *Initial Reports Deep Sea Drilling Project* 79, K. Hinz, E. L. Winterer *et al.*, 449–91. Washington, DC: US Government Printing Office.

Stelting, C. E., K. T. Pickering *et al.* & DSDP Leg 96 Shipboard Scientists 1985. Drilling results on the middle Mississippi Fan. In *Submarine fans and related turbidite systems*, A. H. Bouma, W. R. Normark & N. E. Barnes (eds), 283–90. New York: Springer.

Stevens, S. H. & G. F. Moore 1985. Deformational and sedimentary processes in trench slope basins of the western Sunda Arc, Indonesia. *Marine Geology* **69**, 93–112.

Stewart, I. J. 1987. A revised stratigraphic interpretation of the Early Palaeogene of the central North Sea. In *Petroleum geology of north west Europe*, J. Brooks & K. W. Glennie (eds), 557–76. London: Graham & Trotman.

Stommel, H. 1957. A survey of ocean currents theory. *Deep-Sea Res.* **4**, 149–84.

Stommel, H. 1958. The abyssal circulation. *Deep-Sea Res.* **5**, 80–2.

Stommel, H. & A. B. Aarons 1960a. On the abyssal circulation of the World ocean – I. Stationary planetary flow patterns on a sphere. *Deep-Sea Res.* **6**, 140–54.

Stommel, H. & A. B. Aarons 1960b. On the abyssal circulation of the World ocean – II. An idealized model of the circulation pattern and amplitude in oceanic basins. *Deep-Sea Res.* **6**, 217–33.

Stoneley, R. 1962. Marl diapirism near Gisbourne, New Zealand. *New Zealand J. Geology Geophys.* **5**, 630–41.

Stow, D. A. V. 1976. Deep water sands and silts on the Nova Scotian continental margin. *Maritime Sediments* **12**, 81–90.

Stow, D. A. V. 1979. Distinguishing between fine-grained turbidites and contourites on the Nova Scotian deep water margin. *Sedimentology* **26**, 371–87.
Stow, D. A. V. 1981. Laurentian Fan: morphology, sediments, processes and growth patterns. *Bull. Am. Assoc. Petrolm Geol.* **65**, 375–93.
Stow, D. A. V. 1982. Bottom currents and contourites in the North Atlantic. *Bull. Inst. Geol. Bassin d'Aquitaine* **31**, 151–66.
Stow, D. A. V. 1984. Turbidite facies associations and sequences in the southeastern Angola Basin. In *Initial Reports Deep Sea Drilling Project 75*, W. W. Hay, J. C. Sibuet *et al.*, 785–99. Washington, DC: US Government Printing Office.
Stow, D. A. V. 1985. Deep-sea clastics: where are we and where are we going? In *Sedimentology: recent developments and applied aspects*, P. J. Brenchley & B. P. J. Williams (eds), 67–93. Geol Soc. (London) Spec. Publn 18. Oxford: Blackwell Scientific.
Stow, D. A. V., 1986. Deep clastic seas. In *Sedimentary environments and facies* 2nd edn, H. G. Reading (ed.), 399–444. Oxford: Blackwell Scientific.
Stow, D. A. V. & A. J. Bowen 1980. A physical model for the transport and sorting of fine-grained sediments by turbidity currents. *Sedimentology* **27**, 31–46.
Stow, D. A. V. & W. E. Dean 1984. Middle Cretaceous black shales at Site 530 in the south-eastern Angola Basin. In *Initial Reports Deep Sea Drilling Project 75*, W. W. Hay, J. C. Sibuet *et al.*, 809–17. Washington, DC: US Government Printing Office.
Stow, D. A. V. & J. A. Holbrook 1984. North Atlantic contourites: an overview. In *Fine-grained sediments: deep-water processes and facies*, D. A. V. Stow & D. J. W. Piper (eds), 245–56. Geol Soc. (London) Spec. Publn 15. Oxford: Blackwell Scientific.
Stow, D. A. V. & J. P. B. Lovell 1979. Contourites: their recognition in modern and ancient sediments. *Earth Sci. Rev.* **14**, 251–91.
Stow, D. A. V. & D. J. W. Piper (eds) 1984a. *Fine-grained sediments: deep-water processes and facies*. Geol Soc. (London) Spec. Publn 15. Oxford: Blackwell Scientific.
Stow, D. A. V. & D. J. W. Piper 1984b. Deep-water fine-grained sediments: facies models. In *Fine-grained sediments: deep-water processes and facies*, D. A. V. Stow & D. J. W. Piper (eds), 611–46. Geol Soc. (London) Spec. Publn 15. Oxford: Blackwell Scientific.
Stow, D. A. V. & G. Shanmugam 1980. Sequence of structures in fine-grained turbidites: comparison of recent deep-sea and ancient flysch sediments. *Sed. Geology* **25**, 23–42.
Stow, D. A. V., C. D. Bishop & S. J. Mills 1982. Sedimentology of the Brae oilfield, North Sea: fan models and controls. *J. Petrolm Geology* **5**, 129–48.
Stow, D. A. V., D. G. Howell & C. H. Nelson 1984. Sedimentary, tectonic, and sea-level controls on submarine fan and slope-apron turbidite systems. *Geo-Marine Letts* **3**, 57–64.
Stow, D. A. V., D. G. Howell & C. H. Nelson 1985. Sedimentary, tectonic, and sea-level controls. In *Submarine fans and related turbidite systems*, A. H. Bouma, W. R. Normark & N. E. Barnes (eds), 15–22. New York: Springer.
Stride, A. H., R. H. Belderson & N. H. Kenyon 1982. Structural grain, mud volcanoes and other features on the Barbados Ridge Complex revealed by GLORIA long-range side-scan sonar. *Marine Geology* **49**, 187–96.
Strong, P. G. & R. G. Walker 1981. Deposition of the Cambrian continental rise: the St. Roch Formation near St. Jean-Port-Joli, Quebec. *Can. J. Earth Sci.* **18**, 1320–35.
Strong, S. W. S. 1931. Ejection of fault breccia in the Waimata survey district, Gisbourne. *New Zealand J. Sci. Technology* **12**, 257–67.
Stuart, C. J. & C. A. Caughey 1976. Form and composition of the Mississippi Fan. *Trans. Gulf Coast Assoc. Geol Societies* **26**, 333–43.
Stubblefield, W. L., B. A. McGregor, E. Forde, D. N. Lambert & G. F. Merrill 1981. Reconnaissance in DSRV ALVIN of a 'fluvial-like' meander system in Wilmington Canyon Trough (abstract). *Bull. Geol Soc. Am.* **7**, 538–9.
Stubblefield, W. L., B. A. McGregor, E. B. Forde, D. N. Lambert & G. F. Merrill 1982. Reconnaissance in DSRV Alvin of a 'fluvial-like' meander system in Wilmington Canyon and slump features in South Wilmington Canyon. *Geology* **10**, 31–6.
Sugimura, A. & S. Uyeda 1973. *Island arcs: Japan and its environs*. New York: Elsevier.
Sullwold H. H. Jr 1960. Tarzana fan, deep submarine fan of late Miocene age, Los Angeles County, California. *Bull. Am. Assoc. Petrolm Geol.* **44**, 433–57.
Summerhayes, C. P. 1979. *Marine geology of the New Zealand subAntarctic seafloor*. Bull. New Zealand Dep. Sci. Ind. Res. 190.
Summerhayes, C. P. & N. J. Shackleton (eds) 1986. *North Atlantic palaeoceanography*. Geol Soc. (London) Spec. Publn 21. Oxford: Blackwell Scientific.
Sunit, K. A. & H. Kagami 1979. Sedimentation in a closed trough north of the Iberia Abyssal Plain in the northeast Atlantic. *Sedimentology* **26**, 561–76.
Surlyk, F. 1973. The Jurassic-Cretaceous boundary in Jameson Land East Greenland. In *The Boreal Lower Cretaceous, Geol J. Spec. Issue 5*, 81–100. Liverpool: Steelhouse Press.
Surlyk, F. 1975. Block faulting and associated marine sedimentation at the Jurassic-Cretaceous boundary, East Greenland. In *Proceedings of the Jurassic Northern North Sea Symposium* **7**, K. G. Finstad & R. C. Selley (eds), 1–31. Norwegian Petroleum Society.
Surlyk, F. 1978. *Submarine fan sedimentation along fault scarps on tilted fault blocks (Jurassic-Cretaceous boundary, East Greenland)*. Bull. Grønlands Geologiske Unders. 128.
Surlyk, F. 1984. Fan-delta to submarine fan conglomerates of the Volgian-Valanginian Wollaston Foreland Group, East Greenland. In *Sedimentology of gravels and conglomerates*, E. H. Koster & R. J. Steel (eds), 359–82. Can. Soc. Petrolm Geol. Mem. 10.
Surlyk, F. 1987. Slope and deep shelf gully sandstones, Upper Jurassic, East Greenland. *Bull. Am. Assoc. Petrolm Geol.* **71**, 464–75.
Surlyk, F. & J. M. Hurst 1984. The evolution of the early Paleozoic deep-water basin of North Greenland. *Bull. Geol Soc. Am.* **95**, 131–54.
Surlyk, F., J. H. Callomon, R. G. Bromley & T. Birkelund 1973. *Stratigraphy of the Jurassic-Lower Cretaceous sediments of Jameson Land and Scoresby Land, East Greenland*. Bull. Grønlands Geologiske Unders. 105.
Suzuki, T. & S. Hada 1979. Cretaceous tectonic mélange of the Shimanto Belt in Shikoku, Japan. *J. Geol Soc. Japan* **85**, 467–79.
Sverdrup, M. U., M. W. Johnson & R. M. Fleming 1942. *The oceans, their physics, chemistry and general biology*. Englewood Cliffs, NJ: Prentice-Hall.
Swift, S. A., C. D. Hollister & R. S. Chandler 1985. Close-up stereo photographs of abyssal bedforms on the Nova Scotian continental rise. *Marine Geology* **66**, 303–22.

Syvitski, J. P. M. & G. E. Farrow 1988. Fjord sedimentation as an analogue for small hydrocarbon-bearing submarine fans. In *Deltas: sites and traps for fossil fuels*, M. K. G. Whateley & K. T. Pickering (eds). Geol Soc. (London) Spec. Publn 41. Oxford: Blackwell Scientific.

Syvitski, J. P. M., D. C. Burrell & J. M. Skei 1987. *Fjords: processes and products*. New York: Springer.

Taira, A. 1981. The Shimanto Belt of southwest Japan and arc–trench sedimentary tectonics. *Recent Prog. in Natural Sci. Japan* **6**, 147–62.

Taira, A. 1985. Sedimentary evolution of the Shikoku subduction zone: the Shimanto Belt and Nankai Trough. In *Formation of active margins*, N. Nasu *et al.* (eds), 835–51. Tokyo: Terrapub.

Taira, A. & N. Nitsuma 1986. Turbidite sedimentation in Nankai Trough as interpreted from magnetic fabric, grain size and detrital mode analysis. In *Initial Reports Deep Sea Drilling Project 87*, H. Kagan, D. E. Karig, W. T. Coulburn *et al.*, 611–32. Washington, DC: US Government Printing Office.

Taira, A., H. Okada, J. H. McD Whitaker & A. J. Smith 1982. The Shimanto Belt of Japan: Cretaceous–lower Miocene active-margin sedimentation. In *Trench–forearc geology*, J. K. Leggett (ed.), 5–26. Oxford: Blackwell Scientific.

Taira, A., Y. Saito & M. Hashimoto 1983. The role of oblique subduction and strike-slip tectonics in the evolution of Japan. In *Geodynamics of the western Pacific–Indonesian region*, 303–16. Am. Geophys. Union Geodynamics Series II.

Taira, A., M. Tashiro, M. Okamura & J. Katto 1980. The geology of the Shimanto Belt in the Kochi Prefecture, Shikoku, Japan. In *Geology and paleontology of the Shimanto Belt (Cretaceous)*, A. Taira & M. Tashiro (eds), 319–89. Kochi: Rinyakosaikai Press.

Takahashi, T. 1981. Debris flow. *Annual Rev. Fluid Mechanics* **13**, 57–77.

Talwani, K. & W. C. Pitman III (eds) 1977. Island arcs. In *Deep sea trenches and back-arc basins*. Maurice Ewing Series 1. Am. Geophys. Union.

Talwani, T., M. J. C. Mutter, R. Houtz & M. Konig 1979. The crustal structure and evolution of the area underlying the magnetic quiet zone on the margin south of Australia. In *Geological and geophysical investigations of continental margins*, J. S. Watkins, L. Montadert & P. W. Dickerson (eds), 151–75. Am. Assoc. Petrolm Geol. Mem. 29.

Talwani, T., G. Udinstev *et al.* 1976. *Initial Reports Deep Sea Drilling Project 38*. Washington, DC: US Government Printing Office.

Tamano, T., T. Toba & Y. Aoki 1985. Development of fore-arc continental margins and their potential for hydrocarbon accumulation. *Proc. World Petrolm Congress PD3(2)*, 1–11.

Tankard, A. J. & H. J. Welsink 1987. Extensional tectonics and stratigraphy of Hibernia oil field, Grand Banks, Newfoundland. *Bull. Am. Assoc. Petrolm Geol.* **71**, 1210–32.

Tazaki, K. & M. Inomata 1980. Umbers in pillow lava from the Mineoka tectonic belt, Boso Peninsula. *J. Geol Soc. Japan* **86**, 413–16.

Teale, T. 1985. Occurrence and geological significance of olistoliths from the Longobucco Group, Calabria, southern Italy. In *Abstracts and poster abstracts*, 457–60. 6th European Regional Meeting, Lleida, Spain, Int. Assoc. Sedimentologists.

Teraoka, Y. 1979. Provenance of the Shimanto geosynclinal sediments inferred from sandstone compositions. *J. Geol Soc. Japan* **83**, 795–810.

Thompson, A. F. & M. R. Thomasson 1969. Shallow to deep water facies development in the Dimple Limestone (Lower Pennsylvanian), Marathon region, Texas. In *Depositional environments in carbonate rocks*, G. M. Friedman (ed.), 57–78. Soc. Econ. Paleont. Mineral. Spec. Publn 14.

Thompson, C. W. 1878. *The voyage of the* Challenger, Vols I & II. New York: Harper & Bros.

Thornbjarnarson, K. W., C. A. Nittrouer & D. J. DeMaster 1986. Accumulation of modern sediment in Quinault submarine canyon. *Marine Geology* **71**, 107–24.

Thornburg, T. M. & L. D. Kulm 1987. Sedimentation in the Chile Trench: depositional morphologies, lithofacies, and stratigraphy. *Bull. Geol Soc. Am.* **98**, 33–52.

Thorne, J. & A. B. Watts 1984. Seismic reflectors and unconformities at passive continental margins. *Nature* **311**, 365–8.

Thornton, S. E. 1981. Suspended sediment transport in surface waters of the California Current off southern California: 1977–78 floods. *Geo-Marine Letts* **1**, 23–8.

Thornton, S. E. 1984. Basin model for hemipelagic sedimentation in a tectonically active continental margin: Santa Barbara Basin, California Continental Borderland. In *Fine-grained sediments: deep-water processes and facies*, D. A. V. Stow & D. J. W. Piper (eds), 377–94. Geol Soc. (London) Spec. Publn 15. Oxford: Blackwell Scientific.

Tiffin, D. L., B. E. B. Cameron & J. W. Murray 1972. Tectonics and depositional history of the continental margin off Vancouver Island, British Columbia. *Can. J. Earth Sci.* **9**, 280–96.

Tillman, R. W. & S. A. Ali (eds) 1982. Deep water canyons, fans and facies: models for stratigraphic trap exploration. *Reprint Ser.* **26**. Am. Assoc. Petrolm. Geol.

Tolmazin, D. 1985. *Elements of dynamic oceanography*. London: Allen & Unwin.

Tucholke, B. E. 1979. Furrows and focussed echoes on the Blake Outer Ridge. *Marine Geology* **31**, M13–M20.

Tucholke, B. E., C. D. Hollister, P. E. Biscaye & W. D. Gardner 1985. Abyssal current character determined from sediment bedforms on the Nova Scotian continental rise. *Marine Geology* **66**, 43–57.

Turner, C. C., J. M. Cohen, E. R. Connell & D. M. Cooper 1987. A depositional model for the South Brae oilfield. In *Petroleum geology of north west Europe*, J. Brooks & K. W. Glennie (eds), 853–64. London: Graham & Trotman.

Twichell, D. C. & D. G. Roberts 1982. Morphology, distribution, and development of submarine canyons on the United States Atlantic continental slope between Hudson and Baltimore Canyons. *Geology* **10**, 408–12.

Tyler, J. E. & N. H. Woodcock 1987. The Bailey Hill Formation: Ludlow Series turbidites in the Welsh Borderland reinterpreted as distal storm deposits. *Geol J.* **22**, 73–86.

Uchupi, E. & D. G. Aubrey 1988. Suspect terranes in the North American margins and relative sea-level trends. *J. Geology* **96**, 79–90.

Underwood, M. B. 1986. Transverse infilling of the Central Aleutian Trench by unconfined turbidity currents. *Geo-Marine Letts* **6**, 7–13.

Underwood, M. B. & S. B. Bachman 1982. Sedimentary facies associations within subduction complexes. In *Trench–forearc geology*, J. K. Leggett (ed.), 537–50. Geol Soc. (London) Spec. Publn 10. Oxford: Blackwell Scientific.

Underwood, M. B. & D. E. Karig 1980. Role of submarine canyons in trench and trench-slope sedimentation. *Geology* **8**, 432–6.

Unrug, R. 1963. Istebna Beds – a fluxoturbidite formation in the

Carpathian flysch. *Ann. Soc. Geol. Pologne* **33**, 49–92.

Uyeda, S. 1974. Northwest Pacific Trench Margins. In *The geology of continental margins*, C. A. Burk & C. L. Drake (eds), 473–91. New York: Springer.

Vail, P. R. & R. M. Mitchum Jr 1979. Global cycles of relative changes of sea level from seismic stratigraphy. In *Geological and geophysical investigations of continental margins*, J. S. Watkins, L. Montadert & P. W. Dickerson (eds), 469–72. Am. Assoc. Petrolm Geol. Mem. 29.

Vail, P. R. & R. G. Todd 1981. Northern North Sea Jurassic unconformities, chronostratigraphy and sea-level changes from seismic stratigraphy. In *Petroleum geology of the continental shelf of northwest Europe*, L. V. Illing & G. D. Hobson (eds), 216–35. London: Heyden.

Vail, P. R., J. Hardenbol & R. G. Todd 1984. Jurassic unconformities, chronostratigraphy, and sea-level changes from seismic stratigraphy and biostratigraphy. In *Inter-regional unconformities and hydrocarbon accumulation*, J. S. Schlee (ed.), 129–44. Am. Assoc. Petrolm Geol. Mem. 36.

Vail, P. R., R. M. Mitchum & S. Thompson III 1977. Global cycles of relative changes of sea level. In *Seismic stratigraphy–applications to hydrocarbon exploration*, C. E. Payton (ed.), 83–98. Am. Assoc. Petrolm Geol. Mem. 27.

Valloni, R. & G. Mazzadri 1984. Compositional suites of terrigenous deep-sea sands of the present continental margins. *Sedimentology* **31**, 353–64.

Van Andel, Tj. H. 1964. *Recent marine sediments of Gulf of California: a symposium*. Am. Assoc. Petrolm Geol. Mem. 3, 216–310.

Van Andel, Tj. H. & P. D. Komar 1969. Ponded sediments of the Mid-Atlantic Ridge between 22° and 23° north latitude. *Bull. Geol Soc. Am.* **80**, 1163–90.

Van Hinte, J. E. 1978a. A Cretaceous time scale. In *Contributions to the geologic time scale*, G. V. Cohee, M. F. Glaessner & H. D. Hedberg (eds), 269–87. Am. Assoc. Petrolm Geol. Studies in Geology 6.

Van Hinte, J. E. 1978b. A Jurassic time scale. In *Contributions to the geologic time scale*, G. V. Cohee, M. F. Glaessner & H. D. Hedberg (eds), 289–97. Am. Assoc. Petrolm Geol. Studies in Geology 6.

Van Hinte, J. E., S. W. Wise Jr, B. N. M. Biart, J. M. Covington, D. A. Dunn, J. A. Haggerty, M. W. Johns, P. A. Meyers, M. R. Moullade, J. P. Muza, J. G. Ogg, M. Okamura, M. Sarti & U. von Rad 1985a. DSDP Site 603: first deep (>1000 m) penetration of the continental rise along the passive margin of eastern North America. *Geology* **13**, 392–6.

Van Hinte, J. E., S. W. Wise Jr, B. N. M. Biart, J. M. Covington, D. A. Dunn, J. A. Haggerty, M. W. Johns, P. A. Meyers, M. R. Moullade, J. P. Muza, J. G. Ogg, M. Okamura, M. Sarti & U. von Rad 1985b. Deep-sea drilling on the upper continental rise of New Jersey, DSDP Sites 604 and 605. *Geology* **13**, 397–400.

Van Hinte, J. E., S. W. Wise *et al.* 1987. *Initial Reports Deep Sea Drilling Project 93.* Washington, DC: US Government Printing Office.

Van Hoorn, B. 1970. Sedimentology and paleography of an Upper Cretaceous turbidite basin in the south-central Pyrenees, Spain. *Leidse Geol. Meded.* **45**, 73–154.

Van der Kamp, P. C. & B. E. Leake 1985. Petrography and geochemistry of feldspathic and mafic sediments of the northeastern Pacific margin. *Trans. R. Soc. Edinburgh: Earth Sci.* **76**, 411–49.

Van der Lingen, G. J. 1969. The turbidite problem. *New Zealand J. Geology Geophys.* **12**, 7–50.

Van der Lingen, G. J. 1982. Development of the North Island subduction system, New Zealand. In *Trench–forearc geology*, J. K. Leggett (ed.), 259–72. Geol Soc. (London) Spec. Publn 10. Oxford: Blackwell Scientific.

Vanney, J.-R. & D. J. Stanley 1983. Shelfbreak physiography: an overview In *The shelfbreak: critical interface on continental margins*, D. J. Stanley & G. T. Moore (eds), 1–24. Soc. Econ. Paleont. Mineral. Spec. Publn 33.

Vassoevitch, N. B. 1948. *Flysch i Metodika jevo Izuchenija. Vses. Neft. Geol. Razved. Nauchn. Issled. Inst. Moskva* (French translation: Le Flysch et les Methodes de son Etude. Bur. Rech. Geol. Minières, Paris).

Veevers, J. J. *et al.* 1974. *Initial Reports Deep Sea Drilling Project 27.* Washington, DC: US Government Printing Office.

Visser, J. N. J. 1983. Submarine debris flow deposits from the Upper Carboniferous Dwyka Tillite Formation in the Kalahari Basin, South Africa. *Sedimentology* **30**, 511–23.

Visser, J. N. J. & J. C. Loock 1987. Ice margin influence on glaciomarine sedimentation in the Permo-Carboniferous Dwyka Formation from the southwestern Karoo, South Africa. *Sedimentology*, **34**, 929–42.

Von Huene, R. & M. A. Arthur 1982. Sedimentation across the Japan Trench off northern Honshu Island. In *Trench–forearc geology*, J. K. Legget (ed.), 27–48. Geol Soc. (London) Spec. Publn 10. Oxford: Blackwell Scientific.

Von Huene, R., J. Aubouin *et al.* 1982. A summary of Deep Sea Drilling Project Leg 67 shipboard results from the Mid-America Trench transect off Guatemala. In *Trench–forearc geology*, J. K. Leggett (ed.), 121–47. Geol Soc. (London) Spec. Publn 10. Oxford: Blackwell Scientific.

Von Huene, R., J. Aubouin *et al.* 1985. *Initial Reports Deep Sea Drilling Project 84.* Washington, DC: US Government Printing Office.

Von Huene, R., J. Miller, D. Taylor & D. Blackman 1985. A study of geophysical data along the Deep Sea Drilling Project active margin transect off Guatemala. In *Initial Reports Deep Sea Drilling Project 89*, R. von Huene, J. Aubouin *et al.*, 895–909. Washington, DC: US Government Printing Office.

Von Rad, U. & N. F. Exon 1983. Mesozoic–Cenozoic sedimentary and volcanic evolution of the starved passive continental margin off northwest Australia. In *Studies in continental margin geology*, J. S. Watkins & C. L. Drake (eds), 253–81. Am. Assoc. Petrolm Geol. Mem. 34.

Wagner, H. 1900. *Lehrbuch der Geographie*. Hanover: Hahn.

Walcott, R. I. 1970. Flexural rigidity, thickness and viscosity of the lithosphere. *J. Geophys. Res.* **75**, 3941–54.

Walcott, R. I. 1978. Geodetic strains and large earthquakes in the axial tectonic belt of North Island, New Zealand. *J. Geophys. Res.* **83**, 4419–29.

Waldron, J. W. F. 1987. A statistical test for significance of thinning- and thickening-upward cycles in turbidites. *Sed. Geology* **54**, 137–46.

Walker, K. R., G. Shanmugam & S. C. Ruppel 1983. A model for carbonate to terrigenous clastic sequences. *Bull. Geol Soc. Am.* **94**, 700–12.

Walker, R. G. 1965. The origin and significance of the internal sedimentary structures of turbidites. *Proc. Yorkshire Geol Soc.* **35**, 1–32.

Walker, R. G. 1966. Deep channels in turbidite-bearing formations. *Bull. Am. Assoc. Petrolm Geol.* **50**, 1899–917.

Walker, R. G. 1967a. Upper flow regime bed forms in turbidites of the Hatch Formation, Devonian of New York State. *J. Sed. Petrol.* **37**, 1052–8.
Walker, R. G. 1967b. Turbidite sedimentary structures and their relationship to proximal and distal depositional environments. *J. Sed. Petrol.* **37**, 25–43.
Walker, R. G. 1970. Review of the geometry and facies organisation of turbidites and turbidite-bearing basins. In *Flysch sedimentology in North America*, J. Lajoie (ed.), 219–51. Geol Assoc. Can. Spec. Pap. 7. Toronto: Business & Economic Service.
Walker, R. G. 1975a. Generalized facies model for resedimented conglomerates of turbidite association. *Bull. Geol Soc. Am.* **86**, 737–48.
Walker, R. G. 1975b. Upper Cretaceous resedimented conglomerates at Wheeler Gorge, California: description and field guide. *J. Sed. Petrol.* **45**, 105–12.
Walker, R. G. 1976. Facies models 2. Turbidites and associated coarse clastic deposits. *Geoscience Canada* **3**, 25–36.
Walker, R. G. 1977. Deposition of Upper Mesozoic resedimented conglomerates and associated turbidites in southwestern Oregon. *Bull. Geol Soc. Am.* **88**, 273–85.
Walker, R. G. 1978. Deep water sandstone facies and ancient submarine fans: models for exploration for stratigraphic traps. *Bull. Am. Assoc. Petrolm Geol.* **62**, 932–66.
Walker, R. G. 1984. Turbidites and associated coarse clastic deposits. In *Facies models*, 2nd edn, R. G. Walker (ed.), 171–88. Geoscience Canada Reprint Ser. 1. Kitchener, Ontario: Ainsworth Press.
Walker, R. G. & E. Mutti 1973. Turbidite facies and facies associations. In *Turbidites and deep-water sedimentation*, G. V. Middleton & A. H. Bouma (eds), 119–57. Soc. Econ. Paleont. Mineral. Pacific Section Short Course Notes, Anaheim.
Walther, J. 1894. *Einleitung in die Geologie als historische Wissenschaft.* Jena Verlag von Gustav Fischer. 3 vols.
Walton, E. K. 1967. The sequence of internal structures in turbidites. *Scottish J. Geology* **3**, 306–17.
Wang, Y., D. J. W. Piper & G. Vilks 1982. Surface textures of turbidite sand grains, Laurentian Fan and Sohm Abyssal Plain. *Sedimentology* **29**, 727–36.
Warme, J. E., R. G. Douglas & E. L. Winterer (eds) 1981. *The Deep Sea Drilling Project: a decade of progress.* Soc. Econ. Paleont. Mineral. Spec. Publn 32.
Watkins, J. S. & C. L. Drake (eds) 1983. *Studies in continental margin geology.* Am. Assoc. Petrolm Geol. Mem. 34.
Watkins, J. S., J. W. Ladd, R. T. Buffler, F. J. Shaub, M. H. Houston & J. L. Worzel 1982. Occurrence and evolution of salt in the deep Gulf of Mexico. In *Framework, facies, and oil-trapping characteristics of the upper continental margin*, A. H. Bouma, G. T. Moore & J. M. Coleman (eds), 43–65. Am. Assoc. Petrolm Geol Studies in Geology 7.
Watson, M. P. 1981. Submarine fan deposits of the Upper Ordovician–Lower Silurian Milliners Arm Formation, New World Island, Newfoundland. D. Phil thesis, Oxford University.
Watts, A. B. & M. S. Steckler 1979. Subsidence and eustacy at the continental margin of eastern North America. In *Deep sea drilling results in the Atlantic Ocean: continental margins and palaeoenvironments*, M. Talwani, W. W. Hay & W. B. F. Ryan (eds), 273–310. Am. Geophys. Union, Maurice Ewing Symposium 3.
Watts, A. B. & J. Thorne 1984. Tectonic, global changes in sea level, and their relationship to stratigraphical sequences at the US Atlantic continental margin. *Marine Petrolm Geology* **1**, 319–39.
Watts, A. B., G. D. Karner & M. S. Steckler 1982. Lithospheric flexure and the evolution of sedimentary basins. *Phil. Trans. R. Soc. London (A)***305**, 249–81.
Watts, K. F. & R. E. Garrison 1986. Sumeini Group, Oman – evolution of a Mesozoic carbonate slope on a South Tethyan continental margin. *Sed. Geology* **48**, 107–68.
Weaver, P. P. E. & A. Kuijpers 1983. Climatic control of turbidite deposition on the Madeira Abyssal Plain. *Nature* **306**, 360–3.
Weaver, P. P. E. & R. G. Rothwell 1987. Sedimentation on the Madeira Abyssal Plain over the last 300 000 years. In *Geology and geochemistry of abyssal plains*, P. P. E. Weaver & J. Thomson (eds), 71–86. Geol Soc. (London) Spec. Publn 31. Oxford: Blackwell Scientific.
Weaver, P. P. E. & J. Thomson (eds) 1987. *Geology and geochemistry of abyssal plains.* Geol Soc. (London) Spec. Publn 31. Oxford: Blackwell Scientific.
Weaver, P. P. E., R. C. Searle & A. Kuijpers 1986. Turbidite deposition and the origin of the Madeira Abyssal Plain. In *North Atlantic palaeoceanography*, C. P. Summerhayes & N. J. Shackleton (eds), 131–43. Geol Soc. (London) Spec. Publn 21. Oxford: Blackwell Scientific.
Weaver, P. P. E., J. Thomson & P. M. Hunter 1987. Introduction. In *Geology and geochemistry of abyssal plains*, P. P. E. Weaver & J. Thomson (eds), vii–xii. Geol Soc. (London) Spec. Publn 31. Oxford: Blackwell Scientific.
Weber, K. J. 1971. Sedimentological aspects of oil fields in the Niger delta. *Geol. Mijnb.* **50**, 559–76.
Weedon, G. P. 1986. Hemipelagic shelf sedimentation and climatic cycles: the basal Jurassic (Blue Lias) of South Britain. *Earth & Planet. Sci. Letts* **76**, 321–35.
Weedon, G. P. 1989. The detection and illustration of regular sedimentary cycles using Walsh power spectra and filtering, with examples from the Lias of Switzerland. *J. Geol Soc.* (London) **146**, 133–44.
Weeks, L. A. & R. K. Lattimore 1971. Continental terrace and deep plain offshore central California. *Marine Geophys. Res.* **1**, 145–61.
Weirich, F. H. 1988. Field evidence for hydraulic jumps in subaqueous sediment gravity flows. *Nature* **332**, 626–9.
Weissert, H. 1981. The environment of deposition of black shales in the Early Cretaceous: an ongoing controversy. In *The deep sea drilling project: a decade of progress*, J. E. Warme, R. G. Douglas & E. L. Winterer (eds), 547–60. Soc. Econ. Paleont. Mineral. Spec. Publn 32.
Wentworth, C. M. 1967. Dish structure, a primary sedimentary structure in coarse turbidites (abstract). *Bull. Am. Assoc. Petrolm Geol.* **51**, 485.
Werner, F. & A. Wetzel 1982. Interpretation of biogenic structures in oceanic sediments. *Bull. Inst. Geol. Bassin d'Aquitaine* **31**, 275–88.
Wernicke, B. 1985. Uniform-sense normal simple shear of the continental lithosphere. *Can. J. Earth Sci.* **22**, 108–25.
Wescott, W. A. & F. G. Ethridge 1982. Bathymetry and sediment dispersal dynamics along the Yallahs fan delta front, Jamaica. *Marine Geology* **46**, 245–60.
Westbrook, G. K. 1982. The Barbados Ridge Complex: tectonics of a mature forearc system. In *Trench–forearc geology*, J. K. Leggett (ed.), 275–90. Geol Soc. (London) Spec. Publn 10. Oxford: Blackwell Scientific.
Westbrook, G. K. & M. J. Smith 1983. Long decollements and mud volcanoes: evidence from the Barbados Ridge Complex

for the role of high pore fluid pressure in the development of an accretionary complex. *Geology* **11**, 279–83.

Westbrook, G. K., A. Mascle & B. Biju-Duval 1984. Geophysics and the structure of the Lesser Antilles Forearc. In *Initial Reports Deep Sea Drilling Project 78A*, B. Biju-Duval, J. C. Moore *et al.*, 23–38. Washington, DC: US Government Printing Office.

Wetzel, A. 1983. Biogenic structures in modern slope to deep-sea sediments in the Sulu Sea Basin (Philippines). *Palaeogeog. Palaeoclimatol. Palaeoecol.* **42**, 285–304.

Wetzel, A. 1984. Bioturbation in deep-sea fine-grained sediments: influence of sediment texture, turbidite frequency and rates of environmental change. In *Fine-grained sediments: deep-water processes and facies*, D. A. V. Stow & D. J. W. Piper (eds), 595–608. Geol Soc. (London) Spec. Publn 15. Oxford: Blackwell Scientific.

Weyl, P. 1968. The role of the oceans in climatic change: a theory of the ice ages. *Meteorological Monographs* **8**, 37–62.

Whateley, M. K. G. & K. T. Pickering (eds) 1989. *Deltas: sites and traps for fossil fuels.* Geol Soc. (London) Spec. Publn 41. Oxford: Blackwell Scientific.

Whitaker, J. H. McD. 1962. The geology of the area around Leintwardine, Herefordshire. *Quarterly J. Geol Soc. (London)* **118**, 319–51.

Whitaker, J. H. McD. 1974. Ancient submarine canyons and fan valleys. In *Modern and ancient geosynclinal sedimentation*, R. H. Dott Jr & R. H. Shaver (eds), 106–25. Soc. Econ. Paleont. Mineral. Spec. Publn 19.

White, R. S. 1982. Deformation of the Makran accretionary sediment prism in the Gulf of Oman (north-west Indian Ocean). In *Trench–forearc geology*, J. K. Leggett (ed.), 357–72. Geol Soc. (London) Spec. Publn 10. Oxford: Blackwell Scientific.

White, R. S. & K. E. Louden 1983. The Makran continental margin: structure of a thickly sedimented convergent plate boundary. In *The geology of continental margins*, J. S. Watkins & C. L. Drake (eds), 499–518. Am. Assoc. Petrolm Geol. 34.

Williams, B. G. & R. J. Hubbard 1984. Seismic stratigraphic framework and depositional sequences in the Santos Basin, Brazil. *Marine Petrolm Geology* **1**, 90–104.

Williams, H. 1984. Miogeoclines and suspect terranes of the Caledonian–Appalachian Orogen: tectonic patterns in the North Atlantic region. *Can. J. Earth Sci.* **21**, 887–901.

Williams, H. 1985. Paleozoic miogeoclines and suspect terranes of the North Atlantic region: Cordilleran comparisons. In *Tectonostratigraphic terranes of the circum-Pacific region*, D. G. Howell (ed.), 71–5. Houston, Texas: Circum-Pacific Council for Energy & Mineral Resources.

Williams, H. & R. D. Hatcher Jr 1982. Suspect terranes and accretionary history of the Appalachian orogen. *Geology* **10**, 530–6.

Williams, H. & R. D. Hatcher Jr 1983. Appalachian suspect terranes. In *Contributions to the tectonics and geophysics of mountain chains*, R. D. Hatcher Jr, H. Williams & I. Zietz (eds), 33–53. Geol Soc. Am. Mem. 158.

Williams, H. & R. K. Stevens 1974. The ancient continental margin of eastern North America. In *The geology of continental margins*, C. A. Burk & C. L. Drake (eds), 781–96. New York: Springer.

Williams, P. R., C. J. Pigram & D. B. Dow 1984. Mélange production and the importance of shale diapirism in accretionary terrains. *Nature* **309**, 145–6.

Wilson, D. S. 1986. A kinematic model for the Gorda deformation zone as a diffuse southern boundary for the Juan de Fuca plate. *J. Geophys. Res.* **91**, 259–70.

Wilson, D. S., R. N. Hey & C. Nishimura 1984. Propagation as a mechanism of reorientation of the Juan de Fuca Ridge. *J. Geophys. Res.* **89**, 9215–25.

Wilson, J. T. 1966. Did the Atlantic close and then re-open? *Nature* **210**, 678–81.

Winker, C. D. & M. B. Edwards 1983. Unstable progradational clastic shelf margins. In *The shelfbreak: critical interface on continental margins*, D. J. Stanley & G. T. Moore (eds), 139–57. Soc. Econ. Paleont. Mineral. Spec. Publn 33.

Winn, R. D. Jr, R. J. Bailes & K. I. Lu 1981. Debris flow, turbidites and lead-zinc sulfides along a Devonian submarine fault scarp, Jason prospect, Yukon Territory. In *Deep-water clastic sediments, a core workshop*, C. T. Seimers, R. W. Tillman & C. R. Williamson (eds), 396–416. Soc. Econ. Paleont. Mineral. Core Workshop No. 2.

Winn, R. D. Jr & R. J. Bailes 1987. Stratiform lead-zinc sulfides, mudflows, turbidites: Devonian sedimentation along a submarine fault scarp of extensional origin, Jason deposit, Yukon Territory, Canada. *Bull. Geol Soc. Am.* **98**, 528–39.

Winn, R. D. Jr & R. H. Dott Jr 1977. Large-scale traction-produced structures in deep-water fan-channel conglomerates in southern Chile. *Geology* **5**, 41–4.

Winn, R. D. Jr & R. H. Dott Jr 1978. Submarine-fan turbidites and residimented conglomerates in a Mesozoic arc–rear marginal basin in southern South America. In *Sedimentation in submarine canyons, fans, and trenches*, D. J. Stanley & G. Kelling (eds), 362–76. Stroudsburg, PA: Dowden, Hutchinson & Ross.

Winn, R. D. Jr & R. H. Dott Jr 1979. Deep-water fan-channel conglomerates of Late Cretaceous age, southern Chile. *Sedimentology* **26**, 203–28.

Winterer, E. L. & A. Bosellini 1981. Subsidence and sedimentation on Jurassic passive continental margin, Southern Alps, Italy. *Bull. Am. Assoc. Petrolm Geol.* **65**, 394–421.

Winterer, E. L. & K. Hinz 1984. The evolution of the Mazagan continental margin: a synthesis of geophysical and geological data with results of drilling during Deep Sea Drilling Project Leg 79. In *Initial Reports Deep Sea Drilling Project 79*, K. Hinz, E. L. Winterer *et al.* 79, 893–919. Washington, DC: US Government Printing Office.

Wise, S. W. & J. E. van Hinte 1987. Mesozoic–Cenozoic depositional environments revealed by Deep Sea Drilling Project Leg 93 drilling on the continental rise off the Eastern United States: cruise summary. In *Initial Reports Deep Sea Drilling Project 93*, J. E. van Hinte, S. W. Wise *et al.*, 1367–423. Washington, DC: US Government Printing Office.

Wood, A. & A. J. Smith 1959. The sedimentation and sedimentary history of the Aberystwyth Grits (upper Llandoverian). *Quart. J. Geol Soc. (London)* **114**, 163–95.

Wood, A. W. 1981. Extensional tectonics and the birth of the Lagonegro Basin (southern Italian Apennines), *N. Jb. Geol. Palaeont. Abh.* **161**, 93–131.

Woodcock, N. H. 1976a. Structural style in slump sheets: Ludlow Series, Powys, Wales. *J. Geol Soc. (London)* **132**, 399–415.

Woodcock, N. H. 1976b. Ludlow Series slumps and turbidites and the form of the Montgomery Trough, Powys, Wales. *Proc. Geol. Assoc.* **87**, 169–82.

Woodcock, N. H. 1979a. Sizes of submarine slides and their significance. *J. Structural Geology* **1**, 137–42.

Woodcock, N. H. 1979b. The use of slump structures as palaeoslope orientation estimators. *Sedimentology* **26**, 83–99.

Woodside, J. M. 1977. Tectonic elements and crust of the eastern Mediterranean Sea. *Marine Geophys. Res.* **3**, 317–54.
Worthington, L. V. 1968. Genesis and evolution of water masses. *Meteorological Monographs* **8**, 63–7.
Worthington, L. V. 1976. On the North American circulation. Baltimore: Johns Hopkins University Press.
Worzel, J. L. *et al.* 1973. *Initial Reports Deep Sea Drilling Project 10*. Washington, DC: US Government Printing Office.

Yeats, R. S., B. U. Haq *et al.* 1981. *Initial Reports Deep Sea Drilling Project 63*. Washington, DC: US Government Printing Office.
Yerkes, R. F., D. S. Gorsline & G. A. Rusnak 1967. Origin of Redondo submarine canyon, southern California. *US Geol Survey Research Prof. Paper 575-C*, 97–105.

Ziegler, P. A. 1986a. Late Caledonian framework of western and central Europe. In *The Caledonide Orogen – Scandinavia and related areas*, D. G. Gee & B. A. Sturt (eds), 3–18. London: Wiley & Sons.
Ziegler, P. A. 1986b. Caledonian, Acadian – Ligurian, Bretonian, and Variscan orogens – is a clear distinction justified ? In *The Caledonide Orogen–Scandinavia and related areas*, D. G. Gee & B. A. Sturt (eds), 1241–8. London: Wiley.
Ziegler, P. A. 1986c. Geodynamic model for the Palaeozoic crustal consolidation of Western and Central Europe. *Tectonophysics* **126**, 303–28.
Zonenshain, L. P. & X. Le Pichon 1986. Deep basins of the Black Sea and Caspian Sea as remnants of Mesozoic back-arc basins. *Tectonophysics* **123**, 181–211.

Additional references

Anderton, R. 1985. Clastic facies models and facies analysis. In *Sedimentology. Recent developments and applied aspects*, P. J. Brenchley & B. P. J. Williams (eds), 31–47. London: Blackwell Scientific.

Ballance, P. F. 1964. Streaked out mud ripples below Miocene turbidites, Puriri Formation, New Zealand. *J. Sed. Petrol.*, **34**, 91–101.
Barron, E. J., M. A. Arthur & E. G. Kauffman 1985. Cretaceous rhythmic bedding sequences: a plausible link between orbital variation and climate. *Earth Planet. Sci. Letts* **72**, 327–40.

Damuth, J. E., R. D. Flood, R. O. Kowsmann, R. H. Belderson & M. A. Gorini 1988. Anatomy and growth pattern of Amazon deep-sea fan as revealed by long-range side-scan sonar (GLORIA) and high-resolution seismic studies. *Bull. Am. Assoc. Petrolm Geol.* **72**, 885–911.
Fischer, A. G. 1986. Climatic rhythms recorded in strata. *Ann. Rev. Earth Planet. Sci.* **14**, 351–76.

Imbrie, J. 1985. A theoretical framework for the Pleistocene ice ages. *J. Geol Soc. (London)* **142**, 417–32.
Imbrie, J., J. D. Hays, D. G. Martinsen, A. McIntyre, A. C. Mix, J. J. Morley, N. G. Pisias, W. L. Prell & N. J. Shackleton 1984. The orbital theory of Pleistocene climate: support from a revised chronology of the marine ^{18}O record. In *Milankovitch and climate*, A. Berger, J. Imbrie, G. Kukla & B. Saltzman (eds), 269–305. Dordrecht: D. Reidel.

Manley, P. L. & R. D. Flood 1988. Cyclic sediment deposition within Amazon Deep-Sea Fan. *Bull. Am. Assoc. Petrolm Geol.* **72**, 912–25.

Siedlecka, A., K. T. Pickering & M. B. Edwards 1989. Upper Proterozoic passive margin deltaic complex, Finnmark, N Norway. In *Deltas: Sites and traps for fossil fuels*, M. K. G. Whateley & K. T. Pickering (eds). Geol Soc. (London) Spec. Publn 41. Oxford: Blackwell Scientific.
Syvitski, J. P. M., J. N. Smith, E. A. Calabrese & B. P. Boudreau 1988. Basin sedimentation and the growth of prograding deltas. *J. Geophys. Res.* 93, 6895–908.

Watkins, J. S., L. Montadert & P. W. Dickerson (eds) 1979. Geological and geophysical investigations of continental margins. *Am. Assoc. Petrol. Geol. Memoir* **29**.

Index